2nd edition
biochemistry

The National Medical Series for Independent Study

2nd edition
biochemistry

Ian D. K. Halkerston, Ph.D.

formerly, Associate Professor
  of Biochemistry
University of Massachusetts
  Medical Center
Worcester, Massachusetts

A WILEY MEDICAL PUBLICATION
JOHN WILEY & SONS
New York • Chichester • Brisbane • Toronto • Singapore

Harwal Publishing Company, Media, Pennsylvania

Library of Congress Cataloging-in-Publication Data

Halkerston, Ian D. K.
  Biochemistry.

  (The National medical series for independent study)
  Includes index.
  1. Biochemistry—Outlines, syllabi, etc.
2. Biochemistry—Examinations, questions, etc.
I. Title. II. Series. [DNLM: 1. Biochemistry—
examination questions. 2. Biochemistry—outlines.
QU 18 H173b]
QP518.3.H35 1988   591.19′2   87-29112
ISBN 0-471-85554-5

10 9 8 7 6 5 4 3 2 1

Contents

Preface to the Second Edition

The first edition of this text was developed from the medical biochemistry course taught to first-year medical students at the University of Massachusetts Medical School up until 1983. The present edition expands and strengthens the chapters covering information transfer, in recognition of the very rapid expansion of knowledge in this area, which is of particular importance to modern medicine. The two-part organization of the text has been retained with the first part giving the essentials of information transfer, protein structure–function, and enzymes, and the second part dealing with the metabolism of carbohydrates, fats, and proteins (and other nitrogen-containing compounds) together with major mechanisms of integration and control of metabolism.

Ian D. K. Halkerston

Preface to
the First Edition

This review of biochemistry was developed from the medical biochemistry course taught to first-year medical students at the University of Massachusetts Medical School. It is presented in two parts: the first gives the essentials of information transfer, protein structure–function, and enzymes, and the second deals with the metabolism of carbohydrates, fats, and proteins (and other nitrogen-containing compounds) together with major mechanisms of integration and control of metabolism.

Ian D. K. Halkerston

Acknowledgments

The medical biochemistry course for first-year medical students at the University of Massachusetts Medical School is taught by many individuals who are members of the Department of Biochemistry. These individuals have been most considerate and helpful to me in this endeavor. I would like to thank in particular Dr. Michael Czech, Department Chairman, for his understanding and support, and with gratitude thank Drs. Barber, Carruthers, Burstein, Chlapowski, Cimbala, Flatt, Melchior, Miller, and Nemecek for numerous discussions and assistance.

I wish also to acknowledge the most skillful editing on the part of Jane Edwards and Gloria Hamilton and the artwork of Wieslawa B. Langenfeld.

Publisher's Note

The objective of the *National Medical Series* is to present an extraordinarily large amount of information in an easily retrievable form. The outline format was selected for this purpose of reducing to the essentials the medical information needed by today's student and practitioner.

While the concept of an outline format was well received by the authors and publisher, the difficulties inherent in working with this style were not initially apparent. That the series has been published and received enthusiastically is a tribute to the authors who worked long and diligently to produce books that are stylistically consistent and comprehensive in content.

The task of producing the *National Medical Series* required more than the efforts of the authors, however, and the missing elements have been supplied by highly competent and dedicated developmental editors and support staff. Editors, compositors, proofreaders, and layout and design staff have all polished the outline to a fine form. It is with deep appreciation that I thank all who have participated, in particular, the staff at Harwal—Debra L. Dreger, Jane Edwards, Gloria Hamilton, Judy Johnson, Susan Kelly, Wieslawa B. Langenfeld, Keith LaSala, June Sangiorgio Mash, Jane Velker, and Elizabeth Waddington.

The Publisher

Introduction

Biochemistry is one of ten basic science review books in the *National Medical Series for Independent Study*. This series has been designed to provide students and house officers, as well as physicians, with a concise but comprehensive instrument for self-evaluation and review within the basic sciences. Although *Biochemistry* would be most useful for students preparing for the National Board of Medical Examiners examinations (Part I and FLEX) as well as FMGEMS, it should also be useful for students studying for course examinations. These books are not intended to replace the standard basic science texts, but, rather, to complement them.

The books in this series present the core content of each basic science area using an outline format and featuring 500 study questions. The questions are distributed throughout the book at the end of each chapter and in a challenge exam at the end of the text. In addition, each question is accompanied by the correct answer, a paragraph-length explanation of the correct answer, and specific reference to the outline points under which the information necessary to answer the question can be found.

We have chosen an outline format to allow maximum ease in retrieving information, assuming that the time available to the reader is limited. Considerable editorial time has been spent to ensure that the information required by all medical school curricula has been included and that the question format parallels that of the National Board examinations. We feel that the combination of the outline and the board-type study questions provides a unique teaching device.

We hope you will find this series interesting, relevant, and challenging. The authors and the staff at Harwal/Wiley Medical welcome your comments and suggestions.

Part I
Information, Proteins,
and Enzymes

DNA: Genetic Material

I. COVALENT STRUCTURE OF NUCLEIC ACIDS

A. Nucleic acids are polymers of nucleoside monophosphates (nucleotides).

1. Each monomeric unit consists of
   a. A nitrogenous base, which may be a **purine** or a **pyrimidine**
   b. A pentose sugar, **ribose** or **deoxyribose**, in furanose (ring) form
   c. A **phosphate group** esterified to the sugar (Fig. 1-1)

2. The coupling of base to sugar yields a **nucleoside**, which becomes a **nucleotide** when it is phosphorylated.

B. Nucleotides can be linked together by phosphodiester bonds between the $3'$-hydroxyl on the sugar of one nucleotide and the $5'$-phosphate on the sugar of another nucleotide. Figure 1-1 illustrates the structure of such a **dinucleotide**.

1. The addition of another nucleotide at the $3'$-hydroxyl produces a **trinucleotide**, and with many nucleotides added, a **polynucleotide** exists.

2. These compounds have **polarity**. One end of the molecule is different from the other in the sense that the pyrimidine nucleotide (cytosine) in Figure 1-1 has a free $3'$-hydroxyl group on its sugar moiety, whereas the purine nucleotide (guanine) does not.

3. In long chains formed from nucleotides in this manner, one speaks of the $5'$ end of the molecule and of the $3'$ end (free hydroxyl).

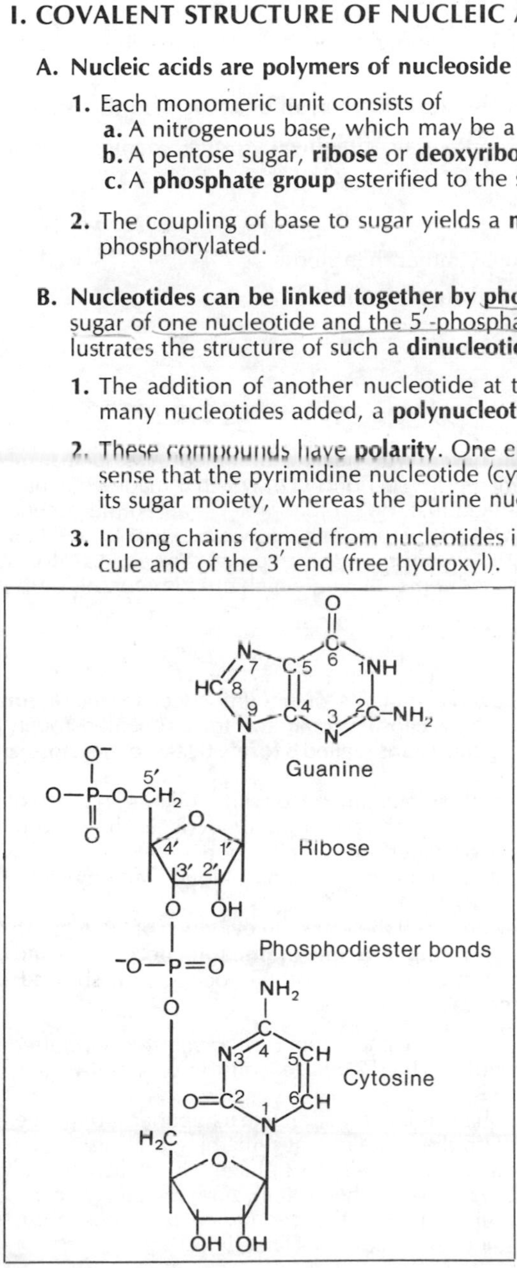

Figure 1-1. Structure of a dinucleotide. The bases here are guanine and cytosine. A base plus a sugar comprise a nucleoside, and a nucleoside plus a phosphate comprise a nucleotide. Because the sugar is ribose, this is a diribonucleotide. If the sugar were deoxyribose, it would be a dideoxyribonucleotide.

II. NOMENCLATURE OF NUCLEIC ACIDS (Figure 1-2 illustrates the structures of the five common bases found in nucleic acids.)

A. Polynucleotides

1. The dinucleotide illustrated in Figure 1-1 can be represented in an abbreviated form as pGpC or pGC.

   a. There is a phosphate group at the 5′ end, which by convention is always placed at the left-hand end of the representation.

   b. The 5′-nucleotide is guanosine phosphate (guanidylic acid).

   c. There is no phosphate on the 3′ end.

2. If the dinucleotide illustrated in Figure 1-1 were modified in the 2′ position of each sugar moiety, by replacing the hydroxyl group (OH) with hydrogen (H) to give deoxyribose, the result would be a **deoxyribonucleotide**. Naturally occurring polydeoxyribonucleotides are called **deoxyribonucleic acid (DNA)**.

B. Polyribonucleotides. Polynucleotides containing ribose, as opposed to deoxyribose, are called polyribonucleotides.

1. Naturally occurring polyribonucleotides also are known as **ribonucleic acid (RNA)**.

2. With a few exceptions, the bases found in RNA are **guanine**, **cytosine**, **adenine**, and **uracil** (see Chapter 3, section II B 2 c).

C. Primary structure of DNA and RNA

1. DNA has 2′-deoxyribose as its sugar moiety rather than ribose.

2. DNA has one different base—thymine (5-methyl uracil).

3. RNA has one different base—uracil.

III. DNA AS THE GENETIC MATERIAL

A. Genes of eukaryotes
(organisms whose cells have a true nucleus bounded by a nuclear membrane; see section VI A) are present in chromatin (see section VI D), which is made up of protein and nucleic acid. For a long time, it was thought that the protein moiety carried the genetic information and that the role of nucleic acid was a structural one. However, DNA is now known to be the carrier of genetic information in all organisms except certain viruses (e.g., tobacco mosaic virus, poliovirus, and influenza virus) and some phages, which use RNA for this purpose.

B. Evidence for DNA as the genetic material

1. Early work with pneumococci

   a. In 1928, Griffith found that nonpathogenic mutants of pneumococci (called **R forms** because the mutant strain could not form a capsular coat and thus appeared rough on plates) could be transformed into pathogenic forms (called **S forms** because the capsulated organisms appeared shiny on plates).

   (1) This transformation was accomplished by injecting mice with a mixture of heat-killed S-form (pathogenic) and live R-form (nonpathogenic) pneumococci. The mice died, and live S-form pneumococci were recovered from their blood.

   (2) When either live R-form or heat-killed S-form pneumococci alone were injected, the mice lived.

   b. In 1944, Avery, McLeod, and McCarthy provided the first solid evidence of DNA as the genetic material in their work on the **transforming principle of pneumococci**. Avery and co-workers purified extracts made from heat-killed S-form pneumococci and showed unequivocally that the transforming factor was DNA.

2. Studies of bacteriophage T2, a virus that infects the bacterium *Escherichia coli*, supported the conclusion that DNA is the genetic material. The T2 phage consists of a **DNA core** surrounded by a **protein coat**.

   a. In 1951, Herriott suggested that the T2 phage might function like a hypodermic syringe, injecting the nucleic acid into the cell, with the protein coat remaining outside.

   b. In 1952, Hershey and Chase tested this idea by labeling T2 phage with radioactive phosphorus in the nucleic acid moiety and with sulfur in the protein. After infecting *E. coli* cells with the labeled T2 phage, they were able to show that the radioactive phosphorus remained with the cells, whereas the radioactive sulfur was largely lost.

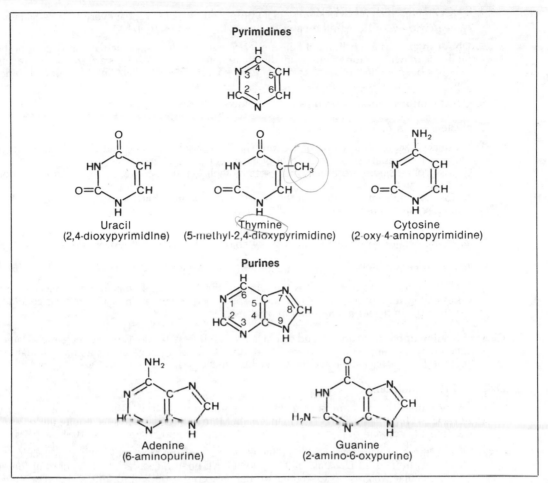

Figure 1-2. The five common pyrimidine and purine bases in nucleic acids.

IV. STRUCTURAL CHARACTERISTICS OF DNA

A. Basic structure of DNA. In 1953, Watson and Crick, using x-ray diffraction data of Franklin and Wilkins, proposed a structure for DNA that has become known as the **double helix**. Using this structure, Watson and Crick proposed a mechanism for the replication of DNA that is still accurate today.

1. **Two antiparallel polydeoxyribonucleotide chains** are wound around each other with the purine and pyrimidine bases on the inside of the helix and the deoxyribose and phosphates on the outside.

2. **Stacking of the hydrophobic rings** results when the planes of the bases are perpendicular to the axis of the helix.

3. **Hydrogen bonding** occurs between pairs of bases, one base on one strand pairing with another base on the antiparallel strand.
   a. **Adenine (A) always pairs with thymine (T), and guanine (G) always pairs with cytosine (C).**
   b. The above explains **Chargaff's rules**, which state that in nearly all DNA studied, the content of A equals the content of T, and the content of G equals the content of C (see section IV B).

4. **The helix is 20 μ in diameter.**
   a. In crystalline DNA, the bases are separated by 3.4 Å along the helix axis, and each base is rotated 36° in relation to the previous base. This means that in this type of DNA, the helical structure repeats at intervals of 34 Å (i.e., every 10 base-pairs).

      b. In solution, the helical structure of DNA probably does not repeat every 10 base-pairs but nearer to every 11 with the bases inclined 20° from horizontal.

  5. DNA can exist as a left-handed helix under certain conditions, particularly in regions of the molecule in which purines and pyrimidines alternate. The DNA adopts a zigzag structure and thus has been named **Z-DNA**. Evidence for the existence of Z-DNA in eukaryotic chromosomes is being actively explored.

  6. Genetic information is stored as the sequence of bases along the chain.

B. Equivalence of A to T and G to C

  1. The mole fraction of any one of the four standard bases (A, T, G, and C) in double helical (duplex) DNA establishes the content of each of the other three.
    a. Base compositions are usually expressed by their G plus C content (i.e., the mole fraction of GC pairs).
    b. The complementary nature of the two strands dictates that the relative GC content of duplex DNA is equal to that of either strand.

  2. The base composition can be determined by
    a. Sedimentation to equilibrium in cesium chloride, because the buoyant density measured is a direct function of GC content
    b. Measurement of the melting temperature (T_m) [see section V A 3]

  3. The GC content (mole fraction) is the same for the DNA of all cells in a given species but varies widely from one species to another, especially in bacteria. The GC content is generally less than 0.5 in higher animals (0.4 in humans).

C. Modification of bases. In animal and plant DNAs, as much as 10% of C residues are methylated. Modified bases, or analogues, also are found in bacterial and phage DNAs and are thought to protect the DNA against cleavage by intracellular nucleases.

D. Size of DNA molecules

  1. A very simple chromosome such as that of the simian virus 40 (SV40) has only about 5 or 6 genes (5243 base-pairs) and is 1.7 μm in length. More complex organisms require more DNA, and thus the genome of *E. coli* is more than 1 mm long and contains 4×10^6 base-pairs.

  2. Measurement of very long DNA is difficult because of the tendency for the molecules to shear, but stretches of DNA about 1 mm long have been measured. Human chromosomes with molecular weights up to 160×10^9 daltons, corresponding to 240×10^6 base-pairs, would, if continuous, measure up to 8.2 cm.

  3. Studies of the **viscoelastic properties** of DNA have revealed the lengths of different DNA molecules in the chromosomes of many cells; and there is strong agreement between the length of DNA in *Drosophila* chromosomes, as measured by this technique, and the DNA mass in *Drosophila* determined cytologically.

  4. DNA molecules run the full length of a chromosome without discontinuity at the centromere, and the continuity is conserved throughout the cell cycle. **These data provide the most compelling evidence that a chromosome of a higher organism is composed of a single molecule of DNA.**

E. Shape of DNA molecules

  1. Chromosomal DNA may be linear or circular (a closed loop). DNA molecules do not exist in an extended form in either case. Rather, both rods and circles are twisted (supercoiled) into complexes that result in a several thousandfold condensation of the DNA in a chromosome or the head of a virus. (The contour length of the *E. coli* genome is 1000 times the length of the bacterium.)

  2. DNA is complexed with histones in the eukaryotic chromosome (see section VI D 1, E), which profoundly influence the structure of the chromosome. Nonhistone proteins and possibly RNA also may affect the packaging of DNA. When a single molecule of DNA is complexed with histones, it forms a chromatin thread, which is a beaded string coiled upon itself. In DNA viruses, DNA condensation is achieved by regular spool-like windings assisted by polyamines and metal ions.

  3. Supercoiling of DNA is essential for stages of replication, transcription, and recombination. It occurs only when there is restraint upon the rotation of the ends of the DNA chains, such as occurs in circular DNA or in linear DNA when the ends are too far apart to allow rapid spinning (see Chapter 2, section II C).

V. DISSOCIATION AND REASSOCIATION OF THE DOUBLE HELICAL CHAINS OF DNA.
Double helical chains of DNA have a remarkable ability to dissociate from one another and to reassociate again. This behavior is essential to the processes of **replication** (see Chapter 2, section IV B) and **transcription** (see Chapter 3, section I C 1 c).

- **A. Denaturation.** Rupture of the hydrogen bonds between the bases, resulting from increasing temperature or the alteration of the hydrogen ion concentration, causes the two strands to come apart.

    1. **Increasing pH** deprotonates ring nitrogens of guanine and thymine; **decreasing pH** protonates ring nitrogens of adenine, guanine, and cytosine.

    2. **Increasing acidity** can cause the rupture of purine glycosidic bonds, and at high temperatures, phosphodiester bonds may be broken. Thus, **alkali is the method of choice for denaturing DNA.** In contrast to the stability of DNA chains in alkali, RNA chains are hydrolyzed.

    3. **Increasing temperature** has been employed by many studies to denature DNA's double helical chains. The dissociation of the two chains occurs at a definite temperature, and thus the process is referred to as **melting.**
        a. T_m is the temperature at which 50% of the double helix is unwound.
        b. The melting of DNA can be studied by measuring the absorption of light at 260 nm; melting causes unstacking of the base-pairs, which results in increased absorbance (**hyperchromic effect**).
        c. T_m is strongly influenced by the base composition of the DNA.
            (1) DNA rich in GC pairs has a higher T_m than DNA with a high proportion of AT pairs.
            (2) The T_m of DNA extracted from different species and measured at pH 7 in an isotonic salt solution varies linearly with GC content with synthetic poly-AT having a T_m of about 65°C and synthetic poly-GC having a T_m of 105°C (Fig. 1-3).
            (3) Mammalian DNA, which is about 40% GC pairs, has a T_m of about 87°C.

- **B. Renaturation (annealing)**

    1. If the temperature of melted (dissociated) duplex DNA is rapidly reduced, the original double helical structure does not reform (anneal).

    2. If, however, the temperature is held at a value of about 20°C to 25°C below the T_m, the original double helical structure reforms.

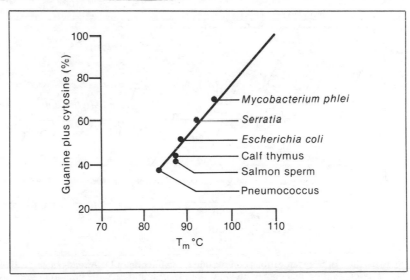

Figure 1-3. Relationship between the content of guanine (G) plus cytosine (C) and melting temperature (T_m) in DNAs from various sources. (Reprinted with permission from Doty P: Inside nucleic acids. *Harvey Lectures* 55:125, 1961, Orlando, FL, Academic Press.)

3. The rate at which the structure reforms is related to the concentration of complementary sequences.
   a. A measure of the complexity of duplex DNA genomes can be obtained from annealing studies in which the fraction (F) of single-stranded DNA that reanneals is plotted against the logarithm of the product of the initial concentration (Co) of single-stranded DNA in mol/L and the time (t) in seconds. The plots of F against log Cot form sigmoidal curves (Fig. 1-4).
   b. The genome of the phage T4 reanneals more rapidly than the larger genome of *E. coli*. The much larger, nonrepetitive fraction of calf thymus reanneals about 3 orders of magnitude more slowly than that of *E. coli*.
   c. Eukaryotic DNA contains many sequences of DNA that reanneal very rapidly (see section VII B).

VI. EUKARYOTIC CHROMOSOMES

A. Eukaryotic cells

1. Eukaryotic cells comprise all cells in multicellular organisms and unicellular organisms other than bacteria and blue-green algae, which are prokaryotes. They are about 20 μ in diameter, have a distinct nuclear membrane and multiple chromosomes, and have DNA that is confined to the nucleus except for small amounts found in mitochondria or chloroplasts.

2. In comparison, prokaryotic cells, such as *E. coli*, are cylindrical, measuring about 1 μ in diameter and 2 μ in length. They have a single chromosome, which lacks a distinct topologic barrier to distinguish it from the cytoplasm.

B. Eukaryotic DNA

1. DNA is packaged into unit structures called **chromosomes**. The number of chromosomes depends upon the species: The fruit fly (*Drosophila*) has 4 chromosomes; humans have 46.
   a. **Somatic cells** (i.e., all cells except gametes) have paired (homologous) chromosomes, one derived from each parent. This quality is known as **diploidy**.
   b. Each chromosome at the metaphase stage of mitosis contains two complete and identical molecules (i.e., four strands) of DNA because the DNA has undergone replication during the preceding S phase of the cell cycle.

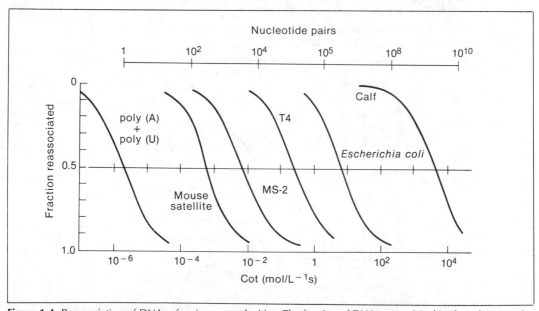

Figure 1-4. Reassociation of DNAs of various complexities. The fraction of DNA reassociated is plotted against the logarithm of the product of the initial concentration (Co) of DNA in mol/L × the time (t) in seconds (log Cot). The correlation between Cot and sequence complexity is indicated by the points where the sigmoidal Cot curves cross the 0.5 reassociation line. Sample DNAs include poly (A) + poly (U) [a synthetic double-stranded polynucleotide], a mouse satellite DNA, phages MS-2 and T4, *Escherichia coli*, and a nonrepetitive fraction of calf DNA. (Reprinted with permission from Britten RJ, Kohne DE: Repeated sequences in DNA. *Science* 161:529–540, 1968. ©1968 by the American Association for the Advancement of Science.)

(handwritten annotations: "At metaphase of mitosis 2 complete + identical molecules", "(2 molecules but 4 strands)", "DNA replication", "DNA synthesis", "G₁ or G₀")

Figure 1-5. The life cycle of a eukaryotic cell. *M* = mitosis; G_1 = gap 1 (preceding DNA synthesis); S = period of DNA synthesis; G_2 = gap 2 (preceding mitosis). (After Stryer L: *Biochemistry*, 2nd ed. New York, WH Freeman, 1981, p 693.)

2. The human diploid cell contains about 6.4×10^9 base-pairs of DNA, about 1000 times as much as in *E. coli*.
   a. The 6.4×10^9 base-pairs of DNA are distributed among the 46 chromosomes, and although the chromosomes vary in size, each contains approximately 1.5×10^8 base-pairs.
   b. If stretched out, 1.5×10^8 base-pairs would have a contour length of 47 mm. Obviously, DNA occurs in a highly condensed form.

C. **Chronologic divisions of a eukaryotic cell cycle** (Fig. 1-5)

   1. During the **metaphase** stage of mitosis, the chromosomes are visible under the light microscope as individual entities.

   2. Metaphase is followed by a period called **gap 1 (G_1)** in the daughter cells. This is a gap in time that precedes DNA synthesis. *(handwritten: paired chromosomes)*
      a. A cell in G_1 is diploid: It contains two copies of each chromosome.
      b. When tissues are not actively growing, the cell cycle stays at G_1; thus, most cells in adult multicellular organisms (including humans) are usually diploid. Cells in this state are said to be in the **G_0 phase**.

   3. The period of **DNA synthesis (S phase)** is next during which the bulk of the DNA is replicated. After the S phase, there is twice as much DNA as in G_1, and each chromosome has become two identical sister chromatids.

   4. Lastly, **G_2** is the gap between the S phase and the next metaphase.

D. **Chromatin.** The combination of DNA and proteins is called chromatin because it can be stained by a number of characteristic microscopy stains. Chromatin contains about equal amounts (by weight) of DNA and protein. Small amounts of RNA are also present. Some is presumably RNA in the process of being transcribed (see Chapter 3, section I A), but a portion of the RNA appears to be associated with chromosome function.

   1. **Histones.** The DNA in eukaryotes (but not prokaryotes) is associated with basic proteins called histones and with other proteins called **nonhistone chromosomal proteins**. These are noncovalent associations.
      a. Histones are small proteins that carry a considerable positive charge.
      b. They are divided into five classes—H1, H2A, H2B, H3, and H4—which vary from 11.3 kdal to 21 kdal in mass (Table 1-1).
      c. These basic proteins have amino acid sequences that have been conserved for more than an eon (1.2×10^9 years).
         (1) The amino acid sequences of histone H4 from the cow and the pea, for example, differ

Table 1-1. Types of Chromosomal Histones, Their Molecular Weights, and Their Location in the Nucleosome

Type	Molecular Weight (daltons)	Location in Nucleosome
H1	21,000	Near linker DNA
H2A	14,500	Core
H2B	13,800	Core
H3	15,300	Core
H4	11,300	Core

by only 2 out of 102 residues with both amino acid substitutions being **conservative** (i.e., a lysine residue for a histidine and a valine for an alanine).

 (2) The extreme conservation of the histone structure strongly implies that the histones play a very specific role in the structure of eukaryotic DNA.

 d. Chemical analysis of chromatin reveals that
   (1) For every 200 base-pairs of DNA, there are two molecules each of H2A, H2B, H3, and H4 and usually one molecule of H1.
   (2) H3 and H4 spontaneously form a tetramer with structure (H3)$_2$ (H4)$_2$. Similarly, H2A and H2B form the tetramer (H2A)$_2$ (H2B)$_2$.

 2. **Nucleosomes.** Electron microscopy shows that chromatin is made up of a series of repeating structures called **nucleosomes**, which resemble a chain of beads. Each bead is made up of two molecules each of H2A, H2B, H3, and H4.
   a. The nucleosomes have a **core** that consists of a histone octamer (made up of the two tetramers) noncovalently bonded to 140 base-pairs of DNA, which are wrapped around the histone in a left-handed superhelix.
   b. The remainder (about 60) of the DNA base-pairs form the chain that links the beads.
   c. A single H1 molecule may be present on the outside of the core near the linker DNA.
   d. The nucleosome represents the first stage in the condensation of DNA.
     (1) The tightly wrapped DNA of the nucleosome is 1/7 the length of fully extended DNA; that is, it has a **packing ratio** (degree of condensation) of 7.
     (2) In comparison, the DNA in a human metaphase chromosome is about 1/10,000 the length of fully extended DNA; here the packing ratio is about 10^4.
     (3) Thus, the condensation of DNA in metaphase must involve several levels of organization above that of the nucleosome.

 3. **Euchromatin and heterochromatin.** Two types of chromatin have been found in cytologic studies.
   a. **Euchromatin**, which does not stain well, represents dispersed chromatin. The DNA in euchromatin probably can be transcribed into RNA.
   b. **Heterochromatin** stains darkly and is found mainly in the centromere region of chromosomes. The DNA in heterochromatin probably cannot be easily transcribed. H1 is associated with heterochromatin.

E. **Cytoplasmic DNA in eukaryotic organelles.** Mitochondria and chloroplasts of eukaryotic cells contain DNA that differs from the nuclear DNA.

 1. Neither mitochondrial nor chloroplast DNA is associated with histones.

 2. Mitochondrial DNA of animal cells is
   a. Double-stranded
   b. Circular
   c. About 15,000 base-pairs in length

 3. Mitochondrial DNA sequences code for only about 5% of the protein components of mitochondrial structure and function. The bulk of the information for mitochondrial protein synthesis is stored in nuclear DNA.

VII. **HETEROGENEITY OF EUKARYOTIC DNA.** Studies of the renaturation of eukaryotic DNA (see section V B) show that different parts of a genome reanneal at different rates, indicating that several types of DNA exist. Four classes of DNA sequences have been recognized.

A. **Unique sequences (single-copy genes)**

 1. About 70% of the DNA of many eukaryotes consist of unique sequences that reanneal very slowly.
   a. These unique sequences code for the major proteins required by the cell.
   b. They are present only in a single copy (or only in a few copies) per genome.
   c. They are interspersed with moderately repetitive sequences, which **are not transcribed** (see section VII C 4).

 2. The coding sequences, or expressed regions (**exons**), of unique sequences are split up by intervening sequences (**introns**), which **are transcribed** but are then removed by **splicing enzymes** and degraded.

B. **Highly repetitive sequences**

 1. Highly repetitive sequences, known as **satellite DNA**, are 2 to 10 base-pairs long and repeat in tandem 10^5 to 10^7 times in the genome.

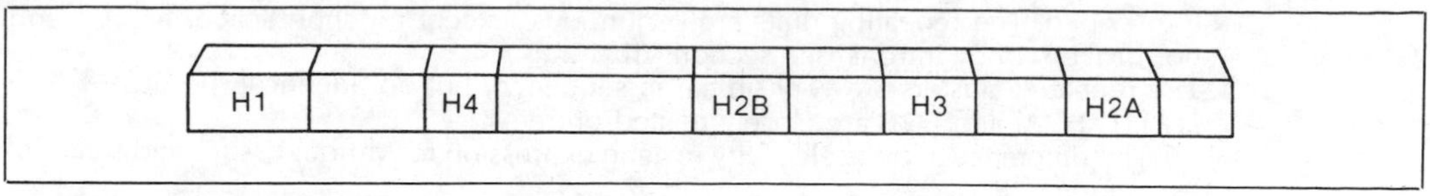

| H1 | | H4 | | | H2B | | H3 | | H2A | |

Figure 1-6. Arrangement of histone genes in a sea urchin. The unlettered blocks are spacer regions that are not transcribed.

   a. They make up about 15% of the genome.
   b. They are not transcribed.
   c. They are concentrated in the centromere region of the chromosome.
   2. Highly repetitive sequences reanneal very rapidly (see Fig. 1-4).

C. Moderately repetitive sequences. Some gene products are coded for by DNA sequences that are repeated **in tandem**, often more than 100 times; for example:

   1. Ribosomal RNA (rRNA) genes in areas of the chromosome associated with nucleoli (see Chapter 3, section II C 2 b)

   2. Transfer RNA (tRNA) [see Chapter 3, section II B]

   3. Histone genes
      a. Sea urchin histone genes are clustered in a repeating unit 7 kb long (Fig. 1-6). The five coding regions correspond to the five histones, and the five intercalated spacer regions are not

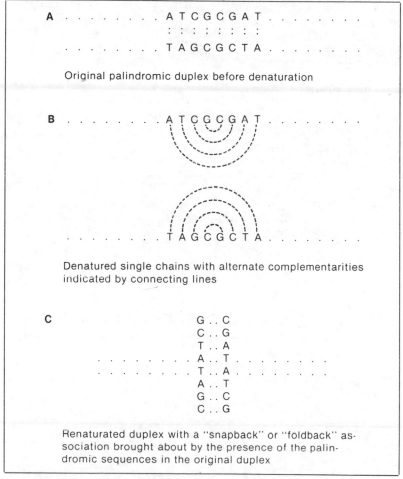

Figure 1-7. Formation of a "foldback" or "snapback" sequence from a palindromic section of DNA.

transcribed. The repeating units are continuous, except for short spacer blocks, and are not broken up by introns (see section VII A 2).

   b. The repeated clusters are very similar in sequence, but not identical, because

   (1) H1, H2A, and H2B are closely related proteins.

   (2) The difference allows selectivity in gene expression for various tissues and stages of development.

   4. Untranscribed DNA sequences. There are a great many moderately repetitive untranscribed DNA sequences of about 300 base-pairs in length; these lie between the unique sequences of single-copy genes that code for cellular proteins.

D. Inverted repetitive sequences make up about 6% of the human genome, but their function is unknown.

   1. These sequences vary in length up to 1200 base-pairs.

   2. They have an axis of twofold rotational symmetry; that is, like the word "madam," they are palindromes, as the same message can be read off each strand in opposite directions.

   3. They reanneal rapidly, forming "foldback" or "snapback" sequences (Fig. 1-7).

STUDY QUESTIONS

Directions: Each question below contains five suggested answers. Choose the **one best** response to each question.

1. If the base sequence of a segment of DNA is

    $5'$pCpApGpTpTpApGpC-3'-OH,

    which of the following sequences is complementary?

    (A) pGpTpCpApApTpCpG-3'-OH
    (B) pGpGpCpTpApApCpTpG-3'-OH
    (C) 5'-OH-GpTpCpApApTpCpGp
    (D) 5'-OH-GpCpTpApApCpTpGp
    (E) 5'-OH-GpCpTpApApCpTpG-3'-OH

2. Which of the following findings first gave a strong indication that DNA was the genetic material?

    (A) RNA synthesis depends on a DNA template
    (B) DNA is in the nuclei of all eukaryotic cells
    (C) Viral nucleic acid can be DNA
    (D) Transforming activity is due to DNA
    (E) Duplex DNA contains two complementary chains

3. Which of the following statements is true of duplex DNA?

    (A) The planes of the bases lie parallel to the helix axis
    (B) The chains have a backbone of linked glycosides
    (C) Unless the DNA is circular, the 3' hydroxyl groups of each chain are at opposite ends of the molecule
    (D) The duplex structure is stabilized only by hydrogen bonding between bases
    (E) Although they are associated in antiparallel fashion, the two chains have an identical base sequence

4. Chargaff's rules state that

    (A) in RNA, A = U, and in DNA, A = T
    (B) G = C in both RNA and DNA
    (C) $\dfrac{(A + T)}{(G + C)} = 1$
    (D) $\dfrac{A}{T} = \dfrac{G}{C}$
    (E) A = G = T = C

5. If a sample of DNA is found to have the base composition (mole ratios) of adenine, 40; thymine, 22; guanine, 19; and cytosine, 19, what conclusion can be drawn?

    (A) The DNA is a circular duplex
    (B) The DNA is a linear duplex
    (C) The DNA is single-stranded
    (D) The DNA has highly repetitive sequences
    (E) The DNA has a high melting point

6. If the cytosine content of duplex DNA is 20% of the total bases, the adenine content would be

    (A) 10%
    (B) 20%
    (C) 30%
    (D) 40%
    (E) 50%

7. The hyperchromic effect refers to

    (A) a change in the optical rotatory dispersion (ORD) of a DNA solution upon heating
    (B) a maximum rate of denaturation versus temperature for duplex DNA
    (C) an increase in the absorbance of light at 260 nm upon denaturation of DNA
    (D) an increase in the absorbance of light at 260 nm when DNA-RNA hybrids are annealed
    (E) none of the above

8. The melting temperature (T_m) of DNA is

    (A) directly proportional to the length of the DNA chain
    (B) directly proportional to the GC content
    (C) directly proportional to the AT content
    (D) not related to the base composition
    (E) the same for all eukaryotes

9. Which of the following statements concerning characteristics of histones is true?

(A) They are acidic proteins found in the nucleus
(B) They are covalently linked to single-stranded DNA
(C) They are very highly conserved proteins
(D) They are large polymeric proteins
(E) They are associated with both prokaryotic and eukaryotic DNA

10. All of the following statements regarding nucleosomes are true EXCEPT

(A) they are beaded structures, each with a histone core wrapped in a left-handed super-helix of DNA
(B) the beads are linked by a DNA chain
(C) each contains two molecules of histones H2A, H2B, H3, and H4
(D) the DNA has a packing ratio of about 7 in the nucleosome
(E) the primary structure of the histones is a characteristic of the cell type and the species

11. Highly repetitive sequences in DNA are characterized by

(A) 2 to 10 base-pairs repeated in tandem 10^5 to 10^7 times in a genome
(B) a slowness to reanneal as compared to unique sequences
(C) unique sequences transcribed together
(D) uniform distribution in the chromosome
(E) their absence in human DNA

Directions: Each question below contains four suggested answers of which **one or more** is correct. Choose the answer

A   if **1, 2, and 3** are correct
B   if **1 and 3** are correct
C   if **2 and 4** are correct
D   if **4** is correct
E   if **1, 2, 3, and 4** are correct

12. Correct statements regarding nucleotide structure include which of the following?

(1) Removal of the 5′-phosphate group from guanylic acid (GMP) yields guanine
(2) Removal of the 5′-phosphate group from adenylic acid (AMP) yields adenosine
(3) Removal of the ribose 5-phosphate from thymidylic acid (TMP) yields uracil
(4) Removal of the ribose 5-phosphate from cytidylic acid (CMP) yields cytidine

13. Correct statements concerning heterochromatin include which of the following?

(1) It is not easily transcribed
(2) It occurs in the centromeres of chromosomes
(3) It is associated with histone H1
(4) It is relatively dispersed and stains poorly

Directions: The group of questions below consists of lettered choices followed by several numbered items. For each numbered item, select the one lettered choice with which it is most closely associated. Each lettered choice may be used once, more than once, or not at all. Choose the answer

A   if the item is associated with **(A) only**
B   if the item is associated with **(B) only**
C   if the item is associated with **both (A) and (B)**
D   if the item is associated with **neither (A) nor (B)**

Question 14–17

For each structural characteristic listed below, select the type of DNA in which it occurs.

(A) Genomic prokaryote DNA
(B) Genomic eukaryote DNA
(C) Both
(D) Neither

14. Complexed with histones
15. Often circular (i.e., without free ends)
16. Duplex, consisting of two antiparallel chains
17. Duplex, consisting of two parallel chains

ANSWERS AND EXPLANATIONS

1. The answer is B. (*II A 1 a; IV A 3 a*) Adenine (A) always pairs with thymine (T), and guanine (G) always pairs with cytosine (C). By convention, the base sequence of the segment of DNA is written in the $5'$ to $3'$ direction. The chain complementary to the one in question can be written in the $3'$ to $5'$ direction as

$$3'\text{-OH-GpTpCpApApTpCpGp}\cdots 5'$$

and then rewritten in the $5'$ to $3'$ direction,

$$\text{pGpCpTpApApCpTpG-}3'\text{-OH.}$$

2. The answer is D. (*III B 1 a, b*) The first solid evidence that DNA is the genetic material came in 1944 from the work of Avery, McLeod, and McCarthy on the transforming principle of pneumococci. These scientists transformed nonpathogenic pneumococci (R forms) into pathogenic forms (S forms) using purified extracts obtained from heat-killed pathogenic organisms. They then showed unequivocally that the transforming activity was due to DNA.

3. The answer is C. (*IV A*) Duplex DNA molecules are formed from two antiparallel, complementary, polydeoxyribonucleotide chains wound around one another with the purine and pyrimidine bases on the inside of the helix and the deoxyribose and phosphates on the outside. Because linear DNA molecules are antiparallel, the $3'$-hydroxyl groups lie at opposite ends of each molecule. The base sequences complement one another in terms of Watson-Crick base-pairing (adenine with thymine and guanine with cytosine); therefore, they are not identical in sequence. Duplex formation is made possible by hydrogen bonding between base-pairs, but the stability of the double helix owes much to the hydrophobic bonds between the stacked bases that lie perpendicular to the helix axis, which is composed of deoxyribose-phosphate chains. The bases are attached to the deoxyribose by glycosidic bonds.

4. The answer is D. (*IV A 3*) Chargaff's rules apply to the base composition of duplex DNA and state that in nearly all duplex DNA studied, the content of A equals the content of T and the content of G equals the content of C. This is not true for RNA, nor is it true for single-stranded DNA. As A = T and G = C, then A/T = G/C, but all four bases are not necessarily present in equal proportions; therefore, (A + T)/(G + C) = 1 and A= G = T = C are not generally true for DNA.

5. The answer is C. (*IV A 3, B 1; V A 3*) The mole fraction of any one of the four standard bases (A, G, T, C) in duplex DNA establishes the content of each of the other three. Because A = T and G = C in all duplex DNAs, the sample cannot be a duplex molecule. The DNA could have highly repetitive sequences, but this conclusion cannot be drawn from the data given. Melting of DNA refers to the dissociation of duplex DNA and does not apply to single-stranded DNA.

6. The answer is C. (*IV A 3, B 1*) Since the mole fraction of C equals the mole fraction of G in duplex DNA, G + C = 40%. The remaining 60% must be divided equally between A and T because the mole fraction of A equals the mole fraction of T in duplex DNA (Chargaff's rules). Thus, the adenine content must be 30% of the total bases.

7. The answer is C. (*V A 3 b*) The bases in duplex DNA are sequestered inside the double helix away from water. Interaction between the stacked, planar ring structures reduces their absorbance of ultraviolet light at 260 nm. Upon denaturation, the bases lose interaction with one another and the absorbance of light at 260 nm increases. This is the hyperchromic effect of denaturation.

8. The answer is B. (*V A 3 c*) Dissociation of the two chains of duplex DNA occurs at a definite increased temperature and is thus referred to as melting. The melting temperature (T_m) of DNA, if measured under standard conditions of pH and salt concentration, varies linearly with GC content; it is directly proportional to GC content, whereas it varies inversely with AT content, which declines as GC increases (because A + T + G + C = 1). The T_m of DNA is not directly proportional to the length of the DNA chain, nor is it the same for all eukaryotes.

9. The answer is C. (*VI D 1 c*) Histones are small, highly basic proteins. They have amino acid sequences that are the most conserved known. They are associated with eukaryotic DNA with which they form complexes by noncovalent bonds in structures called nucleosomes. Prokaryotic DNA is not associated with histones.

10. The answer is E. (*VI D 2*) Nucleosomes consist of a histone core with 140 base-pairs of DNA wrapped around it in a left-handed superhelix. The histone core is made up of two molecules each of the highly conserved histones H2A, H2B, H3, and H4. Nucleosomes are joined by the linker DNA, that is, the parts of the DNA chain not wound around the nucleosome core. The histone H1 is associated with the linker DNA in heterochromatin. The nucleosome represents the first step in the condensation of DNA and shows a packing ratio of about 7. Eventually the condensation has to reach a packing ratio of about 10^4, as seen in the chromosome.

11. The answer is A. (*VII B*) Eukaryotic genomes contain several classes of DNA. One DNA class, which reanneals at a much faster rate than the others, shows sequences that are highly repetitive. Generally 2 to 10 base-pairs long, these sequences are repeated in tandem 10^5 to 10^7 times per genome. They are clustered in the centromere region of the chromosome and are not transcribed.

12. The answer is C (2, 4). (*II; Figure 1-1*) Guanylic, adenylic, thymidylic, and cytidylic acids are nucleotides and consist of a base covalently bonded to the C-1 of ribose and a phosphate group bound to the C-5 of ribose. Removal of the phosphate group yields the base + sugar, which is a nucleoside. The nucleosides corresponding to the above four nucleotides are guanosine, adenosine, thymidine, and cytidine. Removal of the ribose 5-phosphate from a nucleotide leaves the free base. For the four nucleotides above, the corresponding bases are guanine, adenine, thymine, and cytosine.

13. The answer is A (1, 2, 3). (*VI D 3 b*) Heterochromatin stains darkly and is found mainly in the centromere region of the chromosome. Histone H1 is associated with heterochromatin. The DNA in heterochromatin cannot be easily transcribed. Euchromatin, which is relatively dispersed and stains poorly, is probably transcribed into RNA.

14–17. The answers are: 14-B, 15-A, 16-C, 17-D. (*IV A 1, E 1, 2; VI A 2*) The DNA of the genome is always complexed with histone in eukaryotic cells but not in prokaryotic cells. Both eukaryotic and prokaryotic DNAs are duplex, consisting of two antiparallel complementary strands. Duplex DNA cannot be formed by two parallel strands because in polymerization, the 5′-phosphate of the incoming nucleotide reacts with the 3′-hydroxyl of the previously positioned nucleotide.

2
Replication of DNA

I. SEMICONSERVATIVE REPLICATION

A. Watson-Crick proposal. The Watson Crick model of DNA suggests that because one strand of DNA is the complement of the other (due to specific base-pairing between A and T and between G and C), upon unwinding of the double helix, each strand acts as a template for the formation of a new strand.

  1. Semiconservative replication. The process of unwinding of the double helix results in two double-helical daughter molecules, each of which is composed of a parental strand and a newly synthesized strand, formed from the complementary strands. This type of replication is called semiconservative.

  2. Conservative replication. In conservative replication, the parental strands never completely separate. Thus, after one round of replication, one daughter duplex contains only parental strands and the other, only daughter strands.

B. Meselson-Stahl demonstration. Meselson and Stahl demonstrated experimentally that the replication of *Escherichia coli* DNA was indeed semiconservative.

  1. *E. coli* was grown in a medium containing a nitrogen (N) source with an atomic mass of 15 instead of the usual mass of 14.

  2. The *E. coli* was grown for several generations until all of the nitrogen in the DNA was ^{15}N, and the presence of the isotope of high atomic mass had made the DNA of the bacteria significantly denser than normal.

  3. The ^{15}N medium then was replaced with usual medium containing ^{14}N, and the *E. coli* was grown for several more generations. During the growth of the ^{15}N *E. coli* in the ^{14}N medium, cells were harvested and their DNA extracted and centrifuged in cesium chloride to provide a density gradient that separated DNAs of different densities.
     a. After the first generation in the ^{14}N medium, **all** of the DNA was the same hybrid density (i.e., of medium density halfway between ^{14}N and ^{15}N DNA).
     b. After the second generation, 50% was hybrid-density DNA, and 50% was ^{14}N (light) DNA.
     c. After the third generation, 25% was hybrid-density DNA, and 75% was light DNA.
     d. This pattern proved that replication was semiconservative (Fig. 2-1).

C. Applicability to single-stranded DNA and RNA

  1. Single-stranded DNA viruses (e.g., phage φX174, which infects *E. coli*) are single-stranded for only part of their life cycle. They have a replicative form that is double-stranded and that serves as the template for the biosynthesis of the DNA of the progeny virus.

  2. Single-stranded RNA viruses also duplicate by using a replicative form that is double-stranded.

II. GEOMETRIC PROBLEMS IN SEMICONSERVATIVE REPLICATION

A. Theta replication. Most prokaryotic DNA replicates when it is in a circular form (i.e., in a molecule with no free ends). Such replication is called theta replication (Fig. 2-2). The replicating circular DNA is sometimes referred to as a theta structure or a **Cairns molecule** (after John Cairns who first described it).

B. Eukaryotic chromosomal DNA. DNA in eukaryotic chromosomes probably exists in looped domains similar to the situation in some bacterial DNA. Here, as in circular DNA, the ends of the domain are fixed (Fig. 2-3).

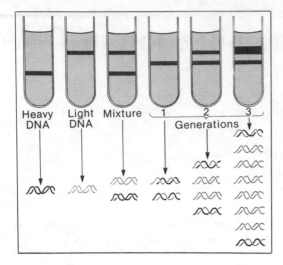

Figure 2-1. The results of the Meselson-Stahl experiment demonstrating semiconservative replication of *Escherichia coli* DNA are represented schematically. The tubes in which the equilibrium centrifugation conditions have been established are represented at the *top* and contain the bands of DNA with indicated densities. The parent DNA strands containing the heavy isotope of nitrogen (^{15}N) and the strands isolated from daughter cells grown in the presence of the naturally occurring light isotope (^{14}N) are represented in the *lower part* of the figure. A band of DNA with an intermediate density and its persistence through three generations with the subsequent appearance of totally light DNA confirm the semiconservative nature of DNA replication in prokaryotes. (Reprinted with permission from Lehninger AL: *Biochemistry*, 2nd ed. New York, Worth Publishers, 1975, p 893.)

C. Double-stranded (duplex) DNA replication. The double helix of DNA has to be unwound into single-stranded DNA in order to replicate. An understanding of the events taking place in the unwinding of duplex DNA requires an understanding of **supercoiling (superhelicity)** [Fig. 2-4].

1. Naturally occurring DNA is **underwound** (see Fig. 2-4*B*).

2. DNA replication promotes positive supercoiling (see Fig. 2-4C) on the strands of the DNA molecule as the replication fork opens (see section IV B 1).

3. The supercoiling constraints are relieved by enzymes called **DNA topoisomerases**, which remove superhelicity and produce **negative superhelicity** (negative supercoiling or underwinding). **DNA gyrase** (Eco topoisomerase II) is the enzyme responsible for relieving supercoiling in *E. coli* DNA replication.

III. MAJOR ENZYME ACTIVITIES INVOLVED IN DNA REPLICATION.
While the formation of a sister DNA strand (i.e., a complementary copy) on an existing DNA strand (i.e., the template) is the province of DNA polymerase enzymes, the replication of the DNA genome in cells requires the participation of other enzymes, including RNA polymerases to form RNA primers, a DNA ligase to join duplex DNA strands, and enzymes that are needed in the unwinding of duplex DNA helices.

A. DNA polymerases

1. **Prokaryotic DNA polymerase** exists in multiple forms: **DNA pol I**, **DNA pol II**, and **DNA pol III**.

   a. **DNA pol I**

   (1) DNA pol I, also known as **Kornberg's enzyme**, was the first DNA polymerase to be discovered. To carry out polymerization, DNA pol I requires

   (a) A template to copy

   (b) A primer of RNA or DNA, which is H-bonded to the template

   (c) A free 3'-hydroxyl (OH) on the growing strand

   (d) All four nucleoside triphosphates

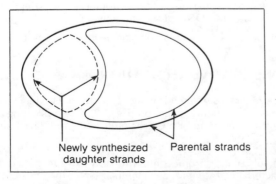

Newly synthesized daughter strands

Parental strands

Figure 2-2. Diagram of the *Escherichia coli* circular chromosome during replication showing the theta structure. (After Stryer L: *Biochemistry*, 2nd ed. New York, WH Freeman, 1981, p 582.)

(2) DNA pol I can copy long stretches of primed single-stranded DNA, which DNA pol III has difficulty doing.

(3) DNA pol I is a single-chain (109 kdal) **multifunctional protein**, performing several distinct enzymatic activities in addition to its polymerizing function.

    (a) Editing function. DNA pol I catalyzes the hydrolytic removal of 3′-terminal nucleotides (i.e., it has 3′- to 5′-exonuclease activity).

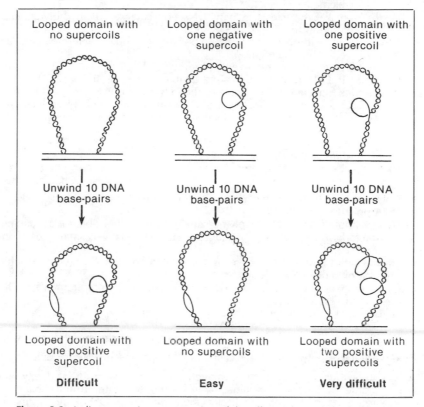

Figure 2-3. A diagrammatic representation of the effects of superhelical tension on a helix opening in a looped domain in bacterial DNA. If the looped domains are negatively supercoiled, the initiation of RNA synthesis is easier than it would be if the domains were not negatively supercoiled and much easier than if they were positively supercoiled. Similar looped domains are believed to be present in eukaryotic DNA. (Reprinted with permission from Alberts B, et al: *Molecular Biology of the Cell*. New York, Garland, 1983, p 447.)

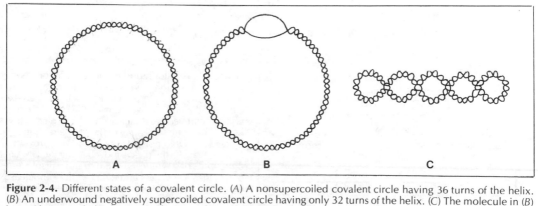

Figure 2-4. Different states of a covalent circle. (A) A nonsupercoiled covalent circle having 36 turns of the helix. (B) An underwound negatively supercoiled covalent circle having only 32 turns of the helix. (C) The molecule in (B) but with 4 superhelical turns to eliminate the underwinding. In solution, (B) and (C) would be in equilibrium; the equilibrium would shift toward (B) with increasing temperature. (Reprinted with permission from Freifelder D: *Molecular Biology*. Boston, Jones and Bartlett, 1983, p 132.)

        (i) It will add on an incoming nucleotide only if the base will pair with the template base in a Watson-Crick base-pair relationship. An incorrectly paired nucleotide is hydrolyzed off by the 3′- to 5′-exonuclease activity.

        (ii) Analogues of purine and pyrimidine nucleotides can be incorporated into DNA only if they form Watson-Crick base-pairs with the template.

        (iii) The cleaved nucleotide must have a free 3′-OH but must not be part of a double helix.

    (b) 5′- to 3′-Exonuclease activity. DNA pol I catalyzes the hydrolytic removal of nucleotides from the 5′ end of DNA chains.

        (i) The hydrolyzed bond has to be part of a duplex DNA molecule, and it requires a 5′-phosphate group.

        (ii) It functions to remove RNA primers in DNA replication (see section IV E 1) and in DNA repair systems [see Chapter 5, section II A 1 b, 2 b (3)].

    (c) Nick translation. At a single-stranded break, or **nick**, DNA pol I can function as an exonuclease and a polymerase at the same time. As a 5′-phosphate nucleotide is removed, it can be replaced by the polymerase activity. However, DNA pol I cannot link a 3′-OH group with a 5′-monophosphate, so the nick is moved along the DNA in the direction of synthesis (Fig. 2-5).

    (d) Endonuclease cutting

        (i) DNA pol I will cut the base-pair that follows a 5′-phosphate unpaired segment.

        (ii) It can function this way in excision and repair (see Chapter 5, section II A).

  b. DNA pol II. A biologic function for DNA pol II has not been elucidated.

  c. DNA pol III

    (1) DNA pol III is the major polymerizing enzyme responsible for advancing the replication fork (see section IV F 3). It functions to fill in relatively small gaps of single-stranded DNA.

    (2) The active form of the enzyme is called DNA pol III holoenzyme, which consists of at least nine different proteins.

    (3) To carry out polymerization, DNA pol III has all the requirements of DNA pol I [see section III A 1 a (1)].

    (4) DNA pol III has polymerase and 3′- to 5′-exonuclease activities but no 5′- to 3′-exonuclease activity [see section III A 1 a (3) (b)].

    (5) DNA pol III also performs an editing function [see section III A 1 a (3) (a)].

2. Eukaryotic DNA polymerases are called α, β, γ and σ (see section V D).

3. All DNA polymerases catalyze the formation of DNA chains by adding deoxyribonucleotides, one at a time, to the 3′-OH terminus of an existing chain (the **primer chain**).

  a. This reaction is a nucleophilic displacement with the 3′-OH of the terminal nucleotide at the growing end of the DNA chain attacking the electrophilic α-phosphorus (inner) atom of the incoming nucleoside triphosphate. Pyrophosphate is the leaving group, and a phosphodiester link is formed (Fig. 2-6).

  b. The pyrophosphate product is rapidly hydrolyzed by cellular pyrophosphatase, which drives the reaction forward.

  c. As the incoming nucleotides are added onto the 3′-OH group of the existing chain, the polymerization proceeds in the 5′ to 3′ direction.

B. DNA ligase

  1. DNA polymerases cannot connect two DNA chains, nor can they close a single chain, pro-

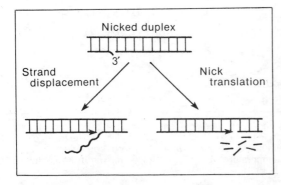

Nicked duplex

Strand displacement

Nick translation

3′

Figure 2-5. An illustration of the function of DNA pol I in nick translation and strand displacement. In the former instance, DNA pol I functions both as an exonuclease and a polymerase. As each 5′-phosphate nucleotide is removed (exonuclease), it is replaced by the nucleotide complementary to the template strand (polymerase). As DNA pol I cannot link a 3′-hydroxyl with a 5′-phosphate so the nick is moved along the DNA in the direction of synthesis. In strand displacement, polymerization occurs at the nick with exonuclease action, and the growing strand displaces the parental strand, a process that may be important in genetic recombination. (Reprinted with permission from Freifelder D: *Molecular Biology.* Boston, Jones and Bartlett, 1983, p 273.)

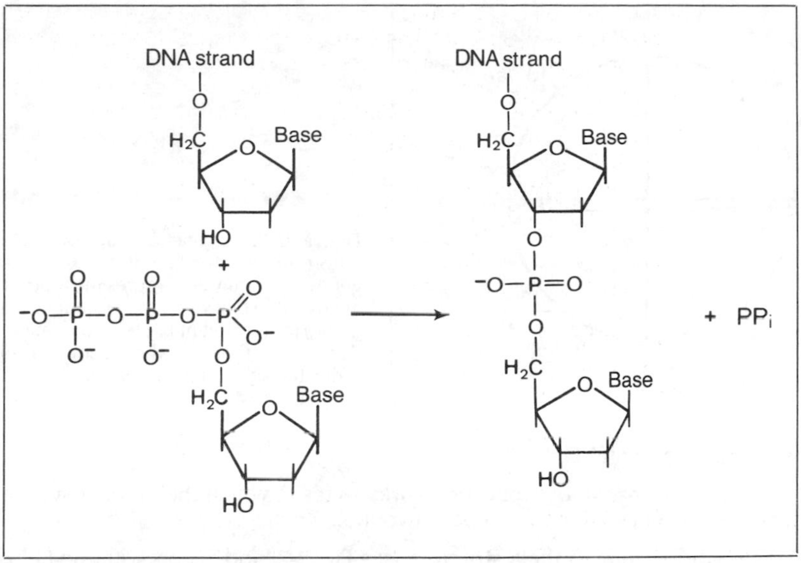

Figure 2-6. DNA polymerase catalyzes the formation of DNA chains.

cesses that must occur in order to form circular (closed-loop) DNA. This function is instead carried out by DNA ligase. To perform this function, ligase requires
  a. A free 3′-OH terminus on one chain and a 5′-phosphate at the end of the other chain
  b. That the chains it connects be part of duplex rather than single-stranded DNA molecules

  2. DNA ligase forms a phosphodiester bond between two chains in an endergonic reaction that requires a source of energy.
  a. In bacteria, the energy source is the hydrolysis of the pyrophosphate bond in nicotin amide-adenine dinucleotide (NAD^+).
  b. In animal cells and phages, the energy source is the hydrolysis of the high-energy phosphate bond in adenosine triphosphate (ATP).
  c. DNA ligase reacts with either NAD^+ or ATP to form a complex with adenosine monophosphate (AMP). The AMP is covalently bonded to the ε-amino group of a lysine residue on the enzyme (a **phosphoamide bond**).
  d. The AMP moiety activates the 5′-phosphate terminal of the DNA chain, which is then subjected to a nucleophilic attack by the 3′-OH group of the other DNA chain.
  e. A phosphodiester bond is formed, and AMP is released.
  f. As in polymerization by DNA polymerases, the reaction is driven by the hydrolysis of the pyrophosphate, which is derived from the formation of the enzyme–adenylate complex.

  C. Primase is a specific RNA polymerase that synthesizes short strands of RNA (4 to 10 bases long) on the **lagging strand** of duplex DNA (see section IV D 2 b). It does not require a primer but does require duplex DNA and all four ribonucleoside triphosphates. RNA polymerase (see Chapter 3, section I B) synthesizes short stretches of RNA primers on the **leading strand**.

  D. Other proteins. In addition to DNA pol I and III, DNA ligase, primase, and RNA polymerase, approximately 10 other proteins are needed to carry out DNA replication in *E. coli*.

IV. REPLICATION PROCESS IN *E. COLI*

  A. Initiation point. Replication begins at a specific initiation point.

  1. This is a unique sequence of bases (called **ori**) for each organism.

  2. It is normally a site where (or near where) the DNA comes into close proximity with an infolding of the cell wall (the replication process is synchronized with events preparing for cell fission).

  3. The initiation point is known with great precision in terms of the genetic map of the *E. coli* chromosome. The process starts at 74′, a site near the *ilv* gene, and proceeds at equal speed in both directions, finishing near the *trp* gene opposite the initiation site (Fig. 2-7).

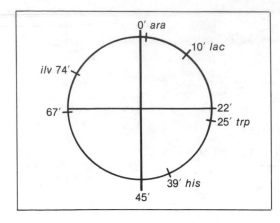

Figure 2-7. A genetic map of the *Escherichia coli* chromosome showing the position of a few of the genes. The numbers represent time in minutes for the genes transferred from a genetic donor bacterium (male) to a recipient bacterium (female) during conjugation. Total time for the genome is 90–100 minutes. OriC, the specific initiation point in *E. coli*, lies near the *ilv* gene at 74'.

B. Unwinding of parental DNA

1. Prior to the formation of **replication forks**—sites at which the nascent synthesis of DNA occurs—the parental DNA has to be unwound.

2. The unwinding forms a **theta structure** (see Fig. 2-2) and occurs with the help of **rep protein**, a helicase, which requires the energy of hydrolysis of two ATP molecules per base-pair broken.

3. Another protein, the **single-strand binding protein**, binds to the unwound DNA to prevent reannealing.

C. Polymerization

1. At the replication forks, daughter strands are formed on the parental strands with each of the daughter strands keeping pace with the replication fork (Fig. 2-8).

2. The parental strands run in antiparallel fashion; therefore, the direction for overall synthesis for one daughter strand would be expected to occur in the 3' to 5' direction, while for the other daughter strand, synthesis would be expected to occur in the 5' to 3' direction.

3. However, the fact that no DNA polymerase synthesizes in the 3' to 5' direction raised a problem, which was eventually resolved when Okazaki discovered that a considerable part of newly synthesized DNA is made in the form of discontinuous small fragments (**Okazaki fragments**).
   a. Lagging strand. These fragments are synthesized in the 5' to 3' direction and are then joined together by DNA ligase to form one daughter strand, referred to as the lagging strand.
   b. Leading strand. The other daughter strand is synthesized more or less continuously in the 5' to 3' direction and is referred to as the leading strand.

D. Synthesis of primer

1. The primer for DNA synthesis in *E. coli* is a small (60 bases) piece of RNA, not DNA.

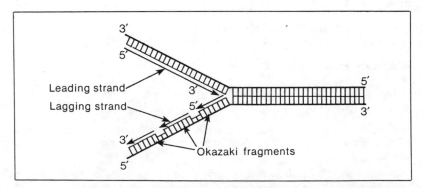

Figure 2-8. Diagram of the replication fork. (After Alberts B et al: *Molecular Biology of the Cell*. New York, Garland, 1983, p 224.)

2. Two enzymes synthesize the primers:
   a. RNA polymerase (see Chapter 3, section I B) on the leading strand
   b. Primase, the product of the *dnaG* gene, on the lagging strand

3. The protein product of the *dnaB* gene binds to the lagging strand, allowing primase to synthesize the primer.

E. **Joining precursor strands** is mainly a problem in the lagging strand synthesis. There are three steps:

   1. Removal of the RNA primer by DNA pol I

   2. Replacement of the primer by DNA pol I via **nick translation** [see section III A 1 a (3) (c)]

   3. Ligation of the replacement DNA to the copied DNA by DNA ligase

F. **Summary of events at the replication fork** (Fig. 2-9)

   1. **Helicase** unwinds the helix, using ATP hydrolysis as a source of energy.

   2. The **single-strand binding protein** binds to the single-stranded DNA to prevent reannealing.

   3. The leading strand advances by **DNA pol III holoenzyme** action, using an RNA primer laid down by RNA polymerase.

   4. The ***dnaB* gene product** binds to the lagging strand, allowing **primase** to synthesize an RNA primer on the lagging strand.

   5. Using the RNA primer on the lagging strand, DNA pol III holoenzyme synthesizes the daughter strand.

   6. DNA pol III holoenzyme stops when it hits the previous RNA primer.

   7. DNA pol I, via nick translation, removes the previous RNA primer and replaces it with DNA.

   8. The nick in the strand just synthesized by DNA pol I is repaired by DNA ligase to produce a continuous piece of DNA.

V. REPLICATION OF DNA IN EUKARYOTES

A. **Semiconservative replication.** Eukaryotic DNA is replicated semiconservatively and is distributed semiconservatively between sister chromatids.

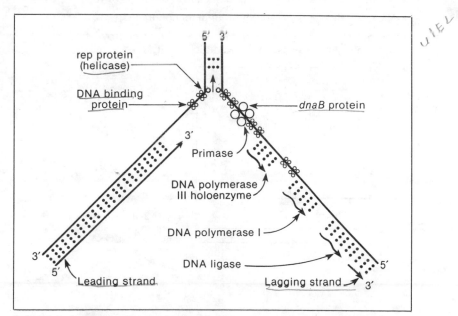

Figure 2-9. Schematic diagram of the enzymatic events at the replication fork of *Escherichia coli*. The events shown include duplex unwinding, chain initiation, elongation, and ligation. (Reprinted with permission from Stryer L: *Biochemistry*, 2nd ed. New York, WH Freeman, 1981, p 586.)

B. Multiple replication forks

1. Replication is achieved by the bidirectional movement of a large number of replication forks (Fig. 2-10).

2. A great many replication forks are needed because the eukaryotic DNA sequences are long and the replication forks move more slowly than those of prokaryotes.

3. At the replication forks, the leading strand is synthesized continuously, and the lagging strand is synthesized discontinuously, as in prokaryotes.

C. Histone separation

1. Histone separation from DNA, as well as unwinding of the duplex DNA, must occur in eukaryotic DNA replication.

2. During replication, the existing histones become associated with the duplex containing the leading daughter strand.

3. Newly synthesized histones become associated with the duplex containing the lagging daughter strand.

D. Eukaryotic DNA polymerase. Eukaryotic cells contain four different types of DNA polymerase: α, β, γ, and σ. Eukaryotic DNA polymerases, with the exception of DNA pol σ, have no exonuclease activities. Proofreading functions, such as the 3′- to 5′-exonuclease action, are carried out by accessory proteins that are specific for each form of the polymerase.

1. DNA pol α

a. DNA pol α functions in replication of nuclear DNA.
b. It may exist in several forms: α_1, α_2, and α_3.
c. It is induced maximally in the mid–S-phase of the cell cycle and can also be induced by small viral DNAs such as polyoma or simian virus 40 (SV40).
d. It is specifically inhibited by **aphidicolin**. As inhibition of DNA pol α prevents DNA replication, this inhibitor is a potential antitumor growth agent.
e. DNA pol α has many accessory proteins that enhance, and thus regulate, its activity. These proteins are tissue-specific.
f. Another accessory protein is a binding protein for Ap4A (adenosine-5′$_{pppp}$-5′-adenosine). During cell proliferation, Ap4A increases more than 1000 times and may have important primer or regulatory functions. (Either RNA or DNA can prime eukaryotic DNA synthesis.)

2. DNA pol β functions in repair of DNA and is constitutively expressed.

3. DNA pol γ functions in replication of mitochondrial DNA.

4. DNA pol σ seems to be expressed only in bone marrow. It is the only eukaryotic DNA polymerase that also has a deoxyribonuclease (DNase) activity.

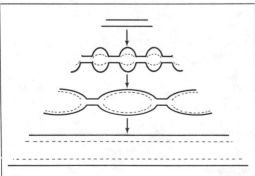

Figure 2-10. Schematic representation of multiple replication forks in eukaryotic DNA. The *dotted lines* represent daughter DNA strands.

STUDY QUESTIONS

Directions: Each question below contains five suggested answers. Choose the **one best** response to each question.

1. *Escherichia coli* cells that had been cultured in a light medium containing ^{14}N-ammonium chloride as the sole nitrogen source were transferred to a heavy medium containing ^{15}N-ammonium chloride. After three generations of growth in the heavy medium, what will be the proportion of dual light strands (LL), dual heavy strands (HH), and hybrid light-plus-heavy strands (LH) in the DNA duplexes?

(A) 3 LH to 1 HH
(B) 3 HH to 1 LH
(C) 3 HH to 1 LH/4 LL
(D) 15 LL to 1 LH
(E) 7 HH to 1 LH

2. Which of the following statements regarding the supercoiling of duplex DNA is correct?

(A) DNA replication promotes positive supercoiling downstream from the replication fork
(B) DNA polymerase III can replicate positively supercoiled duplex DNA
(C) DNA topoisomerases unwind duplex DNA into single-stranded DNA chains
(D) DNA topoisomerases unwind positively supercoiled duplex DNA without cutting the phosphosugar backbone
(E) Negatively supercoiled DNA is very unstable

3. All of the following statements characterize DNA replication EXCEPT

(A) RNA and DNA chains are linked covalently
(B) growth of the new DNA chain occurs in the 5′ to 3′ direction
(C) growth of the DNA chain is discontinuous
(D) overall growth of the DNA chain is bidirectional
(E) DNA is synthesized in a 3′ to 5′ direction on one parental strand and in a 5′ to 3′ direction on the other

4. Which of the following statements about DNA polymerase I is correct?

(A) It functions as a DNA repair enzyme but is not involved in the DNA replication process
(B) It requires a template and a primer to polymerize deoxyribonucleoside triphosphates
(C) It joins together Okazaki fragments to complete the lagging strand during DNA replication
(D) It produces Okazaki fragments linked to RNA primer chains
(E) Its presence is not necessary for normal growth in some bacterial mutants

5. DNA ligase of *E. coli* requires which of the following cofactors?

(A) Flavin-adenine dinucleotide as an electron acceptor
(B) Nicotinamide-adenine dinucleotide phosphate as a phosphate donor
(C) Nicotinamide adenine dinucleotide (NAD$^+$) to form an active adenyl enzyme
(D) NAD$^+$ as an electron acceptor
(E) None of the above

6. Short chains of nucleic acid can be isolated from cells in which DNA is undergoing replication. These segments, known as Okazaki fragments, have which of the following properties?

(A) They are double-stranded
(B) They contain covalently linked RNA and DNA
(C) They are DNA-RNA hybrids
(D) They arise from the nicking of the sugar-phosphate backbone of the parental DNA chain
(E) They are removed by nuclease activity

7. Which of the sequences listed below best describes the order in which the following enzymes participate in the replication of DNA in bacteria?

> 1 = DNA polymerase I
> 2 = 5′-Exonuclease
> 3 = DNA polymerase III
> 4 = DNA ligase
> 5 = RNA polymerase (primase)

(A) 5,3,1,2,4
(B) 3,2,1,5,4
(C) 5,3,4,2,1
(D) 5,3,2,1,4
(E) 3,2,5,1,4

8. Which of the following activities is typical of histones during replication of eukaryotic DNA?

(A) They remain bound to the DNA throughout replication
(B) They undergo proteolysis and are replaced by newly synthesized protein on each daughter strand
(C) They are both conserved and associated with the leading daughter strand and newly synthesized and associated with the lagging daughter strand
(D) They are not synthesized for cell division but are retained from generation to generation
(E) They do not become reassociated with newly synthesized DNA until metaphase

Directions: The question below contains four suggested answers of which **one or more** is correct. Choose the answer

A   if **1, 2, and 3** are correct
B   if **1 and 3** are correct
C   if **2 and 4** are correct
D   if **4** is correct
E   if **1, 2, 3, and 4** are correct

9. Correct statements describing DNA polymerase I include which of the following?

(1) It can add deoxyribonucleoside triphosphates to the 3′-hydroxyl end of a DNA primer chain
(2) It can act as an endonuclease in DNA repair
(3) It can act as a 5′- to 3′-exonuclease
(4) It is the major polymerizing enzyme in the replication process

ANSWERS AND EXPLANATIONS

1. The answer is B. (*I A, B*) The proportion of 3 HH to 1 LH is the basis of the Meselson-Stahl demonstration that DNA replication in bacteria is semiconservative; however, the order for growing the organism in light and heavy media is reversed from the original study. Semiconservative replication means that at each replicative event the new duplex genome contains one strand of parental DNA (conserved) and one strand of newly synthesized DNA.

2. The answer is A. (*II C*) Naturally occurring DNA is underwound, but as replication proceeds, the duplex DNA becomes positively supercoiled downstream from the replication fork. This would stop replication if the constraints were not removed. DNA topoisomerases relieve positive supercoiling and produce negative superhelicity by cutting, uncoiling, and repairing the phosphosugar backbone ahead of the replication fork.

3. The answer is E. (*III; V D*) DNA is replicated by a polymerase complex, which adds deoxyribonucleotide triphosphates to an RNA primer chain, selecting for insertion only the base-pairing nucleotide complementary to that in the parental DNA chain. The first deoxyribonucleotide inserted is bonded covalently to the primer nucleotide by a phosphodiester link. Growth of the chain is always in the 5′ to 3′ direction on both strands. Replication starts at a specific point and proceeds bidirectionally.

4. The answer is B. (*III A 1 a*) DNA polymerase I (DNA pol I) requires both a template and a primer in order to function as a polymerizing enzyme. Both 3′- and 5′-exonuclease activities are also integral parts of this single-peptide protein. It functions in normal DNA replication, hydrolyzing the stretches of primer chain RNA (5′- to 3′-exonuclease activity) and replacing them with template-directed deoxyribonucleotides (polymerizing function). DNA pol I is also involved in the repair of DNA. For example, it participates in the excision of thymine dimers, which is a 5′- to 3′-exonuclease activity. During normal DNA replication, DNA pol I exhibits 3′-exonuclease activity, excising incorrectly introduced deoxyribonucleotide residues. Okazaki fragments are produced by DNA pol III complex, not DNA pol I, and after the removal of the RNA primer chains, the fragments are linked by DNA ligase. DNA pol II, the function of which is unknown, is absent in some bacterial mutants, which, however, show normal growth characteristics. DNA pol I is required for normal growth.

5. The answer is C. (*III B*) DNA ligase can join DNA strands that have been nicked or close single DNA chains to form circular DNA molecules. The enzyme requires a free 3′-hydroxyl at the end of the one chain and a 5′-phosphate at the end of the other chain with both chains part of a duplex DNA molecule. The reaction is endergonic, and a source of energy is required. In *E. coli*, and in other bacteria, the energy source is the pyrophosphate bond in nicotinamide-adenine dinucleotide (NAD⁺). An enzyme–adenosine monophosphate (AMP) complex is formed with the release of the nicotinamide monophosphate moiety of NAD⁺, which activates the 5′-phosphate terminus of the DNA chain. The phosphodiester bond is formed and AMP released.

6. The answer is B. (*IV C 3*) Both complementary strands of DNA can be replicated by a mechanism that polymerizes only in the 5′ to 3′ direction if short stretches of DNA are synthesized on both strands in a 5′ to 3′ direction and are then joined into continuous chains. Okazaki first isolated newly formed DNA from cells undergoing replication and found it to consist of short pieces (Okazaki fragments). Using labeled precursors, Okazaki isolated labeled DNA fragments after denaturation (i.e., chain separation) of the cellular DNA. The fragments are thus single-stranded, and because DNA polymerase III complex uses RNA as a primer, the fragments can consist of both RNA and DNA, which are linked covalently. The RNA and DNA are not hybrids arising from complementarity in base sequence, nor do they arise by fragmentation of the parental chains. Far from being removed by nucleases, they are, in combination, the newly formed daughter chains.

7. The answer is D. (*IV C–F*) The RNA primer is first laid down on the separated DNA strands, where primase carries out this function on the lagging strand and RNA polymerase carries out this function on the leading strand. Short stretches of DNA (Okazaki fragments) are then formed on the template by the DNA polymerase III (DNA pol III) holoenzyme. The primer RNA is removed from both daughter strands by the 5′-exonuclease activity of DNA polymerase I (DNA pol I) and is replaced with deoxyribonucleotides by its polymerizing activity. Finally, the fragments are joined by the DNA ligase.

8. The answer is C. (*V C*) Histone separation from the DNA has to precede DNA replication. However, the old histone is not degraded but becomes associated with the leading daughter strand. Newly synthesized histone becomes associated with the lagging daughter strand.

9. The answer is B (1, 3). (*III A*) DNA polymerase I (DNA pol I) is a trifunctional enzyme with three enzymatic activities residing in a single polypeptide chain. It can add deoxyribonucleoside triphosphates to the 3′-OH of a DNA primer chain. It cannot nick the backbone of a DNA chain, as it is not an endonuclease, but it can act as an exonuclease, cleaving nucleotides from either the 3′ or the 5′ ends; that is, it has both 3′-and 5′-exonuclease activity. In DNA repair systems, DNA pol I functions as a 5′-exonuclease, removing the damaged sequence, and as a polymerase, forming the newly repaired sequence. In normal DNA replication, DNA pol I acts as a 3′-exonuclease in the removal of the RNA primer chains and again as a polymerase in the template-directed replacement of the RNA by DNA.

3

Transcription of DNA: RNA Synthesis

I. TRANSCRIPTION CYCLE

A. Definition. Transcription is the process by which the synthesis of RNA molecules is initiated and terminated.

B. RNA polymerase

1. In prokaryotes, a single complex enzyme, **RNA polymerase**, synthesizes all cellular RNA on DNA templates.

2. Requirements for this synthesis include
   a. A template of double-stranded DNA (or occasionally of single-stranded DNA)
   b. All four ribonucleoside triphosphates [i.e., adenosine triphosphate (ATP), guanosine triphosphate (GTP), uridine triphosphate (UTP), and cytidine triphosphate (CTP)]
   c. Mg^{2+} or Mn^{2+}

3. The first base laid down is in the triphosphate form (5′-PPP). The other bases lose a pyrophosphate (PP_i) group.

4. The reaction catalyzed by RNA polymerase is represented by

$$(RNA)_{n \text{ bases}} + \text{ribonucleoside triphosphate} \rightarrow (RNA)_{n+1} \text{ bases} + PP_i$$

   a. The hydrolysis of PP_i by pyrophosphatases drives the reaction to the right, as in the DNA polymerase reaction.
   b. Also, as in DNA polymerization, the 3′-hydroxyl (OH) group of the terminal nucleotide makes a nucleophilic attack on the α-phosphorus of the incoming nucleoside triphosphate (NTP).
   c. Synthesis is in the 5′ to 3′ direction.

5. In *Escherichia coli* one enzyme polymerizes all types of cellular RNA, but in mammalian cells several different enzymes are required to synthesize the different types of RNA.

6. The RNA polymerase of *E. coli* (465 kdal) is a complex enzyme.
   a. There are five subunits in the holoenzyme:
      (1) Two called α (36,500 kdal) coded for by the gene *rpoA*
      (2) One called β (150,000 kdal) coded for by the gene *rpoB*
      (3) One called β′ (160,000 kdal) coded for by the gene *rpoC*
      (4) One called σ (82,000 kdal) coded for by the gene *rpoD*
   b. The σ subunit specifically dissociates at one stage in transcription.
   c. The **core enzyme** refers to RNA polymerase without the σ subunit (i.e., $\alpha_2\beta\beta'$).

C. Stages of transcription. Transcription can be divided into four stages: **binding of polymerase to the template at specific sites**, **initiation of polymerization**, **chain elongation**, and **chain termination/release**.

1. **Binding of RNA polymerase to the DNA template**
   a. Promoters are specific regions contained in DNA that are recognized by RNA polymerase. The size of the promoter region is variable. In prokaryotes, the region ranges from 20–200 bases.
   b. The core enzyme ($\alpha_2\beta\beta'$) alone cannot recognize the promoter regions. The σ subunit is essential for this function.
   c. During binding of RNA polymerase to the template, the following sequence of events occurs.

(1) The σ subunit of RNA polymerase recognizes the promoter sequence.

(2) RNA polymerase attaches to the promoter region.

(3) RNA polymerase melts the helical structure and opens the DNA.

(4) RNA polymerase initiates RNA synthesis on the denatured single-stranded DNA template.

d. Characteristics of the promoter sequence

(1) **The Pribnow box** is a sequence contained within all prokaryotic promoter regions. It is located 5–10 bases to the left (upstream) from the first base that will be copied into RNA. It orients RNA polymerase as to the direction and start of synthesis (Fig. 3-1).

(a) All Pribnow boxes are variants of TATAATG.

(b) The T at position 6 (**conserved T**) is present in every promoter region.

(2) **The "−35" sequence** is a second recognition site in many promoter regions upstream from the Pribnow box. It is thought to be the initial site of σ subunit binding. Typically, it contains nine bases.

(a) RNA polymerase binds first to the upstream side of the "−35" sequence via a site recognized by the σ subunit.

(b) Because of its huge size, the RNA polymerase then comes into contact with the Pribnow box.

(c) Once bound to the Pribnow box, RNA polymerase dissociates from the initial recognition site (Fig. 3-2).

e. The open–promoter complex is an active intermediate in RNA chain initiation. In this complex, localized melting of the DNA has occurred, starting about 10 bases from the upstream side of the Pribnow box and extending to the first transcribed base.

2. Initiation of polymerization

a. The first base copied is always within six to nine bases of the conserved T of the Pribnow box.

b. RNA polymerase contains two specific sites for the binding of NTPs.

c. The first incoming NTP binds to RNA polymerase at the **initiation site** and H-bonds to the complementary base on the DNA within the open–promoter complex. This site binds only purine NTPs (most transcripts of DNA start with an A).

d. The second incoming NTP binds to the **elongation site** on the polymerase. The NTP is chosen on the basis of its ability to H-bond with the complementary base on the DNA.

e. RNA polymerase covalently bonds the first and second bases.

f. The first base dissociates from the initiation site, and initiation is then complete.

g. Inhibitors of initiation

(1) The antibiotic **rifamycin SV** binds to the β subunit of RNA polymerase, preventing the incoming NTP from binding to the initiation site.

(2) A related compound, rifampicin, allows the first dinucleotide to form but blocks the movement of the polymerase along the DNA. If, however, a third nucleotide has been added to the chain, rifampicin is without further effect.

(3) Both rifamycin and rifampicin are thus inhibitors of chain initiation but not of chain growth.

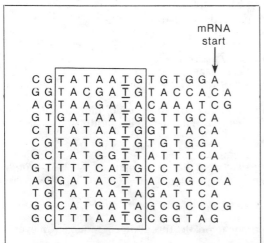

Figure 3-1. Segments of the noncoding strand from various genes showing the common sequence of seven bases (*boxed area*) known as the Pribnow box. The starting point for mRNA synthesis is marked with an *arrow*. The conserved T is *underlined*. (Reprinted with permission from Freifelder D: *Molecular Biology.* Boston, Jones and Bartlett, 1983, p 376.)

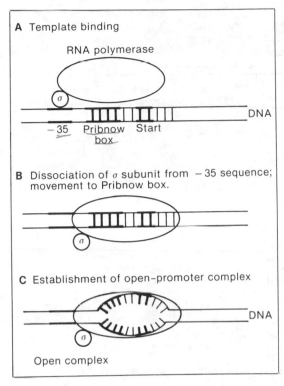

A Template binding

RNA polymerase

σ

DNA

−35   Pribnow   Start
      box

B Dissociation of σ subunit from −35 sequence; movement to Pribnow box.

σ

C Establishment of open-promoter complex

DNA

σ

Open complex

Figure 3-2. A proposed scheme for the binding of RNA polymerase to a promoter to form an open–promoter complex. Regions of the DNA molecule important for binding are indicated by *boxing lines*. The shape of the RNA polymerase is idealized for schematic purposes. The enzyme covers the region from bases − 45 to + 15, and the unpaired region in (C) extends from about − 12 to + 2. The enzyme is shown in contact with both strands because the strands are actually wrapped around one another in a helical array; however, true binding occurs only to bases in the coding strand. (Reprinted with permission from Freitelder D: *Molecular Biology*. Boston, Jones and Bartlett, 1983, p 379.)

3. **Chain elongation**
    a. Most of the **elongation** of the RNA chain is carried out by the **core enzyme**. The σ subunit dissociates after about eight bases have been polymerized.
    b. Elongation does not occur at a constant rate, as the synthesis slows down (**pauses**) during the transcription of certain regions of DNA molecules. These pauses tend to occur in GC-rich regions of the DNA.
    c. The DNA helix recloses after RNA polymerase transcribes through it, and the growing RNA chain dissociates from the DNA.

4. **Termination of synthesis**
    a. Termination of transcription is determined by specific sequences in the DNA.
    b. **Termination sites** contain three important regions.
        (1) An inverted-repeat base sequence contains a central nonrepeating segment (XYZ), for example: ABCDEF-XYZ-F'E'D'C'B'A' (the prime signs indicate the complementary base). This sequence can form a **stem-and-loop configuration** (Fig. 3-3).
        (2) A sequence high in GC is near the loop.
        (3) A sequence of As in the DNA that codes for six to eight U residues in the RNA is usually present. The U residues are followed by an A.
        (4) There is no unique base where transcription stops (e.g., for a given promoter some RNAs will end with 5 Us, some with 6 Us, some with 6 Us + 1 A).
    c. **Role of the rho protein.** At specific termination sites, the new RNA chain is released without the presence of a special protein. However, at other sites, the **rho protein** must participate in chain termination.
        (1) The rho protein binds very tightly to the RNA (not to the RNA polymerase, and only weakly to DNA), and in this bound state, it becomes an **ATPase**.
        (2) The rho protein probably has multiple functions in relation to termination of synthesis and to release of the RNA and RNA polymerase from the DNA. Its actions, however, are not well understood.

    5. **Summary of the transcription cycle in *E. coli*** is illustrated in Figure 3-4.

D. **Eukaryotic RNA polymerases.** Eukaryotic cells contain three classes of RNA polymerase. Each class synthesizes a different type of RNA.

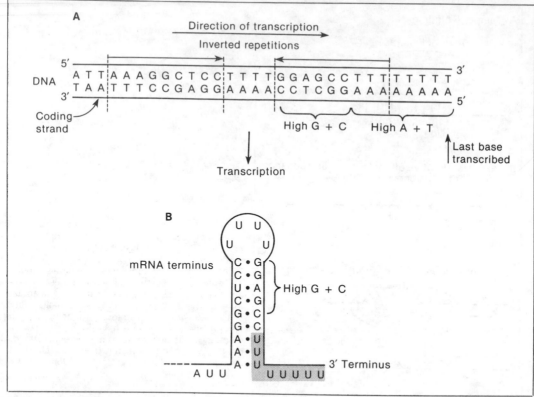

Figure 3-3. Base sequence of the DNA of *Escherichia coli trp* operon at which transcription termination occurs (A) and of the 3′ terminus of the mRNA molecule (B). The inverted-repeat sequence is indicated by the *reversed arrows*. The mRNA molecule is folded to form a stem and loop structure. The relevant regions are *bracketed* and labeled. The Us in the terminal sequence of the mRNA are *shaded*. (Reprinted with permission from Freifelder D: *Molecular Biology*. Boston, Jones and Bartlett, 1983 p 385.)

1. **RNA polymerase I (RNA pol I)**
   a. Is localized in the nucleolus
   b. Transcribes **ribosomal RNA** (rRNA, see section II C) genes

2. **RNA polymerase II (RNA pol II)**
   a. Is localized in the nucleoplasm
   b. Transcribes **messenger RNA** (mRNA, see section II A) genes
   c. Is strongly inhibited by α-amanitin (a mushroom poison)

3. **RNA polymerase III (RNA pol III)**
   a. Is localized in the nucleoplasm
   b. Transcribes **transfer RNA** (tRNA, see section II B) and 5S RNA genes
   c. Is less sensitive to inhibition by α-amanitin than pol II

II. CLASSES OF RNA MOLECULES

A. mRNA

1. **Introduction**
   a. DNA base sequences specify the amino acid sequence of every polypeptide in the cell.
   b. A multistep process allows the conversion of the information in the DNA into the amino acid sequence of a polypeptide.
   c. Transcription of the **coding (sense) strand** of the DNA into an RNA molecule (mRNA) is the first step.
   d. The **antisense strand** is the complementary DNA strand that is not read.
   e. A **cistron** is a DNA sequence that codes for a polypeptide chain together with start and stop signals for the protein synthetic machinery.

2. **Prokaryotic mRNA is polycistronic**; that is, it is a template for several polypeptide chains.

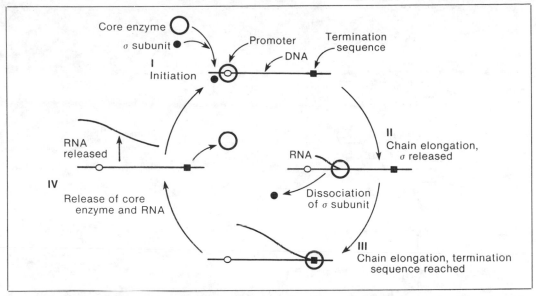

Figure 3-4. The transcription cycle of *Escherichia coli* RNA polymerase. *I*, The σ subunit must be present for initiation to occur. *II*, The σ subunit dissociates shortly after chain elongation begins. *III*, The core enzyme dissociates during termination. *IV*, The core enzyme and RNA are released until *I*, the holoenzyme, is reformed from the core enzyme and the σ subunit. Initiation of a second round of transcription does not use the same core enzyme and σ subunit used in the first round. (Reprinted from Freifelder D: *Molecular Biology*. Boston, Jones and Bartlett, 1983, p 388.)

  a. In polycistronic mRNA, the individual cistrons may be separated by intercistronic sequences called **spacers**.
  b. The 5′ end of the mRNA may contain a sequence that is never translated into protein. These are called **leader sequences** or **5′ untranslated regions**.
  c. A similar sequence, called the **3′ untranslated sequence**, may occur at the 3′ end of the mRNA.
  d. The lifetime of prokaryotic mRNA is short compared to that of eukaryotic mRNA and is often stable only for a few minutes.

 3. Eukaryotic mRNA
  a. Eukaryotic mRNA is **monocistronic**; that is, it is a template for the synthesis of a single polypeptide chain.
  b. Eukaryotic mRNAs are formed (processed) from nuclear RNA precursors, which are known as **heterogeneous nuclear RNA (hnRNA)** because they vary widely in size.
  c. The cap structure
   (1) The 5′ end of all eukaryotic mRNA is attached to a 7-methylguanosine by a 5′-5′-PP linkage (Fig. 3-5), a structure referred to as a **cap**. The adjacent riboses may be methylated at their 2′-OH positions.
   (2) Capping occurs before the RNA pol II leaves the initiation site of transcription and precedes all splicing events.
   (3) The functions of capping may be
    (a) Enhancement of the translational ability of mRNA
    (b) Protection of the mRNA from ribonuclease (RNase) action
  d. Poly A tail
   (1) Most eukaryotic mRNA has a **polyadenylic acid (poly A) tail**, consisting of 20–200 adenine nucleotides at the 3′ end.
   (2) Transcription usually reads through and passes a polyadenylation site before terminating. An endonuclease (an enzyme that cleaves phosphodiester bonds located in the interior of polynucleotides) then cleaves the mRNA 10–25 bases from an AAUAAA sequence. This sequence is thought to be necessary for polyadenylation to occur.
    (a) Polyadenylation occurs after capping but before splicing (see section II A 3 e).
    (b) Polyadenylation may increase the stability of mRNA and may enhance the attachment of mRNA to intracellular membranes.
   (3) Some mRNAs function without a poly A tail (e.g., histone mRNA), and some mRNAs have more than one polyadenylation site.

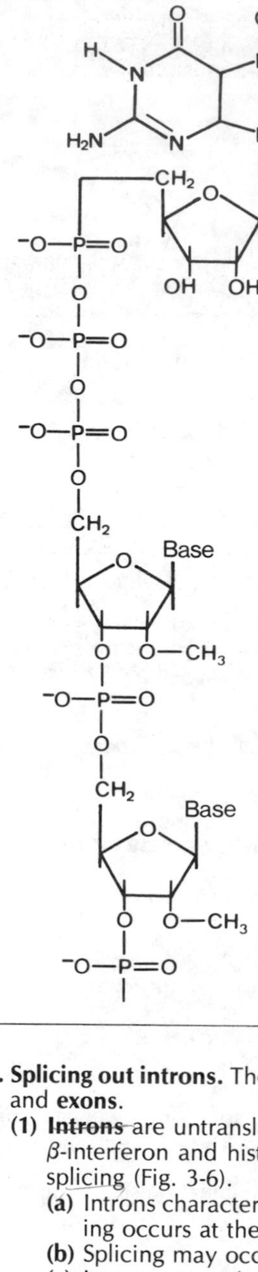

Figure 3-5. The cap structure found on the 5′ end of eukaryotic mRNA. The cap is a 5′-5′ pyrophosphate linkage of the 5′ end of eukaryotic mRNA to a 7-methylguanosine.

    e. Splicing out introns. The eukaryotic primary mRNA transcripts can be divided into **introns** and **exons**.

        (1) Introns are untranslated regions found in almost all eukaryotic mRNA genes (except β-interferon and histone genes). They have no known function and are removed by splicing (Fig. 3-6).

            (a) Introns characteristically have a GU at their 5′ end and an AG at their 3′ end. Splicing occurs at these sites in the primary transcripts.

            (b) Splicing may occur with the aid of a **small nuclear RNA** (snRNA) [Fig. 3-7].

            (c) Introns are excised one at a time, and the presence in the nucleus of all the splicing precursors accounts for the size diversity of hnRNA.

            (d) Transport of mRNA from the nucleus to the cytoplasm will not occur until all the splicing is complete.

        (2) Exons. The final product of splicing contains all the exons joined together. An exon is any part of the primary transcript that is retained in the final mRNA molecule. Exons contain the coding sequences of the spliced genes.

B. tRNA

    1. Introduction

        a. Transfer RNAs are adapter molecules designed to recognize both a specific amino acid and a specific three-base sequence on mRNA (see codon–anticodon interaction, Chapter 4, section V).

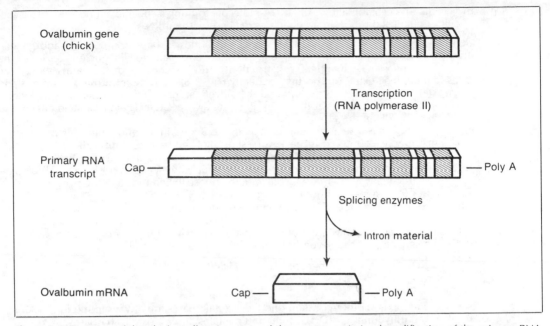

Figure 3-6. Structure of the chick ovalbumin gene and the post-transcriptional modification of the primary RNA transcript to give the functional ovalbumin mRNA. The *shaded regions* of the primary RNA transcript are those removed by splicing enzymes and are called introns. The *clear regions* (exons) are combined into the mature mRNA.

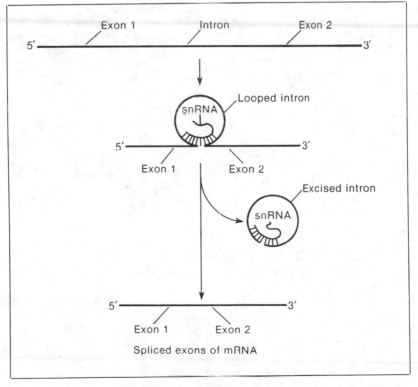

Figure 3-7. Function of a small nuclear RNA (snRNA) in splicing exons and excising introns. The ends of the intron base-pair with a segment of snRNA. The splicing reaction is accompanied by excision of the intron. (Reprinted with permission from Lehninger AL: *Principles of Biochemistry*. New York, Worth, 1983, p 861.)

b. This dual recognition allows the tRNA molecule to bring a specific amino acid into a specific sequential position in a peptide as defined by the mRNA.

2. General structure. Virtually all tRNA shares similarities in base composition and general structure. Both prokaryotic and eukaryotic tRNAs are approximately 80 nucleotides long. The basic two-dimensional **cloverleaf construction** of tRNAs consists of four arms with three loops, although an extra arm occurs frequently (Fig. 3-8). Starting from the 3′ terminus and progressing in a 3′ to 5′ direction around the cloverleaf, other common features include:

a. The last three nucleotides at the 3′ terminus are always CCA. The amino acid attaches to the 2′- or 3′-OH of the terminal A.

b. The next region is a segment that pairs with the 5′ end in a double helical structure.

c. The first loop encountered is the TψC (ψ stands for the base **pseudouridine**).

    (1) This base is formed after transcription, as are all the unusual bases in RNA.

    (2) Although pseudouridine is very like uridine in its chemical composition and base-pairing, structurally it is 5-ribosyluracil with the ribose linked to C-5 of the pyrimidine ring, rather than to N-1 as in uridine (Fig. 3-9).

    (3) Next to the pseudouridine is a thymine residue (ribothymine, not deoxyribothymine).

d. Continuing in the 3′ to 5′ direction, the next arm of the tRNA contains the **anticodon loop**.

    (1) The anticodon is the triplet of bases that pairs with the triplet codon on mRNA.

    (2) The sequence of the seven bases in the anticodon loop is always

$$5'\text{-pyr-pyr-X-Y-Z-modified pu-variable base-}3'$$

anticodon

e. The next loop is another constant one (i.e., certain bases are always found in tRNA). It is called the **dihydrouracil (DHU) loop** and contains at least two DHU bases.

f. The 5′ end of the tRNA molecule is usually a G, which is phosphorylated.

3. Structure of alanine and phenylalanine tRNA

a. Yeast alanine tRNA has the covalent structure shown in Figure 3-10.

    (1) There are a number of unusual bases in addition to pseudouridine (ψ), ribothymine (T), and DHU.

    (2) The anticodon triplet contains inosine (I), a common participant in anticodons, which has its own specific base-pairing properties (see Chapter 4, section V).

b. Yeast phenylalanine tRNA has the three-dimensional structure shown in Figure 3-11.

    (1) The molecule is **L-shaped**.

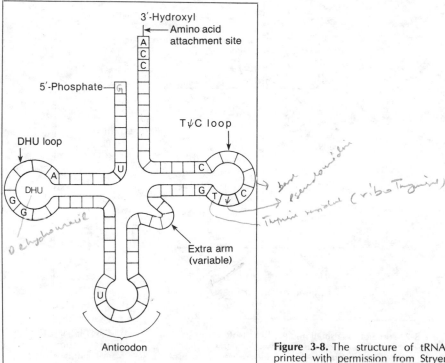

Figure 3-8. The structure of tRNA molecules. (Reprinted with permission from Stryer L: *Biochemistry*, 2nd ed. New York, WH Freeman, 1981, p 647.)

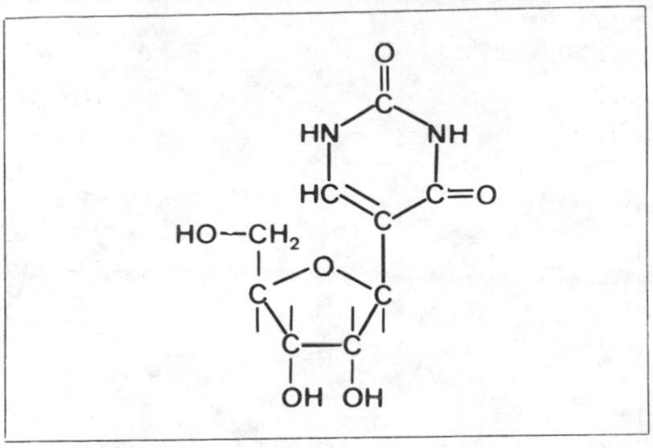

Figure 3-9. Structure of pseudouridine (ψ).

(2) The L-shape is achieved by folding over the TψC loop close to the DHU loop.
(3) The two loops interact strongly with each other through tertiary forces. These interactions keep the structure of tRNA in the L-shape.
(4) The L-shape is the conformation required for tRNAs to interact with aminoacyl-tRNA synthetases (see Chapter 4, section IV) and ribosomes.

4. **Synthesis of tRNA** begins on a promoter and terminates at a terminator sequence. In both prokaryotes and eukaryotes, a **primary tRNA transcript** may contain one or more tRNA sequences, which are transcribed together. These primary transcripts undergo considerable post-transcriptional modification (**processing**). Processing occurs in several stages (Fig. 3-12).
   a. **Formation of the 3$'$-OH terminus**
      (1) An endonuclease recognizes the hairpin loop 1 and makes a cut.
      (2) RNase D (an exonuclease, an enzyme that removes terminal nucleotides) removes seven bases from this terminus and stops two bases before a CCA sequence.
   b. **Formation of the 5$'$-phosphate terminus.**
      (1) RNase P (an endonuclease) removes the excess RNA from the 5$'$ end.
      (2) RNase P seems to recognize the three-dimensional shape of the tRNA rather than a specific sequence.
   c. **Final modification of the 3$'$ end.** RNase D now removes the two bases before the CCA and thus generates the correct amino acid acceptor site.

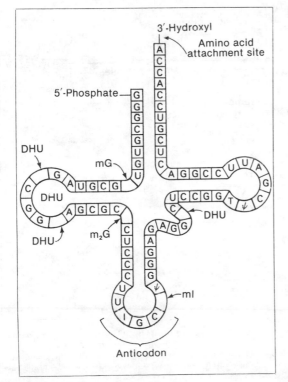

Figure 3-10. The covalent structure of yeast alanine tRNA. (Reprinted with permission from Stryer L: *Biochemistry*, 2nd ed. New York, WH Freeman, 1981, p 645.)

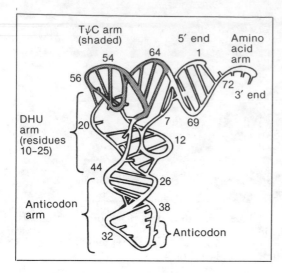

Figure 3-11. Three-dimensional conformation of yeast phenylalanine tRNA deduced from x-ray diffraction analysis at 0.3-nm resolution. (Reprinted with permission from Kim SH, et al: Three-dimensional tertiary structure of yeast phenylalanine transfer RNA. *Science* 185:435–440, 1974. ©1974 by the American Association for the Advancement of Science.)

   d. tRNA is highly and uniquely modified in terms of the unusual bases. In the $tRNA_1{}^{tyr}$ gene, for example:

      (1) Two Us are changed to pseudouridines (ψ)

      (2) Two Us are changed to 4-thiouridines (4tU)

      (3) One G is changed to 2′-O-methyl G (2mG)

      (4) One A is changed to isopentyl A (2ipA)

5. Eukaryotic tRNA

   a. Primary tRNA transcripts (**pre-tRNA transcripts**) may also contain one or more tRNA sequences.

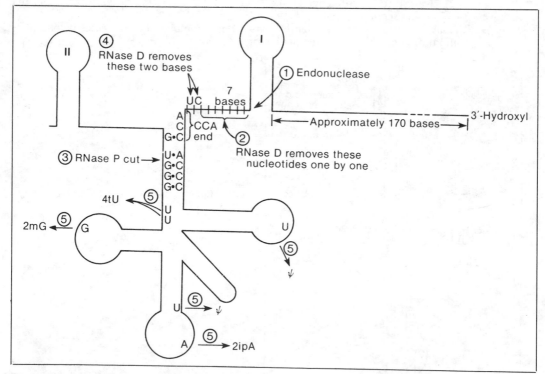

Figure 3-12. The stages in processing of the *Escherichia coli* $tRNA_1{}^{tyr}$ gene transcript. The five stages are given *arabic numbers*. *Step 3* generates the 5′-phosphate end. *Step 4* generates the 3′-hydroxyl end (the CCA end). In *step 5*, six bases, all in or near the loops of the tRNA molecule, are modified to form pseudouridine (ψ), 2-isopentenyl-adenosine (2ipA), 2-O-methylguanosine (2mG), and 4-thiouridine (4tU). (Reprinted with permission from Freifelder D: *Molecular Biology.* Boston, Jones and Bartlett, 1983, p 395.)

b. Many (but not all) eukaryotic genes contain introns.

c. The major difference between eukaryotic tRNA processing and prokaryotic tRNA processing is that in eukaryotic genes the CCA end is not in the primary transcript but is added later.

C. rRNA

1. Prokaryotic rRNA

a. There are three kinds of prokaryotic rRNA:
  (1) 5S rRNA (120 nucleotides)
  (2) 16S rRNA (1541 nucleotides)—for the small ribosomal subunit (see Chapter 4, section VI)
  (3) 23S rRNA (2904 nucleotides)—for the large ribosomal subunit

b. The primary transcript is greater than 5000 bases long and contains one copy each of 5S, 16S, 23S, and a tRNA transcript. (Fig. 3-13).

c. Seven different primary transcripts have been identified, differing in the type of tRNA present and in its location. The presence of the rRNA moieties in each primary transcript ensures that the molar ratio of 5S:16S:23S rRNA remains constant.

d. Processing of prokaryotic rRNA involves RNase III (a nuclease specific for double-stranded RNA) and RNase P (for the tRNA segments).

2. Eukaryotic rRNA

a. Eukaryotic genes contain four types of rRNA:
  (1) 5S rRNA
  (2) 5.8S rRNA
  (3) 18S rRNA
  (4) 28S rRNA

b. Eukaryotic rRNA genes repeat in tandem, often more than 100 times. They are found in areas of the chromosome associated with nucleoli, except for the 5S genes, which are found outside the nucleolus grouped at the chromosome ends.

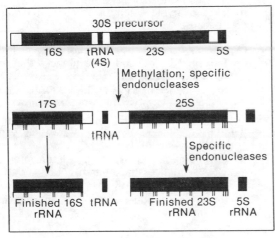

Figure 3-13. Processing of ribosomal RNA transcript in prokaryotes. The final 16S and 23S rRNA products are made from a longer 30S RNA precursor by the action of specific nucleases. Prior to cleavage of 30S RNA, it is methylated at specific bases (*short lines*). From the mid-section, a single tRNA is formed. (Reprinted with permission from Lehninger AL: *Principles of Biochemistry*. New York, Worth 1983, p 858.)

STUDY QUESTIONS

Directions: Each question below contains five suggested answers. Choose the **one best** response to each question.

1. DNA contains sequences of bases known as promoters which

(A) are recognized by the core enzyme of RNA polymerase
(B) occur in the chain just before the message sequence that defines the messenger RNA
(C) are single-stranded regions of the duplex DNA
(D) are transcribed into RNA sequences
(E) are not functional in prokaryotes

2. RNA polymerase recognizes sites on DNA known as promoters. All of the following statements regarding these sites are correct EXCEPT

(A) a sequence of DNA base-pairs known as the Pribnow box is located 5–10 bases upstream from the first base to be copied into RNA
(B) all Pribnow box sequences are variants of the sequence TATAATG
(C) a second sequence, the "– 35" sequence, in many promoters, is located about 35 bases downstream from the Pribnow box
(D) the σ subunit of prokaryotic RNA polymerase is involved in recognition of the promoter sequence
(E) RNA polymerase binds first to the "– 35" sequence region and then to the Pribnow box

3. A segment of a DNA chain with the base sequence TCAGAT would be transcribed by RNA polymerase to an RNA chain with which of the following base sequences?

(A) CUAAGU
(B) AGUCUA
(C) AGTCTA
(D) AUCUGA
(E) ATCTGA

4. Which of the following agents inhibits the initiation of RNA synthesis in prokaryotes by binding to the β subunit of RNA polymerase?

(A) α-Amanitin
(B) Actinomycin D
(C) Rifamycin SV
(D) Puromycin
(E) Streptolydigin

5. Which of the following statements regarding the termination of transcription of DNA is correct?

(A) The rho protein binds tightly to RNA polymerase and inactivates the polymerization activity
(B) Termination sites on DNA contain an inverted-repeat base sequence that results in the RNA formed adopting a "stem and loop" configuration
(C) Each termination site contains a unique base where transcription stops
(D) All prokaryote termination sites contain the "stem and loop" promoting sequence and also require the participation of the rho protein to halt RNA synthesis
(E) None of the above statements is correct

6. The colinearity of gene and product clearly seen in prokaryote systems is obscured in eukaryotes because

(A) the DNA sequences coding for proteins is interrupted by noncoding sequences that are not transcribed
(B) the sequence of bases in the mRNA is modified in the cytoplasm prior to interaction with the ribosomes
(C) ribosomes in eukaryotes only translate exon regions of the mRNA
(D) the coding sequence in DNA is interrupted by DNA sequences that are transcribed and then spliced out of the mRNA
(E) the protein product is modified after leaving the ribosome

7. All of the following statements about tRNA molecules are true EXCEPT that

(A) they have extensive intrachain hydrogen bonding
(B) they contain a number of bases that have been modified post-transcriptionally
(C) the amino acid is attached to the 2' or 3' position of the 3' terminal adenosine
(D) they contain an anticodon triplet of bases at the 5' end of the molecule that is complementary to the triplet codon on mRNA
(E) they have binding sites for a specific amino acid, a specific aminoacyl transferase, the ribosome, and a specific codon on mRNA

Directions: Each question below contains four suggested answers of which **one or more** is correct. Choose the answer

A   if **1, 2, and 3** are correct
B   if **1 and 3** are correct
C   if **2 and 4** are correct
D   if **4** is correct
E   if **1, 2, 3, and 4** are correct

8. DNA-dependent RNA polymerase of *Escherichia coli* is characterized by which of the following properties?

(1) It requires a DNA template, Mg^{2+}, and all four ribonucleoside triphosphates and forms a phosphodiester bond and pyrophosphate
(2) It is a polymeric protein
(3) It requires a protein factor (σ factor) for initiation of transcription
(4) It requires a primer with a free 3'-hydroxyl group

9. Similarities between RNA polymerase and DNA polymerase activities in prokaryotes include

(1) the requirement for a template
(2) the requirement for a primer
(3) synthesis of the nascent chain in the 5' to 3' direction
(4) the 3' to 5' exonuclease editing function

10. Correct statements concerning eukaryotic mRNA include which of the following?

(1) It is formed from nuclear RNA precursors of widely variable size known as heterogeneous nuclear RNA
(2) It has a 7-methylguanosine at the 3' end of the chain
(3) It is prepared from precursor RNA by splicing enzymes, which remove introns and join exons
(4) It has a poly A tail consisting of 150 to 200 nucleotides at the 5' end

11. Post-transcriptional modification of prokaryote RNA molecules includes

(1) cleavage of primary transcripts to form functional molecules of rRNA and tRNA
(2) addition of a CCA 3' terminus to all tRNA molecules
(3) methylation of bases using S-adenosylmethionine as methyl donor
(4) addition of a cap structure to the 5' end of mRNAs

ANSWERS AND EXPLANATIONS

1. The answer is B. (*I C 1*) Initiation of transcription of DNA occurs when RNA polymerase recognizes special sequences (promoter regions), about 40 base-pairs long, that occur just before the message that codes for the mRNA. The RNA polymerase recognizes and binds to these sequences only because of the presence of the σ factor. The core enzyme ($α_2 β β'$) alone cannot recognize the promoter regions. The promoter regions are duplex DNA strands and are vital to replication of DNA in prokaryotes.

2. The answer is C. (*I C 1*) DNA contains sequences of bases known as promoters, which RNA polymerase recognizes. The σ subunit of prokaryotic RNA polymerase is essential for this recognition. One such sequence, found 5–10 bases upstream (to the left) from the first base that will be copied into RNA, is called the Pribnow box. All Pribnow sequences known are variants of TATAATG. A second sequence, the "−35" sequence, is found in many promoters about 35 bases upstream from the Pribnow box. It is thought to be the initial site of σ subunit binding. When this occurs, the huge RNA polymerase molecule is also able to interact with the Pribnow box.

3. The answer is D. (*I C 1*) The complementary strand of RNA formed from the DNA template TCAGAT would be AGUCUA written in the 3′ to 5′ direction with uracil in place of thymine. By convention, the sequence should be written 5′ to 3′ to give AUCUGA.

4. The answer is C. (*I C 2 g, D 2 c; Chapter 4 X A*) The antibiotic rifamycin SV (produced by *Streptomyces mediterranei*) used in the treatment of tuberculosis inhibits prokaryote RNA polymerase by binding to its β subunit and preventing the formation of the first phosphodiester bond, thereby blocking initiation of RNA synthesis. It does not affect elongation of already initiated chains. In contrast, streptolydigin blocks the elongation of RNA chains by prokaryotic RNA polymerase. α-Amanitin, derived from the poisonous mushrooms, *Amanita phalloides*, strongly inhibits eukaryotic RNA polymerase II (synthesis of mRNA) and weakly inhibits RNA polymerase III (synthesis of tRNA). Actinomycin D blocks all DNA-dependent RNA synthesis by binding to guanine-rich regions of duplex DNA. Puromycin, as an analogue of aminoacyl-tRNA, inhibits protein synthesis.

5. The answer is B. (*I C 4*) Termination of transcription is determined by specific sequences in DNA and in some systems, but not all, the presence of a protein called rho is required. Termination sites contain three important regions, one of which is an inverted-repeat base sequence, and the RNA formed from this sequence adopts a "stem and loop" configuration. When required, the rho protein binds very tightly to the RNA, not to the polymerase, and in this state becomes an ATPase, although its action is poorly understood. There does not appear to be a unique base where transcription stops.

6. The answer is D. (*II A 3 e*) In eukaryote genomes, the DNA unique sequences coding for cellular proteins are interrupted by regions that are transcribed but not translated. These so-called exon regions are spliced out of the mRNA during its processing in the nucleus. The sequence of bases in the coding region of the mature mRNA and the gene product are colinear.

7. The answer is D. (*II B 2; Figure 3-8*) Molecules of tRNA contain binding sites for a specific amino acid that is attached to the 2′ or 3′ position of the 3′ terminal adenosine for the amino acid–activating enzyme (aminoacyl-tRNA transferase), and for specific sites on the ribosome; and they contain an anticodon triplet that is borne on the so-called anticodon loop, which lies approximately midway between the 5′ and 3′ ends of the molecule. This triplet of bases is the anticodon and is complementary to the triplet codon on mRNA. The tRNA molecule contains a number of unusual bases, which arise by the post-transcriptional modification of one or another of the four common bases, A, G, U, or C. The three-dimensional structure of the tRNA molecule is important to its interaction with the ribosome and shows considerable intrachain hydrogen bonding (i.e., base-pairing).

8. The answer is A (1, 2, 3). (*I B*) RNA polymerase prefers a duplex DNA template but can use single-stranded DNA. It does not require a primer but does need to have all four ribonucleoside triphosphates available simultaneously and also requires Mg^{2+}. The apoenzyme is a tetramer ($α_2ββ'$), and holoenzyme formation requires the association of a protein factor, the σ factor, for recognition of and binding to the promoter region of DNA. After the first phosphodiester bond is formed, the σ factor leaves the enzyme complex, and the apoenzyme continues polymerization.

9. The answer is B (1, 3). (*I B; Chapter 2 III A*) Both DNA and RNA polymerase require templates: DNA polymerase I (DNA pol I) requires duplex DNA, and RNA polymerase prefers duplex DNA but is also able to use single-stranded DNA as a template. DNA polymerase III requires an RNA primer for the initiation of DNA replication, and DNA pol I adds on to existing DNA chains. RNA polymerase does not require a primer. Both polymerases synthesize nascent chains in the 5′ to 3′ direction. In contrast to

DNA pol I, which has a 3′ to 5′ exonuclease editing function, RNA polymerase does not exhibit exonuclease activity.

10. The answer is B (1, 3). (*II A 3*) Eukaryotic mRNA is formed from nuclear RNA precursors 2 to 20 kb long, known as heterogeneous nuclear RNA. The 5′ ends of all eukaryotic mRNAs have a 7-methylguanosine in a 5′-5′ pyrophosphate linkage (the cap structure). Most eukaryotic mRNAs have a 3′ poly A tail 150 to 200 nucleotides long. The splicing of introns from mRNA precursors is carried out by enzymes exhibiting very high specificities. The introns are not translated.

11. The answer is B (1, 3). (*II B 4, C 1*) Three types of post-transcriptional modification occur in prokaryote RNA. The formation of functional molecules of tRNA and rRNA occurs by the splitting of a large primary transcript that contains sequences for the three types of rRNA of *Escherichia coli*, a tRNA sequence, and spacer sequences that are not part of the final functional molecule. Methylation of bases and other structural changes occur after transcription. The cap structure on the 5′ end of mRNA is a feature of eukaryote post-transcriptional modification of RNA.

4

Translation of mRNA: Protein Synthesis

I. RELATIONSHIP BETWEEN BASE SEQUENCES IN mRNA AND AMINO ACID SEQUENCES IN A POLYPEPTIDE

A. Codons and anticodons

1. The message contained in mRNA is made up of **codons** (base triplets), which are read by **adaptor molecules** carrying specific amino acids. *(tRNA's)*

2. The adaptor molecules are tRNAs, which have
   a. An attachment site for a specific amino acid
   b. A template (mRNA) recognition site—that is, the **anticodon**

3. The amino acid is carried in an activated form, the carboxyl group being esterified to the 3′- or 2′-hydroxyl (OH) of the 3′ terminal ribose of the RNA chain.

B. mRNA coding sequence

1. An amino acid is specified by a group of 3 bases. With 4 bases possible at each of 3 positions, such a code theoretically could specify 64 (4^3) different amino acids. However, the code actually contains 3 stop signals and 61 triplets that specify **only** 20 amino acids. Thus, more than 1 codon can specify the same amino acid, and the code is said to be **degenerate**.

2. The code is nonoverlapping.

3. The code is read sequentially, without spacer bases, from a fixed starting point.

II. DECIPHERING THE GENETIC CODE

A. RNA-directed peptide synthesis

1. **A cell-free protein synthesizing system** can be prepared from *Escherichia coli*.
   a. The cell walls are ruptured and then removed, together with membrane fragments, by centrifugation.
   b. The supernatant mixture obtained contains DNA, tRNA, mRNA, ribosomes, enzymes, and other important factors.
   c. When the mixture is supplemented with adenosine and guanosine triphosphates (ATP and GTP), amino acids, and a polynucleotide that serves as mRNA, the system synthesizes polypeptides.
   d. Using the cell-free protein synthesizing system, Nirenberg found that
   (1) Polyuridylate (poly U) supports the synthesis of polyphenylalanine by acting as a template for peptide synthesis (i.e., as mRNA); the codon for phenylalanine is, therefore, UUU.
   (2) Similarly, polyadenylate supports the synthesis of polylysine, and polycytidylate supports the synthesis of polyproline; hence the code is AAA for lysine and CCC for proline.

2. **Polynucleotide phosphorylase**, which was discovered by Grunberg-Manago and Ochoa, is an enzyme that catalyzes the formation of polyribonucleotides.

$$(RNA)_n + \text{ribonucleoside diphosphate} \rightleftharpoons (RNA)_{n+1} + \text{inorganic phosphate } (P_i)$$

   a. No template is required for this reaction; rather, the composition of the synthesized RNA is determined by the ratios of the nucleoside diphosphates present in the incubation medium.
   b. The reaction uses nucleoside diphosphates rather than triphosphates.
   c. No pyrophosphate (PP_i) product is split in the reaction, which would drive the reaction to the right; in fact, the equilibrium lies to the left.

    d. Using polynucleotide phosphorylase, not only can homopolymers (e.g., poly U) be made, but mixed polymers, such as a copolymer of U and A (with a random sequence of U and A), can be prepared.

  B. Determination of the sequence of bases in codons

    1. Khorana devised a combination of chemical and enzymatic procedures to prepare templates with known base sequences for use in the cell-free protein synthesizing system.

    2. Nirenberg's ribosomal binding assay, in which the ability of trinucleotides to promote the binding of specific tRNA molecules to ribosomes is measured, was used to determine codon sequences. These studies showed that a trinucleotide specifically promotes the binding of the tRNA molecule for which it is the code word.

III. THE CODE

  A. Codon assignments

    1. Of the 64 possible combinations of three bases, 61 have been assigned to the coding of amino acids and 3 to stop or chain termination signals.

    2. The code designations are given in Table 4-1. The triplet codons are represented in the 5' to 3' direction with the first base shown on the left, the second base at the top, and the third base on the right. For example, the codon for tryptophan is UGG only; UUU and UUC both code for phenylalanine.

  B. Degeneracy of the code

    1. Most amino acids are coded for by more than one codon (i.e., the code is degenerate).

    2. Normally, the codon is not ambiguous, because a given codon designates only one amino acid.

    3. Codons that designate the same amino acid are called **synonyms**. Most synonyms differ only in the third base of the codon (see Table 4-1).

  C. Chain termination and start signals

    1. Three codons—UAA, UAG, and UGA—designate chain termination. These codons, which are read by special proteins, signal the end of the translation process (see section VII C 1 c).

Table 4-1. The Genetic Code

First Base	Second Base								Third Base
	U		C		A		G		
U	UUU	Phe	UCU	Ser	UAU	Tyr	UGU	Cys	U
	UUC	Phe	UCC	Ser	UAC	Tyr	UGC	Cys	C
	UUA	Leu	UCA	Ser	UAA	Stop	UGA	Stop	A
	UUG	Leu	UCG	Ser	UAG	Stop	UGG	Try	G
C	CUU	Leu	CCU	Pro	CAU	His	CGU	Arg	U
	CUC	Leu	CCC	Pro	CAC	His	CGC	Arg	C
	CUA	Leu	CCA	Pro	CAA	Gln	CGA	Arg	A
	CUG	Leu	CCG	Pro	CAG	Gln	CGG	Arg	G
A	AUU	Ile	ACU	Thr	AAU	Asn	AGU	Ser	U
	AUC	Ile	ACC	Thr	AAC	Asn	AGC	Ser	C
	AUA	Ile	ACA	Thr	AAA	Lys	AGA	Arg	A
	AUG	Met	ACG	Thr	AAG	Lys	AGG	Arg	G
G	GUU	Val	GCU	Ala	GAU	Asp	GGU	Gly	U
	GUC	Val	GCC	Ala	GAC	Asp	GGC	Gly	C
	GUA	Val	GCA	Ala	GAA	Glu	GGA	Gly	A
	GUG	Val	GCG	Ala	GAG	Glu	AAA	Gly	G

Note.—To find the codon for a given amino acid read the first base (*left-hand column*) for the amino acid, then the second base (*middle columns*) and then the third base (*right-hand column*). Codons are referred to as written in the 5' to 3' direction. Most amino acids are coded for by more than one codon. The codons called stop codons indicate to the RNA polymerase that it is the end of the message. (After Dyson R: *Cell Biology—A Molecular Approach.* Boston, Allyn and Bacon, 1974, p 351.)

2. The start signal in bacteria specifies the use of a specific formylmethionine-tRNA (fmet-tRNA$_f$), which recognizes the codon AUG (or GUG), provided that this codon is preceded by a stretch of mRNA that is part of the initiation signal.

D. Universality of the genetic code

1. The genetic code is the same for all examined species of plants and animals with the exception of the mitochondrial genome where AUA equals met and UGA equals try.

2. Studies of mutations in viruses, bacteria, and higher organisms have established the universality of the genetic code.

3. Most of the amino acid substitutions in proteins can be accounted for by a change of a single DNA base (Table 4-2).

E. Colinearity of gene and product

1. The product of a gene is the peptide specified by the base sequence of the exon regions of the gene.

2. High-resolution genetic mapping techniques have established that there is a linear correspondence in prokaryotes between the base sequence in a gene and the amino acid sequence in the peptide specified by that gene.

3. In eukaryotes, the colinearity of gene base sequence and peptide amino acid sequence is interrupted by intron sequences, which are transcribed and then spliced out (see Chapter 3, section II A 3 e).

F. Overlapping genes

1. The bacteriophage ϕX174 genome codes for more proteins than would be expected, given its known nucleotide content.

2. Sanger's identification of the complete base sequence for this genome revealed that ϕX174 translates regions of its genome in **different reading frames**, producing mRNA that codes for different proteins.

IV. ACTIVATION OF AMINO ACIDS: FORMATION OF AMINOACYL-tRNAs

A. Introduction. An mRNA molecule specifies a specific peptide or, in the case of the polycistronic message of prokaryotes, a group of peptides. In general terms, the **translation** of the message of mRNA into the amino acid sequence of a polypeptide is achieved by the binding of the mRNA

Table 4-2. Mutations in Human Hemoglobin, *Escherichia coli* Tryptophan Synthetase, and Tobacco Mosaic Virus (TMV) Coat Protein

Protein	Amino Acid Substitution	Inferred Codon Change
Hemoglobin	Glu→Val*	GAA→GUA
Hemoglobin	Glu→Lys*	GAA→AAA
Hemoglobin	Glu→Gly	GAA→GGA
Tryptophan synthetase	Gly→Arg	GGA→AGA
Tryptophan synthetase	Gly→Glu	GGA→GAA
Tryptophan synthetase	Glu→Ala	GAA→GCA
TMV coat protein	Leu→Phe	CUU→UUU
TMV coat protein	Glu→Gly	GAA→GGA
TMV coat protein	Pro→Ser	CCC→UCC

Note.—Reprinted with permission from Stryer L: *Biochemistry,* 2nd ed. New York, Freeman, 1981, p 631.
*The substitutions Glu→Val and Glu→Lys occur at position 6 in the β chain of human hemoglobin S and C, respectively.

to ribosomes with the amino acids being delivered to the ribosome attached covalently to tRNAs. This process requires the concerted participation of more than 100 macromolecules.

B. **Aminoacyl-tRNA synthetases** are enzymes that activate amino acids and attach them to the 3′ end of their cognate tRNA chain.

1. There is at least one, although occasionally two, specific enzymes for each amino acid.

2. The enzymes vary in molecular weight, number of subunits, and amino acid composition.

3. They have to be, and are, extremely specific; that is, because once an amino acid is attached to a given tRNA, recognition of a specific codon on mRNA is due entirely to the tRNA, not to the amino acid.
   a. The very high specificity of these enzymes is caused by
      (1) High selectivity in terms of the acceptance of the amino acid to be activated
      (2) High selectivity toward the tRNA to which the activated amino acid will be transferred
   b. Incorrectly activated amino acids are hydrolyzed off the tRNA synthetase rather than transferred to the tRNA.

4. The reaction of an amino acid and tRNA requires ATP and proceeds via the formation of an aminoacyl intermediate, which may or may not be detectable, depending upon the amino acid and tRNA in question.
   a. **Formation of an aminoacyl–adenylate complex** is the first step.

$$\text{Amino acid} + \text{ATP} \rightleftharpoons\ ^+\text{H}_3\text{N}-\underset{\underset{\text{R}}{|}}{\overset{\overset{\text{H}}{|}}{\text{C}}}-\overset{\overset{\text{O}}{\parallel}}{\text{C}}-\text{O}-\underset{\underset{\text{O}^-}{|}}{\overset{\overset{\text{O}}{\parallel}}{\text{P}}}-\text{O}-\text{ribose}-\text{adenine} + \text{PP}_i$$

   b. **Transfer of the aminoacyl group to tRNA** is the next step.

   Aminoacyl–adenosine monophosphate (AMP) + tRNA → aminoacyl-tRNA + AMP

   c. **The sum of the activation and transfer steps** is

   Amino acid + ATP + tRNA → aminoacyl-tRNA + AMP + PP$_i$

   d. **The hydrolysis of pyrophosphate**, which renders the reaction virtually irreversible, drives the reaction to completion as in the formation of polyribonucleic and polydeoxyribonucleic acid polymers. Thus, two high-energy bonds of ATP are expended in the formation of an aminoacyl-tRNA.

V. CODON–ANTICODON RECOGNITION

A. **Antiparallel relationship.** The triplet base codon of mRNA and the triplet base anticodon on tRNA interact in antiparallel relation to one another.

1. The first two bases of the codon and the last two bases of the anticodon pair in the usual Watson-Crick manner.

2. The pairing of the third base of the codon on mRNA with the first base (i.e., the base at the 5′ end) of the anticodon follows less rigid requirements than the first two bases. Thus, some tRNAs recognize and pair with more than one triplet codon on mRNA; this leniency is part of the explanation for the degeneracy of the genetic code.

B. **Crick's wobble hypothesis** addresses the question of the less stringent pairing of the third base in an mRNA codon.

1. When the first base in the anticodon is C or A, the pairing with the third base in the codon is regular (i.e., G and U, respectively).

2. When the first base in the anticodon is U, then the third base in the codon can be either of the purines (i.e., G or A).

3. When the first base in the anticodon is G, then the third base in the codon can be either of the pyrimidines (i.e., U or C).

4. When the first base in the anticodon is I (inosine), the third base in the codon can be A, C, or U.

5. Note that the third base in the codon frequently does not make any difference in the amino acid that is specified (see Table 4-1).

VI. STRUCTURE OF RIBOSOMES

A. Prokaryotic ribosomes

1. In *E. coli*, the ribosome is about 65% RNA and 35% protein.

2. It is a large particle with a sedimentation coefficient of 70S; thus, it is called the 70S particle.

3. It can be split by treatment with ethylenediaminetetraacetic acid (EDTA), which reduces the Mg^{2+} concentration of the medium, to a **50S large subunit** and a **30S small subunit**.
  a. The 50S subunit comprises about 34 proteins (L-proteins) and two molecules of RNA, a 23S and a 5S RNA. The latter is about 200 bases long and has a molecular weight of about 5 kdal.
  b. The 30S component contains about 21 proteins (S-proteins) and a 16S RNA molecule.
  c. Most ribosomal proteins are low molecular weight, basic proteins. The basic charge reflects their ability to interact with negatively charged RNA.
  d. The RNA molecules within the ribosome have defined secondary structures, and they interact with the ribosomal proteins in a defined manner. Prokaryotic ribosomes can be disassembled into purified RNA and protein components, and then reassembled into active functional ribosomes.
  e. With one exception (protein L7/L12), each ribosomal protein is present in a single copy per ribosome.

B. Eukaryotic ribosomes are larger than bacterial ribosomes and have an intact particle with a sedimentation coefficient of 80S.

1. These ribosomes can be dissociated into a **60S large subunit**, which contains about 45 proteins and three RNA components (28S, 5.8S, and 5S), and a **40S small subunit**, which contains about 30 proteins and an 18S RNA component.

2. The ribosomes of mitochondria and chloroplasts resemble those of prokaryotes, both in structure and in their sensitivity to antibiotic inhibitors of translation (see section X).

VII. PROTEIN SYNTHESIS IN PROKARYOTES occurs in three stages: initiation, elongation, and termination.

A. Initiation

1. Formation of a 30S initiation complex (as required by bacteria) consists of
  a. The 30S ribosomal subunit
  b. fmet-tRNA$_f$
  c. Initiation factors (proteins) and GTP
  d. mRNA (Fig. 4-1)

2. Initiation factors and GTP
  a. There are three initiation factors (proteins): IF-1, IF-2, and IF-3.
  b. IF-1 and IF-3 are relatively heat-stable basic proteins of approximately 9000 and 22,000 daltons, respectively. IF-2 is a relatively heat-labile acidic protein of approximately 100,000 daltons.
  c. GTP binding to the 30S particle is also required.

3. Source of the 30S subunit
  a. Following termination of mRNA translation, the 70S ribosomes are released from the polysomal complex and are then in equilibrium with their subunits.

$$30S + 50S \rightleftharpoons 70S$$

At physiologic concentrations of Mg^{2+}, nearly all ribosomes exist as tight 70S couples. Initiation requires a free 30S subunit.
  b. The dissociation of the 70S ribosomal particle is mediated by IF-1 and IF-3 (see Fig. 4-1).

$$70S \underset{k_2}{\overset{k_1}{\rightleftharpoons}} 50S + 30S \overset{IF\text{-}3}{\underset{(IF\text{-}1)}{\longrightarrow}} 30S \cdot IF\text{-}3 + 50S$$

  c. Function of initiation factors
    (1) IF-3 acts primarily as an antiassociation factor.

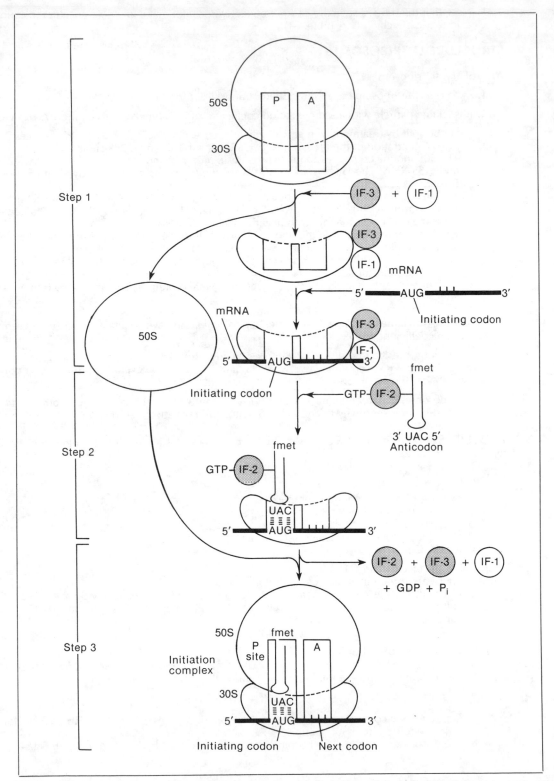

Figure 4-1. Formation of the initiation complex in three steps at the expense of the hydrolysis of guanosine triphosphate (GTP) to guanosine diphosphate (GDP) and P_i. IF-1, IF-2, and IF-3 are the initiation factors. P designates the peptidyl site and A the aminoacyl site. (Reprinted with permission from Lehninger AL: *Principles of Biochemistry*. New York, Worth, 1982, p 884.)

      (2) IF-1 acts directly to increase k_1. It may also aid in the binding of IF-3 to the 30S particle.

4. Source of fmet-tRNA$_f$
    a. A specific transfer RNA molecule, called the **initiator tRNA (tRNA$_f$)**, brings fmet to the 30S ribosomal subunit (to initiate protein synthesis).
    b. This tRNA$_f$ is different from the tRNA that inserts methionine in internal positions in the peptide chain (tRNA$_m$).
    c. The same aminoacyl-tRNA synthetase links both of these tRNAs to methionine, but a specific enzyme (a formyltransferase) adds a formyl group to the amino group of met that is attached to the tRNA$_f$ to form N-fmet-tRNA$_f$. The formyl group is transferred from N^{10}-formyltetrahydrofolate (N^{10}-THFA) [Fig. 4-2].

5. mRNA. In prokaryotes, mRNA is polycistronic and for each independently translated gene product, it bears a sequence of bases that indicate initiation sites.

6. Formation of the 30S initiation complex (see Fig. 4-1).
    a. Following binding of the three initiation factors to the 30S ribosomal particles, mRNA and the initiator fmet-tRNA$_f$ bind to the 30S.IF-1.IF-2.IF-3 particle. The order of addition is uncertain.
    b. The 30S.IF-1.IF-2.IF-3 particle binds to mRNA at a site that includes the initiation (start) codon, AUG or GUG.
      (1) The 30S subunit has the ability to specifically recognize initiation sites in natural mRNA, and IF-3 aids in this process.
      (2) Shine, Dalgarno, and others have shown that virtually all ribosome-binding sites in natural mRNA have a 5′-AGGAGGU-3′ sequence (or a close derivative) on the 5′ side of the initiating codon of a cistron.
      (3) This sequence base-pairs with a pyrimidine-rich region at the 3′ end of 16S rRNA chains. The base-pairing positions the initiating AUG codon so that it can bind to the anticodon of an initiator tRNA.
    c. Binding of fmet-tRNA$_f$
      (1) IF-2 is the factor responsible for the binding of the initiator tRNA to the 30S particle. The exact steps in this process remain unclear, although it appears to stabilize the interaction of the formylated initiator tRNA with the 30S particle.
      (2) GTP acts as a steric effector in the formation of the 30S initiation complex, permitting stable association of IF-2 with the 30S particle at a relatively low concentration of the factor.
    d. Overall reaction in the formation of the 30S initiation complex is

$$(30S.IF-1.IF-2.IF-3) + mRNA + GTP + fmet\text{-}tRNA_f \rightarrow$$

$$(30S.IF-1.IF-2.mRNA.GTP.fmet\text{-}tRNA_f) + IF-3$$

IF-3 dissociates from the 30S initiation complex upon the binding of fmet-tRNA$_f$ (see Fig. 4-1).

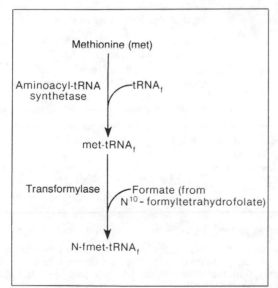

Figure 4-2. Outline of the pathway for the formation of N-formylmethionine-tRNA$_f$ (N-fmet-tRNA$_f$).

7. Formation of the 70S initiation complex and the recycling of IF-2. The 30S initiation complex reacts with the 50S ribosomal subunit in a 1:1 molar ratio to form the **70S initiation complex**.

  a. During the formation of the 70S particle, GTP is hydrolyzed to guanosine diphosphate (GDP) and P_i.

  b. IF-2 (and IF-1), GDP, and P_i are all released from the ribosomal complex (see Fig. 4-1). Thus,

$$(30S.mRNA.IF-1.IF-2.fmet-tRNA_f.GTP) + 50S \rightarrow$$

$$(mRNA.70S.fmet-tRNA_f) + IF-1 + IF-2 + GDP + P_i$$

  c. The hydrolysis of GTP is necessary to convert one steric effector, GTP, to another, GDP. The latter has a low affinity for the ribosomal complex and dissociates from it, allowing IF-1 to effect the release of IF-2.

  d. The P (peptide) and A (amino) sites on the ribosome form by conjunction of S and L ribosomal proteins.

  e. The P site is occupied by fmet-tRNA$_f$ (Fig. 4-3).

B. Elongation

1. **The aminoacyl-tRNA cognate to the next codon after the initiating AUG binds to the ribosome at the A site**, starting the process of elongation.

   a. The process of elongation requires three protein factors, elongation factors EF-Tu, EF-Ts, and EF-G. They are present in the cell at a concentration of 5%–10% of all bacterial protein.

   b. Delivery of an aminoacyl-tRNA to the empty A site on the ribosome is effected by EF-Tu (see Fig. 4-3).

   c. GTP bound to EF-Tu is hydrolyzed at this time.

   d. GDP, from the hydrolysis of GTP, remains associated with EF-Tu until displaced by EF-Ts.

   e. EF-Tu and EF-Ts form a complex that is split by the binding of another GTP, providing an EF-Tu-GTP complex for the delivery of the next aminoacyl-tRNA.

   f. Note that EF-Tu does **not** interact with fmet-tRNA$_f$.

2. **The link between the amino acid and fmet-tRNA$_f$ is an amino acid ester bond**.

   a. It is a high-energy bond with a free energy of hydrolysis similar to that of the terminal phosphate bond in ATP.

   b. It provides more than enough energy for the amino group of the amino acid at the A site to deliver a nucleophilic attack on the carbon of the ester linkage on the amino acid at the P site.

   c. The attack results in two amino acids becoming attached to the tRNA at the A site and a naked tRNA at the P site (Fig. 4-4).

   d. The enzyme peptidyl transferase catalyzes the formation of the peptide bond and is an integral part of the 50S subunit.

3. A very interesting but little understood reaction then occurs.

   a. The ribosome is moved bodily to the right in relation to the mRNA (see Fig. 4-4).

   (1) The dipeptidyl-tRNA is now on the P site.

   (2) The A site is empty.

   (3) The naked tRNA is released.

   b. The third elongation factor, EF-G, controls this process of **translocation**.

   c. EF-G forms a complex with GTP, which is hydrolyzed to GDP and P_1 during translocation. The detailed events are not known.

C. Termination

1. Elongation continues until one of the stop codons (UAA, UAG, or UGA) is reached.

   a. These codons do not specify an amino acid.

   b. They have no tRNAs that pair with them.

   c. They are recognized by certain proteins called **release factors** (RF-1, RF-2, and RF-3).

2. When a termination codon in mRNA is reached, no complementary tRNA molecule can bind to the A site in normal cells.

3. RF-3 promotes either RF-1 or RF-2 binding to the ribosome in a GTP-dependent manner.

4. Binding of the release factors to the termination codon causes hydrolysis of the ester linkage of the peptidyl-tRNA. This hydrolysis is probably effected by the peptidyl transferase of the ribosome working in reverse, as antibiotics, which inhibit transpeptidation (peptide bond formation), also inhibit termination (peptide bond hydrolysis).

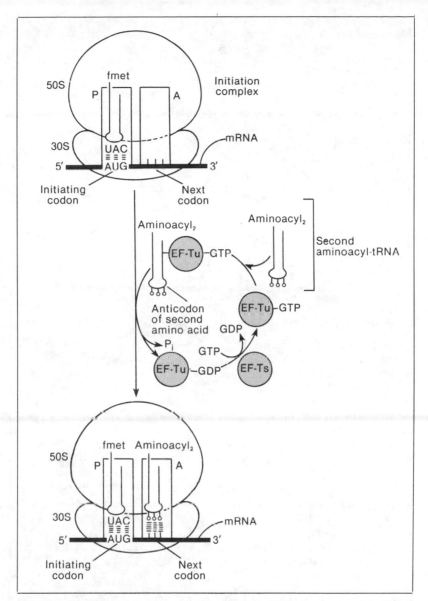

Figure 4-3. The first step in elongation is the binding of the second aminoacyl tRNA. It enters bound to elongation factor Tu (EF-Tu), which also contains guanosine triphosphate (GTP). Binding of the second aminoacyl-tRNA is accompanied by hydrolysis of the bound GTP. The bound guanosine diphosphate (GDP) so formed is displaced by elongation factor Ts (EF-Ts). The nucleotides for the anticodon and codon for subsequent amino acids are symbolized by *small open circles*. (Reprinted with permission from Lehninger AL: *Principles of Biochemistry*. New York, Worth, 1982, p 886.)

5. Upon hydrolysis of the peptidyl-tRNA linkage, the polypeptide leaves the ribosome, the tRNA leaves, and mRNA dissociates.

D. Energy requirements for protein synthesis

    1. Each peptide bond requires at least four high-energy phosphate bonds for its formation.
        a. Acylation of tRNA. ATP is split to adenosine monophosphate (AMP) and PP$_i$, which is the equivalent of using two high-energy phosphate bonds.
        b. Aminoacyl-tRNA binding to the A site. GTP is hydrolyzed to GDP and P$_i$. Thus, one high-energy phosphate bond is used per elongation cycle.

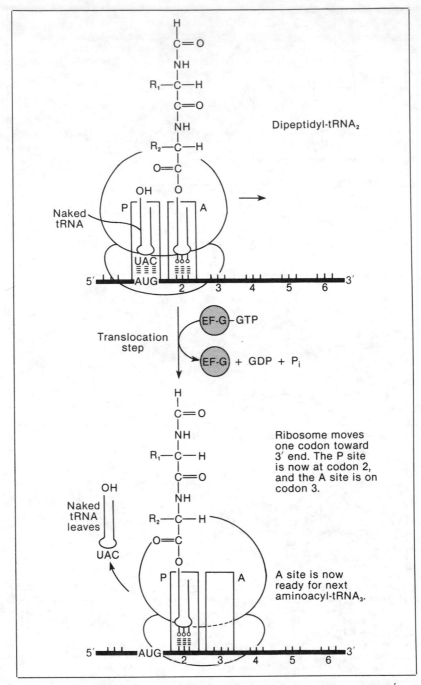

Figure 4-4. The translocation step. The ribosome moves one codon toward the 3′ end of mRNA accompanied by the hydrolysis of guanosine triphosphate (GTP) bound to elongation factor G (EF-G). The dipeptidyl-tRNA₂ is now on the P site, leaving the A site open for the incoming aminoacyl-tRNA₃. (Reprinted with permission from Lehninger AL: *Principles of Biochemistry.* New York, Worth, 1982, p 889.)

    c. Translocation. GTP is hydrolyzed to GDP and P_i. Thus, one high-energy phosphate bond is used per elongation cycle.

    d. Binding of fmet-tRNA$_f$ to the P site requires the hydrolysis of GTP to GDP and P_i. Thus, one high-energy phosphate bond is used per protein synthesized.

2. Roles for the hydrolysis of GTP
   a. The affinity of protein factors for the ribosome appears to be modulated by GTP.
   b. GTP hydrolysis is essential for the release of bound factors.
   c. GTP hydrolysis may provide energy for the translocation step.
   d. GTP hydrolysis is essential for aminoacyl-tRNA binding [see section VII A 6 c (2), B 1 c].

E. Fidelity of translation

   1. Consequences of low fidelity of translation are
   a. Nonconservative replacement of an amino acid can yield a nonfunctional or deleterious protein.
   b. Misreading of termination codons yields abnormal proteins.

   2. Maintenance of fidelity of translation is by
   a. Accuracy of aminoacyl transferases and their proofreading capability
   b. Codon–anticodon recognition

   3. However, the above two mechanisms cannot account for the **in vivo fidelity of protein synthesis**.
   a. Error frequency in *E. coli* translation is estimated to be 1 error per 2000 amino acids.
   b. Codon–anticodon mismatch should yield an error frequency of 1 per 100 amino acids.

   4. Genetic and biochemical evidence suggest that GTP hydrolysis is coupled to a proofreading function that "tests" a codon–anticodon interaction for mismatch.

VIII. PROTEIN SYNTHESIS IN EUKARYOTES

A. Introduction. The mechanism of protein synthesis in eukaryotes is similar to that of prokaryotes. The differences are a consequence of the more complex cellular organization of the eukaryotic cell.

   1. In a prokaryotic cell, translation of an mRNA molecule is initiated while the mRNA is still being transcribed. The mRNA is usually polycistronic with a short half-life of several minutes.

   2. Eukaryotic mRNA molecules must be transported from the nucleus into the cytoplasm prior to initiation. The mRNA is monocistronic with a long half-life. Eukaryotic translation systems will not translate the polycistronic mRNA of a prokaryote.

   3. Gene expression in eukaryotes may be controlled at the level of translation (see Chapter 8, section I B), as well as transcription, while control of gene expression in prokaryotes occurs principally at the level of transcription

B. Initiation. Although initiation in eukaryotes is considerably more complex than in prokaryotes, the same basic events occur.

   1. Initiation codon and initiator tRNA
   a. AUG is the only initiating codon in eukaryotes.
   b. The initiating amino acid is methionine, not fmet.
   c. Eukaryotes contain two tRNA molecules that recognize AUG codons. The initiator tRNA is called met-tRNA$_f$ or met-tRNA$_i$. Internal AUGs are recognized by met-tRNA$_m$.
   d. These two tRNA molecules differ structurally. The most notable difference is in the first loop, where the met-tRNA$_f$ has a GAUC loop, rather than the GTψC loop.

   2. Eukaryotic initiation requires the following components:
   a. Ribosomal subunits, 40S and 60S
   b. met-tRNA$_f$
   c. mRNA (the cap structure on eukaryotic mRNA is important for initiation as the cap analogue, 7-methylguanosine monophosphate, is a potent initiation inhibitor)
   d. GTP
   e. Eukaryotic initiation factors (i.e., eIF-1, eIF-2, and eIF-3) [eukaryotic initiation requires at least 10 identified initiation factors (more than in prokaryotes)]
   f. ATP, which is required for eukaryotic initiation but not for prokaryotic initiation

   3. Stabilization of ribosomal subunits by dissociation and antiassociation factors
   a. Specific initiation factors bind to subunits and prevent interaction of 40S and 60S subunits.
   b. eIF-6 binds to 60S subunits.
   c. eIF-4C binds to 40S subunits as may eIF-3. The role of eIF-3 in the eukaryote system appears similar to that of IF-3 in prokaryotic initiation.

4. Initiator tRNA binding
    a. A ternary complex of eIF-2, GTP, and met-tRNA$_f$ is formed.
    b. This complex binds to the 40S subunit to form a **preinitiation complex**.

$$(40S.eIF\text{-}4C.eIF\text{-}3) + (eIF\text{-}2.GTP.met\text{-}tRNA_f) \rightarrow$$

$$(40S.eIF\text{-}4C.eIF\text{-}3.eIF\text{-}2.GTP.met\text{-}tRNA_f)$$

    c. The function, if not the mechanism, of eIF-2 is analogous to the function of IF-2 in prokaryotes.

5. mRNA binding to the preinitiation complex
    a. The mechanism of mRNA binding to the ribosome is considerably more complex than in prokaryotes, and the exact events and function of initiation factors are not entirely understood.
    b. A **cap binding protein (CBP)**, eIF-4A, eIF-4B, as well as eIF-4C and eIF-3 all function during mRNA binding.
    c. CBP may initiate mRNA binding by interaction with the 5′ cap structure of an mRNA. It also promotes eIF-4A and eIF-4B binding to mRNA.
    d. A **40S initiation complex** is formed.

$$(40S.eIF\text{-}4C.eIF\text{-}3.eIF\text{-}2.GTP.met\text{-}tRNA_f) + (CBP.eIF\text{-}4A.eIF\text{-}4B.mRNA)$$

$$+ \; ATP \rightarrow (40S.eIF\text{-}4C.eIF\text{-}3.eIF\text{-}2.GTP.met\text{-}tRNA_f.mRNA)$$

$$+ \; CBP + eIF\text{-}4A + eIF\text{-}4B + ADP + P_i$$

    e. ATP hydrolysis occurs during mRNA binding. A proposed function for ATP hydrolysis is the unwinding of mRNA secondary structure in the 5′ untranslated region.

6. Recognition of initiation codon
    a. The 5′ untranslated region of eukaryotic mRNA is of variable length, usually between 40 and 80 nucleotides long, although it can be greater than 700 nucleotides long. No sequence of the Shine-Dalgarno type [see section VII A 6 b (2)] is apparent, so the initiation codon must be selected by some alternate mechanism.
    b. Initiation occurs at the first AUG codon in greater than 90% of eukaryotic mRNA sequences.
    c. The most common sequence for an initiation site is AXXAUGG where X is any base. If the first AUG is **not** used, this AUG usually occurs in the sequence YXXAUGY, where Y is either a U or a C.
    d. The scanning mechanism postulate is that the 40S subunit binds initially near the 5′ end of the mRNA and moves in the 3′ direction until the first AUG in the proper context is encountered. ATP hydrolysis may be essential for this scanning process.

7. Formation of the 80S initiation complex
    a. An essential component for joining the 40S initiation complex to 60S subunits is eIF-5.

$$(40S.eIF\text{-}4C.eIF\text{-}3.eIF\text{-}2.GTP.met\text{-}tRNA_f.mRNA) + eIF\text{-}5 + (60S.eIF\text{-}6) \rightarrow$$

$$(80S.met\text{-}tRNA_f.mRNA) + GDP.eIF\text{-}2 + P_i + eIF\text{-}3 + eIF\text{-}4C + eIF\text{-}6$$

    b. The (GDP.eIF-2) complex is converted to (GTP.eIF-2) by the action of another initiation factor, using the GTP/GDP exchange mechanism described for EF-Ts recycling of EF-Tu (see section VII B 1).

8. Gene expression modulation. Gene expression in eukaryotes is modulated by altering the rate of initiation complex formation. Several distinct mechanisms have been described, including
    a. Storage of mRNA as ribonucleoprotein particles in the cytoplasm
    b. Phosphorylation of eIF-2 to alter initiation rate (see Chapter 8, section I B 1 b)
    c. Phosphorylation of protein S6 in the small ribosomal subunit in response to growth factors, which affects the rate of initiation of protein synthesis
    d. Poliovirus, which has uncapped mRNAs (when the virus infects cells, it shuts off host protein synthesis by inactivating CBP, so that host mRNAs initiate very poorly)

C. Elongation

    1. The necessary components, which are analogous to those required by prokaryotes, include
        a. The 80S initiation complex
        b. Aminoacyl-tRNA
        c. GTP
        d. Eukaryotic elongation factors eEF-1$_\alpha$, eEF-1$_\beta$, and eEF-2

2. Aminoacyl-tRNAs are bound to the P site as ternary complexes (eEF-1$_\alpha$.GTP.aminoacyl-tRNA).

3. eEF-1$_\alpha$ is analogous to EF-Tu, and eEF-1$_\beta$ is analogous to EF-Ts, in that it catalyzes GTP/GDP exchange to recycle eEF-1$_\alpha$.

4. Transpeptidation is similar to that in prokaryotes.

5. Translocation requires the eEF-2.GTP complex, which is analogous to the EF-G.GTP complex of prokaryotes.

D. Termination

1. A single release factor recognizes all three termination codons.

2. GTP is required for release factor binding to the ribosome.

3. When release factor binds to the ribosome, termination is effected by the peptidyl transferase of the ribosome.

IX. PROTEIN LOCALIZATION IN EUKARYOTES

A. Introduction. In eukaryotic cells most of the protein synthesis occurs in the cytoplasm. Many proteins have to cross at least one membrane between their cytoplasmic site of synthesis and their final location in the cell. Proteins that require transport across a membrane include:

1. Secretory proteins such as immunoglobulins, peptide hormones, serum albumin, and other plasma proteins secreted from hepatocytes

2. Organelle proteins such as those of mitochondria, of which over 90% have to be imported from the cytoplasm

3. Integral membrane proteins (see Chapter 25, III B 2 d), which have to be inserted into the correct intracellular membrane with the correct orientation to function correctly

B. Protein synthesis on membrane-bound ribosomes

1. **Endoplasmic reticulum** is a membrane system to which eukaryotic ribosomes may be attached. Eukaryotic ribosomes may also be found free in the cytoplasm.
   a. Endoplasmic reticulum that bears ribosomes is called **rough endoplasmic reticulum**. The membrane ribosomes of the rough endoplasmic reticulum are the sites of synthesis for secretory proteins, lysosomal proteins, and many integral membrane proteins.
   b. Ribosome-free endoplasmic reticulum is called **smooth endoplasmic reticulum**.
   c. Only the **product of the ribosome synthesizing activity** distinguishes a ribosome attached to the endoplasmic reticulum from a free one.

2. **Signal hypothesis.** The precursors of proteins that are synthesized on the rough endoplasmic reticulum contain a special amino terminal sequence of approximately 20 amino acids called the **signal sequence**. These sequences contain a stretch of hydrophobic amino acids. Two components of the endoplasmic reticulum translocation apparatus, which are essential for the movement of the protein into the cisternae of the endoplasmic reticulum, have been isolated.
   a. **The signal recognition particle (SRP)** is a ribonucleoprotein particle, which can bind to 80S ribosomes.
      (1) It consists of six nonidentical peptides and a 7S RNA.
      (2) Synthesis of secretory proteins initiates on free ribosomes in the cytoplasm, and upon translation of the amino terminal sequence, SRP binds with high affinity to the ribosome via the signal sequence. This causes an arrest of the elongation step of protein synthesis.
   b. **The SRP receptor, or docking protein**, is the second component isolated. The (SRP-ribosome) complex interacts with the docking protein, which is on the endoplasmic reticulum and releases the elongation arrest by displacing SRP from the ribosome, which is then bound to the endoplasmic reticulum by unknown components.
   c. Translocation of the protein resumes with the nascent chain being translocated across the membrane. It is not known whether this occurs through a protein channel or directly through the lipid bilayer (Fig. 4-5).
   d. The signal sequence is removed from the protein by **signal peptidase** during translocation.
   e. Synthesis of integral membrane proteins is similar, except that a sequence of hydrophobic amino acids (20–25 residues) halts transfer of the protein across the membrane and functions as a membrane-binding sequence.

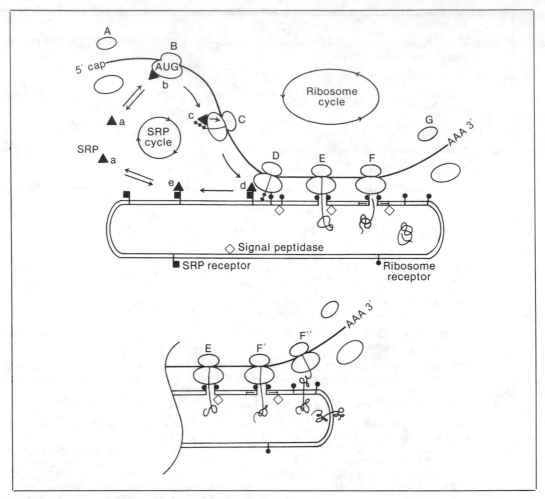

Figure 4-5. Protein translocation across the endoplasmic reticulum. (Reprinted with permission from Walter P et al: Protein translocation across the endoplasmic reticulum. *Cell*, vol 38. Cambridge, MA, MIT Press, 1984, pp 5–8.)

3. **Glycoproteins.** Proteins that are synthesized on membrane-bound ribosomes are glycoproteins that contain covalently attached carbohydrate units.
   a. **Oligosaccharide units** are attached by N-glycosidic bonds to asparagine side chains.
   b. **Sugars** are attached by O-glycosidic bonds to serine or threonine side chains.
   c. **A core oligosaccharide complex** (Fig. 4-6) is linked to the asparagine residues of the growing peptide chain on the luminal side of the endoplasmic reticulum membrane by transfer from an activated lipid carrier, **dolichol phosphate** (Fig. 4-7). The transfer of the oligosaccharide causes the release of dolichol pyrophosphate, which is recovered as dolichol phosphate by the action of a phosphatase.

X. ANTIBIOTIC INHIBITORS OF PROTEIN SYNTHESIS

A. **Puromycin** is an analogue of the aminoacyl–adenosine moiety of an aminoacyl-tRNA (Fig. 4-8).

1. It binds to the ribosome at the A site.
   a. The entry of aminoacyl-tRNAs is blocked.
   b. The amino group forms a peptide bond with the carboxyl group of the peptidyl-tRNA on the P site.
   c. The resulting peptide with a C-terminal puromycin moiety leaves the ribosome even though translation is incomplete.

2. Puromycin inhibits protein synthesis in both prokaryotes and eukaryotes.

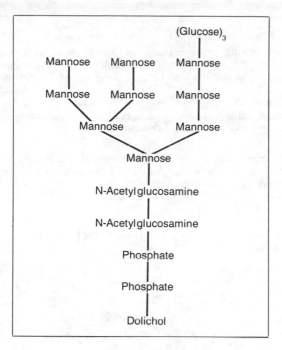

Figure 4-6. Structure of the core oligosaccharide involved in the formation of glycoproteins.

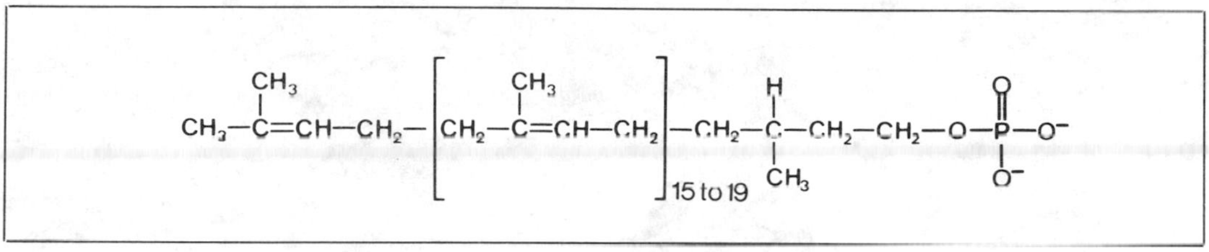

$$CH_3-C=CH-CH_2-\left[CH_2-C=CH-CH_2\right]_{15\,to\,19}-CH_2-C-CH_2-CH_2-O-P-O^-$$

Figure 4-7. Structure of dolichol phosphate.

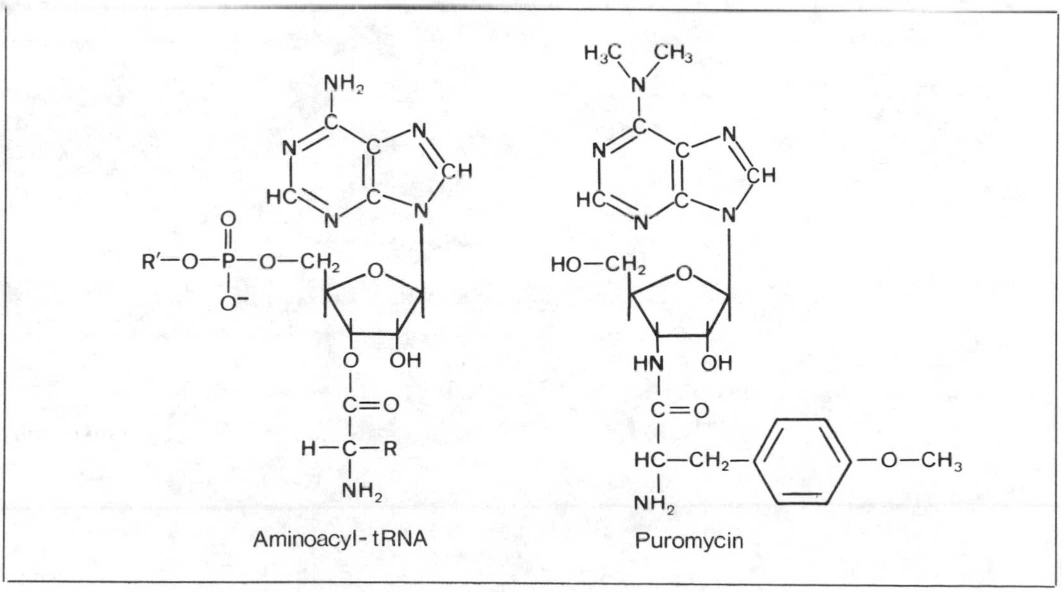

Aminoacyl-tRNA

Puromycin

Figure 4-8. A comparison of the structures of an aminoacyl-tRNA and puromycin. R′ = the remainder of the tRNA molecule.

B. Protein synthesis inhibitors

1. Prokaryotes

  a. Streptomycin is an aminoglycoside that inhibits the initiation of protein synthesis by preventing the binding of fmet-tRNA$_f$ to the P site of the initiation complex. Its effect has been localized to the S12 protein component of the 30S ribosomal subunit. It also causes misreading of the mRNA sequence.

  b. Tetracycline inhibits the binding of aminoacyl-tRNAs to the 30S subunit.

  c. Chloramphenicol inhibits the peptidyl transferase activity of 50S ribosomal subunits.

  d. Erythromycin inhibits translocation by binding to 50S ribosomal subunits.

2. Eukaryotes

  a. Cycloheximide inhibits the peptidyl transferase activity of 60S ribosomal subunits.

  b. Ricin and abrin are toxic lectins. They inhibit protein synthesis by unknown mechanisms, preventing aminoacyl-tRNA from binding to the A site on the ribosome.

  c. α-Sarcin is a toxic RNase that prevents aminoacyl-tRNA binding by cleaving a single phosphodiester bond in 28S rRNA.

  d. Sparsomycin inhibits peptidyl transferase and release factor–dependent termination. It does not prevent termination codon-dependent binding of release factor to the ribosome.

  e. Diphtheria toxin (see Chapter 8, section I B 2).

STUDY QUESTIONS

Directions: Each question below contains five suggested answers. Choose the **one best** response to each question.

1. The genetic code refers to the

(A) number of chromosomes in the diploid cells of a species
(B) nucleotide sequences that correspond to common amino acids
(C) amino acid sequence of cellular proteins
(D) ratios of mendelian inheritance
(E) hierarchy of DNA, RNA, and protein

2. Translation of a synthetic polyribonucleotide containing the repeating sequence CAA in a cell-free protein synthesizing system produced three homopolypeptides: polyglutamine, polyasparagine, and polythreonine. If the codons for glutamine and asparagine are CAA and AAC, respectively, which of the following triplets is a codon for threonine?

(A) AAC
(B) CAA
(C) CAC
(D) CCA
(E) ACA

3. The translation of mRNA into the amino acid sequence of a polypeptide in prokaryotes is terminated at the end of the message by one of three chain termination codons in the mRNA. The stop codon is recognized by

(A) a specific uncharged tRNA
(B) a specific aminoacyl-tRNA
(C) a specific ribosomal RNA
(D) a specific protein
(E) none of the above substances

4. The "wobble" hypothesis refers to the less stringent base-pairing specificity of the

(A) 5'-end base of the codon
(B) 3'-end base of the anticodon
(C) middle base of the anticodon
(D) 5'-end base of the anticodon
(E) middle base of the codon

5. All of the following statements about amino acid activation are correct EXCEPT

(A) aminoacyl-tRNA synthetases transfer the amino acids to the A on the 3' end of tRNA via an aminoacyl–adenosine monophosphate intermediate
(B) the aminoacyl-tRNA synthetase is specific for a given amino acid
(C) there may be more than one kind of tRNA for a given amino acid
(D) the activation step requires the hydrolysis of the terminal phosphate group of guanosine triphosphate
(E) once an amino acid is attached to a given tRNA, the specificity for recognition of the appropriate codon on mRNA is due entirely to the tRNA, and not at all to the amino acid

6. All of the following factors are required to initiate translation of mRNA in prokaryotes EXCEPT

(A) formylmethionine-tRNA$_f$
(B) 30S ribosomal subunits
(C) guanosine triphosphate
(D) mRNA
(E) σ factor

7. Which of the following statements about chain termination in peptide synthesis is true?

(A) The formation of peptide bonds by the ribosomal mRNA complex continues until a stop codon on mRNA is reached
(B) The formation of peptide bonds by the ribosomal mRNA complex continues until a tRNA with an anticodon for UAA, UAG, or UGA interacts with the A site on the ribosome
(C) In prokaryotes, peptide bond formation ceases when a formylmethionine-tRNA$_f$ interacts with the A site on the ribosome
(D) Peptide bond formation ceases when the ribosome reaches the 5' end of the mRNA
(E) Peptide bond formation continues until the ribosome dissociates into large and small subunits

64 *Biochemistry*

8. The signal hypothesis refers to

(A) the hormonal control of eukaryote protein synthesis
(B) the synthesis of secreted proteins by membrane-bound ribosomes
(C) characteristics of special ribosomes involved in the synthesis of proteins for export by the cell
(D) a sequence of amino acids at the carboxyl end of a secretory protein, which is cleaved off after the secretion of protein by the cell
(E) none of the above

9. Which of the following statements about antibiotic inhibitors of protein synthesis is correct?

(A) Chloramphenicol inhibits the peptidyl transferase activity of the large ribosomal subunit in eukaryotes c (prokaryotes)
(B) Cycloheximide inhibits the peptidyl transferase activity of the large ribosomal subunit in eukaryotes
(C) Puromycin blocks protein synthesis in prokaryotes only
(D) Streptomycin inhibits protein synthesis by binding to 30S ribosomal subunits
(E) Erythromycin inhibits translocation in eukaryotes only

Directions: Each question below contains four suggested answers of which **one or more** is correct. Choose the answer

A   if **1, 2, and 3** are correct
B   if **1 and 3** are correct
C   if **2 and 4** are correct
D   if **4** is correct
E   if **1, 2, 3, and 4** are correct

10. Peptide bond synthesis requires an input of energy during

(1) amino acid activation
(2) the formation of the 70S initiation complex of prokaryotes
(3) the binding of the aminoacyl-tRNA to the A site on the ribosome
(4) the movement of the peptidyl-tRNA to the P site and the associated movement of the mRNA

11. If a genetic code is degenerate, it means that

(1) a given base triplet can code for more than one amino acid
(2) there is no punctuation in the code sequences
(3) the third base in a codon is not important in coding
(4) a given amino acid can be coded for by more than one base triplet

12. Characteristics of prokaryote ribosomes include which of the following?

(1) They are composed of two subunit particles with sedimentation coefficients of 30S and 50S
(2) They are composed of about 65% RNA and about 35% protein
(3) They can be split into their component subunits by treatment with a Mg^{2+} chelator
(4) They have all of their protein in the 50S subunit

13. During the synthesis of glycoproteins, a core oligosaccharide complex is linked to amino acid side chains of the growing peptide chain by transfer from an activated lipid carrier. Correct statements regarding this process include which of the following?

(1) This process takes place on free ribosomes in the cytoplasm
(2) The lipid carrier involved is carnitine
(3) The lipid carrier has to cross the endoplasmic reticulum membrane in order to function
(4) The core oligosaccharide complex is linked to asparagine residues on the protein

ANSWERS AND EXPLANATIONS

1. The answer is B. (*I A, B; Chapter 6 II C 1*) The genetic code refers to the nucleotide triplet sequences in mRNA that code for the 20 common amino acids found in proteins and that react in an antiparallel manner with triplet nucleotide sequences on tRNAs. The amino acid sequence of any protein is encoded in a gene in the form of nucleotide triplets. The genetic code refers to these sequences and not to the amino acid sequences in the protein. The genetic code does not refer to chromosome number or behavior during cell division. The hierarchy of DNA, RNA, and protein was an early assumption in the development of molecular genetics, which has now been modified by the finding of RNA-directed DNA synthesis in certain viruses by the enzyme reverse transcriptase.

2. The answer is E. (*I B; III A*) The synthetic polynucleotide sequence of CAACAACAACAA . . . could be read by the in vitro protein synthesizing system starting at the first C, the first A, or the second A. In the first case, the first triplet codon would be CAA, which codes for glutamine. In the second case, the first triplet would be AAC, which codes for asparagine; and in the last case, the triplet would be ACA, which would have to be a codon for threonine.

3. The answer is D. (*III C, VII C 1 c*) In prokaryotes, the three chain termination codons, UAA, UAG, and UGA, do not specify amino acids, and there are no tRNAs whose anticodon recognizes them. They are recognized by the protein release factors, RF-1 and RF-2, which bind to the termination codons. RF-1 recognizes UAA or UAG, and RF-2 recognizes UAA or UGA. The binding of the release factors initiates the hydrolysis of the link between the polypeptide and the tRNA.

4. The answer is D. (*V B*) The codon–anticodon interaction is due to complementary base-pairing between the two triplets, which line up antiparallel to one another. The base-pairing between the 5′-end base of the codon and the 3′-end base of the anticodon is the usual Watson-Crick base-pairing, that is, A with U and G with C, as is the base-pairing of the middle base of the triplet. The base-pairing of the 3′-end base of the codon and the 5′-end base of the anticodon shows more latitude. When the 5′-end base (first base) of the anticodon is C or A, the base-pairing with the codon is regular; that is, the base in the codon is G or U. If the first base of the anticodon is U, then the 3′-end base (third base) of the codon can be either G or A. If the first base of the anticodon is G, then the third base of the codon can be U or C. If the first base of the anticodon is I (inosine), then the third base of the codon can be A, C, or U, but not G.

5. The answer is D. (*IV B*) The reaction of amino acid and tRNA requires adenosine triphosphate and proceeds via the formation of an aminoacyl intermediate, which may or may not be detectable, depending upon the specific amino acid and tRNA in question. The aminoacyl-tRNA synthetase (transferase) must be highly specific for both amino acid and tRNA, as once an amino acid is attached to a tRNA the specificity for recognition of the appropriate codon on mRNA is due entirely to the tRNA, and protein synthesis proceeds without further check on the amino acid carried by the tRNA.

6. The answer is E. (*VII A 1*). Initiation of a round of protein synthesis involves the interaction of the 30S ribosomal subunit, mRNA, formylmethionine-tRNA$_f$, initiation factors, and guanosine triphosphate (GTP) to form the initiation complex. This complex then interacts with the 50S ribosomal subunit, forming the 70S initiation complex; the hydrolysis of the terminal high-energy phosphate bond of GTP is essential for the complex to form. The σ factor is required for RNA polymerase to bind specifically to the promoter region of DNA in the transcription process. It is not involved in translation.

7. The answer is A. (*VII C*) The genetic code contains three stop codons—UAA, UAG, and UGA—for which there are no corresponding tRNA molecules. In prokaryotes, these codons are recognized by two protein release factors, one of which recognizes UAA or UGA, the other recognizing UAA or UAG. The binding of these factors to the stop codons activates the hydrolysis of the bond between the polypeptide and the tRNA.

8. The answer is B. (*IX B 2*) Cells that synthesize proteins for export contain a high concentration of ribosomes bound to the endoplasmic reticulum (i.e., rough endoplasmic reticulum). These ribosomes are no different from free ribosomes. Proteins that are to be secreted contain a special sequence of highly hydrophobic amino acids at their amino end, which is the signal for attachment of the ribosomes to the ribosomal receptor proteins of the endoplasmic reticulum. This signal sequence is removed from the protein after the latter is extruded through the endoplasmic reticulum membrane into the lumen of the endoplasmic reticulum.

9. The answer is B. (*X B 2 a*) Cycloheximide inhibits the peptidyl transferase activity of 50S ribosomal subunits in eukaryotes. This function of the larger subunit is blocked in prokaryotes by chloramphenicol. Erythromycin inhibits the translocation activity of prokaryote ribosomes by binding to the 50S subunit. Puromycin inhibits protein synthesis in both prokaryote and eukaryote cells, as it is an analogue of the aminoacyl–adenosine moiety of an aminoacyl-tRNA. It binds at the A site on the ribosome, and its amino group forms a peptide bond with the carboxyl group of the peptidyl-tRNA on the P site. The resulting peptide with a C-terminal puromycin group leaves the ribosome. Streptomycin inhibits protein synthesis by blocking the binding of formylmethionine-tRNA$_f$ to the P site on the ribosome.

10. The answer is E (all). (*IV B 4; VII A 7, B 1, 3*) The formation of aminoacyl-tRNAs requires the participation of adenosine triphosphate (ATP) and the formation of pyrophosphate. Thus, the activation step costs the expenditure of two high-energy phosphate bonds. The interaction of the 30S initiation complex, which contains guanosine triphosphate (GTP), with the 50S ribosomal subunit requires the hydrolysis of the GTP. The binding of the incoming aminoacyl-tRNA to the A site and, after peptide bond formation, the movement of the peptidyl-tRNA to the P site, together with the coordinated movement of the mRNA, all require the hydrolysis of another terminal phosphate bond of GTP.

11. The answer is D (4). (*III B; V B*) Of the 20 common amino acids, 18 are coded for by more than one base triplet (codon), which is the reason for calling the code degenerate. However, the code is not ambiguous; that is, a given codon only specifies one amino acid. The third base in a codon—that is, the 3'-end base—has less stringent pairing requirements than the other two bases, and many, but not all, of the multiple codons for an amino acid differ only in the 3' base. This latitude has been termed "wobble."

12. The answer is A (1, 2, 3). (*VI A 1, 3*) The *Escherichia coli* ribosome is composed of about 65% RNA and 35% protein. It is a large particle, with a sedimentation coefficient of 70S and can be split by treatment with ethylenediaminetetraacetic acid, a Mg^{2+} chelator, into two particles of 30S and 50S, both of which contain RNA and protein. The 30S component contains about 31 proteins and the 50S component about 34 proteins.

13. The answer is D (4). (*IX B 3*) Oligosaccharide units are attached by N-glycosidic bonds to asparagine residue side chains during the biosynthesis of glycoproteins. This process takes place on the luminal side of the endoplasmic reticulum membrane after the protein has been formed on membrane-bound ribosomes (rough endoplasmic reticulum). The oligosaccharide unit is transferred from the activated carrier dolichol phosphate. Carnitine is the carrier of fatty acids crossing the mitochondrial inner membrane.

5

Mutation and
Repair of DNA

I. MUTATIONS

A. Modification of base sequences. Mutations are caused by changes in the base sequence of DNA. They can also arise from the malfunctioning of DNA polymerase. There are several kinds of mutations.

1. Substitution of one base for another. This is the most common type of mutation; it can be divided into two subtypes:

a. Transition, in which one purine is replaced by another purine or one pyrimidine by another pyrimidine. Watson and Crick suggested a mechanism that could be responsible for the occurrence of **spontaneous transitional mutations.** They hypothesized the following:

(1) On rare occasions, a base would be in its unusual tautomeric form at the time of base-pairing.

(2) The rare tautomeric form would change the normal base-pairing.

(3) The rare imino tautomer of adenine was found to pair with cytosine (Fig. 5-1), resulting in a daughter molecule containing a GC base-pair in place of an AT base-pair.

b. Transversion, in which a purine is replaced by a pyrimidine, or vice versa

2. Deletion of one or more base-pairs. In cases of deletion, the entire sequence following the error will be read incorrectly (i.e., the **reading frame** will be disturbed, and the resulting gene product will probably be nonfunctional).

3. Insertion of one or more base-pairs. In cases of insertion, just as with deletion, the entire sequence following the error will be read incorrectly.

B. Types of mutagens

1. Base analogues, such as 5-bromouracil or 2-aminopurine (see Fig. 5-1), can be incorporated into a DNA chain.

a. 5-Bromouracil, a thymine substitute, usually pairs with adenine. However, because of the 5-bromo substitution, 5-bromouracil has a higher proportion of the rare enol tautomer than thymine, and this tautomer pairs with guanine; the transition produced is AT to GC.

b. 2-Aminopurine usually pairs with thymine. However, its normal tautomer can bond with cytosine (one bond); the transition produced is GC to AT.

2. Chemical mutagens

a. Nonalkylating agents

(1) Formaldehyde (HCHO) has the following characteristics:

(a) It reacts with amine groups.

(b) It eventually crosslinks DNA, RNA, and protein.

(2) Hydroxylamine (NH_2OH) has the following characteristics:

(a) It reacts most readily with single-stranded nucleic acids.

(b) It specifically reacts with cytosine.

(c) The modified C will only pair with A.

(3) Nitrous acid (HNO_2) has the following characteristics:

(a) It oxidatively deaminates C, A, or G.

(b) It changes an amine to a ketone.

(c) Cytosine becomes uracil, adenine becomes hypoxanthine, and guanine becomes xanthine.

(d) The base-pairing is changed.

b. Alkylating agents

(1) Ethylmethane sulfonate (EMS) has the following characteristics:

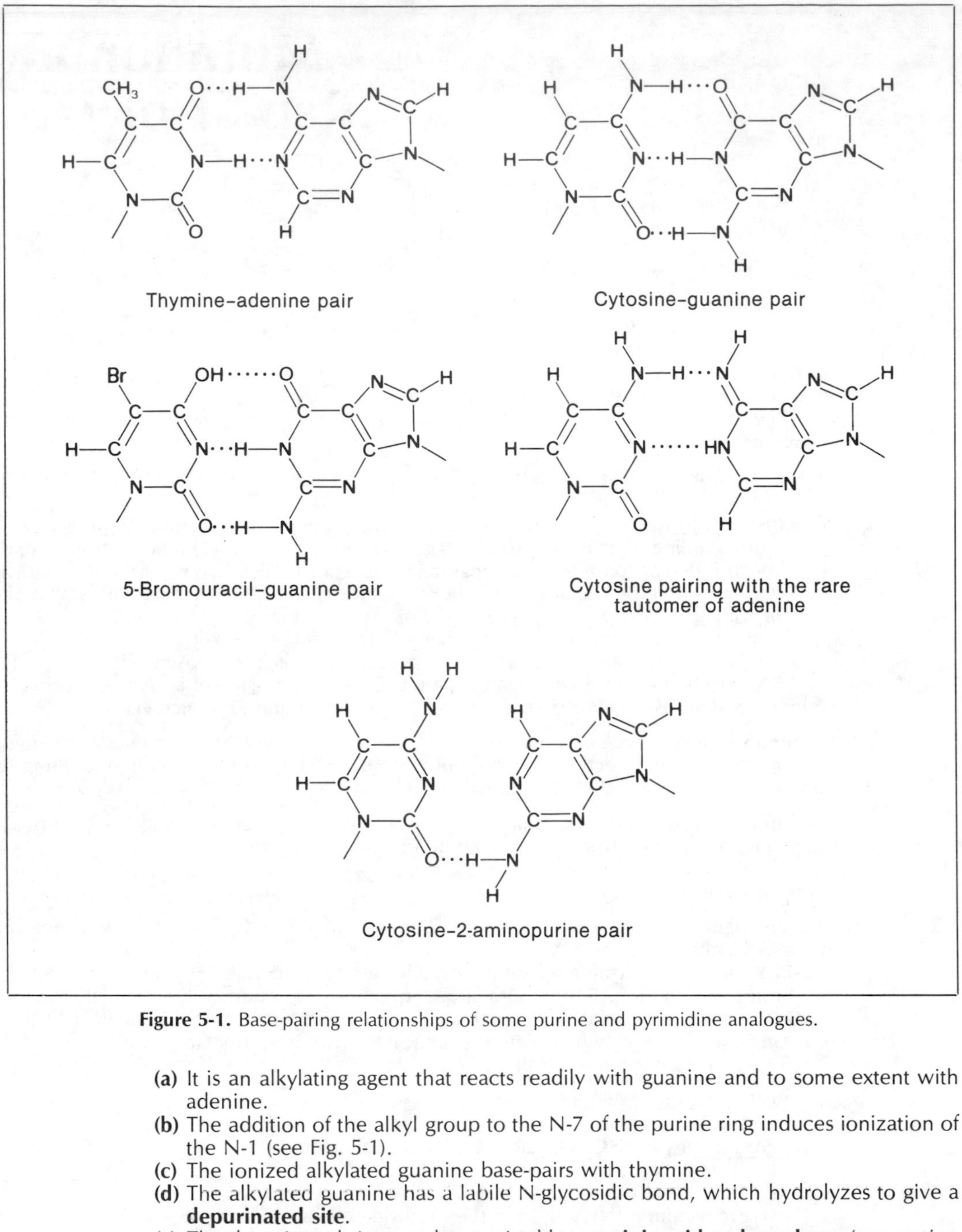

Figure 5-1. Base-pairing relationships of some purine and pyrimidine analogues.

    (a) It is an alkylating agent that reacts readily with guanine and to some extent with adenine.
    (b) The addition of the alkyl group to the N-7 of the purine ring induces ionization of the N-1 (see Fig. 5-1).
    (c) The ionized alkylated guanine base-pairs with thymine.
    (d) The alkylated guanine has a labile N-glycosidic bond, which hydrolyzes to give a **depurinated site**.
    (e) The depurinated site may be repaired by **apurinic acid endonuclease** (see section II A 2 b), but if replication precedes this repair, any base may be inserted.
    (f) After the second round of replication, the original GC pair may become a CG, AT, or TA.
    (g) This is an example of a **transversion mutation**.
  (2) N-methyl-N′-nitro-N-nitrosoguanidine (NNG) has the following characteristics:
    (a) It is an alkylating agent and a very powerful mutagen.
    (b) NNG-induced mutations usually affect several sites in a given gene as well as other genes that are nearby in the genetic map.
    (c) Since NNG exerts its effect mostly in a replication fork, the mutations are grouped in the region of DNA that is being replicated during exposure to the mutagen.

    c. Intercalating agents
      (1) Acridines are flat molecules that can intercalate between the bases of a DNA chain, a process that leads to a shift in the reading frame.
      (2) This insertion mutation (**frame-shift mutation**) produces a nonfunctional gene product.

  3. Ultraviolet (UV) light [200–400 nm]
    a. UV light induces changes in the structure of DNA. In particular, it induces the formation of dimeric adjacent pyrimidine bases, mostly TT dimers (Fig. 5-2).
    b. DNA polymerase cannot replicate through a dimer as the helix is distorted and DNA synthesis stalls.
    c. DNA polymerase III (DNA pol III) stalls because the distortion in the helix due to the TT dimers activates the editing function, so that it repeatedly puts in an A and excises it out again (**"stuttering"**).

C. Ames test

  1. The Ames test assumes that there is a relationship between mutagenesis and carcinogenesis.

  2. It uses a mutant strain of *Salmonella typhimurium*, which requires histidine in the medium in order to grow.   *is mutagen*

  3. Some of these mutants normally revert back to wild type, and the rate of **back conversion** is greatly enhanced by the presence of a mutagen.

  4. The test compares the relative mutagenicity of a substance with the rate of reversion of the *Salmonella*.

  5. There is a 90% agreement between carcinogens defined by animal studies and those found by the Ames test.

II. REPAIR OF DNA

A. Excision repair

  1. Repair of thymine dimers
    a. A UV-specific endonuclease makes a nick in the affected DNA strand, usually on the 5′ side of the dimer (Fig. 5-3), and the defective segment swings out.
    b. DNA polymerase I (DNA pol I) synthesizes new DNA in the 5′ to 3′ direction with the 3′ end of the nicked strand serving as the primer and the intact complementary strand serving as the template.
    c. The 5′- to 3′-exonuclease activity of DNA pol I then removes the damaged sequence.
    d. Finally, DNA ligase seals the gap between the newly synthesized segment and the main chain.
    e. Xeroderma pigmentosum
      (1) This condition, which is transmitted as an autosomal recessive trait, is an example of a defective mechanism for the repair of pyrimidine dimers in DNA.
      (2) Individuals affected by this disease, who are abnormally sensitive to UV light, are more prone to skin cancer than normal subjects.
        (a) In some cases, the defect appears to involve the UV-specific endonuclease required for the recognition of the dimers.
        (b) In other cases, a normal excision repair may occur, but a deficiency in the proofreading process of DNA pol I may be present.

Enzyme is defective Endonuclease

  2. Cytosine deamination to uracil
    a. The deamination of cytosine (2-oxy, 4-aminopyrimidine) to form uracil (2,4-dioxypyrimidine) occurs spontaneously due to the inherent instability of cytosine. On replication, an A will be inserted to pair with the U formed, and this will be a mutation.

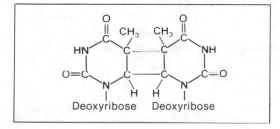

Figure 5-2. Structure of an intrastrand thymine dimer.

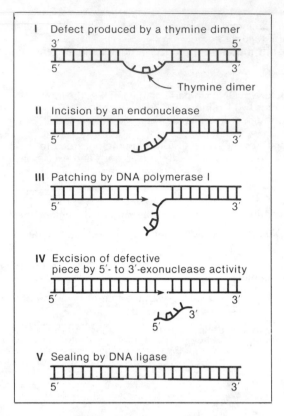

I Defect produced by a thymine dimer

— Thymine dimer

II Incision by an endonuclease

III Patching by DNA polymerase I

IV Excision of defective piece by 5'- to 3'-exonuclease activity

V Sealing by DNA ligase

Figure 5-3. Excision of a thymine dimer by enzymatic processes. (Reprinted with permission from Hanawalt PC: Repair of genetic material in living cells. *Endeavour* 31:84, 1972.)

   b. The repair involves the following:
   (1) Removal of the U by the specific enzyme **uracil N-glycosylase** cleaves the N-glycosidic bond, leaving the deoxyribose phosphate backbone intact.
   (2) Another enzyme, **apurinic acid endonuclease**, cuts the phosphodiester backbone on the 5' side.
   (3) DNA pol I then fills the gap with the correct C and also clips off the deoxyribose residue left over from the excised U.
   (4) DNA ligase seals the break.

B. Photoreactivation, or light-induced repair, is an enzymatic cleavage of thymine dimers activated by visible light (300–600 nm), leading to a restoration of the monomeric condition. The enzyme involved is the **photoreactivating (PR) enzyme**.

   1. This enzyme is found in all organisms.

   2. It binds to NN dimers, including cytosine dimers, in the DNA.

   3. It is activated by **visible light energy**, and then cleaves the CC bonds of the dimer.

   4. It is then released from the DNA.

C. Recombinational repair (sister-strand exchange)

   1. An *Escherichia coli* cell can deal with TT dimers by stopping chain growth at the dimer and reinitiating it on the other side of the dimer (unprimed postdimer initiation). This leaves gaps (Fig. 5-4).

   2. In sister-strand exchange, the unmutated single-stranded segment from homologous DNA is excised from the "good" strand and inserted into the "gap" opposite the dimer.

   3. The recombinational repair system can identify and repair mutations, which, although they might stop synthesis, do not disturb the helix. Excision repair (see section II A 1) would not detect such mutations.

   4. Recombinational repair occurs after the first round of DNA replication, but excision repair occurs before this round.

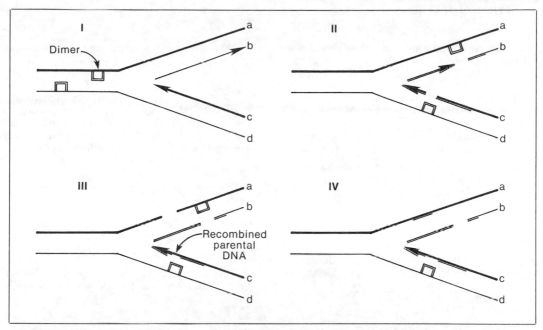

Figure 5-4. Recombinational repair. *I*, A molecule containing two thymine dimers (*boxes*) in strands a and d is being replicated. *II*, By transdimer synthesis, a molecule is formed whose daughter strands b and c have gaps. If repair does not occur, in the next round of replication, strands a and d would yield gapped daughter strands, and strands b and c would again be fragmented. *III*, By sister-strand exchange, a continuous segment of a is excised and inserted into strand c. (In a second round of replication, strand c would be a template for synthesis of functional DNA.) *IV*, The gap in a is next filled. Such a DNA molecule would probably engage in a second sister-strand exchange in which a segment of c would fill the gap in b. The gap in c would then be filled. In this way, strand c also becomes a functional template. *Heavy and thin lines* are used for purposes of identification only. (Reprinted with permission from Freifelder D: *Molecular Biology*. Boston, Jones and Bartlett, 1983, p 329.)

STUDY QUESTIONS

Directions: Each question below contains five suggested answers. Choose the **one best** response to each question.

1. 5-Bromouracil can cause a transitional mutation in the base sequence of DNA because it

(A) bonds irreversibly with A
(B) may resemble C and bond with G
(C) resembles T but cannot bond with A
(D) resembles U and would not be incorporated into DNA
(E) does none of the above

2. Acridines act as mutagens to duplex DNA, causing which of the following mutations?

(A) Transitional mutation
(B) Transversional mutation
(C) Frame-shift mutation
(D) Single amino acid substitution in the gene product
(E) Silent mutation

3. Xeroderma pigmentosum is an autosomal recessive mutation that affects the mechanism for repair of ultraviolet (UV) light-induced thymine dimer formation in DNA. Which of the following enzymes is most likely to be defective?

(A) DNA polymerase I
(B) DNA polymerase II
(C) DNA ligase
(D) Endonuclease
(E) Exonuclease

4. Excision repair can correct DNA modification arising from which of the following causes?

(A) Base deletion
(B) Base insertion
(C) Methylation of bases
(D) Thymine dimer formation
(E) Alkylation of bases

Directions: Each question below contains four suggested answers of which **one or more** is correct. Choose the answer

A   if **1, 2, and 3** are correct
B   if **1 and 3** are correct
C   if **2 and 4** are correct
D   if **4** is correct
E   if **1, 2, 3, and 4** are correct

5. Which of the following abnormalities in DNA can be corrected by excision repair systems?

(1) Intrastrand pyrimidine dimers
(2) Alkylated bases
(3) Cytosine deamination
(4) Missing base-pairs

6. Photoreactivation in DNA repair requires

(1) an endonuclease
(2) pyrimidine dimers
(3) activation by ultraviolet (UV) light (200–400 nm)
(4) an enzyme that cleaves —C—C— bonds

7. The Ames test of mutagenicity is characterized by which of the following statements?

(1) The test uses a mutant strain of bacteria that requires histidine in the medium to grow but that reverts to wild type at a slow rate
(2) The mutant bacterial strain used reverts back to wild type on withdrawal of histidine from the medium
(3) Mutagens markedly enhance the rate at which the bacterial strain reverts to wild type
(4) Mutagenicity (in the Ames test) and carcinogenicity (defined by animal studies) are not correlated

ANSWERS AND EXPLANATIONS

1. The answer is B. (*I B 1 a*) 5-Bromouracil can be incorporated into DNA, where it usually pairs with adenine, as does thymine. However, because of the 5-bromo substitution, the proportion of the rare enol tautomer is higher than for thymine, and this tautomer pairs with guanine. A transitional mutation is one in which one purine is replaced by another, or one pyrimidine is replaced by another. In the case of 5-bromouracil–induced mutations, the change is from an AT pairing to a GC pairing when DNA is replicated.

2. The answer is C. (*I B 2 c*) Acridines are flat molecules that intercalate between the bases in a DNA chain, leading to a shift in the reading frame. In such a mutation, the entire message from the point of insertion to the 3′ end is misread, and usually the product (the peptide specified) is biologically inactive. Transitional and transversional mutations are substitutions of one base for another; transitions occur when one purine is substituted for another or one pyrimidine for another, and transversions occur when there is a substitution of a pyrimidine for a purine or vice versa.

3. The answer is D. (*II A 1 e*) Thymine dimers in DNA chains can form on irradiation with ultraviolet (UV) light. A specific repair system depends upon UV-specific endonuclease that recognizes the thymine dimer and makes a nick in the DNA strand, usually on the 5′ side of the dimer. Removal of the damaged sequence and its replacement by bases that are complementary to the parental DNA strand are carried out by DNA polymerase I. The gap between the main chain and the newly synthesized segment is sealed by DNA ligase. In the genetic abnormality underlying the condition known as xeroderma pigmentosum, the UV-specific endonuclease that recognizes the thymine dimers is missing or much reduced in activity. Thus, the excision repair process does not function.

4. The answer is D. (*II A 1*) Thymine dimers in DNA chains can form on irradiation with ultraviolet (UV) light. A specific repair system depends upon a UV-specific endonuclease, which recognizes the thymine dimer and makes a nick in the DNA strand usually on the 5′ side of the dimer. Removal of the damaged sequence and its replacement by bases, which are complementary to the parental DNA strands, are carried out by DNA polymerase I. The gap between the main chain and the newly synthesized segment is sealed by DNA ligase. Base deletions, insertions, methylations, or alkylations are not recognized by the UV-specific endonuclease as targets for the excision repair system.

5. The answer is B (1, 3). (*II A 1, 2*) Intrastrand pyrimidine dimers can arise in DNA as a result of irradiation by ultraviolet (UV) light (200–400 nm). The dimer prevents the replication of DNA because the resulting distortion of the helix activates the editing function of DNA polymerase III, and the enzyme stalls. The repair system requires a UV-specific endonuclease and the participation of DNA polymerase I (DNA pol I) and DNA ligase. The spontaneous deamination of cytosine to uracil is repaired by the action of the specific enzyme uracil N-glycosylase, which cleaves the N-glycosidic bond, leaving the deoxyribose phosphate backbone intact. Apurinic acid endonuclease then cuts the phosphodiester backbone, DNA pol I fills the gap, and DNA ligase seals the break. Alkylated bases cannot be repaired, and missing base-pairs cannot be replaced.

6. The answer is C (2, 4). (*II B*) Photoreactivation, or light-induced repair, is an enzymatic cleavage of thymine dimers in DNA, which is activated by visible light (300–600 nm). The enzyme involved is the photoreactivating enzyme, and it breaks the —C—C— bonds of the dimer restoring the monomeric condition.

7. The answer is B (1, 3). (*I C*) The Ames test uses a mutant strain of *Salmonella typhimurium*, which requires histidine in the medium. Some of the mutants revert back to wild type, and the rate of conversion is greatly enhanced by the presence of a mutagen. There is a 90% agreement between mutagenicity (defined by the Ames test) and carcinogenicity (defined by animal studies).

histidine

6
Recombinant DNA

I. REARRANGEMENT OF GENES. Different kinds of gene rearrangements have occurred during evolution, a process that is dependent upon the generation of **genetic variation**.

A. Mutation is the first step in the development of genetic variation (see Chapter 5). These variations are enhanced by rearrangements of mutations among the other genetic material.

B. Sexual reproduction requires that the organism has generations of cells containing single sets of chromosomes (**haploid cells**) alternating with generations of cells containing a double set of chromosomes (**diploid cells**). Cycles of haploidy, fusion, diploidy, and meiosis create new combinations of genes.

    1. Fusion of two haploid cells (**gametes**) to form a diploid cell (**zygote or progeny**) involves the mixing (**rearrangement**) of genomes.

    2. Meiosis. Haploid cells arise from diploid cells by **meiosis**, a process in which the genes of diploid cells are sorted into sets of single chromosomes (Fig. 6-1). Meiosis also provides occasion for genetic recombinations between chromosomes to give each cell of the new haploid generation a novel assortment of genes derived from both parents.

    3. Diploidy provides an organism with an important advantage over haploid organisms in that there is an extra copy (**allele**) of each gene that can mutate and form part of a pool of genomic changes, which may be of survival value to the organism.

    4. Haploidy, which is the state of the genome in the gametes after the process of meiosis, allows the mixing of parental genomes as each gamete contains only one copy of each of the homologous chromosomes. As the chromosomes before pairing in meiosis exist as pairs of attached sister chromatids, two nuclear divisions are needed to produce four haploid cells (see Fig. 6-1).

C. Acquisition of genetic information by bacteria. Knowledge of the mechanisms involved in rearrangements of genetic material has come primarily from the study of microorganisms, although the general principles established also apply to higher organisms. However, microorganisms are haploid, and rearrangements between alleles do not normally occur. These organisms acquire new genetic information by obtaining DNA from other cells and incorporating it into their own genome. Several mechanisms for acquiring foreign DNA include

    1. Transformation. A recipient cell takes up DNA that has been released to the environment after a cell death and incorporates this DNA into its genome.

    2. Transduction. A phage acquires DNA from the genome of a cell that it has infected, and on leaving this cell, the combination of phage and foreign DNA may infect another cell. Transduction occurs when the foreign DNA is incorporated into the genome of the recipient cell.

    3. Conjugation. A plasmid, or DNA, from a **high frequency of recombination cell (Hfr cell)** [see section I D 2 c (1) (b)] is transferred from one cell to another by cell-to-cell contact.

D. Mobile genetic elements

    1. Lysogenic phage
        a. Are viral particles that parasitize bacteria
        b. Consist of a single molecule of nucleic acid, which may be
            (1) Single- or double-stranded DNA
            (2) Linear or circular DNA
            (3) Single-stranded linear RNA
        c. May contain one or more proteins in addition to the nucleic acid

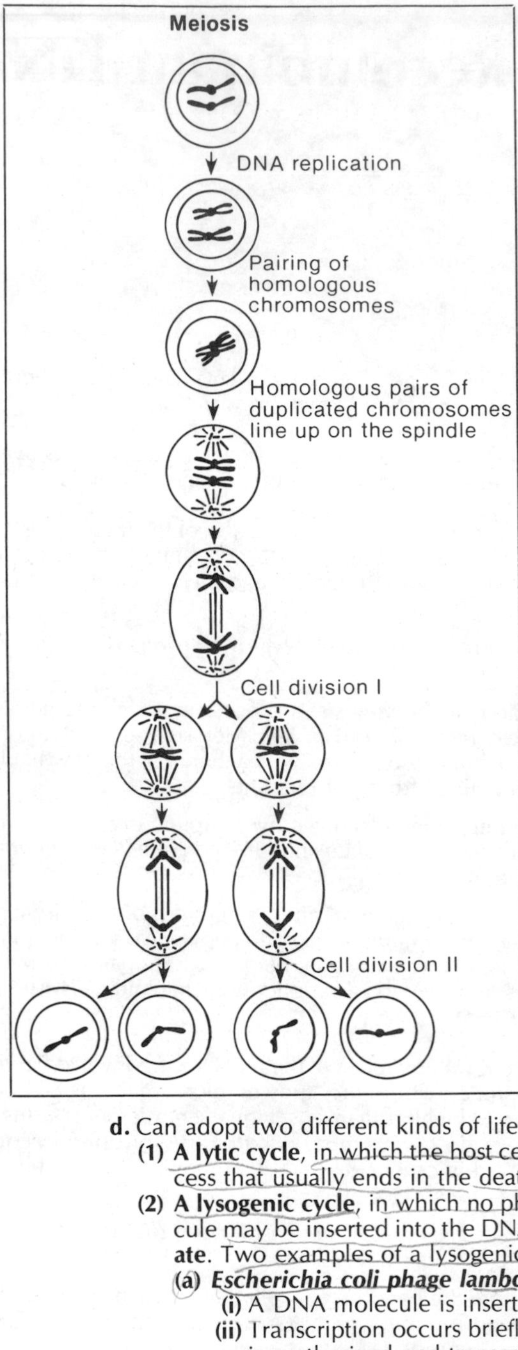

Figure 6-1. Schematic diagram illustrating normal meiosis. The pairing of homologous chromosomes (homologues) is unique to meiosis; because each chromosome is duplicated and exists as attached sister chromatids before this pairing occurs, two nuclear divisions are required to produce the haploid gametes. Each diploid cell that enters meiosis, therefore, produces four haploid cells. (Reprinted with permission from Alberts B, et al: *Molecular Biology of the Cell.* New York, Garland, 1983, p 777.)

   d. Can adopt two different kinds of life cycle

      (1) **A lytic cycle**, in which the host cell is induced to produce many phage progeny, a process that usually ends in the death of the cell. Such phages are said to be **virulent**.

      (2) **A lysogenic cycle**, in which no phage progeny are formed, and the phage DNA molecule may be inserted into the DNA of the host cell. Such phages are said to be **temperate**. Two examples of a lysogenic cycle are

         (a) *Escherichia coli phage lambda*

            (i) A DNA molecule is inserted into a bacterial cell.

            (ii) Transcription occurs briefly. An integration enzyme, coded for by the *int* gene, is synthesized and transcription is shut down by a repressor.

            (iii) A phage molecule is inserted into the genome of the bacterial cell.

            (iv) As the bacterium grows and divides, the phage genes are replicated as part of the bacterial genome.

         (b) *E. coli phage P1*

            (i) There is no integration system.

            (ii) The phage DNA becomes a plasmid—an independently replicating DNA molecule.

2. Plasmids

   a. Definition

      (1) Plasmids are extrachromosomal, circular, duplex, DNA molecules that are found in most species of bacteria and in some eukaryotes.

    (2) They are usually not essential to the host cell, although, in certain circumstances, they may be. For example, plasmids may
        (a) Contain genes for resistance to antibiotics (**R plasmids**) and would thus confer that trait upon the host
        (b) Carry genes from a host cell genome (**F′ plasmids**), which might counteract a mutant host gene

b. General properties
    (1) All plasmids are dependent upon the metabolism of the host cell for their reproduction.
    (2) Each kind of plasmid has, however, unique genes and regulatory sites that determine the time of synthesis and the number of copies per cell of progeny plasmids.
        (a) The **stringent, or low-copy-number, plasmids** produce one or two copies per cell.
        (b) The **high-copy-number plasmids** produce 10 to 100 copies per cell.
    (3) The segregation of plasmid replicas into daughter cells upon division of the parental cell is carefully controlled by an unknown plasmid–chromosome association.

c. Types. Three main types of plasmids have been studied from among the many found in a variety of strains of *E. coli.*
    (1) F, the sex plasmids
        (a) This plasmid transfers a replica of the plasmid from a **donor (F +)** cell to a **recipient (F −)** cell without the F+ cell losing its plasmid.
        (b) The F plasmid DNA can integrate into the chromosome of the recipient cell to produce an Hfr cell. *high frequency of recombination cell*
        (c) Imperfect excision of F may also occur, which produces a plasmid-containing chromosomal gene. This is called an **F′ plasmid.**
    (2) R, the drug resistance plasmids, carry genes conferring resistance to one or more antibiotics and usually can transfer this resistance to an R-free recipient cell.
    (3) Col, the colicinogenic factor plasmids
        (a) Carry genes for the synthesis of proteins called **colicins**, which can kill related strains of bacteria that lack the col plasmid
        (b) Are **non-selftransmissible**—that is, they can prepare their DNA for transfer but do not have the genes necessary for determining effective contact between donor and recipient cells.
    (4) Clinically interesting plasmids include
        (a) Plasmids called *Ent*, which are responsible for traveler's diarrhea and some types of dysentery; the plasmid contains genes that code for an intestinal irritant called **enterotoxin.**
        (b) A plasmid found in some strains of *Bacillus thuringiensis*, which has genes that code for a product that is toxic for gypsy moths and tentworms

3. Transposable elements or transposons are linear segments of DNA, which occur in the genome of a cell and which appear at a second site in the genome of the same cell.
    a. The so-called transposons have no base sequence homology with the DNA at the new site.
    b. Insertion of the transposons into the DNA at the new site is *not* dependent upon the bacterial *recA* gene (see section I E 2 a).
    c. Although the process is known as **transposition**, one copy of the transposable element remains at its original site. Thus, DNA replication is required, in contrast to homologous recombination (see section I E 2 a).

E. Types of recombination mechanisms. The transfer of DNA into the genome of a recipient cell is necessary if the donated DNA is to be maintained through subsequent cycles of cell division, unless the transfer is effected by a plasmid, which can replicate independently of the DNA of the host cell. In all types of recombination events, two DNA molecules (or two portions of the same DNA molecule) interact in an event called **pairing**. Four types of recombination events can be distinguished.

1. Site-specific recombination is
    a. Mediated by the phage lambda *int* gene [see section I D 1 d (2) (a)]
    b. Characterized by an exchange between two identical DNA sequences (15 base-pairs each) that are flanked by four nonidentical sites, which are essential to the specificity of the exchange

2. Homologous recombination. The processes of transformation, transduction, and conjugation require **extensive homology** between the interacting sequences. These interacting regions have hundreds of base-pairs that are identical or nearly identical. Thus, these processes are called homologous recombination.
    a. Some processes of this type require, in addition to homology, the function of a **host gene**; in *E. coli*, it is called *recA*, and the process is termed **recA-dependent.**

    b. The *recA* product is required for the pairing of the DNA strands that will interact.
    c. The proteins involved do not require specific base sequences, but they do require segments having identical, or almost identical, sequences.
    d. Four facts have been established.
        (1) DNA molecules are broken and then rejoined.
        (2) Homologous base-pairing is required for the rejoining to occur.
        (3) "Lap joints," or regions of overlap, are seen during the recombination process.
        (4) DNA replication is **not** necessary for recombination to occur.

  3. Transposition
    a. Involves the movement of **transposons** (see section I D 3) from one part of a DNA molecule to another or from one chromosome to another in a diploid cell
    b. Does not require homology and is *recA*-**independent**
    c. Accounts for a number of deletions and inversions found in bacterial genomes but occurs with a lower frequency than homologous recombination

  4. "Illegitimate" recombination occurs between nonhomologous sequences and requires neither the *recA* product nor transposable elements. There is little understanding of this type of recombination. Table 6-1 summarizes the four types of recombination.

II. CREATION OF RECOMBINANT DNA MOLECULES

  A. Introduction

    1. Recombinant DNA methodology involves the linking of DNA molecules with two basic problems to be solved.
      a. The coupling of the DNA molecule of interest to a DNA molecule that can replicate
      b. The provision of a system that allows replication of the recombinant DNA

    2. Donors. Replication of the genes contained in the donor segment of the recombinant DNA molecule is called **cloning**.

    3. Vectors. The carrier portion of the recombinant DNA molecule is called the vector. Three properties are important for its function.
      a. The ability to enter into a cell
      b. The ability to replicate
      c. Innate properties that allow its presence in a cell to be detected

  B. Restriction endonucleases (or restriction enzymes) are of prime importance in the preparation of both the donor and vector segments of recombinant DNA molecules.

    1. Restriction endonucleases are bacterial enzymes that recognize specific base sequences in double-stranded DNA and make two highly sequence-specific cuts, one in one strand and one in the other. These cuts generate 3'-hyroxyl (OH) and 5'-phosphate (P) ends.

    2. In most cases, the sequence recognized has a **twofold axis of symmetry**. For example, the enzyme *EcoRI* recognizes the following duplex sequence:

$$5'\text{-GAA.TTC-}3'$$
$$3'\text{-CTT.AAG-}5'$$

    3. The sequence-specific cuts made by the restriction enzymes are of two kinds.
      a. Blunt ends. For example, the enzyme *HpaI* makes cuts as shown in Figure 6-2A, breaking the phosphate backbones of the two chains next to complementary bases.
      b. Staggered ends. For example, *EcoRI* breaks the phosphate backbones in such a way that short single-stranded ends are produced (sticky ends) at opposite ends of the duplex fragment, producing the staggered ends depicted in Figure 6-2B.

Table 6-1. Characteristics of Four Types of Genetic Recombination

Type	Homology Required	RecA Product Required	Sequence-Specific Enzyme Required
Site-specific	Yes	No	Yes
Homologous	Yes	Yes	No
Transposition	No	No	Yes
Illegitimate	No	No	?

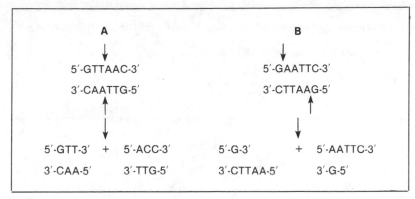

Figure 6-2. Restriction enzymes make two different kinds of cuts in duplex DNA molecules. *A,* One kind of cut is illustrated where both breaks are at the center of symmetry, producing fragments with blunt ends. *B,* The other kind of cut is illustrated where the two cuts are symmetrically positioned about the line of symmetry, producing fragments with staggered ends. These are also called cohesive ends as the two single-stranded segments are complementary.

C. Synthesis of cDNA

1. A double-stranded DNA molecule, complementary in base sequence to an mRNA molecule, can be prepared using the enzyme **reverse transcriptase**.
   a. This enzyme, which is prepared from RNA tumor viruses, can use RNA as a template to synthesize an **RNA–DNA hybrid molecule**.
   b. It requires a primer, which is provided by hybridizing a short chain of oligo dT to the 3′-poly A tail of the mRNA (Fig. 6-3).

2. The DNA strand has a hairpin loop at its 3′ end, which provides a primer for the synthesis of a second DNA strand by **DNA polymerase I** (DNA pol I), which uses the first DNA strand as a template.

3. The double-stranded DNA molecule has a connecting hairpin loop, which can be cleaved by **S1 nuclease**, a single-strand–specific nuclease.

4. The cDNA does not contain **introns**, which are present in the genomic DNA sequence, as these have been spliced out during cellular processing of the mRNA from the primary RNA transcript (see Chapter 3, section II A 3 e).

5. The source of the mRNA is important, as the tissue that is extracted to obtain the mRNA should express adequate amounts of the gene in question.

D. Cloning of cDNA

1. **"Safe" hosts.** The development of methods for the cloning of cDNA required the development of disabled bacterial hosts that had no significant possibility of surviving outside the laboratory.

2. **"Safe" vectors.** Similarly, the plasmid vectors required to insert the cDNA into the bacterial hosts had to be disabled in the sense that they could not move from cell to cell by a sexual process.

3. **Drug resistance plasmids**
   a. The use of vectors containing genes for antibiotic resistance allows the detection of the entry of the plasmids into the previously plasmid-free host, which then becomes resistant to the antibiotic.
   b. The plasmid **pBR322** carries genes for resistance against both ampicillin and tetracycline (Fig. 6-4). It also contains sites for several restriction enzymes, allowing the insertion of foreign DNA into a variety of positions.
      (1) The cleavage of the genome by a restriction enzyme within a gene sequence inactivates that gene but leaves the other drug resistance gene still active.
      (2) Knowledge of the genetic map of pBR322 and the specificity of the restriction enzyme employed allows prediction of the antibiotic resistance pattern of the bacterial cell that is carrying the plasmid.

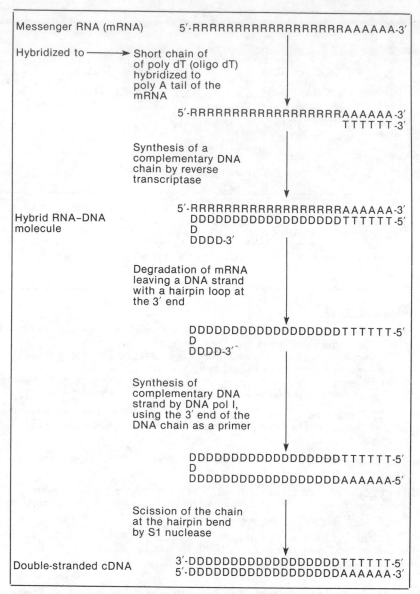

Figure 6-3. Synthesis of double-stranded cDNA from mRNA. The enzyme reverse transcriptase requires a primer, which is provided by hybridizing a short chain of poly dT (oligo dT) to the poly A tail of the messenger RNA. Under in vitro conditions, the reverse transcriptase appears to "turn the corner" at the end of the newly formed DNA chain and start to form a complementary strand. This forms the hairpin bend. The 3′ end of the hairpin bend forms a primer for DNA polymerase I to complete the synthesis of the complementary strand. The hairpin bend is then removed by the action of S1 nuclease, which is specific for single-stranded DNA.

4. **Integration of cDNA with vector DNA.** Insertion of the cDNA into the plasmid can be accomplished by either one of two procedures.
    a. **Tailing with terminal transferase**
        (1) pBR322 is cleaved by a restriction endonuclease that makes blunt cuts, and terminal transferase and deoxyadenosine triphosphate (dATP) are employed to add poly A tails to the 3′ single-stranded ends.
        (2) Terminal transferase and deoxythymidine triphosphate (dTTP) are then employed to attach poly T tails to the cDNA (Fig. 6-5).
    b. **Attaching artificial restriction enzyme sites ("linkers") onto the ends of the cDNA.** The linkers are
        (1) Six to ten base-pair oligonucleotides that are synthesized chemically

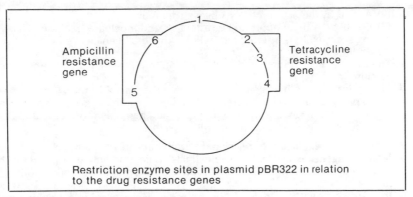

Figure 6-4. The plasmid pBR322 has several restriction enzyme sites, which are indicated in the figure by numerals. *EcoRI* cuts at site 1, *HindIII* at 2, *BamHI* at 3, *SalI* at 4, *PstI* at 5, and *XorII* at 6. Five of these sites are in one or other of the two drug resistance genes in the plasmid. Sites 2, 3, and 4 are in the gene for resistance to tetracycline, and sites 5 and 6 are in the ampicillin resistance gene. (After Watson JD, et al: *Recombinant DNA*. New York, Scientific American Books, WH Freeman, 1983, p 74.)

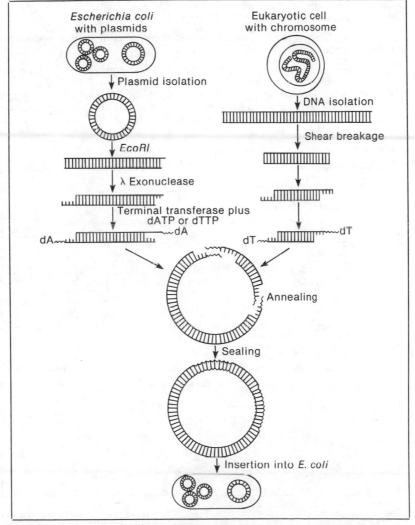

Figure 6-5. Construction of a recombinant plasmid, using the dA-dT tailing technique. (Reprinted with permission from Klein J: *Immunology: The Science of Self–Nonself Discrimination*. New York, John Wiley, 1982, p 254.)

(2) Added by the action of **DNA ligase** (Fig. 6-6)
(3) Cut open with an appropriate restriction endonuclease, and the cDNA, now containing sticky ends, is inserted into pBR322 that has been cleaved with the same restriction endonuclease

c. The plasmid containing the cDNA is put into the host cells by treatment of the cells with calcium ion (Ca^{2+}).

d. The majority of the host cells do not take up the plasmids; therefore, those cells that contain a plasmid have to be separated from those that do not. This is accomplished by growing the host cells on ampicillin or tetracycline; then only those cells that contain the plasmid with its drug resistance genes will be viable.

5. Preparation of genetically pure, clonal populations of recipient cells
   a. **Nonhomogeneous cDNAs.** Tissue extracts containing only one species of mRNA are difficult to obtain, and the cDNA that is prepared from mRNA preparations is usually a mixture of cDNAs.
   b. **Infecting cells with a single cDNA molecule.** The ratio of infecting plasmids to recipient cells can be adjusted so that each recipient bacterial cell receives only one cDNA molecule. Thus, the descendants of this cell will only contain one kind of cDNA.
   c. **Clonal populations of cells** are prepared by selecting single cells and growing them into homogeneous cultures containing the same kind of cDNA in each cell.

E. Construction of chromosomal clones

1. **Genomic "libraries."** If total genomic DNA of a cell is extracted, fragmented, and the fragments cloned, a so-called genomic library may be obtained.

2. **Extraction of DNA** is usually accomplished using **phenol** with precipitation by **ethanol**. **Blood leukocytes** are a convenient source of DNA.

3. **Fragmentation of DNA**, using a variety of restriction endonucleases under conditions that result in incomplete digestion, results in
   a. A mixture of fragments in which there are not too many very small fragments
   b. Any particular sequence will be contained in a number of different fragments
   c. A high probability that a particular gene will be found intact on at least one fragment

4. **Cloning of DNA fragments**
   a. **Plasmids** may be used for the smaller fragments.

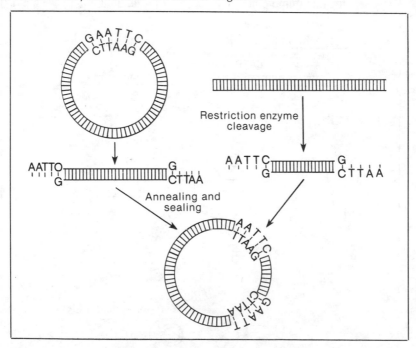

Figure 6-6. Construction of a recombinant plasmid by restriction-enzyme cleavage, resulting in "sticky ends." (Reprinted with permission from Klein J: *Immunology: The Science of Self–Nonself Discrimination*. New York, John Wiley, 1982, p 254.)

   b. Phage lambda [see section I D 1 d (2) (a)] is a more useful vector for chromosomal DNA as
      (1) It has a much larger DNA than plasmids and can accommodate a larger fragment of foreign DNA.
      (2) The recombinant DNA will replicate several thousandfold in the host cell to form complete viral particles that infect surrounding bacteria, providing a cycle of amplification (Fig. 6-7).

   c. Cosmids
      (1) Are synthetic hybrids of plasmid and phage DNA, which will accommodate even larger inserts of foreign DNA than phage lambda
      (2) Contain two plasmid elements, an antibiotic resistance gene, and a sequence of DNA needed to start DNA replication in the bacterial host cell, plus the sequence of DNA to be cloned

 5. Chromosomal walking. Analysis of eukaryotic genomes requires observation of several hundred kilobases of contiguous DNA sequences. However, it is not possible to incorporate this amount of DNA into a single phage or cosmid. Thus, one recombinant phage or cosmid is used to isolate a different recombinant that contains overlapping sequences from the genome—a technique known as "chromosomal walking." The process is as follows:
   a. A segment of DNA from one end of the first recombinant is isolated and then used as a probe to screen the phage or cosmid library again.
   b. The recombinant that results contains the small piece of DNA and the next section of the genome.
   c. This recombinant is then used to obtain a third. With repetition of the process, a set of overlapping fragments can be obtained.

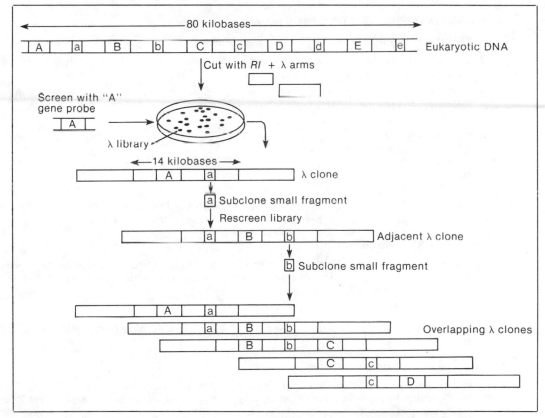

Figure 6-7. Chromosomal walking. One recombinant λ phage obtained from a library of a eukaryotic genome can be used to isolate another recombinant containing the neighboring segment of eukaryotic DNA. The first recombinant is subcloned into small pieces in pBR322. A subcloned DNA fragment that is a single-copy sequence in the genome must be identified; this fragment must be different from the probe that was used to isolate the original recombinant. The subclone is used to rescreen the λ library to obtain a recombinant that has partial overlap with the first λ clone, but that also contains neighboring DNA. This second recombinant is subcloned, and a suitable fragment is used to rescreen the λ library to obtain another partially overlapping clone. (Reprinted with permission from Watson JD, et al: *Recombinant DNA*. New York, Scientific American Books, WH Freeman, 1983, p 81.)

F. Identification of specific clones by hybridization. The cDNA is extracted from the pure clonal cultures, denatured so that the strands separate, and the single-strand molecules bind to nitrocellulose fibers, which prevents the strands from reannealing.

  1. Hybridization with mRNA
  a. Mixtures of mRNAs can be passed through the nitrocellulose filters, which contain bound cDNA, when only the mRNA molecules that are complementary to the cDNA are retained. The noncomplementary mRNAs are washed through the filter of nitrocellulose.
  b. The bound mRNAs can be washed off the nitrocellulose filters and translated in a cell-free system into protein (see Chapter 4, section II A 1), which can often be identified.

  2. Hybridization with cloned cDNA. A cDNA that has been cloned can be used as a probe to screen bacterial colonies for other recombinant, or chimerical, plasmids, which are carrying the same or closely related gene sequences. The nitrocellulose filter technique is used in a manner similar to that employed with mRNA probes.

  3. Hybridization with synthetic oligonucleotides
  a. If the amino acid sequence of a particular protein has been determined, a sequence of five or six amino acids in the protein can be used to predict all the possible mRNA sequences that may code for this sequence of the peptide.
  b. A set of complementary oligonucleotides may then be synthesized chemically and used as probes to screen a cDNA, or a genomic library.
  c. A further use for the set of complementary nucleotides is as a primer for reverse transcriptase in the synthesis of a cDNA that should be enriched in the specific sequence of interest.

III. ANALYSIS OF DNA AND RNA

A. Analysis of chromosomal DNA (Southern blot)

  1. Cleavage. DNA is cleaved with restriction enzymes.

  2. Electrophoresis. The DNA fragments are separated by agarose gel electrophoresis [see Chapter 9, section IV B 3 c (2)].

  3. Blotting. The DNA fragments are transferred to a sheet of nitrocellulose by a flow of buffer (blotting). The fragments bind to the nitrocellulose, creating a replica of the pattern of DNA fragments.

  4. Hybridization. A labeled probe with homology with the gene of interest is hybridized to the filter. The probe may be
  a. Purified RNA
  b. cDNA
  c. A segment of cloned DNA

  5. Autoradiography. The pattern of bands that contain the DNA fragment, or fragments that contain the gene, are visualized by virtue of the radiation from the probe.

B. Analysis of RNA (northern blot)

  1. Electrophoresis. Total cellular RNA, or isolated mRNA, is subjected to agarose gel electrophoresis in the presence of a denaturing agent to remove secondary structure constraints.

  2. Blotting. The gel is blotted onto paper that has been treated chemically so that it will covalently bind RNA.

  3. Hybridization. A labeled probe is used, which allows visualization of the RNA species that are complementary to the probe.

C. Determination of base sequences

  1. The essential process in the analysis of long sequences of DNA is the formation of small fragments of the polymer, in a reproducible way, using the highly sequence-specific restriction endonucleases (see section II B).

  2. Analysis of restriction fragments of DNA was first achieved by synthesizing complementary RNA chains, using RNA polymerase, and by determining the RNA sequence, using procedures developed by **Sanger**.

  3. The first direct procedure for determining the sequence of DNA fragments was called the "plus-minus" method and was an enzymatic method also developed by Sanger.

4. A method based on the chemical degradation of DNA chains, equally as useful as the Sanger procedure, was developed by **Maxam and Gilbert**.

5. Sanger later introduced a second procedure for sequencing DNA, using specific terminators of DNA chain elongation. These were the $2',3'$-dideoxyribonucleotide triphosphates.

STUDY QUESTIONS

Directions: Each question below contains five suggested answers. Choose the **one best** response to each question.

1. Plasmids are characterized best by which of the following statements?

(A) They are the specialized regions of a bacterial chromosome

(B) They are extrachromosomal, circular, duplex, DNA molecules

(C) They are that part of a bacterial genome that is transferred from a donor cell to a recipient cell during conjugation

(D) They are groups of bacterial genes that code for bacterial drug resistance

(E) They are the infoldings of the bacterial cell membrane

2. All of the following statements regarding genetic recombination events are correct EXCEPT

(A) site-specific recombination is characterized by an exchange between two identical DNA sequences

(B) the processes of transformation and transduction require extensive homology between the interacting sequences

(C) *recA*-dependent recombination requires homology between interacting segments as well as the *recA* gene product

(D) DNA replication is required for genetic recombination to occur

(E) during genetic recombination events, the DNA molecules are broken and then rejoined

3. Which of the following statements gives a correct explanation for the use of vectors containing drug resistance genes in the cloning of recombinant DNA (cDNA) molecules?

(A) The products of the drug resistance genes protect the cDNA from destruction by the host cells

(B) The drug resistance genes provide additional base sequences that enable the vector to accommodate larger inserts of cDNA

(C) Entry of the vector containing the cDNA and the drug resistance genes into the host cell renders the latter identifiable as it is now resistant to antibiotic drugs

(D) The cloned cDNA imparts drug resistance upon any cellular system with which it is used

(E) None of the above provide an adequate explanation

4. An analysis of chromosomal DNA, using the Southern blot technique, involves the following five major steps:
1. Autoradiography
2. Blotting
3. Cleavage
4. Electrophoresis
5. Hybridization
Which of the following sequences of steps best illustrates this technique?

(A) 1 2 3 4 5
(B) 1 3 2 4 5
(C) 3 5 2 4 1
(D) 3 2 5 4 1
(E) 3 4 2 5 1

Directions: Each question below contains four suggested answers of which **one or more** is correct. Choose the answer

- **A** if **1, 2, and 3** are correct
- **B** if **1 and 3** are correct
- **C** if **2 and 4** are correct
- **D** if **4** is correct
- **E** if **1, 2, 3, and 4** are correct

5. Microorganisms are haploid, and rearrangements between alleles do not normally occur. These organisms may, however, obtain new genetic information by

(1) taking up DNA from the environment
(2) incorporating into their genome foreign DNA obtained from an infecting phage
(3) receiving a part of the chromosome of another cell by the process of conjugation
(4) rearrangement of the genes in the existing genome

6. Restriction endonucleases are enzymes that

(1) cleave the 5′ terminal nucleotides from duplex DNA molecules
(2) make sequence-specific cuts in both strands of duplex DNA molecules
(3) promote circularization of the duplex DNA molecule by removal of the 5′ terminal nucleotides
(4) generate 3′-hydroxyl and 5′-phosphate ends in the cut DNA stands.

7. Correct statements regarding the enzyme reverse transcriptase include which of the following?

(1) It requires a primer
(2) It requires an RNA template
(3) It synthesizes an RNA–DNA hybrid molecule
(4) It can be extracted from RNA tumor viruses

ANSWERS AND EXPLANATIONS

1. The answer is B. (*I D 2*) Plasmids are extrachromosomal, circular, duplex DNA molecules that are found in most species of bacteria and in some eukaryotes. They may contain: (1) genes for resistance to antibiotics (R plasmids); (2) colicinogenic factors (colicins), which can kill related strains of bacteria that lack the col plasmid; or (3) genes that code for substances toxic to man such as the enterotoxin of *Ent* plasmids.

2. The answer is D. (*I E 1, 2*) In all types of recombination events, two DNA molecules (or two portions of the same DNA molecule) interact in an event called pairing. Several types of recombination events can be distinguished. Site-specific recombination, mediated by the phage lambda *int* gene, is characterized by an exchange between two identical DNA sequences, which are flanked by four nonidentical sites that are essential to the specificity of the exchange. The processes of transformation and transduction require extensive homology between the interacting sequences; thus, these processes are called homologous recombination. Some of the homologous recombination processes require, in addition to homology, the function of a host gene; in *Escherichia coli*, it is called *recA*, and the process is termed *recA*-dependent. The *recA* product is required for the pairing of the DNA strands that will interact. During recombination events, DNA molecules are broken down and then rejoined. DNA replication is *not* necessary for recombination to occur.

3. The answer is C. (*II D 3*) The use of plasmid vectors containing genes for antibiotic resistance allows the detection of the entry of the plasmids into the previously plasmid-free host, which then becomes resistant to the antibiotic.

4. The answer is E. (*III A*) An analysis of chromosomal DNA using the Southern blot technique starts with the cleavage of the DNA with restriction enzymes followed by electrophoresis on agarose gel, which separates the DNA fragments by fragment size. The DNA fragments are transferred to a sheet of nitrocellulose by a flow of buffer (blotting). The fragments bind to the nitrocellulose creating a replica of the pattern of DNA fragments on the gel, which is visualized by autoradiography.

5. The answer is A (1, 2, 3). (*I C 1–3*) Microorganisms cannot sort allelic chromosomes and obtain a new assortment of genes, because they normally do not have alleles. There are, however, three ways in which bacteria can acquire new genetic information. They can take up DNA from the environment and integrate this alien DNA into their genome. This material may be derived from other cells, which die in the environment; the process is known as transformation. Bacteria may acquire DNA by transduction, a process by which a phage acquires DNA from the genome of a cell that it has infected, and on leaving this cell, the combination of phage and foreign DNA may infect another cell. Transduction occurs when foreign DNA is incorporated into the genome of the recipient cell. The third way in which a bacterial cell can obtain new DNA is by conjugation. In this transfer, a plasmid, or DNA, from a high frequency of recombination cell is transferred from one cell to another by cell-to-cell contact.

6. The answer is C (2, 4). (*II B 1*) Restriction endonucleases are bacterial enzymes that recognize specific base sequences in duplex DNA and make two highly sequence-specific cuts, one in one strand and one in the other. These cuts generate 3′-hydroxyl and 5′-phosphate ends. They do not remove 5′ (or 3′) terminal nucleotides from duplex DNA molecules as they are endonucleases, not exonucleases, and removal of a terminal nucleotide does not necessarily provide chains with sticky ends.

7. The answer is E (all). (*II C 1*) Reverse transcriptase is an enzyme found in RNA tumor viruses, which can use an RNA template to synthesize an RNA–DNA hybrid molecule. It requires a primer chain.

7

Regulation of Gene Expression in Prokaryotes

I. OPERON MODEL FOR THE REGULATION OF GENE EXPRESSION

A. Induction. Based on data from genetic studies, Jacob and Monod proposed the operon model in 1961 to explain the phenomenon of induction.* Among the observations that led to the model was the finding that there were two kinds of genes.

1. **Inducible genes.** Some proteins produced by *Escherichia coli*, such as β-galactosidase, are said to be inducible because they are only produced in significant amounts when a specific inducing substance (**inducer**) is present. In the case of β-galactosidase, synthesis of the enzyme is induced by the presence of its substrate, lactose, in the medium.

2. **Constitutive genes.** Most *E. coli* genes maintain a constant rate of synthesis for the protein that they specify; such genes are called constitutive genes, and their expression is a **constitutive expression**.

B. Use of β-galactosides by *E. coli*. The enzyme system in *E. coli* that allows the bacterium to use β-galactosides (e.g., lactose) as a source of carbon and energy was extensively studied by Jacob and Monod in the development of their theory of gene control.

1. **β-Galactosidase.** The key enzyme in the use of lactose by *E. coli* is β-galactosidase, which catalyzes the following reaction:

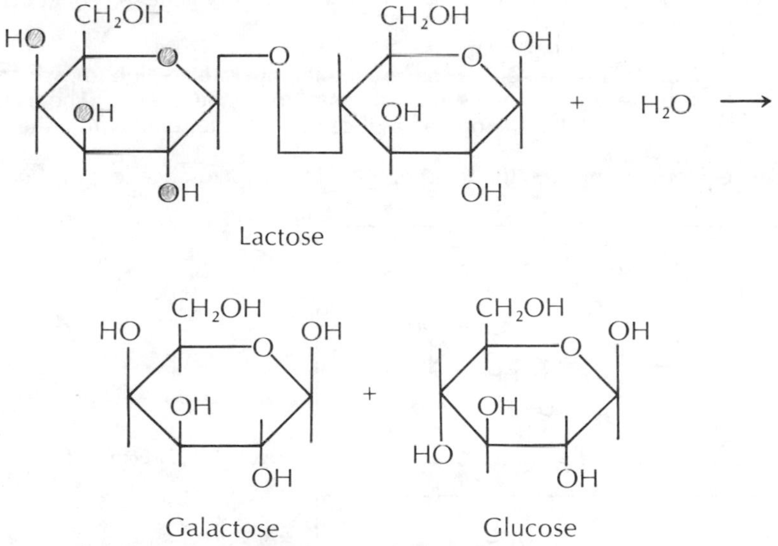

Lactose

Galactose   Glucose

2. **Baseline catalysis by β-galactosidase** occurs when *E. coli* is grown in a medium that is devoid of lactose. Under these conditions, there are fewer than 10 molecules of β-galactosidase per cell.

---

*Jacob F, Monod J: Genetic regulatory mechanisms in the synthesis of proteins. *J Mol Biol* 3:318–356, 1961.

3. **Induction of β-galactosidase by lactose**. If lactose is supplied and other energy sources are not available, the number of β-galactosidase molecules per cell can increase to several thousand.

4. **Induction of associated enzymes.** As the concentration of β-galactosidase rises in the cell, two other enzymes also increase in concentration.
   a. **Galactoside permease**, which is involved in the entry of galactosides into the cell
   b. **Thiogalactosidase transacetylase**, which catalyzes the transfer of an acetyl group from acetyl coenzyme A (acetyl CoA) to the C-6 hydroxyl of a thiogalactoside in vitro but whose function in vivo is unknown

5. **Inducer of β-galactosidase.** Because the addition of lactose to a medium induces the synthesis of enzymes that metabolize lactose, lactose itself was originally thought to be the inducer; however, the natural inducer is **allolactose** (Fig. 7-1), which is produced from lactose by the few β-galactosidase molecules present in the cell during the noninduced state.

6. **Gratuitous inducers.** Some galactosides, such as **isopropylthiogalactoside** (Fig. 7-2), are strong inducers of β-galactosidase but do not act as substrates for the enzyme. Such galactosides are known as **nonmetabolizable** (or gratuitous) **inducers**.

C. Lactose (*lac*) operon

1. **Definition.** The operon of Jacob and Monod was envisaged as a complex of elements, which act in concert to control effectively the expression of a group of genes, the gene products of which catalyze a multienzyme metabolic pathway (Fig. 7-3).

2. **Nature.** Specifically, the *lac* operon is a region of the DNA in the genome in which can be found
   a. **Three structural genes**, z, y, and a, which code for
      (1) β-Galactosidase
      (2) Galactosidase permease
      (3) Thiogalactoside transacetylase
   b. A **promoter region (p)**, which is the site where RNA polymerase is bound (see Chapter 3, section I C 1)
   c. An **operator region (o)**, to which the *lac* repressor binds (see section I D)

D. *Lac* repressor

1. The operon concept also requires that a regulatory gene (*i*) should code for a gene product (*lac* repressor) that
   a. Is a diffusible repressor
   b. Binds to a region of the DNA called the **operator region (o)**, which, along with the promoter regions and the structural genes, forms the operon. (The regulatory gene is not regarded as part of the operon as it can be at a site remote from the structural genes that it regulates.)
   c. Blocks (represses) the transcription of the structural genes z, y, and a

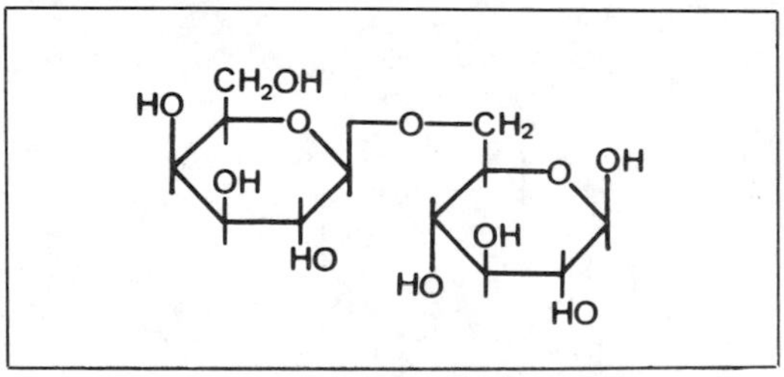

Figure 7-1. Structure of allolactose.

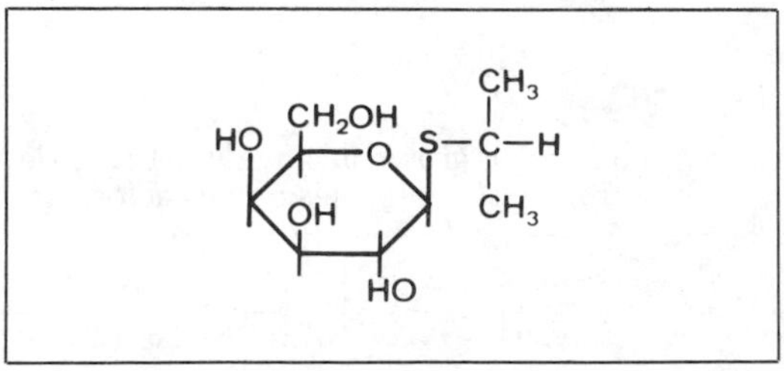

Figure 7-2. Structure of isopropylthiogalactoside.

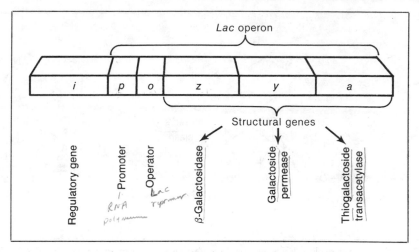

Figure 7-3. Diagram of the lactose (*lac*) operon.

2. The *lac* repressor is always being transcribed because the regulatory gene is a constitutive gene.
   a. The repressor is a tetrameric protein that binds to DNA containing the *lac* operon but not to other DNA.
   b. Each subunit of the *lac* repressor has one binding site for an inducer.
   c. In the absence of an inducer, the *lac* repressor binds tightly to the operator ($K_d = 10^{-13}$M), which prevents transcription of the structural genes.
   d. The *lac* operator region, the region of the DNA to which the *lac* repressor binds, has a symmetric base sequence (Fig. 7-4). The operator is probably recognized by the protein repressor by virtue of a matching symmetry in the protein.
   e. When **inducer molecules** bind to the repressor, they can overcome the strong repressor–operator binding, allowing structural gene transcription and the consequent production of
      (1) β-Galactosidase
      (2) Galactoside permease
      (3) Thiogalactoside transacetylase

II. CONTROL OF OPERONS IN GENERAL. The *lac* operon (described in section I C) is **inducible** and **negatively controlled**. The control of other operons may be different, as for example, when the controlling metabolite acts to prevent transcription. These latter operons are said to be **repressible**, and the controlling metabolite is called the **corepressor**. Again, in contrast to the *lac* operon, operons in which the regulatory protein acts to promote transcription are said to be **positively controlled**.

A. **Positive control of operons.** There is a general positive control system that overrides the transcriptional control of many operons related to sugar metabolism.

   1. **Catabolite repression.** When *E. coli* is grown in a medium containing glucose, the organism contains very low levels of **catabolic enzymes**, such as β-galactosidase, galactokinase, and arabinose isomerase, a phenomenon known as **catabolite repression**.
      a. The presence of glucose suppresses transcription of the structural genes that code for these enzymes by lowering the intracellular level of cyclic adenosine monophosphate (cAMP).

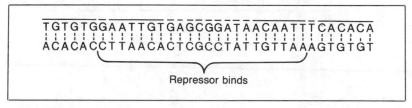

Figure 7-4. Nucleotide sequence of the lactose (*lac*) operator region showing the bases covered by the binding of the *lac* repressor. (After Stryer L: *Biochemistry*, 2nd ed. New York, WH Freeman, 1981, p 673.)

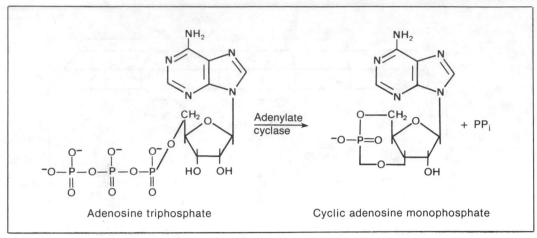

Figure 7-5. Formation of cyclic adenosine monophosphate (cAMP) from adenosine triphosphate (ATP). PP_i = inorganic pyrophosphate.

> **b.** In the absence of glucose, cAMP, formed from adenosine triphosphate (ATP) by adenylate cyclase (Fig. 7-5), binds to a protein called the **catabolite activator protein (CAP)**, a dimer of 22-kdal subunits.
>
> **c.** The cAMP–CAP complex stimulates transcription of several operons, including the *lac* operon.
>
> > **(1)** The complex binds to the promoter sites of these operons.
> >
> > **(2)** In the case of the *lac* operon, the cAMP–CAP complex binds next to the RNA polymerase attachment site, probably by recognizing the twofold symmetry of the DNA base sequence there.
> >
> > **(3)** In the presence of glucose, the cellular levels of cAMP are too low for the cAMP–CAP complex to form, and transcription is not activated because CAP alone cannot recognize the promoter regions.

2. Positive control of the inducible arabinose (*ara*) operon

> **a.** The *ara* operon is induced by one protein in two forms: one form that **inhibits** transcription and the other that **activates** transcription.
>
> **b.** Arabinose is used as an energy and carbon source by bacteria, which convert it to xylulose phosphate, a pentose sugar phosphate. The following three enzymes are involved in this conversion.
>
> > **(1) Arabinose isomerase** is encoded for by the structural gene *araA*.
> >
> > **(2) Ribulose kinase** is encoded for by *araB*.
> >
> > **(3) Ribulose 5-phosphate epimerase** is encoded for by *araD*.
>
> **c.** A map of the *ara* operon shown in Figure 7-6 places the three structural genes in relation to a promoter (*I*), the operator (*o*), and a regulatory gene (*araC*).

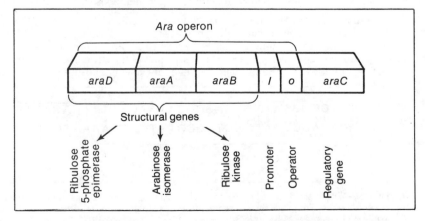

Figure 7-6. Diagram of the arabinose (*ara*) operon.

   d. The gene *araC* codes for a protein that exists in two forms.

     (1) *P1* acts as a repressor and binds to the operator, **blocking** transcription (**negative control**).

     (2) *P2*, in conjunction with the cAMP–CAP complex (see section II A 1), binds to the promoter sites, **initiating** transcription (**positive control**).

   e. Arabinose functions in this system by releasing *P1* from the operator and changing the conformation of *P1* to create *P2*.

B. Negative control of repressible operons in which the controlling metabolite acts to prevent transcription

   1. Bacteria can synthesize all 20 amino acids. However, their synthesis is often a complex time- and energy-consuming process. Therefore, it is economical for the cell to cut off the synthesis when an amino acid is available in the medium.

   2. The enzymes that control the synthesis of many of the amino acids are governed by **negatively controlled repressible operons** with the amino acid functioning as a corepressor.

   3. The **tryptophan (*trp*) operon**, which is an example of a negatively controlled repressible operon, is, in fact, regulated in two ways, by an operator and by an attenuator.

    a. Operator control

     (1) Figure 7-7 is a diagram of the *trp* operon. The structural genes *A*, *B*, *C*, *D*, and *E* code for the five enzymes involved in the synthesis of tryptophan from chorismic acid.

     (2) The ***trp* repressor**, a 58-kdal protein, is coded for by the *trpR* gene, which lies at a site remote from the *trp* operon.

      (a) The *trp* repressor forms a complex with tryptophan, the corepressor.

      (b) The complex binds to the operator site, overlapping the promoter site in the process.

      (c) The complex thus prevents the binding of RNA polymerase at the promoter site, thereby forestalling transcription of the structural genes.

    b. Attenuator control is related to the level of tryptophan available to the cell.

     (1) When enough tryptophan is available, initiation of transcription of the *trp* operon is prevented by the binding of the tryptophan-repressor complex to the operator, as just described.

     (2) If the level of tryptophan in the cell decreases, this repression is relieved and transcription occurs.

      (a) During the transcription process, some RNA polymerase molecules become detached from the DNA at a control point called the **attenuator**. This point is located in a stretch of DNA known as the **leader sequence**, which lies between the operator and the *trpE* gene (see Fig. 7-7).

      (b) The lower the concentration of tryptophan in the cell, the greater the number of RNA polymerase molecules that can travel beyond the attenuator and reach the promoter.

     (3) **Sensing of the cellular tryptophan concentration** by the *trp* operon is due to the presence of *trp* codons in the leader sequence.

      (a) If tryptophan is in ample supply, the entire leader sequence is transcribed, and the resulting mRNA has a secondary structure that enables it to terminate transcription at the attenuator region before the structural genes are transcribed.

      (b) If tryptophan is scarce, the ribosome cannot pass the *trp* codons in the leader mRNA sequence, and it comes to a halt. The arrested ribosome alters the secondary structure of the mRNA so that it cannot terminate transcription at the attenuator region, and RNA polymerase can continue on to the DNA sequences of the structural genes.

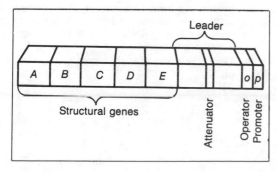

Figure 7-7. Diagram of the tryptophan (*trp*) operon.

STUDY QUESTIONS

Directions: Each question below contains five suggested answers. Choose the **one best** response to each question.

1. Which of the following compounds is a gratu-itous inducer of β-galactosidase in *Escherichia coli*?

(A) Glucose
(B) Allolactose
(C) Isopropylthiogalactoside
(D) Lactose
(E) Fructose

2. All of the following statements about the re-pressor of the lactose (*lac*) operon of *Escherichia coli* are true EXCEPT

(A) it is the gene product of the regulatory gene
(B) it binds to the operator region of the DNA of the *lac* operon
(C) it blocks the transcription of structural genes
(D) it combines with allolactose, and the com-bination will not bind to the operator
(E) it binds allolactose, preventing the sugar de-rivative from activating transcription

3. Cyclic adenosine monophosphate regulates the lactose (*lac*) operon by

(A) binding to the operator to turn on transcrip-tion
(B) binding to the *lac* repressor to prevent tran-scription
(C) combining with the catabolite activator pro-tein (CAP) to form a complex that turns on transcription by binding to the promoter
(D) combining with the CAP to remove the latter's inhibition of transcription
(E) none of the above

Directions: Each question below contains four suggested answers of which **one or more** is correct. Choose the answer

> A   if **1, 2, and 3** are correct
> B   if **1 and 3** are correct
> C   if **2 and 4** are correct
> D   if **4** is correct
> E   if **1, 2, 3, and 4** are correct

4. The products of constitutive genes may be

(1) involved in catabolic pathways
(2) involved in anabolic pathways
(3) continuously synthesized
(4) involved in the synthesis of structural ele-ments in the cell

5. The arabinose (*ara*) operon regulates tran-scription of the structural genes by which of the following processes?

(1) The product of the regulatory gene binds to the operator, blocking transcription
(2) A complex of the catabolite activator protein and cyclic adenosine monophosphate binds to the promoter, initiating transcription
(3) Arabinose modifies the conformation of the protein product of the regulatory gene
(4) Arabinose releases the protein product of the regulatory gene from the operon

6. True statements about the bacterial trypto-phan (*trp*) operon include which of the following?

(1) The *trp* operon has two kinds of regulation, one by an operator and the other by an atten-uator
(2) The repressor complexes with the corepres-sor and on binding to the operator blocks transcription
(3) When tryptophan is in short supply, the leader sequence containing the attenuator cannot be passed, and the stalled ribosome modifies the mRNA so that the structural genes are transcribed
(4) The corepressor is the catabolite activator protein

ANSWERS AND EXPLANATIONS

1. The answer is C. (*I B 6*) Some galactosides, such as isopropylthiogalactoside, are strong inducers of β-galactosidase in *Escherichia coli* but do not act as substrates for the enzyme. Such galactosides are known as nonmetabolizable, or gratuitous, inducers. Lactose and allolactose are substrates for β-galactosidase; glucose and fructose are not. *And inducers also*

2. The answer is E. (*I D*) The lactose (*lac*) repressor is the gene product of the regulatory gene *i*, which is a constitutive gene—that is, it is always turned on. The repressor is a tetrameric protein that binds specifically to the DNA in the operator region of the *lac* operon, thereby blocking transcription of the inducible structural genes. The repressor can form a complex with allolactose (the inducer), and because the complex cannot bind to the operator region of the DNA, transcription of the genes coding for β-galactosidase, galactoside permease, and thiogalactoside transacetylase will occur. Allolactose arises from lactose by the action of β-galactosidase.

3. The answer is C. (*II A 1*) The presence of glucose suppresses the transcription of the structural genes of the lactose (*lac*) operon. It does so by lowering cellular levels of cyclic adenosine monophosphate (cAMP) by an unknown mechanism. In the absence of glucose, cAMP binds to the catabolite activator protein (CAP), and the complex stimulates transcription in several operons, including the *lac* operon, by binding to the operon promoter sites. In the presence of glucose, cAMP levels are too low for the cAMP–CAP complex to form, and transcription is not activated.

4. The answer is E (all). (*I A*) Most of the genes in a cell such as *Escherichia coli* allow the synthesis of their protein product at a more or less constant rate. Such genes are called constitutive, and their products are called constitutive proteins. The constitutive proteins may be enzymes involved in either anabolic or catabolic pathways, or protein regulators, or enzymes involved in the synthesis of structural elements in the cell. The constitutive proteins are in contrast to the inducible proteins, which may be induced (increased synthesis) or repressed (decreased synthesis) at the level of transcription.

5. The answer is E (all). (*II A 2*) The arabinose (*ara*) operon is inducible with the same protein product of the regulatory gene serving to inhibit transcription when the protein is in one form and activating transcription when it is in another form. In the *P1* form, the protein acts as a repressor, blocking transcription by binding to the operator. Arabinose releases *P1* from the operator and changes its conformation to the *P2* form. As *P2*, the protein combines with the catabolite activating protein, and this complex binds to the promoter to initiate transcription.

6. The answer is A (1, 2, 3). (*II B 3*) The tryptophan (*trp*) operon of bacteria has two kinds of regulation, one by an operator and one by an attenuator. In the operator system, a repressor protein with tryptophan as corepressor binds to the operator in such a manner that RNA polymerase cannot transcribe the structural genes. This is the negative control. In addition, as tryptophan availability decreases, positive control is exerted by an attenuator system. In the presence of tryptophan, RNA polymerase molecules become disengaged from the DNA at a control point called the attenuator in the leader sequence between the operator and the first structural gene. As the concentration of tryptophan in the cell decreases, the number of RNA polymerase molecules that can pass the attenuator increases. The cellular tryptophan concentration is sensed by tryptophan codons in the leader sequence. If these cannot be translated (due to lack of tryptophan), the ribosome stalls. The stalled ribosome modifies the mRNA so that transcription can continue, including that of the structural genes.

8

Regulation of Gene Function in Eukaryotes

I. TRANSLATIONAL AND TRANSCRIPTIONAL CONTROL

A. Introduction. Because of the complexity of both the genome and the cellular architecture, eukaryotic gene expression can be controlled at many levels. Thus, control may be exerted at **transcription** or at **translation**, in contrast to prokaryotes where gene control is largely at the level of transcription. Other forms of control involve a variety of **gene rearrangements**, which fall into three classes:

1. **Deletion**, or loss of genomic information (e.g., the deletions that are necessary to activate the immunoglobulin genes)

2. **Translocation**, or change in the relative position of a gene

3. **Amplification**, or increase in the number of copies of a specific gene

B. Regulation of translation

1. **Protein synthesis in reticulocytes**
   a. In the formation of hemoglobin, an adequate supply of heme ensures the rapid synthesis of globin. If the supply of heme is inadequate, globin synthesis is turned off by an inhibitor of protein synthesis.
   b. The control of hemoglobin synthesis is an example of a **protein kinase cascade** (see Chapter 30, section II).
      (1) The **eukaryotic initiation factor (eIF-2)** [see Chapter 4, section VIII B 2 e] is inactivated when it is phosphorylated by a protein kinase, which transfers the terminal phosphate of adenosine triphosphate (ATP) to the eIF-2 (Fig. 8-1).
      (2) The eIF-2 protein kinase exists in active and inactive forms that are interconverted by the action of a second protein kinase, which is cyclic adenosine 3′,5′-monophosphate–dependent (cAMP-dependent).
      (3) The cAMP-dependent protein kinase also exists in active and inactive forms. The inactive form is a tetramer containing two identical catalytic (C_2) and two identical regulatory (R_2) subunits.
         (a) cAMP binds to the regulatory subunits and liberates active monomeric catalytic subunits.

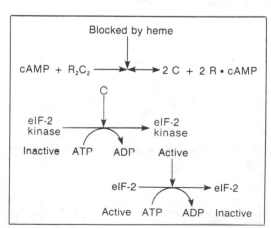

Figure 8-1. Protein kinase cascade in inactivation of the eukaryotic initiation factor (eIF-2). C_2 and R_2 = catalytic and regulatory subunits of a cyclic adenosine monophosphate–dependent (cAMP-dependent) protein kinase; ATP = adenosine triphosphate; and ADP = adenosine diphosphate. (After Stryer L: *Biochemistry*, 2nd ed. New York, WH Freeman, 1981, p 711.)

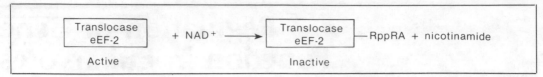

Figure 8-2. Inactivation of the eukaryotic elongation factor 2 (eEF-2, translocase) by covalent modification. RppRA = adenosine diphosphate ribose.

    (b) These in turn activate the eIF-2 kinase, which phosphorylates eIF-2 and prevents globin synthesis.

    c. The entire cascade is interrupted by heme, which blocks the dissociation of the cAMP-dependent protein kinase (see Fig. 8-1) and thus permits the synthesis of globin.

  2. Blockade of translation by diphtheria toxin

    a. The diphtheria exotoxin is produced by some strains of *Cornynebacterium diphtheriae,* a bacillus that lodges in the nasopharynx. The potent toxin is encoded by a lysogenic phage (see Chapter 6, section I D 1).

    b. The toxin blocks protein synthesis in eukaryotes by inactivating the **eukaryotic elongation factor 2 (eEF-2)** [see Chapter 4, section VIII C].

      (1) Normally, eEF-2 is the translocase associated with the guanosine triphosphate–dependent (GTP-dependent) transfer of peptidyl-tRNA from the A site to the P site on the ribosome (see Chapter 4, section VIII C).

      (2) The toxin inactivates eEF-2 by catalyzing the transfer of adenosine diphosphate ribose (ADP-ribose) from nicotinamide-adenine dinucleotide (NAD^+) to eEF-2 with the release of nicotinamide (Fig. 8-2). The covalently modified eEF-2 is inactive.

C. Regulation of transcription

  1. Cellular response to steroid hormones. Some insight into the regulation of gene expression in eukaryotes has been obtained from steroid hormones, which alter gene expression without changing copy number or other genomic rearrangements.

  2. Steroid hormones, which must enter the cell to fulfill their function, are

    a. Cortisol
    b. Aldosterone
    c. Estradiol
    d. Progesterone
    e. Testosterone
    f. 1,25-Dihydroxycholecalciferol (the hormonal derivative of **vitamin D**)

  3. Target cells for several steroids contain a cytoplasmic protein–receptor molecule to which the hormones bind very tightly.

    a. The cytoplasmic receptor for estrogen is a 4S protein, which is modified after binding the hormone, forming a 5S protein–hormone complex that is able to enter the nucleus.

    b. The hormone–receptor complex interacts with chromatin to start the transcription of DNA that codes for a number of proteins involved in the cellular response.

    c. Figure 8-3 illustrates the activation of a target cell by estradiol.

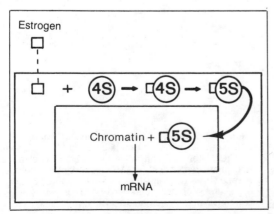

Figure 8-3. The action of estradiol on a target cell, such as the chick oviduct.

    d. In the case of progesterone, the hormone–receptor complex is a dimer that binds to a non-histone chromosomal protein associated with the DNA. One of the subunits is then released to bind directly with the DNA in the region where the hormone–receptor complex is associated with the DNA. Under these conditions, RNA polymerase can initiate transcription.

II. GENE REARRANGEMENTS

A. Structure of the immunoglobulin genes. (Before reading about the structure of immunoglobulin genes, it is important to review the structure of the immunoglobulin proteins in Chapter 10, section V.)

    1. Gene families. Immunoglobulin genes are arranged in three different families with each family consisting of a complex of genes, which are resident on a different chromosome.
        a. The kappa ($\varkappa$) light [L]-chain gene family is found in one locus on chromosome 2.
        b. The lambda (λ) light-chain gene family is found in another locus on chromosome 22.
        c. All of the heavy (H)-chain immunoglobulin classes are found on one locus on chromosome 14.

    2. Separation of genomic information. In each of the loci, the DNA sequences coding for the variable (**V**) region are very far removed from the sequences coding for the rest of the immunoglobulin molecule (Fig. 8-4).

    3. Constant regions. Each constant (**C**) region is encoded for by a unique gene—that is, the genes for the constant regions of λ, $\varkappa$, γ, μ, α, δ and ϵ, are all different.

    4. Variable regions. The genome contains hundreds of variable regions for the heavy chains (V_H) and for the variable regions of the kappa chains ($V_\varkappa$). In mice, there are 100–300 $V_\varkappa$ genes and 100–300 V_H genes. The V genes are clustered in tandem, separated by about 15,000 base-pairs. There are two genes for the variable regions of the lambda chains [V_λ].

B. Translocation and rearrangement of the antibody genes. Immunoglobulins are produced in the B cells, and during differentiation of the antibody-producing B cell, there are two functionally and mechanistically distinct rearrangements of the immunoglobulin gene loci. The first of these is V gene translocation, and the second is heavy-chain class switching.

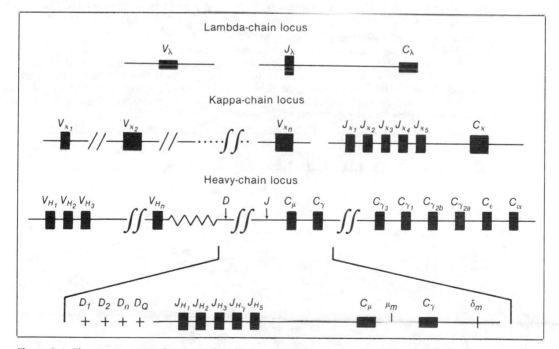

Figure 8-4. The arrangement of immunoglobulin genes in mouse germline DNA. The kappa- and lambda-chain gene families are drawn to the same scale, whereas the heavy-chain (H-chain) gene family is on a twentyfold larger scale. The region of the heavy-chain locus shown is enlarged and has been expanded tenfold. (Reprinted with permission from Gough N: *Trends in Biochemical Sciences*, vol 6. New York, Elsevier, 1981, p 203.)

1. V gene translocation

a. V gene selection. One specific V gene is selected from the many V_L or V_H that exist.

(1) The DNA for the selected V gene is translocated to the $J_\varkappa$ or D_H gene for that locus.

(2) The association of the V gene with its C gene generates a **functional immunoglobulin gene** and determines the **antigen specificity** of that **lymphocyte** and its **progeny**.

(3) This rearrangement occurs independently of contact with antigen.

(4) Germline DNA lying between the selected V gene and the J or D segment may be lost, but this is not certain.

(5) In any one cell, only $\varkappa$ or λ will be expressed.

b. V_H translocations

(1) In V gene translocation for the heavy chain, one V is joined to the D region (which precedes both J and C).

(2) In heavy-chain synthesis, a complete V_H gene is then formed by joining the VD segment to a selected J region.

(3) Formation of a complete V_H gene involves two joining events (both involving the deletion of any intervening DNA).

(a) V/D joining

(b) VD/J joining

$$V_1 \ V_2........V_n............D_1 \ D_2....D_n \ J_1 \ J_2 \ J_3 \ J_4 \ C_\mu$$
$$\downarrow$$
$$V_1 \ D_2.....D_n \ J_1 \ J_2 \ J_3 \ J_4 \ C_\mu$$
$$\downarrow$$
$$V_1 \ D_2 \ J_4 \ C_\mu$$

c. V_L translocation

(1) In V gene translocation for the light chain, V is actually joined to the J region, which precedes the C gene.

(2) In light-chain synthesis, the last 13 amino acids of the V chain are actually encoded by the J region.

(3) In the mouse, there is only one J_λ gene but five $J_\varkappa$ genes.

d. The J region plays a role in two functions:

(1) V/J joining

(2) VJ/C endonuclease recognition of the intron that exists between the J and C segments (Fig. 8-5)

2. Heavy-chain class switching

a. Definition. Heavy-chain class switching refers to an event in which the selected V_H gene segment reassociates with different classes of C_H genes during the course of lymphocyte maturation. This results in a change in the class of antibody that is expressed (Fig. 8-6).

b. Retention of the selected V_H gene. The same V_H gene is retained throughout all these switches. It is also not surprising that IgM is the class of immunoglobulins first synthesized since the gene is 5′ to all the other C_H genes.

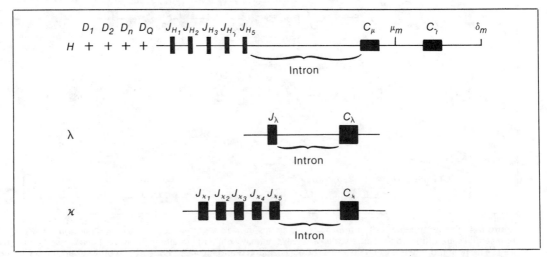

Figure 8-5. Recognition of the introns that lie between the *J* and *C* segments by a VJ/C endonuclease.

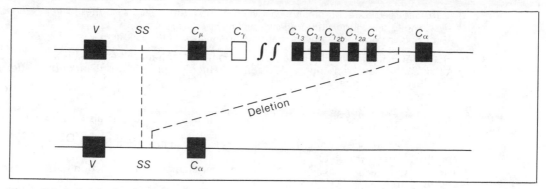

Figure 8-6. Switch in the class of heavy chain expressed; a change from μ to α. SS = DNA segments called switch sites. (Reprinted with permission from Gough N: *Trends in Biochemical Sciences,* vol 6. New York, Elsevier, 1981, p 203.)

    c. Switch sites (SS). The DNA segments involved in directing this relocation are composed of tandemly repeated homologous sequences termed switch sites. They are different from the sites joining V/D/J segments and are denoted by S_μ, S_γ, S_α, and so on.
    (1) Each class of heavy chain has its own set of unique switch sites.
    (2) Deleting a specific switch site will prevent the lymphocyte from secreting that class of immunoglobulin and may "freeze" it forever in IgM production.
    (3) A model for the mechanism of class switching has been proposed; for example, the switching of C_μ to C_α (Fig. 8-7).
        (a) Proteins that are specific for each switch sequence (P_μ and P_α in this example) bind to the S_μ and S_α DNA.

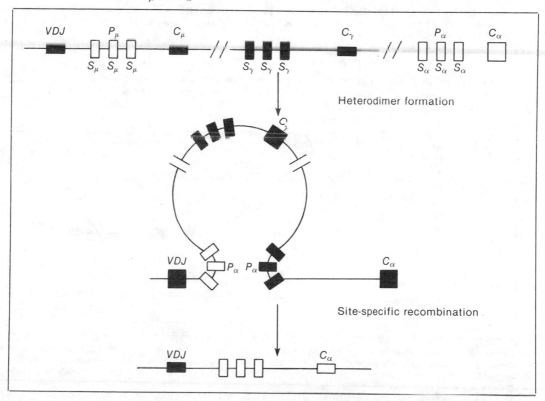

Figure 8-7. Model for class-specific heavy-chain (H-chain) class switching. The *small boxes* represent switching sequences (S) that bind to specific proteins that mediate class-switching. These coding proteins are denoted P_α and P_μ for the S_α and S_μ genes, respectively. Site-specific recombination between S_α and S_μ genes moves the C_α gene up to the VDJ coding sequence for the heavy chain. (Reprinted with permission from Davis MM, Kim SK, Hood LE: DNA sequences mediating class switching in α-immunoglobulins. *Science* 1209:1360–1365, 1980. © 1980 by the American Association for the Advancement of Science.)

(b) Site-specific recombination between homologous regions on these *S* genes moves the C_α gene up to the VDJ segment.
(c) Splicing produces a gene that now connects a $VDJC_\alpha$ sequence (see Fig. 8-7).

3. Regulation of immunoglobulin gene transcription. An "**enhancer**" element is a segment of DNA that can activate transcription of a gene whether it is present upstream or downstream from the 5' end of that gene.
   a. Enhancers can be tissue- or species-specific.
   b. They have no polarity (i.e., they can be inverted).
   c. They are required for the transcription of a particular gene but will also work with other heterologous genes.
   d. The heavy-chain locus contains an enhancer element in the intron between the J segment and the S_μ region (Fig. 8-8).
   e. The light-chain locus may also have an enhancer element at the junction of the C_L exon and the intron that precedes it.

4. Summary of B-cell ontogeny (development)
   a. Cells that are committed to become B cells undergo a series of transitions (Fig. 8-9).
      (1) They differentiate to a stage called the pre-B cell, which produces H_μ chains but no light chains.
      (2) Pre-B cells become B cells when they start to synthesize light chains and when complete immunoglobulin molecules appear in the cell membrane.
   b. A B cell can switch from making a **plasma membrane-bound antibody** to a **secreted form** of the same antibody.
      (1) It does so by altering the RNA transcripts formed during splicing of the heavy-chain RNA primary transcript by an unknown mechanism.
      (2) There are two forms of the heavy chain produced, which differ only in their carboxyl terminals—the membrane-bound form having a hydrophobic tail and the secreted form having a hydrophilic tail.

C. Translocation and change in the relative position of a gene. A chromosomal translocation may result in the removal of a gene from its normal environment with relocation in a new environment, which permits its expression. Some chromosomal translocations are so characteristic of some **human lymphomas** (e.g., **Burkitt's lymphoma**—a B-cell lymphoma) that their occurrence is thought to be critical to the malignant transformation of the cell.

   1. In human Burkitt's lymphoma and in murine plasmacytoma cells, the chromosomal translocations involve chromosomes that carry the immunoglobulin genes.

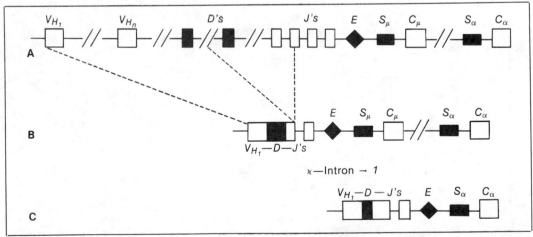

Figure 8-8. Enhancer element in heavy-chain (H-chain) gene. The germline arrangement of the heavy-chain gene is shown in diagram A. The enhancer element (E) is between the segments coding for the J portion of the variable region and the switch (S) region for the first constant region exon (C_μ). Diagram B shows the arrangement after joining of the V, D, and J segments. This brings the control sequences to the left of the V region exon much closer to the enhancer, thus allowing transcription to produce mRNA for the heavy chain. The large intron is spliced out of the original transcript before the messenger is translated into protein structure. The switch from one immunoglobulin class to another involves the deletion of the C region exons preceding the one to be used, a rearrangement, which is mediated by the switch region. As shown in diagram C, the enhancer remains in the appropriate location to facilitate transcription after the class switch. (Reprinted with permission from Marx JL: Immunoglobulin genes have enhancers. *Science* 221:735, 1981. © 1981 by the American Association for the Advancement of Science.)

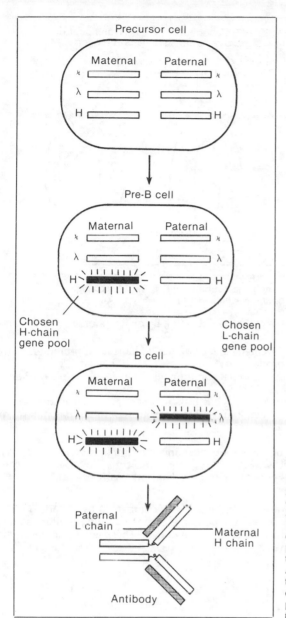

Figure 8-9. The sequential choices in immunoglobulin gene activation that developing B cells must make in order to produce antibodies with only one type of antigen-binding site. Each cell must choose one of four light-chain (L-chain) gene pools and one of two heavy-chain (H-chain) gene pools. During development, a precursor cell first activates one heavy-chain gene pool to become a pre-B cell, making free μ heavy chains. After a period of extensive proliferation, a pre-B cell activates one $\varkappa$ or λ light-chain gene pool to become a B cell that makes a unique IgM molecule. (Reprinted with permission from Alberts B, et al: *Molecular Biology of the Cell*. New York, Garland, 1983, p 904.)

2. In 90% of the patients with Burkitt's lymphoma, there is an exchange of genetic material (DNA) between chromosomes 8 and 14 (Fig. 8-10). Chromosome 14, in humans, bears the genes of the heavy-chain locus.

3. In 10% of the patients with Burkitt's lymphoma, there is an exchange between chromosomes 8 and 2 or 22, which, in the human genome, carry the kappa and lambda immunoglobulin genes, respectively.

4. The chromosomal breakpoint for chromosome 8 is the site where the oncogene called *myc* sits.
   a. The *myc* oncogene (myelocytomatosis virus strain MC29) was originally detected as the oncogene of a virus that caused a blood cell cancer in chickens (v-*myc*).
   b. Injection of the c-*myc* gene (the cellular gene, as opposed to the oncogene) does not transform cells. The v-*myc* gene has acquired this ability from its association with the virus.
   c. The possibility that *myc* gene expression is involved in the development of lymphomas is heightened by results from studies of another virus, the **avian leukosis virus (ALV)**.

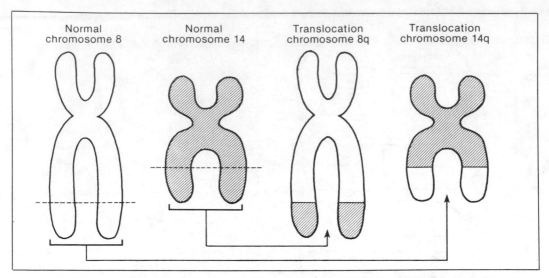

Figure 8-10. Chromosome translocation in Burkitt's lymphoma. Chromosomes 8 and 14 exchange portions of their long arms (designated q) to form the abnormal chromosomes 8q and 14q.

  (1) ALV causes a variety of tumors in chickens, including B-cell lymphomas, but does not carry a v-*myc* sequence.

  (2) Instead, ALV inserts near the c-*myc* gene and using ALV enhancers as promoters, turns on production of c-*myc* gene product.

 d. After translocation, the *myc* gene inserts into the immunoglobulin gene complex, where it usually replaces the VDJ region (Fig. 8-11).

 e. It is possible that c-*myc* translocation may occur by a mechanism similar to the normal mechanism of V/J recombination. In fact, it may be that a "misuse" of this mechanism results in the chromosomal translocation.

D. Amplification of genes. Amplification of DNA sequences involves multiple duplications, via replication, and can occur rapidly, that is, within one generation. It may involve a localized burst of DNA replication, which escapes normal control mechanisms. Many examples of amplification are known.

 1. **Oogenesis in *Xenopus laevis*** (the African clawed toad), is accompanied by a marked selective gene amplification. The number of rRNA gene copies rises to about 2 million, so that these genes make up about 75% of the total DNA in the oocyte. These genes, however, are amplified as **extrachromosomal elements**.

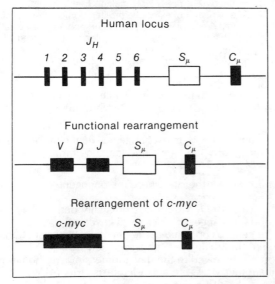

Figure 8-11. A model that would account for the observed genomic rearrangements following insertion of c-*myc*.

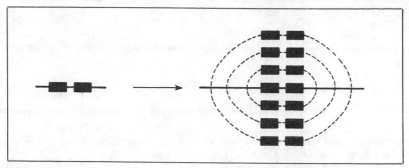

Figure 8-12. The amplified gene locus of the chorion gene of insects has been likened to an "onion skin" in the distribution of the individual amplified genes.

2. **The dihydrofolate reductase gene** (see Chapter 27, section III B and Chapter 29, section VIII F) is amplified in response to methotrexate (an inhibitor of the dihydrofolate reductase enzyme) until 200 copies of the gene accumulate.

3. **The chorion gene**, which codes for the thick, tough, covering of insect eggs, consists of 90 base-pairs, which are amplified. The result is a gene locus that resembles an "onion skin" layering of DNA (Fig. 8-12).

4. **The metallothionein genes** are also amplified.
    a. The gene products are a family of low-molecular-weight proteins that bind heavy metals such as copper, zinc, and mercury.
    b. They are thought to function in heavy metal detoxification and in zinc and copper homeostasis.
    c. Amplification of metallothionein genes in DNA by heavy metals increases the copy number of the genes tenfold. In addition, 25 kilobases of DNA flanking the gene (on both sides) are also increased tenfold.

STUDY QUESTIONS

Directions: Each question below contains five suggested answers. Choose the **one best** response to each question.

1. Hemoglobin synthesis in reticulocytes is characterized by all of the following statements EXCEPT

(A) globin synthesis continues at a rapid rate as long as the supply of heme is adequate
(B) eukaryotic initiation factor 2 (eIF-2) is activated by phosphorylation by a protein kinase
(C) cyclic adenosine monophosphate–dependent (cAMP-dependent) protein kinase regulates the concentration of the active form of the protein kinase, which phosphorylates eIF-2
(D) cAMP binds to the regulatory subunits of the cAMP-dependent protein kinase, liberating active monomeric catalytic subunits
(E) heme regulates the synthesis of globin by blocking the dissociation of the cAMP-dependent protein kinase

2. Amplification of genes involves

(A) removal of histones from the DNA to allow transcription of the gene
(B) multiple duplications of the gene via replication
(C) multiplication of extrachromosomal elements only
(D) invertebrate genomes only
(E) none of the above

Directions: Each question below contains four suggested answers of which **one or more** is correct. Choose the answer

A if **1, 2, and 3** are correct
B if **1 and 3** are correct
C if **2 and 4** are correct
D if **4** is correct
E if **1, 2, 3, and 4** are correct

3. Diphtheria toxin inhibits translation by

(1) activating a protein kinase, which phosphorylates the eukaryotic initiation factor-2 (eIF-2)
(2) covalently modifying the eukaryotic elongation factor 2 (eEF-2)
(3) inhibiting the eIF-2
(4) transferring the adenosine diphosphate ribose moiety of nicotinamide-adenine dinucleotide to eEF-2

4. Correct statements about immunoglobulin genes include which of the following?

(1) Immunoglobulin genes are arranged in families with each family consisting of a complex of genes resident on a different chromosome
(2) The constant regions of the immunoglobulin chains are encoded for by genes that are unique for each class of immunoglobulin
(3) The DNA sequences coding for the variable regions of the immunoglobulin chains are located far away from sequences coding for the rest of the immunoglobulin molecule
(4) Maturation of immunoglobulins involves the translocation and rearrangement of immunoglobulin genes

5. During differentiation of the immunoglobulin-producing B cell, there are translocations of genes coding for variable regions of the immunoglobulin chains (*V* genes). Which of the following statements accurately describe these translocations?

(1) Selection of a *V* gene of either a light or heavy chain and its translocation to the *J* (light) or *D* (heavy) gene for that locus
(2) Amplification (i.e., repeated replication) of the translocated *V* gene together with its *J* or *D* gene
(3) Association of the *V* gene with its constant region gene (*C* gene)
(4) Translocation of the functional immunoglobulin gene back to the locus of the original *V* gene

6. Enhancer elements are segments of DNA that

(1) are required for the transcription of certain genes
(2) activate transcription of a gene irrespective of their relative position in the DNA chain
(3) have no polarity (i.e., they can be inverted)
(4) are nonspecific for tissue or species

ANSWERS AND EXPLANATIONS

1. The answer is B. (*I B 1*) In the formation of hemoglobin, an adequate supply of heme ensures the rapid synthesis of globin. If the supply of heme is inadequate, globin synthesis is turned off. In the reticulocyte, protein synthesis requires the active (dephosphorylated) form of the eukaryotic initiation factor 2(eIF-2). The protein kinase is itself activated by phosphorylation of the cyclic adenosine monophosphate–dependent (cAMP-dependent) protein kinase. The latter is converted to its active form by cAMP, which releases active monomers of the catalytic units of cAMP-dependent protein kinase. Excess heme blocks the dissociation of cAMP-dependent protein kinase; thus it blocks the activation of the protein kinase required for the activation of eIF-2 and the onset of protein (hemoglobin) synthesis.

2. The answer is B. (*II D*) The amplification of DNA sequences involves multiple duplications via replication and can occur rapidly, that is, within one generation. Thus, they are one form of control in times of need, for example, when there is an increased rate of protein synthesis, the number of genes coding for ribosomal RNA (rRNA) is greatly increased. In the case of *Xenopus laevis* (the African clawed toad), the increase in rRNA genes is a cytoplasmic event, not a nuclear event. In other cases of gene amplification, the nuclear DNA is amplified.

3. The answer is C (2, 4). (*I B 2*) The potent toxin of some strains of *Cornynebacterium diphtheriae* is encoded by a lysogenic phage and blocks protein synthesis in infected cells by inactivating the eukaryotic elongation factor 2 (eEF-2). This factor is the translocase associated with the guanosine triphosphate–dependent transfer of peptidyl-tRNA from the A site to the P site on the ribosome. The toxin inactivates eEF-2 by catalyzing the transfer of adenosine diphosphate ribose (not nicotinamide) from nicotinamide-adenine dinucleotide to eEF-2 with the release of nicotinamide. The covalently modified eEF-2 is inactive.

4. The answer is E (all). (*II A, B*) Immunoglobulin genes are arranged in three different families with each family consisting of a complex of genes, which are resident on a different chromosome. The kappa gene family (one form of the genes encoding the sequences found in the constant regions of the light chains) is found on one locus; the lambda gene family (another form coding for constant regions of light chains) is found at another locus; and at a third locus (on a third chromosome) are found all the heavy-chain genes. There is wide separation of genomic information in this system, initially, and the DNA sequences coding for the variable regions (V genes) of immunoglobulin chains are very far removed from the sequences coding for the rest of the immunoglobulin molecule. Maturation of immunoglobulins to form functional molecules involves translocations of the V genes to a constant region gene (C gene) and also switching events (heavy-chain class switching) in which the selected V gene segment reassociates with different classes of C genes during the course of lymphocyte maturation. This results in a change in the class of antibody that is expressed.

5. The answer is B (1, 3). (*II B 1*) During differentiation of the immunoglobulin-producing B cell, there are two functionally and mechanistically distinct rearrangements of the immunoglobulin gene loci. One of these is variable (V) gene translocation. The first event is the selection of one V gene from the many V_L or V_H that exist. The DNA of the selected gene is translocated to the J_x or D_H gene (depending upon whether it is a V_L or a V_H gene) for that locus. There is no amplification of the V gene at this stage, rather it associates with its constant region (C) gene to form a functional immunoglobulin gene and determines the antigen specificity of that lymphocyte and its progeny. The functional immunoglobulin gene is not transferred back to the original locus of the V gene.

6. The answer is A (1, 2, 3). (*II B 3*) An "enhancer" element is a segment of DNA in a genome that can activate transcription of a gene whether it is present upstream or downstream from the 5' end of that gene. They are required for the transcription of a particular gene but will also work with other heterologous genes. They have no polarity (i.e., they can be inverted). They can be tissue- or species-specific.

9

Protein Structure

I. NATURE OF PROTEINS

A. Structure and size

1. Proteins are linear, unbranched polymers constructed from 20 different α-amino acids that are encoded in the DNA of the genome (see Chapter 4, section III).

2. All living organisms use the same 20 amino acids and the same genetic code (see Chapter 4, section III).

3. Proteins are large molecules with molecular weights ranging from 10 to 50 kdal for single-chain proteins. Multichain (oligomeric) proteins of 150 to 200 kdal are frequently encountered.

B. Function. Proteins serve a wide range of functions in living organisms. They are involved in the following:

1. **Enzymatic catalysis.** All known enzymes are proteins.

2. **Transport and storage** of small molecules and ions.

3. **Systematic movements.** Both striated and smooth muscle are composed chiefly of protein, as are structures involved in the motility of certain free-living cells (e.g., sperm).

4. **The structure of skin and bone. Collagen** (see Chapter 10, section III), the most abundant protein in the body, gives these structures high tensile strength.

5. **The immune defense system.** Antibodies are specialized proteins recognizing self and nonself.

6. **Hormonal regulation**
   a. Some hormones are proteins [e.g., somatotropin (pituitary growth hormone) and insulin].
   b. The cellular receptors that recognize hormones and neurotransmitters are proteins.

7. **Control of genetic expression**
   a. Repressor molecules in bacteria are proteins that suppress certain DNA sequences.
   b. Protein initiation and termination factors serve in the transcription and translation phases of gene function (see Chapters 3 and 4).

C. Unique conformation

1. Proteins show an exquisite specificity of biologic function—a consequence of the uniqueness of the three-dimensional structural shape, or **conformation**, of each protein.

2. In humans, disease states are often related to the altered function of a protein. This is due to an anomaly in the structure of the protein, which in turn may be due to a deficiency in its synthesis.

II. AMINO ACIDS

A. Composition. The fundamental units of protein polymers are α-**amino acids**.

1. They are composed of an **amino group**, a **carboxyl group**, a hydrogen atom, and a distinctive **side chain**, all bonded to a carbon atom (the α-**carbon**). Table 9-1 lists the 20 amino acids according to their side chains.

2. One of the 20 amino acids, **proline**, is an **imino acid**, not an α-amino acid as are the other 19.

Table 9-1. The 20 Amino Acids Used for Building Protein Chains

Name	Symbol	Structural Formula
Aliphatic Nonpolar Side Chains		
Glycine	Gly (G)	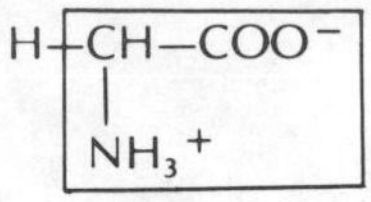 $H-CH-COO^-$ with NH_3^+ on the α-carbon
Alanine	Ala (A)	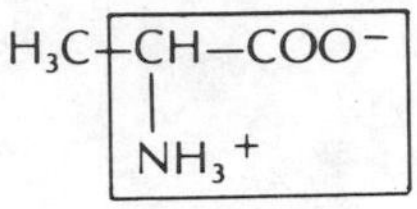 $H_3C-CH-COO^-$ with NH_3^+
Valine	Val (V)	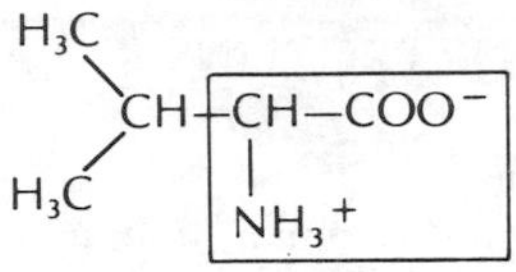 $(H_3C)_2CH-CH-COO^-$ with NH_3^+
Leucine	Leu (L)	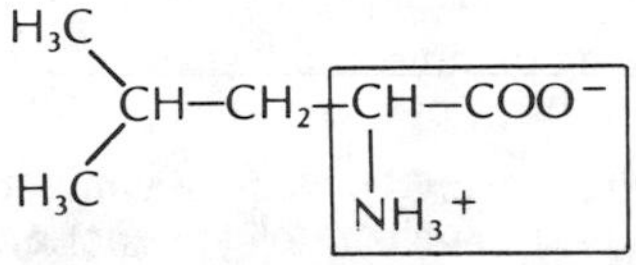 $(H_3C)_2CH-CH_2-CH-COO^-$ with NH_3^+
Isoleucine	Ile (I)	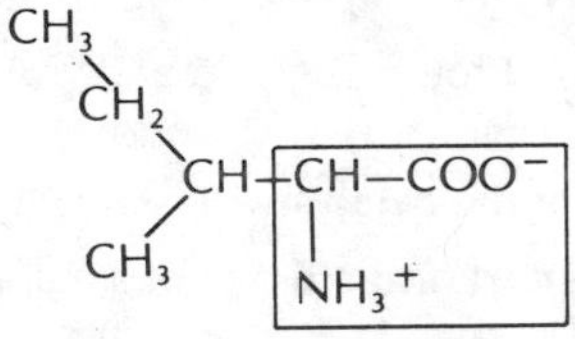 $CH_3-CH_2-CH(CH_3)-CH-COO^-$ with NH_3^+
Aromatic Side Chains		
Phenylalanine	Phe (F)	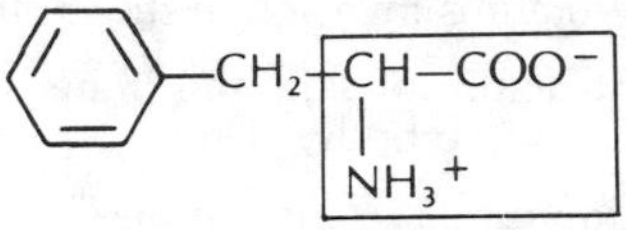 $C_6H_5-CH_2-CH-COO^-$ with NH_3^+
Tyrosine	Tyr (Y)	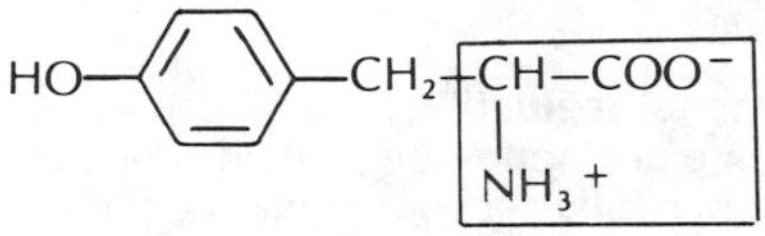 $HO-C_6H_4-CH_2-CH-COO^-$ with NH_3^+
Tryptophan	Trp (W)	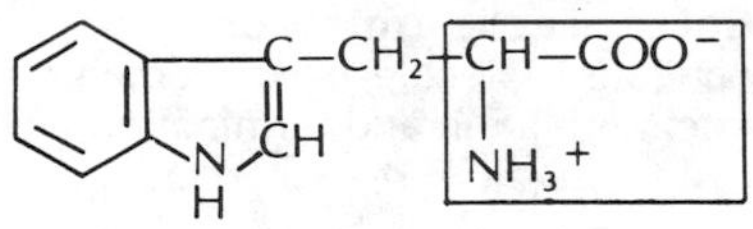 indole ring $-C=CH-CH_2-CH-COO^-$ with NH_3^+ (ring N–H)
Hydroxyl-Containing Side Chains		
Serine	Ser (S)	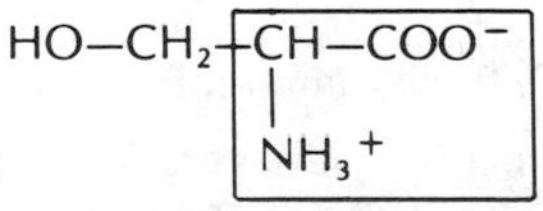 $HO-CH_2-CH-COO^-$ with NH_3^+
Threonine	Thr (T)	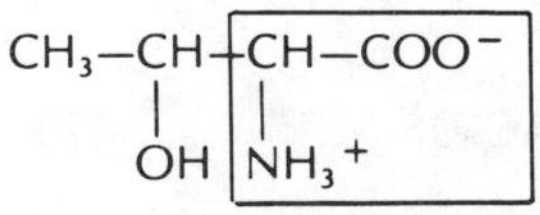 $CH_3-CH(OH)-CH-COO^-$ with NH_3^+
Acidic Side Chains		
Aspartate	Asp (D)	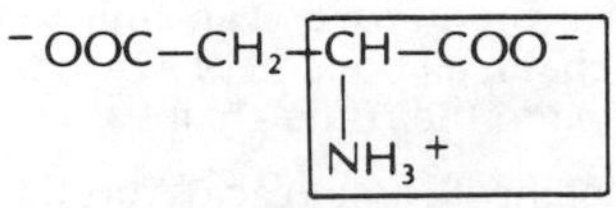 $^-OOC-CH_2-CH-COO^-$ with NH_3^+

Table 9-1. Continued

Name	Symbol	Structural Formula
Glutamate	Glu (E)	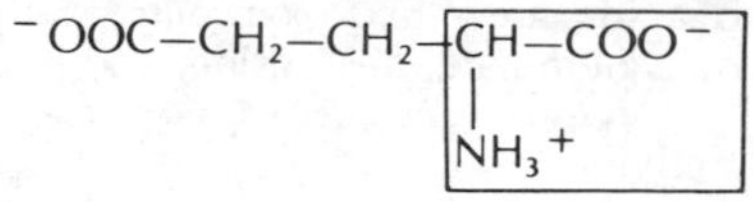 $^-OOC-CH_2-CH_2-CH-COO^-$ $\quad NH_3^+$

Amidic Amino Acids

Name	Symbol	Structural Formula
Asparagine	Asn (N)	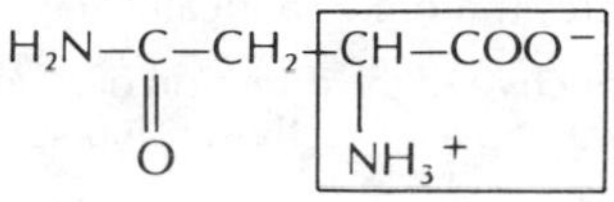 $H_2N-\overset{\displaystyle \,}{\underset{\displaystyle O}{C}}-CH_2-CH-COO^-$ $\quad NH_3^+$
Glutamine	Gln (Q)	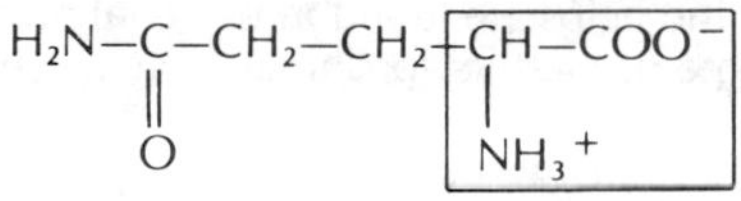 $H_2N-\overset{\displaystyle \,}{\underset{\displaystyle O}{C}}-CH_2-CH_2-CH-COO^-$ $\quad NH_3^+$

Basic Side Chains

Name	Symbol	Structural Formula
Lysine	Lys (K)	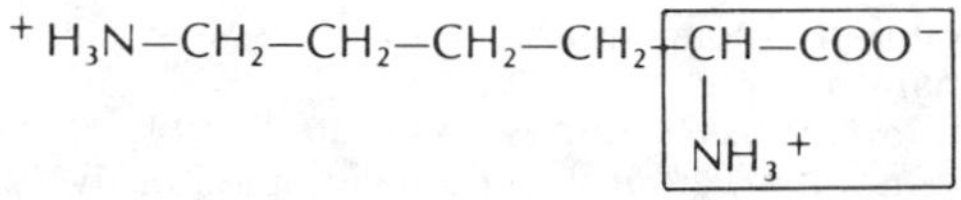 $^+H_3N-CH_2-CH_2-CH_2-CH_2-CH-COO^-$ $\quad NH_3^+$
Arginine	Arg (R)	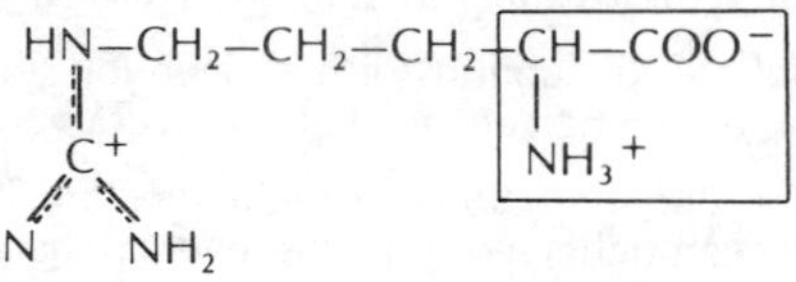 $HN-CH_2-CH_2-CH_2-CH-COO^-$ $\quad\overset{\displaystyle \,}{C^+}\qquad\qquad NH_3^+$ $H_2N\quad NH_2$
Histidine	His (H)	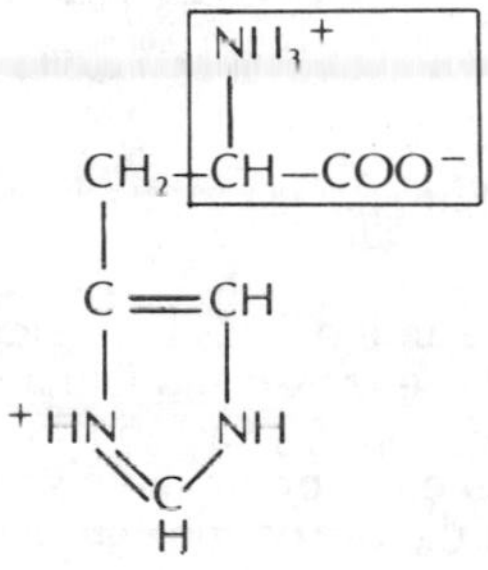 NH_3^+ $CH_2-CH-COO^-$ $C=CH$ $^+HN\quad NH$ $\underset{\displaystyle H}{C}$

Sulfur-Containing Side Chains

Name	Symbol	Structural Formula
Cysteine	Cys (C)	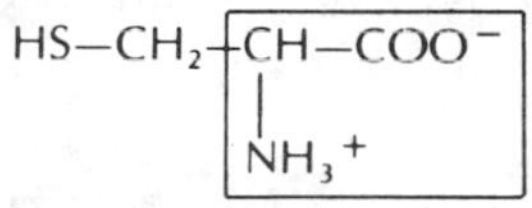 $HS-CH_2-CH-COO^-$ $\quad NH_3^+$
Methionine	Met (M)	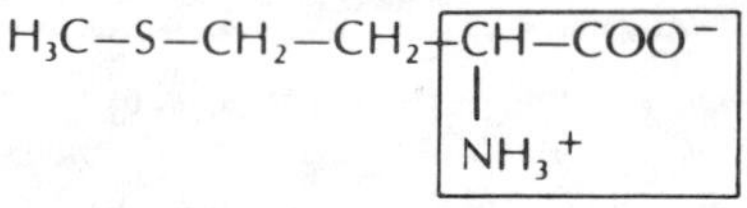 $H_3C-S-CH_2-CH_2-CH-COO^-$ $\quad NH_3^+$

Imino Acid

Name	Symbol	Structural Formula
Proline	Pro (P)	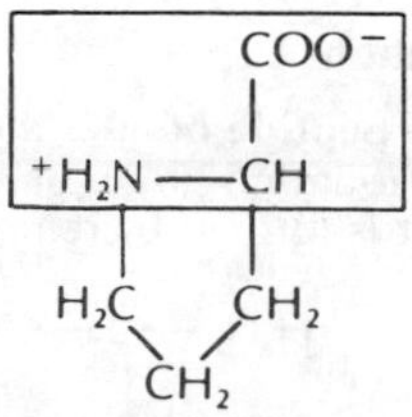 COO^- $^+H_2N-CH$ $H_2C\quad CH_2$ CH_2

3. A few other amino acids are found in a number of proteins but are not coded for in DNA; they are derived from one or another of the 20 fundamental amino acids after these have been incorporated into the protein chain (**post-translational modification**). The **derived amino acids** are 4-hydroxyproline, 5-hydroxylysine, ϵ-N-methyl-lysine, 3-methylhistidine, γ-carboxyglutamate, desmosine, and isodesmosine (Table 9-2).

B. Optical activity

1. With the exception of glycine, all amino acids contain at least one asymmetric carbon atom and are, therefore, optically active.

2. Irrespective of the direction of rotation of plane polarized light, which can be levo- or dextro-, the only optically active amino acids that are incorporated into proteins are of the L-configuration.

3. D-Amino acids are found in bacterial products (e.g., in cell walls) but are not incorporated into proteins via the ribosomal protein synthesizing system.

C. Amphoteric properties

1. Amino acids are **amphoteric molecules**—that is, they have both basic and acidic groups.

2. Monoamino–monocarboxylic acids exist in aqueous solution as **dipolar molecules (zwitterions)**.
   a. The α-carboxyl group is dissociated and negatively charged.
   b. The α-amino group is protonated and positively charged.
   c. Thus, the overall molecule is electrically neutral.

3. At high concentrations of hydrogen ion (low pH), the carboxyl group accepts a proton and becomes uncharged, so that the overall charge on the molecule is positive.

4. At low concentrations of hydrogen ion (high pH), the amino group loses its proton and becomes uncharged; thus the overall charge on the molecule is negative.

$$H_3N^+\text{—CH—COOH} \rightleftharpoons H_3N^+\text{—CH—COO}^- \rightleftharpoons H_2N\text{—CH—COO}^-$$
$$\qquad\qquad |\qquad\qquad\qquad\qquad |\qquad\qquad\qquad\qquad |$$
$$\qquad\qquad R\qquad\qquad\qquad\qquad R\qquad\qquad\qquad\qquad R$$

Overall charge   1 +   0   1 −

5. Some amino acids have **side chains** containing **dissociating groups**.
   a. Those of Asp and Glu are acidic; those of His, Lys, and Arg are basic.
   b. Two others, Cys and Tyr, have a negative charge on the side chain when dissociated.
   c. Whether or not these groups are dissociated depends upon the prevailing pH and the pK_a of the dissociating groups.
   d. These dissociating amino acids also exist in solution as zwitterions; for example, for Glu, the effect of a changing pH is as follows:

$$\begin{array}{llll}
H_3N^+\text{—CH—COOH} & H_3N^+\text{—CH—COO}^- & H_3N^+\text{—CH—COO}^- & H_2N\text{—CH—COO}^- \\
\quad\quad | & \quad\quad | & \quad\quad | & \quad\quad | \\
\quad\quad CH_2 \quad pK_{a_1}' = 2.19 & \quad\quad CH_2 \quad pK_{a_2}' = 4.25 & \quad\quad CH_2 \quad pK_{a_3}' = 9.67 & \quad\quad CH_2 \\
\quad\quad | \quad\quad \rightleftharpoons & \quad\quad | \quad\quad \rightleftharpoons & \quad\quad | \quad\quad \rightleftharpoons & \quad\quad | \\
\quad\quad CH_2 & \quad\quad CH_2 & \quad\quad CH_2 & \quad\quad CH_2 \\
\quad\quad | & \quad\quad | & \quad\quad | & \quad\quad | \\
\quad\quad COOH & \quad\quad COOH & \quad\quad COO^- & \quad\quad COO^-
\end{array}$$

Overall charge 1 +   0   1 −   2 −

III. PEPTIDES AND POLYPEPTIDES

A. Formation

1. **The peptide bond** is the bond formed between the α-carboxyl group of one amino acid and the α-amino group of another.
   a. It is formed by removal of the elements of water.

$$H_3N^+\text{—CH—COO}^- + H_3N^+\text{—CH—COO}^- \rightarrow H_3N^+\text{—CH—CO—NH—CH—COO}^-$$
$$\qquad\quad |\qquad\qquad\qquad\qquad\quad |\qquad\qquad\qquad\qquad\quad |\qquad\qquad\qquad |$$
$$\qquad\quad R_1\qquad\qquad\qquad\qquad\quad R_2\qquad\qquad\qquad\qquad\quad R_1\qquad\qquad\qquad R_2$$

Table 9-2. Structure and Source of Some Amino Acids Found in Proteins and Formed Post-translationally

Name	Structure	Source
4-Hydroxyproline	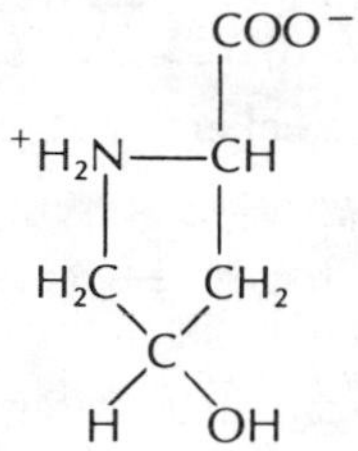 COO^- on $^+H_2N{-}CH$, ring with H_2C, CH_2, C bearing H and OH	Collagen and gelatin
5-Hydroxylysine	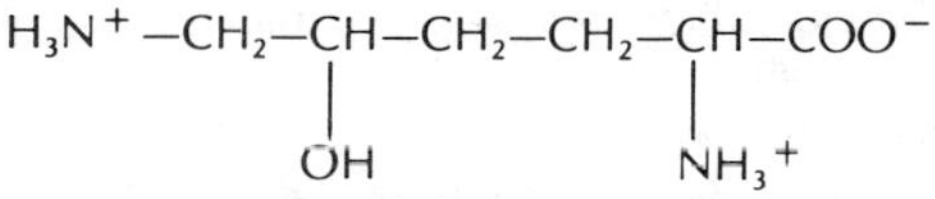 $H_3N^+{-}CH_2{-}CH{-}CH_2{-}CH_2{-}CH{-}COO^-$ with OH on third carbon and NH_3^+ on the α-carbon	Collagen and gelatin
ϵ-N-Methyl-lysine	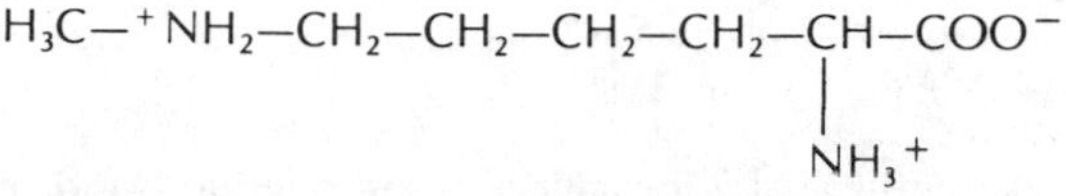 $H_3C{-}^+NH_2{-}CH_2{-}CH_2{-}CH_2{-}CH_2{-}CH{-}COO^-$ with NH_3^+ on the α-carbon	Fast muscle myosin
3-Methylhistidine	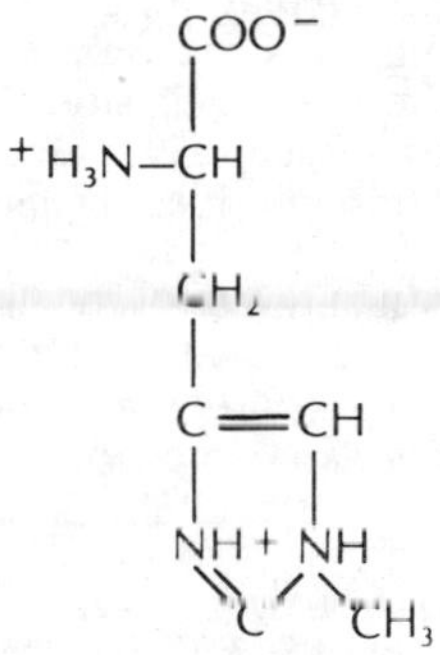 COO^- on $^+H_3N{-}CH$, CH_2, imidazole ring $C{=}CH$, NH^+ NH, C, CH_3	Muscle myosin
γ-Carboxyglutamate	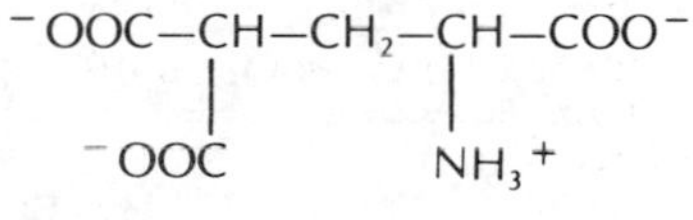 $^-OOC{-}CH{-}CH_2{-}CH{-}COO^-$ with ^-OOC and NH_3^+	Prothrombin and bone protein
Desmosine and isodesmosine	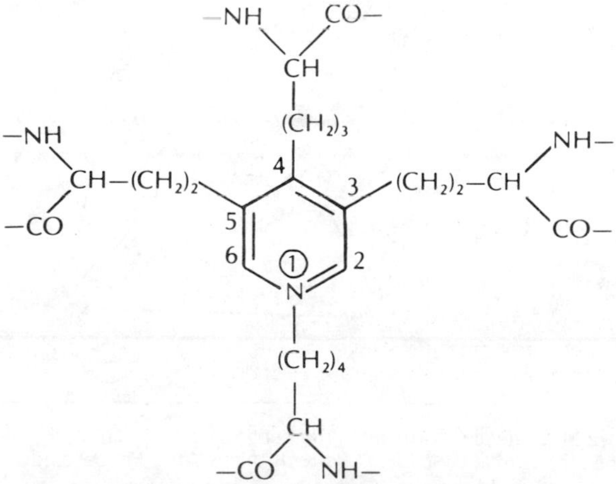 pyridinium ring structure with substituents: $-NH$ $CO-$ on CH, $(CH_2)_3$ at position 4; $-NH$ $CH{-}(CH_2)_2$ $-CO$ at position 5; $NH-$ $(CH_2)_2{-}CH$ $CO-$ at position 3; ring positions numbered 1–6 with N at position 1; $(CH_2)_4$ CH $-CO$ $NH-$ off nitrogen	Elastin

      b. The process is highly **endergonic** and requires the concomitant hydrolysis of high-energy phosphate bonds.

      c. The peptide bond is a **planar** structure with the two adjacent α-carbons, a carbonyl oxygen, and α-amino-N and its associated H atom, and the carbonyl carbon all lying in the same plane (Fig. 9-1). The —CN— bond has a partial double-bond character that prevents rotation about the bond axis.

  2. Polypeptide chains. The linking together of many amino acids by peptide bonds produces polypeptide chains.

      a. Amino acids, when in polypeptide chains, are customarily referred to as **residues**.

      b. Protein polypeptide chains are typically more than 100 amino acid residues long. The recurring sequence

$$C \ - \ C \ - \ N \ - \ C$$

                  α-carbon—carbonyl carbon—α-amino-N—α-carbon

    along the chain is referred to as the "backbone" of the chain.

      c. Smaller peptides, however, are common and often have important biologic roles; for example, the hormones vasopressin (9 residues), glucagon (29 residues), and luteotropin-releasing hormone (3 residues).

  3. Peptide structures. By convention, peptide structures are written from left to right, starting with the amino acid residue having a free α-amino group (the so-called **N-terminal amino acid**) and ending with the residue having a free α-carboxyl group (the **C-terminal**). Either the three-letter abbreviations or (for long peptides) the single-letter abbreviations, given in Table 9-1, are used (e.g., Ala-Glu-Lys).

B. Amphoteric properties. The formation of the peptide bond removes two dissociating groups, one α-amino and one α-carboxyl, per residue. Although the N-terminal and C-terminal α-amino and α-carboxyl groups can play important roles in the formation of protein structures and thus in protein function, the amphoteric properties of a polypeptide are mainly governed by the **dissociable groups** on the amino acid **side chains**. These properties of proteins are not only important in terms of protein structure and function but are also useful in a number of analytic procedures for the purification and identification of proteins.

C. Optical activity. Polypeptides are optically active by virtue of their component amino acids. For short peptides, the optical activity is the algebraic sum of the optical rotations of the individual amino acids, but this rule soon breaks down with increasing chain length, when the optical activity may become more an expression of the **secondary ordering** of the polypeptide chain (see section V B 2).

IV. PURIFICATION AND ANALYSIS OF PROTEINS

A. General considerations

  1. A protein in biologic fluids may require some degree of purification before it can be specifi-

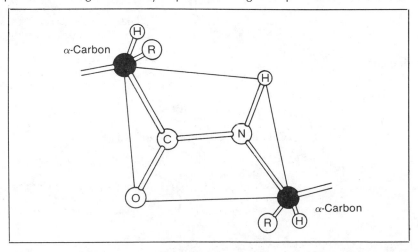

Figure 9-1. The planar nature of the peptide bond. (Adapted from Dickerson RE, Geiss J: *The Structure and Action of Proteins*. New York, Harper and Row, 1969, p 13.)

cally measured. These purification procedures were developed initially to obtain proteins pure enough to allow structural determinations to be carried out.

2. Separation of a protein from other proteins or from smaller molecules is achieved by applying a combination of several methods based on properties such as solubility, molecular size, molecular charge, or specific binding of the protein to a specific substance.

3. Many of the techniques used in the determination of protein structure have found wide application in clinical laboratory practice. Plasma protein patterns, for example, are routinely examined by gel electrophoresis, and a wide range of affinity binding assays (including radioimmunoassays) for hormones and drugs makes use of the specific binding of one substance to another.

B. Separation procedures

1. Protein solubility is influenced by the **salt concentration** of the solution.
   a. Salting out. Adding divalent salts, such as ammonium sulfate, to a solution of a protein mixture will precipitate some proteins at a given salt concentration, but not others. This type of separation is carried out to obtain an enrichment of a given protein in a fraction of a highly complex mixture, such as a tissue homogenate or a sample of blood plasma.
   b. Salting in. Some proteins require inorganic ions for water solubility. Exhaustive **dialysis** (see section IV B 2 a) may, therefore, cause certain proteins from a protein mixture to precipitate out of solution.

2. Separation on the basis of molecular size
   a. Dialysis
      (1) Small molecules can be removed from protein solutions by dialysis through a semipermeable membrane.
      (2) The protein solution is put in a cellophane bag, and the bag is immersed in water. The cellophane contains pores that allow water and small molecules to pass through, but not the protein molecules.
      (3) Most dialysis tubing restricts the flow of molecules larger than 15 kdal.
   b. Gel filtration (molecular exclusion chromatography, molecular sieving)
      (1) Gel filtration uses a column of insoluble, but highly hydrated, carbohydrate polymer made up in the form of 100 μm porous beads. Small molecules can enter the pores, but large molecules cannot. Therefore, the volume of solvent available (the distribution volume) for the small molecules is greater than for the large molecules, and so the small molecules flow through the column more slowly.
      (2) For estimating the molecular weight of a protein, the gel column is standardized with a protein of known molecular weight.
      (3) The shape of a molecule influences its distribution volume and therefore its rate of passage through the column, so that the procedure is most suitable for globular proteins.
   c. Ultracentrifugation
      (1) For analytic purposes, rather than for preparative work, a protein solution can be examined for multiple components by centrifugation at high speed in an ultracentrifuge.
      (2) The molecular weight of a protein can be estimated by optical procedures if the shape and density of the protein molecule are known.
      (3) The data from the ultracentrifugation study are expressed in terms of **Svedberg units** (**S**). These are derived from the **sedimentation constant** (**s**), which is obtained from the rate of sedimentation of the protein in the centrifugal field. One Svedberg unit = s × 10^{13}. Values of S for proteins range from 1 to 200.
   d. Polyacrylamide gel electrophoresis, done on a cross-linked polyacrylamide gel in the presence of a **detergent**, such as sodium dodecyl sulfate (SDS), separates proteins on the basis of their molecular weight.
      (1) The SDS denatures the protein, thereby minimizing the effects of the protein's shape on the molecular weight determination. The SDS dissociates quaternary structures (see section V B 4) into monomers.
      (2) Because the SDS forms negatively charged micellar particles with the protein, the effect of protein charge is lost.
      (3) The SDS-protein micelles migrate to the positive pole with the cross-linked polyacrylamide acting as a molecular sieve.

3. Separation on the basis of molecular charge
   a. Ion-exchange chromatography
      (1) A column of insoluble ion-exchange material carrying **carboxylate groups** is used. At a

neutral pH, these groups are negatively charged and will bind protein molecules carrying a net positive charge.

   (2) The bound proteins are retained on the column material or retarded in flow rate. They may be eluted from the exchanger by washing with a solution containing positive ions (e.g., Na^+ salts), which will exchange places with the positively charged protein bound to the carboxylate groups.

   (3) This technique is used in the automated analysis of amino acid mixtures, which are obtained from either proteins by hydrolytic methods or existing as free amino acids in body fluids such as urine or plasma.

   b. Electrophoresis

   (1) In an electrical field, proteins migrate in a direction determined by the net charge on the molecule. The **net charge on a protein** is determined by the nature of the ionizing groups on the protein and the prevailing pH (see section II C 5).

   (2) For each protein, there is a pH, called the **isoelectric point (pI)**, at which the molecule has no net charge and will *not* move in an electrical field.

   (3) At pH values **more acid** than the pI, the protein will bear a **net positive charge** and, behaving as a **cation**, will move toward the negatively charged pole (the **cathode***).

   (4) At **pH values above** the pI, the protein will have a **net negative charge** and will behave as an **anion**, moving toward the positively charged pole (the **anode**).

   (5) The migration of a protein in an electrical field is defined by its **electrophoretic mobility** (μ), which is the ratio of the velocity of migration (v) to the field strength (E), or $\mu = v/E$, measured in cm^2 per volt-sec. Proteins migrate much more slowly than simple ions because they have a much smaller **ratio of charge to mass**.

   c. Electrophoretic procedures

   (1) **Moving boundary (free) electrophoresis.** The original electrophoretic cell of Tiselius was based on this principle. The procedure measures the rate of movement of a protein from a solution of the protein into a layer of protein-free buffer. The moving boundary of the protein is visualized by following changes in the refractive index. No actual separation of proteins is achieved.

   (2) **Zone electrophoresis** uses paper, starch, or gel blocks saturated with buffer to separate proteins with different electrophoretic mobilities. This type of electrophoresis is often used to fractionate plasma proteins for diagnostic purposes.

   (3) **Isoelectric focusing.** In isoelectric focusing, polyamino–polycarboxylic acids (amphoteric molecules) with known pI values are used to set up a pH gradient in an electrical field. A protein will migrate to the part of the gradient that has the same pI as the protein. This technique is probably most effective for resolving proteins that have very close pI values.

4. Separation by specific affinity binding

   a. Affinity (adsorption) chromatography is based upon the property of some proteins—that of binding strongly to another molecule (called the **ligand**) by specific, noncovalent bonding.

   (1) The ligand is covalently bound to the surface of large, hydrated particles of a porous material in order to make a chromatographic column.

   (2) If a solution containing several proteins is poured down the column, the protein to be selectively adsorbed will bind tightly to the ligand molecules, whereas the other proteins will run through the column unhindered.

   (3) After traces of the other proteins are washed out, the adsorbed protein is eluted by adding a strong solution of pure ligand, which competes for the protein with the bound ligand.

   b. Precipitation by antibodies

   (1) Antibodies to specific proteins can often be prepared and can be used to react with the desired protein in a mixture of proteins (such as a tissue extract or body fluid).

   (2) The interaction of protein and antibody may produce an antigen–antibody complex large enough to be centrifuged out of solution, allowing recovery of the protein. However, it is often necessary to create a large complex by first adding rabbit anti-gamma globulin (anti-IgG) to the antibody–protein mixture and then recovering the triple complex.

V. CONFORMATION OF PROTEINS. Every protein in its **native state** has a unique **three-dimensional structure**, which is referred to as its **conformation**.

   A. Protein classes. There are two broad classes of proteins:

---

*In a galvanic (energy-producing) cell, the negative pole is termed the *anode*. In *electrolysis*, in a discharge tube, or in *electrophoresis*, the negative pole is called the *cathode*.

1. **Globular proteins**, in which one or more polypeptide chains are folded into tight globular forms. Globular proteins are usually soluble and motile.

2. **Fibrous proteins**, in which the straight polypeptide chains lie parallel (or antiparallel) to one another along a single axis, forming fibers or sheets. Fibrous proteins constitute the structural elements of connective and supporting tissues. They are usually insoluble and nonmotile.

B. **Levels of organization.** Protein structure can be classified into four levels of organization.

1. **Primary structure** is the covalent "backbone" of the polypeptide formed by the **specific amino acid sequence**. This sequence is encoded for in DNA and determines the final three-dimensional form adopted by the protein in its native state.

2. **Secondary structure** is the spatial relationships of neighboring amino acid residues.
   a. **Ordering of polypeptide chains.** The secondary structure is formed by the ordering of the polypeptide chain as dictated by the DNA-encoded primary sequence and may begin as the peptide chain comes off the ribosome.
   b. **Hydrogen bonds.** An important characteristic of the ordering is the formation of hydrogen bonds (H bonds) between the —CO group of one peptide bond and the —NH of another nearby peptide bond.
      (1) If the H bonds form between peptide bonds in the *same* chain, **helical** structures develop, such as the α-**helix**.
      (2) If the H bonds form between peptide bonds in *different* chains, **extended** structures form, such as the β-**configuration**, or β-**pleated sheet**.
   c. **The α-helix** is a rod-like structure with the peptide bonds coiled tightly inside and the side chains of the residues protruding outward (Fig. 9-2).

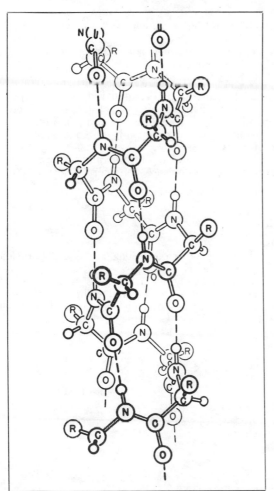

Figure 9-2. An α-helix. The *dashed lines* are the H bonds. (Reprinted with permission from Pauling L: *The Nature of the Chemical Bond*, 3rd ed. Ithaca, NY, Cornell University Press, 1960, p 500.)

     (1) Each —CO group is H-bonded to the —NH of a peptide bond that is **four residues away** from it along the **same chain**.

     (2) There are 3.6 amino acid residues per turn of the helix, and the helix is right-handed (i.e., the coils turn in a clockwise fashion around the axis).

     (3) Helical structures in proteins were predicted by Linus Pauling from his studies of fibrous proteins. However, the α-helix can also be important in the structure of globular proteins, although the chains are much shorter than in the fibrous proteins. The total length of α-helix in a globular protein can vary from almost 0 to more than 75% of the total chain length.

    d. β-**Pleated sheet structures** are found in many proteins, including some globular, soluble proteins as well as some fibrous proteins, such as silk fibroin.

     (1) They are more extended structures than the α-helix and are "pleated" because the carbon–carbon bonds are tetrahedral and cannot be in straight lines.

     (2) The chains lie side by side, with **H bonds** forming between the —CO group of one peptide bond and the —NH of another peptide bond in the **neighboring chain**.

     (3) The chains may run in the same direction, forming a **parallel β-sheet**, or run in opposite directions, as they would do in a globular protein in which an extended chain is folded back on itself, forming an **antiparallel β-structure**.

3. Tertiary structure refers to the spatial relationships of more distant residues.

    a. The secondarily ordered polypeptide chains of soluble proteins tend to fold into globular structures with the hydrophobic side chains in the interior of the structure away from water and the hydrophilic side chains on the outside in contact with water.

    b. This folding is due to associations between segments of α-helix, or extended β-chains, or other secondary structures and represents a state of lowest energy (i.e., of **greatest stability** for the protein in question).

    c. The conformation results from

     (1) H-bonding within a chain or between chains

     (2) The **flexibility** of the chain **at points of instability**, allowing water to obtain maximum entropy and thus govern the structure to some extent

     (3) The formation of **other noncovalent bonds** between side-chain groups, such as salt linkages, or π-electron interactions of aromatic rings

     (4) The sites and numbers of **disulfide bridges** within the chain

4. Quaternary structure refers to the spatial relationships between individual polypeptide chains in a multichain protein; that is, the characteristic noncovalent interactions between the chains that form the native conformation of the protein (see section V B 3 c).

    a. Most proteins larger than 50 kdal have more than one chain and are said to be **oligomeric**, with individual chains known as **protomers**.

    b. Most oligomeric proteins are composed of different kinds of **functional subunits** [e.g., the regulatory and catalytic subunits of regulatory proteins (see Chapter 11, section VII H)].

VI. ANALYSIS OF PROTEIN CONFORMATION

A. Primary structure

    1. Determining the amino acid composition

    a. Hot acid hydrolysis (at 110°C in 6 N hydrochloric acid in a sealed tube for 24 hr) cleaves the peptide bonds, liberating the free amino acids.

    b. Ion-exchange chromatography on sulfonated polystyrene columns separates the free amino acids (see section IV B 3 a). The amino acids are detected as they are eluted from the column by the **ninhydrin reaction**, which turns blue with α-amino acids and yellow with imino acids, such as proline.

$+ CO_2 + R—CHO$

Blue product (yellow with proline)

$$^+H_3N-\overset{\overset{\displaystyle COO^-}{|}}{\underset{\underset{\displaystyle R}{|}}{C}}-H \;+\; 2 \qquad \text{(Ninhydrin structure)} \qquad \longrightarrow$$

α-Amino acid   Ninhydrin

   c. The N-terminal residue can be identified by using a reagent that bonds covalently with its α-amino group. Because the bond is stable to hot acid hydrolysis, the derivative of the N-terminal residue can be identified by chromatographic procedures after the protein has been hydrolyzed. Two reagents are commonly used.

     (1) Fluorodinitrobenzene (FDNB), also called **Sanger's reagent**, was named after the investigator who used it to determine the primary structure of the small protein, insulin, the first time that the primary structure of a protein was determined. The reaction with FDNB is as follows:

$$O_2N-\!\!\left\langle \text{(benzene ring)} \right\rangle\!\!-F \;+\; H_2N-\overset{\overset{\displaystyle }{|}}{\underset{\underset{\displaystyle R}{|}}{CH}}-COO^- \xrightarrow{\text{alkali}}$$

$$O_2N-\!\!\left\langle \text{(benzene ring)} \right\rangle\!\!-\overset{\overset{\displaystyle H}{|}}{N}-\overset{\overset{\displaystyle }{|}}{\underset{\underset{\displaystyle R}{|}}{CH}}-COO^- \;+\; HF$$

     (2) Dansyl chloride, another N-terminal residue reagent, has the advantage of forming a fluorescent N-terminal derivative that can be detected and quantitated in very low concentration.

   d. The number of peptide chains in the protein can be determined by N-terminal amino acid analysis. The appearance of two or more kinds of N-terminal amino acids, for example, indicates a multiplicity of chains.

   e. The analysis of amino acids has been automated, allowing the rapid measurement of amino acids in body fluids as a diagnostic aid.

 2. Identifying the amino acid sequence

   a. Edman reaction

     (1) The reagent phenylisothiocyanate was developed for a technique that allows **repetitive sequencing** of a peptide.

     (2) Under mildly acid conditions, a phenylthiocarbamyl derivative is formed by the reaction of the reagent with the N-terminal amino group and is cleaved off the peptide. It cyclizes to form a phenylthiohydantoin derivative that can be identified chromatographically (Fig. 9-3).

     (3) The remainder of the peptide is left intact with the next residue in the chain bearing a free amino group. This can in turn be reacted with fresh reagent.

     (4) The procedure has been automated and can be used to determine the amino acid sequences of peptides 50 to 60 residues in length.

   b. Sequencing of peptide fragments

     (1) The direct sequencing of long peptides, such as those found in proteins (100 to 200 residues), is a difficult matter, and the analysis is considerably aided by first breaking the chains into short lengths. This **cleavage** is carried out by using reagents (chemical or enzymatic) that cleave the peptide chain at specific points. The sequence of these shorter

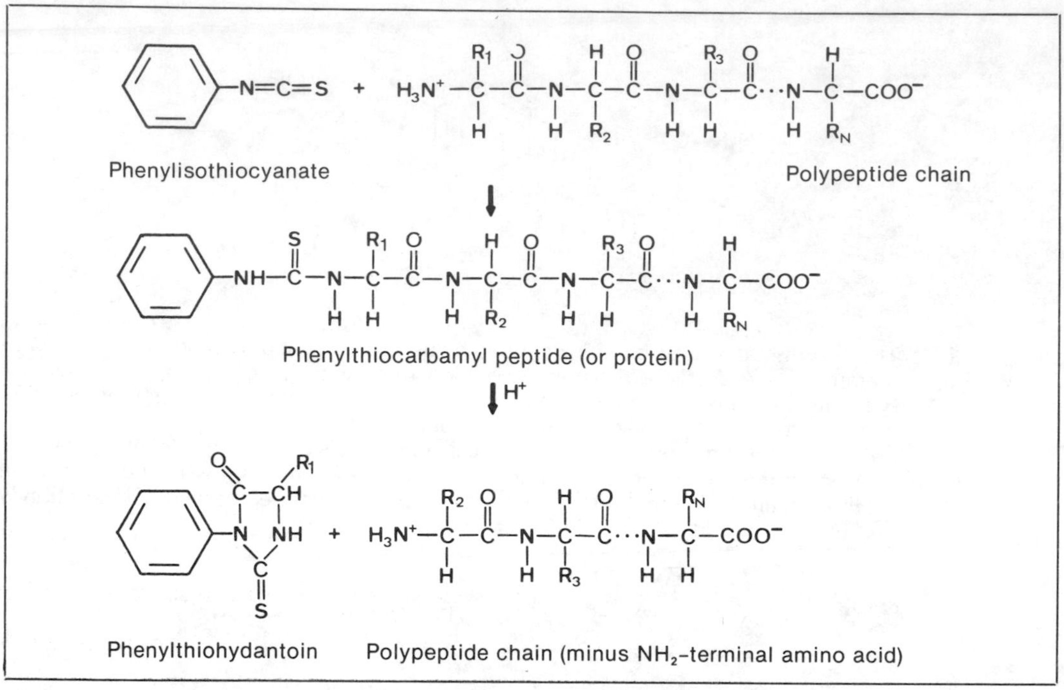

Figure 9-3. The Edman reaction.

peptides is then determined by the Edman technique, using the automated Sequenator.

(2) The problem of **arranging the fragment peptides in their true order** is overcome by fragmenting the original protein with several reagents that cut at different points. Knowing where these reagents cleave the peptide fragments and knowledge of the N-terminal residue allow the determination of the entire sequence of the original peptide.

(3) The reagents used to obtain peptide fragments include proteolytic enzymes and chemical reagents.

    (a) Proteolytic enzymes

        (i) Trypsin cleaves peptide bonds on the carboxyl side of basic amino acids (Arg and Lys).

        (ii) Chymotrypsin cleaves on the carboxyl side of aromatic amino acids (Phe, Tyr, and Trp).

        (iii) Staphlococcal protease cleaves on the carboxyl side of acidic amino acids (Asp and Glu).

    (b) Chemical reagents

        (i) Cyanogen bromide cleaves on the carboxyl side of met residues.

        (ii) Hydroxylamine cleaves at Asn–Gly bonds.

3. Locating the disulfide bonds. Disulfide bonds between cysteinyl residues may cross-link two peptide chains (Fig. 9-4). These bonds are usually oxidized by performic acid to cysteic acid residues, and the chains are separated by chromatography or electrophoresis before sequence analysis. The position of the **cysteic acid residues** in the sequence indicates where the disulfide cross-links occur in the peptide.

B. Secondary and tertiary structures

1. Studies on crystallized proteins

    a. X-ray crystallography has been a powerful technique for gaining insight into the three-dimensional conformation of proteins. For this procedure, the protein must be crystallized with the addition of a heavy metal or must have an elongated structure with repeating units, as in the fibrous proteins.

    b. X-ray diffraction measurements are made in which the heavy metal acts as a reference for the measurement of the phase angles of the diffracted beams. This heavy metal must not modify the conformation of the protein.

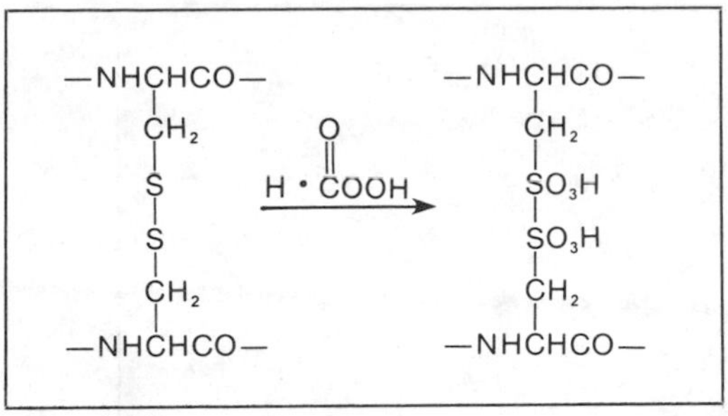

—NHCHCO—   —NHCHCO—
$\quad$ | $\quad$ |
$\quad$ CH_2 $\quad$ CH_2
$\quad$ | $\quad$ |
$\quad$ S $\quad$ O $\quad$ SO_3H
$\quad$ | $\quad$ ‖ $\quad$ $H \cdot COOH$ → $\quad$ |
$\quad$ S $\quad$ COOH $\quad$ SO_3H
$\quad$ | $\quad$ |
$\quad$ CH_2 $\quad$ CH_2
$\quad$ | $\quad$ |
—NHCHCO—   —NHCHCO—

Figure 9-4. The oxidation of disulfide bonds to cysteic acid residues by performic acid.

    c. The picture that emerges portrays a "frozen state" of a dynamic system, often in an inactive form. However, much evidence from the physical measurement of proteins in solution supports the general conclusions that have been drawn from the results of x-ray crystallographic studies.

2. Studies on proteins in solution
    a. Studies of optical behavior
        (1) Optical rotatory dispersion (ORD) and **circular dichroism (CD)** are changes in optical behavior produced by clockwise and counterclockwise vectors of plane polarized light. ORD refers to differences in the refractive index; CD refers to differences in light absorption.
        (2) Helical segments can form either right-handed or left-handed spirals, which can be detected by ORD or CD measurements (Fig. 9-5).
        (3) The large asymmetric side chains of aromatic amino acids also give characteristic ORD and CD patterns.
    b. Ultraviolet light spectroscopy (UV spectroscopy)
        (1) Proteins absorb ultraviolet light by virtue of their tyrosine, phenylalanine, tryptophan, and cysteine content and also because of absorption by the peptide bond (Fig. 9-6).
        (2) In **helical structures**, the interaction between peptide bonds of the same chain results

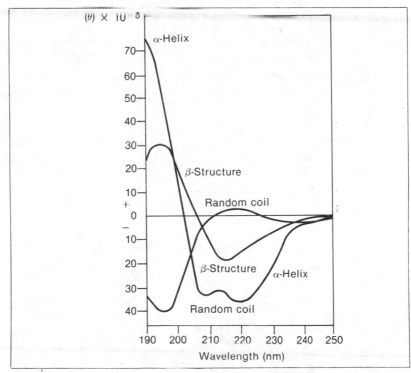

Figure 9-5. Circular dichroism spectra for polypeptide chains in α-helical, β-structure, and random-coil configurations. θ = ellipticity in degree $cm^2/0.1$ mol. (Reprinted with permission from Bull AT, et al: *Companion to Biochemistry.* London, Longman, 1970, p 190.)

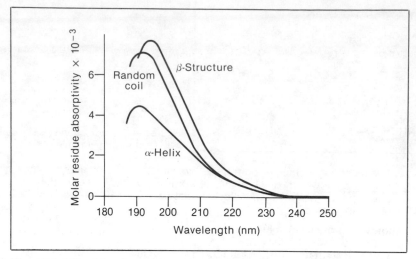

Figure 9-6. Ultraviolet absorption of the peptide bonds of a polypeptide chain in α-helical, random-coil, and antiparallel β-structure configurations. (Reprinted with permission from Bull AT, et al: *Companion to Biochemistry*. London, Longman, 1970, p 175.)

in a shift of the absorption maximum. Thus, UV spectroscopy can be used to monitor changes in protein conformation.

   (3) The absorption spectrum of **aromatic side chains** of the amino acids Tyr, Phe, and Trp is modified by changes in conformation, particularly if the change alters the degree of hydrophobicity in the microenvironment of the chromophore group.

c. Nuclear magnetic resonance (NMR)

   (1) The nuclei of certain isotopes, those with nonzero spin (e.g., ^{13}C, ^{1}H, ^{15}N), are magnetic, and when placed in a homogeneous magnetic field, they absorb electromagnetic energy (radio frequencies) at sharply defined frequencies.

   (2) The nuclear magnetic resonance of a sample is measured by placing the sample in a homogeneous magnetic field and applying electromagnetic energy over a range of frequencies.

   (3) The use of the NMR in probing protein structure rests on the fact that the resonance frequency of a nucleus (e.g., a proton) is influenced by the surrounding molecular structure.

STUDY QUESTIONS

Directions: Each question below contains five suggested answers. Choose the **one best** response to each question.

1. Which of the following α-amino acids is a diamino-monocarboxylic acid?

(A) Leucine
(B) Lysine
(C) γ-Carboxyglutamic acid
(D) Glycine
(E) Proline

[handwritten: typically inactive — No asymmetric carbon]

2. With the exception of glycine, all amino acids found in proteins are

(A) dextrorotatory
(B) of the D-configuration
(C) levorotatory
(D) of the L-configuration
(E) optically inactive

[handwritten: while]

3. The peptide bond has a "backbone" of atoms in which of the following sequences?

(A) C-N-N-C
(B) C-C-C-N
(C) C-C-N-C
(D) N-C-C-C
(E) C-O-C-N

4. Which of the following electrophoretic analytic procedures does not depend on the charge of the protein? *[handwritten: (but on molecular weight)]*

(A) Moving boundary electrophoresis
(B) Zone electrophoresis
(C) Polyacrylamide gel electrophoresis in sodium dodecyl sulfate
(D) Isoelectric focusing
(E) None of the above

5. Which of the following statements about protein structure is correct?

(A) The extended β-configuration is not found in globular proteins
(B) The stability of the α-helix is mainly due to hydrophobic interactions
(C) Globular proteins tend to fold into configurations that keep hydrophobic side chains in the interior of the molecule
(D) The protomers of polymeric proteins are linked by covalent bonds
(E) The primary structure of a peptide does not influence the formation of the native three-dimensional configuration

[handwritten: SANGER REAGENT]

6. A protein treated with fluorodinitrobenzene (FDNB) and subjected to acid hydrolysis yielded two different amino acids with their α-amino groups linked to DNB. A reasonable explanation for this result is that

(A) the protein contained more than one free amino group
(B) the protein contained more than one N-terminal amino acid
(C) the protein contained two basic amino acid residues
(D) the protein contained amide groups
(E) none of the above

7. The information on protein structure that can be obtained from circular dichroism spectra is the

(A) proportion of optically active amino acids in a protein
(B) number of subunits in a polymeric protein
(C) number of disulfide bonds in a protein
(D) presence or absence of prosthetic groups
(E) relative proportion of α-helical and nonhelical segments

Directions: Each question below contains four suggested answers of which **one or more** is correct. Choose the answer

A   if **1, 2, and 3** are correct
B   if **1 and 3** are correct
C   if **2 and 4** are correct
D   if **4** is correct
E   if **1, 2, 3, and 4** are correct

8.   Amino acids found in proteins that are formed by post-translational modification of one of the common amino acids (derived amino acids) include which of the following?

(1) Isoleucine
(2) γ-Carboxyglutamate
(3) Threonine
(4) 4-Hydroxyproline

9.   Which of the following ionic species of glutamic acid would be the prevalent at pH 10?

(1) $HOOC-CH-(CH_2)_2-COOH$
                    |
                 $^+NH_3$

(2) $^-OOC-CH-(CH_2)_2-COOH$
                    |
                 $^+NH_3$

(3) $^-OOC-CH-(CH_2)_2-COO^-$
                    |
                 $^+NH_3$

(4) $^-OOC-CH-(CH_2)_2-COO^-$
                    |
                 NH_2

10.   Separation of one protein from other proteins on the basis of molecular size can be achieved by

(1) polyacrylamide gel electrophoresis
(2) dialysis
(3) gel filtration (molecular exclusion chromatography)
(4) ion-exchange chromatography

11.   Correct statements concerning the Edman reaction include which of the following?

(1) The peptide does not have to be hydrolyzed for recovery of the reacted N-terminal amino acid
(2) The derivation of the N-terminal amino acid is identified by its fluorescent properties
(3) The reaction can be carried out repeatedly on the same sample of the peptide for sequence determination
(4) The reaction can be used in reverse for the synthesis of peptides

ANSWERS AND EXPLANATIONS

1. The answer is B. (*Table 9-1*) Leucine and glycine are α-amino acids; specifically, they are mono-amino-monocarboxylic acids. Proline is an amino acid with one secondary amino group and one carboxyl. γ-Carboxyglutamic acid has one α-amino group, two γ-carboxyls, and one α-carboxyl. It is formed post-transcriptionally from glutamate. Lysine has two amino groups, one α- and ε-carbons and one α-carboxyl. It is, therefore, a diamino-monocarboxylic acid.

2. The answer is D. (*II B*) With the exception of the optically inactive glycine, all amino acids derived from proteins by hydrolysis are of the L-configuration, having the same steric configuration as L-glyceraldehyde. A given amino acid may be either levorotatory or dextrorotatory. Glycine is not optically active because it does not have an asymmetric carbon atom.

3. The answer is C. (*III A*) The atoms forming the peptide bond all lie in the same plane with the carboxyl oxygen and the amide hydrogen being *trans* to one another. The "backbone" of the bond is formed by the two adjacent α-carbons, the carbonyl carbon, and the amide nitrogen. They are in the sequence C-C-N-C when written in the conventional manner with the N-terminal amino acid on the left.

4. The answer is C. (*IV B 2 d*) Electrophoresis carried out on a cross-linked polyacrylamide gel in the presence of a detergent, such as sodium dodecyl sulfate (SDS), separates proteins on the basis of their molecular weight rather than their charge. The SDS forms negatively charged micellar particles with the protein so that the effect of the protein charge is lost. The SDS-protein micelles migrate to the positive pole with the cross-linked polyacrylamide gel acting as a molecular sieve; the smaller the protein, the faster its migration rate. Moving boundary electrophoresis can provide information on the relative amount of proteins present in a mixture but does not actually separate them in a preparative sense. However, it is a procedure based on the charge carried by the protein molecules, as is the case with the other two procedures, zone electrophoresis and isoelectric focusing.

5. The answer is C. (*V B*) The primary structure of a protein (i.e., its amino acid sequence) dictates the final three-dimensional form adopted by the protein. Globular proteins may contain peptide chains, which in part may be extended β-configurations. Some of these might fold back on themselves, forming antiparallel β-sheet conformations in part of the protein. The α-helix is stabilized by intrachain hydrogen bonds that form between peptide bonds four residues apart. Globular proteins fold so that the hydrophobic side chains are kept away from water in the interior of the molecule, and the polar side chains are on the outside in contact with water. The individual chains, or protomers, of polymeric proteins associate by noncovalent bonding, which is frequently largely hydrophobic in nature.

6. The answer is B. (*VI A 1 c, d*) The Sanger reagent, fluorodinitrobenzene (FDNB), reacts with free amino groups on proteins. It thus reacts with the N-terminal amino acid and with ε-amino groups on lysine residues. The derivative DNB–amino acid formed is stable to acid hydrolysis and can be identified by chromatography. The appearance of more than one amino acid with DNB linked to the α-amino group indicates a multiplicity of peptide chains.

7. The answer is E. (*VI B 2 a*) Circular dichroism (CD) arises because clockwise and counterclockwise vectors of plane polarized light have different effects on optical behavior. CD measurements follow changes in light absorbance produced by helical structures and thus provide information on the relative number of helices in a protein. All but one of the amino acids are optically active, but in peptide chains, the optical activity of the individual amino acids is overwhelmed by the effect on light absorbance, or on refraction, by the asymmetric secondary structures. CD spectra do not give useful information on the number of subunits, the number of disulfide bonds, or the presence or absence of prosthetic groups in a protein.

8. The answer is C (2, 4). (*II A 3; Table 9-2*) Some amino acids are found in proteins although they are not coded for in DNA. They are derived from one or another of the 20 common amino acids after the latter has been incorporated into the protein chain (post-translational modification). The derived amino acids are: 4-hydroxyproline, 5-hydroxylsine, ε-N-methyl-lysine, desmosine, isodesmosine, 3-methylhistidine, and γ-carboxyglutamate.

9. The answer is D (4). (*II C 5*) Glutamic acid has three dissociable protons, two with pK_a' values well below pH 7 and one with a pK_a' value well above pH 11. At pH 10, both the α-carboxyl (pK_a' = 2.19) and the γ-carboxyl (pK_a' = 4.25) will have lost a proton and will carry a negative charge. The α-amino group (pK_a' = 9.67) will also have lost a proton and will be uncharged.

10. The answer is B (1, 3). *(IV B 2, 3)* Dialysis may remove small contaminating molecules from a protein, but it is not suitable for the separation of one protein from others. Ion-exchange chromatography is based upon the amphoteric properties of proteins. Gel filtration separates proteins on the basis of their molecular size by virtue of the pore size of the beads of hydrated carbohydrate polymer used to make the filtration column. Small proteins can enter the pores but not large proteins, and because the distribution volume of a small protein is greater than that of a large protein, the small ones flow through the column more slowly. In general, electrophoretic procedures separate proteins on the basis of molecular charge. However, electrophoresis in cross-linked polyacrylamide gels in the presence of the detergent sodium dodecyl sulfate (SDS) separates proteins on the basis of molecular size: The SDS forms negatively charged micellar particles with the protein so that the effect of protein charge is lost, and the cross-linked gel acts as a molecular sieve.

11. The answer is B (1, 3). *(VI A 2)* The Edman reagent, phenylisothiocyanate, reacts with the α-amino group of proteins to form a phenylthiocarbamyl derivative, which is cleaved off the protein by mild acid conditions, leaving the remainder of the peptide intact. The resulting phenylthiohydantoin derivative can be identified chromatographically. It is not fluorescent as are the dansyl derivatives formed by reacting the α-amino group with dansyl chloride. Also, unlike the dansyl reagent, the peptide need not be hydrolyzed in order to identify the amino acid. The Edman reaction can be repeatedly carried out in an automated technique for the sequencing of peptides 30 to 40 amino acids in length. The reaction cannot be reversed for the synthesis of peptides.

10

Structure–Function Relationships in Proteins

I. OXYGEN TRANSPORT PROTEINS

A. Oxygen transport molecule

1. Unicellular organisms and very small metazoans can get oxygen from the aqueous medium by simple molecular diffusion. Increasing size presents a problem met by the development of **circulatory systems** containing an **oxygen-binding molecule**. At first free in solution, the circulating oxygen binder in higher animals is packaged into specialized cells, the **red blood cells**.

2. Oxygen transport in higher animals is mediated by two protein molecules.
   a. **Myoglobin** is found in muscle, where it acts as an oxygen repository and transport molecule.
   b. **Hemoglobin**, found in red blood cells, functions to carry oxygen from the lungs to the tissues and carbon dioxide (CO_2) from the tissues to the lungs.

3. The packaging of hemoglobin into the red blood cells allows high concentrations of carrier without problems of osmotic pressure and viscosity. In humans, there are about 5 billion red blood cells per milliliter of blood, each red cell containing 280 million hemoglobin molecules of 64 kdal each.

B. Myoglobin

1. **Structure**
   a. **Myoglobin is a single peptide chain**, 153 residues in length, which is folded into a very compact structure ($45 \times 35 \times 25$ Å) with little space inside (Fig. 10-1).
   b. **Eight major α-helix segments** (lettered A through H in Figure 10-1) comprise about 75% of the chain.
      (1) The polar (hydrophilic) side chains are on the outside of the molecule.
      (2) Almost all of the nonpolar (hydrophobic) side chains are on the inside.
   c. **Nonpolar residues** (except for two histidines) line a pocket in the molecule into which the heme prosthetic group fits (Fig. 10-2). The nonpolypeptide heme group is composed of protoporphyrin IX (see Chapter 28, section II C 4) bound to an iron atom; the atom in functional myoglobin is in the ferrous state.
   d. **The oxygen-binding site** in myoglobin is on the heme prosthetic group.
      (1) Oxygen is bound to the heme prosthetic group when the latter is associated with the globin.
      (2) The iron atom in heme can form six **ligand bonds**.
         (a) Four of these are in the plane of the heme and are bound to the pyrrole ring nitrogens.
         (b) The fifth bond, to histidine-F8 (His-F8, **proximal histidine**), is on one side of this plane.
         (c) The sixth bond, to His-E7 (**distal histidine**), is on the other side.
         (d) The oxygen molecule is interposed between the iron atom and His-E7.
   e. **The nonpolar environment** of the heme group is important.
      (1) Isolated heme, dissociated from its apoprotein, binds oxygen only transiently in an aqueous environment, because the **ferrous** iron in the heme is rapidly oxidized to the **ferric** form, which cannot bind oxygen.
      (2) Experiments with model systems show that heme, when protected by a nonpolar environment, can bind oxygen well. In the nonpolar environment, it is much harder to strip an electron away from the ferrous iron than it is in water.

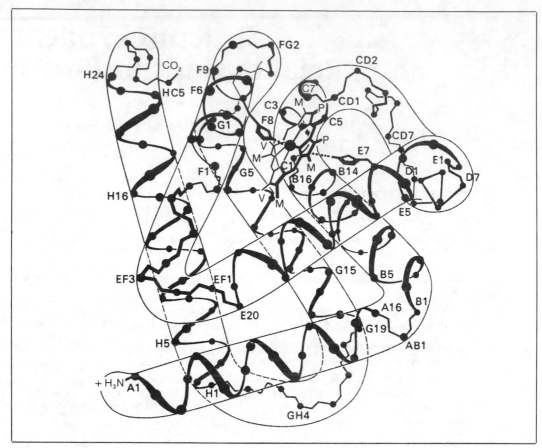

Figure 10-1. Model of myoglobin at high resolution. (Reprinted with permission from Devlin TM: *A Textbook of Biochemistry with Clinical Correlations*, 2nd ed. New York, John Wiley, 1986, p 81.)

2. Function

a. To serve its oxygen repository and transport functions, myoglobin must be able to bind oxygen well at the relatively low oxygen tensions that obtain in the tissues where the carrier (hemoglobin) is giving up oxygen.

b. The oxygen-binding curve for myoglobin is a rectangular hyperbola (Fig. 10-3) with the binding sites 90% saturated at the oxygen pressure of 40 torr that obtains in muscle tissue.

c. This relationship can be expressed as

$$Y = \frac{(MbO_2)}{(Mb) + (MbO_2)} \quad \text{or} \quad Y = \frac{PO_2}{P_{50} + PO_2},$$

where Y is the fraction of myoglobin saturated with oxygen, (MbO_2) is the concentration of oxygenated myoglobin, (Mb) is the concentration of deoxygenated myoglobin, PO_2 is the partial pressure of oxygen, and P_{50} is the partial pressure of oxygen at which myoglobin is 50% saturated with oxygen.

C. Hemoglobin

1. Structure

a. Hemoglobin is an oligomeric conjugated protein with four peptide chains joined by non-covalent bonds (**tetramer**).

b. Several different kinds of hemoglobin are normally found in humans, and these vary in the primary structure of their subunits.

   (1) Hemoglobin A_1, the major circulating form in human adults, has two identical α chains and two identical β chains (abbreviated $\alpha_2\beta_2$). The α and β chains are composed of 141 and 146 residues, respectively.

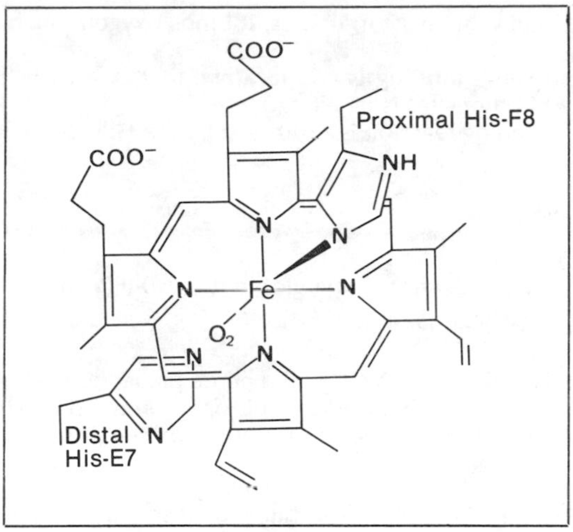

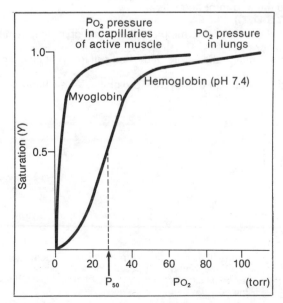

Figure 10-2. The ligand bonds of the ferrous atom in the heme prosthetic group. (Reprinted with permission from Devlin TM: *A Textbook of Biochemistry with Clinical Correlations*, 2nd ed. New York, John Wiley, 1986, p 78.)

(2) Hemoglobin A_2, which accounts for 2% of adult human hemoglobin, has the subunit structure $\alpha_2\delta_2$.

(3) Fetal hemoglobins are different from the adult forms. Three appearing early in fetal life have subunit structures of $\xi_2\zeta_2$, $\xi_2\gamma_2$, and $\delta_2\epsilon_2$. These are followed by hemoglobin F, which is $\alpha_2\gamma_2$.

c. The α and β subunits of hemoglobin have a conformation very much like that of myoglobin.

d. Each hemoglobin subunit contains a pocket for the heme that provides the nonpolar environment needed by the heme-iron in order to bind oxygen reversibly.

e. The tertiary structures of the subunits are very similar but differ enough to provide quaternary structures with different functions, ranging from that of hemoglobin A_1 to hemoglobin F.

2. Function

a. The oxygen carrier in the blood must have a high affinity for oxygen at the relatively high oxygen tension that obtains in the lungs and must be able to release this oxygen at the relatively low oxygen tensions in the tissues.

b. The carrier in the blood also needs to carry CO_2 from the tissues to the lungs.

c. The oxygen-binding curve for hemoglobin is sigmoidal in shape, rather than being a rectangular hyperbola (see Fig. 10-3).

(1) From this relationship, it can be seen that hemoglobin binds oxygen very efficiently at

Figure 10-3. Oxygen-binding curves for myoglobin and hemoglobin. (Reprinted with permission from Devlin TM: *A Textbook of Biochemistry with Clinical Correlations*, 2nd ed. New York, John Wiley, 1986, p 83.)

high oxygen tension, but at PO_2 values that exist in tissues, it binds oxygen much less tightly than myoglobin.

(2) The kinetics of binding indicate that the **binding is cooperative**; that is, as oxygen is bound to some sites, the binding to other sites is facilitated.

(3) The degree of cooperativity in the binding reaction is expressed by the **Hill coefficient**, n, in the expression

$$\text{Saturation } Y = \frac{(PO_2)^n}{(PO_2)^n + P_{50}}$$

For hemoglobin, the value of n is 2.8, whereas for myoglobin it is 1.0 (Fig. 10-4).

3. Structural basis of physiologic functions of hemoglobin

a. Cooperative binding of oxygen

(1) In deoxyhemoglobin, the iron atom is about 0.6 Å out of the heme plane, due in part to steric repulsion between the proximal histidine and the nitrogen atoms of the porphyrin ring. On oxygenation, the iron atom moves into the plane of the porphyrin so that it can form a strong bond with oxygen, the radius of the iron atom being reduced as the sixth ligand bond forms.

(2) The movement of the iron into the plane of the heme pulls the His-F8 residue toward the porphyrin and moves the F helix in relation to the C helix of the adjacent subunit.

(3) These movements of the helical segments occur at the $\alpha_1\beta_2$ and $\alpha_2\beta_1$ interfaces, destabilizing the **deoxyhemoglobin** conformation (**tight or T form**) and switching it to a new tertiary conformation (**relaxed or R oxyhemoglobin form**).

(4) The R form has an affinity constant for the binding of oxygen that is two to three times greater than that of the T form. Hence, the binding of the first oxygen cooperatively enhances the binding of subsequent oxygens to the other heme groups in the hemoglobin molecule.

b. Bohr effect

(1) Oxygen binding by hemoglobin decreases with increasing hydrogen ion (H^+) concentration. An increasing PCO_2, at constant pH, also decreases this affinity and shifts the oxygen-dissociation curve to the right. Figure 10-5 shows the displacement of the oxygen-binding curve to the right by these and other factors.

(2) Since an increase in PCO_2 and a decrease in pH are both characteristic of actively metabolizing cells, these cells promote the release of oxygen from hemoglobin.

(3) Both H^+ and CO_2 bind to hemoglobin.

    (a) The protons bind to the deoxy (T) form because it is a weaker acid, whereas at pH 7.4 (e.g., in the lungs), the change to the R form is enhanced by the loss of 2.8 Eq of protons per mole of hemoglobin.

    (b) The CO_2 binds to hemoglobin, as carbamate, on reaction with deprotonized terminal NH_2 groups. As this tends to stabilize the T form, the binding of CO_2 encourages the release of oxygen. On oxygenation in the lungs, the CO_2 is released because it binds less tightly to the R form of hemoglobin.

c. Role of 2,3-diphosphoglycerate (2,3-DPG)

(1) Blood stored for transfusion use becomes with time more like myoglobin than hemoglobin in terms of oxygen binding.

(2) The increased affinity for oxygen is due to the failure of the red cell to maintain a high level of 2,3-DPG.

(3) 2,3-DPG is required in catalytic amounts in the glycolytic pathway (see Chapter 14,

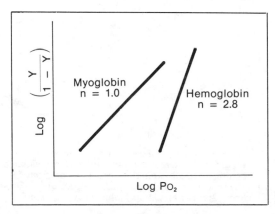

Figure 10-4. Hill plots for myoglobin and hemoglobin HbA₁. (Reprinted with permission from Devlin TM: *A Textbook of Biochemistry with Clinical Correlations,* 2nd ed. New York, John Wiley, 1986, p 83.)

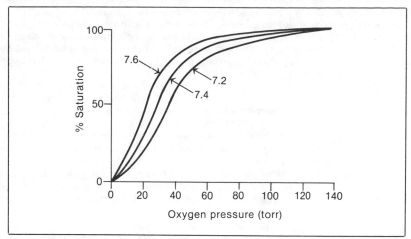

Figure 10-5. Shift of the oxygen–hemoglobin dissociation curve to the right by decreasing pH (i.e., increasing the concentration of hydrogen). Similar shifts of the dissociation curve to the right are brought about by increasing the concentration of carbon dioxide, temperature, or the concentration of 2,3-diphosphoglycerate.

    section III G 4) but is normally present in red cells at about 5 mM, which is almost equimolar with hemoglobin.

(4) 2,3-DPG binds to hemoglobin, stabilizing the deoxy form and promoting the release of oxygen. This is the reason that the Hill coefficient for hemoglobin is 2.8 in the red cell, whereas it is nearer 1 when the protein is isolated (see Fig. 10-4).

(5) 2,3-DPG binds to hemoglobin by interacting with three positively charged groups on each β chain, effectively cross-linking these subunits.

(6) The red cell can adapt to conditions of tissue hypoxia (e.g., from anemia, high altitudes, or pulmonary dysfunction) by increasing the intracellular level of 2,3-DPG, which promotes further release of oxygen.

d. Fetal hemoglobin

(1) The hemoglobin of the fetus must pick up oxygen at the placental junction from the maternal hemoglobin. Therefore, in its oxygen-binding characteristics, fetal hemoglobin must be more like myoglobin than is maternal hemoglobin.

(2) Hemoglobin F, the oxygen carrier formed late in fetal life, has two γ subunits that do not bind 2,3-DPG as strongly as do the β subunits of adult hemoglobin A. Hence, the oxygen-dissociation curve of fetal hemoglobin is shifted to the left compared to hemoglobin A (Fig. 10-6).

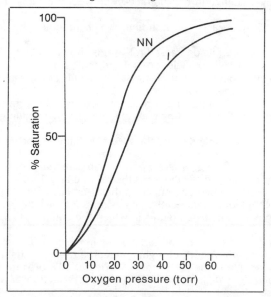

Figure 10-6. Oxygen–hemoglobin dissociation curves in the neonate at birth (*NN*) and in the infant at 3 months (*I*). The shift to the right at 3 months is due to the increased levels of adult hemoglobin (HbA$_1$) and to the decline in the production of fetal hemoglobin (HbF).

4. Modification of hemoglobin structure and disease

a. Glucosylation of hemoglobin A₁ in diabetes mellitus

(1) Hemoglobin A_1 reacts spontaneously with glucose to form a derivative known as **HbA₁c**. The reaction involves the nonenzymatic formation of a Schiff's base between glucose and the N-terminal amino groups of the β chains (valines). The labile Schiff's base then becomes the more stable amino ketone.

(2) Normally the concentration of HbA_{1c} in blood is very low, but in diabetes mellitus, where blood sugar levels may be high, the concentration of HbA_{1c} may reach 12% or more of the total hemoglobin.

(3) The later complications of diabetes mellitus (e.g., angiopathies, neuropathies, nephropathies, and retinopathies) may arise because of prolonged periods of high blood glucose despite insulin therapy, and the blood levels of HbA_{1c} are a valuable guide to the efficacy of insulin therapy.

b. Hemoglobinopathies of genetic origin. Abnormal hemoglobins have arisen by random mutations, which persist if the functional impairment is compatible with life.

(1) **Hemoglobin S** variant in **sickle cell anemia** is formed when valine replaces glutamic acid in the sixth position of the β chains. The α chains are normal.

    (a) On deoxygenation, hemoglobin S polymerizes within the red cell, forming long insoluble fibers that distort the red cell. This leads to the characteristic **sickle-shaped cell**, which impairs circulation, causing painful crises due to vaso-occlusive phenomena. The rate of destruction of the misshapen "sickled" cells is much higher than that of normal cells.

    (b) In other respects, hemoglobin S appears to be normal, the cooperative binding of oxygen and the protection of the ferrous iron by the apoprotein being unaffected.

    (c) Homozygotes have sickle cell *anemia*; heterozygotes are usually symptom-free and are said to have sickle cell *trait*.

    (d) Persons with sickle cell trait show an increased resistance to malaria, and the world distribution of sickle cell anemia closely parallels the distribution of the malarial mosquito.

(2) **Hemoglobin C** is another variant in which lysine replaces glutamic acid in the sixth position of the β chain. In this condition, the cells do not sickle, but they are abnormally susceptible to cell lysis, and anemia develops.

(3) **Hemoglobin M** is a rare abnormal hemoglobin variant in which a mutation modifies the oxygen-binding capability.

    (a) The His-F8 is replaced by tyrosine, which reacts with ferric iron in the heme to stabilize the methemoglobin form in which the iron is in the ferric form and which is incapable of binding oxygen.

    (b) Mutations producing hemoglobin M are known in both α and β subunits, the affected hemoglobins in a given patient carrying only two oxygens. Only heterozygotes are known, as the homozygous form is incompatible with life.

(4) **Hemoglobin Rainier** is an example of a mutant hemoglobin in which the affinity for oxygen is enhanced, in this case by a change in the interaction between the $\alpha_1\beta_2$ contacts.

    (a) In this mutation, tyrosine 145 is replaced by histidine, which modifies the salt bridges of the deoxy form.

    (b) The increased affinity for oxygen (i.e., its decreased release) is shown by a P_{50} for oxygen of 12.9 (normal, 27) and a Hill coefficient of 1.1 (normal, 2.8).

    (c) Polycythemia (an increase in erythrocytes and in hemoglobin) occurs as a means of compensating for the increased oxygen affinity.

(5) **Thalassemias** are hemolytic anemias, which occur as a result of mutations that prevent globin synthesis with some of the mutations affecting control elements involved in genetic expression.

    (a) Each haploid genome has two closely linked genes for the α subchain on one chromosome and a cluster of four genes for other subchains on a different chromosome. This cluster is in the order $^G\gamma\text{-}^A\gamma\text{-}\delta\text{-}\beta$, where $^G\gamma$ and $^A\gamma$ are subchains of fetal hemoglobin F with glycine or alanine, respectively, as the residue at γ-136.

    (b) Several different gene mutations have been identified in thalassemias, including

        (i) The absence of one or both α-globin genes with homozygotes for the latter state exhibiting a syndrome known as **hydrops fetalis** in which the patients die before or soon after birth

        (ii) An instability in mRNA, as in hemoglobin Constant Spring, in which a UAA stop codon has mutated to a CAA, which codes for glutamine; the α-globin chain is 31 residues longer than normal, and the longer mRNA is degraded more rapidly

        (iii) A decrease in the production of β-globin mRNA, causing a reduction or absence of β-globin synthesis

II. GLYCOPROTEINS

A. Nature. Glycoproteins are a class of conjugated proteins with carbohydrate moieties covalently bonded to the peptide chains. They have a number of important physiologic functions.

   1. As components of cell membranes, they are involved in cell–cell recognition and other membrane phenomena.

   2. As components of mucus secreted by specialized epithelial cells, they serve in the protection of tissues that line the ducts of the body.

   3. As proteins exported from cells, they function in extracellular compartments; for example, most plasma proteins are glycoproteins (with the exception of albumin) as are some protein hormones, such as the gonadotropins and thyrotropin.

B. Carbohydrate moiety of glycoproteins

   1. The percentage by weight of carbohydrate in glycoproteins varies over a wide range. Low-carbohydrate glycoproteins typically contain about 4% carbohydrate, whereas high-carbohydrate glycoproteins have from 60% to 80%.

   2. The pattern of distribution of carbohydrate moieties is also variable. They may be evenly distributed or restricted to particular regions of the peptide.

   3. The carbohydrate units may consist of a single sugar (e.g., N-acetyl-β-D-galactosamine) and rarely are composed of more than 12 to 15 residues.

C. Protein–carbohydate linkages

   1. Type I linkages are formed between the amide group of asparagine residues and N-acetyl-β-D-glucosamine (for the synthesis of this sugar derivative see Chapter 19, section V C). They are formed only when the Asn residue is in the sequence Asn-X-Thr (**sequeon**), where X = any amino acid. Type I linkages are found in the plasma protein orosomucoid.

   2. Type II linkages are formed between serine or threonine alcohol groups and the sugar. They are found in the glycoproteins of mucus secretions, in the anticoagulant heparin, and in connective tissue proteoglycans.

   3. Type III linkages are formed between 5-hydroxylysine residues and the carbohydrate moiety. They are found in some collagens and in complement component C1q.

   4. Type IV linkages, formed between the carbohydrate moiety and 4-hydroxyproline residues, are found only in higher plants.

   5. Type V linkages, formed between the carbohydrate and cysteine residues, are found in red blood cell membranes.

   6. Type VI linkages, formed between the carbohydrate and N-terminal amino groups, are found in HbA_{1c} (see section I E 1).

III. COLLAGEN: A FIBROUS PROTEIN

A. Characteristics. Fibrous proteins have a rod-like shape and are insoluble in water. They play a structural role in the organism. Collagen is formed by specialized cells: fibroblasts in connective tissue, osteoblasts in bone, chondroblasts in embryonic cartilage, and odontoblasts in teeth.

   1. Framework protein. Collagen, the most abundant human protein, is found in all tissues and organs, where it serves as the framework protein. It forms long microfibrils by cross-linking staggered arrays of a precursor molecule, **tropocollagen**. These microfibrils associate to form the larger fibers of collagenous tissues. A 70-kg human contains about 12 to 14 kg of protein of which 6 to 7 kg is collagen.

   2. Amino acid composition and sequence
      a. A unique distribution of amino acid residues distinguishes collagen with about 33% of the total residues being glycine (Gly), 10% proline (Pro), 10% hydroxyproline (Hyp), and 1% hydroxylysine (Hyl).
      b. The fundamental unit of collagen structure is **tropocollagen**, a molecule with three peptide chains (α chains) of about 1000 residues each.
      c. Every third amino acid in the α chains is glycine, and the sequences Gly-Pro-X_1, and Gly-X_2-Hyp (where X_1 and X_2 are any amino acids) are each repeated about 100 times; that is, they form 60% of the chains.

3. Structure of tropocollagen

a. The proline-rich chains of tropocollagen form a **helical structure** quite different from the α-helix in that it is more open, there is no intrachain peptide H-bonding, and it is a left-handed (counterclockwise) helix. The synthetic homopolymer polyproline forms a similar structure, and it is believed that these helices are stabilized by steric repulsion of the pyrrolidine rings of proline residues.

b. The three α chains of tropocollagen wind snugly around each other in a right-handed (clockwise) superhelix with about one twist every three residues.

c. The closeness of the three helical chains in tropocollagen is made possible by the high glycine content: every third residue must be a glycine, and its single hydrogen-atom side chain is small enough to fit inside the structure, allowing the tight superhelix to form.

d. Each of the three helical strands initially has nonhelical regions at both the N-terminal and C-terminal ends. These regions are hydrolyzed off during the formation of mature collagen after secretion of the protein from the cell.

B. Biosynthesis (Fig. 10-7)

1. In the formation of mature collagen, there are several stages.

a. The synthesis of an early precursor, **preprocollagen**, takes place on ribosomes attached to the endoplasmic reticulum. Preprocollagen contains **signal peptide sequences**, which allow the molecule to cross the membrane of the endoplasmic reticulum and enter its cisternae.

b. In the cisternae of the endoplasmic reticulum, a series of reactions occur.
  (1) Hydrolysis of the signal peptide forms procollagen.
  (2) Oxidation of cysteine residues in the nonhelical terminal regions forms disulfide bonds.

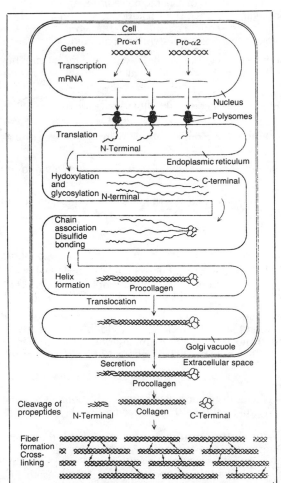

Figure 10-7. Collagen biosynthesis, showing transcription (only two genes are depicted), translation in the endoplasmic reticulum, various and extensive post-translational modifications, helix formation, secretion into the extracellular space, cleavage of propeptides, and cross-linking to give fibrils. (Reprinted with permission from Prockop DJ, Guzman NA: Collagen diseases and the biosynthesis of collagen. *Hosp Pract* 12:61, Dec, 1977.)

These bonds are *interchain* cross-links at the C-terminal end and *intrachain* links at the amino terminal.

(3) Hydroxylation of prolyl and lysyl residues, by the action of three enzymes, forms 4-hydroxyproline, 3-hydroxyproline, and 5-hydroxylysine.

 (a) The hydroxylation requires Fe^{2+}, ascorbate, molecular oxygen, and α-ketoglutarate in the reaction

$$\text{AA residue} + O_2 + \alpha\text{-ketoglutarate} \rightarrow \text{HO—AA residue} + \text{succinate}$$

 (b) The AA residue must be in a nonhelical region and in a particular sequence: (X-Lys-Gly) for lysyl hydroxylase, (X-Pro-Gly) for prolyl-4-hydroxylase, and (Pro-OH-Pro-Gly) for prolyl-3-hydroxylase.

(4) In addition to the hydroxylation reactions, the formation of a stable triple helix requires glycosylation.

 (a) Galactose residues are added to specific hydroxylysyl residues, through the action of galactosyl transferase, to transfer galactosyl residues from uridine diphosphate galactose to the hydroxy group on a hydroxylysyl residue.

 (b) Glucose is then added to some of these new residues by the action of glucosyl transferase to transfer the glucosyl moiety from uridine diphosphate glucose to the galactose.

2. **Triple helix formation** is completed by the end of the hydroxylation and glycosylation program, and the tropocollagen with disulfide bonds still present passes into the Golgi vacuole and is secreted by the cell.

3. **Collagen microfibril formation** occurs spontaneously after secretion when the N-terminal and C-terminal regions containing the disulfide bonds are cleaved off by peptidases. The kind of fibril that forms is dependent upon the amino acid sequence of the chains making up the triple helix.

4. **Collagen fiber formation.** Unit assemblies of 4 or 5 collagen molecules, staggered successively by one repeat unit (about 700 Å), are subsequently overlapped to form continuous filaments. These filaments, called **limiting microfibrils**, are packed together with repeat periods in phase to form collagen fibers.

5. **Maturation of collagen.** Intermolecular cross-links between tropocollagen helices are formed over a considerable period of time during the slow maturation of collagen. These cross-links, which are not disulfide bonds, give mature collagen some of its tensile strength.

 a. Several types of cross-links develop, but only one enzymatic reaction is involved, namely, the oxidative deamination of lysyl and hydroxylysyl residues to form aldehydes. The enzyme is lysyl oxidase, and the reaction requires copper and molecular oxygen.

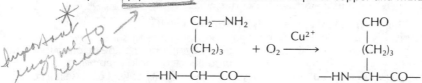

Important to enzyme to recall

 b. The linkages form spontaneously (i.e., nonenzymatically). For example, aldol condensation occurs between two aldehyde groups, usually on the same chain, or a Schiff's base may be formed between an aldehyde on one chain and an ϵ-amino of lysine, hydroxylysine, or glycosylated hydroxylysine residues on neighboring chains. The Schiff's base linkages then rearrange to form more stable linkages.

6. **Classification of collagen types**

 a. Several forms of collagen exist, differing in the amino acid composition of the tropocollagen chains. These variations provide the differences in physical properties that are required by different tissues.

 b. Six types of collagen are recognized; a summary of their structure and function is given in Table 10-1.

C. Diseases related to the metabolism of collagen

1. **Scurvy**

 a. Scurvy is a nutritional disorder caused by a deficiency of ascorbic acid (vitamin C).

 b. Ascorbic acid is required by proline hydroxylase, possibly in order to keep the ferrous iron in the reduced state. A deficiency of the vitamin leads to inadequate formation of hydroxyproline and, in severe deficiency, to a decreased rate of procollagen synthesis.

Table 10-1. Classification of Collagen Types

Type	Chain Designations	Tissue	Characteristics
I	$[\alpha1(I)]_2\alpha2(I)$	Bone, skin, tendons, scar tissue, heart valve, and intestinal and uterine wall	Low carbohydrate; <10 hydroxylysines per chain; two types of polypeptide chains
II	$[\alpha1(II)]_3$	Cartilage and vitreous	10% carbohydrate; >20 hydroxylysines per chain
III	$[\alpha1(III)]_3$	Blood vessels, newborn skin, scar tissue, and intestinal and uterine wall	Low carbohydrate; high hydroxyproline and Gly; contains Cys
IV	$[\alpha1(IV)]_3$ $[\alpha2(IV)]_3$	Basement membrane and lens capsule	High 3-hydroxyproline; >40 hydroxylysines per chain; low Ala and Arg; contains Cys; high carbohydrate (15%)
V	$[\alpha1(V)]_2\alpha2(V)$ $[\alpha1(V)]_3$ $[\alpha1(V)\alpha2(V)\alpha3(V)]$	Cell surfaces or exocytoskeleton; widely distributed in low amounts	High carbohydrate, relatively high Gly, and hydroxylysine
VI	. . .	Aortic intima, placenta, kidney, and skin in low amounts	Relatively large globular domains in telopeptide region; high Cys and Tyr; mol wt relatively low ($\sim 160{,}000$); equimolar amounts of hydroxylysine and hydroxyproline

Note.—Reprinted with permission from Devlin TM: *A Textbook of Biochemistry with Clinical Correlations*, 2nd ed. New York, John Wiley, 1986, p 108.

    c. Symptoms include abnormal bone development in children, poor wound healing, and hemorrhages due to fragility of skin capillaries.

2. Ehlers-Danlos syndrome type VI
    a. In this hereditary disease, the enzyme lysyl hydroxylase is deficient, and collagen with a reduced hydroxylysine content is formed. This reduces the degree of cross-linking and results in less stable collagen.
    b. Clinical symptoms include musculoskeletal deformities, particularly hypermobility of the joints, hyperelastic skin, and poor wound healing.
    c. Some patients have a mutant form of lysyl hydroxylase with an increased K_m for ascorbic acid, and they respond to therapy with ascorbic acid.

3. Diseases resulting from the inhibition of lysyl oxidase activity
    a. **Osteolathyrism** is due to poisoning by β-aminopropionitrile, a substance in sweet pea seeds (*Lathyrus odoratus*) that inhibits lysyl oxidase. Symptoms include deformation of the spine, demineralization of bone, dislocation of joints, and aortic aneurysm.
    b. **Copper deficiency in pigs** causes reduced lysyl oxidase activity as lysyl oxidase is a copper-requiring enzyme system and a consequent cross-linking defect. Symptoms are similar to those of osteolathyrism.
    c. **Therapy with D-penicillamine**, a chelator of copper and other heavy metals, is used in Wilson's disease (in which excess copper occurs from a hereditary lack of ceruloplasmin, the copper transport protein), in mercury and lead poisoning, and in hereditary cystinuria (see Chapter 27, section IV A 2). The penicillamine inhibits lysyl oxidase because the enzyme is a copper-requiring system. Symptoms in humans include extravasation of blood into the skin over the elbows and knees.

4. Homocystinuria. In this condition, there is a failure to metabolize the amino acid homocysteine, and the amino acid, a metabolite of S-adenosylmethionine, accumulates. Homocysteine reacts with the lysyl semialdehydes to block cross-link formation in collagen. Symptoms similar to those in osteolathyrism occur.

5. Ehlers-Danlos syndrome type VII
    a. The proteolytic cleavage of the N-terminal propeptide on procollagen does not occur in this genetic disorder.
    b. This leads to **dermatosparaxis** in cattle and sheep, in which the skin is so fragile that it is incompatible with life.
    c. In humans, the skin is hyperelastic and bruises easily, and bilateral hip dislocation is common.

Table 10-2. Some Characteristics of Plasma Lipoproteins Classified by Hydrated Density

Class	Density (g/ml)	Diameter (nm)
HDL	1.063–1.210	5–13
LDL	1.019–1.063	20–28
IDL	1.006–1.019	25
VLDL	0.95–1.006	25–75
Chylomicrons	<0.95	100–1000

HDL = High-density lipoproteins; LDL = low-density lipoproteins; IDL = intermediate-density lipoproteins; VLDL = very low-density lipoproteins.

IV. LIPOPROTEINS

A. Characteristics. Lipoproteins are complexes of protein and lipids held together by noncovalent bonds. Clinically, the most important are the lipoproteins of plasma, which function as major transporters of lipids.

B. Classification. Two systems for classifying plasma lipoproteins are in use.

   1. Density. One system is based on the density (d) of the lipoprotein particles as determined by their flotation rate on centrifugation in 1.063 d sodium chloride solution. Five fractions are distinguished, and their characteristics are shown in Table 10-2.
   a. High-density lipoproteins (HDL)
   b. Low-density lipoproteins (LDL)
   c. Intermediate-density lipoproteins (IDL)
   d. Very low-density lipoproteins (VLDL)
   e. Chylomicrons

   2. Electrophoretic mobility. The other classification is based on electrophoretic mobility, the plasma lipoproteins migrating with the α- and β-globulins of plasma. The correspondence between the density classification and the electrophoretic behavior is shown in Figure 10-8.

C. Composition

   1. Lipoprotein components do not have an exact stoichiometric relationship, but the different classes have characteristic lipid:protein ratios (Table 10-3), and characteristic lipid classes are associated with the apoproteins (Table 10-4).

   2. When the lipids are extracted from a lipoprotein by organic solvents, the **apoproteins** remain. These distinct proteins were initially classified as apoproteins A and B, but now eight different proteins are known, as shown in Table 10-4.

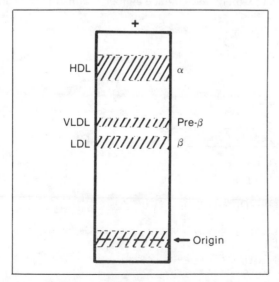

Figure 10-8. Separation of plasma lipoproteins by gel electrophoresis into β, pre-β, and α bands. β-Lipoproteins are mainly low-density lipoproteins (LDL); pre-β-lipoproteins, very low-density lipoproteins (VLDL); and α-lipoproteins, high-density lipoproteins (HDL). Chylomicrons remain at the origin.

Table 10-3. Chemical Composition of the Major Plasma Lipoproteins

Class	Protein (%)	Lipid (%)	Cholesterol (Free)	Cholesterol (Ester)	Triglycerides	Phospholipids
			(As percent of lipid fraction)			
HDL	50	50	3–4	12	3	20–25
LDL	20–25	75–80	7–10	35–40	7–10	15–20
IDL	15–20	80–85	8	22	30	22
VLDL	12	88	5–10	10–15	50–65	15–20
Chylo-microns	0.5–2.5	97.5–99.5	1–3	3–5	84–89	7–9

HDL = high-density lipoproteins; LDL = low-density lipoproteins; IDL = intermediate-density lipoproteins; and VLDL = very low-density lipoproteins.

    3. Structurally, the apoproteins have been shown by circular dichroism and optical rotatory dispersion to be significantly α-helical in nature.

D. Structure

    1. Volume calculations suggest that the lipoprotein complex (Fig. 10-9) can be thought of as a sphere with the protein plus the amphipathic lipid (phosphatidylcholine and unesterified cholesterol) forming an outer shell 21.5 Å in thickness. The apolar segments of the shell are directed inward, and the polar segments face the water outside. The nonpolar lipids (triglycerides and esterified cholesterol) form the inside of the sphere.

    2. From the model illustrated in Figure 10-9, it is easy to see why a decrease in the nonpolar lipid:protein ratio occurs as the size of the lipoprotein particle decreases.

E. Diseases associated with abnormal lipoprotein metabolism are discussed in Chapter 24, section V.

V. IMMUNOGLOBULINS

A. Definitions

    1. Immunoglobulins, or **antibodies**, are proteins produced by the body in response to the presence of foreign compounds (e.g., proteins, complex carbohydrates, and nucleic acid polymers). The human body can produce about 10^6 different kinds of antibody molecules, but all of these molecules have a similar basic structure.

    2. Antigens. Antibodies react with foreign substances, called antigens, to initiate a process whereby the invaders are eliminated.

    3. Haptens are small molecules that cannot by themselves induce antibody formation but can do so when covalently linked to larger molecules.

B. Primary structure of immunoglobulins

    1. Immunogloblins of the G class (IgG) have four peptide chains, two of them identical **heavy**

Table 10-4. Apoproteins of Human Plasma Lipoproteins and the Density Classes in Which They Occur

Apoprotein	Density Class
A-I	HDL
A-II	HDL
B	LDL, VLDL
C-I	
C-II	HDL, LDL, VLDL
C-III	
D	HDL
E	HDL, LDL, VLDL

HDL = high-density lipoproteins; LDL = low-density lipoproteins; and VLDL = very low-density lipoproteins.

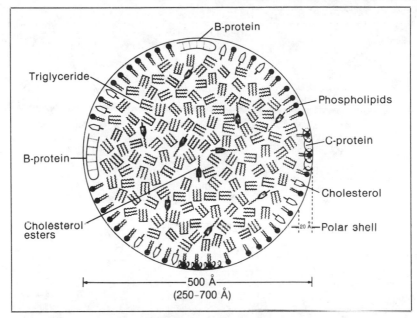

Figure 10-9. Generalized structure for the lipoproteins. The structure shown is very low-density lipoproteins. Protein, phospholipids, and unesterified cholesterol form a shell of 20 Å on the outside. (Reprinted with permission from Morrisett JD, et al: Lipid-protein interactions in the plasma lipoproteins. *Biochim Biophys Acta* 472:93, 1977.)

(H) chains of 50 kdal each, and two identical **light (L) chains** of about 25 kdal each, covalently attached by disulfide bonds (Fig. 10-10).

   2. Immunoglobulins are glycoproteins: There is a small amount of carbohydrate (2% to 12%) associated with the heavy chains.

C. Three-dimensional structure of immunoglobulins

   1. In IgG (see Fig. 10-10), each heavy chain is paired with a light chain. The N-terminal ends of the heavy and light chains are close together, but the light chains are much shorter than the heavy chains, so that the C-terminal halves of the heavy chains are aligned with one another.

   2. The heavy and light chains are joined by **disulfide bonds**.

   3. IgA and IgM form polymers; their monomeric forms are similar to IgG but not the same.

D. Variable and constant regions

   1. Variable (V) regions
      a. The N-terminal ends of the heavy and light chains vary considerably in their amino acid sequences. These variable regions take up one-half the length of each light chain and one-quarter of each heavy chain.
      b. Within the variable regions, hypervariable segments can be distinguished.

   2. Constant (C) regions are segments showing an amino acid sequence that is almost constant for each class of immunoglobulins. The constant regions, encompassing one-half the length of the light chains and three-quarters of the heavy chains, are at the C-terminal ends.

   3. Diversity (D) regions are small segments involved in the maturation of the variable region on the heavy chain [see Chapter 8, section II B 1 b (1)].

   4. Joining (J) chain is a glycopeptide chain that joins the variable regions to the constant regions.

E. Classes of immunoglobulins

   1. Immunoglobulins were originally divided into five classes—**IgG, IgA, IgD, IgE,** and **IgM**—on the basis of their sedimentation coefficients (Table 10-5). It was later found that the constant

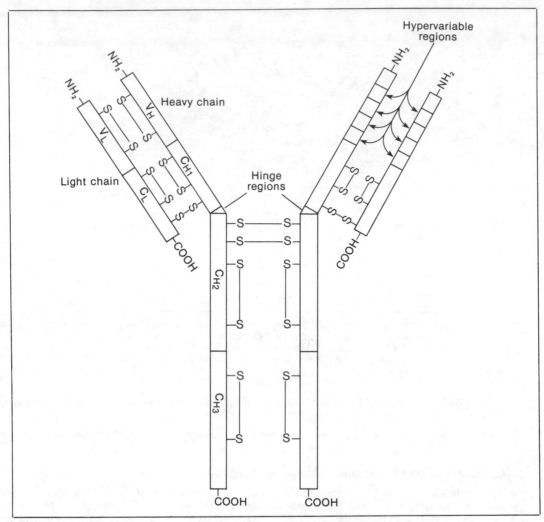

Figure 10-10. Diagrammatic structure of IgG. There are two light chains (*L*) and two heavy chains (*H*) linked by disulfide bridges (*S—S*). The L chains have two domains, one a region of variable amino acid sequence (V_L) and one of constant amino acid sequence (C_L). The heavy chains also have specialized domains, V_H, and three constant regions, C_{H1}, C_{H2}, and C_{H3}. The antigen-binding sites are V_H and V_L. NH$_2$ = amino terminal; COOH = carboxyl terminal.

regions of the heavy chains are homologous within a given class of immunoglobulins and differ from one class to another.

   a. IgG is the predominant immunoglobulin in plasma, but its synthesis is relatively slow (about 10 days).

   b. IgM, which is rapidly synthesized in response to an antigen, serves as a first line of defense.

2. The constant regions of the heavy chains give the five immunoglobulin classes their different physical and biologic characteristics.

3. The heavy chains of the different immunoglobulin classes have been given specific isotype designations (see Table 10-5).

4. Light chains have two types of constant regions, both of which are found in all five immunoglobulin classes. The two types of light chains are termed **kappa** (*ϰ*) and **lambda** (λ) chains.

F. Antigen-binding sites on immunoglobulins

1. The antigen-binding area of the immunoglobulin is formed by the variable regions at the N-terminal ends of the light- and heavy-chain pairs (see Fig. 10-10). Thus, there are two binding sites per immunoglobulin molecule, a property known as **divalency**.

Table 10-5. Immunoglobulin Classes

Classes of Immuno-globulin	Approximate Molecular Weight	Heavy Chain Isotype (and Molecular Weight)	Carbohydrate by Weight (%)	Concentration in Serum (mg/100 ml)
IgG	150,000	γ (53,000)	2–3	600–1800
IgA	170,000–720,000*	α (64,000)	7–12	90–420
IgD	160,000	δ (58,000)	. . .	0.3–40
IgE	190,000	ϵ (75,000)	10–12	0.01–0.10
IgM	950,000*	μ (70,000)	10–12	50–190

Note.—Reprinted with permission from Devlin TM: *A Textbook of Biochemistry with Clinical Correlations*, 2nd ed. New York, John Wiley, 1986, p 92.
*Forms polymer structures of basic structural unit.

2. The antigen-binding sites were identified by enzymatic studies using primarily two proteolytic (peptide bond-splitting) enzymes.
   a. **Papain** hydrolyzes the immunoglobulin molecule at the hinge region of each heavy chain (see Fig. 10-10), resulting in the formation of three fragments.
      (1) Two identical fragments each contain the N-terminal segment of a heavy chain and a complete light chain. Because each fragment holds the antigen-binding region, these fragments are called **Fab** (Fragment **a**ntigen-**b**inding) fragments.
      (2) In the C-terminal portion of the heavy chain, a single covalent fraction is called the **Fc** (Fragment **c**rystallizable) fragment. The Fc fragment cannot bind antigen.
   b. **Pepsin** treatment of the molecule results in digestion of most of the Fc fragment. One large fragment remains, called the **F(ab′)₂** fragment, which consists of two Fab fragments joined by covalent bonds.

3. Being divalent, antibodies can combine with two antigens at a time and thus are able to form interconnected matrices of antigens and antibodies, which facilitate the agglutination and precipitation of antigen molecules (Fig. 10-11).

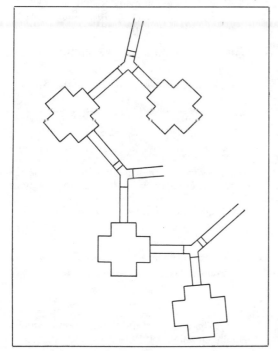

Figure 10-11. Diagrammatic representation of the formation of interconnected matrices by antigen molecules (*crosses*) and the divalent antibodies (*Y*s).

STUDY QUESTIONS

Directions: Each question below contains five suggested answers. Choose the **one best** response to each question.

1. Which of the following statements about oxygen binding and release by hemoglobin is correct?

(A) On binding oxygen, the iron of the heme prosthetic group is oxidized to the ferric state

(B) Lowering the pH accelerates the release of oxygen from oxyhemoglobin

(C) A high concentration of 2,3-diphosphoglycerate in the erythrocyte enhances the binding of oxygen by hemoglobin

(D) Oxygen binding by any one of the four heme groups occurs independently of the other three

(E) None of the above statements is correct

2. Under conditions of repeated episodes of hyperglycemia, hemoglobin A_1 may be modified chemically. What is the effect of this chemical modification?

(A) Hemoglobin loses its ability to show cooperativity in the binding of oxygen

(B) Hemoglobin A_1 becomes less soluble, forming long rod-like structures, which deform the erythrocyte

(C) A stable amino ketone derivative of hemoglobin A_1 appears in the blood and may amount to over 10% of the total hemoglobin

(D) The synthesis of hemoglobin A_1 is depressed in favor of the synthesis of hemoglobin F

(E) The affinity of hemoglobin for oxygen is increased, inhibiting the release of oxygen

3. The substitution of Val for Glu in the sixth position on the β chain of hemoglobin produces sickle cell anemia in the homozygote because

(A) the cooperative binding of oxygen is lost because of the missing Glu-6 residue

(B) the apoprotein can no longer protect the heme iron from becoming oxidized owing to the presence of the hydrophobic Val-6

(C) the Val-6 variant loses solubility upon deoxygenation

(D) the Val-6 variant tends to retain oxygen when the oxygen tension is low

(E) the red cell is unable to sequester high enough concentrations of the Val-6 variant hemoglobin

4. True statements about the structure of collagen include all of the following EXCEPT

(A) tropocollagen chains are helices that are more open than the α-helix; they lack intrachain peptide H bonds and are stabilized by the steric repulsion of proline residue pyrrolidone rings

(B) the close approach of the three helical chains in tropocollagen is made possible by the high glycine content

(C) each of the three helical strands of tropocollagen has nonhelical regions spaced at regular intervals along the chain

(D) hydroxylation of prolyl and lysyl residues occurs in the cisternae of the endoplasmic reticulum

(E) formation of a stable triple helix requires glycosylation of specific hydroxylysyl residues

5. Which of the following cofactors derived from a vitamin is involved in the formation of hydroxyproline during collagen synthesis?

(A) Pyridoxal phosphate (vitamin B_6)
(B) Biotin
(C) Thiamine pyrophosphate (vitamin B_1)
(D) Ascorbic acid (vitamin C)
(E) Methylcobalamin (vitamin B_{12})

6. A deficiency of copper affects the formation of normal collagen by reducing the activity of

(A) glucosyl transferase
(B) galactosyl transferase
(C) prolyl hydroxylase
(D) lysyl hydroxylase
(E) lysyl oxidase

7. Collagen occurs in different forms, which are usually classified on the basis of the

(A) type of carbohydrate present
(B) cysteine content
(C) hydroxyproline and hydroxylysine content
(D) types of peptide chains present
(E) glycine content

8. The lipoproteins with the highest electrophoretic mobility and the lowest triglyceride content are

(A) very low-density lipoproteins
(B) low-density lipoproteins
(C) intermediate-density lipoproteins
(D) high-density lipoproteins
(E) chylomicrons

9. The structure of an antibody molecule is characterized by all of the following EXCEPT

(A) regions of variable amino acid sequences are found in the N-terminal half of the light chains and the N-terminal one-fourth of the heavy chains
(B) regions of constant amino acid composition on the heavy chains are homologous within a given class of immunoglobulin
(C) antigen-binding sites of the immunoglobulin molecule are the N-terminal variable regions of the light- and heavy-chain pairs
(D) digestion with the proteolytic enzyme papain dissociates the immunoglobulin molecule into its separate light and heavy chains
(F) Joining (J) chains are associated with polymeric immunoglobulins

Directions: Each question below contains four suggested answers of which **one or more** is correct. Choose the answer

A   if **1, 2, and 3** are correct
B   if **1 and 3** are correct
C   if **2 and 4** are correct
D   if **4** is correct
E   if **1, 2, 3, and 4** are correct

10. True statements about myoglobin include which of the following?

(1) It is a monomeric protein
(2) It contains four protoheme IX prosthetic groups
(3) The molecule has eight major helical segments
(4) Oxygen binding exhibits cooperative kinetics

11. Fetal hemoglobin (HbF) takes up oxygen well under conditions in which the maternal hemoglobin (HbA) is releasing oxygen because

(1) hemoglobin F binds oxygen more tightly than hemoglobin A
(2) there is no cooperativity in the binding of oxygen by hemoglobin F
(3) the γ chains of hemoglobin F react less strongly with 2,3-diphosphoglycerate than do the β chains of hemoglobin A
(4) the concentration of hemoglobin F in the fetal red blood cell is much higher than the concentration of hemoglobin A in the maternal red blood cell

12. Protein–carbohydrate linkages in the glycoproteins involve sugar residues and which of the following amino acids?

(1) Asparagine
(2) Serine
(3) 5-Hydroxylysine
(4) Cysteine

ANSWERS AND EXPLANATIONS

1. The answer is B. (*I C 3 a, b*) The iron in heme must be in the ferrous state in order to bind oxygen reversibly. On binding oxygen, the iron does not change its valence. A reduction in pH, such as occurs at the tissues relative to the lung, enhances the release of oxygen from oxyhemoglobin. This is the Bohr effect. The binding of 2,3-diphosphoglycerate to hemoglobin increases the stability of the deoxy form, thus promoting the unloading of oxygen. Oxygen binding by hemoglobin exhibits cooperativity with the binding of oxygen to the first heme iron, facilitating the binding of oxygen by the other three.

2. The answer is C. (*I C 4 a*) Hemoglobin A_1, the major circulating form in adult humans, reacts spontaneously with glucose to form a derivative known as HbA_{1c}. The reaction involves the nonenzymatic formation of a Schiff's base between glucose and the N-terminal amino acids of the β chains (valines). The labile Schiff's base then becomes the more stable amino ketone. Normally the level of HbA_{1c} is very low, but in diabetes mellitus, where blood sugar levels may be periodically high, the concentration of HbA_{1c} may reach 12% or more of the total hemoglobin. Vigorous control of blood sugar levels in diabetic patients may reduce blood levels of HbA_{1c} to more normal values.

3. The answer is C. (*I C 4 b*) Hemoglobin S, the form of hemoglobin found in sickle cell anemia, is a mutant molecule with valine substituted for glutamic acid in the sixth position on the β chains. In the homozygote form, deoxyhemoglobin S polymerizes within the red cells. The long insoluble polymer fibers distort the erythrocyte, leading to the characteristic sickle-shaped cells. These misshapen cells are destroyed by the spleen, leading to anemia. In other respects, hemoglobin S appears normal; the cooperative binding of oxygen and the protection of the ferrous iron by the apoprotein heme pocket are not affected.

4. The answer is C. (*III A 3, B*) The proline-rich chains of tropocollagen form a helical structure quite different from that of the α-helix, in that it is more open, there is no intrachain peptide H-bonding, and the helix is stabilized by steric repulsion of the proline residue pyrrolidone rings. It is a left-handed helix. The unit structure of collagen is the triple helix of tropocollagen, which forms because the structure of glycine allows the close approach of the three chains. At first, each of the three helical strands has nonhelical regions at both the N-terminal and C-terminal ends. Preprocollagen is transferred into the cisternae of the endoplasmic reticulum, where signal peptide sequences are removed, prolyl and lysyl residues are hydroxylated, and glycosylation of specific hydroxylysyl residues occurs.

5. The answer is D. (*III B 1 b*) The enzyme prolyl hydroxylase forms hydroxyproline from proline residues in procollagen molecules as they pass through the cisternae of the endoplasmic reticulum during synthesis. Ascorbic acid (vitamin C) is required by this hydroxylase, possibly to keep the essential iron component of the system in the reduced state. The reaction also requires molecular oxygen and α-ketoglutarate, but none of the other vitamin-derived cofactors listed as choices are involved.

6. The answer is E. (*III B 5, C 3*) After tropocollagen is secreted from the cell and the nonhelical N- and C-terminal regions are proteolytically removed, a number of intermolecular cross-links are gradually formed to produce mature collagen. Only one enzymatic reaction is involved, the oxidative deamination of lysyl and hydroxylysyl residues to yield aldehydes. The enzyme involved is lysyl oxidase, which requires copper and molecular oxygen. A deficit of lysyl oxidase activity can result from poisoning with β-aminopropionitrile (in sweet pea seeds); from a copper-deficient diet; or from therapy with D-penicillamine, a chelator of copper. Copper deficiencies do not influence the activity of the glucosyl or galactosyl transferases or the prolyl or lysyl hydroxylases, all of which act on tropocollagen while it is still intracellular, in the cisternae of the endoplasmic reticulum, whereas lysyl oxidase activity is extracellular.

7. The answer is D. (*III B 6 b; Table 10-1*) Collagen types are classified by the nature of the peptide chains, which form the triple helix of procollagen. Six different α_1 chains are recognized, which vary in primary structure; these are labeled $\alpha1$(I), $\alpha1$(II), $\alpha1$(III), $\alpha1$(IV), $\alpha1$(V), and $\alpha1$(VI). The collagen of bone, skin, tendons, and scar tissue contains triple helices made up of two $\alpha1$(I) chains and one $\alpha2$(I) chain. Collagen of cartilage and vitreous contains three $\alpha1$(II) chains. Other collagens are made up of three chains of $\alpha1$(III), three $\alpha1$(IV), and three $\alpha2$(IV) chains, or a combination of $\alpha1$(V), $\alpha2$(V), and $\alpha3$(V) chains.

8. The answer is D. (*IV B; Table 10-3; Figure 10-8*) Two systems for classifying plasma lipoproteins are in use: One is based on the density (d) of the lipoprotein particle as determined by its flotation on centrifugation in 1.063 d sodium chloride solution, and the other is based on their electrophoretic mobility, the plasma lipoproteins migrating with the α- and β-globulins of plasma. Chylomicrons have the highest triglyceride content and remain at the origin on electrophoresis. High-density lipoproteins have the lowest triglyceride content and the highest electrophoretic mobility, migrating with the α globulins.

9. The answer is D. (*V D–F; Figure 10-10*) Immunoglobulins of the G class (IgG) have two identical heavy chains and two identical light chains covalently attached by disulfide bonds. Each heavy chain is aligned with a light chain so that their N-terminal ends are close together. The amino acid sequences of the N-terminal halves of the light chains and the N-terminal one-fourths of the heavy chains are highly variable. The C-terminal three-fourths of the heavy chains and one-half of the light chains are regions of almost constant amino acid sequence. The constant regions of the heavy chains are homologous within a given class of immunoglobulins. The two types of constant regions on the light chains, ϰ and λ, occur in all five immunoglobulin classes. Digestion with the proteolytic enzyme papain splits the molecule into three fragments: two identical fragments, which contain the N-terminal segments of the heavy chain plus the full light chain (Fab fragments), and a single fragment made up of the covalently bonded C-terminal halves of the heavy chains (Fc fragment). Joining (J) chain is a glycopeptide chain that is associated with polymeric forms of immunoglobulins (i.e., those that contain two or more basic units). IgA and IgM have a polymeric structure.

10. The answer is B (1, 3). (*I B*) Myoglobin has a compact structure formed by the folding of a single peptide chain containing eight helical segments into a globular shape. There is one prosthetic group of protoheme IX embedded in a hydrophobic pocket of the apoprotein. The plot of oxygen binding versus oxygen tension is a rectangular hyperbola, and the binding does not display cooperativity.

11. The answer is B (1, 3). (*I C 3 c*) The major fetal hemoglobin, HbF, has two γ chains in place of the two β chains of maternal hemoglobin, HbA. The γ chains bind 2,3-diphosphoglycerate less strongly than the β chains, so that the deoxy form is less favored with HbF than with HbA. HbF displays cooperativity of oxygen binding, and the concentration of HbF in the red blood cell is similar to that of HbA. It is doubtful if significantly more hemoglobin of any kind could be packaged into a red blood cell than is normally present.

12. The answer is E (all). (*II C 1–6*) Type I linkages in glycoproteins occur between the amide group of asparagine and N-acetyl-D-glucosamine. Type II linkages form between serine or threonine hydroxyls and the sugar. Type III linkages form between 5-hydroxylysine residues and the carbohydrate moiety. Type V linkages occur between cysteine residues and the carbohydrate.

I. GENERAL CHARACTERISTICS OF ENZYMES

A. Biologic catalysts

1. Enzymes *share* some of the properties of chemical catalysts.
   a. They are neither consumed nor produced during the course of a reaction.
   b. They do not cause reactions to take place; they speed up reactions that would ordinarily proceed, but at a much slower rate, in their absence. In other words, they do not alter the equilibrium constants of reactions that they catalyze.

2. Enzymes *differ* from chemical catalysts in several ways.
   a. They are invariably proteins.
   b. They are highly specific for the reactions they catalyze and produce only the expected products from the given reactants, or **substrates** (i.e., there are no side reactions).
   c. They often show a high specificity toward one substrate, although some enzymes have a broader specificity, using more than one substrate.
   d. They function within a moderate pH and temperature range.

B. Genetic expression

1. Enzymes are the major means for genetic expression.

2. Cells contain a unique set of enzymes, which determine the cell's function.

3. The set of enzymes in each cell is genetically determined.

C. Activity

1. **Requirements for enzyme activity**
   a. If the enzyme is a simple protein, only the native conformation of the protein is required for activity.
   b. If the enzyme is a conjugated protein, activity will depend upon both the protein's conformation and the availability of **cofactors**.

$$\text{Apoenzyme} + \text{cofactor} = \text{holoenzyme}$$

   (1) Some enzymes require metals, particularly transition elements, as cofactors.
   (2) A **coenzyme** such as oxidized nicotinamide-adenine dinucleotide (NAD^+) or oxidized NAD phosphate ($NADP^+$) may be required as a cofactor for activity.
   (3) In some enzyme systems, the cofactor is tightly bound to the enzyme protein, as in the case of oxidized flavin-adenine dinucleotide (FAD). In such cases, the cofactor is called a **prosthetic group**.

2. **Measures of enzyme activity**
   a. **The unit of enzyme activity** is the amount of enzyme causing transformation of 1 μmol of substrate per minute at 25°C under optimal conditions of measurement.
   b. **The specific activity** is the number of units of enzyme activity per milligram of enzyme protein.
   c. **The katal (kat)** is the amount of enzyme activity that transforms 1 mol of substrate per second.

II. ENZYME NOMENCLATURE. There are several ways of naming enzymes.

A. **Trivial names** (e.g., trypsin, pepsin) are commonly used, which are often the names of the substrates with the suffix *-ase* added (e.g., maltase).

B. **Systematic names** for each enzyme have been assigned in an **international classification of enzymes**.* This systematic scheme classifies enzymes on the basis of the reaction catalyzed.

1. **Oxidoreductases** are involved in oxidation and reduction.

2. **Transferases** transfer functional groups (e.g., amino or phosphate groups) between donors and acceptors.

3. **Hydrolases** transfer water; that is, they catalyze the hydrolysis of a substrate.

4. **Lyases** add (or remove) the elements of water, ammonia, or carbon dioxide (CO_2) to (or from) double bonds.

5. **Isomerases** catalyze changes within one molecule; they include racemases and epimerases, *cis–trans* isomerases, intramolecular oxidoreductases, intramolecular transferases, and intramolecular lyases.

6. **Ligases (synthetases)** join two molecules together at the expense of a high-energy phosphate bond of adenosine triphosphate (ATP).

III. MECHANISM OF ENZYME CATALYSIS

A. **Chemical reactions**

1. **Free-energy changes.** Figure 11-1 illustrates the free-energy changes that occur during a chemical reaction when it is catalyzed (lower curve) and uncatalyzed (upper curve). The **initial state** is the free energy of the substrate at the start of the reaction (only changes in free energy are measurable).

2. **Energy of activation**
   a. A chemical reaction occurs when a certain proportion of the substrate molecules are sufficiently energized to reach a so-called **transition state**, in which there is a high probability that a chemical bond will be made or broken to form the product.
   b. The effect of catalysts (enzymes) is to decrease the energy of activation.

B. **Specificity**

1. The specificity of an enzyme is determined by
   a. The functional groups of the substrate (or product)
   b. The functional groups of the enzyme and its cofactors
   c. The physical proximity of these various functional groups

2. Two theories have been proposed to explain the specificity of enzyme action.
   a. **The lock and key theory.** The active site of the enzyme is complementary in conformation to the substrate, so that enzyme and substrate "recognize" one another.
   b. **The induced-fit theory.** The enzyme changes shape upon binding the substrate, so that the conformation of substrate and enzyme protein are only complementary after the binding reaction.

IV. ENZYME KINETICS

A. **Quantitation of enzyme activity**

1. The rate at which the substrate changes to the product must be directly proportional to time.

2. The rate at which the substrate changes to the product must be directly proportional to the enzyme concentration.

B. **Classification of chemical reactions by kinetic order**

1. **In zero-order reactions**, the rate, or velocity (v), is constant and is independent of the reactant concentration ([A]).

$$v = k[A]^0$$

2. **In first-order reactions**, the rate is proportional to the reactant concentration.

$$v = k[A]^1$$

---

*Enzyme Nomenclature, 1978: *Recommendations of the Nomenclature Committee of the International Union of Biochemistry on the Nomenclature and Classification of Enzymes.* Orlando, FL, Academic Press, 1979.

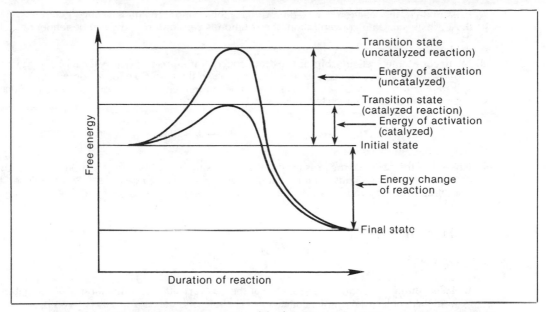

Figure 11-1. Diagrammatic representation of the free energy of activation of a chemical reaction.

3. In second-order reactions, the rate is proportional to the product of the concentrations of the reactants.

$$v = k([A] \times [A]) = k[A]^2$$

C. Michaelis-Menten kinetic theory of enzyme action

1. **Effect of enzyme concentration on reaction velocity.** If the substrate concentration is held constant, the velocity of the reaction is proportional to the enzyme concentration.

2. **Effect of substrate concentration on reaction velocity**
   a. Figure 11-2 illustrates the effect of the substrate concentration ([S]) on the reaction velocity (v) for a typical enzyme-catalyzed reaction.
   b. When the substrate concentration is low, the reaction is first-order with $v \sim [S]$.

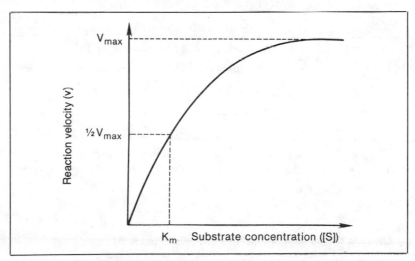

Figure 11-2. The rectangular hyperbola obtained by plotting reaction velocity (v) against substrate concentration ([S]) for a typical enzyme-catalyzed reaction. K_m is the substrate concentration at $\frac{1}{2} V_{max}$. (Adapted from Lehninger AL: *Biochemistry*, 2nd ed. New York, Worth, 1975, p 153.)

c. At mid-[S], the reaction is mixed-order (i.e., the proportionality is changing).
d. At a high substrate concentration, the reaction is zero-order, and v is independent of [S].

3. **Michaelis-Menten relationship of substrate concentration to velocity.** Michaelis and Menten proposed that an enzyme-catalyzed reaction involved the reversible formation of an enzyme–substrate complex, which then broke down to form free enzyme and one or more products. Their postulate can be depicted thus

$$E + S \underset{k_1}{\overset{k_2}{\rightleftharpoons}} ES \overset{k_3}{\rightarrow} E + P,$$

where E is the free enzyme, S is the substrate, ES is the enzyme–substrate complex, P is the product, k_1 is the rate constant for the formation of ES, k_2 is the rate constant for the dissociation of ES to E + S, and k_3 is the rate constant for the dissociation of ES to E + P.

a. Assumption 1. [S] is very large compared to [E], so that when all E is bound in the form ES, there is still an excess of S. By the **law of mass action**, the rate of formation of ES from E + S is proportional to [E] • [S], or

$$\frac{d[ES]}{dt} = k_1[E][S].$$

b. Assumption 2. Conditions are such that there is very little accumulation of P, so that the formation of ES from E + P is negligible. The rate of breakdown of ES is given by

$$\frac{-d[ES]}{dt} = k_2[ES] + k_3[ES].$$

c. Assumption 3. The rate of breakdown of ES very rapidly equals the rate of formation of ES. This is the steady-state assumption:

$$\frac{d[ES]}{dt} = \frac{-d[ES]}{dt}$$

or

$$k_1[E][S] = k_2[ES] + k_3[ES] = (k_2 + k_3)[ES].$$

But [E] = [e] − [ES], where [e] is the total enzyme concentration; therefore,

$$k_1([e] - [ES])[S] = (k_2 + k_3)[ES].$$

Rearranging this equation yields, successively,

$$k_1([e][S] - [ES][S]) = (k_2 + k_3)[ES],$$
$$k_1[e][S] - k_1[ES][S] = (k_2 + k_3)[ES],$$
$$k_1[e][S] = k_1[ES][S] + (k_2 + k_3)[ES],$$
$$k_1[e][S] = [ES](k_2 + k_3 + k_1[S]),$$

or, finally,

$$\frac{k_1[e][S]}{k_2 + k_3 + k_1[S]} = [ES].$$

The rate of formation of the product is always equal to $k_3[ES]$ (see section IV C 3); therefore,

$$v = k_3[ES] = \frac{k_3 k_1[e][S]}{k_2 + k_3 + k_1[S]}.$$

Dividing the numerator and the denominator by $k_1[S]$ yields

$$v = \frac{k_3[e]}{\dfrac{k_2}{k_1[S]} + \dfrac{k_3}{k_1[S]} + 1} = \frac{k_3[e]}{\dfrac{k_2 + k_3}{k_1[S]} + 1}.$$

d. Michaelis constant (K_m). Because k_1, k_2, and k_3 are constants, the expression $(k_2 + k_3)/k_1$ is a constant, which may be written as K_m. Therefore,

$$v = \frac{k_3[e]}{1 + \dfrac{K_m}{[S]}} ,$$

where v is the rate of the enzyme-catalyzed reaction.

e. Michaelis-Menten equation. Maximum velocity (V_{max}) will be attained when the concentration of ES is maximal. This will be the case when all of the enzyme is involved in the enzyme–substrate complex—that is, when $[ES] = [e]$; therefore,

$$V_{max} = k_3[e].$$

Thus, the expression for the velocity of reaction may be rewritten as

$$v = \frac{V_{max}}{1 + \dfrac{K_m}{[S]}}$$

or

$$v = \frac{V_{max}\,[S]}{[S] + K_m} ,$$

which is known as the Michaelis-Menten equation.

D. Using the Michaelis-Menten equation

1. Significance of K_m

    a. When $v = \tfrac{1}{2}\,V_{max}$, then

$$v = \frac{V_{max}}{2} = \boxed{\dfrac{V_{max}}{1 + \dfrac{K_m}{[S]}}}.$$

    b. Rearranging the expression yields

$$1 + \frac{K_m}{[S]} = 2,$$

$$\frac{K_m}{[S]} = 1,$$

and, finally,

$$K_m = [S].$$

    c. Therefore, **K_m is equal to the substrate concentration at which the velocity is half maximal**.

    d. The Michaelis constant, K_m, is not a true dissociation constant, but it does provide a measure of **the affinity of an enzyme for its substrate**. The lower the value of K_m, the greater the affinity of the enzyme for substrate–enzyme complex formation.

2. Linear transforms. Because it is difficult to estimate V_{max} from the position of an asymptote, as in the plot of a rectangular hyperbola (see Fig. 11-2), linear transforms of the Michaelis-Menten equation are often used.

    a. Lineweaver-Burk transform is written as

$$\frac{1}{v} = \frac{1}{V_{max}} + \frac{K_m}{V_{max}} \cdot \frac{1}{[S]} ,$$

$$(y\ =\quad b\quad +\quad mx)$$

where m is the slope and b is the y intercept of the regression of y on x. Figure 11-3 shows the straight-line graph obtained by plotting $1/v$ against $1/[S]$, where the y intercept $= 1/V_{max}$, the x intercept $= -\,1/K_m$, and the slope $= K_m/V_{max}$.

    b. Eadie-Hofstee transform is used to avoid the "bunching" of values that occurs about the

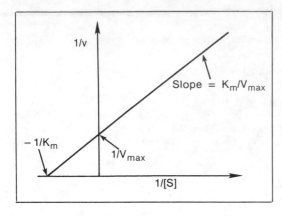

Figure 11-3. A plot of *1/v* against *1[S]* for an enzyme-catalyzed reaction. This straight-line representation is the Lineweaver-Burk transform of the Michaelis-Menten equation.

lower end of the double-reciprocal (Lineweaver-Burk) plots. The Eadie-Hofstee transform can be written as

$$v = V_{max} - K_m \cdot \frac{v}{[S]} \ .$$

$$(y = \quad b \quad + \quad mx)$$

Figure 11-4 shows the straight-line graph obtained by plotting v against $v/[S]$, where the y intercept = V_{max}, the x intercept = V_{max}/K_m, and the slope = $- K_m$.

V. ENZYME INHIBITION

A. Competitive inhibition

1. Competition occurs between the substrate and an inhibitor for binding to the active site of an enzyme.
   a. Such inhibition of enzyme activity is reversible by increasing the substrate concentration relative to that of the inhibitor.
   b. Competitive inhibition changes the K_m of the enzyme but not the V_{max} (Fig. 11-5).
   c. The inhibitor can form an enzyme–inhibitor complex, EI, that is equivalent to ES.

$$E + I = EI$$

$$K_I = \frac{[E] \cdot [I]}{[EI]}$$

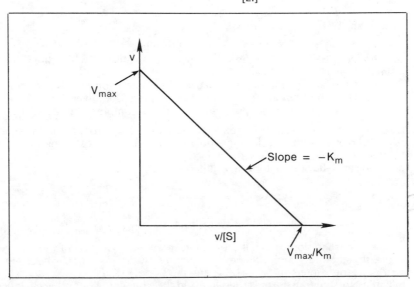

Figure 11-4. A plot of the Eadie-Hofstee transform of the Michaelis-Menten equation.

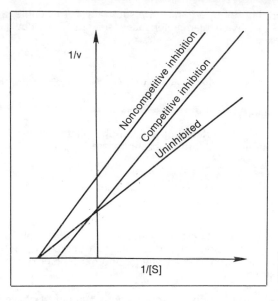

Figure 11-5. Lineweaver-Burk plots showing the effect of competitive and noncompetitive inhibitors on the kinetics of enzyme-catalyzed reactions.

(1) The Michaelis-Menten form of the inhibitor equation is

$$v = \frac{V_{max}}{1 + \dfrac{K_m}{[S]}\left(1 + \dfrac{[I]}{K_I}\right)}.$$

(2) The Lineweaver-Burk transform of the Michaelis-Menten form of the inhibitor equation is

$$\frac{1}{v} = \frac{1}{V_{max}} + \frac{K_m}{V_{max}}\left(1 + \frac{[I]}{K_I}\right)\cdot\frac{1}{[S]}.$$

$$(y = b + mx)$$

2. Examples of competitive inhibitors include malonate inhibition of succinate dehydrogenase (see Chapter 15, section III F) and the action of sulfanilamide on bacterial synthesis of folic acid (see Chapter 27, section III B 3).

B. Noncompetitive reversible inhibition

1. The inhibitor binds at a site on the enzyme in such a manner that the binding of substrate to enzyme is not affected.

$$E + I \rightleftharpoons EI$$

$$ES + I \rightleftharpoons ESI$$

2. However, the binding of the inhibitor to the enzyme molecule blocks its catalytic activity; thus, the amount of enzyme appears to decrease and the V_{max} is lowered. The K_m is not affected (see Fig. 11-5).

a. The Michaelis-Menten form of the inhibitor equation is

$$v = \frac{V_{max}}{\left(1 + \dfrac{K_m}{[S]}\right)\left(1 + \dfrac{[I]}{K_I}\right)}.$$

b. The Lineweaver-Burk transform of the Michaelis-Menten form of the inhibitor equation is

$$\frac{1}{v} = \frac{1}{V_{max}}\left(1 + \frac{[I]}{K_I}\right) + \frac{K_m}{V_{max}}\left(1 + \frac{[I]}{K_I}\right)\cdot\frac{1}{[S]}.$$

$$(y = \qquad b \qquad + \qquad mx)$$

3. This type of inhibition can be reversed, but not by adding more substrate. For example, the inhibition of a Mg^{2+}-requiring enzyme by ethylenediaminetetraacetic acid (EDTA), a chelator of Mg^{2+}, can be reversed by adding excess Mg^{2+}.

C. Noncompetitive irreversible inhibition

1. The inhibitor binds irreversibly to the enzyme and effectively removes the enzyme from the reaction.

2. The V_{max} is lowered, but the K_m is unchanged.

3. For example, the organophosphorus compound diisopropyl fluorophosphate (DFP) reacts with the active site on the enzyme cholinesterase to form a phosphorylated enzyme that is inactive.

D. Mixed inhibitors combine with the enzyme to change both K_m and V_{max}.

VI. ORGANIZATION OF MULTIENZYME SYSTEMS (Fig. 11-6)

A. Least organized. Among multienzyme systems in which a number of enzymes are involved in the stepwise transformation of a substance, the least organized is a solution of enzymes in the cell sap with intermediate substrates that diffuse from one enzyme to another. The enzymes of the glycolytic pathway (see Chapter 14) are an example.

B. Organized. Multienzyme complexes are more highly organized. Here, a number of enzymes that act in sequence are physically bound to one another by noncovalent bonds, so that intermediate products can be passed directly from one enzyme to the next in the reaction sequence. An example is the pyruvate dehydrogenase complex (see Chapter 15, section II).

C. Highly organized. Insoluble enzymes that form part of a membrane structure show a high level of organization. These enzymes provide the means for the very rapid vectorial flow of products from enzyme to enzyme. The electron transport system of the inner mitochondrial membrane (see Chapter 16) is an example.

D. Compartmentalized. An enzyme pathway can be protected from competing substrates and cofactors by enclosure within an intracellular organelle or space. For example, the enzymes of fatty acid oxidation are sequestered within the mitochondrial matrix space, and the enzymes of fatty acid synthesis are confined to the cytosol (see Chapter 20, section III B 1 b).

VII. REGULATION OF ENZYMES

A. pH. A change in pH can alter the rates of enzyme-catalyzed reactions with many enzymes exhibiting a bell-shaped curve when enzyme activity is plotted against pH (Fig. 11-7).

1. A change in pH can alter the **ionization state** of the substrate or of the enzyme-binding site for substrate or cofactor.

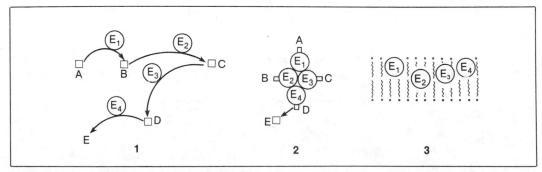

Figure 11-6. Different kinds of multienzyme systems. *(1)* A soluble multienzyme system in which there are no permanent associations between the enzymes E_1, E_2, E_3, and E_4. The substrates A, B, C, D, and E are free to diffuse from one enzyme to another. *(2)* A multienzyme system in which the enzymes are bound to one another by noncovalent bonds and in which the products of the reactions often remain covalently bound to an enzyme until the last reaction in the sequence. *(3)* A membrane-bound multienzyme system in which the enzymes are associated with lipid bilayers in cellular or organellar membranes.

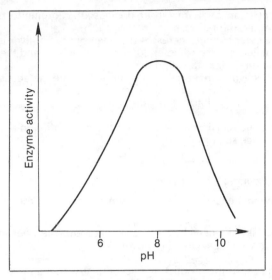

Figure 11-7. A typical pH-activity plot of an enzyme-catalyzed reaction.

2. It can also alter the ionization state at the catalytic site on the enzyme or affect the protein molecule so that its conformation and catalytic activity change.

B. **Temperature** can also affect the rate of enzyme-catalyzed reactions. Most enzymes are thermolabile.

1. The rate of an enzyme-catalyzed reaction usually increases with increasing temperature up to an optimum point (Fig. 11-8).

2. The apparent optimum point shown in Figure 11-8 is the result of an initial increase in reaction rate due to a rise in temperature, followed by a decrease in enzymatic activity as the temperature continues upward with increasing thermal denaturation of the proteinaceous enzyme molecule.

3. The rate by which the velocity of a reaction changes with a 10°C rise in temperature is known as the Q_{10}, or **temperature coefficient**.

C. **Amount of enzyme present** obviously affects the reaction.

1. The amount of enzyme that is present in a cell is determined by its **rate of synthesis** (k_s) and its **rate of degradation** (k_d), the combined process constituting **enzyme turnover** (see Chapter 26, section III).

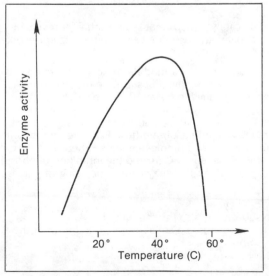

Figure 11-8. Effect of temperature on the activity of an enzyme.

2. The amount of enzyme present in a cell can be increased or decreased according to the body's needs, the process being known as **adaptation**.
   a. For example, an increase in dietary carbohydrate increases the level of several liver enzymes involved in the process of lipogenesis (see Chapter 20, section III E 1 b), and these same enzymes decline in concentration during starvation.
   b. Barbiturates **induce the synthesis** of enzymes that are involved in barbiturate metabolism and that also metabolize other drugs. Thus, an optimal dose level of a drug given with a barbiturate can become an overdose if the barbiturate is withdrawn.

D. Stoichiometric control. The rate of an enzyme reaction depends upon the concentration of reactants and products. In an enzyme sequence such as

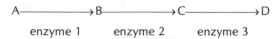

   A————————→B————————→C————————→D
   enzyme 1   enzyme 2   enzyme 3

1. The rate catalyzed by enzyme 1 will be influenced by the rate at which enzyme 2 removes product B.

2. Similarly, the rate catalyzed by enzyme 2 is influenced by the rate at which enzyme 1 delivers the product B from A, and by the rate at which enzyme 3 removes C.

E. Product inhibition. In some enzymatic reactions the product, if it accumulates, can inhibit the enzyme. This form of control will limit the rate of formation of the product when the product is underused. For example, the hepatic formation of the prohormone 25-hydroxycholecalciferol from cholecalciferol (vitamin D) is strongly product-inhibited, which serves to limit the production of the active form of the hormone (see Chapter 23, section IV B 3).

F. Acceptor control. In reactions where there is a transfer of a group from one substrate to another, as in the reaction

   1,3-diphosphoglycerate + adenosine diphosphate (ADP) → 3-phosphoglycerate + ATP

(see Chapter 14, section III F), the reaction rate will be limited by the prevailing concentration of the **acceptor**, ADP.

G. Covalent modification

1. Phosphorylation of serine groups can modify enzyme activity.

2. In some cases, the phosphorylated form is more active, as with glycogen phosphorylase; in other instances, as with glycogen synthetase or pyruvate dehydrogenase, it is less active (see Chapter 18, section IV A, B, and Chapter 15, section II E 3).

3. Phosphorylated forms are returned to their nonphosphorylated forms by phosphatases. Thus, the phosphorylation and the change in function of the enzyme can be reversed.

H. Regulatory or allosteric enzymes

1. Regulatory or allosteric enzymes generally catalyze steps that are essentially irreversible, often the first irreversible step in a metabolic pathway. They are usually large enzymes containing more than one subunit.

2. The term **allosteric** means "other site," indicating that a **modulator** acts at a site on the enzyme molecule that is distinct from the active site involved in substrate modification.
   a. Modulators are **positive** if they enhance the catalytic reaction and **negative** if they attenuate the reaction rate.
   b. Negative modulation by the end product of a metabolic pathway that acts as the first irreversible step in the pathway is referred to as **feedback inhibition**. Examples are the inhibition of the first enzyme in porphyrin synthesis, δ-aminolevulinate synthetase, by the oxidized end product, hemin (see Chapter 28, section II C 5), and the inhibition of phosphofructokinase by ATP or citrate, both products of the enzymatic processing of glucose in cells (see Chapter 14, section III C 3).

3. There are two major classes of regulatory enzymes:
   a. Homotropic enzymes, in which the substrate is also the modulator
   b. Heterotropic enzymes, in which the modulator is a small molecule other than the substrate, often the final product of the pathway

STUDY QUESTIONS

Directions: Each question below contains five suggested answers. Choose the **one best** response to each question.

1. The specific activity of an enzyme is

(A) the amount of enzyme that produces 1 mol of product per second under standard conditions
(B) the activity of an enzyme in relation to a standard preparation of the enzyme
(C) the number of enzyme units per milligram of enzyme protein
(D) the amount of enzyme causing transformation of 1 μmol of substrate per minute under standard conditions
(E) the activity of an enzyme in the presence of its preferred substrate

2. An enzyme that catalyzes the conversion of an aldose sugar to a ketose sugar would be classified as one of the

(A) oxidoreductases
(B) transferases
(C) hydrolases
(D) isomerases
(E) lyases

3. Enzymes increase the rates of reactions by

(A) increasing the free energy of activation
(B) increasing the free energy change of the reaction
(C) changing the equilibrium constant of the reaction
(D) decreasing the energy of activation
(E) decreasing the free-energy change of the reaction

4. What is meant by the steady-state assumption that underlies the Michaelis-Menten relationship between substrate concentration and reaction velocity?

(A) The reaction velocity is linearly related to substrate concentration
(B) The reaction velocity is independent of substrate concentration
(C) The rate of breakdown of the enzyme–substrate complex equals the rate of formation of the complex
(D) The rate of formation of product equals the rate of disappearance of substrate
(E) The amount of enzyme remains constant

5. If the substrate concentration in an enzyme-catalyzed reaction is equal to $\frac{1}{2}$ K_m, the initial reaction velocity will be

(A) 0.25 V_{max}
(B) 0.33 V_{max}
(C) 0.50 V_{max}
(D) 0.67 V_{max}
(E) 0.75 V_{max}

6. The Michaelis-Menten constant (K_m) is

(A) numerically equal to $V_{max}/2$
(B) the equilibrium constant for the dissociation of ES to E + P
(C) increased in value with increasing affinity of the enzyme for its substrate
(D) the substrate concentration at $\frac{1}{2}$ V_{max}
(E) the intercept on the $1/v$ axis of a Lineweaver-Burk transform

7. Which of the following regulatory actions involves a reversible covalent modification of an enzyme?

(A) Allosteric modulation
(B) Competitive inhibition
(C) Conversion of zymogen to active enzyme
(D) Association of apoenzyme with a cofactor
(E) Phosphorylation of a serine hydroxyl on the enzyme

8. An allosteric modulator influences enzyme activity by

(A) competing for the catalytic site with the substrate
(B) binding to a site on the enzyme molecule distinct from the catalytic site
(C) changing the nature of the product formed
(D) changing the specificity of the enzyme for its substrate
(E) none of the above

Directions: Each question below contains four suggested answers of which **one or more** is correct. Choose the answer

A if **1, 2, and 3** are correct
B if **1 and 3** are correct
C if **2 and 4** are correct
D if **4** is correct
E if **1, 2, 3, and 4** are correct

9. The effects of a competitive inhibitor on the kinetics of an enzyme reaction include which of the following?

(1) The V_{max} is not changed
(2) Increased concentrations of substrate reverse the inhibition
(3) The K_m is increased
(4) The inhibitor binds to a site on the enzyme other than the catalytic site

10. The effects of pH on enzyme-catalyzed reactions include which of the following?

(1) The direction of the reaction may be influenced by the $[H^+]$
(2) The ionization state of dissociating groups on the enzyme may be modified
(3) The ionization state of the substrate may be modified
(4) The protein may be denatured at certain pH values

ANSWERS AND EXPLANATIONS

1. The answer is C. (*I C 2*) A unit of enzyme activity is the amount of enzyme causing the transformation of 1 μmol of substrate per minute at 25°C under optimal conditions of measurement. The specific activity of an enzyme is the number of enzyme units per milligram of enzyme protein.

2. The answer is D. (*II B*) The choices listed in the question—oxidoreductases, transferases, hydrolases, isomerases, and lyases—are five of the six groups used by the International Union of Biochemistry to classify all enzymes; the missing group is the ligases. The aldose and ketose sugars are isomers, and an enzyme that catalyzes their interconversion would thus be classified as an isomerase.

3. The answer is D. (*III A 2*) A chemical reaction occurs when a fraction of the population of substrate molecules becomes sufficiently energized to reach a so-called transition state in which there is a high probability that a chemical bond will be formed or broken to form a product. This is the energy of activation, and the effect of a catalyst (i.e., an enzyme) is to decrease the energy of activation. Enzymes do not alter the equilibrium or the overall free-energy change of a reaction.

4. The answer is C. (*IV C 3 c*) In deriving the Michaelis-Menten relationship between substrate concentration and reaction velocity, several assumptions are made. The first is that the concentration of substrate, [S], is very large compared to the concentration of enzyme, [E]. It is also assumed that there is very little accumulation of product, [P], so that the formation of enzyme–substrate complex from enzyme + P is negligible. The third assumption is the so-called steady-state assumption, which is that the rate of breakdown of the enzyme–substrate complex very rapidly equals the rate of formation of the complex, that is, [ES] is in a steady state.

5. The answer is B. (*IV C 3 e*) The Michaelis-Menten equation relates the velocity of an enzyme-catalyzed reaction to the maximum velocity (V_{max}), the substrate concentration, and the Michaelis-Menten constant (K_m). That is,

$$v = \frac{V_{max}}{1 + \frac{K_m}{[S]}} .$$

If [S] = ½ K_m, then

$$v = \frac{V_{max}}{1 + \frac{K_m}{0.5\,K_m}} = \frac{V_{max}}{1 + \frac{1}{0.5}} = \frac{V_{max}}{1 + 2} = \frac{V_{max}}{3} = 0.33\,V_{max}.$$

6. The answer is D. (*IV D*) The Michaelis-Menten constant (K_m) is not a true dissociation constant, as it is derived from the three rate constants describing the formation and dissociation of the enzyme–substrate complex. It is approximately equal to the equilibrium constant for the dissociation of ES to E + S. As such, it does provide a measure of the affinity of an enzyme for its substrate: The lower the value of the K_m, the greater the affinity of the enzyme for enzyme–substrate complex formation. K_m can be estimated from the 1/[S] intercept (X-axis intercept) in a plot of 1/v against 1/[S] (the Lineweaver-Burk transform).

7. The answer is E. (*VII G*) Some enzymes are modulated by an adenosine triphosphate–dependent phosphorylation of a serine, a threonine, or in some cases a tyrosine residue on the enzyme. The phosphorylation may cause activation of the enzyme in some cases and inactivation in other cases. The modulation of the enzyme can be reversed by the action of a phosphatase, which removes the phosphate group. Competitive inhibitors and allosteric modulators interact with the enzyme by noncovalent bonding forces. The conversion of zymogen to an active enzyme involves an irreversible covalent modification, usually a proteolytic cleavage of the enzyme protein. The association of an apoenzyme with its cofactor is often a noncovalent interaction, but even if covalent bonds are involved it would not represent a regulatory mechanism as the enzyme is incomplete without its associated cofactor.

8. The answer is B. (*VII H*) The term allosteric means "other site," indicating that the modulator of an allosteric enzyme acts as a site distinct from the active or catalytic site involved in substrate modification. The substrate specificity need not be changed, and usually is not, nor is the product formed necessarily different in the presence of the modulator. There are two major classes of allosteric enzymes, homotropic, in which the substrate is also the modulator, and heterotropic, in which the modulator is a small molecule other than the substrate, often the final product of a multienzyme pathway.

9. The answer is A (1, 2, 3). (*V A*) A competitive inhibitor binds reversibly to the active site, preventing the binding of substrate. It can be displaced by increasing substrate concentrations. V_{max} can be achieved by increasing the substrate concentration, but the substrate concentration for ½ V_{max} (i.e., the K_m) is now increased.

10. The answer is E (all). (*VII A*) The hydrogen ion concentration may modify the rate of an enzyme-catalyzed reaction in a number of ways: As a product of a reaction, the [H^+] can influence the rate of reaction; it can change the ionization state of active groups on the enzyme or on the substrate; and by modifying the native configuration of the protein, it can induce an inactive state of the protein.

Part II
Metabolism

12

Introduction to Metabolism

I. METABOLISM

A. Overview. Living organisms maintain their complex order in a dynamic steady state by importing food and energy from their surroundings. They have the machinery needed to liberate and store chemical energy from foods and to create complex molecules from simpler ones for the building of new structure. These processes collectively are spoken of as **metabolism** and involve the transformation by enzyme-catalyzed reactions of both **matter** and **energy**.

B. Metabolic pathways. The transformation of matter or energy often involves a number of reactions that are catalyzed by a sequence of enzymes. This sequence of enzymatic reactions collectively constitutes a **metabolic pathway**. In such sequences, the product of one enzyme reaction becomes the substrate for the next reaction in the sequence, the successive products of the reactions being known as **metabolites**, or **metabolic intermediates**.

C. Two aspects of metabolism

1. **Catabolism** encompasses the degradative processes whereby complex molecules are broken down into simpler ones.
   a. One important function of catabolism is to transform the molecules derived from ingested food materials into simpler "building block" compounds.
   b. In catabolic processes, there can be an attendant release of the free energy of the complex molecules.
   c. Some of this free energy can be conserved by coupled enzymatic reactions and stored as adenosine triphosphate (ATP).

2. **Anabolism** encompasses the biosynthetic aspects of metabolism, which are concerned with combining building block compounds into the complex macromolecules required by the organism. Anabolic processes require energy inputs, which can be supplied in two ways:
   a. By ATP transferred from the catabolic pathways
   b. In some cases, by high-energy hydrogen in the form of reduced nicotinamide-adenine dinucleotide phosphate (NADPH)

II. BIOENERGETICS

A. Standard state. When considering biochemical systems, a so-called standard state is presumed to exist.

1. **The standard state** is defined as one in which
   a. The pH is 7 (see Appendix)
   b. The temperature is 25°C (288°K)
   c. All solutes are at 1 molar concentration
   d. All gases are at 1 atm pressure

2. **Thermodynamic concepts** in the area of bioenergetics allow a quantitative approach to metabolism. Three fundamental concepts are important.
   a. **Enthalpy (H)**
   b. **Entropy (S)**
   c. **Free energy (G)** (This concept is the most useful in biochemistry.)

3. **The change in H, S, or G** (written as ΔH, ΔS, or ΔG), rather than absolute values, is usually considered. In the standard state, the three functions are written as
   a. $\Delta H^{\circ'}$

b. $\Delta S^{\circ'}$
c. $\Delta G^{\circ'}$

B. Enthalpy (H)

1. Definition. Enthalpy (H) is the **heat content** of a body and is given by

$$H = E + PV,$$

where E is the internal energy, and PV is the product of pressure times volume.

2. Heat of formation, or enthalpy of formation (H_F), of a chemical compound is the heat absorbed or evolved during the synthesis of 1 mol of the compound from its elements, all components being in their standard state. For compound X, the heat of formation is written as $H_F^{\circ'}(X)$.

3. The enthalpy change in the reaction

$$A + B \leftrightarrow C + D$$

is given by

$$\Delta H^{\circ'} = \Delta H_F^{\circ'}(C) + \Delta H_F^{\circ'}(D) - \Delta H_F^{\circ'}(A) - \Delta H_F^{\circ'}(B).$$

C. Entropy (S)

1. Definition. Entropy (S) is the **degree of randomness**, or disorder, of a system. The greater the degree of disorder, the higher the value of S.

2. Entropy changes cannot be measured directly; thus, ΔS is calculated from ΔH, ΔG, and ΔT, where ΔT is the temperature change.

D. Free energy (G)

1. Definition. Free energy (G) is the **amount of useful work** that can be obtained from a system that is at constant temperature, pressure, and volume. Energy is needed to maintain the entropy level ($T\Delta S$). Because this energy cannot be made to provide useful work, then

$$\Delta G = \Delta H - T\Delta S.$$

2. Standard free energy of formation of a compound is defined as the **change in free energy during the synthesis** of 1 mol of the compound from its constituent elements under standard conditions. For the compound X, it is written as $\Delta G_F^{\circ'}(X)$.

3. Standard free-energy change of a reaction can be calculated by
a. Subtracting the $\Delta G_F^{\circ'}$ values of reactants from those of the products
b. Deriving it from the equilibrium constant (K_{eq}) for the reaction
(1) For the reaction

$$A + B \leftrightarrow C + D,$$

the actual free-energy change (ΔG) is given by

$$\Delta G = \Delta G^{\circ'} + RT \ln \frac{[C][D]}{[A][B]},$$

where R is the gas constant (1.987×10^{-3} kcal $\times$ deg^{-1} $\times$ mol^{-1}), T is the temperature in degrees kelvin ($^{\circ}C + 273$), and ln indicates the natural logarithm, which can be converted to $\log_{10}$ by multiplying by 2.303.
(2) At equilibrium, when the rate of the forward reaction equals the rate of the backward reaction, there is no change in free energy; that is, $\Delta G = 0$. Also, [C][D]/[A][B] is equal to the K_{eq} for the reaction.
(3) Thus, for a temperature of 37°C

$$0 = \Delta G^{\circ'} + 1.987 \times 10^{-3} \times (37 + 273) \times 2.303 \log_{10}K_{eq},$$

or

$$\Delta G^{\circ'} = -1.418 \log_{10}K_{eq}.$$

(4) If $\Delta G^{\circ'}$ is known, the actual free-energy change (ΔG) in a reaction can be calculated by substituting the actual temperatures and concentrations of reactants and products in the equation at section II D 3 b (1).

    4. Spontaneity of a reaction is measured by the value of ΔG.
        a. Exergonic reaction. If the change in free energy is less than zero ($\Delta G < 0$), the reaction can proceed spontaneously with the **release** of energy.
        b. Endergonic reaction. If the change in free energy is more than zero ($\Delta G > 0$), the reaction cannot proceed spontaneously unless there is an **input** of energy to drive the reaction forward.

  E. Coupled reactions. A thermodynamically unfavorable reaction can be driven by being coupled to a thermodynamically favorable reaction.

    1. For example, the formation of glucose 6-phosphate (G6P) from glucose and inorganic phosphate is an endergonic reaction with a $\Delta G^{\circ\prime}$ of + 3.3 kcal/mol. It is, therefore, a highly unfavorable reaction, thermodynamically speaking.

    2. However, when coupled to the hydrolysis of ATP to adenosine diphosphate (ADP) and inorganic phosphate (P_i), an exergonic reaction with a $\Delta G^{\circ\prime}$ of -7.3 kcal/mol, the phosphorylation of glucose can occur with a net $\Delta G^{\circ\prime}$ of -4.0 kcal/mol.
      a. The coupled reactions are

$$\text{Glucose} + P_i \rightarrow \text{G6P} + H_2O \qquad\qquad \Delta G^{\circ\prime} = +3.3\,\text{kcal/mol}$$

$$\text{ATP} + H_2O \rightarrow \text{ADP} + P_i \qquad\qquad \Delta G^{\circ\prime} = -7.3\,\text{kcal/mol}$$

      b. The net reaction is

$$\text{Glucose} + \text{ATP} \rightarrow \text{G6P} + \text{ADP} \qquad\qquad \Delta G^{\circ\prime} = -4.0\,\text{kcal/mol}$$

III. HIGH-ENERGY PHOSPHATE COMPOUNDS

  A. High group transfer potential. High-energy phosphate compounds have a highly negative $\Delta G^{\circ\prime}$ for hydrolysis of the phosphate group. They exhibit a high group transfer potential for phosphate. For example, in the coupled reaction described in section II E 2 above, ATP is the high-energy phosphate compound with a highly negative $\Delta G^{\circ\prime}$ for the hydrolysis of the terminal (γ) phosphate group. In this reaction, the γ-phosphate group is readily transferred to the hydroxyl on carbon 6 of glucose (Fig. 12-1).

  B. High-energy versus low-energy phosphate compounds

    1. The terminal phosphate group of ADP (the β-phosphate group) is likewise a high-energy

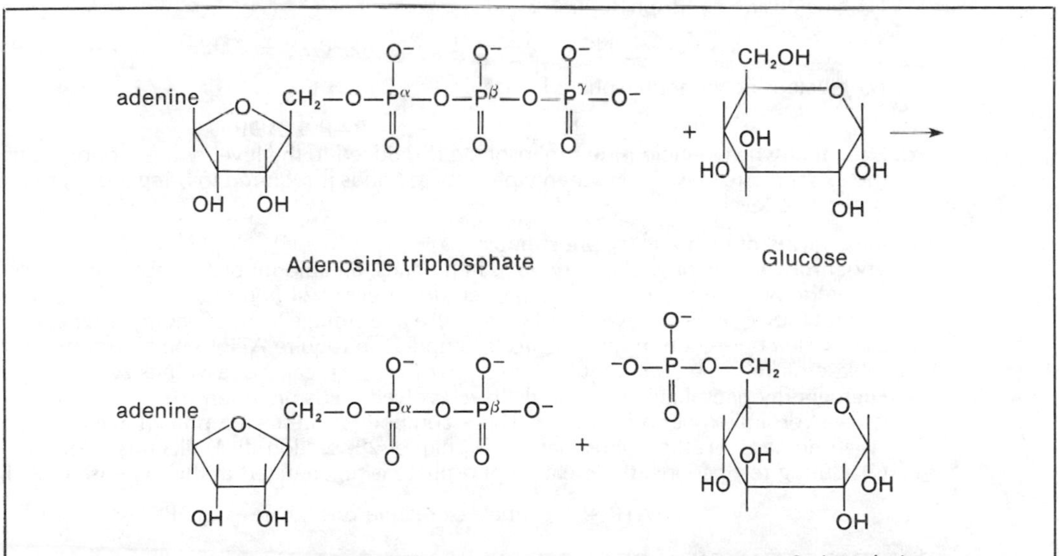

Figure 12-1. Phosphorylation of glucose. The terminal phosphate group is readily transferred to the hydroxyl on carbon 6 of glucose because adenosine triphosphate is a high-energy phosphate compound with a highly negative standard free-energy exchange for the hydrolysis of the terminal phosphate group.

phosphate group, but the α-phosphate of adenosine monophosphate (AMP) and the phosphate on G6P are not, as their free energies of hydrolysis are only -3.4 and -3.3 kcal/mol, respectively.

2. Catabolic reactions can yield compounds that have higher group transfer potentials than ATP, and the hydrolysis of these compounds can be coupled to the formation of ATP by the transfer of phosphate to ADP.

   a. For example, the $\Delta G^{\circ\prime}$ for the hydrolysis of the phosphate group of phosphoenolpyruvate is -14.8 kcal/mol. In the glycolytic pathway (see Chapter 14, section III), the phosphate group is transferred to ADP to form ATP, and the enol group is converted to a ketone to yield pyruvate.

$$\begin{array}{ccc}
\text{COO}^- & & \text{COO}^- \\
| & & | \\
\text{C}-\text{OPO}_3{}^{2-} + \text{ADP}^{3-} \longrightarrow & & \text{C}=\text{O} + \text{ATP}^{4-} \\
\| & & | \\
\text{CH}_2 & & \text{CH}_3
\end{array}$$

Phosphoenolpyruvate   Pyruvate

   b. As discussed in Chapter 14, section III I, this reaction is one of two steps in which ATP is formed during the catabolism of glucose in the glycolytic pathway.

3. As the free energy of hydrolysis of ATP is intermediate in value between compounds with high-phosphate transfer potential (such as phosphoenolpyruvate) and those with a lower transfer potential (such as G6P), ATP can function as an energy carrier between catabolic pathways (where it is formed) and anabolic pathways (where it is used). Thus, in a sense, ATP can act as a "universal currency" for energy in cells.

C. Phosphate pool

1. Nucleoside triphosphates other than that of adenine are also required in metabolic processes.
   a. Guanosine triphosphate (GTP) is used to supply energy in protein synthesis (**as well as act as a modulator of protein conformation**); cytidine triphosphate (CTP), in lipid synthesis; and uridine triphosphate (UTP), in polysaccharide synthesis.
   b. The nucleoside diphosphates or monophosphates that are formed in these reactions are dependent upon ATP for the resynthesis of the triphosphate forms.
   c. Two nucleoside kinases are active in the regeneration of the nucleoside triphosphates
      (1) Nucleoside diphosphate kinase

$$\text{N}-\text{P}-\text{P} + \text{ATP} \rightarrow \text{N}-\text{P}-\text{P}-\text{P} + \text{ADP}$$

      (2) Nucleoside monophosphate kinase

$$\text{N}-\text{P} + \text{ATP} \rightarrow \text{N}-\text{P}-\text{P} + \text{ADP}$$

   d. Note that when a nucleoside triphosphate is reduced to the level of the monophosphate, the expenditure of two high-energy phosphate bonds is required to bring it back to the triphosphate level.

2. **Storage forms of high-energy phosphates**
   a. The turnover rate for ATP is very high. In humans, an amount of ATP appoximately equal to the body weight is formed and broken down every 24 hours.
   b. The rapidity of ATP turnover precludes its use as a storage form of energy. Processes such as muscle contraction in humans, for example, can require ATP to be hydrolyzed for the release of free energy at a rate much higher than ATP can be resynthesized.
   c. High-energy phosphate compounds have evolved that store energy for use during heavy work over limited periods of time. These compounds are called **phosphagens**.
      (1) In humans, creatine phosphate (see Chapter 28, section III A) fills this role.
      (2) During rest periods, the creatine phosphate is regenerated at the expense of ATP.

$$\text{ATP} + \text{creatine} \rightarrow \text{creatine phosphate} + \text{ADP}$$

IV. OXIDATION–REDUCTION REACTIONS

A. Definitions

1. **Oxidation** constitutes a loss of electrons.

2. **Reduction** constitutes a gain of electrons. For example, Fe^{2+} is more reduced than Fe^{3+}.

3. **Oxidation state of a carbon atom** depends upon the electronegativity of the atoms bound to it, that is, upon their tendency to attract the bonding pair of electrons.
   a. In methane (CH_4), the carbon atom is in its most reduced state, as the electrons are shared equally by carbon and hydrogen.
   b. In methanol (CH_3OH), the carbon is more oxidized because the hydroxyl group is more electronegative than carbon. The C—OH bond is, therefore, mildly polar because the carbon has a slight positive charge, having given up an electron to the oxygen.
   c. In formaldehyde (HCHO), the carbonyl bond is more polarized than the C—OH bond.
   d. In formic acid (HCOOH), the carbon is further oxidized.
   e. In carbon dioxide (CO_2), the carbon is in its most oxidized state. (Note that CO_2 cannot be reduced by animal cells, but only by cells of green plants.)

B. Oxidation–reduction potential

1. **Oxidation–reduction reaction** involves **electron transfers**, and each reaction consists of two coupled half-reactions. For example, in the mitochondrial electron transport system (see Chapter 16, section I A), the oxidation of NADH to $NAD^+ + H^+ + 2 e^-$, which is one half-reaction, is coupled to the formation of water from $2 H^+$, $\frac{1}{2} O_2$, and $2 e^-$, which is the other half-reaction.

2. **Standard reduction potential** for the half-reactions is the **electrical potential (E_0)** in volts, measured during the reaction under standard conditions (25°C, all solutes at 1 molar concentration, and all gases at 1 atm pressure). The $H_2{:}H^+$ electrode is used as the reference half-cell with an arbitrarily assigned zero potential. In biochemistry, the standard state also assumes a pH of 7 and is written as E_0'. The E_0' of the $H_2{:}H^+$ electrode is now -0.42 volts.
   a. A compound that is more easily oxidized than H_2 will have a negative E_0'.
   b. In the mitochondrial electron transport system referred to in section IV B 1, the standard reduction potential for the reduction of NAD^+ to NADH is -0.320 volts; therefore, the oxidation of NADH has an E_0' of $+0.320$ volts.
   c. The standard reduction potential for water formation from $\frac{1}{2} O_2$, $2 H^+$, and $2 e^-$ is $+0.816$ volts.

3. **Net reaction potential.** The difference between the standard reduction potentials, $\Delta E_0'$, is the net reaction potential. To calculate $\Delta E_0'$ the **oxidation potential of the oxidation half-reaction** is added to the **reduction potential of the reduction half-reaction**.
   a. The two half-reactions and their potentials are

   $$NADH \rightleftharpoons NAD^+ + H^+ \, 2 e^- \qquad\qquad E = E_0' = +0.320 \text{ volts}$$

   $$\frac{1}{2} O_2 + 2 H^+ + 2 e^- \rightleftharpoons H_2O \qquad\qquad E = E_0' = +0.816 \text{ volts}$$

   b. Adding these together gives the net reaction and the $\Delta E_0'$.

   $$NADH + \frac{1}{2} O_2 + H^+ \rightleftharpoons NAD^+ + H_2O \qquad\qquad \Delta E = \Delta E_0' = +1.140 \text{ volts}$$

   c. When ΔE is positive, a reaction can proceed spontaneously. When ΔE is zero, the reaction is at equilibrium.

4. **Standard free-energy change.** Knowing the $\Delta E_0'$ of an oxidation–reduction reaction and the number of electrons transferred (n), the standard free-energy change can be calculated

   $$\Delta G^{\circ\prime} = -nF\Delta E_0'$$

   where F is the Faraday constant ($23.061 \text{ kcal} \times \text{volt}^{-1} \times \text{equiv}^{-1}$)

C. Electron carriers

1. In aerobic organisms, the ultimate acceptor of electrons derived from fuel molecules is molecular oxygen.

2. The electrons are first transferred from the fuel molecules to specialized electron carriers.
   a. Two compounds accept electrons from fuel molecules and serve as carriers, **NAD^+** and **flavin-adenine dinucleotide (FAD)**. (For the biosynthesis of these compounds see Chapter 29, section VII.)
      (1) The active portion of the NAD^+ molecule is the **nicotinamide ring**, which accepts a proton and two electrons (equivalent to a hydride ion, H^-) from the substrate (Fig. 12-2).
      (2) The active portion of the FAD molecule is the **alloxazine ring**, which accepts two protons and two electrons (equivalent to 2 H) from the substrate (Fig. 12-3).
   b. The electrons from the carriers reach molecular oxygen via the mitochondrial electron transport system (see Chapter 16, section I A).

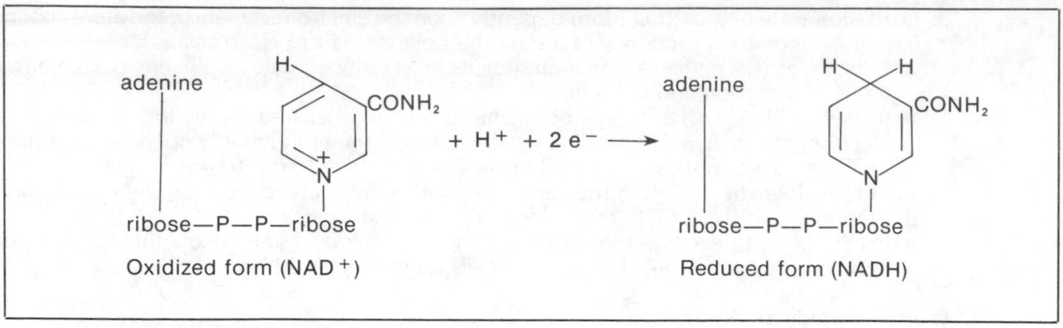

Figure 12-2. Conversion of the oxidized form of nicotinamide-adenine dinucleotide (NAD$^+$) to the reduced form (NADH).

    c. Both NAD$^+$ and FAD are cofactors for the dehydrogenase enzymes involved in the oxidation of fuel molecules.
      (1) In dehydrogenases, using FAD as the electron acceptor, the cofactor is firmly bound to the enzyme.
      (2) NAD$^+$ is more mobile and can move between dehydrogenases.
    d. NADPH serves as the donor of hydrogen and electrons in **reductive syntheses**, where the precursors are more oxidized than the products.

V. COMPARTMENTALIZATION OF METABOLIC PATHWAYS IN CELLS

    A. Eukaryote cell structure. Eukaryote cells have a complex architecture, including the

      1. Nucleus, which is a membrane-bounded area enclosing the chromosomes

      2. Membrane-bounded organelles, such as the **mitochondria** (see Chapter 16, Fig. 16-1), which enclose the major enzyme systems concerned with the final oxidation of foodstuffs and the trapping of free energy in the form of ATP

      3. Endoplasmic reticulum, a continuous network of membrane-bounded vesicles

      4. Golgi apparatus, which is concerned with the manufacture and export of cellular products

      5. Lysosomes, which contain hydrolytic enzymes for the catabolism of cellular macromolecules

    B. Localization of metabolic pathways

      1. The enzymes catalyzing metabolic pathways are frequently localized in a specific intracellular compartment or organelle.
      a. For example, the entire sequence of enzymes involved in the conversion of glucose to lactate (the glycolytic pathway—see Chapter 14) is found in the soluble sap, or cytosol, of the cell; that is, it is not enclosed within an intracellular organelle.
      b. In contrast, the enzymes of the tricarboxylic acid (TCA) cycle (see Chapter 15), which are

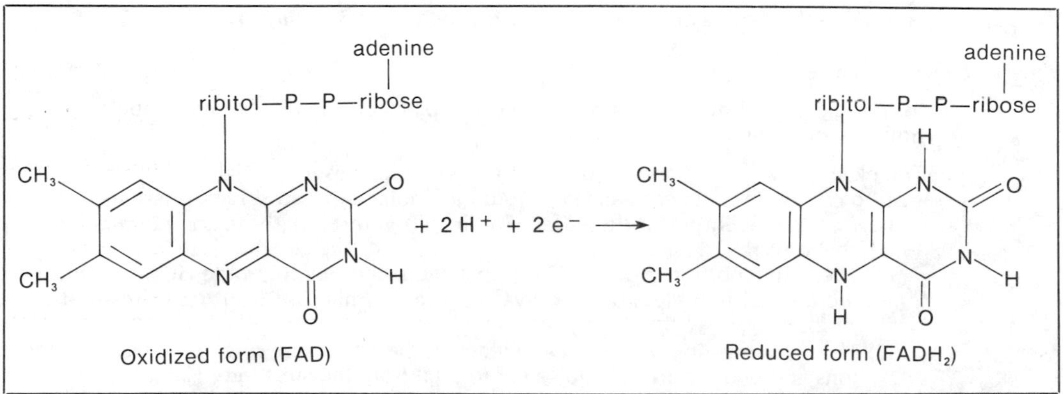

Figure 12-3. Conversion of the oxidized form of flavin-adenine dinucleotide (FAD) to the reduced form (FADH$_2$).

concerned with the final oxidation of two-carbon units to CO_2 and water, are contained within the inner membrane of the double-membrane–bounded mitochondria.

  2. The distribution of enzymes within cells is studied by using the technique of **differential centrifugation**.

    a. This technique provides cellular fractions that are defined operationally, in terms of the major cell component present; for example, the "nuclear fraction," the "mitochondrial fraction," or the "soluble fraction" (the cytosol).

    b. In differential centrifugation, a homogenate of the appropriate tissue (in isotonic sucrose) is subjected to a centrifugal field of increasing force. The usual fractions, obtained as pellets of sedimented material or as nonsedimented "soluble" material, are given in Table 12-1.

C. Biologic advantage of compartmentalization. The compartmentalization of metabolic pathways allows the separation of processes that proceed in opposite directions and may otherwise interfere with one another. For example:

  1. The anabolic processes involved in the *biosynthesis* of fatty acids *from* acetyl coenzyme A (acetyl CoA) are confined to the cytosol fraction.

  2. The catabolic processes concerned with the *oxidation* of fatty acids *to* acetyl CoA are contained within the mitochondrial matrix (see Chapter 20).

VI. STAGES IN THE CATABOLISM OF FUEL MOLECULES. The use of complex molecules in food for the extraction of free energy by humans was depicted by Hans Krebs as occurring in three stages.

A. Stage 1

  1. The complex macromolecules of starch, protein, and triacylglycerols are broken down into smaller units, such as monosaccharides, amino acids, glycerol, and fatty acids.

  2. During this stage, little or no free energy is trapped.

B. Stage 2

  1. The simple molecules of different kinds are catabolized to a few molecules that can be oxidized to CO_2 and water along a common pathway.

  2. In this stage, some free energy is trapped as ATP.

C. Stage 3 is the **final common pathway**.

  1. It consists of
    a. The tricarboxylic acid (TCA) cycle
    b. The electron transport system
    c. Oxidative phosphorylation

  2. Together, these processes oxidize acetyl CoA to CO_2 and water, and trap the available free energy as ATP.

  3. The fact that there is a common pathway for the final oxidation of all metabolic fuels means that these fuels must compete for the necessary enzymes and cofactors (see Chapter 32, section III D).

Table 12-1. Major Operationally Defined Fractions Obtained by Differential Centrifugation of a Tissue Homogenate Prepared in 0.25 M Sucrose

Fraction	Operational Definition	Major Cell Components Present
600-g pellet	Nuclear fraction	Nuclei and unbroken cells
15,000-g pellet	Mitochondrial fraction	Mitochondria and lysosomes
100,000-g pellet	Microsomal fraction	Ribosomes and fragments of endoplasmic reticulum
100,000-g supernatant	Soluble fraction	Soluble enzymes (cytosol)

STUDY QUESTIONS

Directions: Each question below contains five suggested answers. Choose the **one best** response to each question.

1. If a reaction is at equilibrium, the free-energy change (ΔG) is

(A) equal to $- RT \times \ln K_{eq}$
(B) equal to $- nF \times \Delta E_0'$
(C) equal to the ΔG under standard conditions
(D) equal to zero
(E) none of the above

2. The normal concentrations of glucose 6-phosphate (G6P) and fructose 6-phosphate (F6P) in human erythrocytes are 1×10^{-5} M and 1×10^{-6} M, respectively. If the standard free-energy change ($\Delta G°'$) for the reaction G6P $\rightarrow$ F6P is $+ 0.4$ kcal/mol, which of the following statements is correct?

(A) The equilibrium constant for the reaction G6P $\rightarrow$ F6P is 1
(B) The free-energy change (ΔG) is about $- 1.0$ kcal/mol
(C) The ΔG for the reverse reaction is $- 0.4$ kcal/mol
(D) The reaction as written cannot occur in the erythrocyte
(E) None of the above are correct

3. The standard free-energy change ($\Delta G°'$) for the conversion of compound X to compound Y is $+ 1.4$ kcal/mol at 37°C. Which of the following most closely approximates the equilibrium ratio of the concentration of X to that of Y?

(A) 100:1
(B) 1:100
(C) 10:1
(D) 1:10
(E) 1:1

4. If an enzymatic reaction has a standard free-energy change ($\Delta G°'$) of $- 5$ kcal/mol, the equilibrium constant is

(A) greater than one
(B) less than one
(C) zero
(D) dependent upon enzyme concentration
(E) not determinable from the value of $\Delta G°'$

5. The values of the standard free-energy change ($\Delta G°'$) for the hydrolysis of adenosine triphosphate (ATP) and creatine phosphate are $- 7.4$ kcal/mol and $- 10.2$ kcal/mol, respectively. How then can you account for the formation of creatine phosphate from ATP and creatine, catalyzed by creatine kinase?

(A) Two molecules of ATP are necessary to form one molecule of creatine phosphate
(B) Because ATP contains two high-energy phosphate bonds, there is sufficient free energy available to drive the reaction
(C) The reaction involves the use of another high-energy phosphate compound
(D) The formation of creatine phosphate from ATP and creatine will proceed if the concentration of creatine kinase is high enough
(E) The free-energy change (ΔG) of the reaction is dependent upon the concentration of reactants as well as on the $\Delta G°'$ values of hydrolysis

6. Coupled reactions involved in the trapping of free energy from fuel molecules consist of

(A) two endergonic reactions, one of which has a higher standard free-energy change ($\Delta G°'$) than the other
(B) two exergonic reactions, one of which has a lower $\Delta G°'$ than the other
(C) one endergonic reaction coupled to an exergonic reaction, which has a lower $\Delta G°'$ than the endergonic reaction
(D) one exergonic reaction coupled to an endergonic reaction, which has a lower $\Delta G°'$ than the exergonic reaction
(E) two exergonic reactions, both of which have the same $\Delta G°'$

7. Which of the following compounds has a higher group transfer potential for phosphate than adenosine triphosphate?

(A) Glucose 6-phosphate
(B) Fructose 1,6-diphosphate
(C) Ribose 5-phosphate
(D) 2-Phosphoenolpyruvate
(E) Pyruvate

8.   Which of the following parameters can be determined from a knowledge of the standard oxidation–reduction potential for an oxidation–reduction reaction?

(A) The standard free-energy change of the reaction

(B) The activation energy of the reaction

(C) The concentrations of oxidant and reductant in a nonequilibrium mixture

(D) The hydrogen ion concentration in the reaction mixture

(E) None of the above

9.   If an oxidation–reduction reaction with a two-electron transfer has a standard reduction potential of +0.3 volts, what is the free-energy change under standard conditions?

(A) +6.9 kcal/mol

(B) –13.8 kcal/mol

(C) +46.1 kcal/mol

(D) +13.8 kcal/mol

(E) –6.9 kcal/mol

ANSWERS AND EXPLANATIONS

1. The answer is D. (*II D 3*) At equilibrium, the free-energy change of a reaction (ΔG) is zero. The standard free-energy change ($\Delta G^{\circ\prime}$) is equal to $- RT \ln K_{eq}$ at equilibrium, and for an oxidation–reduction reaction, it is equal to $-nF\Delta E_0$.

2. The answer is B. (*II D 3*) The free-energy change (ΔG) for the conversion of glucose 6-phosphate (G6P) to fructose 6-phosphate (F6P) can be found from

$$\Delta G = \Delta G^{\circ\prime} + RT \ln \frac{[F6P]}{[G6P]} = +0.4 + 1.4 \log \frac{10^{-5}}{10^{-6}} = -1.$$

The equilibrium constant for the reaction as written (G6P → F6P) can be found from

$$\Delta G^{\circ\prime} = -1.4 \log K_{eq}, \text{ or, } \log K_{eq} = \frac{-0.4}{1.4}.$$

Thus, K_{eq} is less than 1. The $\Delta G^{\circ\prime}$ for the reverse reaction is -0.4 kcal/mol, but not the ΔG, which is $+1$ kcal/mol. The reaction as written can occur in erythrocytes because the ΔG calculated from the concentration of G6P and F6P is negative.

3. The answer is C. (*II D 3 b*) The ratio of the concentration of the product (Y) to the concentration of the reactant (X) at equilibrium is the value for the equilibrium constant (K_{eq}). Given the standard free-energy change ($\Delta G^{\circ\prime}$), the value for the equilibrium constant can be calculated from

$$\Delta G^{\circ\prime} = -RT \ln K_{eq},$$
$$\Delta G^{\circ\prime} = -RT \times 2.303 \log_{10} K_{eq}, \text{ and}$$
$$\log_{10} K_{eq} = \frac{\Delta G^{\circ\prime}}{-RT \times 2.303}.$$

Note that at 25°C to 37°C, the value for RT $\times$ 2.303 is approximately 1400.

4. The answer is A. (*II D 3 b*) At equilibrium, the free-energy change (ΔG) for a reaction is zero, and the standard free-energy change ($\Delta G^{\circ\prime}$) is given by

$$\Delta G^{\circ\prime} = -RT \ln K_{eq},$$

where K_{eq} is the equilibrium constant for the reaction, R is the gas constant (1.98×10^{-3} kcal), and T is the temperature in degrees kelvin ($^\circ K = {}^\circ C + 273$). Rearranging the equation gives

$$\ln K_{eq} = \Delta G^{\circ\prime} \div -RT.$$

Note that $\ln$ ($\log_e$) = 2.303 $\log_{10}$. For a $\Delta G^{\circ\prime}$ of -5 kcal/mol, the value for the logarithm of the equilibrium constant at 37°C would be given by

$$\log_{10} K_{eq} = 5 \div (1.98 \times 10^{-3} \times 310 \times 2.303) = 3.537.$$

Therefore, K_{eq} is 3.44×10^3, which is greater than one. The equilibrium constant for a reaction is not changed by the presence of the enzyme.

5. The answer is E. (*II E*) The standard free-energy change ($\Delta G^{\circ\prime}$) associated with the formation of adenosine triphosphate (ATP) from creatine phosphate and adenosine diphosphate (ADP) is -3.0 kcal/mol, a value that could be predicted from the greater $\Delta G^{\circ\prime}$ for the hydrolysis of creatine phosphate in relation to that for ATP. Thus, the $\Delta G^{\circ\prime}$ for the reverse reaction, that is, the formation of creatine phosphate from ATP and creatine, is $+3.0$ kcal/mol. However, the spontaneity of a reaction is indicated by the value for the change in free energy (ΔG), which depends not only upon the $\Delta G^{\circ\prime}$ but also on the concentration of reactants and products. The actual ΔG for the formation of creatine phosphate from creatine and ATP, catalyzed by creatine kinase, is given by

$$\Delta G = +3.0 + RT \ln \frac{[\text{creatine P}][\text{ADP}]}{[\text{creatine}][\text{ATP}]}.$$

If the concentrations of ATP and creatine are high enough, creatine phosphate will be formed.

6. The answer is D. (*II E*) The catabolism of fuel molecules must include reactions in which the standard free-energy change is sufficient to support the phosphorylation of adenosine diphosphate (ADP) to adenosine triphosphate. These reactions are exergonic and have a standard free-energy change greater than the -7.3 kcal/mol required for the endergonic phosphorylation of ADP.

7. The answer is D. (*III B*) The free energy of hydrolysis of 2-phosphoenolpyruvate is -14.8 kcal/mol, a high value, giving it a greater group transfer potential than that of adenosine triphosphate, which has a free energy of hydrolysis of -7.3 kcal/mol. None of the other phosphates listed are high-energy phosphate compounds, and pyruvate does not have a phosphate group.

8. The answer is A. (*IV B*) The standard free-energy change ($\Delta G^{\circ\prime}$) of an oxidation–reduction reaction is given by $\Delta G^{\circ\prime} = -nF\Delta E_0{}^{\prime}$, where $\Delta E_0{}^{\prime}$ is the change in the oxidation–reduction potential under standard conditions. If the number of electrons transferred is known, then $\Delta G^{\circ\prime}$ can be calculated from $\Delta E_0{}^{\prime}$. The concentrations of oxidant and reductant in a nonequilibrium mixture cannot be calculated from the standard reduction potential alone but only can if the actual observed reduction potential is determined. The activation energy is the free energy required to convert the reactants into their reactive states and cannot be determined from standard oxidation–reduction potentials alone. A standard oxidation–reduction potential, $E_0{}^{\prime}$, is, as the prime indicates, determined at pH 7; that is, at a $[H^+]$ of 10^{-7}M. In a nonequilibrium reaction, the $[H^+]$ could not be calculated from $\Delta E_0{}^{\prime}$ alone.

9. The answer is B. (*IV B 4*) Given the standard reduction potential change ($\Delta E_0{}^{\prime}$) for an oxidation–reduction reaction, the standard free-energy change ($\Delta G^{\circ\prime}$) can be calculated from

$$\Delta G^{\circ\prime} = -nF\Delta E_0{}^{\prime},$$

where n is the number of electrons transferred, and where F is the Faraday constant (23.061 kcal per volt • equivalents). Thus,

$$\Delta G^{\circ\prime} = -2 \times 23.061 \times 0.3 = +13.8 \text{ kcal/mol}.$$

13
Carbohydrate Chemistry

I. GENERAL CHARACTERISTICS OF CARBOHYDRATES

A. Definition. Carbohydrates are polyhydroxy aldehydes, polyhydroxy ketones, or compounds that can be hydrolyzed to them.

B. Classification

1. **Monosaccharides** are simple sugars that cannot be broken down into smaller molecules by hydrolysis.

2. **Disaccharides** can be hydrolyzed to give two monosaccharides.

3. **Oligosaccharides** are polymers made up of two to ten monosaccharide units.

4. **Polysaccharides** are polymers with many monosaccharide units.

C. Nomenclature

1. Monosaccharides can be named by a system that is based on the number of carbons with the suffix **-ose** added. They have a general formula $C_nH_{2n}O_n$ where **n** is the number of carbons.
   a. Triose = three carbons
   b. Tetrose = four carbons
   c. Pentose = five carbons
   d. Hexose = six carbons

2. Nomenclature can also indicate the reactive groups.
   a. Aldoses are monosaccharides with an aldehyde group.
   b. Ketoses are monosaccharides containing a ketone group.

3. Monosaccharide and reactive-group nomenclature can be combined in designating compounds. Thus, the sugar **glucose** is an **aldohexose**; that is, it is a six-carbon monosaccharide (-hexose) containing an aldehyde group (aldo-).

4. **System for numbering the carbons.** The carbons are numbered sequentially with the aldehyde or ketone group being on the carbon with the lowest possible number.

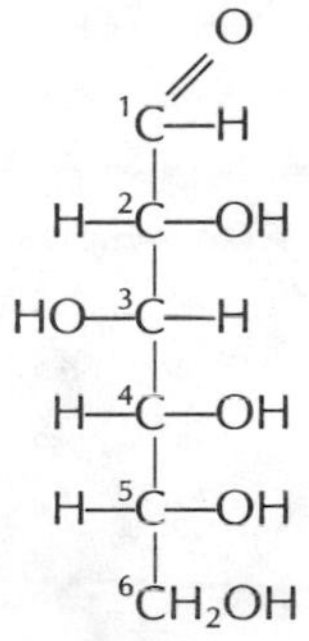

Correct 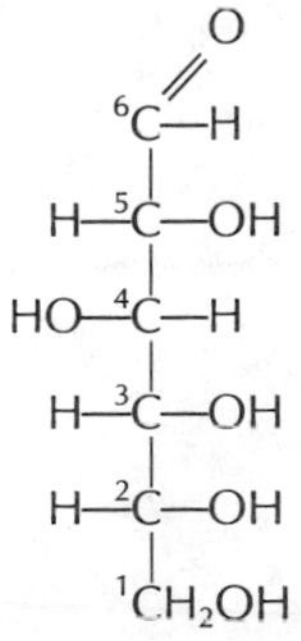   Incorrect

II. CYCLIC FORMS OF SUGAR.
Study of the chemical and physical properties of many sugars has shown that cyclic forms predominate over open-chain structures, both in solution and in the solid state.

A. The cyclic form of glucose is a six-membered ring; such sugars are called **pyranoses** because they resemble pyran.

B. Fructose forms a five-membered ring called a **furanose**.

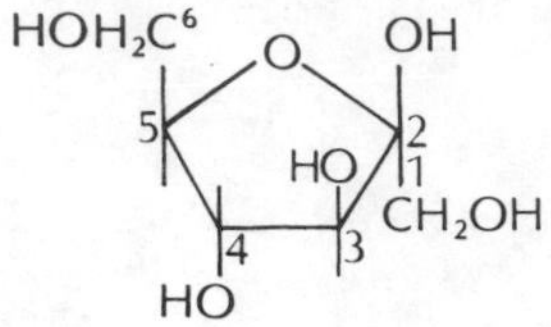

$$\text{HOH}_2\text{C}^6 \quad \text{O} \quad \text{OH}$$

C. The **intermolecular hemiacetals** occur when an alcohol reacts with an aldehyde. These are unstable compounds.

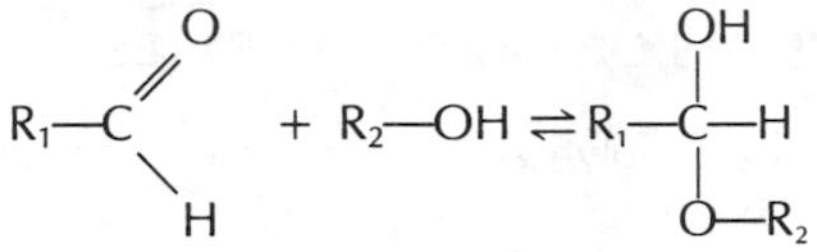

$$R_1\!-\!\overset{\displaystyle O}{\overset{\|}{C}}\diagdown_{\!H} \;+\; R_2\!-\!OH \;\rightleftharpoons\; R_1\!-\!\overset{\displaystyle OH}{\underset{O\!-\!R_2}{\overset{|}{C}}}\!-\!H$$

     Aldehyde   Alcohol   Hemiacetal

D. Cyclized forms of hemiacetals can be formed by similar **intramolecular reactions**. In glucose, the —OH on carbon 5 (C-5) can react intramolecularly with the carbonyl group (on carbon 1) to form a stable, **cyclic hemiacetal**.

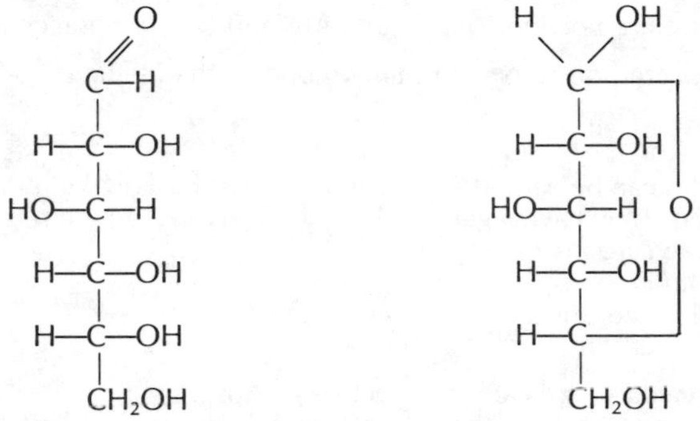

     Open-chain form   Cyclic form

III. OPTICAL ISOMERISM

A. Asymmetric carbons. Carbohydrates contain asymmetric carbons; namely, those bonded to four different atoms or groups of atoms. Because of this carbon asymmetry, carbohydrates are optically active, rotating the plane of polarized light. The carbons below labeled with an asterisk are asymmetric.

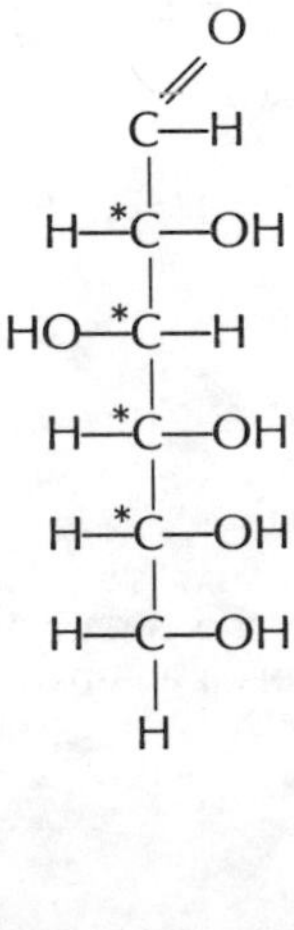

B. Specific rotation

1. If plane polarized light is rotated to the *right* (clockwise), the compound is **dextrorotatory** (indicated by *d* or +).

2. If plane polarized light is rotated to the *left* (counterclockwise), the compound is **levorotatory** (indicated by *l* or −).

C. Configuration

1. The simplest carbohydrates are the monosaccharide trioses; for example, glyceraldehyde, which has two optically active forms, designated L and D.

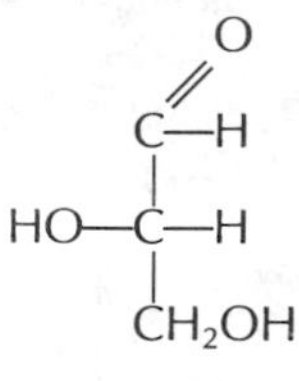

L-Glyceraldehyde 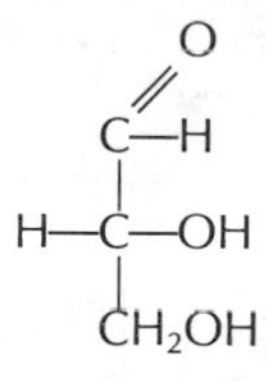   D-Glyceraldehyde

2. For the purposes of nomenclature, other sugars are considered to be derived from glyceraldehyde. Thus, a D sugar is one that matches the configuration of D-glyceraldehyde about the asymmetric carbon that is *farthest from the aldehyde or ketone group*; an L-sugar correspondingly matches L-glyceraldehyde.

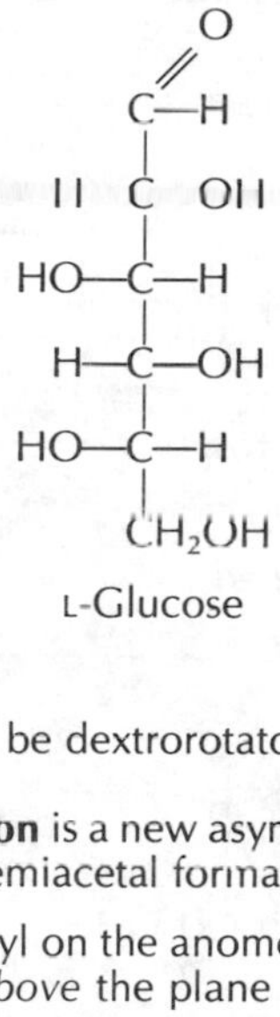

L-Glucose 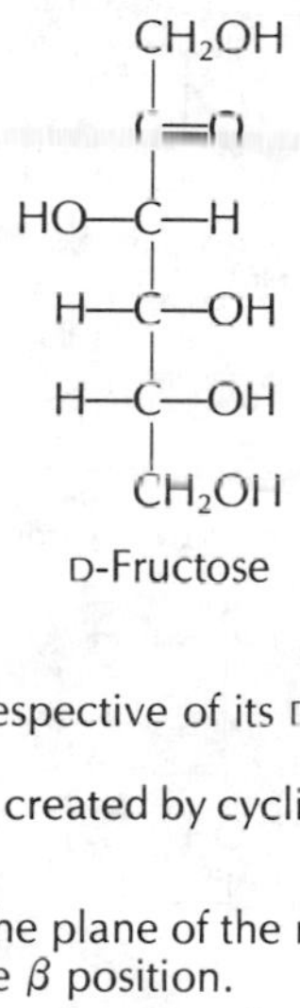   D-Fructose

3. A sugar may be dextrorotatory or levorotatory irrespective of its D- or L-configuration.

D. Anomeric carbon is a new asymmetric carbon that is created by cyclization at the carbon bound to oxygen in hemiacetal formation.

1. If the hydroxyl on the anomeric carbon is *below* the plane of the ring, it is said to be in the α position; if *above* the plane of the ring, it is in the β position.

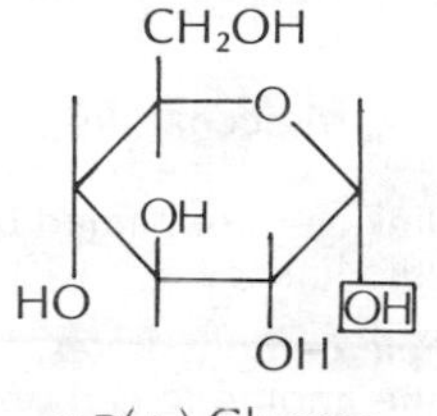

α-D(+)-Glucose   β-D(+)-Glucose

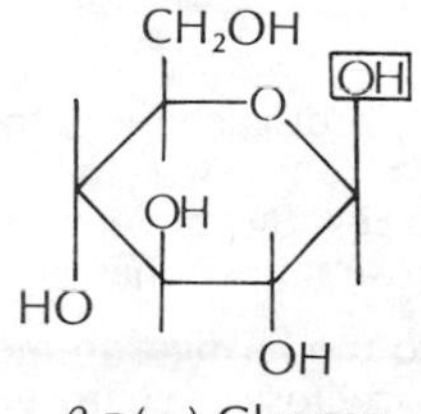

2. In solution, α- and β-sugars slowly change into an equilibrium mixture of both. This process is known as **mutarotation**.

IV. GLYCOSIDIC LINK

A. **Acetals.** Hemiacetals react with alcohols to form acetals.

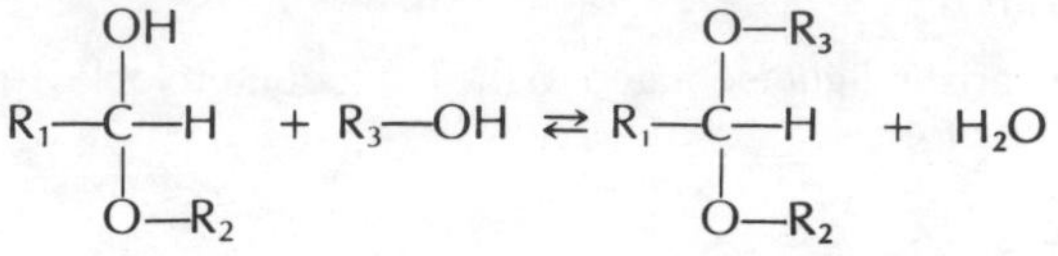

$$R_1\!-\!\underset{\underset{O-R_2}{|}}{\overset{\overset{OH}{|}}{C}}\!-\!H \quad + \quad R_3\!-\!OH \quad \rightleftarrows \quad R_1\!-\!\underset{\underset{O-R_2}{|}}{\overset{\overset{O-R_3}{|}}{C}}\!-\!H \quad + \quad H_2O$$

       Hemiacetal   Alcohol   Acetal

B. **Glycosides.** A sugar can react with an alcohol to form an acetal known as a glycoside.

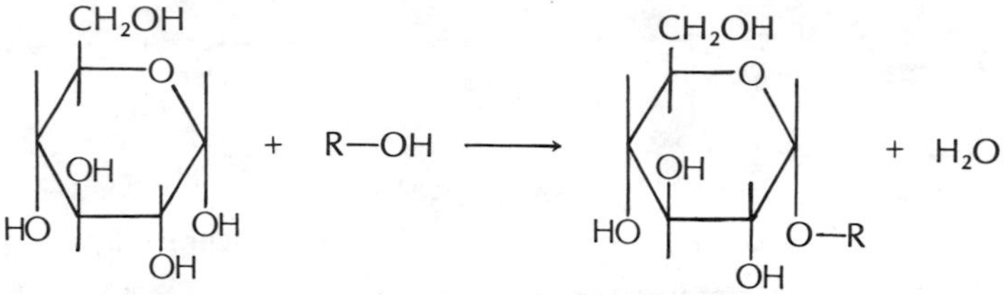

    α-Sugar   α-Glycoside

1. If the sugar residue is glucose, the derivative is a glucoside; if fructose, it is a fructoside; if galactose, it is a galactoside.

2. When R is another sugar, the glycoside is a **disaccharide**.

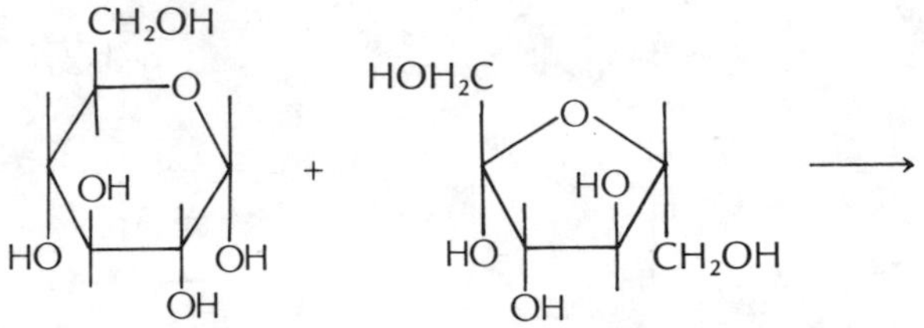

    α-D-Glucose   β-D-Fructose

(Note that the fructose representation has been flipped over.)

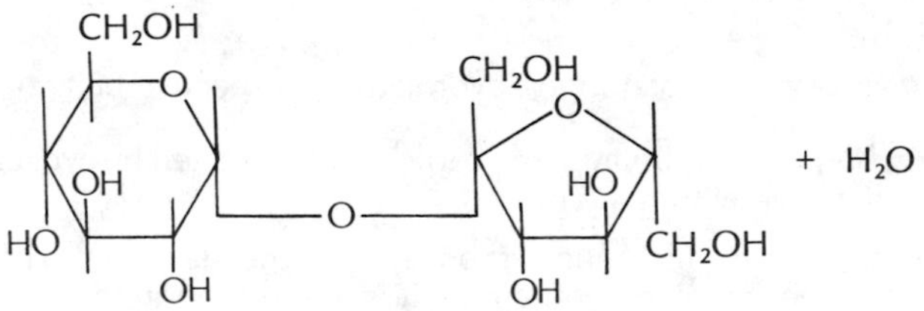

Sucrose (α-D-glucopyranosyl-β-D-fructofuranoside)

3. If R is already a disaccharide, the glycoside is a trisaccharide.

C. **Glycosidic linkages.** By convention, glycosidic linkages are named by reading from left to right. Therefore, sucrose (above) has an α-1,2-glycosidic linkage.

1. In contrast to the linkages in most other simple carbohydrates, the oxygen bridge between glucose and fructose in sucrose is between the anomeric carbons.
   a. Consequently, there is no free aldehyde or ketone group in sucrose.
   b. Therefore, this disaccharide is **not a reducing sugar**. For example, it will not reduce an alkaline copper reagent such as Fehling's solution.

2. The disaccharide **maltose** possesses an *unattached* anomeric carbon atom, which may have

either the α or β configuration. However, the glycosidic linkage in maltose must be in the α-1,4-glycosidic configuration.

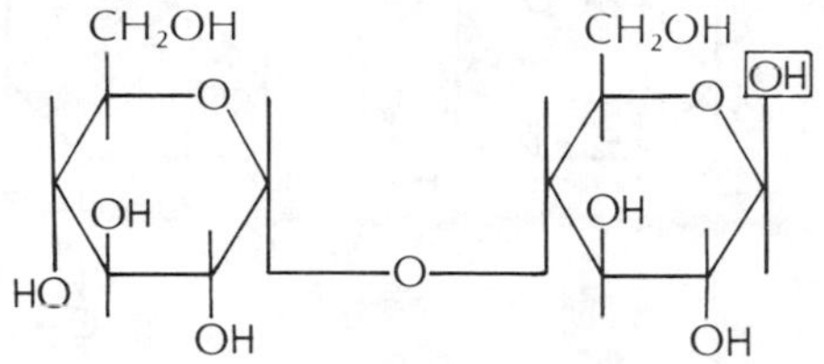

β-Maltose

V. POLYSACCHARIDES

A. Amylose is a linear unbranched polymer of α-D-glucose units in a repeating sequence of α-1,4-glycosidic linkages.

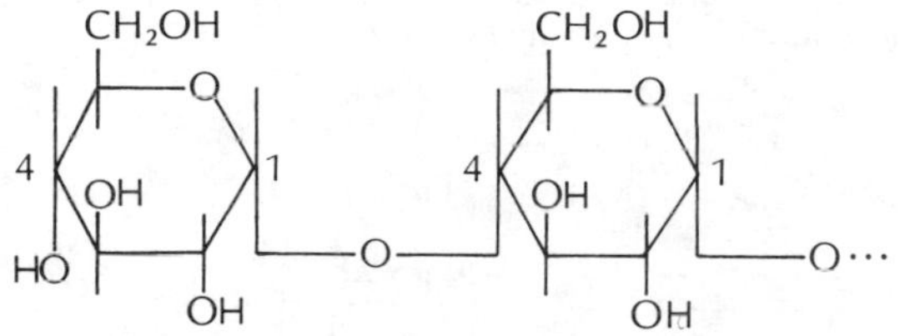

Amylose

B. Amylopectin is a branched polymer of α-D-glucose with α-1,4-glycosidic linkages and with α-1,6 branching points that occur at intervals of approximately 25 to 30 α-D-glucose residues.

C. Starch is a mixture of amylose and amylopectin. It is the storage form for glucose in plants.

D. Glycogen is the major storage form of carbohydrate in animals, found mostly in liver and muscle. It is a highly branched form of amylopectin: α-1,6 branching points occur every eight to ten D-glucose residues, and the latter are in α-1,4 linkages.

E. Cellulose is composed of chains of D-glucose units joined by β-1,4-glycosidic linkages. The chains are exclusively linear (i.e., unbranched).

   1. Cellulose is a structural polysaccharide of plant cells.

   2. Although cellulose forms a part of the human diet (e.g., in vegetables and fruit), it is not hydrolyzed by human enzyme systems.

VI. CARBOHYDRATE DERIVATIVES

A. Phosphoric acid esters of monosaccharides

   1. Phosphorylation is the initial step in the metabolism of sugars.

   2. Phosphorylated sugars such as D-glucose-1-phosphate are metabolic intermediates.

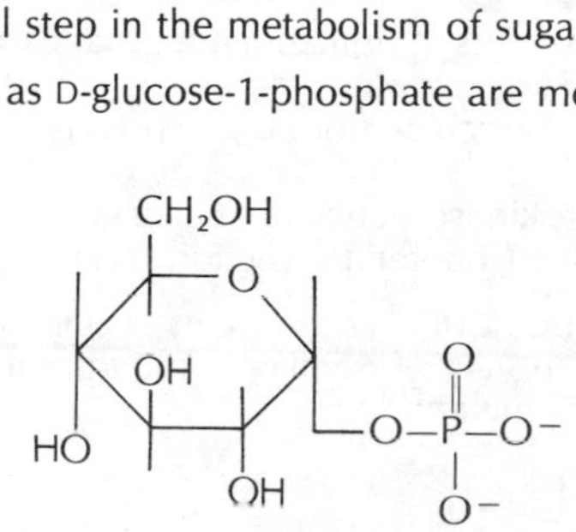

D-Glucose-1-phosphate

B. Amino sugars. In these, a hydroxyl group is replaced by an **amino** or an **acetylamino** group.

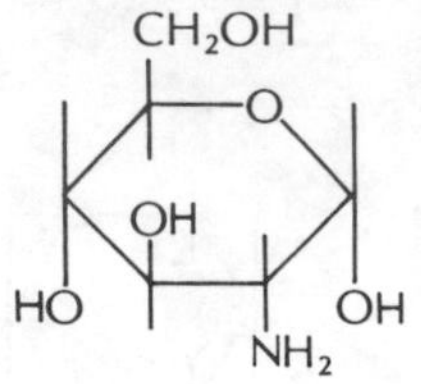

D-Glucosamine 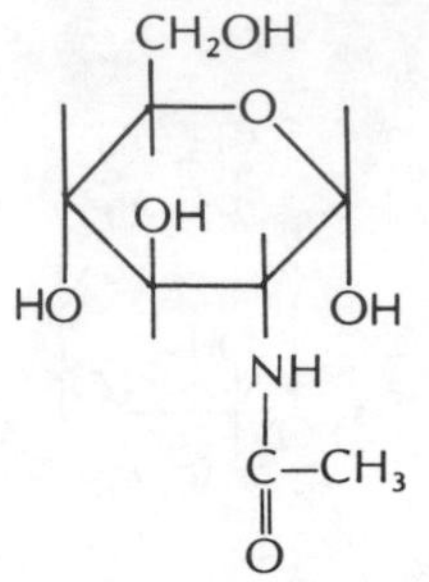   D-Acetylglucosamine

1. **Glucosamine** is the product of the hydrolysis of chitin, the major polysaccharide of the shells of insects and crustaceans.

2. **Galactosamine** is found in the polysaccharide of cartilage, chondroitin sulfate.

C. Sugar acids are produced by oxidation of the aldehydic carbon, the hydroxyl carbon, or both. Ascorbic acid (vitamin C) is a sugar acid.

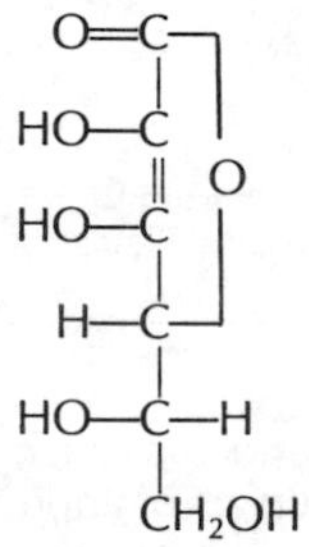

Ascorbic acid

D. Deoxy sugars include 2-deoxyribose, found in DNA (see Chapter 1, section I A 1).

E. Sugar alcohols

1. Monosaccharides, both aldoses and ketoses, may be reduced at the carbonyl carbon, to the corresponding polyhydroxy alcohols (sugar alcohols).

2. Aldoses yield the corresponding alcohol, while ketoses form two alcohols, because a new asymmetric carbon is formed in the process.

3. Commonly occurring aldoses and ketoses form the following sugar alcohols:
   a. D-Glucose yields D-sorbitol.
   b. D-Mannose yields D-mannitol.
   c. D-Galactose yields dulcitol.
   d. D-Fructose, a ketose, yields D-mannitol and D-sorbitol.

4. There are enzyme systems in human tissues that can reduce monosaccharides to their corresponding alcohols. For example:
   a. The formation of D-fructose from D-glucose via D-sorbitol in seminal vesicles (see Chapter 19, section III B 1)
   b. In cases of galactokinase deficiency, excess galactose may be reduced to dulcitol in the lens of the eye [see Chapter 19, section IV D 2 a (2)].

5. Sugar alcohols are also normal constituents of cell components, as for example, ribitol, which is the carbohydrate moiety of riboflavin, found in the electron carriers riboflavin phosphate and flavin-adenine dinucleotide.

STUDY QUESTIONS

Directions: Each question below contains five suggested answers. Choose the **one best** response to each question.

1. Which of the following statements best characterizes glucose?

   (A) It usually exists in the furanose form
   (B) It is a ketose
   (C) Carbon 2 is the anomeric carbon atom
   (D) It forms part of the disaccharide sucrose
   (E) Only the L-isomer can be transported into mammalian cells

2. The sugar residues of amylose are

   (A) in β-1,4 linkages
   (B) in α-1,4 linkages
   (C) galactose units only
   (D) fructose units only
   (E) both galactose and fructose units

Directions: Each question below contains four suggested answers of which **one or more** is correct. Choose the answer

   A if **1, 2, and 3** are correct
   B if **1 and 3** are correct
   C if **2 and 4** are correct
   D if **4** is correct
   E if **1, 2, 3, and 4** are correct

 3. Which of the following are reducing sugars?

   (1) Sucrose
   (2) Lactose
   (3) Xylitol
   (4) Ribose

4. Correct statements regarding the structure of polysaccharides include which of the following?

   (1) Amylose is a linear unbranched polymer of D-glucose in a repeating sequence of α-1,4-glycosidic linkages
   (2) Amylopectin is a branched polymer of D-glucose with α-1,4-glycosidic linkages and with α-1,6 branching points every 25 to 30 D-glucose residues
   (3) Glycogen is a branched polymer of D-glucose with α-1,4-glycosidic linkages and with α-1,6 branching points every 8 to 10 D-glucose residues
   (4) Cellulose is composed of linear chains of D-glucose units joined by α-1,4-glycosidic linkages

ANSWERS AND EXPLANATIONS

1. The answer is D. (*I C 3; II A; III D 2; IV B 2*) Glucose is an aldose sugar and forms a six-membered ring structure, pyranose form, which is similar to the six-member pyran ring compound. Carbon 1 is the anomeric carbon atom, and solutions of α-glucose or β-glucose gradually change into a mixture of both anomers (mutarotation). The D-form of glucose is the naturally occurring form and is the isomer handled by the sugar transport system in mammalian cells. Sucrose is α-D-glucopyranosyl-β-D-fructofuranoside, a disaccharide of glucose and fructose.

2. The answer is B. (*V A*) Amylose is a linear unbranched polymer of α-D-glucose units in a repeating sequence of α-1,4-glycosidic linkages.

3. The answer is C (2, 4). (*IV C 1*) Ribose is a reducing sugar; that is, it will reduce alkaline copper reagents by virtue of its aldehyde group on carbon 1. The disaccharide lactose is also a reducing sugar because the linkage between glucose and galactose is β-1,4, which leaves the reducing group of glucose free to react. Sucrose, on the other hand, is a disaccharide of glucose and fructose in α-1,2 linkage. As this link involves both anomeric carbon atoms, neither is free to react. Xylitol is a sugar alcohol derived by reduction of L-xylulose and does not contain a reducing group.

4. The answer is A (1, 2, 3). (*V A–E*) Amylose and cellulose are linear unbranched polymers of D-glucose, joined in amylose by α-1,4 linkages and in cellulose by β-1,4 linkages. Amylopectin and glycogen are branched polymers of D-glucose with main chain units joined by α-1,4 linkages and the branch sites by α-1,6 linkages. Glycogen is highly branched with branch chains arising every 8 to 10 D-glucose units, whereas they arise every 20 to 30 D-glucose units in amylopectin.

14
Glycolysis

I. CARBOHYDRATES: SOURCE OF CARBON AND ENERGY

A. Physiology

1. Carbohydrate sources
   a. Starch is a major nutrient derived from ingested plant cells.
   b. Lactose, a disaccharide of **glucose** and **galactose**, is the major carbohydrate component of milk.
   c. Fructose, a ketose isomer of glucose, is ingested as such in fruits and is also an important carbohydrate source when the intake of sucrose (a glucose–fructose disaccharide) is high.

2. Glucose, the product of starch digestion, is the major form in which carbohydrate is presented to the cells of the body.

3. Glycogen, the major form of carbohydrate storage in animals, is a highly branched polymer of glucose.

B. Digestion

1. α-Amylase is responsible for several activities.
   a. It hydrolyzes hydrated starch and glycogen in saliva (although an isozyme of α-amylase in pancreatic juice is more important). Because starch becomes hydrated when it is heated, cooking is important for its digestion and absorption.
   b. It attacks α-1,4 linkages, except for those of glucose units that serve as branch points.
   c. It does not attack α-1,6 linkages.
   d. Its action produces
      (1) Maltose [α-glucose (1,4) glucose]
      (2) Maltotriose [α-glucose (1,4) α-glucose (1,4) glucose]
      (3) Limit dextrins (highly branched molecules composed of about eight glucose units joined by one or more α-1,6 bonds)

2. Intestinal enzymes
   a. Oligosaccharidases complete the hydrolysis of disaccharides and oligosaccharides on the surface of epithelial cells in the small intestine. These enzymes remove successive units from the nonreducing ends (the ends opposite the aldehyde or ketone groups).
   b. α-Glucosidases of the small intestine in humans occur in excess, as do the enzymes that hydrolyze sucrose. In contrast, **β-galactosidases** (required for the hydrolysis of lactose) can be rate-limiting.
   c. If disaccharides, oligosaccharides, and polysaccharides are not hydrolyzed by α-amylase and intestinal enzymes, they cannot be absorbed. In this case, they may be metabolized by intestinal bacteria into short-chain fatty acids, lactate, and gases [H_2, methane, and carbon dioxide (CO_2)], causing fluid secretion, increased intestinal mobility, and cramps.

3. Major monosaccharide products of carbohydrate digestion are
   a. D-Glucose
   b. D-Galactose
   c. D-Fructose

C. Active (carrier-mediated) transport of hexoses

1. Two known transport systems are involved in the uptake of monosaccharides by the epithelial cells of the intestinal lumen.
   a. A Na^+–monosaccharide cotransport system is highly specific for D-glucose and D-galactose and actively transports these sugars.

b. A Na^+-independent facilitated diffusion is specific for D-fructose.

2. The monosaccharides are transported out of the intestinal cells and into the circulation by a Na^+-independent facilitated diffusion of D-glucose and D-galactose in the contraluminal plasma membrane of the intestinal epithelial cells; D-fructose apparently leaves the intestinal cells by passive diffusion.

3. The Na^+–monosaccharide transport is inhibited by the plant glycoside **phlorhizin**, and the Na^+-independent systems by **cytochalasin B**. These inhibitors are not specific for intestinal hexose transport but also affect transport in renal tubules, erythrocytes, adipocytes, muscle cells, and other cells.

II. GLYCOLYTIC (EMBDEN-MEYERHOF) PATHWAY

A. Definition. Glycolysis is a process in which glucose is enzymatically split into two molecules of pyruvate.

1. It is the primary sequence in the metabolism of glucose by all cells.

2. It is an oxidative pathway that does not require oxygen.
 a. Anaerobic glycolysis is a process that functions in the absence of oxygen.
 b. Aerobic glycolysis is a process that functions when oxygen is available.

3. Whether anaerobic or aerobic, it results in the net production of adenosine triphosphate (ATP).

B. Stages. Two stages that can be distinguished are

1. The conversion of hexose to triose phosphate. This stage involves a series of reactions that requires the expenditure of two molecules of ATP for each molecule of hexose that is split.

2. The conversion of triose phosphate to pyruvate. During this stage, two molecules of ATP are produced for each molecule of triose phosphate converted to pyruvate, or four molecules of ATP per molecule of hexose used.

III. ENZYMATIC PATHWAY OF GLYCOLYSIS

A. Glucose to glucose 6-phosphate (G6P) $\rightarrow F6P \rightarrow F1,6DP$

1. Glucose (and other hexoses) are phosphorylated immediately upon entry into the cell. In most cells, except perhaps liver cells, the concentration of free glucose is very low.

2. Hexokinase is the enzyme catalyzing the phosphorylation of glucose at the expense of ATP.

β-D-Glucose + ATP $\longrightarrow$ D-Glucose 6-phosphate + ADP

a. The reaction is essentially irreversible, and glucose is efficiently trapped inside the cell, as phosphorylated intermediates do not readily pass through cell membranes.
b. Hexokinase is an allosteric enzyme that is strongly inhibited by its product, G6P.
c. Several isoenzymes of hexokinase are found that exhibit different Michaelis constant (K_m) values for glucose and different specificities for the hexose substrates.
 (1) Most hexokinases have a low K_m for glucose and readily take up glucose and other sugars from the blood. The hexokinase isoenzyme in the brain has a particularly low K_m for glucose.
 (2) In contrast, the liver is unique in that its major enzyme for phosphorylating glucose is **glucokinase**. This enzyme is specific for glucose and has a high K_m, which enables it to handle the high concentration of glucose that is present in the portal venous blood after a meal; it is an inducible enzyme, increasing its synthesis probably in response to insulin secretion.

3. G6P is an intermediate of pivotal importance: It not only is an intermediate of the glycolytic pathway but also serves as a precursor for several other metabolic pathways, both anabolic and catabolic.

B. G6P to fructose 6-phosphate (F6P)

1. The reaction is catalyzed by **phosphoglucoisomerase**.

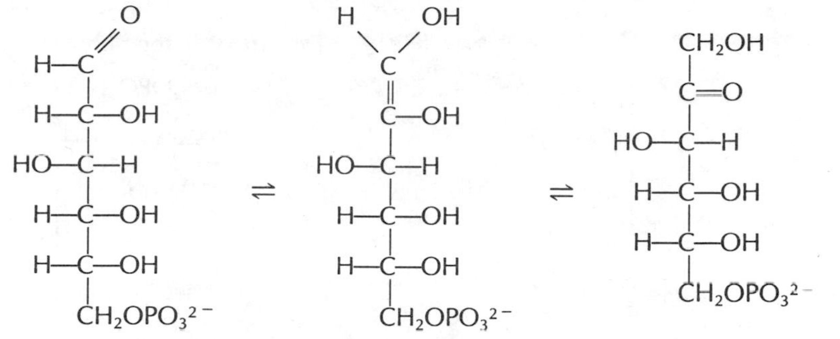

D-Glucose 6-phosphate   Enediol   D-Fructose 6-phosphate
                                  (enzyme-bound)

2. This enzymatic step prepares the first carbon (C-1) for phosphorylation.

3. It is a freely reversible reaction controlled by substrate–product levels (stoichiometric control).

C. F6P to fructose 1,6-diphosphate (F1,6DP)

1. The reaction is catalyzed by **phosphofructokinase (PFK)**.

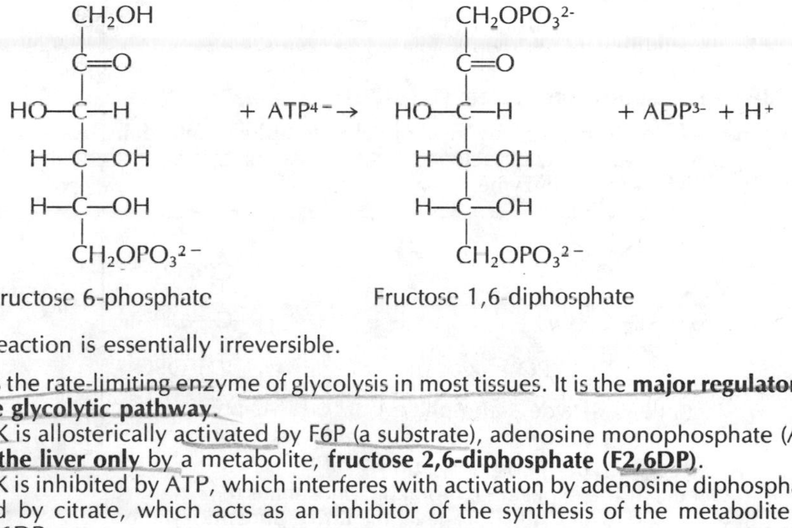

Fructose 6-phosphate   Fructose 1,6-diphosphate

2. The reaction is essentially irreversible.

3. PFK is the rate-limiting enzyme of glycolysis in most tissues. It is the **major regulatory enzyme of the glycolytic pathway.**

 a. PFK is allosterically activated by F6P (a substrate), adenosine monophosphate (AMP), and **in the liver only** by a metabolite, **fructose 2,6-diphosphate (F2,6DP)**.

 b. PFK is inhibited by ATP, which interferes with activation by adenosine diphosphate (ADP), and by citrate, which acts as an inhibitor of the synthesis of the metabolite activator, F2,6DP.

 c. The activation of PFK by AMP is logical in the sense that an active PFK will enhance the flow of metabolic fuel through glycolysis and thus provide more pyruvate for the generation of ATP by the mitochondria (see Chapter 16, section I B).

 d. Similarly, the inhibition of PFK by ATP is logical in the sense that a high cellular ATP indicates satiety of energy reserves and calls for a slowdown of glycolysis. Inhibition of the rate-limiting step of PFK achieves that.

 e. Citrate, which is formed in the mitochondria during the oxidation of 2-carbon fragments from pyruvate (see Chapter 15, section III A), becomes available in the cytoplasm to influence enzymes of the glycolytic pathway because a high level of ATP prevents its further metabolism (see Chapter 15, section III C 5).

 f. The kinase (phosphofructokinase-2) that synthesizes the PFK activator F2,6DP is itself a regulatory enzyme. It is inhibited by citrate (see section III C 3 b) and by ATP, both of which signal a reduction of glycolysis. The kinase is also inhibited by phosphorylation of the enzyme by cyclic AMP–dependent (cAMP-dependent) protein kinase.

  g. The fructose 2,6 diphosphatase (F2,6DPase) kinase is activated by F1,6DP, a product of PFK, and by dephosphorylation of the kinase.

D. F1,6DP to dihydroxyacetone phosphate (DHAP) and glyceraldehyde 3-phosphate (G3P)

  1. The first stage of glycolysis (hexose to triose phosphate) is completed by this reaction. It is catalyzed by **aldolase**.

Dihydroxyacetone phosphate

$$CH_2OPO_3^{2-}$$
$$|$$
$$C=O$$
$$|$$
$$CH_2-OH$$
$$+$$
$$H-C=O$$
$$|$$
$$H-C-OH$$
$$|$$
$$CH_2OPO_3^{2-}$$

triose phosphate isomerase

$$CH_2OPO_3^{2-}$$
$$|$$
$$C=O$$
$$|$$
$$HO-C-H \rightleftharpoons$$
$$|$$
$$H-C-OH$$
$$|$$
$$H-C-OH$$
$$|$$
$$CH_2OPO_3^{2-}$$

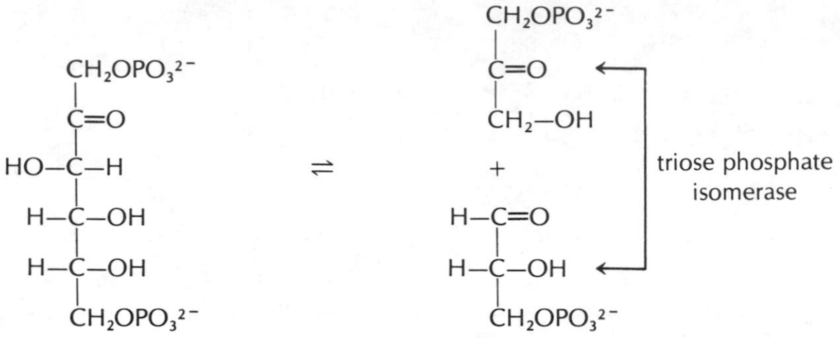

Fructose 1,6-diphosphate   Glyceraldehyde 3-phosphate

  2. It is an energetically unfavorable reaction in the direction written with a standard free-energy change ($\Delta G^{\circ\prime}$) of + 5.73 kcal, but the rapid conversion of G3P to pyruvate drives the reaction.

  3. All of the carbons of glucose can end up as pyruvate because of the equilibrium between DHAP and G3P, catalyzed by **triose phosphate isomerase**. As G3P is used by subsequent reactions of glycolysis, carbon is drawn from DHAP to form G3P.

  4. A number of isoenzymes of aldolase exist. In the rabbit, for example, muscle, liver, and brain tissues each have their own aldolase isoenzyme.

E. G3P to 1,3-diphosphoglycerate (1,3-DPG)

  1. The reaction, catalyzed by **glyceraldehyde 3-phosphate dehydrogenase (G3PD)**, requires nicotinamide-adenine dinucleotide (NAD) as an electron carrier. In its oxidized form, NAD$^+$ binds tightly to the enzyme.

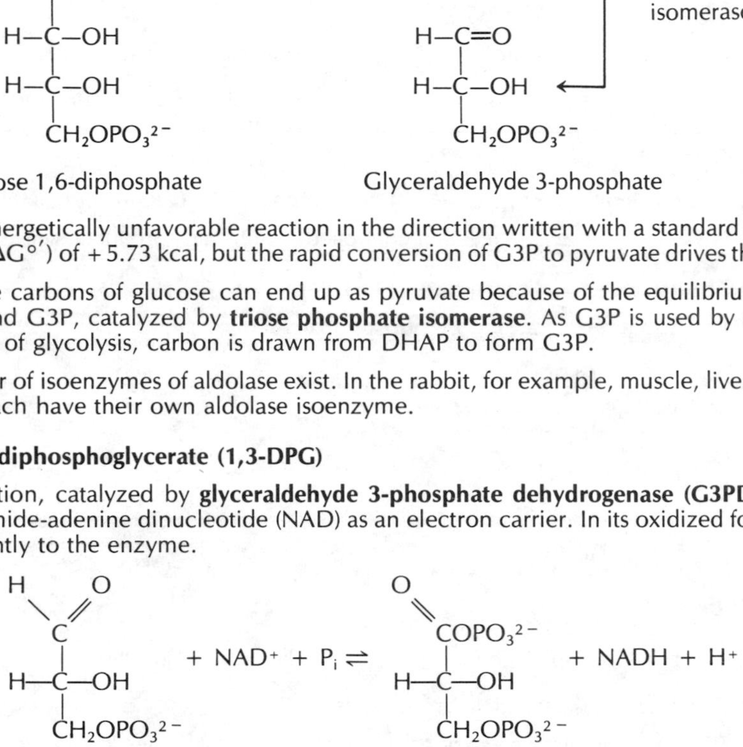

$$\begin{array}{c} H \quad O \\ \diagdown \diagup \\ C \\ | \\ H-C-OH \\ | \\ CH_2OPO_3^{2-} \end{array} + NAD^+ + P_i \rightleftharpoons \begin{array}{c} O \\ \diagup\diagup \\ COPO_3^{2-} \\ | \\ H-C-OH \\ | \\ CH_2OPO_3^{2-} \end{array} + NADH + H^+$$

Glyceraldehyde 3-phosphate   1,3-Diphosphoglycerate

  2. In the reaction, the phosphorylation
    a. Occurs at the expense of inorganic phosphate (P_i)
    b. Is an example of **substrate-level oxidative phosphorylation**

  3. The reaction generates a **high-energy phosphate bond** in 1,3-DPG, which is a mixed anhydride of phosphoric acid and a carboxylic acid (i.e., it is an **acyl phosphate**). Because of this, 1,3-DPG has a high group transfer potential.

  4. Action of G3PD

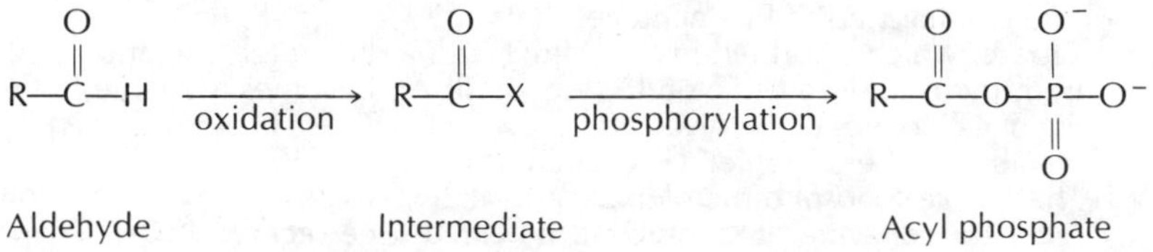

$$\begin{array}{c} O \\ \| \\ R-C-H \end{array} \xrightarrow[\text{oxidation}]{} \begin{array}{c} O \\ \| \\ R-C-X \end{array} \xrightarrow[\text{phosphorylation}]{} \begin{array}{c} O \quad O^- \\ \| \quad | \\ R-C-O-P-O^- \\ \| \\ O \end{array}$$

Aldehyde   Intermediate   Acyl phosphate

   a. The aldehyde (G3P) is first oxidized.
   - **(1)** The oxidation step involves the removal of a hydride ion ($:H^-$); this is facilitated by the addition of a nucleophile (X in the equation above).
   - **(2)** The nucleophile, X, is the sulfhydryl group of a cysteine residue on the enzyme G3PD.
   - **(3)** The acceptor for the hydride ion is NAD^+, which is tightly bound to the enzyme until its reduction to NADH, when it leaves the enzyme, to be replaced there by another molecule of NAD^+.

   b. A thioester is formed during the transfer of the hydride ion. This thioester is next phosphorylated by P_i to form 1,3-DPG.

   c. Arsenate ($AsO_4{}^{3-}$) can substitute for P_i in the phosphorylation reaction by combining with the energy-rich thioester intermediate to form an unstable compound, 1-arseno-3-phosphoglycerate. This compound breaks down to yield 3-phosphoglycerate (3-PG), which is an intermediate of the glycolytic pathway but is not phosphorylated. Thus, $AsO_4{}^{3-}$ uncouples oxidation and phosphorylation at this step.

5. Replacing the NADH formed in the G3PD reaction by a molecule of NAD^+ requires the continuous reoxidation of NADH in order to keep glycolysis going.

   a. Under anaerobic conditions, this is accomplished by the conversion of pyruvate to lactate (see section III J)

   b. Under aerobic conditions, oxidation of the NADH and the generation of ATP by mitochondrial systems are achieved via the glycerol 3-phosphate shuttle (see section IV B 3).

F. 1,3-DPG to 3-PG

1. The reaction is catalyzed by **phosphoglycerate kinase**.

$$\text{1,3-Diphosphoglycerate} + ADP^{3-} \rightleftharpoons \text{3-Phosphoglycerate} + ATP^{4-}$$

2. This is the first step in glycolysis that generates ATP. It is another example of a substrate level oxidative phosphorylation.

   a. In the prior steps, two molecules of 1,3-DPG were formed from each molecule of glucose. Therefore, two ATP molecules are now formed per original molecule of glucose.

   b. Because up to triose formation (section III D), two molecules of ATP have been used per molecule of glucose consumed, the balance sheet for ATP use and formation is even at this step.

G. 3-PG to 2-PG

1. The reaction is catalyzed by **phosphoglycerate mutase**.

3-Phosphoglycerate $\underset{Mg^{2+}}{\longleftrightarrow}$ 2-Phosphoglycerate

2. This reversible reaction has a $\Delta G^{\circ\prime}$ of $+1.1$.

3. The reaction requires trace concentrations of **2,3-diphosphoglycerate (2,3-DPG)** as a cofactor.

   a. In most cells, 2,3-DPG is present in low concentration.

   b. By contrast, the concentration of 2,3-DPG is about 4 mM (equal in molarity to hemoglobin) in red blood cells, where 2,3-DPG acts as a regulator of oxygen transport, stabilizing the deoxygenated form of hemoglobin (see Chapter 10, section I C 3 c).

4. Formation and degradation of 2,3-DPG in the red cell

   a. 2,3-DPG is formed from 1,3-DPG by a mutase and is degraded to 3-PG by a phosphatase.

$$
\begin{array}{ccc}
\overset{\displaystyle O}{\underset{\displaystyle |}{\overset{\displaystyle \|}{C}OPO_3{}^{2-}}} & \overset{\displaystyle COO^-}{\underset{\displaystyle |}{|}} & \overset{\displaystyle COO^-}{\underset{\displaystyle |}{|}} \\
H-\overset{|}{C}-OH \xrightarrow{\ \text{mutase}\ } & H-\overset{|}{C}-OPO_3{}^{2-} \xrightarrow{\ \text{phosphatase}\ } & H-\overset{|}{C}-OH \\
\overset{|}{C}H_2OPO_3{}^{2-} & \overset{|}{C}H_2OPO_3{}^{2-} & \overset{|}{C}H_2OPO_3{}^{2-}
\end{array}
$$

1,3-Diphosphoglycerate   2,3-Diphosphoglycerate   3-Phosphoglycerate

    b. The mutase reaction requires the presence of 3-PG for the formation of a ternary com-plex—1,3-DPG, 3-PG, and enzyme—in which the transfer of the phosphoryl group occurs.
    c. 2,3-DPG is a strong competitive inhibitor of its own synthesis by the mutase.

H. 2-PG to phosphoenolpyruvate (PEP)

    1. The reaction is catalyzed by **enolase**.

$$
\begin{array}{cc}
COO^- & COO^- \\
| & | \\
H-\overset{|}{C}-OPO_3{}^{2-} \rightleftharpoons & \overset{\|}{C}-OPO_3{}^{\cdot-} + H_2O \\
| & \| \\
CH_2OH & CH_2
\end{array}
$$

2-Phosphoglycerate   Phosphoenolpyruvate

    2. It is a reversible reaction with a relatively small $\Delta G^{\circ\prime}$ but a large change in the distribution of energy.

    3. The phosphoenol bond is a **high-energy phosphate bond**.

I. PEP to pyruvate

    1. The reaction is catalyzed by **pyruvate kinase**.

$$
\begin{array}{cc}
COO^- & COO^- \\
| & | \\
\overset{\|}{C}-OPO_3{}^{2-} + ADP^{3-} \rightarrow & \overset{\|}{C}=O + ATP^{4-} \\
\| & | \\
CH_2 & CH_3
\end{array}
$$

    Phosphoenolpyruvate   Pyruvate

    2. The formation of ATP from ADP at the expense of the high-energy phosphoenol bond of PEP is another example of substrate-level oxidative phosphorylation.

    3. Two molecules of pyruvate are formed per original molecule of glucose. Thus, the net pro-duction of ATP per molecule of glucose used in the formation of pyruvate is $+2$ ($-2 + 2 + 2$).

    4. The pyruvate kinase reaction is essentially irreversible under cellular conditions ($\Delta G^{\circ\prime} = -7.5$ kcal).

    5. Pyruvate kinase is an allosteric enzyme.
      a. It is inhibited by ATP, alanine, fatty acids, and acetyl CoA.
      b. The isoenzyme found in liver is strongly activated by F1,6DP.
      c. The liver enzyme is also regulated by covalent modification (see Chapter 11, section VII G). It is phosphorylated by cAMP-dependent protein kinase and dephosphorylated by a phosphatase. It is active in the dephosphorylated form and inactive when phosphorylated.
      d. The liver enzyme is also an inducible enzyme, increasing in concentration with high car-bohydrate intake and high insulin levels.

J. Pyruvate to lactate

    1. The reaction is catalyzed by **lactate dehydrogenase**.

$$
\begin{array}{cc}
COO^- & COO^- \\
| & | \\
\overset{\|}{C}=O + NADH + H^+ \rightleftharpoons & H-\overset{|}{C}-OH + NAD^+ \\
| & | \\
CH_3 & CH_3
\end{array}
$$

    Pyruvate   Lactate

    2. The reaction is an essential step in anaerobic glycolysis, as it is the means for reoxidizing the reduced form of NAD^+, NADH, formed in the G3PD step.

IV. HIGH-ENERGY PHOSPHATE (ATP) PRODUCTION

A. Anaerobic glycolysis

    1. Two moles of ATP per mole of glucose consumed are *used* to form F1,6DP.

    2. Two moles of ATP per mole of glucose are *produced* by the phosphoglycerate kinase reaction.

    3. Two moles of ATP per mole of glucose are *produced* by the pyruvate kinase reaction.

    4. Thus, two *net* moles of ATP per mole of glucose are *produced* in anaerobic glycolysis.

B. Aerobic glycolysis

    1. Two net moles of ATP per mole of glucose are produced via substrate-level oxidative phosphorylations, as in the anaerobic state.

    2. Four moles of ATP per mole of glucose are produced via the NADH that is formed in the G3PD reaction and oxidized by means of the glycerol 3-phosphate shuttle.

    3. The **glycerol 3-phosphate shuttle** overcomes the problem of the impermeability of the inner mitochondrial membrane to NADH.
        a. In the cytosol, G3PD is an NAD^+-linked enzyme, and it catalyzes the reduction of DHAP to glycerol 3-phosphate with transfer of electrons from NADH.
        b. The glycerol 3-phosphate enters the mitochondria and is oxidized back to DHAP by the mitochondrial G3PD.
        c. The mitochondrial dehydrogenase is linked to flavin-adenine dinucleotide (FAD) as the electron acceptor, and on oxidation of the reduced FAD ($FADH_2$) formed in the reaction, the mitochondrial electron transport chain and oxidative phosphorylation can support the synthesis of two ATPs.
        d. The DHAP formed then leaves the mitochondria to continue the shuttle.

    4. Overall, 6 mol of ATP are formed per mole of glucose in aerobic glycolysis.

V. SUMMARY OF ENZYMATIC CONTROL OF GLYCOLYSIS. Three regulatory enzymes take part in glycolysis.

A. Hexokinase

    1. Is allosterically inhibited by G6P

    2. Limits glucose phosphorylation if G6P is not used

B. Phosphofructokinase

    1. Is the major regulatory enzyme in most tissues

    2. Is activated by F6P, AMP, and in the liver only, by F2,6DP

    3. Is inhibited by ATP, citrate, and H^+
        a. Glycolysis is limited by a high-cell charge or excess citrate over that used in fatty acid synthesis (see Chapter 20, section III B 3).
        b. Inhibition of PFK is relieved by ADP, AMP, and P_i.

C. Phosphofructokinase-2

    1. Is allosterically activated by F1,6DP

    2. Is inhibited, allosterically, by ATP and by citrate. It is also inhibited by phosphorylation of the enzyme by cAMP-dependent protein kinase.

D. F2,6DPase is activated by phosphorylation of the enzyme by cAMP-dependent protein kinase.

E. Pyruvate kinase

    1. Is inhibited by ATP, alanine, fatty acids, and acetyl CoA

    2. Is present in the liver as an isoenzyme which
        a. Is strongly activated by F1,6DP

      b. Is regulated via phosphorylation and dephosphorylation, by cAMP-dependent protein kinase and a phosphatase, respectively

      c. Is inactive when phosphorylated

VI. DISEASES ASSOCIATED WITH IMPAIRED GLYCOLYSIS

A. Hexokinase deficiency

1. In patients with inherited defects of hexokinase activity, the red blood cells contain low concentrations of the glycolytic intermediates, including the precursor of 2,3-DPG.

2. In consequence, the hemoglobin of these patients has an abnormally high oxygen affinity.

3. The oxygen saturation curves of red blood cells from a patient with hexokinase deficiency are shifted to the left, which indicates that oxygen is less available for the tissues.

B. Pyruvate kinase deficiency (hemolytic anemia)

1. All red blood cells are completely dependent upon glycolytic activity for ATP production.

2. Failure of the pyruvate kinase reaction (step I of the glycolytic pathway) drastically impedes the production of ATP.

3. Inadequate production of ATP reduces the activity of the Na^+- and K^+-stimulated ATPase ion pump, which maintains the shape of the red blood cell membrane. In consequence, the cells swell and lyse, and this excess red blood cell destruction results in hemolytic anemia.

C. Lactic acidosis

1. Blood levels of lactic acid are normally less than 1.2 mM. In lactic acidosis, the values for blood lactate may be 5 mM or more.

2. The high concentration of lactate results in lowered blood pH and bicarbonate levels.

3. High blood lactate levels can result from increased formation or decreased utilization of lactate.

4. A common cause of hyperlactidemia is anoxia, in which
    a. The formation of lactate is increased because the shortage of oxygen reduces mitochondrial production of ATP with the consequent activation of PFK, causing increased glycolysis and lactate production.
    b. The use of lactate by the tissues is reduced because the use of lactate as an energy source requires oxygen (for the mitochondrial oxidation of acetyl CoA, derived from lactate, to CO_2 and water).

5. Tissue anoxia may occur in shock and other conditions that impair blood flow, in respiratory disorders, and in severe anemia.

STUDY QUESTIONS

Directions: Each question below contains five suggested answers. Choose the **one best** response to each question.

1. All of the following statements apply to the digestive enzyme α-amylase EXCEPT

(A) glycogen and hydrated starch are normal substrates
(B) the form of α-amylase in the human pancreas is the most important isozyme
(C) it catalyzes the hydrolysis of α-1,4-glycosidic linkages except those of glucose units that serve as branch points
(D) glucose is the major product of α-amylase action on starch
(E) cellulose, a plant carbohydrate, is not hydrolyzed by α-amylase in man

2. In anaerobic glycolysis, 2 mol of inorganic phosphate are used per mole of glucose consumed. Which of the following enzymes catalyzes the uptake of inorganic phosphate?

(A) Hexokinase
(B) Phosphofructokinase
(C) Pyruvate kinase
(D) Glyceraldehyde 3-phosphate dehydrogenase
(E) Enolase

3. Which of the following statements best describes 2,3-diphosphoglycerate (2,3-DPG)?

(A) It is only found in red blood cells
(B) It is present in red blood cells at about 4–5 mM
(C) It is formed in the liver and exported to red blood cells
(D) It is formed by the phosphorylation of 2-phosphoglycerate
(E) It increases the affinity of hemoglobin for oxygen

4. Which of the following enzyme-catalyzed reactions has a product containing a newly formed high-energy phosphate bond?

(A) The phosphorylation of glucose
(B) 2-Phosphoglycerate to phosphoenolpyruvate
(C) 3-Phosphoglycerate to 2-phosphoglycerate
(D) Dihydroxyacetone phosphate to glyceraldehyde 3-phosphate
(E) Fructose 1,6-diphosphate to glyceraldehyde 3-phosphate and dihydroxyacetone phosphate.

5. The oxidation of 1 mol of glucose by anaerobic glycolysis yields a net of

(A) 2 mol of lactate and 2 mol of adenosine triphosphate (ATP)
(B) 2 mol of lactate, 2 mol of reduced nicotinamide-adenine dinucleotide (NADH), and 2 mol of ATP
(C) 2 mol of lactate, 2 mol of NAD^+, and 6 mol of ATP
(D) 2 mol of pyruvate and 2 mol of ATP
(E) 2 mol of pyruvate, 2 mol of NADH, and 2 mol of ATP

6. What is the net production of high-energy phosphate bonds in aerobic glycolysis?

(A) 4
(B) 6
(C) 8
(D) 10
(E) 12

7. The glycolytic pathway requires which of the following as allosteric regulatory enzymes?

(A) Glucokinase, phosphofructokinase, and pyruvate kinase
(B) Hexokinase, aldolase, and pyruvate kinase
(C) Hexokinase, glyceraldehyde 3-phosphate dehydrogenase, and enolase
(D) Phosphofructokinase, enolase, and pyruvate kinase
(E) Hexokinase, phosphofructokinase, and pyruvate kinase

Directions: Each question below contains four suggested answers of which **one or more** is correct. Choose the answer

A   if **1, 2, and 3** are correct
B   if **1 and 3** are correct
C   if **2 and 4** are correct
D   if **4** is correct
E   if **1, 2, 3, and 4** are correct

8. The transport of D-glucose into the epithelial cells of the small intestine involves

(1) Adenosine triphosphate (ATP)–linked extrusion of Na^+ across the serosal membrane
(2) phosphorylation of glucose by hexokinase
(3) Na^+ on the mucosal surface for cotransport
(4) ATP activation of the system by its binding to a carrier protein for D-glucose

9. Glucokinase is characterized by which of the following statements?

(1) It has a much higher Michaelis constant (K_m) for glucose than hexokinase
(2) It phosphorylates most hexoses
(3) It is an inducible enzyme
(4) It is found in most cells

10. Phosphofructokinase is characterized by which of the following statements?

(1) It is a major regulatory enzyme in glycolysis
(2) Adenosine triphosphate (ATP) is a substrate of the enzyme
(3) ATP is a negative modulator of the enzyme
(4) Citrate is a positive modulator of the enzyme

ANSWERS AND EXPLANATIONS

1. The answer is D. (*I B 1; Chapter 13 V E*) α-Amylase hydrolyzes α-1,4-glycosidic linkages of hydrated starch and glycogen except those of glucose units that serve as branch points. (Cooking enhances the hydration, and thus the hydrolysis, of starch.) The products of α-amylase action are maltose, maltotriose, and limit dextrins (composed of about eight glucose units with one or more α-1,6 linkages). The isozyme found in the human pancreas is more important in carbohydrate digestion than the salivary enzyme. Cellulose, as found in vegetables and other plant foods, is not digested by human enzyme systems.

2. The answer is D. (*III E*) The oxidation of glyceraldehyde 3-phosphate by oxidized nicotinamide-adenine dinucleotide–linked glyceraldehyde 3-phosphate dehydrogenase involves a concomitant phosphorylation to form 1,3-diphosphoglycerate. This phosphorylation is carried out at the expense of inorganic phosphate (P_i). As 2 mol of triose are formed per mole of glucose used in the glycolytic pathway, 2 mol of P_i are consumed per mole of glucose used.

3. The answer is B. (*III G 3–4*) 2,3-Diphosphoglycerate is required in trace amounts for glycolysis in all cells. It is a cofactor in the conversion of 3-phosphoglycerate to 2-phosphoglycerate catalyzed by phosphoglycerate mutase. In the red blood cell, it is present in high concentration (4–5 mM) and acts as a negative modulator of the oxygen-carrying capacity of hemoglobin. It is formed from 1,3-diphosphoglycerate by a mutase.

4. The answer is B. (*III H*) High-energy phosphate bonds are formed at two points in the glycolytic pathway. The first compound formed with a new high-energy phosphate bond is 1,3-diphosphoglycerate, a mixed anhydride and a product of glyceraldehyde 3-phosphate dehydrogenase catalysis. The second new high-energy phosphate bond is found in phosphoenolpyruvate, a product of enolase catalysis. The phosphate bonds in glucose 6-phosphate, 2- and 3-phosphoglycerates, glyceraldehyde 3-phosphate, and dihydroxyacetone phosphate are not high-energy phosphate bonds and do not exhibit a large negative free-energy change upon hydrolysis.

5. The answer is A. (*IV A*) Under anaerobic conditions, the pyruvate formed from glucose by glycolysis is converted to lactate. There is concomitant oxidation of reduced nicotinamide-adenine dinucleotide formed at the glyceraldehyde 3-phosphate dehydrogenase step, to nicotinamide-adenine dinucleotide (NAD^+), thus permitting the continuation of the pathway. Under these conditions, 1 mol of glucose yields 2 mol of lactate and a net of 2 mol of adenosine triphosphate.

6. The answer is B. (*IV B*) In aerobic glycolysis, the reduced nicotinamide-adenine dinucleotide (NADH) formed at the glyceraldehyde 3-phosphate dehydrogenase step is oxidized by the glycerol 3-phosphate shuttle. In this reaction, the NADH reduces dihydroxyacetone phosphate (DHAP) in the cytoplasm to glycerol 3-phosphate, which can enter the mitochondria. Inside the mitochondria, there is another dehydrogenase, which is flavin-adenine dinucleotide (FAD)–linked and which converts the glycerol 3-phosphate back to DHAP, which then leaves the mitochondria. The mitochondrial $FADH_2$ formed can support the formation of 2 mol of adenosine triphosphate (ATP) per mole of DHAP formed, or 4 mol per mole of glucose used. As the two substrate-level phosphorylations of glycolysis also take place aerobically, a total of 6 mol of ATP are formed per mole of glucose used.

7. The answer is E. (*V*) There are three allosteric regulatory enzymes in the glycolytic pathway. Hexokinase is inhibited by glucose 6-phosphate, but the main liver enzyme with the same function, glucokinase, is not a regulatory enzyme. Phosphofructokinase is allosterically inhibited by adenosine triphosphate, citrate, or H^+, and activated by adenosine diphosphate and monophosphate or inorganic phosphate. Pyruvate kinase is activated by fructose 1,6-diphosphate.

8. The answer is B (1, 3). (*I C*) D-Glucose is transported into the epithelial cells of the small intestine (and into other cells) by a Na^+-monosaccharide cotransport system. Na^+ enters from the mucosal surface, down the concentration gradient, and there is an obligatory cotransport of D-glucose (or D-galactose) into the cell. Na^+ is pumped out of the cell on the serosal side by the adenosine triphosphate–dependent Na^+ pump.

9. The answer is B (1, 3). (*III A 2 c*) Glucokinase and hexokinase both catalyze the phosphorylation of glucose to glucose 6-phosphate. Most hexokinases have a low Michaelis constant (K_m) for glucose and are not specific for glucose. On the other hand, glucokinase has a high K_m for glucose and is very specific for that sugar. The high K_m allows glucokinase to handle the high blood glucose levels found in the portal vein following a meal. Glucokinase is an inducible enzyme, increasing its concentration probably in response to insulin, and is only found in the liver.

10. The answer is A (1, 2, 3). *(III C; V B)* Phosphofructokinase is an allosteric enzyme catalyzing a virtually irreversible step in glycolysis and is the major control enzyme of this pathway in most tissues. Adenosine triphosphate (ATP) is one of its substrates, fructose 6-phosphate being the other. ATP is also a negative modulator of the enzyme, as is citrate.

15

Tricarboxylic Acid Cycle

I. OVERVIEW

A. Description

1. The tricarboxylic acid (TCA) cycle (Fig. 15-1), also known as the **citric acid cycle** or the **Krebs cycle**, is a cyclic series of enzymatically catalyzed reactions carried out by a multienzyme system.

2. It oxidizes the acetyl group of acetyl coenzyme A (acetyl CoA) to carbon dioxide (CO_2), reduced nicotinamide-adenine dinucleotide (NADH), hydrogen ion (H^+), and reduced flavin-adenine dinucleotide ($FADH_2$), using oxidized nicotinamide-adenine dinucleotide (NAD^+) and oxidized flavin-adenine dinucleotide (FAD) as electron acceptors.

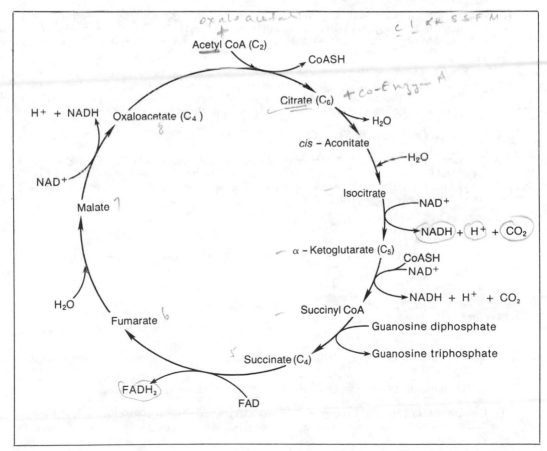

Figure 15-1. The tricarboxylic acid cycle (TCA, citric acid, or Krebs cycle). The overall reaction catalyzed by the cycle is $CH_3COOH + 2\ H_2O \rightarrow 2\ CO_2 + 8\ H^+$.

B. Functions

1. The TCA cycle is an **amphibolic** pathway; that is, it is involved in both anabolic and catabolic processes.
   a. **Anabolic process.** The TCA cycle provides routes for the synthesis of biosynthetic precursors.

   α-Ketoglutarate + alanine $\rightleftharpoons$ glutamate + pyruvate
   Oxaloacetate + alanine $\rightleftharpoons$ aspartate + pyruvate
   Succinyl CoA $\rightarrow$ heme $\rightarrow$ hemoglobin, myoglobin, and cytochromes

   b. **Catabolic process.** The TCA cycle provides the degradation of two-carbon acetyl residues derived from carbohydrate, fatty acids, and amino acids.

2. The TCA cycle is a component of respiration, the energy-trapping system of aerobic cells. It forms a common pathway for the **final oxidation of all metabolic fuels** (carbohydrates, free fatty acids, ketone bodies, and amino acids), which compete for available CoA, NAD^+, and FAD.

C. Stoichiometry

1. The TCA cycle accomplishes
   a. The complete oxidation of both carbons in acetate to CO_2
   b. The generation of three molecules of NADH, three of H^+, one of $FADH_2$, and one of guanosine triphosphate (GTP); and the transfer of four pairs of electrons

2. The cycle does not produce, nor consume, any intermediate of the cycle.

D. Mitochondrial locations of the enzymes

1. The reactions of the TCA cycle occur within the inner mitochondrial membrane, in the so-called **mitochondrial matrix** (see Chapter 16, Fig. 16-1). They are thus sequestered from the reactions of glycolysis and fatty acid synthesis, which are cytosolic systems but are in competition with the β-oxidation of fatty acids.

2. Electrons from the TCA cycle and from the β-oxidation of fatty acids (see Chapter 20, section VI) are transferred to acceptors in the **electron transport chain**, which is also located within the inner mitochondrial membrane, as are the enzymes of **oxidative phosphorylation** [the formation of adenosine triphosphate (ATP) via the electron transport chain; see Chapter 16].

II. OXIDATION OF PYRUVATE TO ACETYL CoA

A. Pyruvate dehydrogenase (PDH) enzyme complex.
Pyruvate, whether formed by glycolysis, from lactate, or by transamination of alanine, can be oxidized to acetyl CoA and CO_2 by the **PDH enzyme complex**, a highly integrated multienzyme complex. This irreversible step provides a link between glycolysis and the TCA cycle.

$$
\begin{array}{l}
COO^- \\
| \\
C{=}O + NAD^+ + CoASH \rightarrow \\
| \\
CH_3
\end{array}
\qquad
\begin{array}{l}
O \\
\| \\
C{-}S{-}CoA + CO_2 + NADH + H^+ \\
| \\
CH_3
\end{array}
$$

Pyruvate   Acetyl coenzyme A

1. Note that although this reaction takes place within the mitochondrial matrix, it is not part of the TCA cycle.

2. The standard free-energy change ($\Delta G^{\circ\prime}$) of the reaction is -8.0 kcal/mol.

B. Components of the PDH complex.
There are four enzymatic activities concerned with the formation of acetyl CoA from pyruvate. Figure 15-2 summarizes the reactions catalyzed by the PDH complex.

1. **Pyruvate decarboxylase**
   a. The reaction requires thiamine pyrophosphate (TPP) as coenzyme. TPP is formed from vitamin B_1 (thiamine) by transfer of a pyrophosphate group (PP_i) from ATP.

b. Pyruvate decarboxylase (E_1 in the displayed equation below) catalyzes the decarboxylation of pyruvate with the formation of an α-hydroxyethyl derivative of TPP.

$$E_1\text{-TPP} + \overset{\displaystyle COO^-}{\underset{\displaystyle CH_3}{C}=O} + H^+ \rightarrow E_1\text{-TPP}\overset{\displaystyle OH}{—CH—}CH_3 + CO_2$$

2. PDH next catalyzes the dehydrogenation of the α-hydroxyethyl group to form an acetyl group.

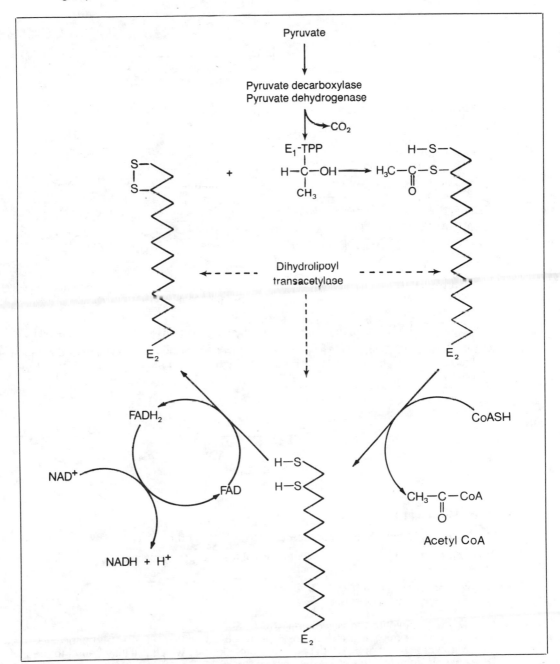

Figure 15-2. Summary of the reactions catalyzed by the pyruvate dehydrogenase (PDH) complex. The "zig-zag" structures represent the carbon chain of lipoamide. E_1 = pyruvate decarboxylase and E_2 = dihydrolipoyl transacetylase.

3. Dihydrolipoyl transacetylase catalyzes the next steps.
   a. The acetyl group is first transferred to the lipoic acid prosthetic group of dihydrolipoyl transacetylase. The prosthetic group, which serves as a coenzyme, is known as **lipoamide**.

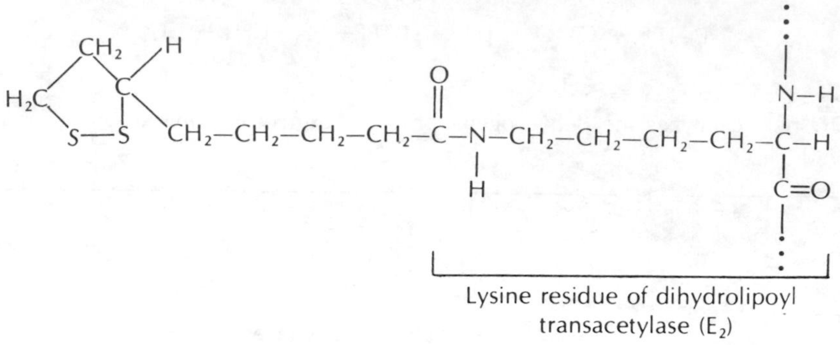

Lysine residue of dihydrolipoyl
transacetylase (E_2)

   b. The reaction yields acetyl lipoamide–E_2.

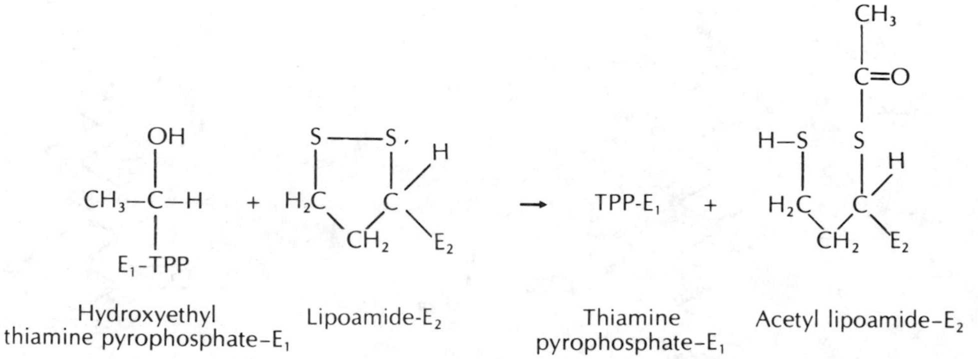

Hydroxyethyl   Lipoamide–E_2 Thiamine   Acetyl lipoamide–E_2
thiamine pyrophosphate–E_1 pyrophosphate–E_1

   c. The acetyl group is then transferred by dihydrolipoyl transacetylase to form acetyl CoA.

4. Dihydrolipoyl dehydrogenase has FAD as its prosthetic group. This enzyme oxidizes the reduced form of lipoamide. $FADH_2$ is oxidized by transfer of the electrons to NAD^+ to form NADH.

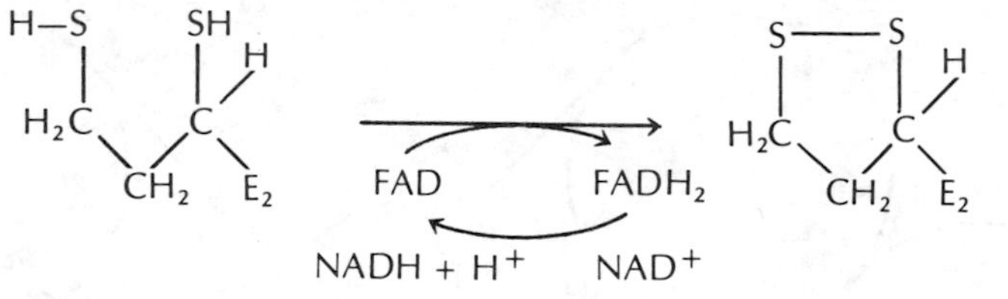

C. Control of the PDH complex

   1. Product inhibition. Both acetyl CoA and NADH inhibit the PDH complex.

   2. Acceptor control. Adequate concentrations of the acceptor molecules CoA and NAD^+ must be present for the complex to function.

   3. Covalent modification (Fig. 15-3)
      a. PDH exists in two forms.
         (1) Inactive, phosphorylated form
         (2) Active, dephosphorylated form
      b. A serine residue on the decarboxylase (the α-subunit) of PDH is phosphorylated by a Mg^{2+}-dependent and ATP-dependent protein kinase, which is tightly bound to the PDH enzyme complex; this yields the inactive form.
      c. Reactivation of the complex is carried out by dephosphorylation by a phosphoprotein phosphatase.

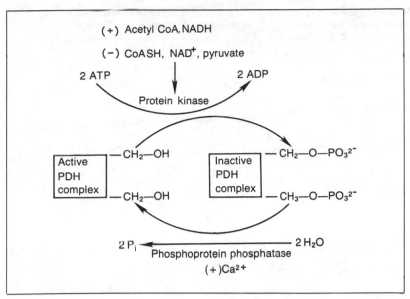

Figure 15-3. Summary of the covalent modification involved in the control of the pyruvate dehydrogenase (PDH) complex.

   d. Regulation of the complex is due to a differential regulation of the kinase and the phosphatase.

     (1) Acetyl CoA and NADH stimulate the protein kinase reaction, while free CoA (CoASH) and NAD^+ inhibit it, as will pyruvate, the substrate of the PDH complex.

     (2) The phosphoprotein phosphatase is activated by increasing calcium concentrations in the mitochondria that occur when the system is short of ATP.

   e. Insulin, by an unknown mechanism, can activate PDH in adipose tissue, as can catecholamines in cardiac muscle.

III. REACTIONS OF THE TCA CYCLE

  A. Citrate synthetase

    1. The TCA cycle proper begins with the formation of citrate via the condensation of a four-carbon unit, **oxaloacetate**, with a two-carbon unit, the acetyl group of **acetyl CoA**.

Acetyl coenzyme A + oxaloacetate → Citrate + Coenzyme A

    2. The reaction has a $\Delta G^{\circ\prime}$ of -7.7 kcal/mol.

    3. Citrate synthetase can react with monofluoroacetyl CoA to form monofluorocitrate, a potent inhibitor of the next step in the cycle.

4. In vitro studies suggest that this enzyme is a regulatory step with ATP, NADH, succinyl CoA, and long-chain fatty acyl CoAs serving as inhibitors, but these effects may not hold true under physiologic conditions.

B. Aconitase

1. Citrate is converted to isocitrate in the following reaction:

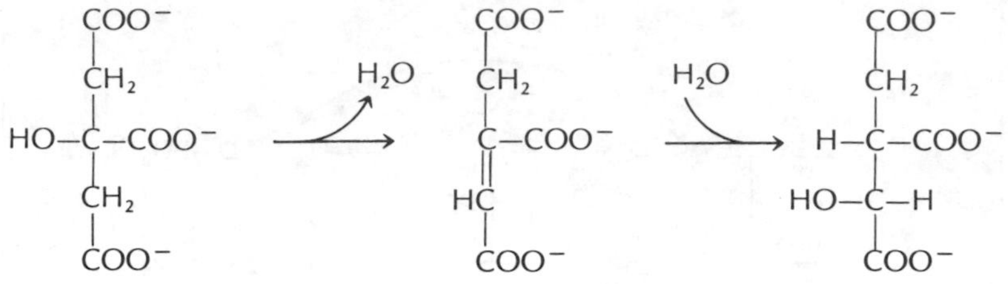

          Citrate   *cis*-Aconitate   Isocitrate

2. Removal of the isocitrate by the next step in the cycle pulls the reaction to the right.

3. The enzyme requires Fe^{2+}.

C. Isocitrate dehydrogenase

1. This step is the first of four oxidation–reduction reactions; in it, the first of two carbons are lost as CO_2, and the first of three NADH molecules are produced.

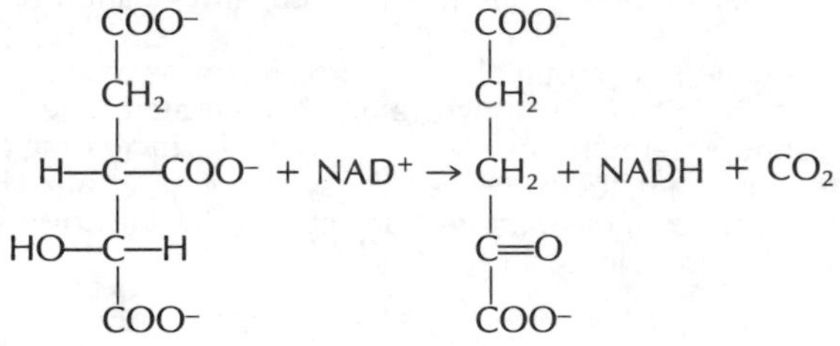

          Isocitrate α-Ketoglutarate

2. The $\Delta G°'$ is -5.0 kcal/mol.

3. As the mitochondrial enzyme involved in the TCA cycle in mammalian tissues, this form of isocitrate dehydrogenase requires NAD^+ as the acceptor of reducing equivalents.

4. Another isocitrate dehydrogenase of mammalian tissues requires $NADP^+$ as the acceptor; found in both mitochondria and cytosol, its role is in biosynthesis auxiliary to the TCA cycle.

5. The NAD^+-linked enzyme is stimulated by adenosine diphosphate (ADP) and is inhibited by ATP and NADH. Thus, under a high-energy charge (i.e., with high ATP:ADP + P_i or NADH:NAD^+ ratios), the isocitrate dehydrogenase is inhibited, and it is stimulated under conditions of low-energy charge.

D. α-Ketoglutarate dehydrogenase

1. The second oxidation–reduction reaction shows the second of two carbons lost as CO_2, and the second of three NADH molecules formed.

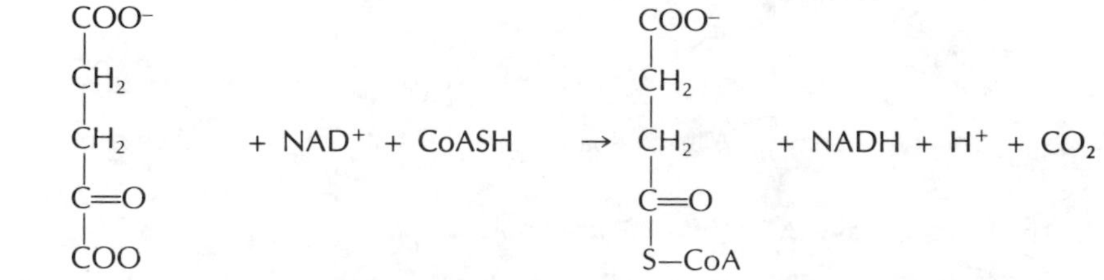

   α-Ketoglutarate   Coenzyme A   Succinyl coenzyme A

2. The $\Delta G^{\circ\prime}$ is -8.0 kcal/mol.

3. With TPP, lipoic acid, CoASH, FAD, and NAD^+ participating in the reaction, this is an enzyme complex very similar to the PDH complex, but with several differences.

   a. Decarboxylase and dehydrogenase activities appear to be on the same polypeptide chain.

   b. The complex is *not* regulated by phosphorylation.

4. The complex is inhibited by ATP, GTP, NADH, succinyl CoA, and Ca^{2+}.

5. The product, succinyl CoA, is an energy-rich thioester compound similar to acetyl CoA.

E. Succinyl CoA synthetase

1. Succinyl CoA is converted to succinate with the energy-rich character of the thioester being preserved by a substrate-level phosphorylation of guanosine diphosphate (GDP) to GTP.

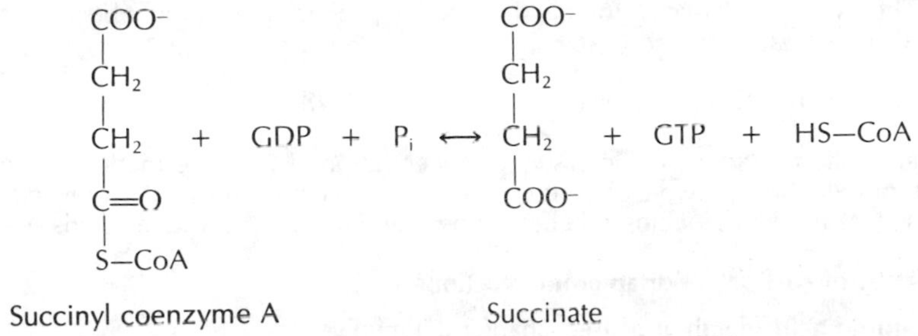

$$
\begin{array}{c}
COO^- \\
| \\
CH_2 \\
| \\
CH_2 \\
| \\
C{=}O \\
| \\
S{-}CoA
\end{array}
\;+\; GDP \;+\; P_i \;\longleftrightarrow\;
\begin{array}{c}
COO^- \\
| \\
CH_2 \\
| \\
CH_2 \\
| \\
COO^-
\end{array}
\;+\; GTP \;+\; HS{-}CoA
$$

Succinyl coenzyme A   Succinate

2. The reaction is freely reversible with a $\Delta G^{\circ\prime}$ of -0.7 kcal/mol.

3. The GTP formed is converted to ATP by nucleoside diphosphate kinase.

F. Succinate dehydrogenase

1. This is the third oxidation–reduction reaction and the site of formation of the only $FADH_2$ produced in the TCA cycle.

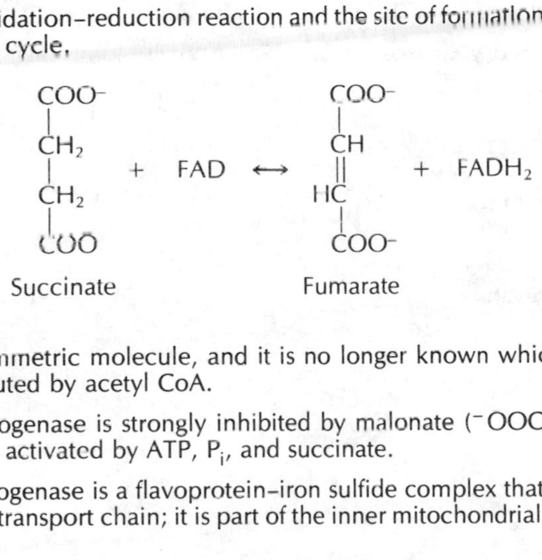

$$
\begin{array}{c}
COO^- \\
| \\
CH_2 \\
| \\
CH_2 \\
| \\
COO
\end{array}
\;+\; FAD \;\longleftrightarrow\;
\begin{array}{c}
COO^- \\
| \\
CH \\
|| \\
HC \\
| \\
COO^-
\end{array}
\;+\; FADH_2
$$

Succinate   Fumarate

2. Succinate is a symmetric molecule, and it is no longer known which carboxyl carbon was originally contributed by acetyl CoA.

3. Succinate dehydrogenase is strongly inhibited by malonate ($^-OOC{-}CH_2{-}COO^-$) and by oxaloacetate. It is activated by ATP, P_i, and succinate.

4. Succinate dehydrogenase is a flavoprotein–iron sulfide complex that is intimately associated with the electron transport chain; it is part of the inner mitochondrial membrane and is not in the matrix.

G. Fumarate hydratase catalyzes the reversible hydration of fumarate to L-malate.

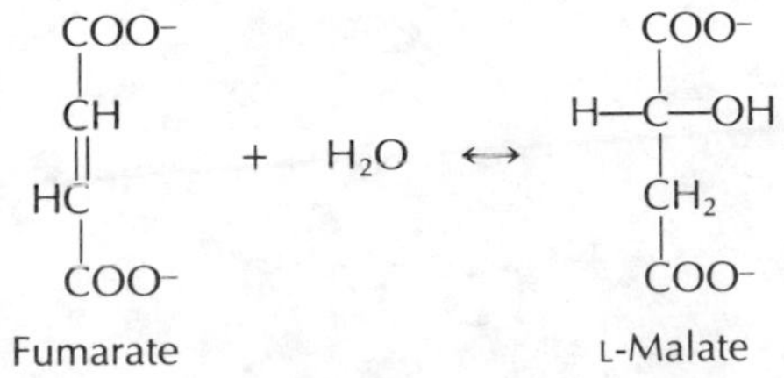

$$
\begin{array}{c}
COO^- \\
| \\
CH \\
|| \\
HC \\
| \\
COO^-
\end{array}
\;+\; H_2O \;\longleftrightarrow\;
\begin{array}{c}
COO^- \\
| \\
H{-}C{-}OH \\
| \\
CH_2 \\
| \\
COO^-
\end{array}
$$

Fumarate   L-Malate

H. NAD$^+$-linked malate dehydrogenase

1. The final reaction of the TCA cycle is the fourth oxidation–reduction reaction and the third formation of an NADH.

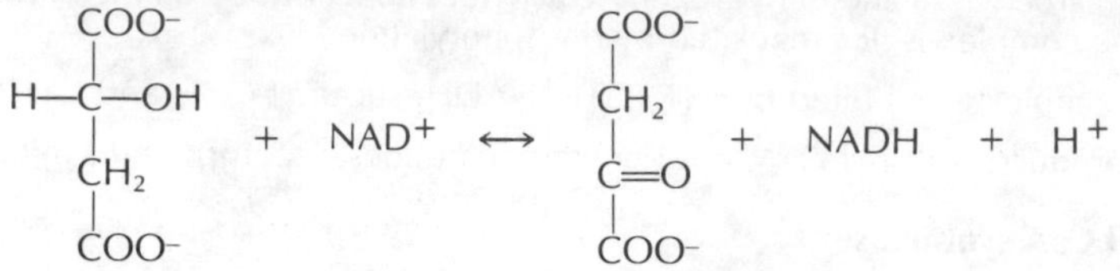

$$
\begin{array}{c}
\text{COO}^- \\
| \\
\text{H—C—OH} \\
| \\
\text{CH}_2 \\
| \\
\text{COO}^-
\end{array}
\; + \; \text{NAD}^+ \;\longleftrightarrow\;
\begin{array}{c}
\text{COO}^- \\
| \\
\text{CH}_2 \\
| \\
\text{C}\!=\!\text{O} \\
| \\
\text{COO}^-
\end{array}
\; + \; \text{NADH} \; + \; \text{H}^+
$$

L-Malate   Oxaloacetate

2. The $\Delta G°'$ is $+7.1$ kcal/mol, but the reaction proceeds to the right because of the rapid removal of oxaloacetate and NADH.

IV. ANAPLEROTIC AND AMPHIBOLIC REACTIONS

A. Anaplerotic reactions can increase the concentration of TCA intermediates, allowing an increased rate of oxidation of two-carbon units. The same reactions may run in reverse, draining off TCA-cycle intermediates for biosynthetic purposes, and in this case, the reactions are called amphibolic.

B. Sources of carbon for anaplerotic reactions

1. Amino acid metabolism (see Chapter 27) involves
   a. Transaminases that form α-ketoglutarate, a TCA-cycle intermediate
   b. Transaminases that form oxaloacetate, also a TCA-cycle intermediate
   c. Glutamate dehydrogenase, which also provides α-ketoglutarate

   $$\text{Glutamate} + \text{NAD}^+ + \text{H}_2\text{O} \rightleftharpoons \alpha\text{-ketoglutarate} + \text{NH}_3 + \text{NADH} + \text{H}^+$$

   d. Succinyl CoA formation from isoleucine, valine, methionine, and threonine, although these reactions would only be concerned with a filling up of the TCA-cycle intermediates

2. Pyruvate carboxylase (see Chapter 17, II C 2) also forms oxaloacetate.

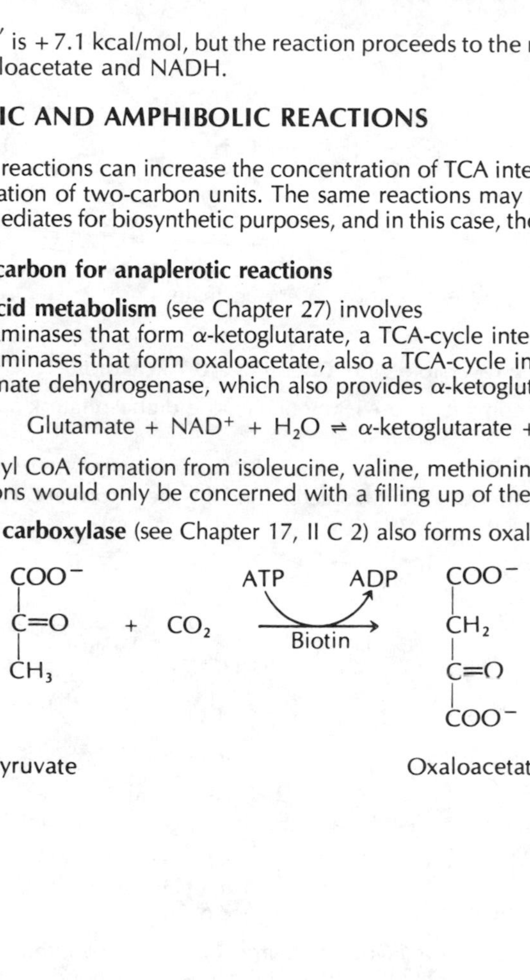

$$
\begin{array}{c}
\text{COO}^- \\
| \\
\text{C}\!=\!\text{O} \\
| \\
\text{CH}_3
\end{array}
\; + \; \text{CO}_2
\quad\xrightarrow[\text{Biotin}]{\text{ATP} \quad \text{ADP}}\quad
\begin{array}{c}
\text{COO}^- \\
| \\
\text{CH}_2 \\
| \\
\text{C}\!=\!\text{O} \\
| \\
\text{COO}^-
\end{array}
$$

Pyruvate   Oxaloacetate

STUDY QUESTIONS

Directions: Each question below contains five suggested answers. Choose the **one best** response to each question.

1. The function of the tricarboxylic acid cycle is characterized by all of the following statements EXCEPT

(A) it generates reduced nicotinamide-adenine dinucleotide and reduced flavin-adenine dinucleotide
(B) it generates guanosine triphosphate
(C) it catalyzes the complete oxidation of acetate to carbon dioxide and water
(D) it provides for the net synthesis of oxaloacetate from acetyl CoA
(E) it cannot function in the absence of oxygen

2. If 1 mol of acetyl coenzyme A labeled with ^{14}C in both carbon positions of the acetate moiety goes through one turn of the tricarboxylic acid cycle, how many moles of $^{14}CO_2$ would be produced?

(A) 0
(B) 1
(C) 2
(D) 4
(E) 6

3. All of the following compounds are intermediates of the tricarboxylic acid cycle EXCEPT for

(A) isocitrate
(B) malate
(C) oxaloacetate
(D) pyruvate
(E) succinate

4. Which of the reactions of the tricarboxylic acid cycle listed below results in the formation of a high-energy phosphate compound?

(A) Isocitrate dehydrogenase
(B) Succinyl coenzyme A synthetase
(C) Succinate dehydrogenase
(D) Citrate synthetase
(E) None of the above reactions

5. All of the enzymes of the tricarboxylic acid cycle are located in the mitochondrial matrix EXCEPT for

(A) citrate synthetase
(B) α-ketoglutarate dehydrogenase
(C) succinate dehydrogenase
(D) fumarase
(E) malate dehydrogenase

6. The enzyme that catalyzes an anaplerotic reaction in the tricarboxylic acid cycle is

(A) succinate dehydrogenase
(B) citrate lyase
(C) citrate synthetase
(D) pyruvate dehydrogenase
(E) pyruvate carboxylase

Directions: Each question below contains four suggested answers of which **one or more** is correct. Choose the answer

A   if **1, 2, and 3** are correct
B   if **1 and 3** are correct
C   if **2 and 4** are correct
D   if **4** is correct
E   if **1, 2, 3, and 4** are correct

7. The pyruvate dehydrogenase complex contains which of the following compounds?

(1) Biotin
(2) Thiamine pyrophosphate
(3) Pyridoxal phosphate
(4) Nicotinamide-adenine dinucleotide

8. Pyruvate dehydrogenase activity is regulated by

(1) covalent modification
(2) acceptor control
(3) product inhibition
(4) insulin

9. The release of carbon dioxide results from which of the following reactions in the tricarboxylic acid cycle?

(1) Isocitrate to α-ketoglutarate
(2) Malate to oxaloacetate
(3) α-Ketoglutarate to succinyl coenzyme A
(4) Succinate to fumarate

10. The isocitrate dehydrogenase–catalyzed reaction in the tricarboxylic acid cycle is

(1) inhibited by increased levels of adenosine triphosphate
(2) inhibited by increased levels of reduced nicotinamide-adenine dinucleotide
(3) stimulated by increased levels of adenosine diphosphate
(4) stimulated by a high-energy charge

ANSWERS AND EXPLANATIONS

1. The answer is D. (*I B; III*) The tricarboxylic acid (TCA) cycle oxidizes the acetyl group of acetyl coenzyme A (acetyl CoA) to carbon dioxide (CO_2) and transfers electrons to nicotinamide-adenine dinucleotide (NAD^+) and flavin-adenine dinucleotide (FAD) to form the reduced forms (NADH and $FADH_2$). At one step in the cycle pathway, at the conversion of succinyl CoA to succinate catalyzed by succinyl CoA synthetase, guanosine diphosphate is phosphorylated to guanosine triphosphate. This is the only step in the cycle where a high-energy phosphate bond is formed. The TCA cycle does not result in a net synthesis of any of the intermediates of the cycle as two carbons are injected at every turn of the cycle, and two carbons are lost as CO_2. The TCA cycle depends upon the reoxidation of the reduced electron carriers in order to continue cycling, and this in turn depends upon the electron transport chain being able to transfer electrons from the reduced electron carriers to molecular oxygen. Without oxygen, the electron transport chain cannot function, and the TCA cycle stops.

2. The answer is A. (*I C*) Citrate synthetase catalyzes the interaction of acetyl coenzyme A (acetyl CoA) and oxaloacetate to form citrate. If both acetate carbons of the acetyl CoA are labeled with ^{14}C, then one carboxyl and an adjacent methylene carbon of the citrate will be labeled with ^{14}C. In one turn of the tricarboxylic acid cycle, neither of the labeled carbons is lost, and when succinate is formed, the symmetry of the molecule ensures that there is no difference between the labeled and unlabeled portions of the citrate molecule in the next turn of the cycle.

3. The answer is D. (*II A; III*) The intermediates of the tricarboxylic acid (TCA) cycle are citrate, isocitrate, α-ketoglutarate, succinyl coenzyme A (succinyl CoA), succinate, fumarate, malate, and oxaloacetate. Pyruvate is not an intermediate of the TCA cycle but is a major source of acetyl CoA derived from carbohydrate and used for oxidation in the TCA cycle.

4. The answer is B. (*III E*) All four enzymes listed catalyze reactions of the tricarboxylic acid (TCA) cycle. Only one reaction in the TCA cycle generates a high-energy phosphate compound, namely, the reaction whereby succinyl coenzyme A (succinyl CoA) is converted to succinate catalyzed by succinyl CoA synthetase. This reaction is coupled to the phosphorylation of guanosine diphosphate to the triphosphate.

5. The answer is C. (*III F 4*) The enzymes of the tricarboxylic acid cycle are located in the mitochondrial matrix (i.e., they are confined within the inner mitochondrial membrane) with the exception of succinate dehydrogenase. This is a flavin-adenine dinucleotide–linked enzyme, which is embedded in the inner mitochondrial membrane.   FAD

6. The answer is E. (*IV B*) Succinate dehydrogenase and citrate synthetase are enzymes of the tricarboxylic acid (TCA) cycle, and in this pathway, there is no reaction providing a net synthesis of TCA intermediates. Reactions that can increase the concentration of TCA-cycle intermediates are called anaplerotic (filling-up) reactions. One of these, catalyzed by pyruvate carboxylase, is the carboxylation of pyruvate to oxaloacetate, an intermediate of the TCA cycle. Citrate lyase is a cytosolic enzyme that cleaves citrate to oxaloacetate and acetyl coenzyme A (acetyl CoA). Pyruvate dehydrogenase provides the TCA cycle with acetyl CoA but does not increase the level of TCA-cycle intermediates.

7. The answer is C (2, 4). (*II B*) The pyruvate dehydrogenase (PDH) complex contains four enzymatic activities and four associated coenzymes: thiamine pyrophosphate, which acts as a carrier of "active acetate"; lipoamide, to which the acetyl group is transferred; flavin-adenine dinucleotide (FAD), which is the electron acceptor for the reoxidation of the reduced form of lipoamide; and nicotinamide-adenine dinucleotide (NAD^+), which is the electron acceptor for the reoxidation of the reduced form of FAD. Biotin is not a cofactor in this enzyme complex but is the carrier of carbon dioxide in carboxylation reactions. Pyridoxal phosphate is also not a cofactor in the PDH complex but is widely used as a cofactor in reactions involving amino acids.

8. The answer is E (all). (*II C*) The pyruvate dehydrogenase (PDH) complex is regulated by a number of mechanisms. The products of the reaction, acetyl coenzyme A (acetyl CoA) and reduced nicotinamide-adenine dinucleotide (NADH), both inhibit the complex (product inhibition). The concentrations of the cofactors CoA and oxidized nicotinamide-adenine dinucleotide limit the activity of the complex, which has to compete with other oxidative systems, such as the β-oxidation of fatty acids, for these cofactors. This form of control is called acceptor control. PDH exists in two forms: an inactive phosphorylated form and an active dephosphorylated form. Regulation of PDH activity is due to a differential regulation of a protein kinase, which creates the phosphorylated form, and a protein phosphatase, which creates the dephosphorylated form. Insulin activates PDH in adipose tissue by an unknown mechanism.

9. The answer is B (1, 3). (*III*) The tricarboxylic acid cycle provides for the oxidation to carbon dioxide (CO_2) of two carbons per turn of the cycle. CO_2 is involved at the isocitrate dehydrogenase step, where isocitrate is converted to α-ketoglutarate, and at the α-ketoglutarate dehydrogenase step, which converts α-ketoglutarate to succinyl coenzyme A. Both of these enzymes are dehydrogenases that transfer electrons from the substrate to oxidized nicotinamide-adenine dinucleotide (NAD^+). Malate dehydrogenase is also NAD^+-linked, but the reaction converting malate to oxaloacetate does not involve a decarboxylation. Succinate dehydrogenase, a flavin-adenine dinucleotide–linked enzyme, converts succinate to fumarate without loss of carbon.

10. The answer is A (1, 2, 3). (*III C 5*) Isocitrate dehydrogenase of the tricarboxylic acid cycle is inhibited by a high-energy charge and deinhibited by a low-energy charge. A state of high-energy charge is associated with high cellular levels of adenosine triphosphate and reduced nicotinamide-adenine dinucleotide, whereas a low-energy charge is associated with high levels of adenosine diphosphate.

16
Mitochondrial Electron Transport and Oxidative Phosphorylation

I. DESCRIPTION

A. **The electron transport chain** is the final common pathway in aerobic cells by which electrons derived from various substrates are transferred to oxygen.

  1. The electron transport chain is a series of highly organized oxidation–reduction enzymes whose reactions can be represented by

$$\text{Reduced A + oxidized B} \leftrightarrow \text{oxidized A + reduced B}$$

  2. It is possible for a variety of substrates to use a common pathway because many of them are oxidized by enzymes that use oxidized nicotinamide-adenine dinucleotide (NAD^+) or oxidized flavin-adenine dinucleotide (FAD) as electron acceptor cofactors.

B. **Oxidative phosphorylation** is the process whereby the free energy that is released when electrons are transferred along the electron transport (respiratory) chain is coupled to the formation of adenosine triphosphate (ATP) from adenosine diphosphate (ADP) and inorganic phosphate (P_i).

  1. In intact mitochondria and in special preparations of submitochondrial particles, the transport of electrons and the phosphorylation of ADP are tightly coupled reactions.

  2. Oxidative phosphorylation is the main source of energy in aerobic cells.

  3. In damaged mitochondria, respiration (i.e., electron transport) may occur unaccompanied by oxidative phosphorylation. The free energy may still be released as the electrons are transferred down the transport chain; however, this energy is not trapped as ATP but appears instead as heat.

II. LOCALIZATION OF THE ELECTRON TRANSPORT CHAIN

A. **Mitochondrial structure** is illustrated in Figure 16-1.

B. **Components of mitochondria**

  1. **The outer membrane** is permeable to most small molecules.

  2. **The intermembrane space** presents no barrier to the passage of intermediates (i.e., substances entering or leaving the mitochondrial matrix).

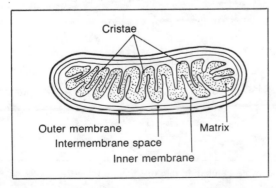

Cristae

Outer membrane

Intermembrane space

Inner membrane

Matrix

Figure 16-1. The structure of a mitochondrion, showing the membrane and intermembrane compartments. The infoldings of the inner membrane are called *cristae*.

   3. The inner membrane shows a highly selective permeability.
     a. It has transport systems only for specific substances.
       (1) ATP, ADP, and P_i
       (2) Pyruvate, succinate, α-ketoglutarate, malate, and citrate
       (3) Cytidine and guanosine triphosphates and diphosphates
     b. The enzymes of the electron transport chain are embedded in the inner membrane in association with the enzymes of oxidative phosphorylation (see section VII).

   4. The matrix is bounded by the inner membrane and contains
     a. The enzymes of the tricarboxylic acid (TCA) cycle with the exception of succinate dehydrogenase, which is embedded in the inner membrane
     b. The enzymes of β-oxidation of fatty acids (see Chapter 20, section VI A 3)
     c. Miscellaneous enzyme systems

C. Sources of electrons

   1. Reduced nicotinamide-adenine dinucleotide (NADH) is derived from NAD^+-linked dehydrogenases, including
     a. The isocitrate, α-ketoglutarate, and malate dehydrogenases of the TCA cycle
     b. Pyruvate dehydrogenase
     c. L-3-Hydroxylacyl coenzyme A (CoA) dehydrogenase of fatty acid oxidation
     d. Miscellaneous NAD^+-linked dehydrogenases

   2. Reduced flavin-adenine dinucleotide (FADH$_2$) is derived from FAD-linked dehydrogenases, including
     a. Succinate dehydrogenase of the TCA cycle
     b. The FAD-linked dehydrogenase of the α-glycerophosphate shuttle
     c. Acyl CoA dehydrogenase of fatty acid oxidation
     d. Miscellaneous FAD-linked dehydrogenases

III. ORGANIZATION OF THE ELECTRON TRANSPORT CHAIN

A. Entry via NADH–coenzyme Q (ubiquinone, often referred to simply as Q) reductase (Fig. 16-2)

   1. Electrons from NADH enter the electron transport chain at the level of an enzyme complex known as NADH-Q reductase (or NADH dehydrogenase), which oxidizes NADH. The enzyme is a flavoprotein with a riboflavin 5-phosphate [flavin mononucleotide (FMN)] prosthetic group.

   2. The prosthetic group accepts two electrons and two protons from NADH, to become $FMNH_2$, the reduced form.

   3. The electrons are then transferred from the reduced $FMNH_2$ to nonheme iron–sulfur–protein (FeS-protein) centers that are also prosthetic groups on the enzyme. (The association of NADH-Q reductase and nonheme FeS-protein is known as **complex I**.)

   4. Finally, the electrons are transferred from the FeS-protein centers to coenzyme Q.

B. Entry of electrons from FADH$_2$

   1. The succinate-Q reductase complex serves as the entry path for electrons from $FADH_2$ into the electron transport system. The complex has two components:
     a. The flavin prosthetic group of succinate dehydrogenase
     b. The FeS-protein center

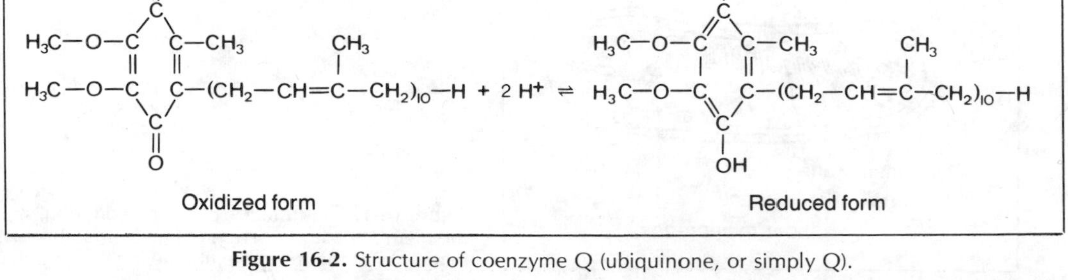

Figure 16-2. Structure of coenzyme Q (ubiquinone, or simply Q).

2. Electrons from $FADH_2$ are transferred via the FeS-protein to Q. The association of succinate-Q reductase and the FeS-protein centers forms **complex II**.

C. Electron transport by cytochromes

1. The cytochromes (see Chapter 28, section II C 4) and the FeS–protein centers are one-electron carriers. NADH, $FADH_2$, and Q are two-electron carriers.

2. Thus, Q transfers its two electrons to two molecules of cytochrome b, the next electron carrier in the electron transport chain.

3. There are six electron carriers lying between Q and molecular oxygen:

$$QH_2 \leftrightarrow FeS \leftrightarrow cyt\ b \leftrightarrow FeS \leftrightarrow cyt\ c_1 \leftrightarrow cyt\ c \leftrightarrow cyt\ a\text{-}a_3 \leftrightarrow O_2$$

4. The **oxidation–reduction potentials** (i.e., the electron affinities) of the cytochromes in the above sequence increase from left to right (see Chapter 12, section IV B).

5. The prosthetic groups of the cytochromes are iron porphyrins (see Chapter 28, section II). Those of cytochromes b, c, and c_1 are iron protoporphyrin IX (heme) and those of cytochrome a-a_3, heme a.

D. QH_2–cytochrome c reductase complex

1. This complex (**complex III**) consists of cytochrome b, an FeS-protein center, and cytochrome c_1; it transfers an electron from reduced coenzyme Q (QH_2) to cytochrome c.
  a. Cytochrome c is a water-soluble enzyme.
  b. It is only loosely associated with the inner surface of the inner mitochondrial membrane.

2. Cytochrome c then transfers its electron to the cytochrome c oxidase complex.

E. Cytochrome c oxidase complex

1. The electron transferred from cytochrome c is accepted by cytochrome a, which forms part of a copper-containing cytochrome complex known as cytochrome a-a_3, or cytochrome oxidase (**complex IV**).

2. Cytochrome oxidase can interact directly with molecular oxygen.

3. The resulting formation of water involves the transfer of four electrons from cytochrome a-a_3 molecules. The mechanism is unknown.

IV. FREE ENERGY AND SITES OF ATP FORMATION

A. The decline in free energy as electrons flow down the electron transport chain to oxygen is 53 kcal/mol of electron pairs. This can be calculated from the standard oxidation–reduction potentials of the NAD^+/NADH and O_2/H_2O oxidation–reduction couples (see Chapter 12, section IV).

B. At three sites (Table 16-1), the free energy released per electron pair transferred is sufficient to support the phosphorylation of ADP to ATP, which requires about 7 kcal/mol.

C. Electrons that enter the transport chain through the NADH-Q reductase complex support the synthesis of 3 mol of ATP per mole of reducing power (or per electron pair). By contrast, electrons injected at the level of Q, as they would be if donated by $FADH_2$, only support the synthesis of 2 mol of ATP per mole of reducing power.

V. INHIBITORS OF ELECTRON TRANSPORT. Selective inhibition of various components of the electron transport chain can be achieved by a variety of substances, some of which are used as

Table 16-1. Oxidation–Reduction Potentials and Free-Energy Changes at Sites in the Electron Transport Chain That Can Support ATP Formation

Site	Oxidation–Reduction Potential (volts)	Free Energy Released (kcal/mol)
At NAD^+-Q	0.27	12.2
At cytochrome b $\rightarrow$ cytochrome c_1	0.22	9.9
At cytochrome a_3 $\rightarrow$ O_2	0.53	23.8

poisons (e.g., insecticides) and some of which are used as drugs. Many of the inhibitors have been used to investigate the sequence of electron carriers in the transport chain.

A. **Rotenone** inhibits the transfer of electrons through the NADH-Q reductase complex; it is used as a fish poison and as an insecticide.

B. **Amobarbital** (Amytal) and **secobarbital** (Seconal) also inhibit electron transport through the NADH-Q reductase complex; there is no direct evidence that their hypnotic action is a result of blocking of the electron transport chain.

C. **Piericidin A**, an antibiotic, blocks the transfer of electrons at the NADH-Q reductase complex by competing with Q. The electrons from the complex are transferred to piericidin A instead of to Q.

D. **Antimycin A**, also an antibiotic, blocks electron transport at the level of the cytochrome b-cytochrome c_1 complex (complex III).

E. **Hydrogen cyanide**

1. Cyanide (CN^-) inhibits the terminal transfer of electrons to oxygen by the cytochrome oxidase complex.
2. CN^- binds strongly to the ferric form of cytochrome oxidase, inhibiting its reduction to the acceptor ferrous form.

F. **Carbon monoxide (CO)** inhibits cytochrome oxidase by combining with the oxygen-binding site. The CO binding to cytochrome oxidase is reversed by illumination with visible light.

VI. JUSTIFICATION OF THE SEQUENCE OF ELECTRON CARRIERS

A. The standard oxidation–reduction potentials of the electron carriers in the transport chain become more positive, going from the NADH-Q reductase complex through the postulated sequence to oxygen.

B. Experimental reconstructions of fragmented electron transport chains (see section VII) have shown a specificity in the reactions of electron carriers; for example, NADH can reduce NADH dehydrogenase but not cytochromes b or c, or cytochrome oxidase.

C. Complexes have been isolated that were postulated to contain functionally related carriers, and the expected components were shown to be present.

D. **Difference spectra**

1. Suspensions of isolated mitochondria are turbid and unsuitable for direct spectrophotometry.
2. Difference spectra can be obtained by simultaneously measuring two aliquots of a suspension of mitochondria.
   a. One in the reduced state with substrate present and oxygen absent
   b. One in the oxidized state without substrate and well oxygenated
3. The difference in light absorption, over a range of wave lengths, between the two aliquots shows that
   a. The electron carriers become increasingly more oxidized as the transport sequence progresses toward oxygen
   b. The postulated sequence matches the sequence revealed by the characteristic absorption spectra seen

VII. MITOCHONDRIAL FRAGMENTATION (Fig. 16-3)

A. **Oxidative phosphorylation**, a mitochondrial process, is studied by isolating and then fragmenting mitochondria.

1. In the first fragmentation step, the outer membrane is removed by treatment with various detergents (e.g., phospholipase and digitonin).
   a. The two particulate fractions that result are
      (1) The outer membrane, either in the form of vesicles or completely solubilized
      (2) The inner membrane plus the mitochondrial matrix enzymes. This fraction contains the enzymes of
         (a) The electron transport chain

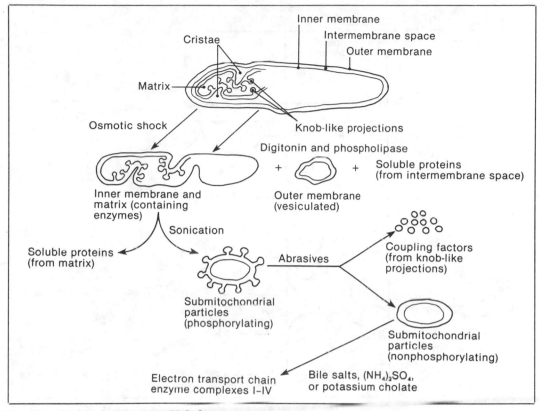

Figure 16-3. Steps in the fragmentation of mitochondria.

    (b) Oxidative phosphorylation
    (c) The TCA cycle
  b. A soluble fraction containing the enzymes from the intermembrane space is also obtained.

  2. The inner membrane is then subjected to mild sonication,
    a. The enzymes present in the matrix are released into the medium, and the inner membrane forms vesicles.
    b. The vesiculated inner membrane can carry out coupled electron transport and oxidative phosphorylation.

  3. If the phosphorylating particles are abraded, knob-like projections containing the so-called **coupling factors** are separated from the membrane, which now cannot carry out phosphorylation of ADP to ATP but can still transport electrons through the transport chain.

  4. Treatment of the transport-chain fraction of the inner membrane with bile salts, $(NH_4)_2 SO_4$, or potassium cholate (often used sequentially) breaks the fraction up into the four complexes concerned with electron transport (see section III). These isolated complexes can be used to study electron transport in reconstitution experiments.

B. Location of the enzymes of the electron transport chain in relation to coupled oxidative phosphorylation is illustrated in Figure 16-4.

  1. Succinate dehydrogenase faces the mitochondrial matrix from which it accepts electrons from succinyl CoA synthetase of the TCA cycle.

  2. The cytochrome a-a_3 complex likewise faces the matrix from which it will obtain oxygen for the final step of the electron transport chain.

VIII. COUPLING OF PHOSPHORYLATION TO RESPIRATION. Oxidative phosphorylation has been studied in isolated mitochondria (or in phosphorylating submitochondrial particles) by incubating them in an oxygen electrode cell (called an **oxygraph cell**), which measures the concentration of oxygen dissolved in the suspension medium and thus indicates the **mitochondrial oxygen uptake. Phosphorylation** is assessed by measuring the rate of disappearance of ADP or P_i.

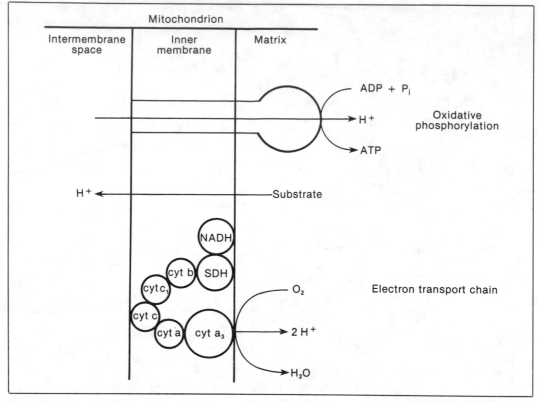

Figure 16-4. Location of the carriers of the electron transport chain (within the inner mitochondrial membrane) in relation to oxidative phosphorylation. SDH = succinate dehydrogenase; NADH = reduced nicotinamide-adenine dinucleotide; cyt = cytochrome; ADP = adenosine diphosphate; P_i = inorganic phosphate; ATP = adenosine triphosphate.

A. An oxygen electrode tracing is illustrated in Figure 16-5.

1. **At point A** on the graph, a substrate, P_i, and Mg^{2+} are all added to the suspension medium. The horizontal tracing indicates that there is no oxygen uptake from the mitochondria-free medium.

2. **At point B**, an aliquot of a mitochondrial suspension is added.
   a. A slow rate of oxygen uptake ensues, as indicated by the rate of disappearance of oxygen from the medium.
   b. This oxygen uptake, or **respiration**, is called **state IV respiration**, and is probably related to the degree of uncoupling of the mitochondria (see section VIII B) rather than to the use of endogenous substrate.

3. **At point C**, a measured amount of ADP is added, producing a more rapid rate of oxygen uptake. This rate, known as **state III respiration**, is limited only by the rate at which the electron transport chain can transfer electrons from substrate to oxygen.

4. **At point D**, all of the added ADP has been phosphorylated to ATP, and the rate of oxygen uptake declines back to the state IV rate. State IV respiration is therefore limited by the concentration of ADP. (**Points E and F** will be discussed below.)

B. Acceptor control ratio (the acceptor being ADP) is defined as

$$\frac{\text{Rate of respiration in state III}}{\text{Rate of respiration in state IV}}$$

It is a measure of the degree to which electron transport and phosphorylation are coupled.

1. Tightly coupled mitochondria have acceptor control ratios of about 10.

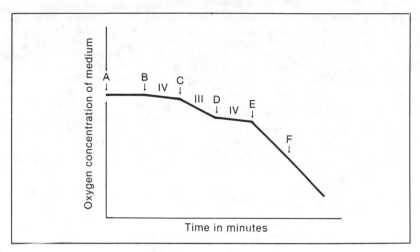

Figure 16-5. An oxygraph tracing showing the change in oxygen uptake by mitochondria on addition of various substances. (See section VIII A, C, and D for a detailed discussion.)

2. Uncoupled mitochondria may have ratios as low as 1, the value produced when state III and state IV respiration rates are equal.

C. **ADP:O (or P:O) ratio** is a measure of how many moles of ATP are formed from ADP by phosphorylation per gram atom of oxygen used. This is usually measured as the number of moles of ADP (or P_i) that disappear per gram atom of oxygen used.

1. If malate was the substrate added at point A in Figure 16-5, the ADP:O (or P·O) ratio would be 3 because of the following steps:
   a. Malate, when it is oxidized, donates its electrons to NAD^+.
   b. When the NADH that is formed enters the electron transport chain, it does so at the level of the NADH-Q reductase complex.
   c. As the electrons pass down the chain, three ATPs are formed per atom of oxygen used.

2. If glycerol 3-phosphate was the substrate added at point A in Figure 16-5, the ADP:O ratio would be 2 because
   a. Some electrons are derived from the FAD-linked dehydrogenase that oxidizes the substrate.
   b. These electrons enter the transport chain at the level of coenzyme Q.
   c. These electrons miss the formation of ATP at the NADH-Q reductase complex.

D. **Uncouplers of oxidative phosphorylation** are compounds that allow mitochondria to use oxygen regardless of whether or not there is any phosphate acceptor (ADP) available.

1. **At point E** on the graph in Figure 16-5, an uncoupler has been added. This causes a marked increase in oxygen uptake, the new rate being greater than the state III rate.

2. Note that no ADP had to be added for the uncoupler to boost the oxygen uptake rate to the new high level. In fact, if ADP is added at **point F**, there will be no further change in the oxygen uptake rate.

3. **Prototype uncouplers**
   a. **2,4-Dinitrophenol** is a classic uncoupler of oxidative phosphorylation. It was once used as a weight-loss drug but was discontinued because of its toxicity.
   b. **Dicumarol** and similar drugs are used clinically as anticoagulants. The closely related **warfarin** is also used as a rat poison, mainly because of its anticoagulant action but also because it uncouples oxidative phosphorylation.
   c. **Calcium** transport into the mitochondria also changes the relationship between electron transport and oxidative phosphorylation.
      (1) The mitochondria of mammalian cells transport calcium against a concentration gradient, and this process is energetically coupled to electron transport.
      (2) The uptake of calcium by mitochondria is also obligatorily coupled to the uptake of a corresponding amount of P_i.

(3) For every pair of electrons that pass from NADH to oxygen along the electron transport chain, approximately six Ca^{2+} accumulate in the mitochondria; that is, <u>two Ca^{2+} per</u> energy-conserving site (or per site of ATP formation) are retained (Fig. 16-6). Note that in the figure, the possibility of a fourth Ca^{2+}-accumulating site is indicated.

(4) When Ca^{2+} is taken up by the mitochondria, electron transport can proceed, but energy is required to pump the Ca^{2+} into the mitochondria; therefore, there is no energy available to be stored as ATP.

(5) If Ca^{2+} is added to an oxygraph cell containing mitochondria, substrate, P_i, and Mg^{2+}, the respiration rate changes from state IV to a very fast rate typical of uncoupled respiration. If ADP is added, there is no additional change in the oxygen uptake rate.

(6) Ca^{2+} does not increase the state IV rate unless P_i is present.

(7) Ca^{2+} and P_i are transported into the mitochondria, and apparently the calcium is stored as a calcium–phosphate complex.

E. Phosphorylation inhibitors. Oligomycin prevents both the stimulation of oxygen uptake by ADP and the phosphorylation of ADP to ATP.

1. If oligomycin is added to a mitochondrial preparation in the presence of substrate, P_i, Mg^{2+}, and ADP, the state III respiration is immediately reduced to the state IV rate.

2. Characteristically, the inhibition of oxygen uptake by oligomycin is relieved by the addition of 2,4-dinitrophenol, which stimulates the usual fast uncoupled rate of oxygen uptake.

3. Oligomycin appears to act by interfering with the ATP synthetase reaction, thus preventing the phosphorylation of ADP.

F. Inhibitors of the ADP–ATP carrier

1. **Atractyloside** is a toxic glycoside from a Mediterranean thistle; **bongkrekic acid** is derived from a mold that grows on coconut "flesh." Both of these compounds block the translocase that is responsible for the movement of ADP and ATP across the inner mitochondrial membrane.

2. The addition of either inhibitor to a mitochondrial preparation incubated in the presence of substrate, P_i, Mg^{2+}, and ADP reduces the state III respiration to state IV.

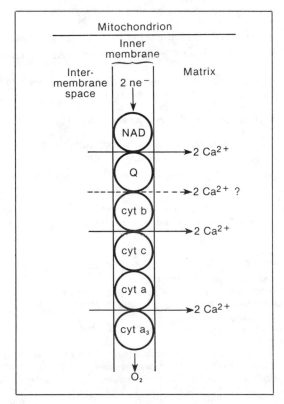

Figure 16-6. Possible sites for calcium transport along the mitochondrial electron transport chain. *2 ne⁻* = electron pairs; cyt = cytochrome.

IX. MECHANISMS OF OXIDATIVE PHOSPHORYLATION. Three major proposals for the mechanism of oxidative phosphorylation have been considered.

A. **The chemiosmotic coupling hypothesis** is probably the most widely accepted of current theories of oxidative phosphorylation.

1. It has been proposed that an electrochemical gradient of protons (H^+) across the mitochondrial inner membrane serves as the means of coupling the energy flow of electron transport to the formation of ATP.

2. The electron carriers are hypothesized to act as pumps, which cause vectorial (directional) pumping of H^+ across the membrane (Fig. 16-7).

3. Because H^+ is a charged particle, the flow of free energy across the inner membrane is due to the combination of a concentration gradient and a charge gradient.

4. In the electron transport chain, H^+ is separated from the electron; thus, according to the chemiosmotic coupling hypothesis, as the electrons move down the chain, H^+ is expelled, traveling from the matrix to the intermembrane space, as shown in Figure 16-7.
   a. Note that the transport sequence shown in Figure 16-7 is different from the sequence given previously in section III C because the chemiosmotic coupling hypothesis requires a reversal of the order of coenzyme Q and cytochrome b in order to allow directional pumping of H^+.
   b. This modification is not unreasonable in view of the difficulty in assigning the lipid soluble Q to a particular position in the electron transport chain.

5. The chemiosmotic coupling hypothesis then proposes that the protons in the intermembrane space pass through the inner membrane and back into the matrix at a special site, or "pore," where ATP synthetase resides. The dissipation of energy that occurs as the protons pass down the concentration gradient to the matrix allows the phosphorylation of ADP to ATP by the synthetase.

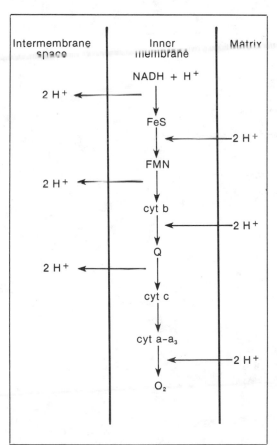

Figure 16-7. The postulated proton pumping during mitochondrial electron transport, according to the chemiosmotic coupling hypothesis. FeS = iron–sulfur–protein center; FMN = flavin mononucleotide; cyt = cytochrome.

6. To account for the actions of 2,4-dinitrophenol, oligomycin, and calcium, the chemiosmotic coupling hypothesis proposes that
   a. 2,4-Dinitrophenol, a lipid-soluble molecule, creates holes in the inner membrane, rendering it permeable to H^+ so that an H^+ gradient cannot form
   b. Oligomycin blocks the ATP synthetase reaction, and electron transport ceases as the gradients back up
   c. Calcium transport acts to break down the proton gradient

B. The chemical coupling hypothesis developed from the concept of a high-energy intermediate common to both electron transport and phosphorylation of ADP.

C. The conformational coupling hypothesis postulates that changes in the architecture of the mitochondrial cristae reflect changes in the different components of the electron chain to one another and that the conformational change represents the formation of a high-energy state. When state III respiration is occurring, the cristae are said to be in a condensed state, in contrast to the orthodox state of the cristae during state IV respiration.

STUDY QUESTIONS

Directions: Each question below contains five suggested answers. Choose the **one best** response to each question.

1. It is possible for a variety of substrates to use a common pathway for the transfer of electrons to oxygen because

(A) the substrates are oxidized in the mitochondria

(B) many of the substrates are oxidized by enzymes linked to oxidized nicotinamide- or flavin-adenine dinucleotide

(C) the substrates are all oxidized by the same enzymes

(D) the electrons from all substrates are transferred to the common acceptor, adenosine triphosphate

(E) protons from all substrates are used to form water

2. The mitochondrial electron transport chain carriers are located

(A) in the inner mitochondrial membrane

(B) in the mitochondrial matrix

(C) in the intermembrane space

(D) on the inner surface of the outer mitochondrial membrane

(E) on the outer surface of the outer mitochondrial membrane

3. All of the following electron carriers are components of the mitochondrial electron transport chain EXCEPT

(A) nicotinamide-adenine dinucleotide

(B) nicotinamide-adenine dinucleotide phosphate

(C) flavin mononucleotide

(D) flavin-adenine dinucleotide

(E) coenzyme Q

4. Electrons from pyruvate enter the mitochondrial electron transport chain at the level of

(A) coenzyme Q (complex II)

(B) NADH-Q reductase (complex I)

(C) QH_2-cytochrome c reductase (complex III)

(D) cytochrome c oxidase (complex IV)

(E) none of the above

5. Some of the free energy released in the mitochondrial electron transport chain can be harnessed to the formation of adenosine triphosphate (ATP). How many moles of ATP can be formed per mole-pair of electrons transferred from reduced nicotinamide-adenine dinucleotide to oxygen?

(A) 0

(B) 1

(C) 2

(D) 3

(E) 4

6. Rotenone, which is used as a fish poison and as an insecticide, blocks mitochondrial electron transport by

(A) inhibiting the interaction between oxygen and the terminal electron carrier

(B) inhibiting the reduction of cytochrome c

(C) inhibiting the transfer of electrons through the NADH-Q reductase complex

(D) inhibiting electron transfer at all nonheme iron–sulfur–protein centers

(E) forming an inactive complex with cytochrome b

7. Carbon monoxide inhibits mitochondrial electron transport by

(A) binding to hemoglobin in the erythrocytes and so blocking the transport of oxygen to tissues

(B) binding to the oxygen-binding site of cytochrome oxidase

(C) blocking electron transport at the level of the cytochrome b–cytochrome c_1 complex (complex III)

(D) combining with coenzyme Q and preventing its interaction with the nonheme iron–sulfur–protein center of complex II

(E) inhibiting the electron transfers of complex I

ADP phosphorylated to ATP when state IV

Dinitrophenol

8. State III respiration in a mitochondrial preparation describes the rate of oxygen uptake

(A) when the ratio of adenosine diphosphate (ADP) to adenosine triphosphate is 1:3
(B) in the presence of Mg^{2+}, substrate, inorganic phosphate, and ADP
(C) in the presence of an uncoupler of oxidative phosphorylation only
(D) in the presence of ADP but in the absence of substrate
(E) in the presence of Mg^{2+} and substrate only

9. The P:O ratio is most closely described by which of the following statements?

(A) The ratio of phosphorus to oxygen atoms in adenosine triphosphate (ATP)
(B) The ratio of moles of phosphate incorporated into ATP to moles of oxygen consumed
(C) The ratio of moles of phosphate incorporated into adenosine diphosphate (ADP) to gram atoms of oxygen consumed
(D) The ratio of moles of ADP phosphorylated to moles of oxygen consumed
(E) The ratio of adenosine monophosphate to ATP present in respiring cells

10. The uncoupling of oxidative phosphorylation in a mitochondrial system describes which of the following actions?

(A) The phosphorylation of adenosine diphosphate (ADP) to adenosine triphosphate accelerates
(B) The phosphorylation of ADP continues but oxygen uptake stops
(C) The phosphorylation of ADP stops but oxygen uptake continues
(D) Oxygen uptake stops
(E) None of the above

11. If both oligomycin and 2,4-dinitrophenol are added to a mitochondrial preparation in the presence of substrate, inorganic phosphate, Mg^{2+}, and adenosine diphosphate (ADP), then

(A) both oxygen uptake and phosphorylation of ADP would cease
(B) oxygen uptake would be reduced to state IV, but phosphorylation of ADP would continue
(C) oxygen uptake would be high, but phosphorylation of ADP would cease
(D) there would be no change in the state III respiration nor of the P:O ratio at the end of the experiment
(E) none of the above

12. The chemiosmotic coupling hypothesis of oxidative phosphorylation proposes that adenosine triphosphate is formed because

(A) of a change in the permeability of the inner mitochondrial membrane toward adenosine diphosphate (ADP)
(B) of the formation of high-energy bonds in mitochondrial proteins
(C) ADP is pumped out of the matrix into the intermembrane space
(D) a proton gradient forms across the inner membrane
(E) none of the above

Directions: Each question below contains four suggested answers of which **one or more** is correct. Choose the answer

A   if **1, 2, and 3** are correct
B   if **1 and 3** are correct
C   if **2 and 4** are correct
D   if **4** is correct
E   if **1, 2, 3, and 4** are correct

13. Which of the following enzymatic reactions donate electrons to the mitochondrial electron transport chain via reduced flavin-adenine dinucleotide?

(1) Mitochondrial glycerophosphate dehydrogenase
(2) Acyl coenzyme A dehydrogenase of the β-oxidation of fatty acids
(3) Succinate dehydrogenase
(4) Pyruvate dehydrogenase

14. In the mitochondrial electron transport chain, carriers with a more positive reduction potential will oxidize carriers with a more negative reduction potential. The reduction potential of cytochrome c is more

(1) positive than that of $2 H^+ + 2 e^- + \frac{1}{2} O_2 \rightarrow H_2O$
(2) negative than that of $2 H^+ + 2 e^- + \frac{1}{2} O_2 \rightarrow H_2O$
(3) negative than that of $NAD^+ \rightarrow NADH + H^+ + 2 e^-$
(4) positive than that of $FAD \rightarrow FADH_2 + 2 e^-$

15. A release of energy in excess of − 7 kcal/mol of transferred electron pairs occurs at which of these sites in the mitochondrial electron transport chain?

(1) NADH-Q reductase
(2) Nonheme iron–sulfur–protein center → cytochrome (cyt) c_1
(3) Cyt a_3 → oxygen
(4) Cyt c_1 → cyt c

ANSWERS AND EXPLANATIONS

1. The answer is B. (*I A 2*) The tricarboxylic acid cycle and the electron transport chain act as a common pathway for the transfer of electrons from a variety of substrates to oxygen because many of the enzymes involved in the oxidation of metabolic fuels are linked to oxidized nicotinamide- or flavin-adenine dinucleotide. These electron carriers pass on their electrons to oxygen via the electron transport chain.

2. The answer is A. (*II B 3 b*) Mitochondria have two membranes, an outer membrane, which is permeable to most small molecules, and an inner membrane, which shows a highly selective permeability. The mitochondrial electron transport chain is localized in the inner mitochondrial membrane in association with the enzymes of oxidative phosphorylation. The enzymes of the electron transport chain accept electrons from nicotinamide-adenine dinucleotide–linked dehydrogenases in the mitochondrial matrix. The flavin-adenine dinucleotide–linked succinate dehydrogenase of the tricarboxylic acid cycle is also in the inner mitochondrial membrane.

3. The answer is B. (*III*) The electron carriers listed in the question are all to be found in the mitochondrial electron transport chain with the exception of nicotinamide-adenine dinucleotide phosphate ($NADP^+$). In general, $NADP^+$ is involved in electron transfer in biosynthetic reactions, such as reductive syntheses, which require the reduced form of the cofactor, NADPH. On the other hand, NAD^+ is, in general, the electron acceptor in catabolic reactions involved in the extraction of energy from metabolic fuels.

4. The answer is B. (*III A*) In the pyruvate dehydrogenase complex, the electrons from pyruvate are transferred to lipoamide and from there to flavin-adenine dinucleotide (FAD). The reduced FAD is re-oxidized by nicotinamide-adenine dinucleotide (NAD^+), and the reduced cofactor (NADH) injects its electrons into the mitochondrial electron transport chain at the level of complex I, the NADH coenzyme Q reductase. The electrons from NADH are first passed to riboflavin phosphate (also known as flavin mononucleotide), which is the prosthetic group of NADH dehydrogenase. The electrons then flow to the nonheme iron–sulfur–protein center and thence to Q. There is a sufficient drop in free energy at complex I to support the phosphorylation of adenosine diphosphate to adenosine triphosphate.

5. The answer is D. (*IV B*) The free energy released in the transfer of electrons from reduced nicotinamide-adenine dinucleotide to oxygen by the mitochondrial electron transport chain is sufficient to support the phosphorylation of adenosine diphosphate to adenosine triphosphate (ATP) at three sites. Thus, for every pair of electrons that pass down the chain, 3 molecules of ATP are formed, or for every mole-pair of electrons transferred, 3 mol of ATP are formed.

6. The answer is C. (*V A*) Rotenone blocks the mitochondrial electron transport chain by interacting with the NADH coenzyme Q reductase complex, complex I.

7. The answer is B. (*V F*) Carbon monoxide inhibits mitochondrial electron transport by combining with the oxygen-binding site of cytochrome oxidase, preventing the electrons passed down the chain from interacting with oxygen. Carbon monoxide also binds to the oxygen-binding site of hemoglobin, reducing the capacity of the blood to carry oxygen.

8. The answer is B. (*VIII A 1–4; Figure 16-5*) A preparation of mitochondria incubated in the presence of Mg^{2+}, substrate, inorganic phosphate, and adenosine diphosphate (ADP) shows an oxygen use rate that is limited by the rate of phosphorylation of ADP to adenosine triphosphate. This is called state III respiration. When all of the ADP has been phosphorylated, the respiration rate will decline to a state IV level.

9. The answer is C. (*VIII C 1, 2*) The P:O (or ADP:O) ratio is a measure of how many moles of adenosine triphosphate (ATP) are formed from adenosine diphosphate (ADP) by phosphorylation per gram atom of oxygen used. In fully coupled mitochondria, reduced nicotinamide-adenine dinucleotide should support a P:O ratio of 3, because there are three sites for the phosphorylation of ADP as electrons pass down the mitochondrial electron transport chain. Reduced flavin-adenine dinucleotide, on the other hand, would only support a P:O ratio of 2, as the ATP formed at complex I is bypassed.

10. The answer is C. (*VIII D*) In fully coupled mitochondria, the passage of electrons down the electron transport chain to oxygen is obligatorily tied to the phosphorylation of adenosine diphosphate (ADP) to adenosine triphosphate (ATP). If ADP is in short supply, then respiration slows down. Uncoupling of oxidative phosphorylation is the undoing of this obligatory tie. In the presence of an uncoupler substance, the respiration rate is no longer governed by the rate of phosphorylation, and the flow of electrons down the electron transport chain is limited only by the capacity of the system. Thus, in the uncoupled state, oxygen uptake continues, and the phosphorylation of ADP to ATP stops.

11. The answer is C. *(VIII E 1–3)* A mitochondrial preparation incubated in the presence of substrate, inorganic phosphate, Mg^{2+}, and adenosine diphosphate (ADP) consumes oxygen at a state III rate. If oligomycin is added to the medium, the phosphorylation of ADP is blocked, and the respiration rate drops to state IV. If 2,4-dinitrophenol, an uncoupler of oxidative phosphorylation, is also added to the preparation, the electron chain is no longer coupled to phosphorylation, and the respiration rate will be higher than state III. Thus, in the presence of both of these inhibitors, phosphorylation stops but oxygen uptake increases. The P:O ratio at the end of the experiment would be zero.

12. The answer is D. *(IX A)* The chemiosmotic coupling hypothesis is probably the most widely accepted of current theories of oxidative phosphorylation. It proposes that an electrochemical gradient of protons (H^+) is established across the mitochondrial membrane by the functioning of the electron carriers of the mitochondrial electron transport chain. The formation of adenosine triphosphate (ATP) is explained by the use of the free energy released as the protons pass down the concentration gradient at a special site where the ATP synthetase is located.

13. The answer is A (1, 2, 3). *(II C 2; Chapter 14 IV B 3; Chapter 20 VI A 4)* Electrons injected into the electron transport chain via reduced flavin-adenine dinucleotide ($FADH_2$) can give rise to the formation of adenosine triphosphate (ATP) at two sites in the electron transport chain en route to oxygen. Substrates linked to FAD miss the NADH coenzyme Q reductase complex (complex I) with its associated site for ATP formation, and interact at the level of Q. Three important substrates use this route: one is succinate of the tricarboxylic acid cycle, another is glycerol 3-phosphate of the glycerophosphate shuttle, and the third is the FAD-linked acyl dehydrogenase of the β-oxidation of fatty acids. The $FADH_2$ of the pyruvate dehydrogenase complex passes its electrons on to nicotinamide-adenine dinucleotide.

14. The answer is C (2, 4). *(III C 4; VI D)* The oxidation–reduction potentials of the electron carriers in the mitochondrial electron transport chain become more positive as the sequence runs from reduced nicotinamide-adenine dinucleotide to cytochrome oxidase. Cytochrome c can oxidize reduced cytochrome c_1; therefore, it has a more positive reduction potential. Any carrier preceding cytochrome c_1 must have a less positive oxidation–reduction potential; thus, the oxidation–reduction potential of cytochrome c is more positive than that of FAD $\rightarrow FADH_2 + 2 e^-$. Similarly, a carrier closer to cytochrome oxidase than cytochrome c will have a more positive oxidation–reduction potential than cytochrome c; thus, cytochrome c has a oxidation–reduction potential more negative than $2 H^+ + 2 e^- + \frac{1}{2} O_2 \rightarrow H_2O$.

15. The answer is B (1, 3). *(Table 16-1; Chapter 12 IV B)* There are three reactions in the mitochondrial electron transport chain where the change in free energy is more negative than -7 kcal per electron pair transferred. The first occurs in complex I, the NADH coenzyme Q reductase, which has an oxidation–reduction potential of 0.27 volts and a standard free-energy change ($\Delta G^{\circ\prime}$) of 12.2 kcal/mol. The second occurs at the cytochrome b–cytochrome c_1 reductase complex, which has an oxidation–reduction potential of 0.22 volts and a $\Delta G^{\circ\prime}$ of 9.9 kcal/mol. The third occurs at the cytochrome a_3 to oxygen transfer (complex IV), which has an oxidation–reduction potential of 0.53 volts and a $\Delta G^{\circ\prime}$ of 23.8 kcal/mol. The free-energy changes associated with the transfer of electrons from the nonheme iron–sulfur–protein center to cytochrome c_1 and from cytochrome c_1 to cytochrome c are not sufficiently negative to support the phosphorylation of adenosine diphosphate to adenosine triphosphate.

17
Gluconeogenesis

I. GENERAL DESCRIPTION

A. Synthesis of "new" glucose

1. Gluconeogenesis is the synthesis of carbohydrate from noncarbohydrate precursors. These precursors include lactate (i.e., pyruvate), glycerol, and some amino acids, termed gluco-genic amino acids. A small amount of glucose can be made from the odd-chain fatty acids.

2. For the most part, gluconeogenesis is confined to the liver.
   a. Under normal circumstances, the liver is responsible for 85% to 95% of the glucose that is made.
   b. During starvation or during metabolic acidosis, the kidney is capable of making glucose and then may contribute up to 50% of the glucose formed, since, in these conditions, the amount contributed by the liver is decreased considerably.
   c. The only other tissue capable of gluconeogenesis is the epithelial cell of the small intestine. The significance of this activity is not clear, and it contributes not more than 5% of the total glucose formation.

B. Functions of gluconeogenesis

1. During starvation or during periods of limited carbohydrate intake, when the levels of liver glycogen are low, gluconeogenesis is important in maintaining adequate blood sugar concentrations.

2. During severe exercise, when high catecholamine levels have mobilized carbohydrate and lipid reserves, the gluconeogenic pathway allows the use of lactate from glycolysis and of glycerol from fat breakdown.

3. During metabolic acidosis, gluconeogenesis in the kidney allows the excretion of an increased number of protons.

4. Gluconeogenesis also allows the use of dietary protein in carbohydrate pathways after disposal of the amino acid nitrogen as urea (see Chapter 27, section I).

II. GLUCONEOGENIC PATHWAY

A. The pathway of gluconeogenesis (the conversion of lactate to glucose) is the reversal of the glycolytic pathway (which converts glucose to lactate; see Chapter 14, section III) except for three irreversible steps in glycolysis that must be bypassed:

1. The phosphorylation of glucose by glucokinase

2. The conversion of fructose 6-phosphate (F6P) to fructose 1,6-diphosphate (F1,6DP) by phosphofructokinase (PFK)

3. The conversion of phosphoenolpyruvate (PEP) to pyruvate by pyruvate kinase

B. Conversion of lactate to pyruvate

1. Lactate is converted to pyruvate by **lactate dehydrogenase (LDH)**, which requires nicotinamide-adenine dinucleotide (NAD^+).

$$\text{Lactate} + NAD^+ \rightarrow \text{pyruvate} + NADH + H^+$$

2. In gluconeogenic tissues, LDH usually runs the reaction in the direction written above, the

critical factors being the ratio of NAD^+ to NADH, the ratio of lactate to pyruvate, and the iso-enzyme of LDH that is present.

C. Pyruvate to PEP. Pyruvate cannot be converted directly to PEP. The conversion requires two reactions that serve to bypass the irreversible pyruvate kinase step of glycolysis. The energy barrier involved in phosphorylating pyruvate to form PEP requires the expenditure of two high-energy phosphate bonds.

1. **The first reaction** for bypassing pyruvate kinase is the conversion of pyruvate to oxaloacetate, catalyzed by **pyruvate carboxylase**.

   Pyruvate + carbon dioxide (CO_2) + adenosine triphosphate (ATP) $\rightarrow$

   oxaloacetate + adenosine diphosphate (ADP)

   a. The prosthetic group of the enzyme is biotin, which carries the CO_2.
   b. The enzyme requires both Mg^{2+} and Mn^{2+}.
   c. There are two steps in the reaction:
      (1) The carboxylation of enzyme-bound biotin, which requires the hydrolysis of ATP and Mg^{2+}
      (2) The transfer of the carboxyl group from the enzyme-bound biotin to pyruvate, forming oxaloacetate, a step that requires Mn^{2+}
   d. Pyruvate carboxylase is a mitochondrial enzyme.
   e. It is an allosteric enzyme with an absolute requirement for acetyl coenzyme A (acetyl CoA), which regulates its activity in a concentration-dependent manner.

2. **The second reaction** for bypassing pyruvate kinase is the conversion of oxaloacetate to PEP, catalyzed by the Mn^{2+}-requiring **phosphoenolpyruvate carboxykinase (PEPCK)**.

   Oxaloacetate + guanosine triphosphate (GTP) $\rightarrow$ PEP + CO_2 + guanosine diphosphate (GDP)

   a. In rat and mouse liver, PEPCK is a cytosolic enzyme, and the mitochondrial inner membrane is impermeable to its substrate oxaloacetate.
      (1) One mechanism for transporting oxaloacetate carbons out of the mitochondria involves the malate shuttle.

      Oxaloacetate + NADH + H^+ $\rightleftharpoons$ malate + NAD^+

      (a) This reaction is catalyzed by *mitochondrial* malate dehydrogenase.
      (b) The malate is transported across the inner membrane. In the cytoplasm

      Malate + NAD^+ $\rightleftharpoons$ oxaloacetate + NADH + H^+

      (c) This reaction is catalyzed by *cytosolic* malate dehydrogenase.
      (2) Another mechanism for moving oxaloacetate carbons out of the mitochondria involves the formation of aspartate from oxaloacetate.

      Oxaloacetate + glutamate $\rightleftharpoons$ aspartate + α-ketoglutarate

      (a) The reaction is catalyzed by aspartate aminotransferase.
      (b) The aspartate can cross the mitochondrial inner membrane and in the cytosol can transaminate with α-ketoglutarate to form glutamate and oxaloacetate.
      (3) A third mechanism for moving oxaloacetate carbons out of the mitochondrial matrix involves the formation of citrate and its transfer across the inner membrane to the cytoplasm.

      Oxaloacetate + acetyl CoA $\rightarrow$ citrate

      (a) In the cytosol

      Citrate + ATP + CoASH $\rightarrow$ oxaloacetate + acetyl CoA + ADP

      (b) The reaction is catalyzed by the citrate cleavage enzyme.
   b. In man, PEPCK is found in both the mitochondria and the cytosol.
   c. In chickens, pigeons, and rabbits, it is purely mitochondrial, and in these species, there is a transport system for moving PEP out into the cytoplasm.

D. PEP to F1,6DP. The conversion of PEP to F1,6DP occurs by way of the reversal of the enzymes of glycolysis, which are controlled strictly by the levels of substrates and products.

1. PEP is converted to 2-phosphoglycerate (2-PG) by **enolase**.
2. 2-PG is converted to 3-phosphoglycerate (3-PG) by **phosphoglycerate 2,3-mutase**.

3. 3-PG is converted to 1,3-diphosphoglycerate (1,3-DPG) by **1,3-phosphoglycerate kinase**, which requires the concomitant expenditure of one high-energy phosphate bond.

4. 1,3-DPG is converted to glyceraldehyde 3-phosphate (G3P) by **glyceraldehyde 3-phosphate dehydrogenase** (G3PD).
   a. This step requires NADH and yields NAD^+ and inorganic phosphate (P_i) as products.

$$1,3\text{-DPG} + NADH + H^+ \rightleftharpoons G3P + NAD^+ + P_i$$

   b. The NADH can come from the lactate-to-pyruvate step if lactate is the substrate for gluconeogenesis. If other substrates are used, the NADH can come from the malate shuttle.

5. G3P is in equilibrium with dihydroxyacetone phosphate (DHAP) through the action of **triose isomerase**, and **aldolase** converts these substrates to F1,6DP.

E. F1,6DP to fructose 6-phosphate (F6P).

   1. This conversion cannot be effected by phosphofructokinase, which is irreversible under cellular conditions.

   2. Cells of the liver, the kidney, and the intestinal epithelium have an enzyme, **fructose 1,6-diphosphatase (F1,6DPase)**, which can cleave F1,6DP to F6P and P_i. This is a slow step in the sequence, but it is not the major rate-limiting step.

   3. FDPase is an allosteric enzyme with ATP acting as a positive modulator and adenosine monophosphate (AMP) as a negative modulator.

F. F6P to glucose

   1. F6P is converted to glucose 6-phosphate (G6P) by the freely reversible phosphoglucoisomerase.

   2. The formation of free glucose from G6P cannot occur by reversal of the glucokinase or hexokinase reactions.
   a. The gluconeogenic tissues contain **glucose 6-phosphatase** (G6Pase), an enzyme on the endoplasmic reticulum, which removes the phosphate from G6P.

$$G6P \rightarrow glucose + P_i$$

   b. This reaction is probably irreversible in the cell.

III. ENERGETICS OF GLUCONEOGENIC PATHWAY

A. Cost in high-energy phosphate bonds

   1. The transformation of lactate to 1,3-DPG costs 3 mol of ATP.
   a. From pyruvate to oxaloacetate   − 1 ATP
   b. From oxaloacetate to PEP   − 1 GTP (= ATP)
   c. From 3-PG to 1,3-DPG   − 1 ATP

   2. Therefore, because 1 mol of glucose is formed from 2 mol of lactate, the total cost is 6 mol of ATP.

B. Efficiency of glucose formation

   1. When 2 mol of pyruvate are oxidized via the tricarboxylic acid (TCA) cycle to CO_2 and H_2O, 30 mol of ATP are formed.

   2. From 2 mol of lactate, 34 mol of ATP can be formed.

   3. Thus, 85% of the lactate can be converted to glucose at the expense of the remaining 15%, which is oxidized to CO_2 and H_2O.

IV. SUBSTRATES FOR GLUCONEOGENESIS

A. Cori cycle

   1. Blood lactate is a substrate for gluconeogenesis. It is formed from glucose by glycolysis in muscle cells (and other cells) from which it is released to become available to the liver for gluconeogenesis. Thus

Plasma glucose → muscle glycolysis → muscle lactate → plasma lactate →

liver gluconeogenesis → plasma glucose

2. Energetics of the Cori cycle

    a. If 20 mol of plasma glucose are used by rapidly exercising muscle, they can provide 40 mol of ATP for the muscle and release 40 mol of lactate into the circulation.

    b. If the 40 mol of lactate are taken up by the liver

        (1) 6 mol of lactate are oxidized to CO_2 and H_2O, providing 102 mol of ATP

        (2) 34 mol of lactate are converted to 17 mol of glucose, which requires 102 mol of ATP

    c. The 17 mol of glucose can return to the circulation.

    d. Thus, the process uses 3 mol (20 − 17) of glucose to produce 40 mol of ATP in muscle, or 13.3 mol of ATP per mole of glucose.

    e. Under aerobic conditions, the oxidation of glucose can provide 36 mol of ATP per mole of glucose used; thus, gluconeogenesis has a relative efficiency of 13.3/36 = 0.37.

B. Alanine cycle

  1. Glutamine and alanine are the major amino acids that enter the liver during fasting and form important substrates for gluconeogenesis (see Chapter 27, I B 3).

  2. Pyruvate formed during glycolysis in muscle can undergo transamination with glutamate to yield alanine.

  3. The alanine is released by the muscle and taken up by the liver.

  4. In the liver, alanine transaminates with α-ketoglutarate to yield glutamate and pyruvate.

  5. The pyruvate is used to produce glucose by the gluconeogenic pathway.

C. Glucose from glycerol

  1. Glycerol is released on lipolysis of triacylglycerols.

  2. Adipose tissue lacks glycerokinase, the enzyme required to phosphorylate glycerol to L-α-glycerophosphate, and the glycerol therefore leaves the adipocyte.

  3. The liver takes up the glycerol and phosphorylates it to L-α-glycerophosphate.

  4. The relatively high NAD^+:NADH ratio in the liver cytosol allows the transformation of the L-α-glycerophosphate into G3P, an intermediate in the gluconeogenic pathway.

V. CONTROL OF GLUCONEOGENESIS

A. Nonhormonal regulation

  1. Substrate concentration

    a. In perfused rat liver, the apparent Michaelis constant (K_m) for gluconeogenesis from lactate is about 3 mM.

    b. In man, blood levels of lactate are normally around 1 mM, an amount that is substantially lower than the probable K_m for gluconeogenesis.

    c. Substrate levels are, therefore, an important factor in the regulation of the gluconeogenic pathway; this is true for pyruvate and alanine as well as for lactate.

    d. Because the rate of glucose production from fructose is much higher than from lactate, there must be a rate-limiting step between lactate and the entry point of fructose into the pathway: This is at G3P.

  2. Identification of rate-limiting steps by crossover plots

    a. To provide baseline measurements, isolated rat livers are first perfused with 10 mM lactate, and the concentrations of all metabolites between lactate and glucose are measured. Each metabolite concentration in this baseline "run" is regarded as 100%.

    b. The livers are then perfused with 20 mM lactate, and the concentration of each metabolite is measured again.

    c. The concentrations of both lactate and pyruvate increase when the substrate is changed from 10 mM to 20 mM, but there is no increase in the concentration of PEP. Therefore, the first rate-limiting step must lie between pyruvate and PEP.

  3. Allosteric regulation

    a. For pyruvate carboxylase, acetyl CoA is a dose-dependent positive regulator.

    b. Pyruvate kinase is inhibited by ATP, alanine, free fatty acids, and acetyl CoA; the inhibition directs the reactions towards gluconeogenesis. In the liver, pyruvate kinase is activated by F1,6DP.

  4. Potential futile cycles exist at several sites.

    a. At pyruvate kinase and pyruvate carboxylase

$$\text{PEP}$$

Inhibited by $\dashrightarrow$ ↓
acetyl CoA

         Pyruvate $\longrightarrow$ Oxaloacetate

             ↑
             |
             Has absolute
             requirement for
             acetyl CoA

     The inverse response of the two enzymes to the acetyl CoA concentration prevents the "futile" cycling of carbon through these enzymes.

  b. At phosphofructokinase and F1,6DP

$$\text{F6P}$$

ATP (−) $\dashrightarrow$ ↓   ↑ $\dashleftarrow$ ATP (+)
AMP (+)   AMP (−)

         F1,6DPase

     The ATP:AMP ratio affects these two enzymes in an inverse manner, again preventing "futile" cycling of carbon.

  c. At glucokinase and G6Pase

$$\text{Glucose}$$

Glucokinase $\dashrightarrow$ ↓   ↑ $\dashleftarrow$ G6Pase

              G6P

    (1) The extent to which G6Pase is regulated by glucose is not known, as the concentration of glucose required to inhibit the enzyme is very high (about 100 mM).

    (2) In diabetes, glucokinase disappears, and glucose is phosphorylated by hexokinase. This enzyme is inhibited allosterically by G6P; therefore, there is no futile cycling at the hexokinase and G6Pase steps under these conditions.

B. Hormonal regulation

  1. Slow effects

    a. Glucagon, epinephine, and glucocorticoids can all cause increased synthesis of the bypass enzymes of gluconeogenesis (i.e., PEPCK, F1,6DPase, and G6Pase). The change in enzyme levels takes 24 to 48 hr.

    b. Insulin suppresses the synthesis of these enzymes and at the same time induces increased synthesis of the three allosteric enzymes of glycolysis (i.e., hexokinase, phosphofructokinase, and pyruvate kinase).

  2. Rapid effects

    a. The rate-limiting steps in the gluconeogenic pathway lie between pyruvate and PEP, and between F1,6DP and F6P, with the pyruvate-to-PEP step the most important.

    b. Glucagon and epinephrine both act to increase the rate of gluconeogenesis, an effect over and above that of substrate control; glucagon is the major physiologic regulator.

    c. The mechanism by which glucagon increases gluconeogenesis probably involves the inactivation of pyruvate kinase by a cyclic adenosine monophosphate–dependent (cAMP-dependent) phosphorylation of the enzyme, and the inhibition of phosphofructokinase by changing levels of intermediates, particularly F2,6DP.

    d. The secretion of glucagon is stimulated by low blood glucose levels, whereas high blood glucose levels shut down glucagon secretion.

    e. A high blood glucose level stimulates insulin secretion, which reduces gluconeogenesis by lowering cAMP levels in the liver. Insulin does not slow down the basal rate of gluconeogenesis, but only the elevated rate due to glucagon action.

    f. The major short-term hormonal control of the gluconeogenic pathway is thus the ratio of glucagon to insulin.

    g. Glucocorticoids have a rapid effect on the pathway by causing peripheral protein breakdown, which provides additional amino acid substrates for gluconeogenesis.

STUDY QUESTIONS

Directions: Each question below contains five suggested answers. Choose the **one best** response to each question.

1. The functions of gluconeogenesis are described by all of the following statements EXCEPT

(A) it maintains blood sugar levels during fasting
(B) it is useful during strenuous exercise
(C) it plays a role in countering metabolic acidosis
(D) it allows the use of acetyl coenzyme A for glucose production
(E) it allows the use of dietary protein for glucose production and as a source of metabolic energy

2. In the sequence of reactions that convert lactate to glucose in cells capable of gluconeogenesis, three reactions are used to bypass enzymes of the glycolytic pathway in which glucose is converted to lactate. All of the following enzymes fall into this category EXCEPT

(A) pyruvate carboxylase
(B) pyruvate dehydrogenase
(C) phosphoenolpyruvate carboxykinase
(D) fructose 1,6-diphosphatase
(E) glucose 6-phosphatase

3. The cost in high-energy phosphate bonds for the formation of 1 mol of glucose from lactate is

(A) 8 mol of adenosine triphosphate (ATP)
(B) 6 mol of ATP
(C) 4 mol of ATP
(D) 2 mol of ATP
(E) none of the above

4. Muscle glycogen cannot be used as a direct source of blood glucose because muscle

(A) uses all of its glucose 6-phosphate for glycolysis
(B) lacks glucose 6-phosphatase
(C) cell membranes transport glucose unidirectionally into the cell
(D) lacks phosphoglucomutase
(E) lacks aldolase

5. What would be the consequence of a genetic absence of fructose 1,6-diphosphatase?

(A) Accumulation of fructose phosphates in the liver
(B) Failure of cells to metabolize glucose 6-phosphate via the glycolytic pathway
(C) Inability to produce glucose from lactate
(D) Pentosuria
(E) None of the above

6. The most important rate-limiting step in gluconeogenesis from lactate or alanine is the

(A) rate of entry of lactate or alanine into hepatocytes
(B) conversion of oxaloacetate to phosphoenolpyruvate
(C) reaction catalyzed by fructose 1,6-diphosphatase
(D) phosphorylation of glycerol by glycerol kinase
(E) phosphorylation of adenosine diphosphate

7. Which of the following compounds is a positive allosteric regulator of the enzyme pyruvate carboxylase?

(A) Adenosine triphosphate
(B) Acetyl coenzyme A
(C) Biotin
(D) Phosphoenolpyruvate
(E) Fructose 1-phosphate

8. The short-term hormonal control of hepatic gluconeogenesis is characterized by all of the following statements EXCEPT

(A) glucagon and epinephrine both act to increase the rate of gluconeogenesis
(B) insulin slows the high rate of gluconeogenesis induced by glucagon
(C) glucocorticoids enhance gluconeogenesis by increasing peripheral protein breakdown
(D) glucagon slows gluconeogenesis by decreasing intracellular levels of cyclic adenosine monophosphate
(E) high blood glucose levels reduce gluconeogenesis by decreasing glucagon secretion and increasing insulin secretion

Directions: Each question below contains four suggested answers of which **one or more** is correct. Choose the answer

A   if **1, 2, and 3** are correct
B   if **1 and 3** are correct
C   if **2 and 4** are correct
D   if **4** is correct
E   if **1, 2, 3, and 4** are correct

9. Cells capable of gluconeogenesis include

(1) hepatocytes
(2) kidney tubule cells
(3) small intestine mucosal cells
(4) pancreatic islet β cells

10. Important substrates for gluconeogenesis by the human liver include which of the following substances?

(1) Serum free fatty acids
(2) Blood lactate
(3) Serum β-hydroxybutyrate
(4) Serum alanine

ANSWERS AND EXPLANATIONS

1. The answer is D. (*I B*) Gluconeogenesis is the formation of glucose from noncarbohydrate precursors. It is an important process in fasting when the blood sugar levels begin to decline, as it provides glucose from the catabolism of amino acids. In strenuous exercise, the glycerol released from adipose tissue as a by-product of fat mobilization can be used to increase blood sugar levels, as can the lactate from rapid muscular activity. Gluconeogenesis can play a role in countering metabolic acidosis, as the ammonia derived from amino acid catabolism can be excreted as NH_4^+, thus taking protons out of the circulation. Acetyl coenzyme A cannot be used by humans to provide a net increase in glucose formation by the liver.

2. The answer is B. (*II*) In the glycolytic pathway, whereby glucose is converted to lactate, three reactions are irreversible under cellular conditions. These reactions are catalyzed by glucokinase (or hexokinase), phosphofructokinase, and pyruvate kinase. The conversion of pyruvate to phosphoenolpyruvate is accomplished by a two-step pathway. Pyruvate is first carboxylated to oxaloacetate by pyruvate carboxylase, and the oxaloacetate is then converted to phosphoenolpyruvate by phosphoenolpyruvate carboxykinase. The next irreversible step in glycolysis is at the phosphofructokinase reaction, which converts fructose 6-phosphate to fructose 1,6-diphosphate. The reverse reaction is catalyzed by a different enzyme, fructose 1,6-diphosphatase. Lastly, the production of glucose from glucose 6-phosphate is catalyzed by glucose 6-phosphatase. Pyruvate dehydrogenase is not an enzyme of glycolysis or gluconeogenesis. It is the main path by which carbohydrate can provide acetyl coenzyme A for the tricarboxylic acid cycle.

3. The answer is B. (*III A*) The conversion of lactate to phosphoenolpyruvate (PEP) requires the expenditure of two high-energy phosphate bonds: the first at the carboxylation of pyruvate to oxaloacetate and the second at the conversion of oxaloacetate to PEP. One additional mole of adenosine triphosphate is required to convert 3-phosphoglycerate to 1,3-diphosphoglycerate. Thus, three high-energy phosphate bonds are required to form one of the substrates for aldolase. To provide both triose substrates for aldolase, and thus 1 mol of glucose, 6 mol of high-energy phosphate bonds have to be expended.

4. The answer is B. (*IV A; Chapter 32 III C 2*) The glycogen that is stored in muscle cannot be used to increase blood glucose directly because muscle lacks glucose 6-phosphatase. However, muscle glycogen can provide glucose indirectly by forming lactate, which is released from muscle and forms a gluconeogenic substrate for liver. This is known as the Cori cycle.

5. The answer is C. (*II E; V B*) The important consequence of a genetic absence of fructose 1,6-diphosphatase would be the inability of the liver to carry out gluconeogenesis from any of the usual substrates, including lactate. Glucose, and fructose phosphates, would still be used by the glycolytic pathway because the reverse of the fructose 1,6-diphosphatase reaction is catalyzed by a different enzyme, phosphofructokinase. There would be no increase in the urinary excretion of pentoses (pentosuria).

6. The answer is B. (*V A 2*) The first rate-limiting step in gluconeogenesis from lactate or alanine is the reaction converting oxaloacetate to phosphoenolpyruvate, catalyzed by phosphoenolpyruvate carboxykinase. Crossover plots carried out on perfused rat liver indicate that in this species the dephosphorylation of fructose 1,6-diphosphate is also a slow step, but it is not as important in regulating gluconeogenesis as the first rate-limiting step.

7. The answer is B. (*V A 3*) Pyruvate carboxylase, which catalyzes the carboxylation of pyruvate to form oxaloacetate, is an allosteric enzyme with an absolute requirement for acetyl coenzyme A, which regulates the activity of the enzyme in a dose-dependent manner. Biotin is the cofactor involved in the carboxylation reaction as the carrier of the carboxyl group. Neither adenosine triphosphate (ATP) nor phosphoenolpyruvate, nor fructose 1-phosphate are allosteric regulators of pyruvate carboxylase, but ATP is required by the carboxylation reaction.

8. The answer is D. (*V B 2*) The most important aspect of the short-term hormonal regulation of hepatic gluconeogenesis is the concentration ratio of circulating insulin to glucagon. Glucagon and epinephrine increase gluconeogenesis, whereas insulin slows the high rate induced by glucagon. A high blood glucose level reduces gluconeogenesis because it induces insulin secretion and suppresses glucagon secretion. Glucocorticoids favor increased protein breakdown, which provides the liver with increased substrate for gluconeogenesis.

9. The answer is A (1, 2, 3). (*I A*) Three tissues—the liver, the kidney, and the small intestine—are capable of forming glucose from noncarbohydrate precursors (gluconeogenesis). Under normal conditions the human liver is the major site of gluconeogenesis, but in starvation or in metabolic acidosis, the rate of gluconeogenesis in the liver declines and that in the kidney increases. After 20 to 30 days of starvation, total glucose production is down, and the contribution by the kidney may be up to 50% of the total. The role of the small intestine in gluconeogenesis is not clear; in any case, it contributes not more than 5% of the total glucose production.

10. The answer is C (2, 4). (*IV A, B*) Circulating lactate and alanine are both major substrates for gluconeogenesis. Lactate arises from glycolysis by skeletal muscle during exercise, and alanine is the form in which much of the amino acids formed during muscle proteolysis reach the liver. Fatty acids cannot give rise to a net synthesis of glucose, and the ketone bodies (β-hydroxybutyrate and acetoacetate) cannot be metabolized by the liver.

18

Glycogen Metabolism

I. GENERAL DESCRIPTION

A. Glycogen structure

1. Glycogen is a highly branched, very large polymer of glucose molecules linked along its main line by α-1,4-glycosidic linkages; branches arise by α-1,6-glycosidic bonds at about every tenth residue.

2. Glycogen occurs in the cytosol as granules, which also contain the enzymes that catalyze its formation and use.

B. Glycogen storage

1. Glycogen is the storage form of glucose. Its polymeric nature allows the sequestering of energy stores with much less of a problem from osmotic effects than glucose would cause.

2. Muscle and liver are the major sites for the storage of glycogen, and although its *concentration* in the liver is higher, the much greater mass of skeletal muscle stores a greater *total* amount of glycogen.

3. Liver can mobilize its glycogen for the release of glucose to the rest of the body, but muscle can only use its glycogen for its own energy needs.

4. The glycogen stores allow us to eat intermittently by providing an immediate source of blood glucose for use as a metabolic fuel.

5. In humans, the liver glycogen stores are only adequate for about 12 hr or less without support from the gluconeogenic pathways (see Chapter 17).

II. GLYCOGEN SYNTHESIS (GLYCOGENESIS) AND USE (GLYCOGENOLYSIS). Glycogenesis and glycogenolysis occur by separate enzyme pathways.

A. Synthesis of the nucleotide precursor

1. **Uridine diphosphate glucose (UDP-glucose),** the precursor of glycogen, is formed from glucose 1-phosphate (G1P) and uridine triphosphate (UTP).

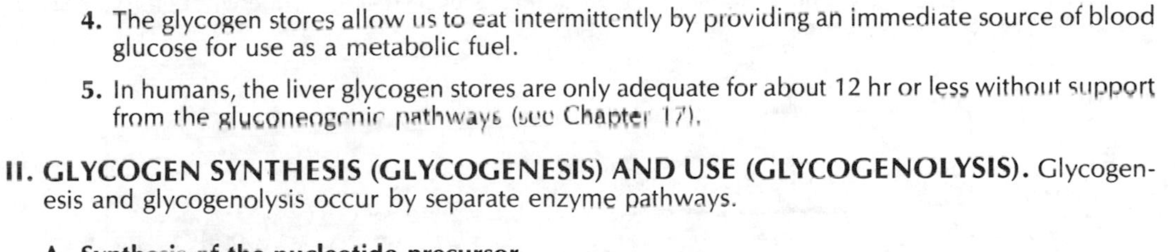

Glucose 1-phosphate   Uridine diphosphate glucose

    a. The reaction is catalyzed by UDP-glucose pyrophosphorylase.
    b. The reaction is reversible, but the hydrolysis of inorganic pyrophosphate (PP_i) by cellular pyrophosphatases renders it essentially irreversible.
    c. UDP-glucose is an activated form of glucose, the C-1 carbon of the glucosyl unit being esterified to the diphosphate moiety of UDP.

2. G1P is derived from glucose 6-phosphate (G6P) by the action of phosphoglucomutase.
   a. The reaction proceeds by way of an intermediate, glucose 1,6-diphosphate (G1,6DP).

$$CH_2OPO_3{}^{2-} \qquad CH_2OPO_3{}^{2-} \qquad CH_2OH$$

Glucose 6-phosphate ⇌ Glucose 1,6-diphosphate ⇌ Glucose 1-phosphate

   b. Catalytic amounts of G1,6DP must be present. [Compare the role of 2,3-diphosphoglycerate (2,3-DPG) in the phosphoglycerate mutase reaction of glycolysis in Chapter 14, section III G.]

B. Synthesis of the glycogen molecule

   1. Formation of amylose chains. The synthesis of new glycogen requires the presence of existing glycogen chains and glucosyl residues from UDP-glucose. The residues are successively transferred to a C-4 terminus of an existing glycogen chain in α-1,4-glycosidic linkages to form a growing chain.

$$\text{UDP-glucose} + (\text{glycogen})_{n\text{ residues}} \rightarrow \text{UDP} + (\text{glycogen})_{n\,+\,\text{residues}}$$

   a. The reaction is catalyzed by glycogen synthetase (UDP-glycogen transferase).
   b. This is the **rate-limiting step** in glycogen synthesis.

   2. Formation of branch chains and further growth (Fig. 18-1). Segments of the amylose chain are transferred onto the C-6 hydroxyl of neighboring chains, forming α-1,6 linkages.
   a. The enzyme responsible is glycosyl-4:6 transferase (**branching enzyme**).
   b. In branch formation, seven-residue segments of the amylose terminal chains are transferred to a C-6 hydroxyl group of a glucosyl residue that is four residues away from an existing branch. A terminal branch must be at least eleven residues in length before a segment is transferred from it.

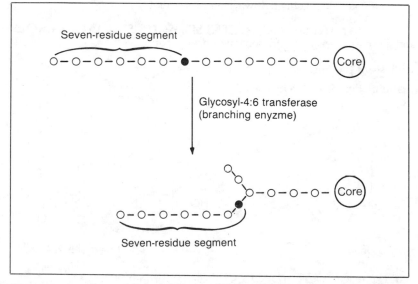

Figure 18-1. Formation of branches in glycogen chains. Glycosyl-4:6 transferase moves a 7-unit segment of α-1,4 residues from a glycogen chain at least 11 residues long onto a segment of the main chain at position 6 of a residue that is at least 4 residues from nearest branch.

C. Glycogenolysis

1. Phosphorolytic cleavage of terminal α-1,4-glycosidic bonds

Glycogen$_{n \text{ residues}}$

Glucose 1-phosphate   Glycogen$_{n-1 \text{ residues}}$

    a. The reaction, catalyzed by glycogen phosphorylase, yields G1P and a gradually shrinking glycogen chain (Fig. 18-2).
    b. This is the **rate-limiting step** in glycogen use.
    c. The reaction is readily reversible in vitro, but under cellular conditions, the concentration of inorganic phosphate (P_i) is always too high for the synthesis of glycogen to occur by this pathway.

2. Removal of branches (see Fig. 18-2)

    a. The reaction is catalyzed by amylo-1,6-glucosidase (**debranching enzyme**).
    b. The enzyme is a single 160-kdal polypeptide, which contains both glucosyl transferase and amylo-6-glucosidase activities.
        (1) As a **glucosyl transferase**, it transfers three glucosyl residues from a branch onto a chain terminus, leaving a single residue on C-6.
        (2) As an **amylo-6-glucosidase**, it removes the single residue on C-6 to yield a **free glucose molecule**.

III. REGULATION OF GLYCOGEN METABOLISM

A. Glycogenesis: glycogen synthetase

    1. The rate-limiting step in glycogen formation is the addition of activated glycosyl units (derived from UDP-glucose) to an existing glycogen chain by glycogen synthetase.

    2. Glycogen synthetase exists in inactive and active forms.
        a. The inactive form is known as the D (dependent) form and the active form as the I (independent) form, based on their relationship with G6P.
            (1) The D form of glycogen synthetase is an allosteric enzyme with high concentrations of G6P acting as a positive modulator.
            (2) Because the enzyme is inactive in the absence of G6P, it is called the dependent form.
        b. Interconversion of the D and I forms takes place by phosphorylation and is catalyzed by cyclic adenosine monophosphate–dependent (cAMP-dependent) protein kinase (see section III C). The phosphorylated form is the inactive (D) form.

B. Glycogenolysis

1. Phosphorylase

    a. The rate-limiting step in glycogen catabolism is the phosphorolysis of glycogen by phosphorylase.
    b. The enzyme is a dimer with an inactive form (phosphorylase b) and an active form (phosphorylase a), and it contains pyridoxal phosphate (see Chapter 31, section II C 3 c).

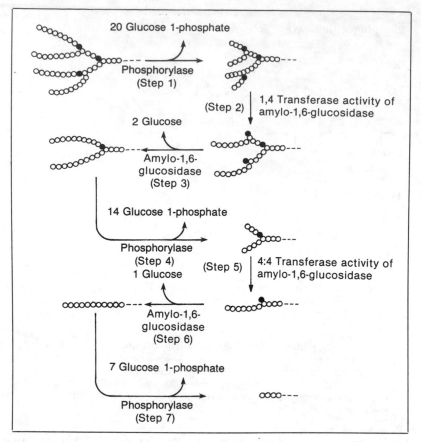

20 Glucose 1-phosphate

Phosphorylase
(Step 1)

(Step 2) | 1,4 Transferase activity of
amylo-1,6-glucosidase

2 Glucose

Amylo-1,6-
glucosidase
(Step 3)

14 Glucose 1-phosphate

Phosphorylase
(Step 4)
1 Glucose

(Step 5) | 4:4 Transferase activity of
amylo-1,6-glucosidase

Amylo-1,6-
glucosidase
(Step 6)

7 Glucose 1-phosphate

Phosphorylase
(Step 7)

Figure 18-2. Summary of the sequence of glycogen breakdown. The diagram shows the fate of four outer chains and the next branches from which they arise. Each chain is shown with nine residues beyond the branch. The residues linked by 1,6 bonds released as free glucose are indicated by *solid circles*. Sum: Of the 44 glucosyl residues removed, 3 have appeared as free glucose and 41 as glucose 1-phosphate. (Reprinted with permission from McGilvery RW: *Biochemistry: A Functional Approach*, 2nd ed. Philadelphia, WB Saunders, 1979, p 491.)

(1) The b form is converted to the a form by phosphorylation of a specific serine group on each subunit (see section III B 2).

(2) Cleavage of the phosphate by a phosphatase reconverts the a form to the inactive b form.

2. Phosphorylase kinase

a. This enzyme catalyzes the phosphorylation of phosphorylase b to give the a form at the expense of adenosine triphosphate (ATP).

b. The enzyme also exists in inactive and active forms, which are interconverted by phosphorylation–dephosphorylation.

(1) Phosphorylation to the active form is catalyzed by cAMP-dependent protein kinase (see section III C).

(2) Ca^{2+} in the range of 10^{-7} M causes partial activation of muscle phosphorylase kinase. This important activation coordinates the contraction of muscle (which is also activated by Ca^{2+}) and the muscle's consequent need for energy, which is then provided via glycolysis.

(3) Phosphorylase kinase is inactivated by dephosphorylation by a phosphatase.

C. Phosphorylation and cAMP-dependent protein kinase

1. The enzyme is a tetramer made up of two catalytic subunits (C) and two regulatory subunits (R).

2. In the absence of cAMP, the tetramer R_2C_2 is inactive.

3. Each of the R subunits binds cAMP, which causes these subunits to dissociate from the C subunits. The free C units are catalytically active and phosphorylate phosphorylase kinase at the expense of ATP.

D. Phosphorylation cascade

1. The enzymes of glycogen metabolism undergo a sequential covalent modification by means of phosphorylation. This process provides a very large amplification of the initial stimulus.

2. As discussed in section III F and in Chapter 30, section II A, B, cAMP is formed when adenylate cyclase is activated by circulating hormone (epinephrine or glucagon, depending on the tissue).

   a. Only a few molecules of hormone are needed to activate adenylate cyclase, which then produces a large number of cAMP molecules, each of which can activate a cAMP-dependent protein kinase enzyme molecule.

   b. This active enzyme in turn phosphorylates a large number of phosphorylase kinase molecules, activating each of them.

   c. Again, each phosphorylase kinase molecule can activate a large number of phosphorylase b molecules.

3. This type of amplification system is called a **cascade**.

E. Antithetic effects of covalent modification

1. With separate systems for the synthesis and degradation of glycogen, and with G1P acting as a common intermediate, the possibility of a futile cycling of glycogen must be considered.

2. The futile cycle is avoided because covalent modification, by phosphorylation, has opposite effects on the enzymes concerned with the synthesis and degradation of glycogen.

   a. An adequate level of cAMP stimulates formation of the inactive D form of glycogen synthetase and the active form of phosphorylase. Thus, glycogenesis is limited and glycogenolysis is increased.

   b. With low levels of cAMP, the active I form of glycogen synthetase predominates and the active form of phosphorylase kinase is low because the common phosphorylating enzyme, cAMP dependent protein kinase, is inactive. Thus, glycogenesis is increased, and glycogenolysis is decreased.

F. Hormonal regulation

1. **In muscle**

   a. **Epinephrine** promotes glycogenolysis and inhibits glycogensis.

      (1) It stimulates the formation of cAMP by activating adenylate cyclase (see Chapter 30, section II A, B).

      (2) Thus, in an emergency, when epinephrine is released and acts upon the muscle cell membrane, glycogenolysis is activated via the phosphorylation cascade, and simultaneously, glycogenesis is retarded.

   b. **Insulin** increases glycogenesis and decreases glycogenolysis.

      (1) It heightens the entry of glucose into the muscle cells.

      (2) It reduces cAMP levels, probably by speeding up the destruction of cAMP by phosphodiesterase (see Chapter 30, section II C).

   a. **Glucagon** activates adenylate cyclase in liver cell membranes and thus turns on glycogenolysis and reduces glycogenesis.

   b. **Insulin** increases glycogenesis in the liver by a mechanism that is not yet clear.

   c. **The glucagon:insulin ratio** appears more important than the absolute level of either hormone, glycogen metabolism being strongly influenced by the predominant hormone.

      (1) Insulin domination provides for the storage of glycogen after a meal.

      (2) Glucagon domination favors mobilization of glycogen stores as the blood glucose level declines.

IV. GENETIC DEFECTS IN GLYCOGEN METABOLISM

A. Type I (von Gierke's disease)

1. A deficiency of glucose 6-phosphatase (see Chapter 17, section II F 2 a) in liver, intestinal mucosa, and kidney causes this glycogen storage disease.

2. It occurs in only 1 person per 200,000 and is transmitted as an autosomal recessive trait.

3. Symptoms include

   a. Fasting hypoglycemia, because the liver cannot release enough glucose by means of glycogenolysis; only the free glucose from debranching enzyme activity is available.

   b. Lactic acidemia, because the liver cannot form glucose from lactate and because the liver responds to the action of glucagon by producing lactate instead of glucose. The increased blood lactate reduces blood pH and the alkali reserve.

   c. Hyperlipidemia, because the lack of hepatic gluconeogenesis (see Chapter 17) results in increased mobilization of fat as a metabolic fuel.

   (1) The increased lipids in the plasma are mainly mobilized free fatty acids.

   (2) The excessive mobilization of fat induces the formation of ketone bodies. While these can provide valuable metabolic fuel, they are acidic and, in conjunction with the increased lactic acid, tend to cause a metabolic acidosis.

   d. Hyperuricemia (with gouty arthritis), due to hyperactivity of the hexose monophosphate shunt pathway, with a concomitantly increased formation of pentose phosphates and phosphoribosyl pyrophosphate (see Chapter 29, section IX B 2 d).

B. Other types. A number of other genetic glycogen storage defects (glycogenoses) have been described. These are summarized in Table 18-1.

Table 18-1. Genetic Defects in Glycogen Metabolism

Type	Enzyme System Affected	Organs Involved	Clinical Symptoms	Eponym
0	Uridine diphosphate glucose–glycogen transferase (glycogen synthetase)	Liver, muscle	Large fatty liver, fasting hypoglycemia	. . .
Ia	Glucose-6-phosphatase	Liver, kidney	Large liver and kidney; growth retardation; severe hypoglycemia; acidosis; hyperlipemia; hyperuricemia	von Gierke's disease
Ib	Glucose-6-phosphatase translocase	Liver, leukocytes	As above but less severe; neutropenia, recurrent gastrointestinal infections	. . .
II	Lysosomal glucosidase (various types)	All organs	Large liver and heart; no abnormal blood chemistry	Pompe's disease
III	Debranching enzyme system	Liver, muscle, heart leukocytes	Enlarged liver, fasting hypoglycemia, variable muscle involvement	Forbes's disease
IV	Branching enzyme system	Liver, muscle, and most tissues	Progressive cirrhosis in juvenile type; myopathy and heart failure in late-onset type	Andersen's disease
V	Muscle phosphorylase	Skeletal muscle	Cramps on exercise with no rise in blood lactate	McArdle's disease
VI	Liver phosphorylase	Liver	Enlarged liver, fasting hypoglycemia but often no symptoms at all	Hers's disease
VII	Phosphofructokinase	Skeletal muscle erythrocytes	Cramps on exercise but no rise in blood lactate, hemolysis	Tarui's disease

VIII, IX, X: Rare disorders involving various components of the liver phosphorylase activating–deactivating cascade

Note.—Reprinted with permission from Berkow R (Ed): *The Merck Manual of Diagnosis and Therapy*, 15th ed. Rahway, New Jersey, Merck and Company, 1987.

STUDY QUESTIONS

Directions: Each question below contains five suggested answers. Choose the **one best** response to each question.

1. What types of linkages are present between the glucose units of glycogen?

(A) β-1,4 linkages only
(B) β-1,4 and β-1,6 linkages
(C) α-1,4 linkages only
(D) α-1,4 and α-1,6 linkages
(E) β-1,6 linkages only

2. Glycogen is stored in the cells as

(A) a component of endoplasmic reticulum membranes
(B) granules, which also contain the enzymes that catalyze its formation and degradation
(C) granules adhering to the inner side of the plasma membrane
(D) a component of the Golgi apparatus, where it is formed
(E) free glycogen in solution in the cytosol

3. The greatest amount of body glycogen can be found in which of the following human tissues?

(A) Liver
(B) Kidney
(C) Skeletal muscle
(D) Cardiac muscle
(E) Adipose tissue

4. Free glucose is formed during glycogenolysis from

(A) glucose residues in 1,4 linkage to the main chain
(B) glucose residues in 1,6 linkage to the main chain
(C) glucose 1-phosphate hydrolysis
(D) glucose 1,6-diphosphate hydrolysis
(E) breakdown of uridine diphosphoglucose

5. Glycogen synthetase is characterized by all of the following statements EXCEPT

(A) the enzyme exists in active and inactive forms
(B) uridine diphosphate glucose is a substrate
(C) it is activated by phosphorylation
(D) it requires a primer strand of glycogen
(E) it is found in association with glycogen granules

6. Glycogen synthetase D is activated allosterically by

(A) glucose 1-phosphate
(B) fructose 6-phosphate
(C) glucose 6-phosphate
(D) fructose 1,6-diphosphate
(E) glucose

7. Which of the following statements regarding the regulation of glycogenolysis is correct?

(A) cyclic adenosine monophosphate (cAMP) enhances glycogenolysis by adenylation of glycogen phosphorylase b
(B) phosphorylase b is activated by phosphorylation
(C) phosphorylase b kinase is inactivated by phosphorylation by cAMP dependent protein kinase
(D) muscle phosphorylase b kinase is inactivated by Ca^{2+}
(E) none of the above

8. A liver biopsy from an infant with a large liver and kidney, dwarfism, hypoglycemia, acidosis, and hyperlipidemia revealed a massive buildup of glycogen with a normal structure. A possible diagnosis would be

(A) α-1,4-glucosidase deficiency (Pompe's disease)
(B) branching enzyme deficiency (Andersen's disease)
(C) debranching enzyme deficiency (Cori's disease)
(D) glucose 6-phosphatase deficiency (von Gierke's disease)
(E) liver phosphorylase deficiency (Hers' disease)

Directions: The question below contains four suggested answers of which **one or more** is correct. Choose the answer

- A if **1, 2, and 3** are correct
- B if **1 and 3** are correct
- C if **2 and 4** are correct
- D if **4** is correct
- E if **1, 2, 3, and 4** are correct

9. Glycogen phosphorylase a is often assayed in vitro by measuring the release of inorganic phosphate from glucose 1-phosphate (G1P); that is, it is assayed in the direction of glycogen synthesis. Glycogen is not synthesized by glycogen phosphorylase in the cell because

(1) this reaction is used by the cell to synthesize glycogen

(2) G1P is an allosteric inactivator of phosphorylase a

(3) under cellular conditions, the concentration of G1P is not sufficiently high

(4) under cellular conditions, the concentration of inorganic phosphate is too high

ANSWERS AND EXPLANATIONS

1. The answer is D. (*I A 1*) Glycogen is a very large polymer of glucose molecules linked along its main line by α-1,4-glycosidic linkages. It is a highly branched molecule with the branches arising from α-1,6-glycosidic bonds at about every tenth residue.

2. The answer is B. (*I A 2*) Glycogen is stored in granules in the cytoplasm. These granules also contain the enzymes that catalyze its formation and degradation.

3. The answer is C. (*I B 2*) Muscle and liver are major sites for glycogen storage, and although the concentration of glycogen is higher in liver than in muscle, the much greater mass of skeletal muscle stores a greater total amount of glycogen. It is important to recognize that liver glycogen can be mobilized for the release of glucose to the rest of the body, but muscle glycogen can only be used to support muscle glycolysis.

4. The answer is B. (*II C*) The rate-limiting step in glycogenolysis is the phosphorylytic cleavage of terminal 1,4-glycosidic residues. This process continues until a glucose residue is reached, which is four residues from a 1,6 branch. The enzyme amylo-1,6-glucosidase (debranching enzyme) then carries out two reactions. In the first, it transfers three glycosyl residues from a branch onto a chain terminus, leaving a single residue on C-6. In the second, the same enzyme removes the single residue on C-6 to yield a free glucose molecule.

5. The answer is C. (*III A 2*) Glycogen synthetase exists in active and inactive forms. Glycogen synthetase D (dependent) is inactive except in the presence of high concentrations of glucose 6-phosphate (G6P), its positive allosteric modulator. Glycogen synthetase I (independent) is active in the absence of G6P. The two forms are interconverted by phosphorylation–dephosphorylation reactions with the phosphorylated form being inactive. The enzyme transfers glucose residues from uridine diphosphate glucose to C-4 terminals on existing glycogen chains, forming α-1,4 linkages.

6. The answer is C. (*III A 2 a*) Glycogen synthetase exists in active and inactive forms. The inactive form is known as glycogen synthetase D (dependent) and the active form as glycogen synthetase I (independent), based on their relationship to glucose 6-phosphate (G6P). Glycogen synthetase D is an allosteric enzyme with high concentrations of G6P acting as a positive modulator. In the absence of G6P, the enzyme is inactive. Glycogen synthetase I is not an allosterically regulated enzyme and is active in the absence of G6P.

7. The answer is B. (*III B*) Phosphorylase b is the inactive form of the rate-limiting enzyme of glycogenolysis. It is converted to the active form, phosphorylase a, by phosphorylation catalyzed by phosphorylase b kinase. Cyclic adenosine monophosphate (cAMP) does not interact directly with phosphorylase b but instead activates the cAMP-dependent protein kinase that activates phosphorylase b kinase by phosphorylation. Muscle phosphorylase b kinase is activated by Ca^{2+} at about 10^{-7} M.

8. The answer is D. (*IV A; Table 18-1*) A deficiency of glucose 6-phosphatase in liver, kidney, and intestinal mucosa causes the glycogen storage disease known as von Gierke's disease. The findings and their causes include the following: (1) fasting hypoglycemia due to the inability of the liver or kidney to form and release free glucose; (2) lactic acidemia because the liver cannot form glucose from lactate; (3) hyperlipidemia, because the lack of hepatic gluconeogenesis results in the increased mobilization of fat for fuel; the excessive mobilization of fat induces the formation of ketone bodies, which exacerbate the acidosis; (4) hyperuricemia, because of increased activity of the hexose monophosphate shunt and the concomitant increased formation of phosphoribosyl pyrophosphate.

9. The answer is D (4). (*II C 1 c*) Glycogen is synthesized from uridine diphosphate glucose and broken down to glucose 1-phosphate (G1P) by separate enzyme systems in the cell. Glycogen phosphorylase a catalyzes the phosphorolysis of glycogen to G1P, a reaction requiring inorganic phosphate as one of the substrates. Although the reaction can be run in reverse in vitro if the phosphate concentration is kept low, it does not do so in the cell because the inorganic phosphate levels are always too high.

19

Alternate Pathways of Carbohydrate Metabolism

I. PENTOSE PHOSPHATE PATHWAY (HEXOSE MONOPHOSPHATE SHUNT)

A. Functional significance

1. The pentose phosphate pathway serves as a source of reduced nicotinamide-adenine dinucleotide phosphate (NADPH) for reductive synthesis.

2. It is a source of pentoses, derived from glucose, for nucleic acid synthesis.

3. It provides a route for the use of pentoses and for their conversion to hexoses and trioses.

B. Enzymatic pathway.
The pathway is most active in the liver, the mammary glands, adipose tissue (in the rat), and the adrenal cortex. In effect, it is most active in fat-synthesizing tissues. The enzymes involved are located in the cytosol.

1. The pathway uses glucose 6-phosphate (G6P) as substrate, and the first two steps are oxidations catalyzed by NADP-linked dehydrogenases. Two moles of NADPH are formed per mole of G6P oxidized by this pathway.

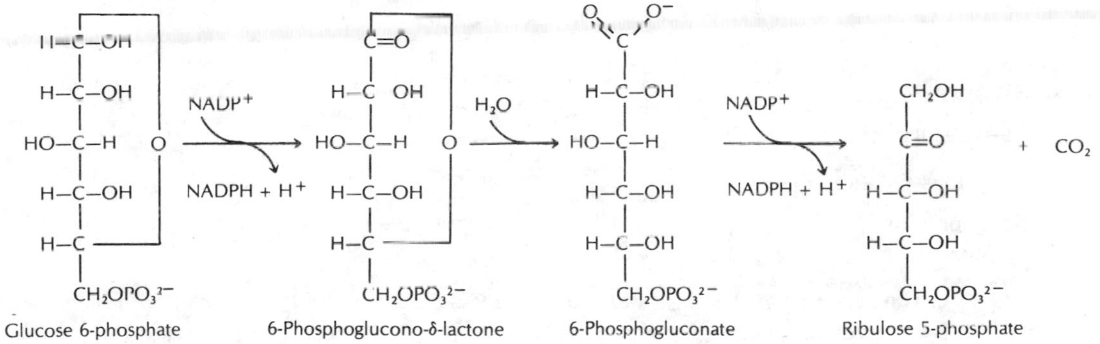

Glucose 6-phosphate   6-Phosphoglucono-δ-lactone   6-Phosphogluconate   Ribulose 5-phosphate

     a. **The first step** is the catalysis of **G6P to 6-phosphoglucono-δ-lactone** by glucose 6-phosphate dehydrogenase (G6PD), an NADP-specific dehydrogenase. The δ-lactone is hydrolyzed enzymatically to 6-phosphogluconate.

     b. **The second step** is the conversion of **6-phosphogluconate to ribulose 5-phosphate** and carbon dioxide (CO_2), catalyzed by 6-phosphogluconate dehydrogenase, which is also an NADP-linked enzyme.

2. **Conversion of ribulose 5-phosphate to ribose 5-phosphate.** In this reaction, the isomerization passes through an enediol intermediate.

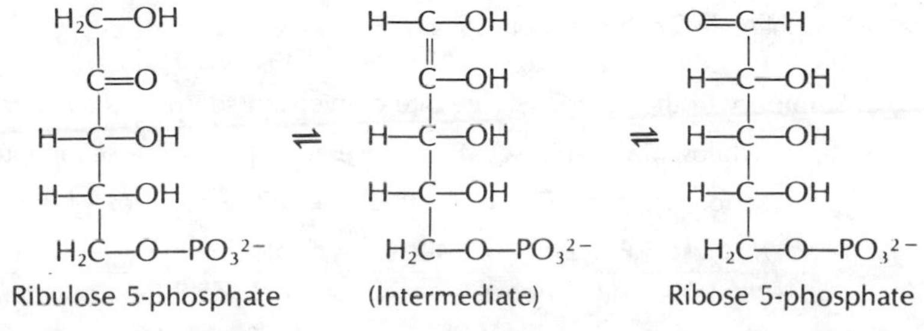

Ribulose 5-phosphate   (Intermediate)   Ribose 5-phosphate

3. **Interconversion of pentoses, hexoses, and trioses.** These steps in the pathway are catalyzed by **transketolase** and **transaldolase**. Transketolase transfers *two-carbon units* from a **ketose** to an **aldose** and is a thiamine pyrophosphate (TPP)–dependent enzyme. Transaldolase transfers *three-carbon units* from a ketose to an aldose and does not require TPP as a cofactor.

   a. The first transketolase step
     (1) In this step, the carrier of the two-carbon moiety is the TPP prosthetic group of the enzyme.
     (2) The xylulose 5-phosphate is derived from ribulose 5-phosphate by the action of an epimerase.

$$
\begin{array}{ccccc}
\text{CH}_2\text{OH} & \text{O=C—H} & & \text{O=C—H} & \text{CH}_2\text{—OH} \\
\text{C=O} & \text{H—C—OH} & & \text{H—C—OH} & \text{C=O} \\
\text{HO—C—H} & \text{H—C—OH} & \rightleftharpoons & \text{H—C—OH} & \text{HO—C—H} \\
\text{H—C—OH} & \text{H—C—OH} & & \text{CH}_2\text{—O—PO}_3^{2-} & \text{H—C—OH} \\
\text{CH}_2\text{—O—PO}_3^{2-} & \text{CH}_2\text{—O—PO}_3^{2-} & & & \text{H—C—OH} \\
& & & & \text{H—C—OH} \\
& & & & \text{CH}_2\text{—O—PO}_3^{2-}
\end{array}
$$

Xylulose 5-phosphate   Ribose 5-phosphate   Glyceraldehyde 3-phosphate   Sedoheptulose 7-phosphate

   b. The transaldolase step. No cofactor is required to act as carrier of the three-carbon unit in this step.

$$
\begin{array}{ccccc}
\text{CH}_2\text{OH} & & & \text{O=C—H} & \text{CH}_2\text{OH} \\
\text{C=O} & & & \text{H—C—OH} & \text{C=O} \\
\text{HO—C—H} & \text{O=C—H} & & \text{H—C—OH} & \text{HO—C—H} \\
\text{H—C—OH} & \text{H—C—OH} & \rightleftharpoons & \text{CH}_2\text{—O—PO}_3^{2-} & \text{H—C—OH} \\
\text{H—C—OH} & \text{CH}_2\text{—O—PO}_3^{2-} & & & \text{H—C—OH} \\
\text{H—C—OH} & & & & \text{CH}_2\text{—O—PO}_3^{2-} \\
\text{CH}_2\text{—O—PO}_3^{2-} & & & &
\end{array}
$$

Sedoheptulose 7-phosphate   Glyceraldehyde 3-phosphate   Erythrose 4-phosphate   Fructose 6-phosphate

   c. The second transketolase step. Again, TPP is the carrier of the two-carbon moiety.

$$
\begin{array}{ccccc}
\text{CH}_2\text{OH} & \text{O=C—H} & & \text{O=C—H} & \text{CH}_2\text{OH} \\
\text{C=O} & \text{H—C—OH} & & \text{H—C—OH} & \text{C=O} \\
\text{HO—C—H} & \text{H—C—OH} & \rightleftharpoons & \text{CH}_2\text{—O—PO}_3^{2-} & \text{HO—C—H} \\
\text{H—C—OH} & \text{CH}_2\text{—O—PO}_3^{2-} & & & \text{H—C—OH} \\
\text{CH}_2\text{—O—PO}_3^{2-} & & & & \text{H—C—OH} \\
& & & & \text{CH}_2\text{—O—PO}_3^{2-}
\end{array}
$$

Xylulose 5-phosphate   Erythrose 4-phosphate   Glyceraldehyde 3-phosphate   Fructose 6-phosphate

   d. Summary of the reactions. The interconversion sequences are summed below.

Xylulose 5-P + ribose 5-P ⇌ glyceraldehyde 3-P + sedoheptulose 7-P

Sedoheptulose 7-P + glyceraldehyde 3-P ⇌ erythrose 4-P + fructose 6-P

Xylulose 5-P + erythrose 4-P ⇌ glyceraldehyde 3-P + fructose 6-P
———————————————————————————————————————
Sum: 2 Xylulose 5-P + ribose 5-P ⇌ glyceraldehyde 3-P + 2 fructose 6-P

C. Regulation of the pathway

1. The cellular concentration of NADP$^+$ is the major controlling factor; its availability regulates the rate-limiting G6PD reaction.

2. A slow regulation occurs by the induction of increased synthesis of G6PD and 6-gluconate dehydrogenase.

3. The control of the interconversion of pentoses and hexoses is not known.

D. G6PD deficiency

1. Red blood cells depend upon the pentose phosphate pathway for the formation of NADPH, which the cells require to maintain glutathione in a reduced state. The reduced glutathione is involved in maintaining the integrity of the red cell membrane.

2. G6PD activity may be genetically absent or may be present as a partially active variant. In affected individuals, many oxidizing substances (e.g., the antimalarial primaquine, aspirin, sulfonamides, nitrofurans, phenacetin, and, in some whites, fava beans) will cause hemolysis of red cells, leading to a hemolytic anemia.

II. URONIC ACID PATHWAY (GLUCURONIC ACID CYCLE)

A. Functional significance

1. The pathway is a source of activated glucuronate for
   a. **The formation of glucuronides.** Many naturally occurring substances, as well as many drugs, are eliminated from the body by the formation of glucuronide derivatives of the substances. This conjugation (detoxification) of waste products allows them to be excreted in the bile and urine as water-soluble compounds. Examples are steroid hormones, bilirubin, morphine, salicylic acid, and menthol. The general reaction is

   Uridine diphosphate glucuronide (UDP-glucuronide) + acceptor → a glucuronide + UDP

   b. **The biosynthesis of certain polysaccharides.** UDP-glucuronate is the donor for the glucuronyl moiety in some polysaccharides, such as heparin.
   c. **The biosynthesis of chondroitin sulfate**, a glycosaminoglycan (see section V B)

2. The pathway is integral to the formation of ascorbic acid (vitamin C) in most animals with the exception of humans, other primates, guinea pigs, and an East Indian fruit bat.

3. It is also needed for the formation of pentoses and the metabolism of nonphosphorylated sugar derivatives.

B. Pathway

1. **G6P to uridine diphosphate glucose (UDP-glucose).** This conversion involves steps that are common to glycogenesis.

Glucose 6-phosphate   Glucose 1-phosphate   Uridine diphosphate glucose

2. UDP-glucose to uridine diphosphate glucuronate (UDP-glucuronate)

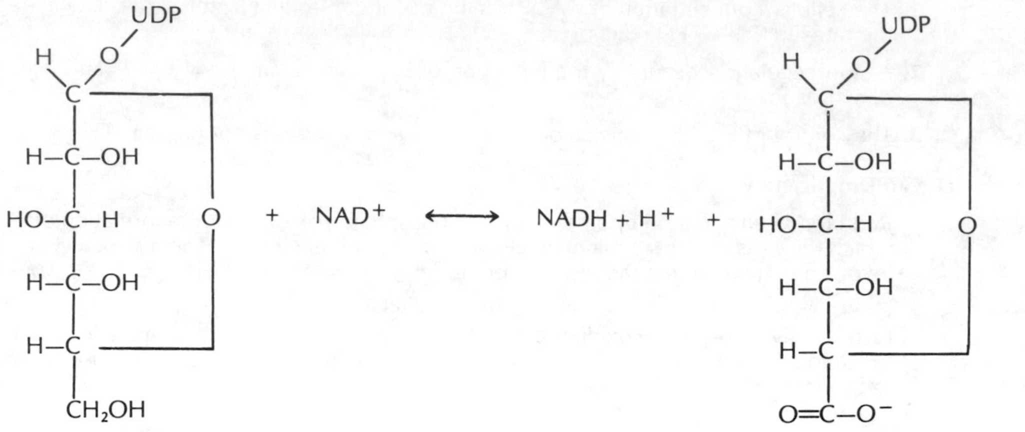

Uridine diphosphate glucose + NAD⁺ ⟷ NADH + H⁺ + Uridine diphosphate glucuronate

3. UDP-glucuronate to L-gulonate

  a. UDP-glucuronate (activated glucuronate) is the donor species in the formation of glucuronides and in the formation of chondroitin sulfate, which contains glucuronate moieties. It also provides glucuronyl for some polysaccharides (see section II A 1).
  b. If UDP-glucuronate is not used as a donor of the glucuronyl moiety, it can be converted to free glucuronate, probably with glucuronate 1-phosphate as an intermediate. The free glucuronate is then reduced to L-gulonate with NADPH as the electron donor.

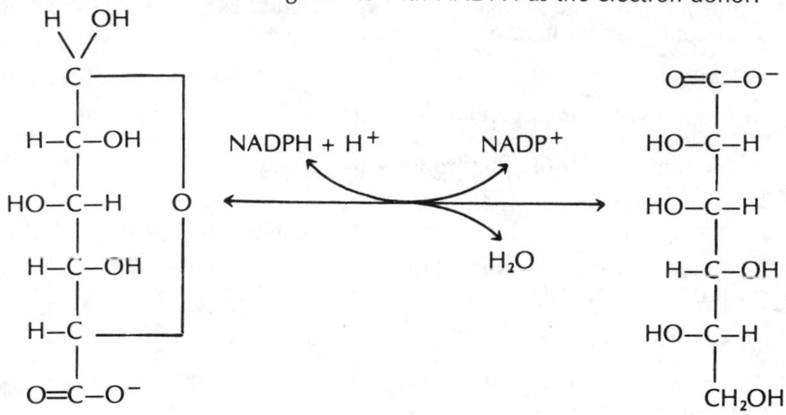

D-Glucuronate   NADPH + H⁺ → NADP⁺ / H₂O   L-Gulonate

  (1) Note the change from D- to L-.
  (2) This occurs because the most highly oxidized carbon in L-gulonate is now C-1, as opposed to C-6 in D-glucuronate.

4. L-Gulonate to ascorbate

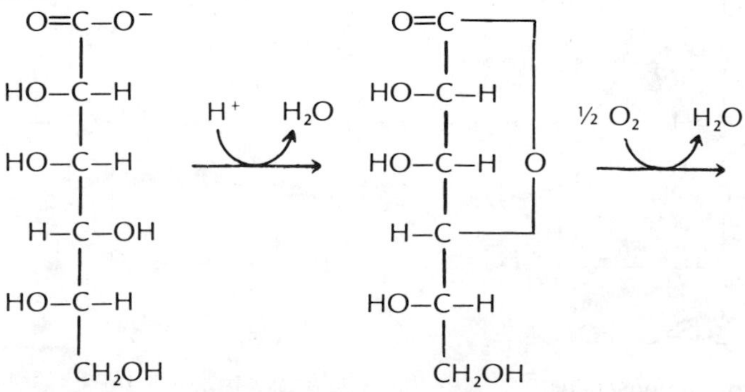

L-Gulonate   H⁺ / H₂O →   L-Gulonolactone   ½ O₂ / H₂O →

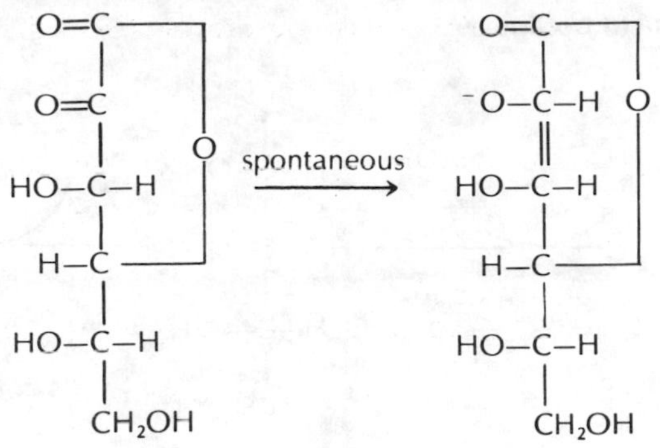

$$\text{2-Keto-L-gulonolactone} \xrightarrow{\text{spontaneous}} \text{L-Ascorbic acid}$$

2-Keto-L-gulonolactone   **L-Ascorbic acid**

a. The formation of 2-keto-L-gulonolactone is the step that is blocked in humans, other primates, and guinea pigs.

b. The oxidation of ascorbic acid occurs readily; it is catalyzed by ascorbic acid oxidase in plants and by heavy metals in animals.

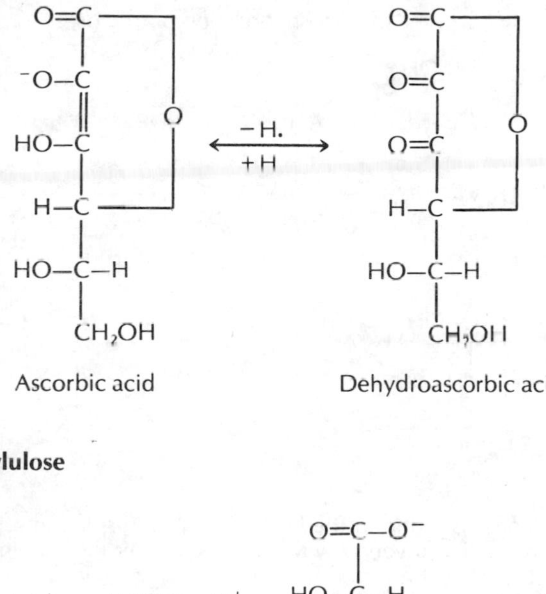

$$\text{Ascorbic acid} \underset{+H}{\overset{-H.}{\rightleftharpoons}} \text{Dehydroascorbic acid}$$

Ascorbic acid   **Dehydroascorbic acid**

5. L-Gulonate to L-xylulose

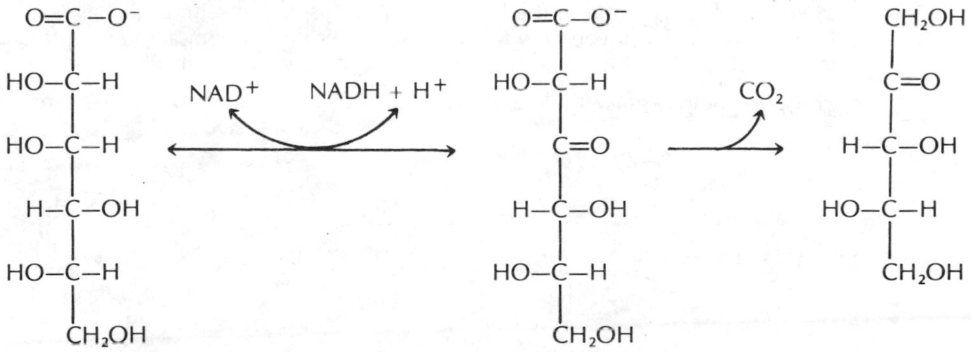

$$\text{L-Gulonate} \underset{NAD^+ \quad NADH + H^+}{\longleftrightarrow} \text{3-Keto-L-gulonate} \xrightarrow{CO_2} \text{L-Xylulose}$$

L-Gulonate   **3-Keto-L-gulonate**   **L-Xylulose**

6. The above reaction yields L-xylulose, but the D- form of this pentose is required by the pentose phosphate pathway.

a. L-Xylulose to D-xylulose

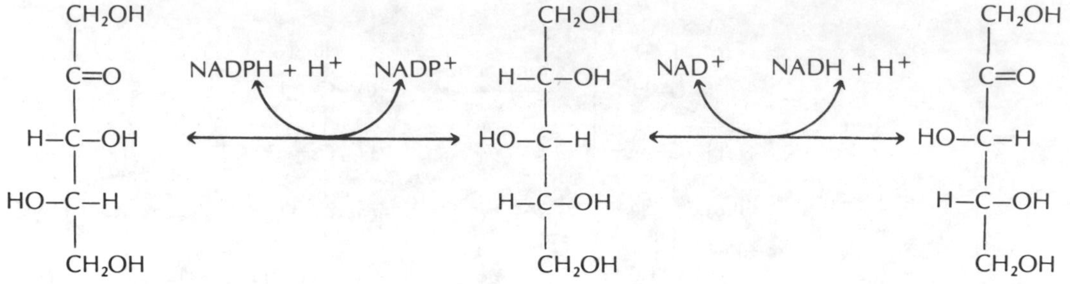

CH₂OH   CH₂OH   CH₂OH

$$
\begin{array}{ccc}
\text{CH}_2\text{OH} & \text{CH}_2\text{OH} & \text{CH}_2\text{OH} \\
| & | & | \\
\text{C}=\text{O} & \text{H}-\text{C}-\text{OH} & \text{C}=\text{O} \\
| & | & | \\
\text{H}-\text{C}-\text{OH} & \text{HO}-\text{C}-\text{H} & \text{HO}-\text{C}-\text{H} \\
| & | & | \\
\text{HO}-\text{C}-\text{H} & \text{H}-\text{C}-\text{OH} & \text{H}-\text{C}-\text{OH} \\
| & | & | \\
\text{CH}_2\text{OH} & \text{CH}_2\text{OH} & \text{CH}_2\text{OH}
\end{array}
$$

L-Xylulose $\xrightarrow{\text{NADPH} + \text{H}^+ \quad \text{NADP}^+}$ Xylitol $\xrightarrow{\text{NAD}^+ \quad \text{NADH} + \text{H}^+}$ D-Xylulose

b. D-Xylulose is phosphorylated to D-xylulose 5-phosphate at the expense of adenosine triphosphate (ATP) and then can be metabolized by the pentose phosphate pathway.

III. METABOLISM OF FRUCTOSE.
Fructose is a product of sucrose hydrolysis and is probably the second most abundant dietary sugar after glucose. Sources include sugar beets, sugar cane, sorghum, maple sugar, pineapple, ripe fruits, and honey.

A. Major pathway of fructose metabolism

1. Fructokinase catalyzes the phosphorylation of fructose.

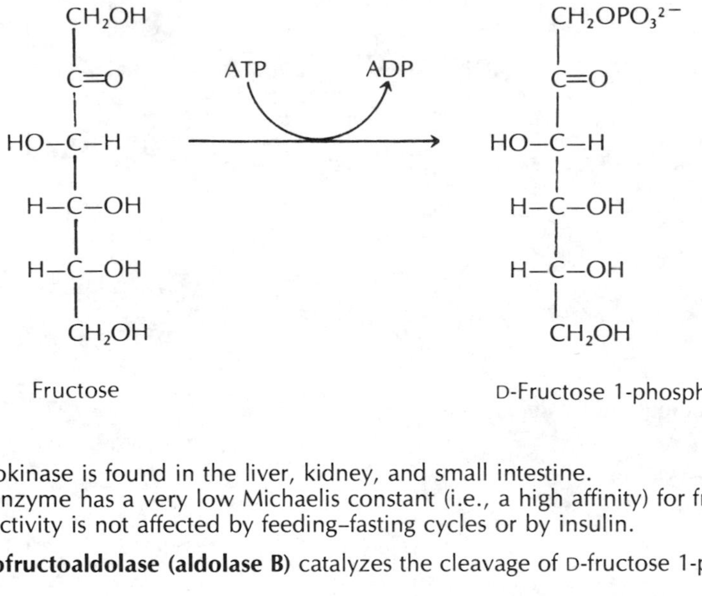

$$
\begin{array}{ccc}
\text{CH}_2\text{OH} & & \text{CH}_2\text{OPO}_3{}^{2-} \\
| & & | \\
\text{C}=\text{O} & \xrightarrow{\text{ATP} \quad \text{ADP}} & \text{C}=\text{O} \\
| & & | \\
\text{HO}-\text{C}-\text{H} & & \text{HO}-\text{C}-\text{H} \\
| & & | \\
\text{H}-\text{C}-\text{OH} & & \text{H}-\text{C}-\text{OH} \\
| & & | \\
\text{H}-\text{C}-\text{OH} & & \text{H}-\text{C}-\text{OH} \\
| & & | \\
\text{CH}_2\text{OH} & & \text{CH}_2\text{OH}
\end{array}
$$

Fructose   D-Fructose 1-phosphate

a. Fructokinase is found in the liver, kidney, and small intestine.
b. The enzyme has a very low Michaelis constant (i.e., a high affinity) for fructose.
c. The activity is not affected by feeding–fasting cycles or by insulin.

2. Phosphofructoaldolase (aldolase B) catalyzes the cleavage of D-fructose 1-phosphate.

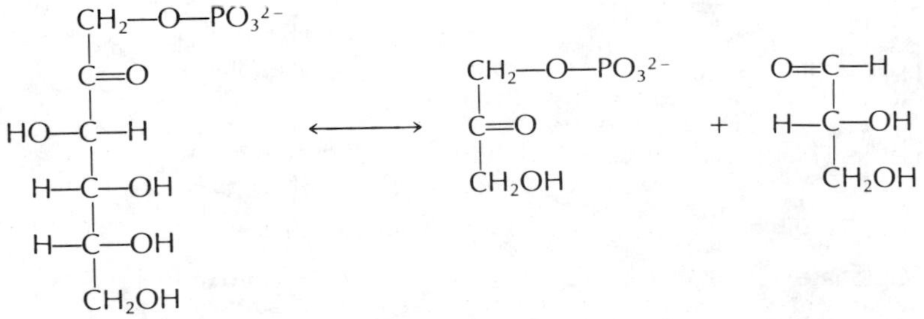

$$
\begin{array}{ccc}
\text{CH}_2-\text{O}-\text{PO}_3{}^{2-} & & \\
| & & \\
\text{C}=\text{O} & & \text{CH}_2-\text{O}-\text{PO}_3{}^{2-} \qquad \text{O}=\text{C}-\text{H} \\
| & & | \qquad\qquad\qquad\qquad\quad | \\
\text{HO}-\text{C}-\text{H} & \longleftrightarrow & \text{C}=\text{O} \qquad + \quad \text{H}-\text{C}-\text{OH} \\
| & & | \qquad\qquad\qquad\qquad\quad | \\
\text{H}-\text{C}-\text{OH} & & \text{CH}_2\text{OH} \qquad\qquad \text{CH}_2\text{OH} \\
| & & \\
\text{H}-\text{C}-\text{OH} & & \\
| & & \\
\text{CH}_2\text{OH} & &
\end{array}
$$

D-Fructose 1-phosphate   Dihydroxyacetone   D-Glyceraldehyde
                                phosphate

    a. Phosphofructoaldolase is an isozyme of the glycolytic pathway aldolase.
    b. It is abundant in the liver.

  3. The fate of D-glyceraldehyde may take several routes.
    a. Phosphorylation to D-glyceraldehyde 3-phosphate (G3P) and metabolism by the glycolytic pathway or the gluconeogenic pathway.
    b. Oxidation to D-glycerate by a nicotinamide-adenine dinucleotide (NAD^+)–linked glyceraldehyde dehydrogenase. D-Glycerate may then be converted to L-serine via hydroxypyruvate [see Chapter 27, section III B 5 a (4)].
    c. Reduction to glycerol by a NADH-linked glycerol dehydrogenase, followed by phosphorylation to glycerol 3-phosphate. The latter may be used for gluconeogenesis or triglyceride synthesis.

B. Specialized pathways of fructose metabolism

  1. Fructose metabolism in spermatozoa
    a. Fructose is the major energy source for spermatozoa and is formed from glucose in the seminal vesicle.
    b. The pathway involves reduction of glucose to D-**sorbitol** and oxidation of the latter to D-**fructose**.

D-Glucose   D-Sorbitol   D-Fructose

    c. The fructose concentration of semen may reach 10 mM. Most of this is available for the spermatozoa because fructose is used sparingly by the other tissues that come in contact with the seminal fluid.
    d. The mitochondria of sperm are the only such organelles to contain lactate dehydrogenase. Because this enzyme is present, the lactate that is formed by fructolysis can be completely oxidized to CO_2 and H_2O without the need for a shuttle system to transport reducing equivalents into the mitochondria.

  2. Sorbitol metabolism in diabetes
    a. The formation of sorbitol from glucose proceeds rapidly in the lens of the eye and in the Schwann cells of the nervous system.
    b. Sorbitol cannot pass through the cell membrane, and in diabetic individuals, sorbitol levels build up in these cells because the rate of oxidation of sorbitol to fructose is decreased.
    c. It is thought that the elevated sorbitol concentration causes an increase in osmotic pressure, which might be a causative factor in the development of the lens cataracts and the neural dysfunction that occur in diabetes.

C. Hereditary fructose intolerance is a genetic disorder characterized by vomiting and hypoglycemia after ingestion of fructose or fructose-producing foods (e.g., sucrose and sorbitol).

  1. It is an autosomal recessive condition involving a deficiency of fructose 1-phosphate aldolase.

  2. Eating a meal containing fructose causes an accumulation of fructose 1-phosphate in tissues, which is believed to underlie the hepatic and renal damage found in this condition.

  3. Several enzymatic activities are inhibited by high cellular levels of fructose 1-phosphate.
    a. Fructokinase. Deficiency of fructokinase leads to fructosemia and fructosuria.

b. Hepatic glycogen phosphorylase. The lack of this enzyme is in part responsible for the postprandial hypoglycemia and glucagon unresponsiveness.

c. Fructose 1,6-diphosphate aldolase. Without this enzyme, gluconeogenesis is blocked, contributing to the postprandial hypoglycemia.

IV. METABOLISM OF LACTOSE. Lactose [β-D-galactosyl-(1→4)-α-D-glucose] is a disaccharide of D-galactose and D-glucose. The biosynthesis of lactose is confined to mammary tissue, and it is found in milk at a concentration of about 5%. Lactose makes up about 40% of the dietary intake in the first year of life.

A. Biosynthesis

1. UDP-glucose is converted to UDP-galactose by an epimerase, UDP-glucose-4-epimerase.

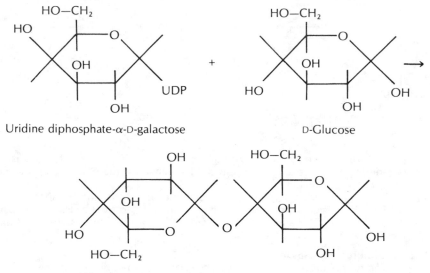

2. Lactose is formed from the reaction of UDP-galactose and D-glucose, which is catalyzed by galactosyl transferase.

Lactose [β-D-galactosyl-(1 → 4)-α-D-glucose]

B. Control of lactose synthesis

1. Galactosyl transferase does not use D-glucose as one of its substrates in tissues other than the mammary gland. Rather, it is involved in the transfer of the galactosyl moiety from UDP-α-D-galactose to **N-acetylglucosamine** to form **N-acetylgalactosamine**.

2. Lactose synthetase has a catalytic subunit, namely galactosyl transferase, and a modifier subunit, α-lactalbumin.

3. In the mammary gland, binding of the α-lactalbumin modifier to the catalytic subunit brings about the transfer of the galactosyl moiety to D-glucose to form lactose.

4. α-Lactalbumin levels in mammary tissue are under hormonal control, and the hormonal changes that follow parturition increase the cellular levels of the modifier subunit.

5. Prolactin increases the rate of synthesis of galactosyl transferase and of α-lactalbumin.

    6. Progesterone inhibits the synthesis of α-lactalbumin. At parturition, the levels of progesterone fall precipitously, and α-lactalbumin is synthesized.

C. Use of galactose for glycolysis

    1. Cleavage by intestinal lactase

      a. Lactose is hydrolyzed to yield galactose and glucose in the small intestine of infants and small children by one form of the enzyme.

      b. An adult form of the enzyme is less active and is missing in many adults, leading to a low tolerance for milk and for milk products containing lactose.

    2. Galactose is phosphorylated by galactokinase to form galactose 1-phosphate.

    3. Hexose 1-phosphate uridyl transferase catalyzes the transfer of UDP from UDP-glucose to galactose 1-phosphate to form UDP-galactose and glucose 1-phosphate (G1P).

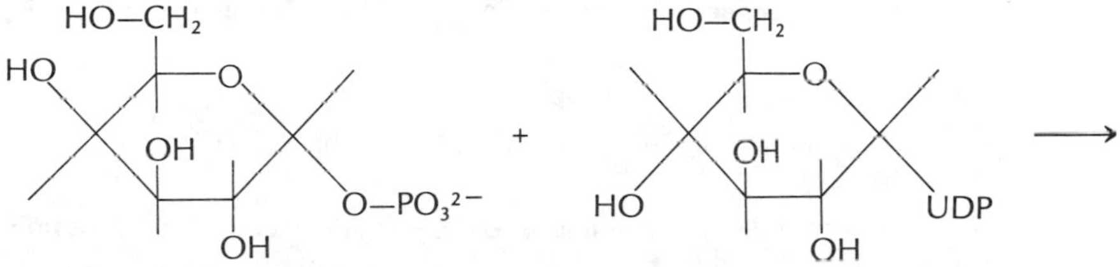

Galactose 1-phosphate   +   Uridine diphosphate-D-glucose   ⟶

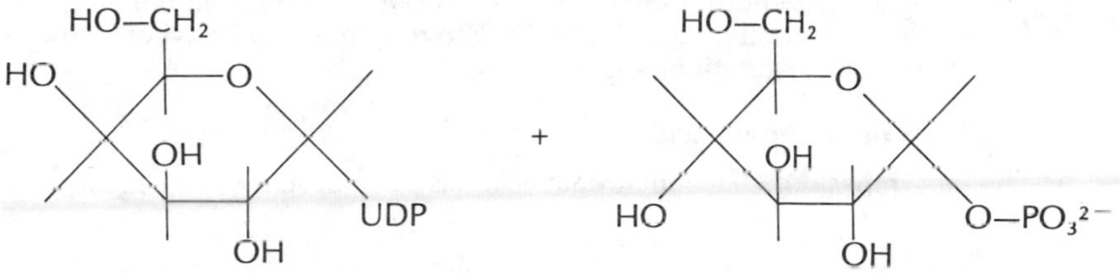

Uridine diphosphate-D-galactose   +   Glucose 1-phosphate

    4. An epimerase converts UDP-galactose to UDP-glucose.

D. Galactosemia

    1. An excessive accumulation of galactose is due to inborn errors of galactose metabolism.

    2. Two types, both autosomal recessive, are known.

      a. Galactokinase deficiency

        (1) In this condition, there is an accumulation of galactose in blood and tissues.

        (2) In the lens of the eye, galactose is reduced by aldose reductase to dulcitol (galacitol), which cannot escape from the cells. The osmotic effect of the sugar alcohol contributes to the development of cataracts.

      b. Galactose 1-phosphate uridyl transferase deficiency

        (1) This condition (also referred to as **classic galactosemia**) leads to the accumulation of both galactose and galactose 1-phosphate in tissues.

        (2) Cataracts also arise, as in galactokinase deficiency.

        (3) It is believed that the mental retardation and liver cirrhosis of classic galactosemia are related to increased cellular levels of galactose 1-phosphate in neural tissues and liver cells.

V. CARBOHYDRATES IN STRUCTURAL ELEMENTS

A. Connective tissues contain considerable quantities of **proteoglycans**, which are compounds of heteropolysaccharides and proteins. Heteropolysaccharides are polymers made up of several different kinds of sugar (or sugar derivative) units. Proteoglycans form the **ground substance**, or **matrix**, of connective tissues.

B. Glycosaminoglycans (GAGs), the heteropolysaccharides of the proteoglycans, are made up of repeating disaccharide units, each of which contains a hexuronic acid (β-D-glucuronic or

L-iduronic acid) and an amino sugar (glucosamine or galactosamine). The structures of the monomeric units are given in Figure 19-1.

C. Biosynthesis of amino sugars

1. **Glucosamine 6-phosphate** is the precursor of all the hexosamine residues in GAGs.

2. Glucosamine 6-phosphate is formed by the transfer of the amide link of glutamine to fructose 6-phosphate, a reaction catalyzed by L-glutamine D-fructose 6-phosphate aminotransferase; there is a simultaneous isomerization of the ketose to the aldose.

$$\text{D-Fructose 6-P + glutamine} \rightarrow \text{D-glucosamine 6-P + glutamate}$$

3. The amino group is then acetylated in a reaction using acetyl coenzyme A (acetyl CoA) as the acetyl donor.

$$\text{D-Glucosamine 6-P + acetyl CoA} \rightarrow \text{N-acetyl-}\beta\text{-D-glucosamine 6-P + CoASH}$$

4. A mutase catalyzes the transfer of the phosphate from the C-6 to the C-1 position to yield N-acetyl-β-D-glucosamine 1-phosphate.

5. A uridyl group is transferred from uridine triphosphate (UTP) to form UDP-N-acetyl-β-D-glucosamine.

$$\text{N-acetyl-}\beta\text{-D-glucosamine + UTP} \rightarrow \text{UDP-N-acetyl-}\beta\text{-D-glucosamine + PP}_i$$

6. UDP-N-acetyl-β-D-galactosamine is formed from UDP-N-acetyl-β-D-glucosamine by a NAD^+-linked epimerase.
   a. In this reaction, the C-4 hydroxyl of the glucosamine moiety is oxidized to a ketone without release of either the product or the NADH formed from the enzyme.
   b. The electrons are then transferred back to reform the hydroxyl (which can be in either the L- or the D-configuration) and NAD^+.

D. Biosynthesis of hexuronic acids

1. The biosynthesis of UDP-glucuronic acid has been described in section II B.

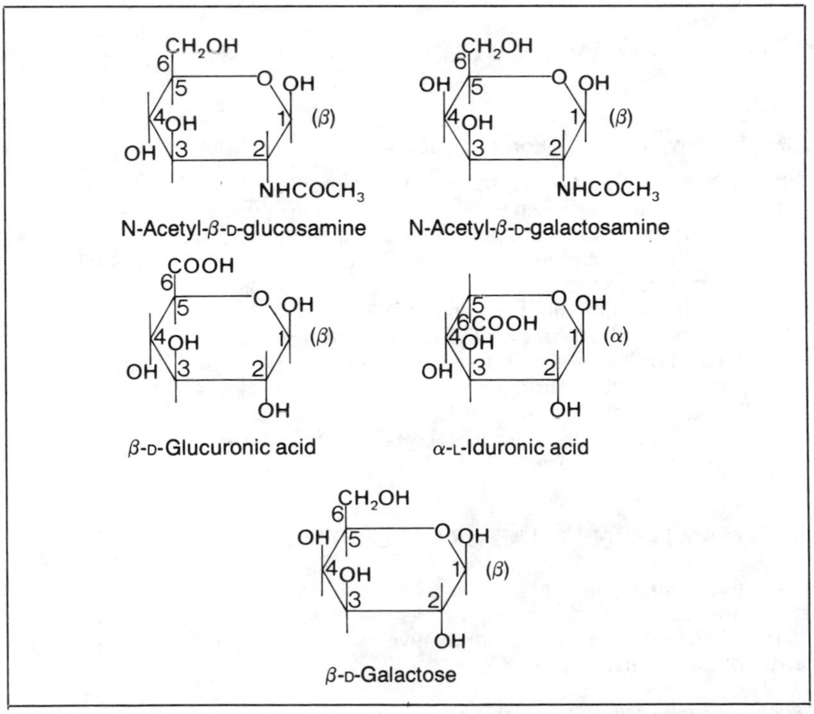

Figure 19-1. Structures of carbohydrate monomers found in the glycosaminoglycans.

2. L-**Iduronic acid** is formed by the action of an NAD$^+$-linked epimerase, which inverts the configuration at the C-5 of UDP-glucuronate to form UDP-L-iduronate.

E. Glycoproteins

1. The carbohydrate chains in the glycoproteins (see Chapter 10, section II) have additional residues as well as those found in the proteoglycans.

   a. Mannose residues are formed from fructose 6-phosphate (note the formation of the guanine nucleotidyl derivative rather than the uridyl form).

$$\text{Fructose 6-P} \rightleftharpoons \text{mannose 6-P}$$

$$\text{Mannose 6-P} \rightleftharpoons \text{mannose 1-P}$$

$$\text{Mannose 1-P} + \text{guanosine triphosphate} \rightarrow \text{guanosine diphosphate (GDP) mannose} + PP_i$$

   b. Fucose residues arise from GDP-mannose by a series of reactions.

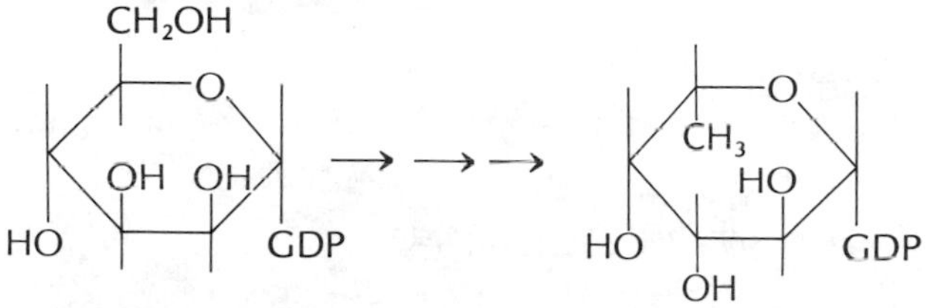

Guanosine diphosphate-D-mannose   Guanosine diphosphate-L-fucose

2. Acetylneuraminic acid

   a. This compound is formed by the introduction of the pyruvate moiety from phosphoenol pyruvate into N-acetylmannosamine 6-phosphate.

   b. The N-acetylmannosamine 6-phosphate is formed from UDP-N-acetylmannosamine by a reaction in which there is an inversion at C-2 and the loss of UDP. The loss of UDP renders the reaction irreversible.

STUDY QUESTIONS

Directions: Each question below contains five suggested answers. Choose the **one best** response to each question.

1. D-Ribose 5-phosphate is formed in the pentose phosphate pathway from

(A) ribulose 5-phosphate by an isomerase
(B) xylulose 5-phosphate by an epimerase
(C) the interaction of sedoheptulose 7-phosphate and erythrose 4-phosphate
(D) the interaction of fructose 6-phosphate and glyceraldehyde 3-phosphate
(E) the interaction of xylulose 5-phosphate and erythrose 4-phosphate

2. Which enzymatic activity in the pentose phosphate pathway requires thiamine pyrophosphate as a cofactor?

(A) Glucose 6-phosphate dehydrogenase
(B) 6-Phosphogluconate dehydrogenase
(C) Transketolase
(D) Transaldolase
(E) None of the above

3. The antimalarial drug primaquine (an oxidant) causes hemolysis of red cells in certain individuals and not in others because

(A) it inhibits glucose 6-phosphate dehydrogenase (G6PD) in the susceptible individuals only
(B) it inhibits the interaction of reduced nicotinamide-adenine dinucleotide phosphate (NADPH) and glutathione, which is necessary for the integrity of the red cell membrane
(C) it damages red cell membranes, which are unprotected by NADPH in individuals with G6PD deficiency
(D) the red cells of susceptible individuals cannot phosphorylate glucose to provide substrate for the pentose phosphate pathway
(E) the red cells of susceptible individuals metabolize primaquine to form an active inhibitor of the glycolytic pathway

4. Which of the following compounds can be formed by the oxidation of D-glucose at C-6?

(A) Gulonic acid
(B) Glucuronic acid
(C) Gluconic acid
(D) Glucitol
(E) Fructose

5. All of the following statements regarding the functional significance of the uronic acid pathway (glucuronic acid cycle) are correct EXCEPT that

(A) it is a source of active glucuronate for glucuronide formation
(B) it is source of active glucose for glycogen synthesis
(C) it is a source of ascorbic acid in man
(D) it allows the metabolism of nonphosphorylated sugars
(E) it provides for the interconversion of hexoses and pentoses

6. Fructose in seminal fluid is the major energy source for spermatozoa and is formed by

(A) isomerization of glucose 6-phosphate to fructose 6-phosphate and dephosphorylation of the latter
(B) reduction of D-glucose to sorbitol and oxidation of the latter to D-fructose
(C) reduction of D-glucose 6-phosphate to sorbitol 6-phosphate and oxidation and dephosphorylation of the latter to D-fructose
(D) dephosphorylation of fructose 1-phosphate
(E) none of the above reactions

7. Hereditary fructose intolerance is a condition caused by a deficiency of

(A) phosphofructokinase
(B) fructokinase
(C) fructose 1-phosphate aldolase
(D) fructose 1,6-diphosphate aldolase
(E) fructose 6-phosphatase

8. What is the active form of mannose in the synthesis of mannose-containing carbohydrate chains in glycoproteins?

(A) Uridine diphosphate mannose
(B) Guanosine diphosphate mannose
(C) Adenosine diphosphate mannose
(D) Cytidine diphosphate mannose
(E) None of the above

9. Glucosamine 6-phosphate, the precursor of all hexosamine residues in glucosaminoglycans, is formed by

(A) the transfer of the amide moiety of glutamine to fructose 6-phosphate
(B) the transfer of the amide moiety of glutamine to glucose 6-phosphate (G6P)
(C) the transfer of the α-amino group of glutamine to fructose 6-phosphate
(D) the transfer of the α-amino group of glutamine to G6P
(E) none of the above

10. Galactose is used for glycolysis in which of the following ways?

(A) It is phosphorylated to galactose 1-phosphate, which interacts with uridine diphosphate glucose (UDP-glucose) to form UDP-galactose and glucose 1-phosphate. An epimerase converts UDP-galactose to UDP-glucose
(B) It is phosphorylated to galactose 6-phosphate, which is converted to glucose 6-phosphate (G6P) by an epimerase
(C) It is converted to glucose by an epimerase, and the latter is phosphorylated by hexokinase to G6P
(D) It is converted by an isomerase to fructose, which is phosphorylated by fructokinase to fructose 1-phosphate. The latter is converted to dihydroxyacetone phosphate and D-glyceraldehyde by aldolase B
(E) It cannot serve as a source of pyruvate

Directions: Each question below contains four suggested answers of which **one or more** is correct. Choose the answer

A   if **1, 2, and 3** are correct
B   if **1 and 3** are correct
C   if **2 and 4** are correct
D   if **4** is correct
E   if **1, 2, 3, and 4** are correct

11. Functions of the pentose phosphate pathway include which of the following?

(1) It serves as a source of reduced nicotinamide-adenine dinucleotide phosphate
(2) It provides a route for the interconversion of pentoses and hexoses
(3) It serves as a source of D-ribose for nucleic acid synthesis
(4) It is only functional in red blood cells

12. Which of the following statements can be correctly applied to fructokinase?

(1) It catalyzes the phosphorylation of fructose to fructose 1-phosphate
(2) It functions at times of high fructose levels because it has a high Michaelis constant for fructose
(3) Its activity is not affected by feeding–fasting cycles
(4) It is only found in the liver

13. Possible fates of D-glyceraldehyde in the liver include

(1) phosphorylation to D-glyceraldehyde 3-phosphate
(2) oxidation to D-glycerate
(3) reduction to glycerol
(4) reaction with dihydroxyacetone phosphate to form fructose 1,6-diphosphate

14. Lactose synthesis is characterized by which of the following statements?

(1) It is confined to mammary tissue
(2) Lactose synthetase has catalytic and modifier subunits
(3) α-Lactalbumin acts as the modifier of lactose synthetase
(4) It arises from the interaction of uridine diphosphate glucose and D-galactose

SUMMARY OF DIRECTIONS

A	B	C	D	E
1, 2, 3 only	1, 3 only	2, 4 only	4 only	All are correct

15. A patient with classic galactosemia is likely to present with

(1) a deficiency of galactose 1-phosphate uridyl transferase
(2) an accumulation of galactose and galactose 1-phosphate in cells
(3) cataracts as a result of the reduction of galactose to dulcitol
(4) increased levels of galactose 1-phosphate, which can cause neural and liver damage

16. Characteristics of glycosaminoglycans include which of the following?

(1) They are polymers of repeating disaccharide units made up of a hexuronic acid and an amino sugar
(2) They are the polysaccharide components of the proteoglycans of ground substance in connective tissues
(3) The hexuronic acid may be β-D-glucuronic acid or L-iduronic acid
(4) The amino sugar may be glucosamine or fructosamine

ANSWERS AND EXPLANATIONS

1. The answer is A. (*I B 2*) Ribose 5-phosphate, required for the synthesis of nucleotides, is formed from ribulose 5-phosphate by an isomerization. Ribulose 5-phosphate arises from glucose 6-phosphate by the sequential action of glucose 6-phosphate dehydrogenase and 6-phosphogluconate dehydrogenase. The interaction of sedoheptulose 7-phosphate and glyceraldehyde 3-phosphate (G3P), catalyzed by transaldolase, produces erythrose 4-phosphate and fructose 6-phosphate. The interaction of fructose 6-phosphate and G3P, catalyzed by transketolase, produces xylulose 5-phosphate and erythrose 4-phosphate.

2. The answer is C. (*I B 3*) Transketolase is an enzyme that transfers two-carbon units from a ketose to an aldose. This transfer requires a cofactor, thiamine pyrophosphate, which is the carrier of the two-carbon unit. Transaldolase, which transfers three-carbon units from a ketose to an aldose, does *not* require a cofactor. The cofactor required by glucose 6-phosphate dehydrogenase and 6-phosphogluconate dehydrogenase is nicotinamide-adenine dinucleotide phosphate, which is needed as an electron acceptor.

3. The answer is C. (*I D*) Reduced glutathione is involved in maintaining the integrity of red cell membranes, particularly as a protection against oxidants in the blood. Glutathione is maintained in its reduced state by a system that uses reduced nicotinamide-adenine dinucleotide phosphate (NADPH) as the reductant. The NADPH is formed in the pentose phosphate pathway. In some individuals, a deficiency of glucose 6-phosphate dehydrogenase (G6PD), which uses $NADP^+$ as an electron acceptor and produces NADPH, leaves the red cell vulnerable to attack by oxidants, such as primaquine, and hemolysis of the red cells ensues.

4. The answer is B. (*II*) If the primary alcohol group on C-6 of glucose is oxidized to a carboxylate group, the compound formed is glucuronic acid. Gulonic acid arises by reduction of glucuronic acid, and in animals other than primates and guinea pigs, it serves as a precursor of ascorbic acid. Gluconic acid is formed by the oxidation of the aldehyde group on C-1 of glucose to the carboxylate and occurs as the 6-phospho- derivative in the pentose shunt pathway. Glucitol is a sugar alcohol formed by reduction of glucose. Fructose is a ketone isomer of glucose.

5. The answer is C. (*II A, B*) The uronic acid pathway includes the formation of uridine diphosphate glucose (UDP-glucose), used for the formation of glycogen, or the transfer of glucosyl residues in polysaccharide synthesis. UDP-glucose is also converted to UDP-glucuronate, the active form of glucuronate, which is used to form glucuronides in detoxification reactions, and in the synthesis of some polysaccharides (e.g., heparin) and chondroitin sulfate, a glycosaminoglycan. In most animals, the pathway is the route of synthesis of ascorbic acid (vitamin C). However, in man, other primates, guinea pigs, and an East Indian fruit bat, the step from L-gulonate to 2-keto-L-gulonolactone, the precursor of ascorbic acid, is blocked. Thus, in man and in these other animals, ascorbic acid is a vitamin. The pathway also allows for the metabolism of some nonphosphorylated sugars and forms a link between hexoses and pentoses, in addition to the pentose phosphate pathway.

6. The answer is B. (*III B 1 b*) Fructose in the seminal vesicle is formed from glucose by reduction to D-sorbitol, followed by oxidation of the latter to D-fructose. The fructose is used for an energy supply by its conversion to lactate via the glycolytic pathway in spermatozoa and by oxidation of the lactate to carbon dioxide and water by spermatozoal mitochondria, which are unique in their complement of the enzyme lactate dehydrogenase.

7. The answer is C. (*III C*) Hereditary fructose intolerance is characterized by vomiting, postprandial hypoglycemia, and damage to liver and kidney cells, following ingestion of fructose or of fructose-containing foods, such as sucrose or sorbitol. It is due to a deficiency of the aldolase that cleaves fructose 1-phosphate, which is formed from fructose by fructokinase to dihydroxyacetone phosphate and D-glyceraldehyde. Because of this deficiency, fructose 1-phosphate accumulates in cells and is believed to underlie the hepatic and renal damage. Fructose 1-phosphate also inhibits several enzyme activities; for example, hepatic glycogen phosphorylase, which accounts in part for the postprandial hypoglycemia, and fructose 1,6-diphosphate aldolase, which also contributes to the hypoglycemia as gluconeogenesis is blocked.

8. The answer is B. (*V E 1 a*) Carbohydrate chains found in the glycoproteins may contain mannose residues. These are formed from fructose 6-phosphate, which is converted to mannose 1-phosphate, and which, by reaction with guanosine triphosphate, forms guanosine diphosphate mannose, the donor of the mannose moiety.

9. The answer is A. (*V C 2*) The precursor of all hexosamine residues in glucosaminoglycans, glucosamine 6-phosphate, is formed by the transfer of the amide moiety of glutamine to fructose 6-phosphate. In this reaction, catalyzed by L-glutamine D-fructose 6-phosphate aminotransferase, there is a simultaneous isomerization of the ketose to the aldose.

10. The answer is A. (*IV C 2, 3*) Galactose is phosphorylated by galactokinase to form galactose 1-phosphate. Hexose 1-phosphate uridyl transferase catalyzes the transfer of the uridine diphosphoryl moiety from uridine diphosphate glucose (UDP-glucose) to galactose 1-phosphate to form UDP-galactose and glucose 1-phosphate. An epimerase converts UDP-galactose to UDP-glucose, which can be metabolized in the glycolytic pathway.

11. The answer is A (1, 2, 3). (*I A, B, D*) The pentose phosphate pathway is most active in the liver, the mammary glands, adipose tissue (in the rat), and the adrenal cortex. It is in general most active in tissues in which the synthesis of fat is important. It is also an important pathway in red blood cells where reduced nicotinamide-adenine dinucleotide phosphate (NADPH) formed is essential for the preservation of cellular integrity. The major functions of the pathway in all tissues are the production of NADPH for reductive synthesis, D-ribose for nucleic acid synthesis, and the interconversion of pentoses and hexoses.

12. The answer is B (1, 3). (*III A 1*) Fructokinase, which catalyzes the phosphorylation of fructose to fructose 1-phosphate, is found in the liver, kidney, and small intestine. It has a very low Michaelis constant for fructose; that is, it has a very high affinity for the substrate, and its activity is not affected by feeding–fasting cycles.

13. The answer is A (1, 2, 3). [*III A 3; Chapter 27 III B 5 a (4)*] D-Glyceraldehyde is formed by the action of phosphofructoaldolase (aldolase B) on fructose 1-phosphate. There are three possible pathways for the metabolism of D-glyceraldehyde. In one, phosphorylation of D-glyceraldehyde would be followed by metabolism of the D-glyceraldehyde 3-phosphate (G3P) in the glycolytic or gluconeogenic pathways. In another, oxidation to D-glycerate by a nicotinamide-adenine dinucleotide–linked dehydrogenase could be followed by conversion to L-serine via hydroxypyruvate. In a third possible pathway, D-glyceraldehyde could be reduced to glycerol, followed by phosphorylation to yield glycerol 3-phosphate. The latter could be used for triglyceride synthesis or, after oxidation to G3P, could serve in the glycolytic or gluconeogenic pathways. The formation of fructose 1,6-diphosphate by aldolase A requires G3P, not glyceraldehyde.

14. The answer is A (1, 2, 3). (*IV A, B*) Lactose synthetase is confined to mammary tissue. The enzyme has a catalytic subunit and a modifier subunit called α-lactalbumin. The adenohypophyseal hormone, prolactin, induces the synthesis of both the catalytic and modifier subunits. Lactose synthetase catalyzes the transfer of galactose from uridine diphosphate galactose to D-glucose to form lactose [β-D-galactosyl-(1→4)-α-D-glucose]. The modifier subunit functions to change the specificity of the catalytic subunit so that it uses D-glucose as an acceptor of the galactosyl moiety instead of N-acetylglucosamine, the acceptor in other tissues.

15. The answer is E (all). (*IV D 2 b*) Classic galactosemia is due to a deficiency of galactose 1-phosphate uridyl transferase. Another genetic disease of galactose metabolism involves a deficiency of galactokinase. In classic galactosemia, both galactose and galactose 1-phosphate accumulate in cells, with the latter being responsible, it is believed, for the mental retardation and liver cirrhosis seen in these cases. The increased levels of galactose contribute to cataract formation in the lens of the eye, probably because of the reduction of the excess galactose to dulcitol.

16. The answer is A (1, 2, 3). (*V A–C*) The glycosaminoglycans are polysaccharides formed from repeating disaccharide units, which are made up of a hexuronic acid and an amino sugar. They are the polysaccharide components of the proteoglycans, complexes of protein and polysaccharide, which make up the matrix, or ground substance, of connective tissues. The hexuronic acids found may be β-D-glucuronic acid or L-iduronic acid, while the amino sugars found may be glucosamine or galactosamine. Both the amino sugars arise from glucosamine 6-phosphate.

20

Fatty Acid and Acylglyceride Metabolism

I. INTRODUCTION TO LIPIDS

A. Nature. Lipids are intracellular compounds that are insoluble or only poorly soluble in water. They are readily soluble in nonpolar solvents such as ether, chloroform, or benzene. The hydrophobic (water-hating) nature of lipids is due to the predominance of hydrocarbon chains ($-CH_2-CH_2-CH_2-$) in their structures.

B. Functions. Four general functions of biologic lipids have been identified.

1. They serve as a storage form of metabolic fuel.

2. They serve as a transport form of metabolic fuel.

3. They provide structural components of membranes.

4. They have protective functions in bacteria, plants, insects, and vertebrates, serving as a part of the outer coating between the body of the organism and the environment.

C. Classification. There are many different ways to classify lipids. In one of the more common, five classes are recognized.

1. Fatty acids and their immediate derivatives (e.g., prostaglandins and leukotrienes).

2. Glycerol esters (e.g., acylglycerols and phosphoglycerides).

3. Sphingolipids (e.g., sphingomyelin and glycosphingolipids).

4. Cholesterol and its derivatives (e.g., cholesterol esters, bile acids, steroid hormones, and vitamin D).

5. Isoprene derivatives (e.g., dolichols, vitamin A, vitamin E, and vitamin K).

II. FATTY ACIDS

A. Nature and nomenclature. Fatty acids are water-insoluble long-chain hydrocarbons with one carboxyl group at the end of the chain. The chains may be saturated or unsaturated. Table 20-1 lists the major fatty acids found in mammalian tissues.

Table 20-1. Predominant Fatty Acids Found in Mammalian Tissues

Common Name	Systematic Name	No. Carbon Atoms	No. Double Bonds	Melting Point (°C)
Lauric	Dodecanoic	12	0	43.5
Myristic	Tetradecanoic	14	0	54.4
Palmitic	Hexadecanoic	16	0	62.8
Stearic	Octadecanoic	18	0	69.6
Pamitoleic	cis-Δ^9-Hexadecenoic	16	1	1.0
Oleic	cis-Δ^9-Octadecenoic	18	1	13.0
Linoleic	all cis-Δ^9,Δ^{12}-Octadecadienoic	18	2	− 11.0
Linolenic	all cis-$\Delta^9,\Delta^{12},\Delta^{15}$-Octadecatrienoic	18	3	− 11.2
Arachidonic	all cis-$\Delta^5,\Delta^0,\Delta^{11},\Delta^{14}$-Eicosatetraenoic	20	4	− 49.5

1. **Saturated fatty acids** have no double bonds in the chain.
   a. Their general formula is

   $$CH_3-(CH_2)_n-COOH$$

   where n specifies the number of methylene groups between the methyl and carboxyl carbons.
   b. The systematic name gives the number of carbons, with the suffix *-anoic* appended. Palmitic acid, for example, has 16 carbons and has the systematic name hexadecanoic acid.
2. **Unsaturated fatty acids** have one or more double bonds.
   a. The most commonly used system for designating the position of double bonds in an unsaturated fatty acid is the delta (Δ) numbering system.
      (1) The terminal carboxyl carbon is designated carbon 1, and the double bond is given the number of the carbon atom on the carboxyl side of the double bond. For example, palmitoleic acid has 16 carbons and has a double bond between carbons 9 and 10. It is designated as $16:1:\Delta^9$, or 16:1:9.
      (2) The systematic name gives the number of carbon atoms, the number of double bonds (unless it has only one), and bears the suffix *-enoic*. Thus, palmitoleic acid is *cis*-Δ^9-hexadecenoic acid; linoleic acid, which has 18 carbons and two double bonds, is all (i.e., all double bonds are *cis*) *cis*-Δ^9,Δ^{12}-octadecadienoic acid.
   b. Double bonds in naturally occurring fatty acids are always in a *cis* as opposed to a *trans* configuration.

   Cis   Trans

B. **Source**

   1. **Nonessential fatty acids.** All nonessential fatty acids can be synthesized from acetyl coenzyme A (acetyl CoA) derived from glucose oxidation. They are nonessential in the sense that they do not have to be obligatorily included in the diet. However, the bulk of the nonessential fatty acids in humans may in fact be obtained from the diet, particularly in the case of the high-fat diets of affluent societies.

   2. **Essential fatty acids.** Fatty acids of the linoleic ($18:2:\Delta^{9,12}$) and linolenic ($18:3:\Delta^{9,12,15}$) families, which are the precursors of the prostaglandins (see Chapter 22), must be obtained from the diet. There are no human enzyme systems that can introduce a double bond beyond the ninth carbon atom (9–10 position) of a fatty acid chain, and all double bonds that are introduced are separated by three-carbon intervals. This rule, combined with the fact that fatty acid elongation (see section III C) only occurs by two-carbon additions, makes it impossible to synthesize de novo certain polyunsaturated fatty acids.

C. **Physical properties**

   1. Fatty acids are detergent-like due to their **amphipathic nature**; that is, they have nonpolar (CH_3) and polar ($-COOH$) ends and, in biphasic systems, will orient with the polar end associated with water and the nonpolar end associated with the hydrophobic phase.

   2. The melting point of fatty acids is related to chain length and degree of unsaturation. The longer the chain length, the higher the melting point, and the greater the number of double bonds, the lower the melting point (see Table 20-1).

III. BIOSYNTHESIS OF FATTY ACIDS

A. **General features**

   1. Fatty acids are synthesized from dietary glucose; the extent to which this process occurs depends upon the diet. With a high-fat diet, fatty acid synthesis is limited.

   2. In humans, the major site of fatty acid synthesis is in the liver.

B. **Cytoplasmic synthesis of saturated fatty acids.** Enzymes that synthesize fatty acids are located in the cytosol, and the system in humans polymerizes two-carbon fragments to form the C-16 saturated fatty acid palmitic acid (hexadecanoic acid). The cytosolic system that synthesizes fatty acids is quite distinct from the mitochondrial system that oxidizes fatty acids. The former requires acetyl CoA, reduced nicotinamide-adenine dinucleotide phosphate (NADPH), bicarbonate (HCO_3^-), and adenosine triphosphate (ATP) as substrates.

1. Source of acetyl CoA

    a. The two-carbon fragments needed for fatty acid synthesis are supplied by acetyl CoA, which in turn is derived from the oxidation of glucose via the glycolytic pathway (Chapter 14) and pyruvate dehydrogenase (Chapter 15, section II).

    b. Acetyl CoA formed in the matrix of mitochondria cannot cross the inner mitochondrial membrane, and an **acetyl CoA shuttle system** is required to transport the two-carbon fragments out of the mitochondria and into the cytosol.

        (1) Within the mitochondria, acetyl CoA reacts with oxaloacetate to form citrate in a reaction catalyzed by citrate synthetase (see Chapter 15, section III A).

        (2) Under conditions where acetyl CoA and ATP are in high concentration (as in the fed state), the citrate leaves the mitochondria via the tricarboxylic acid (TCA) transport system; and in the cytosol, reacts with cytoplasmic CoA and ATP to form acetyl CoA and oxaloacetate in a reaction catalyzed by citrate cleavage enzyme (citrate lyase).

Mitochondrial acetyl CoA + oxaloacetate → citrate + CoASH

Cytosolic citrate + ATP + CoASH → acetyl CoA + oxaloacetate + adenosine diphosphate (ADP) + inorganic phosphate (P_i)

        (3) The oxaloacetate, which cannot cross the mitochondrial inner membrane, is converted to malate by cytosolic malate dehydrogenase, using reduced nicotinamide-adenine dinucleotide (NADH) as the electron donor.

Oxaloacetate + NADH + H^+ ⇌ malate + NAD^+

        (4) The malate can then re-enter the mitochondrial matrix by the dicarboxylic acid transport system to continue the acetyl group shuttle.

2. Sources of NADPH

    a. In adipose tissue, the acetyl CoA shuttle results in the formation of NADPH because the malate formed from oxaloacetate in the cytoplasm serves as a substrate for the $NADP^+$-linked malic enzyme.

Malate + $NADP^+$ → pyruvate + NADPH + H^+ + CO_2

The pyruvate formed is then converted to oxaloacetate by pyruvate carboxylase within the mitochondria to continue the acetyl group shuttle.

    b. The synthesis of a C-16 fatty acid requires the polymerization of eight two-carbon fragments with two molecules of NADPH being oxidized in the linking of each two-carbon fragment pair, so that a total of fourteen NADPH molecules is required. The production of acetyl CoA by the acetyl shuttle provides eight of the fourteen required NADPH molecules. The other six arise from the pentose phosphate pathway (see Chapter 19, section I).

    c. In humans, adipose tissue is not very active in fatty acid synthesis, and most of the fatty acids are formed in the liver. In this organ, most of the NADPH is provided by the pentose phosphate pathway as the malate formed from the acetyl group shuttle re-enters the mitochondria as described above.

3. Activation of acetyl CoA

    a. The union of two acetyl CoA molecules to form a four-carbon molecule (acetoacetyl CoA) requires an expenditure of over 10 kcal/mol. This barrier is overcome by carboxylation of the methyl group of one of the acetyl CoA molecules in a reaction catalyzed by acetyl CoA carboxylase to form malonyl CoA.

$$\underset{\text{Acetyl CoA}}{CH_3-\overset{\overset{\displaystyle O}{\|}}{C}-S-CoA} + ATP + HCO_3^- \rightarrow \underset{\text{Malonyl CoA}}{{}^-OOC-CH_2-\overset{\overset{\displaystyle O}{\|}}{C}-S-CoA} + ADP + P_i$$

    b. Acetyl CoA carboxylase requires biotin as a cofactor. The HCO_3^- required for the carboxylation is only needed in catalytic amounts, as it is returned to the cytosol as HCO_3^- during the polymerization of malonyl CoA with an acetyl CoA.

4. Fatty acid synthetase complex

    a. The fatty acid synthetase of bacteria (*Escherichia coli*) is a seven-enzyme complex.

b. In contrast, fatty acid synthetases of yeast and animals are multifunctional proteins.

    (1) The yeast enzyme contains two nonidentical subunits, which are coded for by two unlinked genes.

        (a) Subunit A exhibits acyl carrier protein (ACP), β-ketoacyl–ACP synthetase, and β-keto– and acyl–ACP reductase activities.

        (b) Subunit B exhibits ACP–acetyl and ACP–malonyl transacylase activities, acyl–ACP dehydratase and enoyl–ACP reductase activities.

    (2) The animal enzymes are dimers of identical subunits, each polypeptide containing six of the enzyme activities required for fatty acid synthetase activity. In association, they combine to catalyze the β-ketoacyl–ACP synthetase activity.

c. The assembly of two-carbon units to form palmitic acid begins with acetyl CoA, followed by the sequential addition of malonyl CoA units and the loss of carbon dioxide (CO_2).

$$CH_3\text{—}\underset{\underset{O}{\|}}{C}\text{—S—CoA} + 7\ {}^-OOC\text{—}CH_2\text{—}\underset{\underset{O}{\|}}{C}\text{—S—CoA} + 14\ NADPH + 14\ H^+ \rightarrow$$

$$CH_3\text{—}(CH_2)_{14}\text{—}COO^- + 7\ CO_2 + 14\ NADP^+ + 8\ CoASH + 6\ H_2O$$

5. Palmitic acid formation: Lynen cycle

    a. The sequence of events is believed to be similar in prokaryote and eukaryote systems, but it is most easily followed by looking at the multienzyme complex of prokaryotes as outlined below.

    b. The initial step, binding of acetyl CoA to a serine or threonine hydroxyl (known as the loading, or B_1, site) on the enzyme complex, is followed by the transfer of the acetyl group to a sulfhydryl group of a cysteine residue on the ACP in a reaction catalyzed by acetyl CoA–ACP transacylase.

    c. The acetyl group is then transferred to another cysteinyl-sulfhydryl group (the B_2 site), which is a residue in the so-called condensing enzyme, β-ketoacyl–ACP synthetase.

    d. The first of seven malonyl CoA molecules now enters at the B_1 site, and the malonyl moiety is transferred to the ACP-sulfhydryl by malonyl CoA–ACP transacylase.

    e. Condensation of the acetyl group on the B_2 site with the malonyl group on ACP is catalyzed by β-ketoacyl–ACP synthetase with the loss of CO_2.

    f. A reduction, dehydration, and another reduction occur while the four-carbon β-ketoacyl compound is bound to the ACP.

    (1) Reduction by β-ketoacyl–ACP reductase with NADPH as electron donor

$$CH_3\text{—}\underset{\underset{O}{\|}}{C}\text{—}CH_2\text{—}\underset{\underset{O}{\|}}{C}\text{—S—ACP} + NADPH + H^+ \rightarrow CH_3\text{—}\underset{\underset{OH}{|}}{CH}\text{—}CH_2\text{—}\underset{\underset{O}{\|}}{C}\text{—S—ACP} + NADP^+$$

    (2) Dehydration by β-hydroxyacyl–ACP dehydratase

$$CH_3\text{—}\underset{\underset{OH}{|}}{CH}\text{—}CH_2\text{—}\underset{\underset{O}{\|}}{C}\text{—S—ACP} \rightarrow CH_3\text{—}CH=CH\text{—}\underset{\underset{O}{\|}}{C}\text{—ACP} + H_2O$$

    (3) Reduction by enoyl–ACP reductase and NADPH

$$CH_3\text{—}CH=CH\text{—}\underset{\underset{O}{\|}}{C}\text{—S—ACP} + NADPH + H^+ \rightarrow CH_3\text{—}CH_2\text{—}CH_2\text{—}\underset{\underset{O}{\|}}{C}\text{—S—ACP} + NADP^+$$

    g. The saturated acyl–ACP thioester now reacts with the B_2 site on the β-ketoacyl–ACP synthetase, which frees the ACP to react with another malonyl group.

    h. After seven turns of this cycle, the palmitic acid that has formed is released by palmitoyl deacylase. Figure 20-1 illustrates the sequence of reactions in each turn of the cycle.

6. Stoichiometry of the synthesis of palmitate (C-16) from acetyl CoA by the Lynen cycle is given by

$$7\ \text{acetyl CoA} + 7\ CO_2 + 7\ ATP + 7\ H_2O \rightarrow 7\ \text{malonyl CoA} + 7\ ADP + 7\ P_i$$

$$\text{acetyl CoA} + 7\ \text{malonyl CoA} + 14\ NADPH + 14\ H^+ \rightarrow$$
$$\text{palmitate} + 7\ CO_2 + 14\ NADP^+ + 8\ CoASH + 6\ H_2O$$

Sum: $8\ \text{acetyl CoA} + 7\ ATP + 14\ NADPH + 14\ H^+ + H_2O \rightarrow$
$$\text{palmitate} + 7\ ADP + 7\ P_i + 8\ CoASH + 14\ NADP^+$$

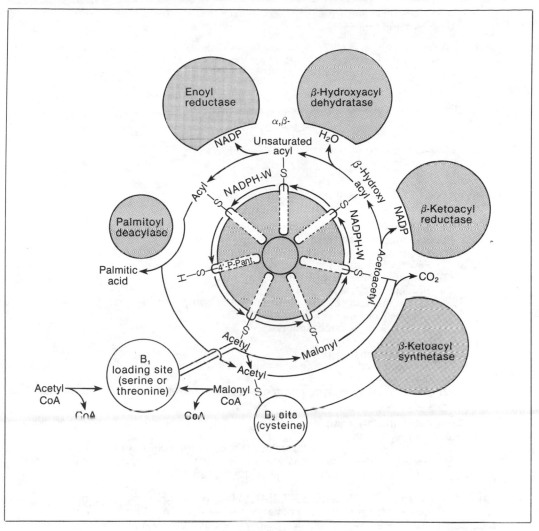

Figure 20-1. The Lynen cycle: A proposal for the mechanism of fatty acid synthesis by the fatty acyl coenzyme A synthetase complex of pigeon liver. (Reprinted with permission from Phillips GT, et al: *Arch Biochem Biophys* 138:390, 1971.)

C. Elongation of fatty acids. The normal product of the cytosolic fatty acid synthetase complex is palmitic acid. Fatty acids longer than 16 carbons can be formed through the addition of two-carbon units by elongation systems. There are two elongation systems.

   1. The most active system is found in the endoplasmic reticulum. It adds malonyl CoA onto palmitate in a manner similar to the action of the action of fatty acid synthetase, except that CoASH is involved rather than the ACP. Stearic acid (18 carbons) is a common product of this elongation system.

   2. A mitochondrial elongation system uses acetyl CoA units, rather than malonyl CoA units, to elongate fatty acids for the synthesis of structural lipids in this organelle.

D. Desaturation of fatty acids

   1. The two most common monounsaturated fatty acids in mammals are palmitoleic acid (16:1:Δ^9) and oleic acid (18:1:Δ^9).

   2. The double bonds are introduced between carbons 9 and 10 by fatty acid oxygenase in the endoplasmic reticulum. The enzyme is a mixed-function oxygenase, which requires molecular oxygen and NADPH.

$$CH_3—(CH_2)_{14}—\overset{O}{\underset{\|}{C}}—S—CoA + NADPH + H^+ + O_2 \rightarrow$$

Palmitoyl CoA

$$CH_3—(CH_2)_7—CH{=}CH—(CH_2)_7—\overset{O}{\underset{\|}{C}}—S—CoA + NADP^+ + 2\ H_2O$$

Palmitoleyl CoA

E. Regulation of fatty acid synthesis

1. **Acetyl CoA carboxylase** catalyzes the rate-limiting step in fatty acid biosynthesis; it is regulated by both rapid- and slow-acting mechanisms.
   a. **Rapid-acting mechanisms**
      (1) Allosteric activation by citrate, which promotes the formation of a very large active polymer form from the inactive monomer form of the enzyme
      (2) Allosteric inhibition by C-16–C-18 acyl CoAs
      (3) Covalent modification of the enzyme
         (a) Inhibition by cyclic adenosine monophosphate–dependent (cAMP-dependent) phosphorylation
         (b) Activation by dephosphorylation
      (4) Stimulation by insulin
   b. **Slow-acting mechanisms**
      (1) Increased synthesis of the enzyme, induced by a high-carbohydrate diet or a fat-free diet
      (2) Decreased enzyme synthesis in response to a high-fat diet, fasting, or elevated plasma glucagon levels

2. **Fatty acid synthetase complex.** The Lynen cycle enzymes also have rapid- and slow-acting regulatory mechanisms.
   a. **Rapid-acting mechanism.** Allosteric activation of the complex by phosphorylated sugars
   b. **Slow-acting mechanism.** The rate of enzyme synthesis (probably the major control) with the inducing factors being the same as for acetyl CoA carboxylase

IV. FORMATION AND STORAGE OF TRIACYLGLYCEROLS. Fatty acids obtained from the diet, or synthesized from glucose, are converted to **triacylglycerols** for transport to tissues and for storage. This conversion involves the acylation of the three hydroxyl groups of glycerol.

A. Activation of the fatty acid occurs first.

1. The formation of the CoA thioester is catalyzed by acyl CoA synthetase in a reaction that requires ATP.

$$CH_3—(CH_2)_{14}—COO^- + CoASH + ATP \longrightarrow CH_3—(CH_2)_{14}—\overset{O}{\underset{\|}{C}}—S—CoA + AMP + PP_i$$

2. The hydrolysis of the pyrophosphate (PP_i) renders the reaction irreversible.

B. Acylation of glycerol

1. **First acylation.** There are two routes for the acylation of the first hydroxyl of glycerol.
   a. One route uses dihydroxyacetone phosphate (DHAP), derived from glucose by the glycolytic pathway, as the acceptor of the acyl moiety from the fatty acyl CoA.

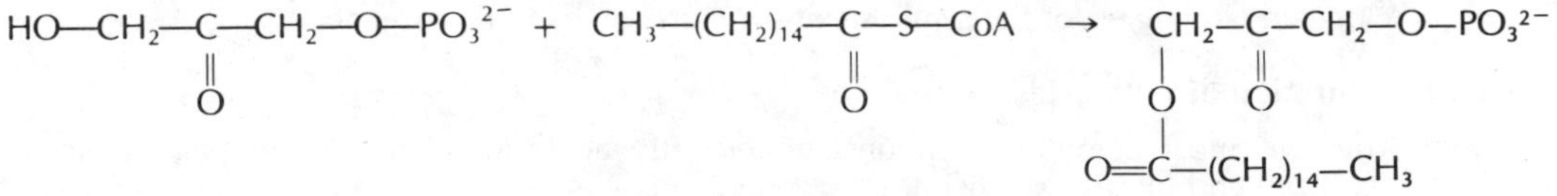

$$HO—CH_2—\overset{O}{\underset{\|}{C}}—CH_2—O—PO_3^{2-} + CH_3—(CH_2)_{14}—\overset{O}{\underset{\|}{C}}—S—CoA \rightarrow \begin{array}{c} CH_2—\overset{O}{\underset{\|}{C}}—CH_2—O—PO_3^{2-} \\ | \\ O \\ | \\ O{=}C—(CH_2)_{14}—CH_3 \end{array}$$

Dihydroxyacetone phosphate   Palmitoyl CoA   1-Palmitoyl dihydroxyacetone 3-phosphate

(1) The initial reaction is followed by a reduction, with NADPH as the electron acceptor, to form **lysophosphatidate**.

$$CH_2\!-\!CH\!-\!CH_2\!-\!O\!-\!PO_3^{2-}$$
$$| \qquad |$$
$$O \quad\; OH$$
$$|$$
$$O\!=\!C\!-\!(CH_2)_{14}\!-\!CH_3$$

(2) The fatty acid preferentially introduced is saturated.

b. The second route gives the same product and shows the same preference for a saturated fatty acid, but the order is reversed and reduction of DHAP to glycerol 3-phosphate occurs before acylation of the C-1 hydroxyl.

2. Second acylation. An unsaturated fatty acyl CoA thioester is introduced to the 2-hydroxyl of lysophosphatidate. An exception occurs in the human mammary gland, where a saturated fatty acyl CoA is used.

3. Third acylation. The phosphate group on C-3 is first removed by a phosphatase, followed by the addition of either a saturated or unsaturated fatty acid to the C-3 hydroxyl.

C. Storage of triacylglycerols in adipose tissue

1. The esterification of fatty acids in adipose tissue to form triacylglycerols is dependent upon ongoing carbohydrate metabolism for the formation of DHAP or glycerol 3-phosphate.

2. Adipose tissue lacks a glycerol kinase and therefore cannot carry out the phosphorylation of glycerol to glycerol 3-phosphate.

3. The only source of glycerol 3-phosphate for triacylglycerol synthesis is from DHAP formed during glycolysis from glucose.

4. The entry of glucose into the adipose tissue cell (adipocyte) is an insulin-dependent process. Thus, insulin is an essential requirement for triacylglycerol synthesis in adipose tissue.

V. LIPOLYSIS OF TRIACYLGLYCEROLS

A. Fatty acid stores as metabolic fuel

1. The main stores of metabolic fuel in humans are the fat deposits in fat cells (adipocytes). A very large proportion of ingested fats are stored as triacylglycerols in the fat droplets of the adipocytes, where they serve long-term needs for metabolic fuel.

2. Triacylglycerols have several advantages over other forms of metabolic fuel.
  a. They are light (less dense than water). They provide a concentrated form of fuel because their complete combustion to CO_2 and water releases 9 kcal/g as opposed to 4 kcal/g for carbohydrate.
  b. As they are water insoluble, they present no osmotic problems to the cell even when stored in large amounts.

3. During normal overnight fasting, the blood levels of **free fatty acids** (also known as **nonesterified fatty acids**) increase to about 0.5 mM from a very low postprandial value.

4. The appearance of fatty acids in the blood during fasting is due to the mobilization of fat stores by the process of **lipolysis**. Lipolysis involves the hydrolysis of triacylglycerols to free glycerol and free fatty acids with both products leaving the adipocyte. The fatty acids are used by most tissues, with the exception of the brain, as a metabolic fuel for energy production or as a source of carbon for biosynthesis. The glycerol, which cannot be used by adipose tissue cells, is picked up by the liver as a substrate for gluconeogenesis (see Chapter 17, section IV C).

B. Enzymatic steps of lipolysis

1. Three different enzymes (lipases) are required to hydrolyze off the fatty acyl moieties from the triacylglycerols.

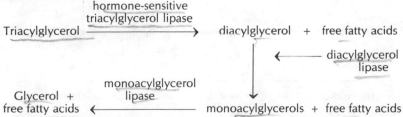

2. The rate-limiting step of lipolysis in adipocytes is the reaction catalyzed by the hormone-sensitive triacylglycerol lipase. The diacyl- and monoacylglycerol lipases are present in excess, so that once the hormone-sensitive triacylglycerol lipase is activated, lipolysis of the triacylglycerols goes to completion.

C. Hormonal control of lipolysis in adipocytes

1. Hormone-sensitive lipase is activated by covalent modification in which phosphorylation of the lipase by cAMP-dependent protein kinase activates the lipase (Fig. 20-2).

2. Epinephrine circulating in the blood in response to an alarm signal, or **norepinephrine** released by neural connections to adipose tissue, activates the cell membrane adenylate cyclase to produce cAMP within the cell (see Chapter 30). Thus, under these conditions (i.e., in stress or when neural signals indicate low levels of metabolic fuel), the hormone-sensitive lipase is activated by the cAMP-dependent protein kinase.

3. Insulin is inhibitory to lipolysis by two mechanisms.
  a. It reduces cAMP levels and thus reduces lipolysis, probably by inhibiting adenylate cyclase activity.
  b. It increases glucose entry into the adipocytes, so that the formation of DHAP and glycerol 3-phosphate is increased. The availability of these products of glycolysis increases the rate of re-esterification of free fatty acids to triacylglycerols, thus reducing the rate of release of the fatty acids from adipocytes.

4. Prostaglandins also inhibit lipolysis by reducing cAMP levels.

D. Lipolysis in other tissues

1. Although adipocytes form the main storage depots for triacylglycerols, other tissues, including muscle and liver, store small amounts of triacylglycerols in the form of intracellular lipid droplets for their own use.

2. These fat stores in cells other than adipocytes appear to be mobilized by the same hormonal controls as are found in adipocytes.

VI. FATTY ACID OXIDATION

A. *β*-Oxidation of fatty acids is the principal pathway for the catabolism of fatty acids; it takes place within the matrix of mitochondria. *β*-Oxidation involves oxidation of the *β*-carbon to form a *β*-keto acid.

$$R—CH_2—CH_2—COO^- + O_2 \rightarrow R—\underset{\underset{O}{\|}}{C}—CH_2—COO^- + H_2O$$

Fatty acid   *β*-keto acid

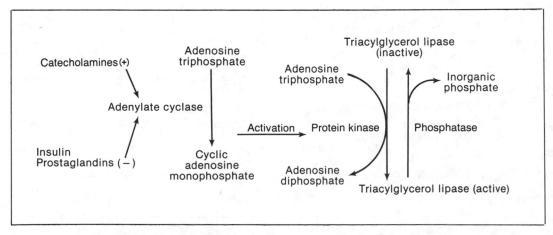

Figure 20-2. Factors in the control of hormone-sensitive lipase in adipocytes.

1. **Role of free fatty acids.** Free fatty acids are taken up from the circulation by cells, and once inside the cell they are activated by formation of acyl CoA derivatives. Two activating systems are present in cells.
   a. An endoplasmic reticulum acyl CoA synthetase (thiokinase) activates long-chain fatty acids (12 or more carbons).

$$\text{Fatty acid} + \text{ATP} + \text{CoASH} \rightarrow \text{acyl CoA} + PP_i + \text{AMP}$$

   b. Inner mitochondrial acyl CoA synthetases activate fatty acids of medium-chain length (4 to 10 carbons) and short-chain length (acetate and propionate). These fatty acids enter the mitochondria freely from the cytoplasm, whereas the fatty acids of long-chain length cannot do so.

2. **Role of carnitine**
   a. Long-chain fatty acyl CoAs cannot freely diffuse across the inner mitochondrial membrane, which is impermeable to CoA.
   b. On the outer surface of the inner mitochondrial membrane, the enzyme carnitine palmitoyl transferase I (CPT I) catalyzes the transfer of the acyl group from CoA to carnitine (β-hydroxy-γ-trimethylammonium butyrate), which is located in the inner mitochondrial membrane. The fatty acyl group is translocated across the membrane to the inner surface, where the enzyme carnitine palmitoyl transferase II (CPT II) catalyzes the transfer of the acyl group to CoA drawn from the matrix CoA pool (Fig. 20-3).

3. **Pathway of β-oxidation**
   a. Long-chain fatty acyl CoAs are subjected to a repeated four-step process, which removes two carbons from the chain successively until the last two-carbon fragment is obtained. The four steps are illustrated in Figure 20-4.
   b. Figure 20-5 shows the first cycle of β-oxidation of palmitoyl CoA, a 16-carbon fatty acyl CoA, with the enzymes involved at each step.

4. **Stoichiometry of β-oxidation**
   a. **β-Oxidation of palmitate (16 carbons)**

$$\text{Palmitoyl CoA} + 7\,\text{CoA} + 7\,\text{FAD} + 7\,H_2O \rightarrow 8\,\text{acetyl CoA} + 7\,FADH_2 + 7\,\text{NADH} + 7\,H^+$$

   (1) A total 7 mol of reduced flavin-adenine dinucleotide ($FADH_2$) yields 14 mol of ATP by electron transport and oxidative phosphorylation.
   (2) A total of 7 mol of NADH yields 21 mol of ATP by electron transport and oxidative phosphorylation.
   (3) Therefore,

$$\text{Palmitoyl CoA} + 7\,\text{CoA} + 7\,O_2 + 35\,P_i + 35\,\text{ADP} \rightarrow$$
$$8\,\text{acetyl CoA} + 35\,\text{ATP} + 42\,H_2O$$

   (4) As it costs 2 mol of ATP to activate the free palmitate, the net yield of energy is $35 - 2 = 33$ mol of ATP per mole of palmitate for β-oxidation.

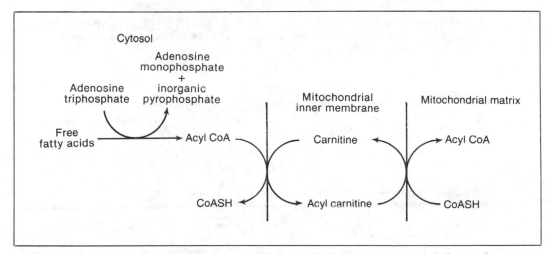

Figure 20-3. Functions of carnitine palmitoyl transferase I and II in the transport of long-chain fatty acyl coenzyme As across the inner mitochondrial membrane.

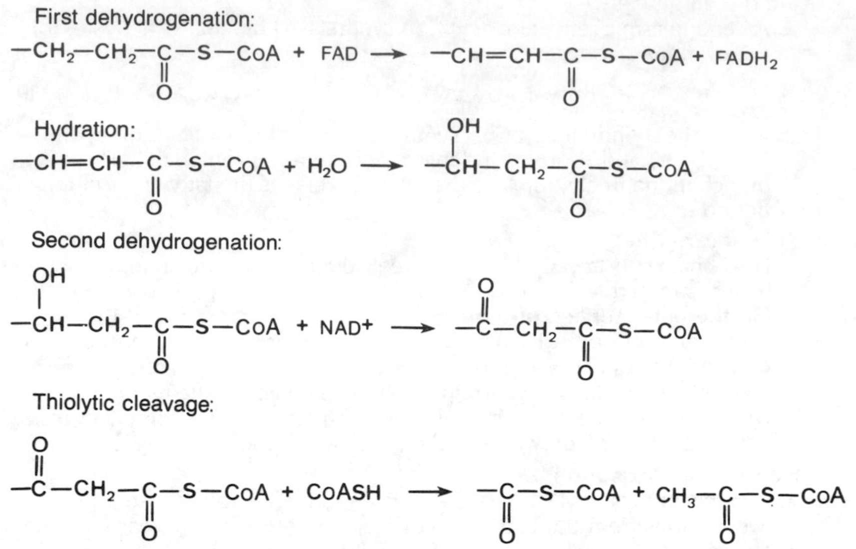

First dehydrogenation:

$$-CH_2-CH_2-\overset{\overset{\text{O}}{\|}}{C}-S-CoA + FAD \longrightarrow -CH=CH-\overset{\overset{\text{O}}{\|}}{C}-S-CoA + FADH_2$$

Hydration:

$$-CH=CH-\overset{\overset{\text{O}}{\|}}{C}-S-CoA + H_2O \longrightarrow -\overset{\overset{\text{OH}}{|}}{CH}-CH_2-\overset{\overset{\text{O}}{\|}}{C}-S-CoA$$

Second dehydrogenation:

$$-\overset{\overset{\text{OH}}{|}}{CH}-CH_2-\overset{\overset{\text{O}}{\|}}{C}-S-CoA + NAD^+ \longrightarrow -\overset{\overset{\text{O}}{\|}}{C}-CH_2-\overset{\overset{\text{O}}{\|}}{C}-S-CoA$$

Thiolytic cleavage:

$$-\overset{\overset{\text{O}}{\|}}{C}-CH_2-\overset{\overset{\text{O}}{\|}}{C}-S-CoA + CoASH \longrightarrow -\overset{\overset{\text{O}}{\|}}{C}-S-CoA + CH_3-\overset{\overset{\text{O}}{\|}}{C}-S-CoA$$

Figure 20-4. The reactions of the repeated four-step process of β-oxidation of long-chain fatty acyl coenzyme As.

b. Total oxidation

(1) The acetyl CoA derived from β-oxidation of fatty acyl CoAs is usually oxidized to CO_2 and H_2O by the TCA cycle.

$$8 \text{ Acetyl CoA} + 16 O_2 + 96 P_i + 96 \text{ ADP} \rightarrow$$
$$8 \text{ CoASH} + 96 \text{ ATP} + 16 CO_2 + 104 H_2O$$

(2) The combined yield of ATP is therefore given by

$$\text{Palmitoyl CoA} + 23 O_2 + 131 P_i + 131 \text{ ADP} \rightarrow$$
$$8 \text{ CoASH} + 131 \text{ ATP} + 16 CO_2 + 146 H_2O$$

or, $131 - 2 = 129$ mol of ATP, corrected for the activation of the fatty acyl CoA by thiokinase.

5. **Respiratory quotient (RQ)** is defined as the moles of CO_2 produced, divided by the moles of oxygen consumed during complete oxidation of a metabolic fuel to CO_2 and H_2O.
   a. For palmitate, the RQ is given by

$$\frac{16 \text{ mol } CO_2 \text{ produced}}{23 \text{ mol } O_2 \text{ consumed}} = 0.7$$

   b. In contrast, the complete oxidation of glucose yields an RQ of 1.0.

$$C_6H_{12}O_6 + 6 O_2 \rightarrow 6 CO_2 + 6 H_2O$$

$$RQ = \frac{6 \text{ mol } CO_2 \text{ produced}}{6 \text{ mol } O_2 \text{ consumed}} = 1.0$$

B. **Oxidation of fatty acids with an odd number of carbon atoms.** Fatty acids that have an odd number of carbon atoms are only a minor species in human tissues, but they do occur.

1. They undergo β-oxidation as acyl CoA derivatives until a three-carbon fragment is reached, which is propionyl CoA.

2. Propionyl CoA is carboxylated to the S form of methylmalonyl CoA by biotin-dependent propionyl CoA carboxylase.

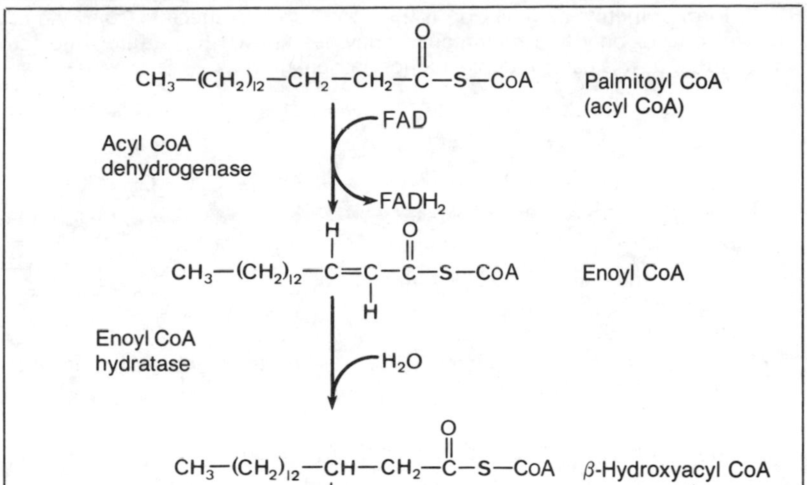

Figure 20-5. Steps in the β-oxidation of the 16-carbon fatty acyl coenzyme A (acyl CoA), palmitoyl CoA, showing the enzymes catalyzing each step.

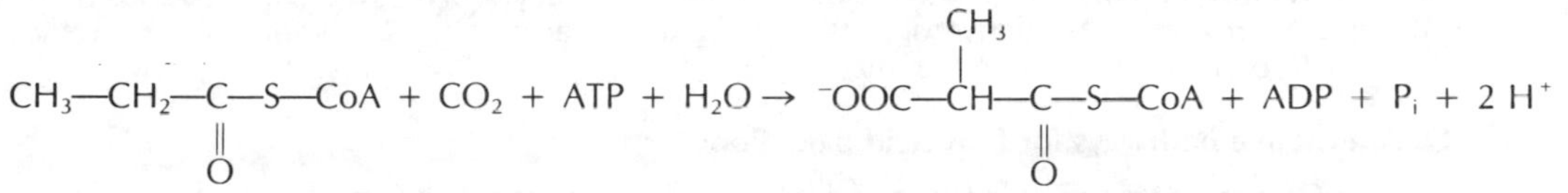

$$CH_3-CH_2-\underset{\underset{O}{\|}}{C}-S-CoA + CO_2 + ATP + H_2O \rightarrow \ {}^-OOC-\underset{\underset{\underset{O}{\|}}{C}}{\overset{\overset{CH_3}{|}}{CH}}-S-CoA + ADP + P_i + 2\,H^+$$

Propionyl CoA   Methylmalonyl CoA (S form)

3. The S form of methylmalonyl CoA is racemized to the R form by methylmalonyl CoA racemase.

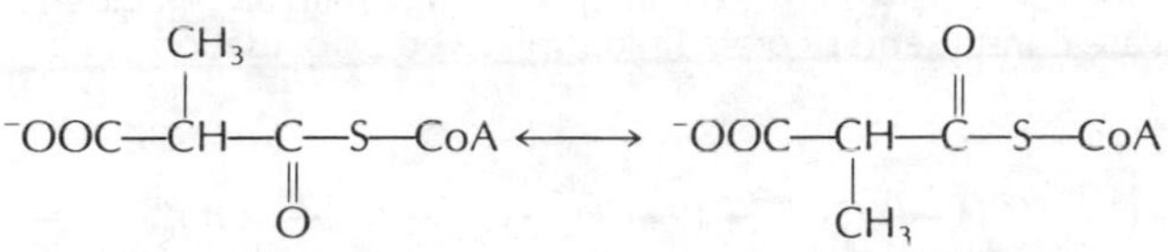

$$ {}^-OOC-\underset{\underset{\underset{O}{\|}}{C}}{\overset{\overset{CH_3}{|}}{CH}}-S-CoA \longleftrightarrow \ {}^-OOC-\underset{\underset{CH_3}{|}}{CH}-\overset{\overset{O}{\|}}{C}-S-CoA$$

S form   R form

4. The R form of methylmalonyl CoA is then converted to succinyl CoA by methylmalonyl CoA mutase, one of only two mammalian enzymes known to require a cofactor derived from vitamin B_{12}, in this case 5′-deoxyadenosyl cobalamin.

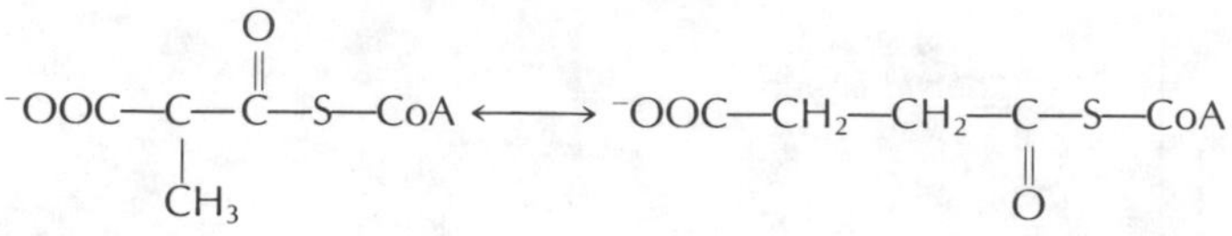

$$^-OOC-\underset{\underset{CH_3}{|}}{C}-\overset{\overset{O}{||}}{C}-S-CoA \longleftrightarrow {}^-OOC-CH_2-CH_2-\overset{\overset{}{}}{C}-S-CoA$$
$$\underset{O}{||}$$

Methylmalonyl CoA (R form)   Succinyl CoA

5. Succinyl CoA can be metabolized via the TCA cycle, of which it is an intermediate.

C. Oxidation of unsaturated fatty acids

1. In the human body, 50% of the fatty acids are unsaturated.

2. A wide range of unsaturated fatty acids can be degraded by the β-oxidation pathway with the assistance of only two other enzymes.

3. The double bonds in the naturally occurring fatty acids are *cis*, and the β-oxidation pathway can only deal with *trans* double bonds as at the enoyl CoA hydratase step (see Fig. 20-5).

4. A Δ^3-*cis*–Δ^2-*trans* enoyl CoA isomerase shifts the double bond to the preferred Δ^2-*trans* configuration. For example, in the β-oxidation of palmitoleate ($16:1:\Delta^9$):

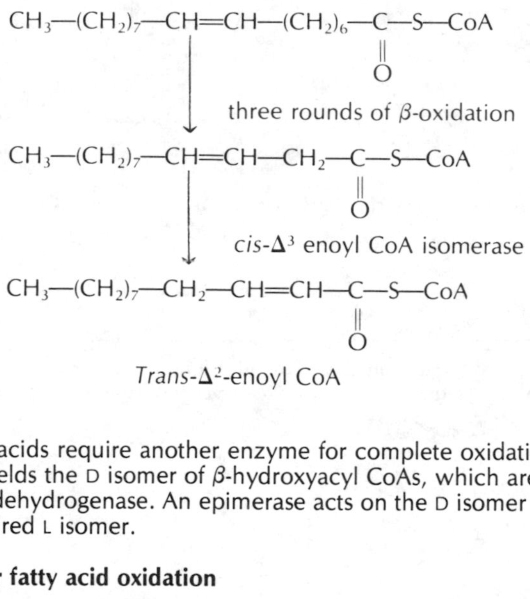

$$CH_3-(CH_2)_7-CH=CH-(CH_2)_6-\overset{\overset{}{}}{C}-S-CoA$$
$$\underset{O}{||}$$

↓ three rounds of β-oxidation

$$CH_3-(CH_2)_7-CH=CH-CH_2-\overset{}{C}-S-CoA$$
$$\underset{O}{||}$$

↓ *cis*-Δ^3 enoyl CoA isomerase

$$CH_3-(CH_2)_7-CH_2-CH=CH-\overset{}{C}-S-CoA$$
$$\underset{O}{||}$$

Trans-Δ^2-enoyl CoA

5. Polyunsaturated fatty acids require another enzyme for complete oxidation. Hydration of a *cis*-Δ^2 double bond yields the D isomer of β-hydroxyacyl CoAs, which are *not* substrates for L-β-hydroxyacyl CoA dehydrogenase. An epimerase acts on the D isomer of a β-hydroxyacyl CoA to yield the required L isomer.

D. Alternative pathways for fatty acid oxidation

1. **ω-Oxidation** involves oxidation of the terminal methyl group to form an ω-hydroxy fatty acid. This is a minor pathway observed with liver microsomal preparations.

2. **α-Oxidation**
   a. Removal of one carbon from the carboxyl end of a fatty acid has been demonstrated in microsomal fractions from brain tissue.
   b. This pathway involves the oxidation of long-chain fatty acids to 2-hydroxy fatty acids, which are constituents of brain lipids, followed by oxidation to a fatty acid with one carbon less.

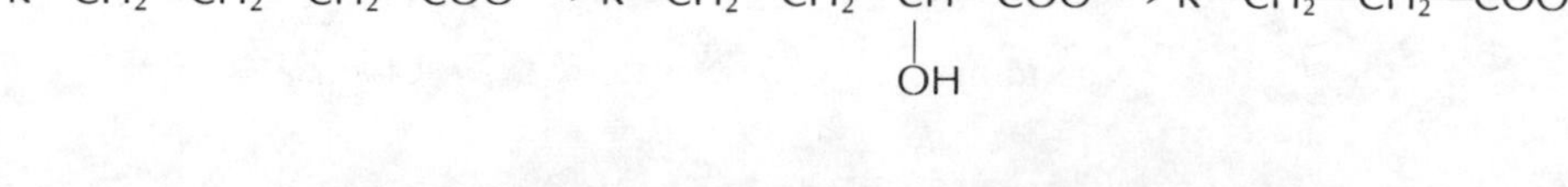

$$R-CH_2-CH_2-CH_2-COO^- \rightarrow R-CH_2-CH_2-\underset{\underset{OH}{|}}{CH}-COO^- \rightarrow R-CH_2-CH_2-COO^-$$

c. In normal individuals, α-oxidation of phytanic acid removes the terminal carboxyl before β-oxidation, allowing the latter pathway to operate, as the β-carbon is now available.

$$
\begin{array}{c}
\text{CH}_3 \\
| \\
-\text{CH}_2-\text{CH}-\text{CH}_2-\text{COO}^- \xrightarrow{\ \alpha\text{-hydroxylation}\ }
\end{array}
\quad
\begin{array}{c}
\text{CH}_3 \\
| \\
-\text{CH}_2-\text{CH}-\text{CH}-\text{COO}^- \xrightarrow{\ \alpha\text{-oxidation}\ } \\
\quad\quad\quad\quad | \\
\quad\quad\quad\quad \text{OH}
\end{array}
$$

$$
\begin{array}{c}
\text{CH}_3 \\
| \\
-\text{CH}_2-\text{CH}-\text{COO}^- + \text{CO}_2 \xrightarrow[\substack{\text{first step of}\\ \beta\text{-oxidation}}]{}
\end{array}
\quad
\begin{array}{c}
\text{CH}_3 \\
| \\
-\text{CH}=\text{C}-\text{COO}^-
\end{array}
$$

(β and α labeled under first carbons)

d. The ability to oxidize fatty acids at the α-carbon is missing in patients with **Refsum's disease** (phytanic acid storage disease).
   (1) In this disorder, there is a buildup of phytanic acid (derived from animal fat and cow's milk and probably originally from cholorophyll). The phytanic acid (Fig. 20-6) cannot be oxidized by the β-oxidation system because the β position is blocked by a methyl group.
   (2) The symptoms include retinitis pigmentosa, failing night vision, peripheral neuropathy, and cerebellar ataxia. The disease has an autosomal recessive pattern of inheritance.

VII. KETONE BODY METABOLISM

A. Nature of ketone bodies

1. Ketone bodies are small, water-soluble potential units of acetate.

2. They are **acetone**, **acetoacetate**, and **β-hydroxybutyrate**.

$$
\underset{\text{Acetone}}{\begin{array}{c}\text{CH}_3\\|\\\text{C}=\text{O}\\|\\\text{CH}_3\end{array}}
\xleftarrow[\substack{\text{spontaneous}\\ \text{CO}_2}]{}
\underset{\text{Acetoacetate}}{\begin{array}{c}\text{CH}_3\\|\\\text{C}=\text{O}\\|\\\text{CH}_2\\|\\\text{COO}^-\end{array}}
\underset{\substack{\beta\text{-hydroxybutyrate}\\ \text{dehydrogenase}}}{\overset{\text{NADH} + \text{H}^+ \quad \text{NAD}^+}{\rightleftharpoons}}
\underset{\beta\text{-Hydroxybutyrate}}{\begin{array}{c}\text{CH}_3\\|\\\text{H}-\text{C}-\text{OH}\\|\\\text{CH}_2\\|\\\text{COO}^-\end{array}}
$$

B. Ketone body formation

1. Ketone bodies are synthesized in liver mitochondria and released from that organ for use as metabolic fuel by other tissues. The liver cannot use ketone bodies as an energy source.

2. Synthesis is limited except under conditions of high rates of fatty acid oxidation and limited carbohydrate intake as, for example, in fasting; in starvation, considerable production will occur.

3. The process uses acetyl CoA derived mainly from β-oxidation of fatty acids, although the catabolism of ketogenic amino acids contributes.

4. **Route of biosynthesis**
   a. The first two steps in ketone body biosynthesis are the same as those involved in cholesterol biosynthesis (see Chapter 23, section II B 1), except that cholesterol biosynthesis occurs in the cytosol and uses excess acetyl CoA derived from carbohydrate metabolism.
   b. In the first step, two acetyl CoA molecules form acetoacetyl CoA in a reaction catalyzed by thiolase, the synthesis being aided by the cleavage of a high-energy thioester bond.

$$
\underset{\text{Acetyl CoA}}{2\ \text{CH}_3-\overset{\overset{\text{O}}{\|}}{\text{C}}-\text{S}-\text{CoA}} \rightarrow \underset{\text{Acetoacetyl CoA}}{\text{CH}_3-\overset{\overset{\text{O}}{\|}}{\text{C}}-\text{CH}_2-\overset{\overset{\text{O}}{\|}}{\text{C}}-\text{S}-\text{CoA}} + \text{CoASH}
$$

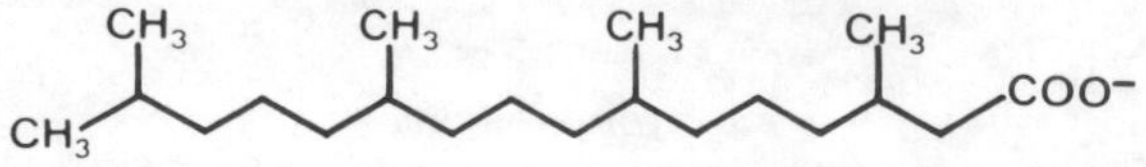

Figure 20-6. The structure of phytanic acid (3,7,11,15-tetramethyl hexadecanoic acid).

c. In the second step, another acetyl CoA reacts with the acetoacetyl CoA to form β-hydroxy-β-methylglutaryl CoA (HMG CoA) in a reaction catalyzed by HMG CoA synthetase.

$$CH_3\text{—}\overset{\overset{\Large O}{\|}}{C}\text{—S—CoA} + CH_3\text{—}\overset{\overset{\Large O}{\|}}{C}\text{—}CH_2\text{—}\overset{\overset{\Large O}{\|}}{C}\text{—S—CoA} + H_2O \rightarrow$$

$$^{-}OOC\text{—}CH_2\text{—}\underset{\underset{\Large OH}{|}}{\overset{\overset{\Large CH_3}{|}}{C}}\text{—}CH_2\text{—}\overset{\overset{\Large O}{\|}}{C}\text{—S—CoA} + \text{CoASH}$$

d. HMG CoA is then cleaved by HMG CoA lyase to acetoacetate and acetyl CoA.

$$^{-}OOC\text{—}CH_2\text{—}\underset{\underset{\Large OH}{|}}{\overset{\overset{\Large CH_3}{|}}{C}}\text{—}CH_2\text{—}\overset{\overset{\Large O}{\|}}{C}\text{—S—CoA} \rightarrow CH_3\text{—}\overset{\overset{\Large O}{\|}}{C}\text{—}CH_2\text{—}COO^{-} + CH_3\text{—}\overset{\overset{\Large O}{\|}}{C}\text{—S—CoA}$$

β-hydroxy-β-methylglutaryl CoA | Acetoacetate | Acetyl CoA

e. β-Hydroxybutyrate is formed from acetoacetate by β-hydroxybutyrate dehydrogenase when the NADH/NAD⁺ ratio is high, as it is in the liver during fasting.

$$CH_3\text{—}\overset{\overset{\Large O}{\|}}{C}\text{—}CH_2\text{—}COO^{-} + NADH + H^{-} \leftrightarrow CH_3\text{—}\overset{\overset{\Large OH}{|}}{CH}\text{—}CH_2\text{—}COO + NAD^{+}$$

Acetoacetate | β-Hydroxybutyrate

f. Acetone is formed spontaneously from a small fraction of the circulating acetoacetate and is lost in expired air from the lungs. In untreated diabetes, the odor of acetone is apparent on the patient's breath, and in this condition, ketone body production can be extremely high.

C. Ketone body oxidation

1. Ketone bodies are the preferred energy substrates of the heart, skeletal muscle, and kidney. If blood levels of β-hydroxybutyrate and acetoacetate increase sufficiently, as they do after about 20 days of starvation, they form a valuable energy substrate for the brain and may account for up to 75% of brain oxidation.
2. The ketone bodies must be activated in order to be used as energy substrates.
3. Activation occurs by the formation of the CoA thioester of acetoacetate, catalyzed by 3-keto acid CoA transferase.

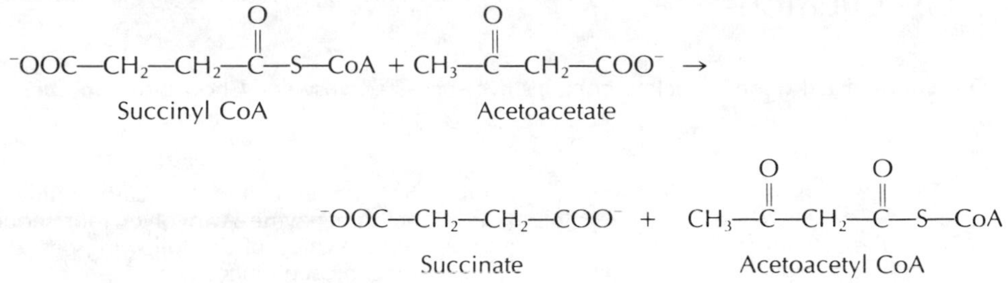

$$^-OOC-CH_2-CH_2-\overset{\overset{\displaystyle O}{\|}}{C}-S-CoA + CH_3-\overset{\overset{\displaystyle O}{\|}}{C}-CH_2-COO^- \rightarrow$$

Succinyl CoA   Acetoacetate

$$^-OOC-CH_2-CH_2-COO^- + CH_3-\overset{\overset{\displaystyle O}{\|}}{C}-CH_2-\overset{\overset{\displaystyle O}{\|}}{C}-S-CoA$$

Succinate   Acetoacetyl CoA

4. Cleavage of the acetoacetyl CoA to two acetyl CoA molecules with the use of CoA provides two-carbon fragments for oxidation by the TCA cycle. This reaction is catalyzed by thiolase.

5. Energy yield from oxidation of ketone bodies
   a. Normally 1 guanosine triphosphate (= 1 ATP) is formed by the conversion of succinyl CoA to succinate in the TCA cycle. This ATP is not formed from succinyl CoA that is a substrate for the 3-keto acid CoA transferase.
   b. Conversion of β-hydroxybutyrate to acetoacetate yields an NADH molecule, which in turn yields three ATP molecules by electron transport and oxidative phosphorylation.
   c. Each mole of acetyl CoA that is formed yields 12 mol of ATP via the TCA cycle, electron transport, and oxidative phosphorylation.
   d. Therefore, oxidation of acetoacetate yields − 1 + 24 = 23 mol of ATP, while oxidation of β-hydroxybutyrate yields − 1 + 3 + 24 = 26 mol of ATP.

6. Ketoacidosis
   a. During starvation, ketone body production increases dramatically, but blood levels rise only slowly, to a maximum of about 7 mM after 20 to 30 days of starvation.
   b. Ketonuria (high output in the urine) accompanies **ketonemia** (high blood levels of ketone bodies); these conditions together are referred to as **ketosis**.
   c. In starvation, the ketonemia and ketonuria are never high enough to precipitate **keto-acidosis** (ketosis accompanied by metabolic acidosis), which results from the swamping of the blood buffering systems by high levels of the acidic ketone bodies. In contrast, in untreated diabetic patients, blood levels of ketone bodies may be extremely high and may well be capable of producing life-threatening ketoacidosis.

STUDY QUESTIONS

Directions: Each question below contains five suggested answers. Choose the **one best** response to each question.

1. The acetyl groups required for cytoplasmic fatty acid synthesis appear in the cytoplasm as a result of the activity of

(A) citrate synthetase
(B) isocitrate dehydrogenase
(C) citrate lyase
(D) thiolase
(E) none of the above enzymes

2. The most important source of reducing equivalents for fatty acid synthesis in the liver is

(A) oxidation of glucuronic acid
(B) oxidation of acetyl coenzyme A
(C) glycolysis
(D) the tricarboxylic acid cycle
(E) the pentose phosphate pathway

3. After the addition of two carbons to the growing fatty acid chain, three sequential reactions occur in the biosynthesis of palmitate by the mammalian cytoplasmic system before another two carbons are added. Which of the following gives the correct sequence for these reactions? (ACP = acyl carrier protein)

(A) Reduction of the enoyl–ACP; reduction of the β-hydroxyacyl–ACP; dehydration of the β-ketoacyl–ACP
(B) Reduction of the β-ketoacyl–ACP; dehydration of the β-hydroxyacyl–ACP; reduction of the enoyl–ACP
(C) Reduction of the β-hydroxyacyl–ACP; reduction of the enoyl–ACP; dehydration of the β-ketoacyl–ACP
(D) Reduction of the enoyl–ACP; dehydration of the β-hydroxyacyl–ACP; oxidation of the β-ketoacyl–ACP
(E) None of the above

4. The biosynthesis of triacylglycerols in adipose tissue cells requires

(A) elevated levels of plasma epinephrine
(B) elevated intracellular levels of cyclic adenosine monophosphate
(C) increased glucose entry into the cells
(D) increased rate of glycerol release from the cells
(E) decreased levels of plasma insulin

5. The removal of two-carbon units from a fatty acyl coenzyme A involves four sequential reactions. Which of the following best describes the reaction sequence?

(A) Oxidation, dehydration, oxidation, cleavage
(B) Reduction, dehydration, reduction, cleavage
(C) Dehydrogenation, hydration, dehydrogenation, cleavage
(D) Hydrogenation, dehydration, hydrogenation, cleavage
(E) Reduction, hydration, dehydrogenation, cleavage

6. The coenzyme A ester of oleic acid (*cis*-Δ^9-octadecenoic acid) can be oxidized completely to carbon dioxide and water by the β-oxidation pathway and the tricarboxylic acid cycle. How many moles of adenosine triphosphate (ATP) are produced per mole of oleate oxidized, not counting ATP required for activation or transport?

(A) 146 mol
(B) 148 mol
(C) 151 mol
(D) 153 mol
(E) 180 mol

18C.
9 × 12 = 108

7. β-Oxidation of phytanic acid, a fatty acid derived from chlorophyll and found in cow's milk, cannot take place without prior oxidation at the α-carbon. Which of the following statements best explains the requirement for α-oxidation?

(A) The chain length of phytanic acid is too long for the mitochondrial β-oxidation system to handle
(B) The β-carbon in phytanic acid is blocked by a methyl group
(C) Phytanic acid cannot be activated by coenzyme A prior to β-oxidation
(D) The mitochondrial acyl carnitine transferase will not transport phytanic acid into the mitochondrial matrix
(E) β-Oxidation of phytanic acid can begin, but the product of the first dehydrogenation reaction is toxic to the β-oxidation system

Directions: Each question below contains four suggested answers of which **one or more** is correct. Choose the answer

    A if **1, 2, and 3** are correct
    B if **1 and 3** are correct
    C if **2 and 4** are correct
    D if **4** is correct
    E if **1, 2, 3, and 4** are correct

8. If an in vitro system is used to synthesize palmitate from acetyl coenzyme A (acetyl CoA), ^{14}C-labeled carbon dioxide and ^{14}C-methylene–labeled malonyl CoA, which carbons in the product will be radioactive?

(1) Carbons 1, 3, 5, 7
(2) Carbons 2, 4, 6, 8
(3) Carbons 3, 6, 9, 12
(4) Carbons 10, 12, 14, 16

9. The cytoplasmic fatty acid synthetase system of mammals is characterized by which of the following statements?

(1) The major product is a C-16 fatty acid
(2) The chain grows by addition of two-carbon fragments donated by a three-carbon compound
(3) The electron donor for the reductive synthesis of fatty acids is reduced nicotinamide-adenine dinucleotide phosphate (NADPH)
(4) The NADPH required by the system is produced by the glycolytic pathway

10. Differences between the systems involved in the β-oxidation of fatty acids and in their biosynthesis include the

(1) intracellular location of the enzymes involved
(2) nature of the thiol acyl carriers
(3) oxidation–reduction state of required pyridine nucleotide cofactors
(4) participation of high-energy phosphate bonds

11. Which of the following fatty acids *could* be a precursor of prostaglandins in humans?

(1) Linoleic acid
(2) $\Delta^8, \Delta^{11}, \Delta^{14}$-Eicosatrienoic acid
(3) Linolenic acid
(4) $\Delta^5, \Delta^8, \Delta^{11}$-Eicosatrienoic acid

ANSWERS AND EXPLANATIONS

1. The answer is C. (*III B 1*) Most of the fatty acid synthesis by humans occurs in the liver during times of plenty; that is, after feeding. The carbon for fatty acid synthesis arises from glucose via glycolysis and the tricarboxylic acid (TCA) cycle, leaving the mitochondria as citrate. The citrate exits from the mitochondria because the adenosine triphosphate and reduced nicotinamide-adenine dinucleotide phosphate levels in the mitochondria are high and effectively block use of the citrate in the TCA cycle by inhibiting isocitrate dehydrogenase. In the cytoplasm, the citrate is cleaved by citrate lyase to oxaloacetate and acetyl coenzyme A (acetyl CoA). The latter is used for the synthesis of fatty acids, and the oxaloacetate is reduced to malate, which, in the liver, enters the mitochondria. Citrate synthetase produces citrate in the mitochondria from acetyl CoA and oxaloacetate, and thiolase catalyzes the reversible cleavage of acetoacetyl CoA to acetyl CoA.

2. The answer is E. (*III B 2*) Reducing equivalents are required during fatty acid synthesis to convert the β-ketoacyl–acyl carrier protein (ACP) intermediate to the saturated acyl–ACP thioester. Reduced nicotinamide-adenine dinucleotide phosphate (NADPH) is used as the electron donor. In the liver, the NADPH arises mainly from the oxidation of glucose 6-phosphate (G6P) via the pentose phosphate pathway (the hexose monophosphate shunt). NADPH is formed at two steps in this pathway: the first at the oxidation of G6P to 6-phosphoglucono-δ-lactone, and the second at the oxidation of 6-phosphogluconate to ribulose 5-phosphate (see Chapter 19, section I B 1 b). Another potential source of NADPH is from the cytoplasmic $NADP^+$-linked malic enzyme, which will oxidize malate to pyruvate and carbon dioxide. The malate arises from the cytoplasmic oxaloacetate formed by citrate cleavage enzyme, a reaction providing the fatty acid synthesizing system with acetyl coenzyme A units (see section III B 1). However, this reaction is of lesser importance in the liver, where malate tends to enter the mitochondria, than in rodent fat tissue, where it is an important source of NADPH for fat synthesis.

3. The answer is B. (*III B 5 f*) In the synthesis of palmitate by the cytoplasmic fatty acid synthetase, a condensation of an acetyl group occurs with a malonyl group on the acyl carrier protein (ACP). There is a loss of carbon dioxide and a ketoacyl–ACP intermediate is formed. Three reactions now take place before the addition of further two-carbon units. These reactions are, in sequence, a reduction, a dehydration, and a second reduction. The first reaction reduces the β-ketoacyl–ACP to a β-hydroxyacyl–ACP, with NADPH as the electron donor. The dehydration reaction then yields the enoyl–ACP intermediate, and the second reduction, also with NADPH as the electron donor, converts this to the saturated acyl–ACP thioester. The latter reacts with the B_2 site on the β-ketoacyl–ACP synthetase, freeing the ACP to react with another malonyl group.

4. The answer is C. (*IV*) The biosynthesis of triacylglycerols involves the esterification of glycerol by fatty acyl coenzyme As (fatty acyl CoAs). Adipose tissue cells (adipocytes) cannot phosphorylate glycerol arising from lipolysis, as they lack a glycerol kinase. Instead, they use dihydroxyacetone phosphate or glycerol 3-phosphate formed during glycolysis. The biosynthesis of triacylglycerols is thus stimulated by increased glucose entry into the cells, which is itself a function of plasma insulin levels. The higher the plasma insulin, the greater the rate of glycolysis, and the more glycerol 3-phosphate available for esterification of fatty acyl CoAs. Elevated plasma levels of epinephrine and of intracellular levels of cyclic adenosine monophosphate stimulate lipolysis, which is the hydrolysis of triacylglycerols, just the reverse of biosynthesis. An increased rate of glycerol release from adipocytes occurs during lipolysis, as the free glycerol formed in this process cannot be used by these cells.

5. The answer is C. (*VI A 3; Figure 20-4*) The four sequential reactions of β-oxidation remove two-carbon units from fatty acyl coenzyme As (fatty acyl CoAs) until, with an even number of carbons, only acetyl CoA is left. With an odd-numbered carbon chain, the final fragment is propionyl CoA. These reactions are, in sequence, a dehydrogenation, a hydration, a dehydrogenation, and a cleavage. The first dehydrogenation converts acyl CoAs to the trans-Δ^2-enoyl CoA derivative; the dehydrogenase requires flavin-adenine dinucleotide as the electron acceptor. Hydration yields the L-β-hydroxyacyl CoA derivative, which is converted by the second dehydrogenation to the β-ketoacyl CoA derivative. The second dehydrogenase requires nicotinamide-adenine dinucleotide as electon acceptor. The last of the four reactions uses CoA to split off two carbons as acetyl CoA, leaving the residual fatty acyl CoA.

6. The answer is A. (*VI A 4, C*) The C-18 unsaturated fatty acid, oleic acid, can be oxidized by β-oxidation to yield nine acetyl coenzyme As (acetyl CoAs), a process that involves eight thiolytic cleavages. In a saturated fatty acid, each cleavage requires one flavin-adenine dinucleotide (FAD) and one nicotinamide-adenine dinucleotide (NAD^+) which, as reduced cofactors, will provide two and three adenosine triphosphates (ATPs), respectively, via the electron transport system. However, oleate already has one double bond, and although it is the Δ^3-cis double bond rather than the required Δ^2-trans, this is isomerized prior to the hydration step. FAD is, therefore, not required in the removal of the two-carbon fragment containing the isomerized double bond. Therefore, the unsaturated oleate provides two ATPs less during β-oxidation than a saturated C-18 fatty acid. The number of ATPs formed by the β-oxidation process from oleate is $8 \times 5 - 2 = 38$. The 9 acetyl CoAs formed by the 8 cuts can be oxidized to carbon dioxide and water in the tricarboxylic acid cycle, yielding $9 \times 12 = 108$ ATPs (see Chapter 15, section I C and Chapter 16, section IV C). The total yield of ATP per mole of oleate oxidized is therefore $108 + 38 = 146$ mol.

7. The answer is B. (*VI D 2*) Phytanic acid is 3,7,11,15-tetramethyl hexadecanoic acid. The methyl group at C-3 is on the β-carbon and effectively blocks the hydroxylation of this carbon. In normal humans, phytanic acid is oxidized at the α-carbon first, a reaction that involves the loss of the terminal carboxyl carbon as carbon dioxide. β-Oxidation can then take place, with alternating release of propionyl coenzyme A (propionyl CoA) and acetyl CoA. The final fragment is isobutyryl CoA.

8. The answer is C (2, 4). (*III B 5*) The rate-limiting step in the de novo synthesis of fatty acids by the cytoplasmic fatty acid synthetase system is the formation of malonyl coenzyme A (malonyl CoA) by the carboxylation of acetyl CoA. The experiment provides the synthesizing system with unlabeled acetyl CoA and ^{14}C-labeled carbon dioxide (CO_2). This will produce malonyl CoA labeled in the carboxyl carbon, which will equilibrate with the ^{14}C-methylene–labeled malonyl CoA added to the system. However, at the condensation step, when malonyl CoA is attached to either the first acetyl CoA or the end of the growing chain, the ^{14}C-carboxyl–labeled carbon is lost as CO_2. Thus, only the ^{14}C-methylene label will appear in the final product. The labeling will thus occur only in even-numbered carbons, as the number one carbon of the product will be the unlabeled carboxyl carbon of the last malonyl CoA.

9. The answer is A (1, 2, 3). (*III B*) The mammalian fatty acid synthetase complex produces the C-16 fatty acid palmitic acid. Chain growth is by the addition of malonyl coenzyme A (malonyl CoA) [three carbons] to the growing chain, which is attached to the acyl carrier protein (ACP). One carbon of the malonyl CoA is lost as carbon dioxide (CO_2). Following the condensation of the two-carbon fragment with the growing chain, three reactions occur before the addition of another malonyl CoA. There is a reduction of the β-ketoacyl–ACP to the β-hydroxyacyl–ACP, dehydration to the enoyl–ACP derivative, and another reduction to the saturated acyl–ACP thioester. Both reductions require reduced nicotinamide-adenine dinucleotide phosphate (NADPH) as electron donor. In adipose tissue, NADPH for the reductive synthesis of fatty acids comes partly from the activity of malic enzyme, which is NADP+-linked and oxidizes malate to pyruvate and CO_2, and partly from the pentose phosphate pathway. In the liver, most of the NADPH comes from the pentose phosphate pathway (see Chapter 19, section I).

10. The answer is E (all). (*III B; VI A*) The principal route for the oxidation of fatty acids is the β-oxidation pathway localized in the mitochondrial matrix. In contrast, the major pathway for the biosynthesis of the C-16 palmitic acid is the cytoplasmic fatty acid synthetase complex. Prior to oxidation, fatty acids are activated to form fatty acyl thioesters of coenzyme A (fatty acyl CoA), which acts as the acyl carrier in this process. During the synthesis of the fatty acid chain, the intermediates are carried on the acyl carrier protein as thiol esters. Oxidation of fatty acids requires the participation of nicotinamide-and flavin-adenine dinucleotides, whereas fatty acid synthesis requires reduced nicotinamide-adenine dinucleotide phosphate. Oxidation of a C-16 fatty acid to carbon dioxide and water via the β-oxidation pathway and the tricarboxylic acid pathway yields a net 129 mol of adenosine triphosphate per mole of fatty acid. Fatty acid synthesis, on the other hand, requires the expenditure of high-energy phosphate bonds for the supply of acetyl CoA from citrate and for malonyl CoA from acetyl CoA.

11. The answer is A (1, 2, 3). (*II B; Table 20-1; Chapter 22 II A*) In human tissues, the immediate precursor of major prostaglandins is arachidonic acid, which cannot be synthesized de novo. Unsaturated fatty acids containing a methyl-end saturated "tail" of six carbons can act as precursors of arachidonic acid and therefore as precursors of prostaglandins. Arachidonic acid has a Δ^{14} double bond, which human tissues cannot duplicate either by direct unsaturation or by elongation two carbons at a time followed by unsaturation.

21

Phosphoglycerides and Sphingolipids: Structures and Metabolism

I. PHOSPHOGLYCERIDES. The phosphoglycerides include **phosphatidylcholine (lecithin)**; the two **cephalins, phosphatidylethanolamine** and **phosphatidylserine**; **phosphatidylinositol**; and **cardiolipin**. Lecithin and the cephalins arise from choline and ethanolamine derived from the diet or from the breakdown of body phospholipids.

A. Characteristics

1. Phosphoglycerides are the most polar of lipids and are **amphipathic**, possessing both hydrophilic and hydrophobic groups. Therefore, they can inhabit transition regions between aqueous and nonaqueous phases.

2. Phosphoglycerides are also **amphoteric**, bearing both negatively charged and positively charged groups.

3. Phospholipids are major constituents of cellular membranes and occur in high concentrations in the lipids of glandular organs, blood plasma, and egg yolk, and in the seeds of legumes. They comprise about 40% of the lipids in the erythrocyte membrane and over 95% in the inner mitochondrial membrane. About 20% of the lipids of the inner mitochondrial membrane are cardiolipin.

B. Biosynthesis

1. Biosynthesis of phosphatidate
   a. The backbone of a phosphoglyceride is glycerol 3-phosphate.
   b. A stereospecific numbering system (sn) is used for glycerol.

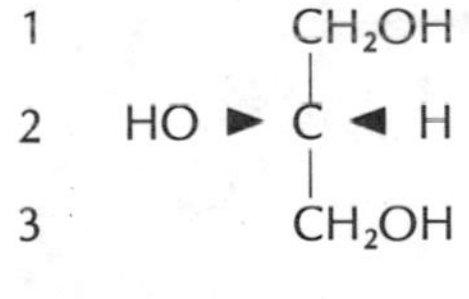

$$
\begin{array}{cc}
1 & CH_2OH \\
2 & HO \blacktriangleright C \blacktriangleleft H \\
3 & CH_2OH
\end{array}
$$

   c. Acylation of the sn-1 position of glycerol 3-phosphate yields **lysophosphatidate**.

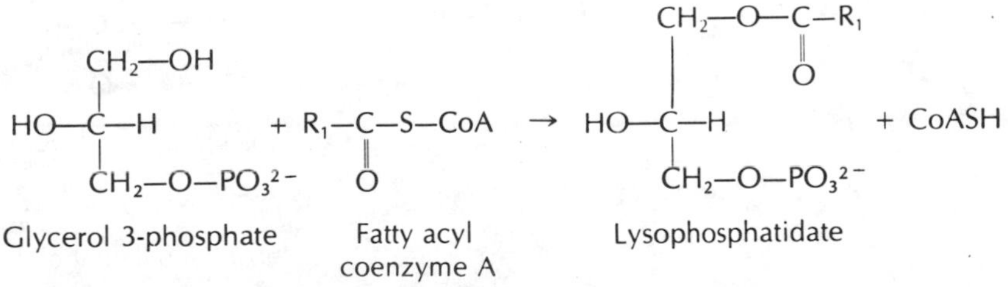

$$
\begin{array}{c}
CH_2-OH \\
HO-C-H \\
CH_2-O-PO_3{}^{2-}
\end{array}
+ R_1-\overset{\displaystyle O}{\underset{\displaystyle \|}{C}}-S-CoA \rightarrow
\begin{array}{c}
CH_2-O-\overset{\displaystyle O}{\underset{\displaystyle \|}{C}}-R_1 \\
HO-C-H \\
CH_2-O-PO_3{}^{2-}
\end{array}
+ CoASH
$$

Glycerol 3-phosphate   Fatty acyl coenzyme A   Lysophosphatidate

   d. Lysophosphatidate can react with another fatty acyl coenzyme A (fatty acyl CoA)

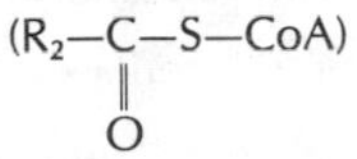

$$(R_2-\overset{\displaystyle O}{\underset{\displaystyle \|}{C}}-S-CoA)$$

to form a phosphatidate.

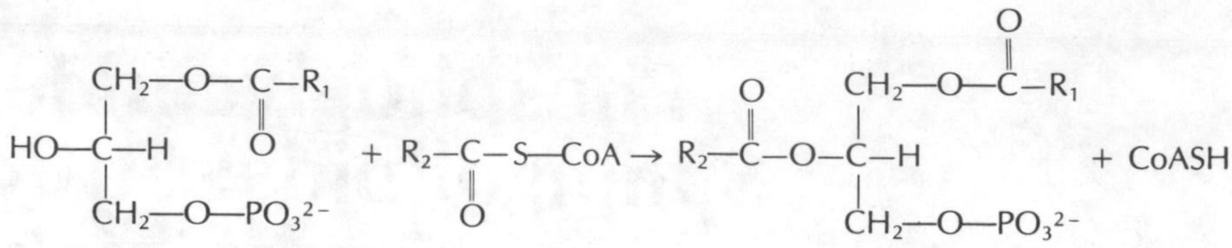

$$\text{Lysophosphatidate} + R_2-C-S-CoA \rightarrow \text{Phosphatidate} + CoASH$$

Lysophosphatidate   Fatty acyl coenzyme A   Phosphatidate

e. A phosphatase removes the phosphate group from a phosphatidate to yield an α,β-diacylglycerol.

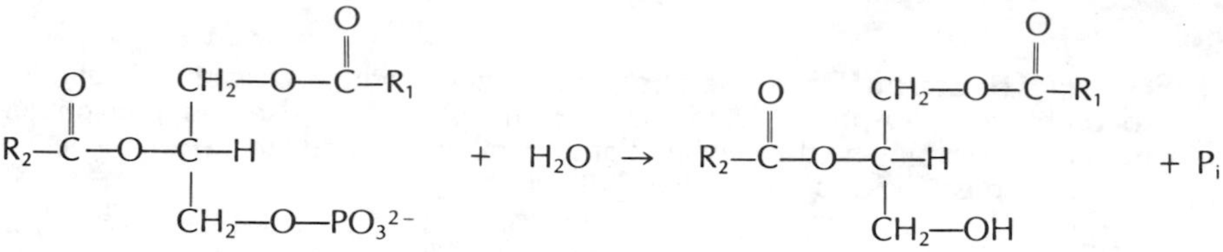

$$\text{Phosphatidate} + H_2O \rightarrow \alpha,\beta\text{-Diacylglycerol} + P_i$$

Phosphatidate α,β-Diacylglycerol

f. The α,β-diacylglycerol can be acylated by another fatty acyl CoA to form a triacylglycerol (see Chapter 20, section IV B 3) or used to form phospholipids (see below).

2. Biosynthesis of phosphatidylethanolamine

a. The first step is the formation of ethanolamine phosphate from ethanolamine and adenosine triphosphate (ATP).

$$^+H_3N-CH_2-CH_2-OH + ATP \rightarrow {}^+H_3N-CH_2-CH_2-O-PO_3^{2-} + PP_i$$

Ethanolamine   Ethanolamine phosphate

b. Ethanolamine phosphate reacts with cytidine triphosphate (CTP) to form cytidine diphosphate ethanolamine (CDP-ethanolamine).

$$^+H_3N-CH_2-CH_2-O-PO_3^{2-} + CTP \rightarrow {}^+H_3N-CH_2-CH_2-O-CDP + PP_i$$

Ethanolamine phosphate   Cytidine diphosphate ethanolamine

c. The reaction of CDP-ethanolamine with an α,β-diacylglycerol yields the phospholipid phosphatidylethanolamine.

$$\alpha,\beta\text{-Diacylglycerol} + {}^+H_3N-CH_2-CH_2-O-CDP \rightarrow$$

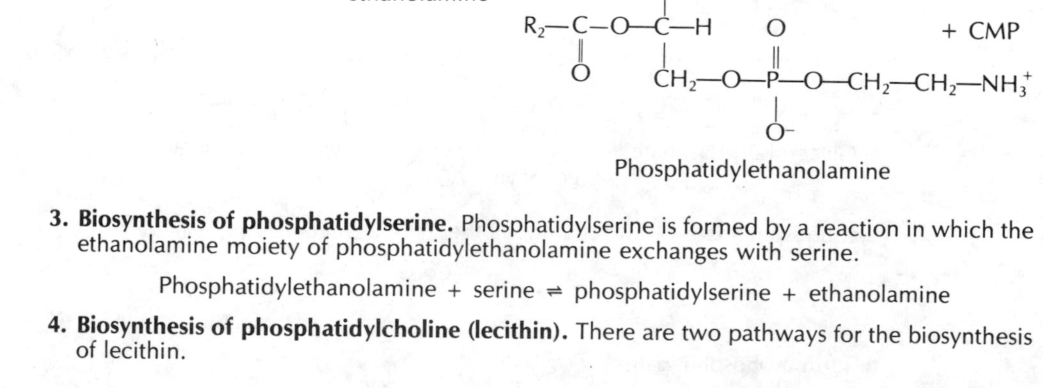

α,β-Diacylglycerol   Cytidine diphosphate ethanolamine $\rightarrow$ Phosphatidylethanolamine $+$ CMP

Phosphatidylethanolamine

3. Biosynthesis of phosphatidylserine.
Phosphatidylserine is formed by a reaction in which the ethanolamine moiety of phosphatidylethanolamine exchanges with serine.

$$\text{Phosphatidylethanolamine} + \text{serine} \rightleftharpoons \text{phosphatidylserine} + \text{ethanolamine}$$

4. Biosynthesis of phosphatidylcholine (lecithin).
There are two pathways for the biosynthesis of lecithin.

a. De novo synthesis from phosphatidylethanolamine takes place by the successive transfer of three methyl groups from the methyl donor S-adenosylmethionine (see Chapter 27, section III C 2).

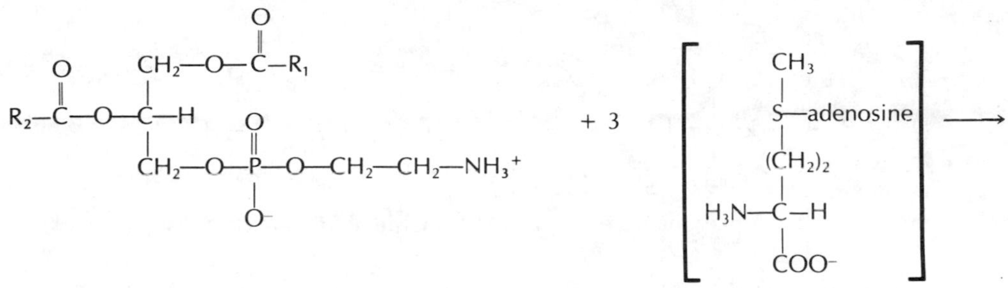

$$R_2-\overset{\overset{\displaystyle O}{\|}}{C}-O-\overset{\overset{\displaystyle CH_2-O-\overset{\overset{\displaystyle O}{\|}}{C}-R_1}{|}}{\underset{\underset{\displaystyle CH_2-O-\overset{\overset{\displaystyle O}{\|}}{\underset{\underset{\displaystyle O^-}{|}}{P}}-O-CH_2-CH_2-NH_3{}^+}{|}}{C}}-H$$

Phosphatidylethanolamine

$$+ 3 \quad \begin{bmatrix} CH_3 \\ | \\ S-adenosine \\ | \\ (CH_2)_2 \\ | \\ H_3N-\overset{\overset{}{|}}{C}-H \\ | \\ COO^- \end{bmatrix} \longrightarrow$$

S-Adenosylmethionine

$$R_2-\overset{\overset{\displaystyle O}{\|}}{C}-O-\overset{\overset{\displaystyle CH_2-O-\overset{\overset{\displaystyle O}{\|}}{C}-R_1}{|}}{\underset{\underset{\displaystyle CH_2-O-\overset{\overset{\displaystyle O}{\|}}{\underset{\underset{\displaystyle O^-}{|}}{P}}-O-CH_2-CH_2-\overset{\overset{\displaystyle CH_3}{|}}{\underset{\underset{\displaystyle CH_3}{|}}{N}}-CH_3}{|}}{C}}-H$$

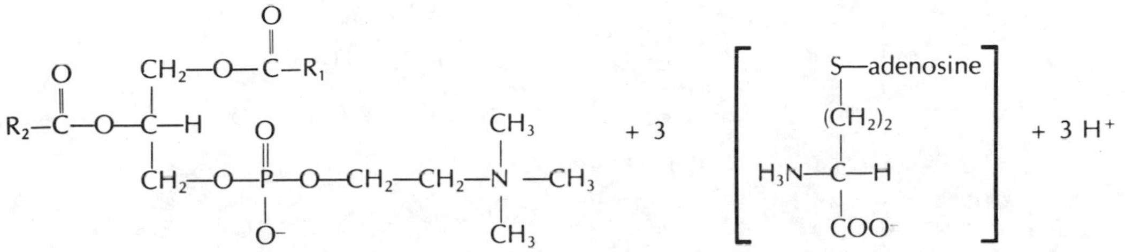

Phosphatidylcholine

$$+ 3 \quad \begin{bmatrix} S-adenosine \\ | \\ (CH_2)_2 \\ | \\ H_3N-\overset{\overset{}{|}}{C}-H \\ | \\ COO^- \end{bmatrix} + 3\ H^+$$

S-Adenosylhomocysteine

b. Salvage pathway synthesis uses choline derived from the degradation of body phospholipids, or from the diet.

(1) Choline is first phosphorylated at the expense of ATP.

$$(CH_3)_3-N^+-CH_2-CH_2-OH + ATP \rightarrow (CH_3)_3-N^+-CH_2-CH_2-O-PO_3{}^{2-} + ADP$$

Choline   Choline phosphate

(2) Choline phosphate next reacts with CTP to form cytidine diphosphate (CDP) choline.

Choline phosphate + CTP → CDP-choline + pyrophosphate (PP$_i$)

(3) CDP-choline then reacts with an α,β-diacylglycerol to form lecithin.

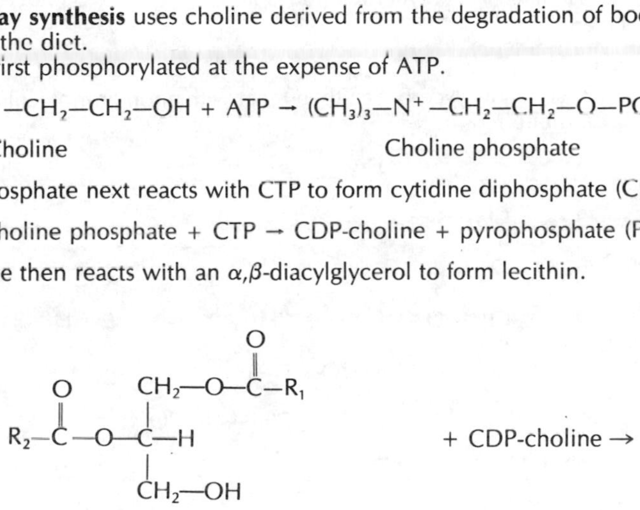

$$R_2-\overset{\overset{\displaystyle O}{\|}}{C}-O-\overset{\overset{\displaystyle CH_2-O-\overset{\overset{\displaystyle O}{\|}}{C}-R_1}{|}}{\underset{\underset{\displaystyle CH_2-OH}{|}}{C}}-H \qquad + \text{ CDP-choline} \rightarrow$$

α,β-Diacylglycerol

$$R_2-\overset{\overset{\displaystyle O}{\|}}{C}-O-\overset{\overset{\displaystyle CH_2-O-\overset{\overset{\displaystyle O}{\|}}{C}-R_1}{|}}{\underset{\underset{\displaystyle CH_2-O-\overset{\overset{\displaystyle O}{\|}}{\underset{\underset{\displaystyle O^-}{|}}{P}}-O-CH_2-CH_2-\overset{\overset{\displaystyle CH_3}{|}}{\underset{\underset{\displaystyle CH_3}{|}}{N^+}}-CH_3}{|}}{C}}-H$$

CH_3 + cytidine monophosphate (CMP)

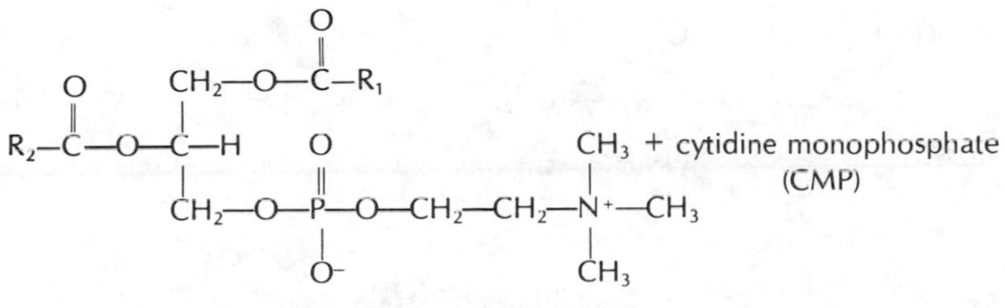

Phosphatidylcholine (lecithin)

5. Biosynthesis of phosphatidylinositol and cardiolipin

a. These glycerol phospholipids are synthesized from phosphatidic acids. The common step in their formation is the reaction between a phosphatidate and CTP to yield **cytidine diphosphate diacylglycerol** (CDP-diacylglycerol).

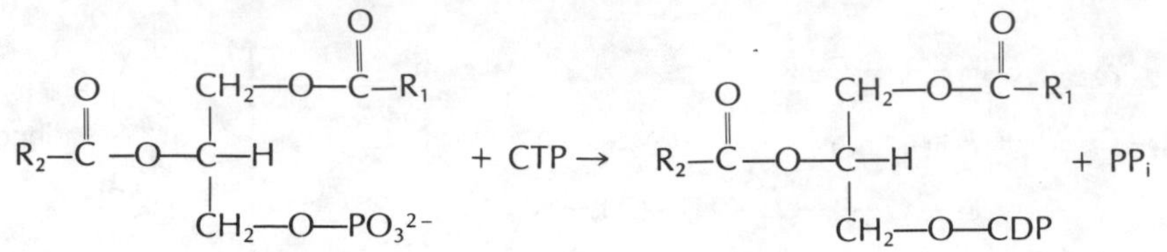

| A phosphatidate | Cytidine diphosphate diacylglycerol |

b. CDP-diacylglycerol reacts with glycerol 3-phosphate to give **phosphatidylglycerol**.

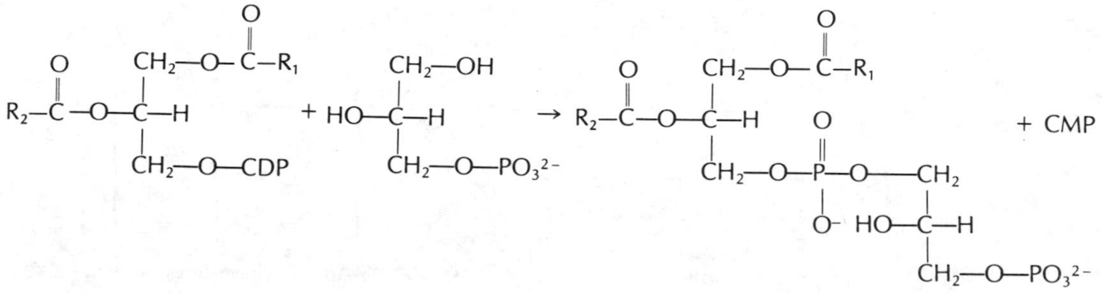

Cytidine diphosphate   Glycerol 3-phosphate   Phosphatidylglycerol
diacylglycerol

c. Phosphatidylglycerol can then react with another molecule of CDP-diacylglycerol to form **cardiolipin**.

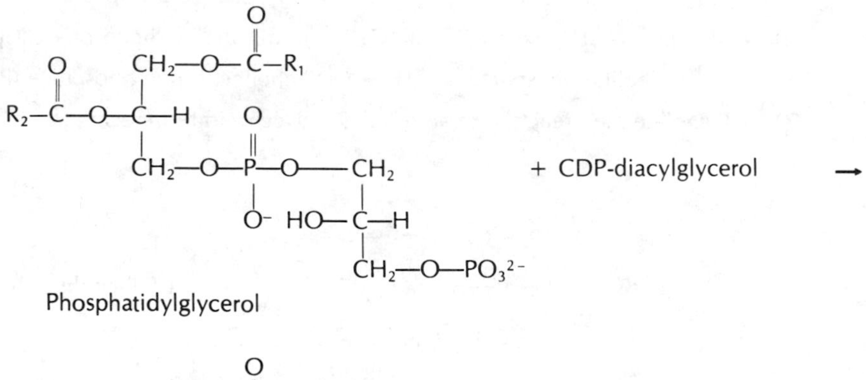

Phosphatidylglycerol

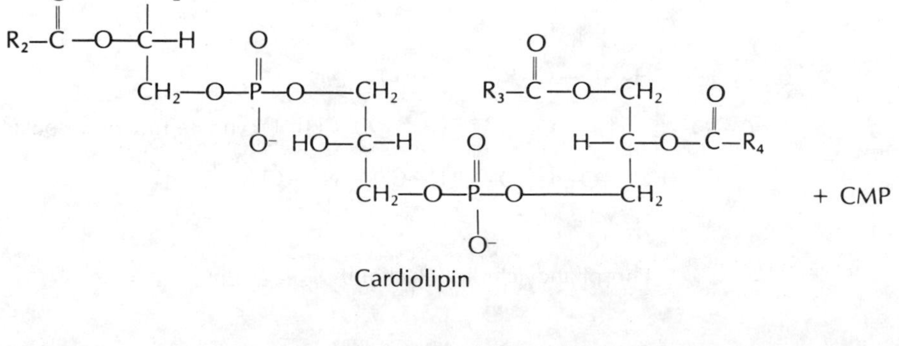

Cardiolipin

d. CDP-diacylglycerol may also react with inositol to form **phosphatidylinositol**.

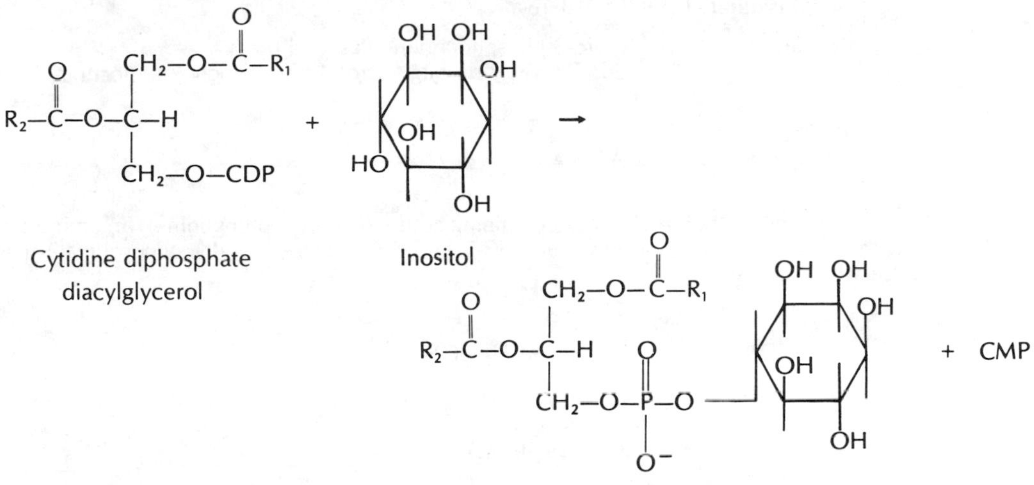

Cytidine diphosphate
diacylglycerol

Inositol

Phosphatidylinositol

C. Modification

1. The fatty acids found in the sn-1 and sn-2 positions of tissue phospholipids (see section I B 1 b) are often not the same fatty acids that were inserted by the synthesizing enzymes. Most fatty acyl CoA transferases and phospholipid synthesizing systems lack the specificity required to account for the asymmetric position, or distribution, found in tissue phospholipids.

2. Tissues adjust the fatty acid composition of their phospholipids by phospholipase excision of the sn-1 or sn-2 fatty acids, followed by reacylation.

   a. Phospholipases. Two hydrolases are important in the modification of phosphoglyceride structures.

   (1) Phospholipase A₁ catalyzes the removal (hydrolysis) of the fatty acid substituent at the sn-1 carbon of the glycerol backbone.

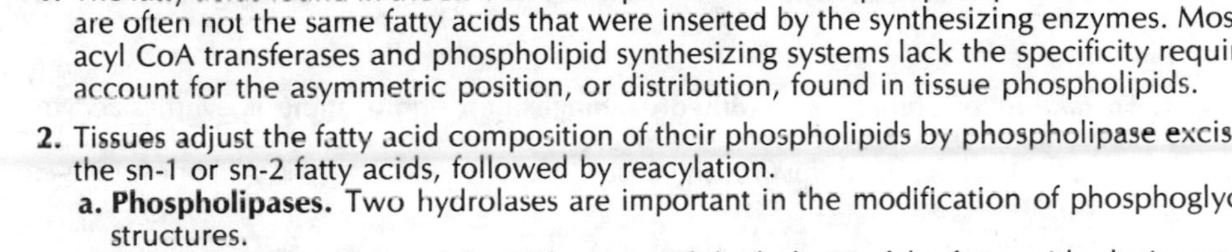

   (2) Phospholipase A₂ catalyzes the removal of the sn-2 fatty acid.

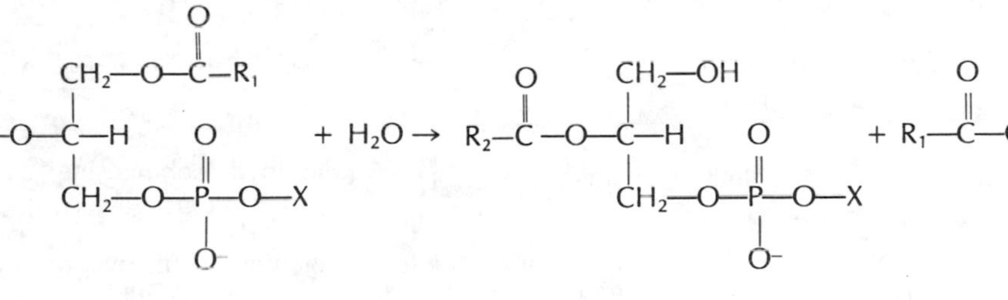

   b. Reacylation can be
   (1) By direct acylation with the appropriate fatty acyl CoA

(2) By exchange-type reactions; for example, by an arachidonic acid–specific acyl CoA acyltransferase, lysolecithin:lecithin acyltransferase (LCAT), which shows a preference for unsaturated acyl CoA derivatives

Acyl CoA + 1-acylglycero-3-phosphocholine (lysolecithin) $\longleftrightarrow$
CoASH + 1,2-diacylglycerol-3-phosphocholine (lecithin)

II. SPHINGOLIPIDS

A. Nature

1. Sphingolipids are complex lipids containing **sphingosine** (4-sphingenine), an amino alcohol with two asymmetric carbon atoms (C-2 and C-3). The naturally occurring form is the D-erythro form.

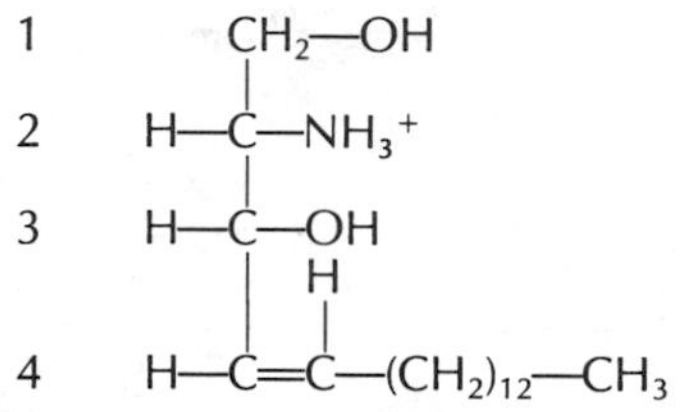

$$
\begin{array}{ll}
1 & CH_2{-}OH \\
2 & H{-}C{-}NH_3{}^+ \\
3 & H{-}C{-}OH \\
  & \quad\quad H \\
4 & H{-}C{=}C{-}(CH_2)_{12}{-}CH_3 \\
\end{array}
$$

Sphingosine

2. The greatest concentration of sphingolipids is found in the central nervous system (CNS), particularly in white matter, although nearly all human tissues contain some.

B. Biosynthesis

1. **Formation of sphinganine (dihydrosphingosine).** Sphinganine is synthesized from serine and palmitoyl CoA by a two-step process.
   a. Formation of 3-ketodihydrosphingosine is catalyzed by a pyridoxal phosphate–dependent enzyme.

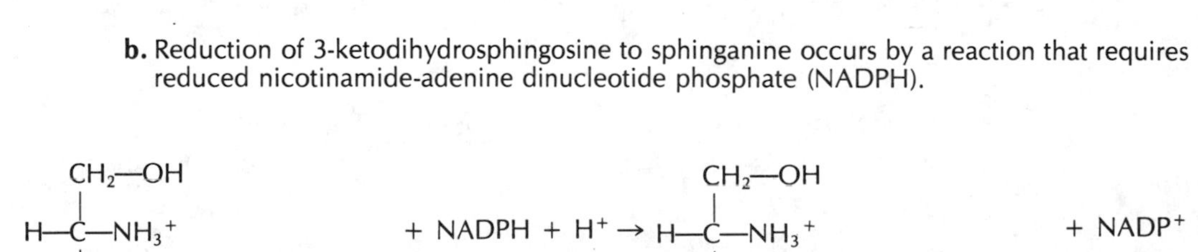

$$
\begin{array}{l}
COO^- \\
H{-}C{-}NH_3{}^+ \\
CH_2{-}OH
\end{array}
\; + \; CoA{-}S{-}\underset{O}{\overset{\|}{C}}{-}(CH_2)_{14}{-}CH_3 \; + H^+ \rightarrow \;
\begin{array}{l}
CH_2{-}OH \\
H{-}C{-}NH_3{}^+ \\
\underset{O}{\overset{\|}{C}}{-}(CH_2)_{14}{-}CH_3
\end{array}
\; + CO_2 + CoASH
$$

Serine Palmitoyl coenzyme A 3-Ketodihydrosphingosine

   b. Reduction of 3-ketodihydrosphingosine to sphinganine occurs by a reaction that requires reduced nicotinamide-adenine dinucleotide phosphate (NADPH).

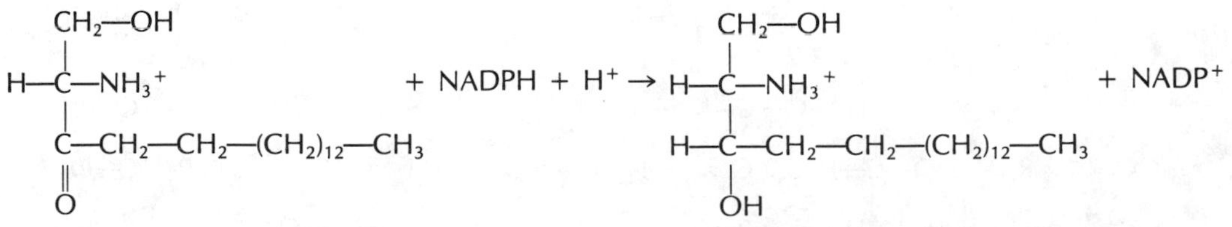

$$
\begin{array}{l}
CH_2{-}OH \\
H{-}C{-}NH_3{}^+ \\
\underset{O}{\overset{\|}{C}}{-}CH_2{-}CH_2{-}(CH_2)_{12}{-}CH_3
\end{array}
\; + NADPH + H^+ \rightarrow \;
\begin{array}{l}
CH_2{-}OH \\
H{-}C{-}NH_3{}^+ \\
H{-}\underset{OH}{C}{-}CH_2{-}CH_2{-}(CH_2)_{12}{-}CH_3
\end{array}
\; + NADP^+
$$

3-Ketodihydrosphingosine Sphinganine (dihydrosphingosine)

2. **Formation of ceramide.** A two-step process involves the introduction of a C-22 fatty acyl moiety and oxidation of the palmitoyl chain.
   a. Formation of dihydroceramide is the first step.

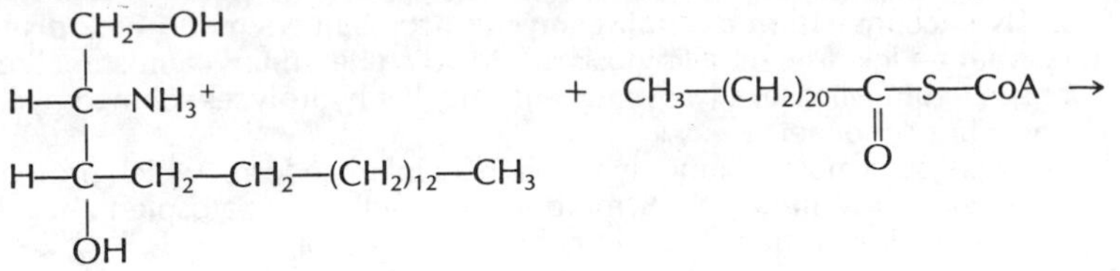

$$\text{Sphinganine} \qquad\qquad + \quad CH_3-(CH_2)_{20}-\overset{O}{\underset{\|}{C}}-S-CoA \quad \rightarrow$$

Sphinganine   Behenyl coenzyme A (docosanoyl CoA)

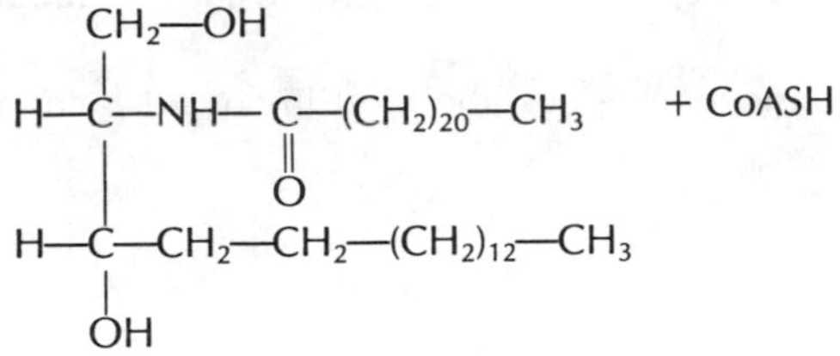

Dihydroceramide   + CoASH

b. Formation of ceramide is the next step. Ceramide forms the **core structure** of naturally oc-curring sphingolipids.

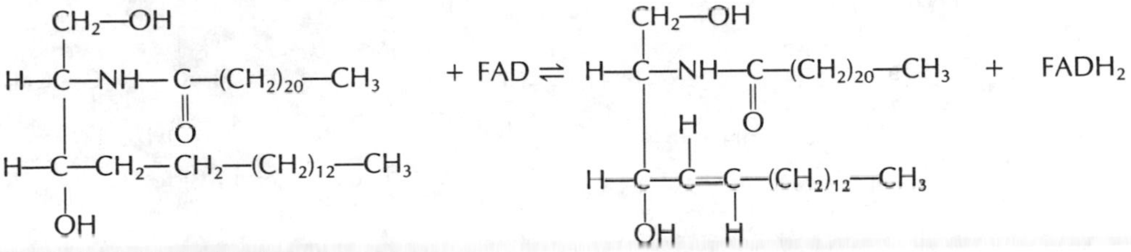

Dihydroceramide + FAD $\rightleftharpoons$ ... + FADH$_2$

Dihydroceramide   Ceramide

C. Sphingomyelin, a major component of membranes in neural tissues, is formed from ceramide and CDP-choline. It is the only sphingolipid that contains phosphate and has no sugar moiety.

1. Synthesis of sphingomyelin is from ceramide.

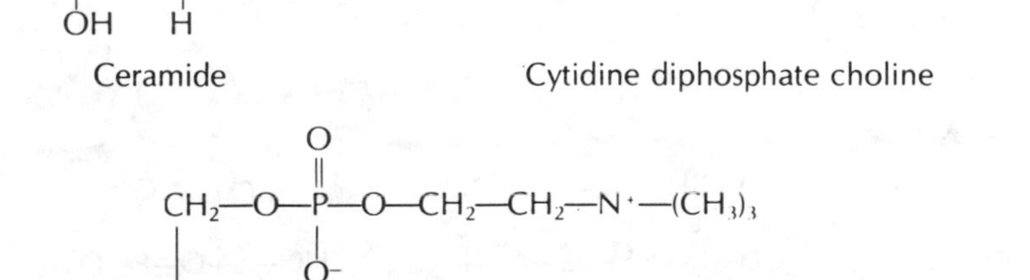

Ceramide + CDP$-$O$-$CH$_2$$-CH_2$$-N^+$$-$(CH$_3$)$_3$ $\rightarrow$

Ceramide   Cytidine diphosphate choline

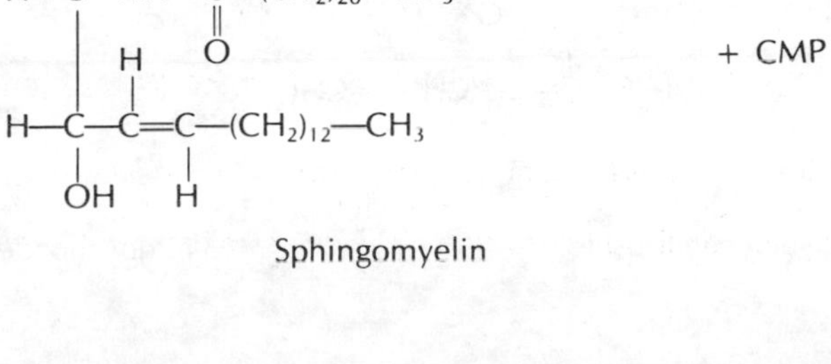

+ CMP

Sphingomyelin

2. **Excessive accumulation of sphingomyelin** occurs in **Niemann-Pick disease**.
   a. Niemann-Pick disease, a lipidosis, is due to an hereditary (autosomal recessive) deficiency of sphingomyelinase, a lysosomal enzyme that hydrolyzes sphingomyelin to ceramide and phosphorylcholine.
   b. The disease is most commonly found in an infantile form, which presents as a neuropathic disorder shortly after birth. Symptoms may include hepatosplenomegaly, failure to thrive, neurologic impairment, and retinal cherry red spots.

D. **Glycosphingolipids.** Sphingolipids that contain carbohydrate moieties are known as glycosphingolipids. At least four classes have been distinguished: **cerebrosides**, **sulfatides**, **globosides**, and **gangliosides**.

   1. **Cerebrosides** are ceramide monohexosides, the most important being **galactocerebroside** and **glucocerebroside**.

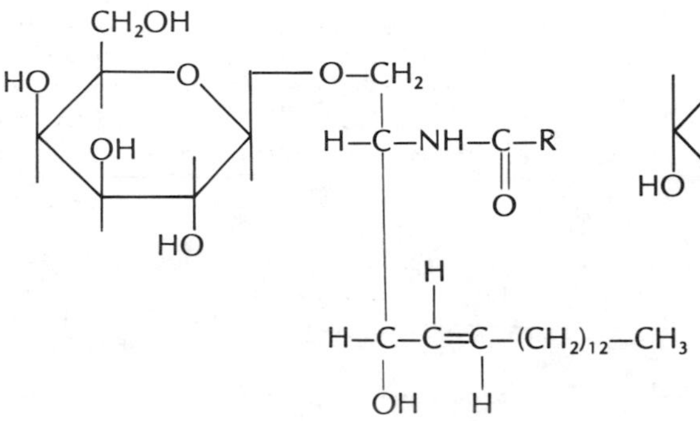

Galactocerebroside 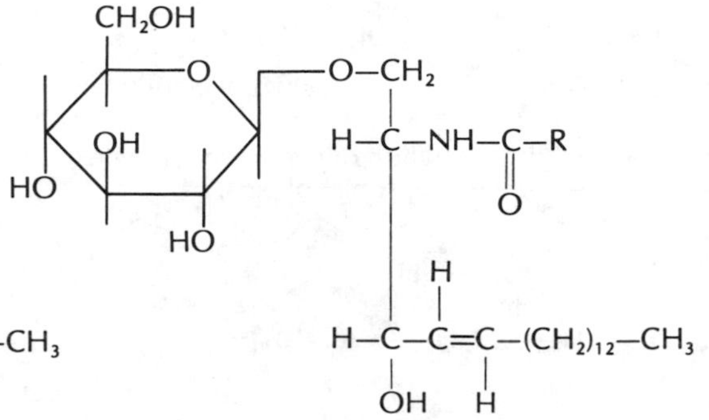   Glucocerebroside

   a. Both compounds can be synthesized from ceramide and uridine diphosphate (UDP) sugars; some tissues can form these compounds by glycosylating sphingosine, followed by fatty acylation.
   b. Glucocerebroside is not normally a component of membranes, but in the inherited (autosomal recessive) **Gaucher's disease**, a deficiency of β-glucocerebrosidase (a lysosomal enzyme hydrolyzing glucocerebrosides to ceramide and glucose) causes a pathologic accumulation of this glycosphingolipid. There are infantile, juvenile, and adult forms of the disease. In the infantile form, there is mental retardation and spasticity that later becomes flaccidity. In the juvenile form, there is ataxia.

   2. **Sulfatides.** Galactocerebroside sulfate, a major brain sulfolipid, accounts for about 15% of the lipids in white matter.
      a. The sulfate group is transferred to galactocerebroside from "active sulfate," 3'-phosphoadenosine-5'-phosphosulfate (PAPS), by a microsomal transferase.

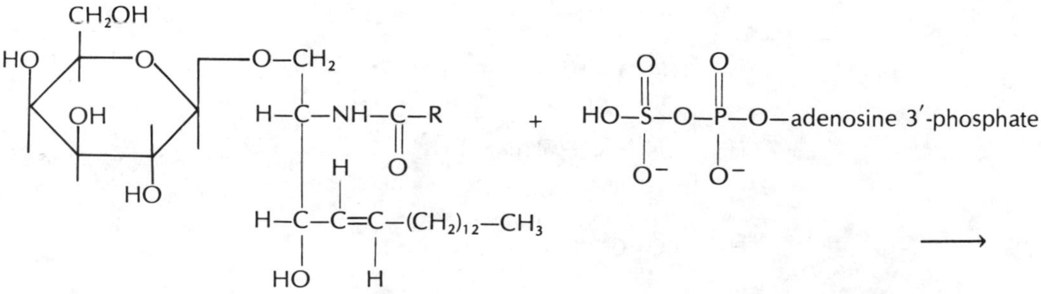

Galactocerebroside   3'-Phosphoadenosine-5'-phosphosulfate

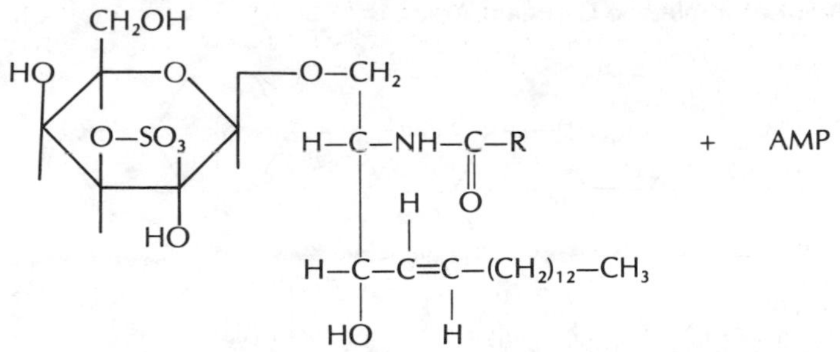

Galactocerebroside sulfate

b. A genetic deficiency involving a lysosomal enzyme that can hydrolyze off the sulfate moiety of galactocerebroside sulfatide results in the accumulation of the galactocerebroside sulfate in tissues, leading to a condition known as **metachromatic leukodystrophy**. The symptoms include mental retardation, leukodystrophy (disturbance of the white matter of the brain), and dementia in adults.

3. Globosides are ceramide oligosaccharides; they contain two or more sugar molecules, most often galactose, glucose, or N-acetylgalactosamine, attached to ceramide.

  a. As an example of a globoside, the structure of **lactosylceramide** [ceramide-βGlc-(4←1)-αGal-(4←1)] is shown below. It is found in erythrocyte membranes.

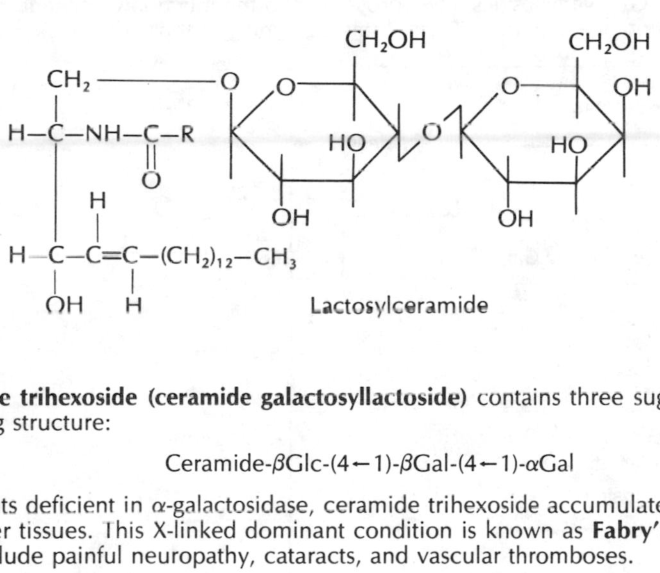

Lactosylceramide

  b. Ceramide trihexoside (ceramide galactosyllactoside) contains three sugars and has the following structure:

$$\text{Ceramide-}\beta\text{Glc-(4←1)-}\beta\text{Gal-(4←1)-}\alpha\text{Gal}$$

In patients deficient in α-galactosidase, ceramide trihexoside accumulates in the kidneys and other tissues. This X-linked dominant condition is known as **Fabry's disease**; symptoms include painful neuropathy, cataracts, and vascular thromboses.

4. Gangliosides

  a. Nature. Gangliosides are glycosphingolipids that contain one or more neuraminic acid residues, usually as the N-acetyl derivative, which is **sialic acid**. They are found in high concentration in ganglion cells of the CNS and in lower concentration in the membranes of most cells.

  b. Nomenclature

    (1) The letter G is used to denote a ganglioside, with subscripts M, D, T, or Q to indicate, respectively, mono-, di-, tri-, or quatro-sialic acid contents, and numerical subscripts to designate the sequence of carbohydrate moieties present.

$$G_{M1} \quad \text{Gal-N-AcGal-Gal-Glc-Cer}$$
$$G_{M2} \quad \text{N-AcGal-Gal-Glc-Cer}$$
$$G_{M3} \quad \text{Gal-Glc-Cer}$$

where Cer = ceramide, Glc = glucose, Gal = galactose, and N-AcGal = N-acetylgalactosamine.

(2) For example, the G_{M2} ganglioside is

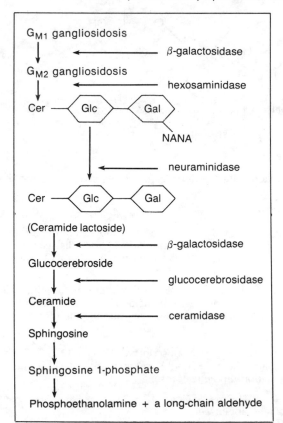

where NANA is N-acetylneuraminic acid (sialic acid).

E. Catabolism

1. The catabolism of sphingolipids occurs by the stepwise hydrolysis of the component moieties, starting at the hydrophilic terminal of the molecule. The enzymes responsible are located in lysosomes. The sequence is summarized in Figure 21-1.

2. Lipidoses due to impaired ganglioside catabolism. A number of inherited defects involve the lysosomal enzymes that degrade gangliosides, the most common of these being Tay-Sachs disease and G_{M1} gangliosidosis.

a. Tay-Sachs disease (G_{M2} gangliosidosis) is due to a deficiency of lysosomal hexosaminidase A, which hydrolyzes off the terminal N-acetylgalactosamine of G_{M2} gangliosides. Its absence leads to the accumulation of the gangliosides in brain tissues. The disorder has an autosomal recessive inheritance pattern, and the symptoms include mental retardation, seizures, blindness, and macrocephaly.

b. G_{M1} gangliosidosis is due to a deficiency of β-galactosidase, leading to the accumulation of G_{M1} gangliosides, glycoproteins, and the mucopolysaccharide keratan sulfate. The inheritance pattern and symptoms are similar to those of Tay-Sachs disease.

Figure 21-1. The sequence of enzymatic steps in the catabolism of sphingolipids. Cer = ceramide; Glc = glucose; Gal = galactose; NANA = N-acetylneuraminic acid.

STUDY QUESTIONS

Directions: Each question below contains five suggested answers. Choose the **one best** response to each question.

1. All of the following statements describe phosphoglycerides EXCEPT

(A) they are both amphipathic and amphoteric
(B) they arise from phosphatidic acid
(C) they are found in cell membranes
(D) they are a major store of metabolic energy
(E) they contain two fatty acid moieties

2. The metabolism of phospholipids is characterized by all of the following statements EXCEPT

(A) phosphatidylethanolamine is formed by the reaction of an α,β-diacylglycerol and cytidine diphosphate (CDP) ethanolamine
(B) phosphatidylserine is formed by a reaction in which the ethanolamine moiety of phosphatidylethanolamine exhanges with serine
(C) phosphatidylcholine is formed by the reaction of choline phosphate and an α,β-diacylglycerol
(D) phosphatidylinositol is formed by the reaction of inositol and CDP-diacylglycerol
(E) phosphatidylglycerol reacts with CDP-diacylglycerol to form cardiolipin

3. Which of the following compounds is a precursor of both phosphatidylcholine and sphingomyelin?

(A) Phosphatidylethanolamine
(B) Acetylcholine
(C) Glycerol 3-phosphate
(D) Uridine diphosphoglucose
(E) Cytidine diphosphocholine

4. The products of hydrolysis of phosphatidylcholine by phospholipase A_2 are

(A) a fatty acid plus 1-acyl lysophosphatidylcholine
(B) a fatty acid plus 2-acyl lysophosphatidylcholine
(C) two fatty acids plus glycerol 3-phosphocholine
(D) two fatty acids plus glycerol 3-phosphate plus choline
(E) two fatty acids plus glycerol plus inorganic phosphate plus choline

Directions: Each question below contains four suggested answers of which **one or more** is correct. Choose the answer

A   if **1, 2, and 3** are correct
B   if **1 and 3** are correct
C   if **2 and 4** are correct
D   if **4** is correct
E   if **1, 2, 3, and 4** are correct

5. Long-chain fatty acid moieties are found in which of the following complex lipids?

(1) G_{M2} gangliosides
(2) Phosphatidylserine
(3) Galactocerebroside
(4) Sphingomyelin

6. Correct statements about inherited sphingolipid disorders include which of the following?

(1) Gaucher's disease is due to a deficiency in the degradation of glucocerebrosides
(2) Tay-Sachs disease is due to a deficiency of lysosomal hexosaminidase A
(3) Niemann-Pick disease is due to a deficiency of sphingomyelinase
(4) All of the known inherited lipidoses are due to deficiencies in biosynthetic pathways

7. Sphingosine formation involves the

(1) reaction of palmitoyl coenzyme A with serine
(2) reaction of ceramide and cytidine diphosphate-choline
(3) action of β-galactosidase on ceramide galactoside
(4) action of ceramidase on ceramide

8. Gangliosides have which of the following characteristics?

(1) They contain sugars and sugar derivatives
(2) They contain sialic acid (N-acetylneuraminic acid)
(3) They are derived from cerebrosides
(4) They are found in abnormally high concentrations in neural tissues of patients with Tay-Sachs disease

ANSWERS AND EXPLANATIONS

1. The answer is D. (*I A, B*) Phosphoglycerides contain both polar and nonpolar moieties and can inhabit transition regions between aqueous and nonaqueous phases. They are also amphoteric, bearing both negatively and positively charged groups. They are constituents of cell membranes; they do *not* play a significant role as stores of metabolic energy. They contain fatty acids esterified to the sn-1 and sn-2 carbons of the glycerol backbone.

2. The answer is C. (*I B*) Phosphatidylcholine cannot be formed from choline phosphate and an α,β-diacylglycerol. An activated form of choline is required, which is met by cytidine diphosphate choline and not by choline phosphate.

3. The answer is E. (*I B 4, II C*) The phosphoglyceride, phosphatidylcholine (lecithin), can be synthesized de novo from phosphatidylethanolamine by three methylation reactions using S-adenosylmethionine as the methyl donor. Phosphatidylcholine can also be formed by salvage reactions, using choline derived from the degradation of body phospholipids, or from the diet, and an α,β-diacylglycerol. The choline is first converted to choline phosphate at the expense of adenosine triphosphate; then the choline phosphate reacts with cytidine triphosphate to form cytidine diphosphate choline (CDP-choline). The latter compound is the donor of the choline group to the α,β-diacylglycerol, forming phosphatidylcholine. CDP-choline is also the choline donor in the formation of sphingomyelin from ceramide. Phosphatidylethanolamine is a precursor of phosphatidylcholine but not of sphingomyelin. Acetylcholine, the neurotransmitter in cholinergic neurons, is not a precursor for either compound. Uridine diphosphoglucose is the activated form of glucose used in carbohydrate polymer biosynthesis and is not a precursor. Glycerol 3-phosphate is a precursor of the glycerol moiety of phosphatidylcholine but is not a precursor of sphingomyelin.

4. The answer is A. (*I C 2 a*) Phospholipase A_2 specifically catalyzes the hydrolysis of the fatty acid substituent at the sn-2 carbon of the glycerol backbone of phosphatidylcholine. The products of hydrolysis are a fatty acid and 1-acyl lysophosphatidylcholine. A fatty acid plus 2-acyl lysophosphatidylcholine would be the products of the action of phospholipase A_1 on phosphatidylcholine. None of the other choices lists the products of a single enzymatic attack on phosphatidylcholine.

5. The answer is E (all). (*I B 3; II C, D 1, 4*) The phosphoglyceride phosphatidylserine contains two long-chain fatty acids esterified to the sn-1 and sn-2 carbons of the glycerol backbone. The other three compounds are all sphingolipids and contain ceramide, which has two long-chain fatty acid moieties in its structure, those of palmitic and behenic acids.

6. The answer is A (1, 2, 3). (*II C–E*) Most of the inherited lipidoses are due to deficiencies of enzymes involved in the degradation of complex lipids. As a result, these complex lipids accumulate, particularly in the nervous system. For example, in Niemann-Pick disease, there is an excessive accumulation of sphingomyelin due to an hereditary deficiency of sphingomyelinase, a lysosomal enzyme that hydrolyzes sphingomyelin to ceramide and phosphocholine. Gaucher's disease is due to a deficiency of β-glucocerebrosidase, a lysosomal enzyme that hydrolyzes glucocerebroside to ceramide and glucose, causing a pathologic accumulation of this glycosphingolipid. Tay-Sachs disease is due to a deficiency of lysosomal hexosaminidase A, which hydrolyzes the terminal N-acetylgalactosamine of G_{M2} gangliosides, leading to the accumulation of gangliosides in brain tissue. Deficiencies in the biosynthesis of complex lipids are not known, probably because such mutations would be lethal in the embryo.

7. The answer is D (4). (*II B, E; Figure 21-1*) In the biosynthesis of sphingolipids, sphinganine (dihydrosphingosine) is first formed from serine and palmitoyl coenzyme A via the formation and then the reduction of 3-ketodihydrosphingosine. The structure of sphingosine does not arise until the reduction of dihydroceramide to ceramide, which contains the sphingosine moiety. During the catabolism of sphingolipids, free sphingosine is released from ceramide upon hydrolysis by ceramidase. β-Galactosidase hydrolyzes ceramide lactoside to yield ceramide. The reaction of ceramide and cytidine diphosphate choline yields sphingomyelin.

8. The answer is E (all). (*II D 1, 4, E 2*) The gangliosides are glycosphingolipids that contain one or more neuraminate residues, usually as the N-acetyl derivative (a sialic acid). They contain sugars and sugar derivatives; for example, a G_{M2} ganglioside contains N-acetylgalactosamine, galactose with N-acetylneuraminic acid on the 3-carbon, glucose, and ceramide. They all contain ceramide and can be regarded as being derived from cerebrosides, which are ceramide monohexosides. G_{M2} gangliosides accumulate in the neural tissues of infants with Tay-Sachs disease, which is an inherited condition in which lysosomal hexosaminidase A is deficient. This enzyme normally hydrolyzes off the terminal N-acetylgalactosamine residue of the G_{M2} ganglioside.

22

Prostaglandins and Related Compounds

I. PRODUCTS OF ARACHIDONIC ACID METABOLISM. The prostaglandins, thromboxanes, and leukotrienes, which may be referred to collectively as **eicosanoids**, are, in humans, mainly products of arachidonic acid (C_{20}-Δ 5, 8, 11, 14) metabolism, although other series of prostaglandins may be derived from dihomo-δ-linoleic acid (C_{20}-Δ 8, 11, 14) or eicosapentaenoic acid (C_{20}-Δ 5, 8, 11, 14, 17). Many of these substances are involved in important physiologic functions, sometimes in a hormone-like manner, although in a number of instances their precise roles have yet to be defined.

A. Prostaglandins

1. The prostaglandins are 20-carbon hydroxy fatty acids containing a 5-membered ring and are derived from linoleic and linolenic acid families.

2. Two general physiologic actions can be distinguished.
   a. They are potent smooth muscle agonists.
   b. They have a modulating action on target cells of adenohypophyseal tropic hormones or on cells with β-adrenergic receptors for the catecholamines.

3. **Prostacyclin (PGI$_2$)** is the main prostaglandin produced by the endothelial cells of the vascular system.
   a. PGI$_2$ is a vasodilator, particularly for coronary arteries.
   b. It antagonizes the aggregation of platelets and their adherence to the endothelial surface.

B. Thromboxane A$_2$ (TXA$_2$)

1. TXA$_2$, a 20-carbon hydroxy fatty acid, has a 6-membered, oxygen-containing ring.

2. It arises from the same precursor as the prostaglandins.

3. It is the main such substance produced by the platelets and has opposite effects from PGI$_2$, contracting arteries and promoting platelet aggregation. It has a very short half-life of about 30 sec.

C. Leukotrienes

1. The leukotrienes are hydroxy fatty acid derivatives of arachidonic acid and do not contain a ring structure.

2. They are involved in chemotaxis, inflammation, and allergic reactions.

3. Leukotriene D$_4$ (LTD$_4$) has been identified as the **slow-reacting substance of anaphylaxis (SRS-A)**, which causes smooth muscle contraction and is about 1000 times more potent than histamine in constricting the pulmonary airways. SRS-A also increases fluid leakage from small blood vessels and constricts coronary arteries.

4. Leukotriene B$_4$ (LTB$_4$) attracts neutrophils and eosinophils, which are found in large numbers at sites of inflammation.

II. NOMENCLATURE AND STRUCTURE

A. Prostaglandins and thromboxanes.
The prostaglandins and thromboxanes are considered to be analogues of prostanoic acid, a 20-carbon fatty acid, which contains a 5-carbon saturated ring.

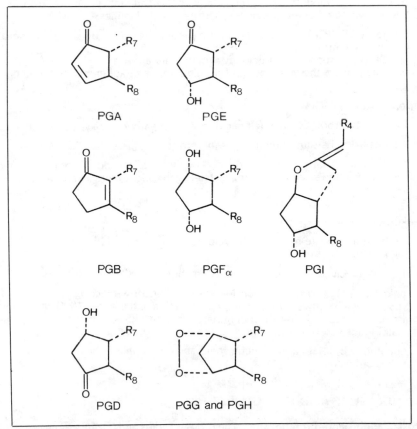

Prostanoic acid

The prostaglandins have a five-membered ring, but the thromboxanes have a six-membered oxygen-containing ring.

1. Prostanoic acid does not occur naturally but is regarded as the parent compound of the prostaglandins for the purpose of nomenclature and carbon numbering.

2. The dotted line between carbons 7 and 8 in the structure shown above indicates that the seven-carbon chain attached to the ring at C-8 projects below the plane of the ring, whereas the eight-carbon chain attached at C-12 projects above the plane.

3. All naturally occurring prostaglandins have a hydroxyl group at C-15 projecting below the plane of the ring.

B. Prostaglandin compounds are abbreviated as **PG**, with an additional capital letter and a numeric subscript: In one case, the subscript also carries a Greek letter.

1. Substitutions on the ring, except in one case, are represented by the capital letters A, B, D, E, F, G, H, and I. Seven kinds of rings are found in naturally occurring prostaglandins, whose structures are shown in Figure 22-1.

  a. PGA and PGB have keto groups at C-9 and a double bond.
   (1) PGA has a 10,11 double bond.
   (2) PGB has an 8,12 double bond.

Figure 22-1. Structures of the seven families of prostaglandins (PGs), showing the different types of ring substitutions found. R_7, R_8, and R_4 refer to the seven-carbon, eight-carbon, and four-carbon chains.

b. PGD has a hydroxyl group at C-9 and a keto group at C-11.

c. PGE has a keto group at C-9 and a hydroxyl group at C-11.

d. PGF has two hydroxyl groups, one at C-9 and the other at C-11.

e. PGG and PGH have the same ring structure (an inconsistency in the nomenclature), a cyclopentane endoperoxide, and differ only in the substitution at C-15 of the eight-carbon chain.

   (1) PGG has a hydroperoxide at C-15.

   (2) PGH has a hydroxyl group at C-15.

f. PGI is a two-ringed structure with oxygen at C-19 linked to C-6 of the seven-carbon side chain.

2. Hydrocarbon structure. The **subscript number** in the prostaglandin nomenclature denotes the **number of unsaturated bonds** that a prostaglandin contains (e.g., PGE_1, PGE_2, and PGE_3). This numeric subscript refers only to double bonds in the **hydrocarbon chains**, but not in the rings. In the 1 series, the double bond is $\Delta^{13,14}$; in the 2 series, they are $\Delta^{5,6}$ and $\Delta^{13,14}$; and in the 3 series, they are $\Delta^{5,6}$, $\Delta^{13,14}$, and $\Delta^{17,18}$.

3. The hydroxyl group at C-9 in the PGF series is represented by the Greek letter in the subscript. Only the F_α series is found naturally; it is the isomer with the hydroxyl above the plane of the ring.

C. Thromboxanes. A similar capital-letter and numeric-subscript nomenclature is used for the thromboxanes, as far as it is applicable. Another inconsistency in nomenclature involves the ring structure found in the thromboxanes. The active form, thromboxane A (TXA), has one oxygen in the ring and another attached to C-9 and C-11.

Thromboxane A

D. Leukotrienes arise by the addition of hydroperoxy groups to arachidonic acid, the products of the lipoxygenase reaction known as hydroperoxyeicosatetraenoic acids (HPETE) [Fig. 22-2].

1. Hydroperoxy substitution of arachidonic acid may occur at positions 5, 8, 9, 11, or 15.

   a. 5-HPETE is the major product of the lipoxygenase reaction in basophils, polymorphonuclear leukocytes, and macrophages.

   b. 12-HPETE occurs in platelets, pancreatic endocrine islet cells, and glomerular cells of the kidney.

   c. 15-HPETE occurs in reticulocytes, eosinophils, and T cells.

2. The HPETE compounds are not as active as hormones themselves but are converted to active compounds by reduction to the analogous hydroxy fatty acid (e.g., LTB_4) or to the leukotrienes.

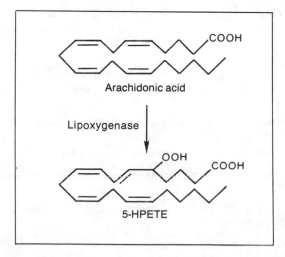

Arachidonic acid

Lipoxygenase

5-HPETE

Figure 22-2. The conversion of arachidonic acid (all *cis*-Δ^5, Δ^8, Δ^{11}, Δ^{14}-eicosatetraenoic acid) into 5 hydroperoxyeicosatetraenoic acid (5-HPETE) by the lipoxygenase system.

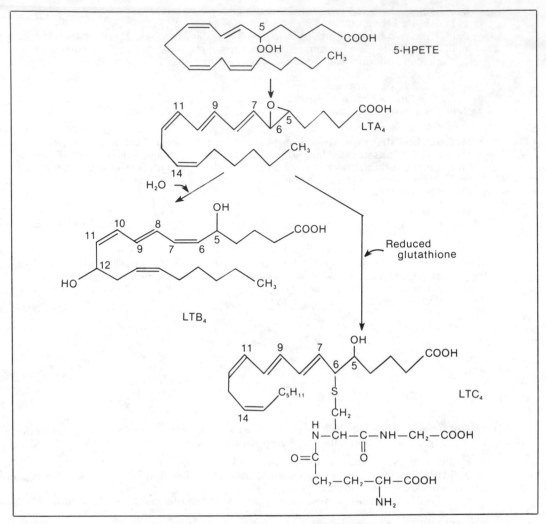

Figure 22-3. Conversion of 5-hydroperoxyeicosatetraenoic acid (5-HPETE) to leukotriene B$_4$ (LTB$_4$) and leukotriene C$_4$ (LTC$_4$) through the intermediate of leukotriene A$_4$. (Reprinted with permission from Devlin TM: *Textbook of Biochemistry with Clinical Correlations*, 2nd ed. New York, John Wiley, 1986, p. 431.)

    a. Leukotrienes are formed from 5-HPETE by the generation of an epoxide, leukotriene A$_4$ (LTA$_4$) [Fig. 22-3].

    b. LTA$_4$ occurs at a branch point and can be converted to either leukotriene B$_4$ (LTB$_4$) or to the leukotrienes C$_4$ and D$_4$ (LTC$_4$ and LTD$_4$).

III. HYDROLYSIS OF PHOSPHOLIPIDS AND THE RELEASE OF ARACHIDONIC ACID

A. Arachidonic acid phospholipids

    1. Arachidonic acid acts as the precursor of the major prostaglandins, thromboxanes, and leukotrienes in humans.

    2. There is little free arachidonic acid in cells. It is stored almost completely as esters of the sn-2 position of the glycerol backbone of cell membrane phospholipids.*

B. Release of arachidonic acid

    1. Because there is little free arachidonic acid in cells, the synthesis of prostaglandins requires its immediate release from membane phospholipids.

---

*For an explanation of the stereospecific numbering system of glycerol (sn), see Chapter 21, section I B 1 b.

2. Stimulation of cells by an appropriate agonist causes the release of arachidonic acid. Agonists have specific target cells; for example, thrombin causes the release of arachidonic acid in platelets and endothelial cells; bradykinin acts similarly in renal tubular cells.

3. Two systems operate to provide free arachidonic acid from membrane phospholipids.
   a. One is the hydrolysis of arachidonate esterified to the sn-2 position of phospholipids by phospholipase A_2. Kidney and platelet phospholipases cleave only arachidonate or eicosatrienoic acid from the sn-2 position.
   b. In platelets, phosphatidylinositol is greatly enriched in arachidonate in the sn-2 position, and arachidonate release is effected by the following pathway:
      (1) A specific phospholipase, phospholipase C, cleaves the inositol from phosphatidylinositol to give an α,β-diacylglycerol rich in arachidonate.
      (2) The α,β-diacylglycerol may yield up its arachidonate by one of two pathways.
         (a) A specific diacylglycerol lipase can hydrolyze off the arachidonate from the sn-2 position.
         (b) Phosphatidic acid may be formed from the α,β-diacylglycerol at the expense of adenosine triphosphate; hydrolysis by phospholipase A_2 then yields arachidonate.

4. The amount of free arachidonic acid that is released determines the rate of synthesis of the product, while the type of product depends upon the cell's enzymatic machinery.

5. Regulation of arachidonate release
   a. The release of arachidonate from membrane phospholipids is regulated by an inhibitory protein whose synthesis is induced by glucocorticoid hormones.
   b. This inhibitory effect on arachidonate release accounts for the so-called **anti-inflammatory action of steroids**, since prostaglandins and prostaglandin-like substances play important roles in the inflammatory reaction.

C. Synthesis of prostaglandins, thromboxanes, and leukotrienes. Once arachidonate is released in the cell, it activates both cyclooxygenase and lipoxygenase enzymes.

1. The cyclooxygenase system converts arachidonate (and other C-20 unsaturated fatty acids) into endoperoxides (Fig. 22-4). These are the key precursor molecules for the thromboxane (TXA_2) and the prostaglandins (PGD_2, PGE_2, PGF_2, and PGI_2 [prostacyclin]) that are of major importance in human tissues.
   a. The first product formed, PGG_2, is largely converted to PGH_2 (see Fig. 22-4) by a peroxidase activity that is part of the cyclooxygenase enzyme.
   b. PGH_2 can then act as the precursor of the important compounds listed above.

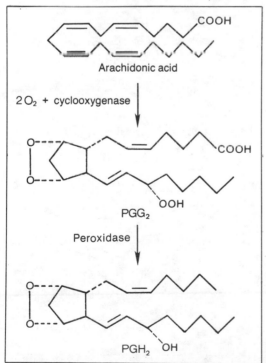

Figure 22-4. The conversion of arachidonic acid into endoperoxides by the cyclooxygenase system.

    c. The cyclooxygenase enzyme is inhibited by the **nonsteroidal anti-inflammatory agents (NSAIDs)**, such as aspirin, indomethacin, or ibuprofen.

    d. The inhibition of platelet cyclooxygenase by aspirin is important as it is irreversible, and because platelets cannot synthesize protein, the permanently inhibited enzyme cannot be replaced.

2. The lipoxygenase system converts free arachidonic acid to 5-HPETE (see Fig. 22-2).

    a. Note that 5-HPETE and the active compounds derived from it do not contain a ring structure.

    b. The lipoxygenase enzyme, like the cyclooxygenase, is substrate activated.

    c. A series of compounds derived from 5-HPETE have been studied, such as

        (1) LTB_4, which is a 5,12-dihydroxyacid

        (2) LTC_4, which has a glutathionyl moiety attached at C-5

        (3) LTD_4, which has lost part of the glutathionyl moiety (see Fig. 22-3)

STUDY QUESTIONS

Directions: Each question below contains five suggested answers. Choose the **one best** response to each question.

1. All of the following functions typify either prostacyclin (PGI$_2$) or thromboxane A$_2$ (TXA$_2$) EXCEPT

(A) PGI$_2$ dilates coronary arteries
(B) PGI$_2$ contracts coronary arteries
(C) PGI$_2$ antagonizes platelet aggregation
(D) TXA$_2$ contracts coronary arteries
(E) TXA$_2$ promotes platelet aggregation

2. The leukotrienes are characterized by all of the following statements EXCEPT

(A) they are obtained from arachidonate by the lipoxygenase system
(B) they contain four double bonds
(C) they may contain amino acids
(D) they all contain a six-membered unsaturated ring
(E) one of them has been identified as the slow-reacting substance of anaphylaxis

3. Correct statements regarding arachidonic acid include all of the following EXCEPT

(A) it is a precursor of thromboxane A$_2$
(B) it activates lipoxygenase
(C) it is a 20-carbon fatty acid with three double bonds
(D) it can be derived from linoleic acid
(E) it is found esterified to the sn-2 position of glycerophospholipids

4. The release of arachidonate by phospholipase A$_2$ from membrane glycerophospholipids is inhibited by

(A) aspirin
(B) linoleic acid
(C) a specific protein induced by glucocorticoids
(D) 2-acyl lysophosphatidylcholine
(E) none of the above

ANSWERS AND EXPLANATIONS

1. The answer is B. (*I A, B*) Prostacyclin is the main prostaglandin formed by the endothelial cells of the vascular system. It is a vasodilator, particularly of coronary arteries, and antagonizes the aggregation of platelets and their adherence to the endothelial surface. Thus, it protects the interests of the vascular system. On the other hand, thromboxane A_2, which is formed in the platelets, promotes contraction of arteries and aggregation of platelets. It helps to repair damaged vascular tissue, a function of platelets.

2. The answer is D. (*I C; II D; III C 2*) The leukotrienes are formed from arachidonate by the action of lipoxygenase, which first forms 5-hydroperoxyeicosatetraenoic acid (5-HPETE). They contain four double bonds, and leukotriene C_4 contains a glutathionyl moiety. Leukotriene D_4, which has lost part of the glutathionyl moiety, has been identified as the slow-reacting substance of anaphylaxis.

3. The answer is C. (*III A, C; Chapter 20 II B 2*) Arachidonic acid is a C-20 fatty acid with four double bonds, namely, all *cis*-Δ^5, Δ^8, Δ^{11}, Δ^{14}-eicosatetraenoic acid. It is stored primarily esterified to the sn-2 position of membrane glycerophospholipids. Upon its release by phospholipase A_2, arachidonate activates both the lipoxygenase and cyclooxygenase systems. The latter system gives rise to the thromboxanes and the former to the leukotrienes. Arachidonate can be derived from linoleate, the conversion requiring both desaturations and elongation of the chain.

4. The answer is C. (*III B 5*) The action of phospholipase A_2, to cleave off the arachidonate moiety from the sn-2 position of membrane glycerophospholipids, is blocked by an inhibitory protein whose synthesis is induced by glucocorticoids. This is the basis for the suppression of inflammation by glucocorticoids. Aspirin and related compounds, such as indomethacin, exert their so-called nonsteroidal anti-inflammatory action by blocking cyclooxygenase, which produces prostaglandins and thromboxanes from arachidonate. Linoleic acid is a precursor of arachidonate, and 2-acyl lysophosphatidylcholine is produced by the action of phospholipase A_1 on phosphatidylcholine.

23

Cholesterol Metabolism

I. STEROID STRUCTURE AND NOMENCLATURE

A. Ring numbering system. Steroids are lipids that contain four carbon rings joined to form the steroid nucleus, **cyclopentanoperhydrophenanthrene**.

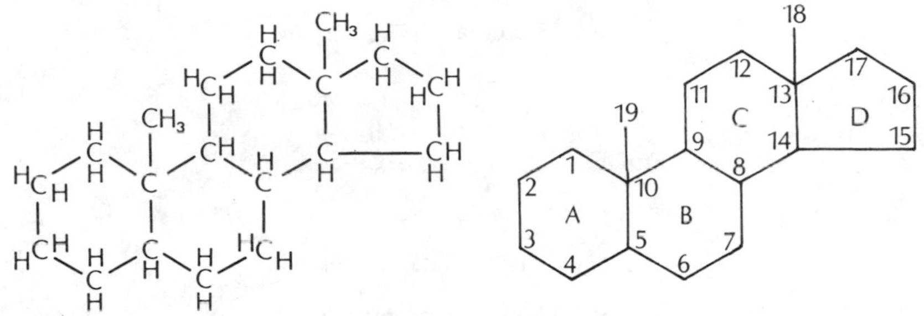

Cyclopentanoperhydrophenanthrene

B. Sterol structure

1. A sterol is a class of steroids characterized by
   a. A hydroxyl group at C-3
   b. An aliphatic chain of at least eight carbons at C-17

2. For example, **cholesterol**, the sterol of vertebrates, has the following structure:

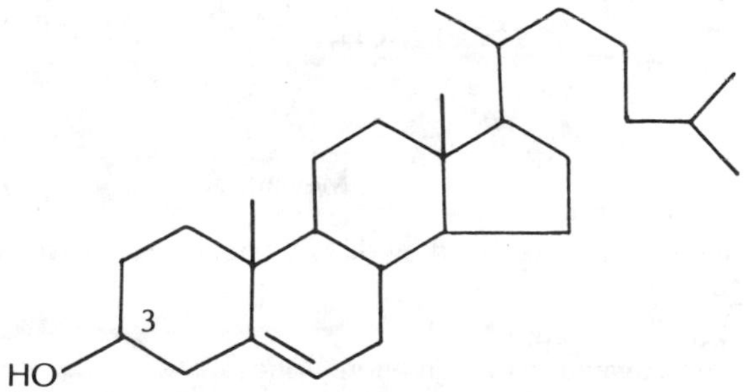

Cholesterol

3. The axial groups at C-18 and C-19 project above the plane of the molecule.
   a. Substituents on the rings that are *cis* in relation to these methyl groups are said to be β-oriented and, as with the HO substituent in the structure of cholesterol above, are indicated by a solid line joining the group to the ring. Thus, cholesterol has a 3-β-hydroxyl group.
   b. Substituents that are *trans* to the axial methyls are said to be α-oriented.

4. Unsaturated bonds are indicated by a delta (Δ) symbol with a superscript that indicates the position of the bond, or bonds. For example, cholesterol has a Δ^5 double bond.

II. CHOLESTEROL BIOSYNTHESIS

A. Site and carbon source

1. The liver is the major site of cholesterol biosynthesis, although other tissues are also active in this regard (e.g., the intestines, adrenals, gonads, skin, neural tissue, and aorta).

2. All 27 carbon atoms of cholesterol are derived from the acetate moiety of acetyl coenzyme A (acetyl CoA), the enzyme system residing in the cytosolic and microsomal fractions.

B. Pathway. Cholesterol biosynthesis can be thought of as occurring in five groups of reactions.

1. Formation of mevalonate from acetyl CoA

a. β-Hydroxy-β-methylglutaryl CoA (HMG CoA) can be formed in the cytosol from acetyl CoA in two steps by **thiolase** and **HMG CoA synthetase**.

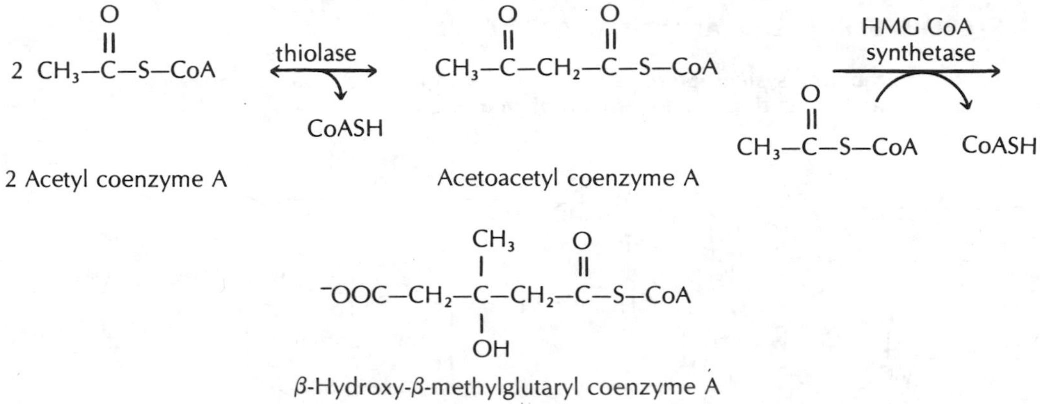

2 Acetyl coenzyme A   Acetoacetyl coenzyme A

β-Hydroxy-β-methylglutaryl coenzyme A

b. The rate-limiting step of cholesterol biosynthesis is the conversion of HMG CoA to **mevalonate** by **HMG CoA reductase**.

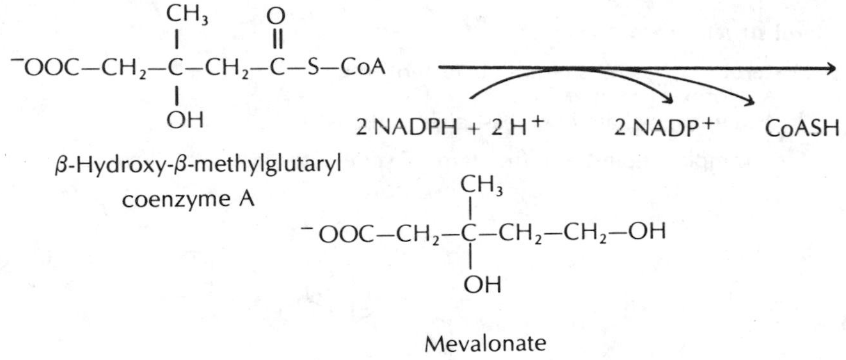

β-Hydroxy-β-methylglutaryl coenzyme A

Mevalonate

c. This rate-limiting step is inhibited by dietary cholesterol as well as by endogenously synthesized cholesterol.

2. Formation of isoprenoid units

a. Mevalonate is activated with high-energy phosphate bonds and then decarboxylated to form the five-carbon isoprenoid isomers, 3,3-dimethylallyl pyrophosphate and isopentenyl pyrophosphate.

(1) Two molecules of adenosine triphosphate (ATP) are first expended to form mevalonate 5-pyrophosphate.

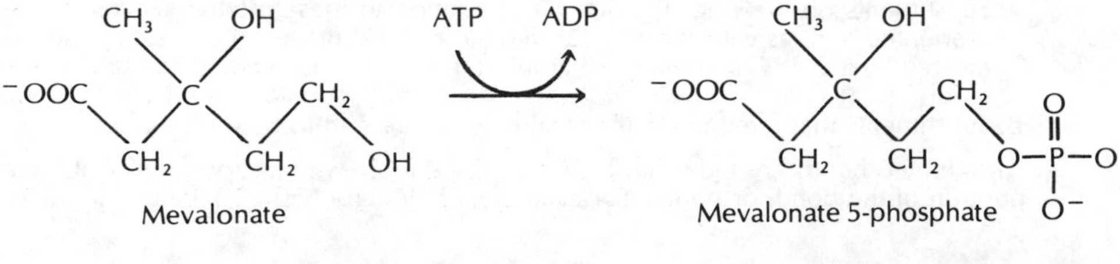

Mevalonate   Mevalonate 5-phosphate

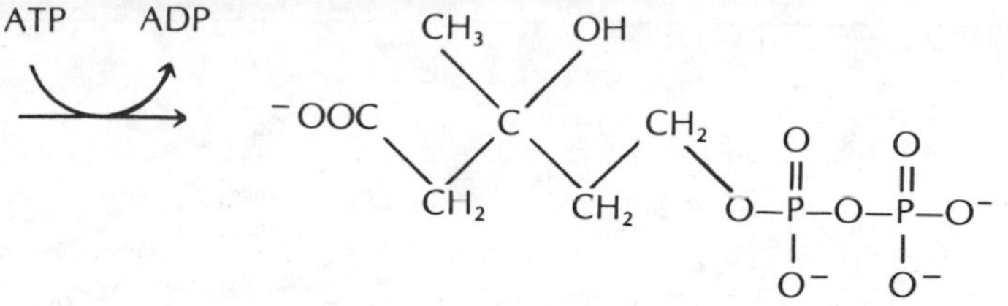

Mevalonate 5-pyrophosphate

(2) The final step in this group of reactions also requires ATP with the probable formation of an unstable intermediate, mevalonate 3-phospho-5-pyrophosphate, which decarboxylates to form **isopentenyl pyrophosphate**.

b. An isomerase governs an equilibrium between isopentenyl pyrophosphate and its isomer, **3,3-dimethylallyl pyrophosphate**.

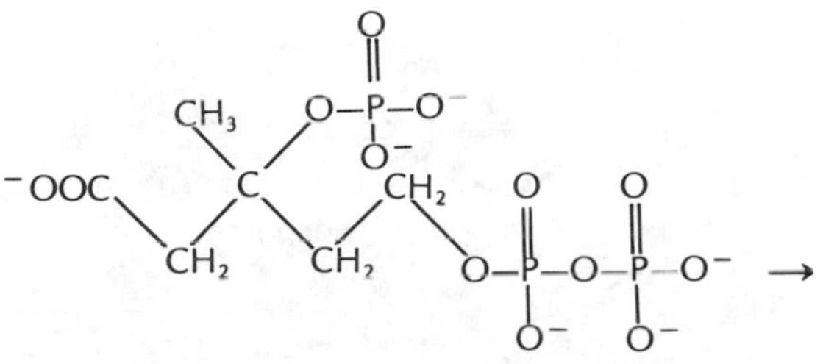

Mevalonate 3-phospho-5-pyrophosphate

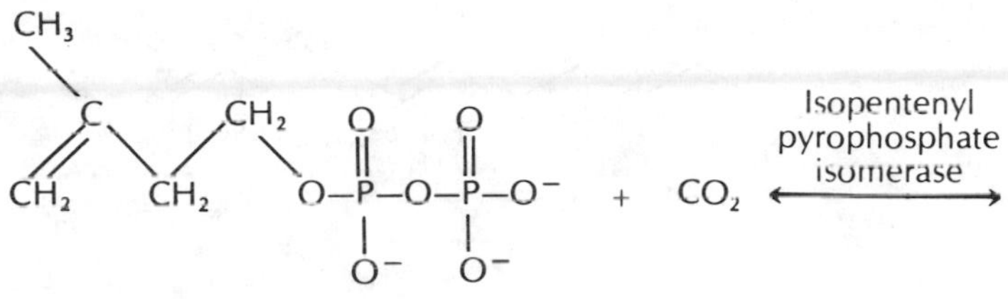

Isopentenyl pyrophosphate isomerase

Isopentenyl pyrophosphate

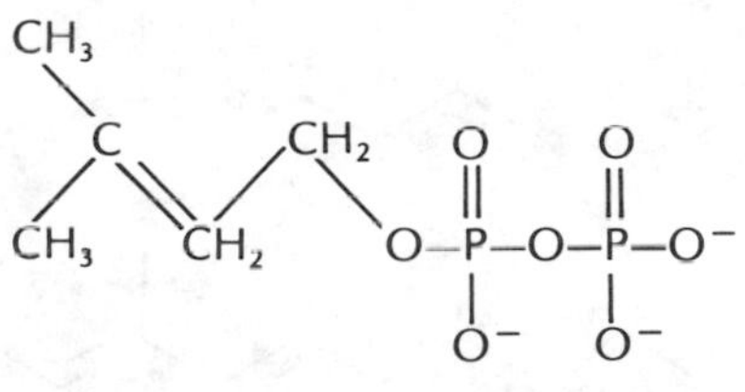

3,3- Dimethylallyl pyrophosphate

3. Formation of squalene

a. The pyrophosphorylated isoprenoid units condense to form a 30-carbon aliphatic chain called **squalene**.

b. The condensation occurs in three steps (Fig. 23-1).

(1) A molecule of 3,3-dimethylallyl pyrophosphate condenses with a molecule of isopentenyl pyrophosphate to form the 10-carbon compound **geranyl pyrophosphate** in a reaction catalyzed by **geranyl pyrophosphate synthetase**.

(2) Another molecule of isopentenyl pyrophosphate reacts with geranyl pyrophosphate to form the 15-carbon compound **farnesyl pyrophosphate**. The reaction is catalyzed by **farnesyl pyrophosphate synthetase**.

(3) Finally, in a reaction that uses reduced nicotinamide-adenine dinucleotide phosphate (NADPH) as an electron donor, two molecules of farnesyl pyrophosphate are linked to form **squalene** in a reaction catalyzed by **squalene synthetase**.

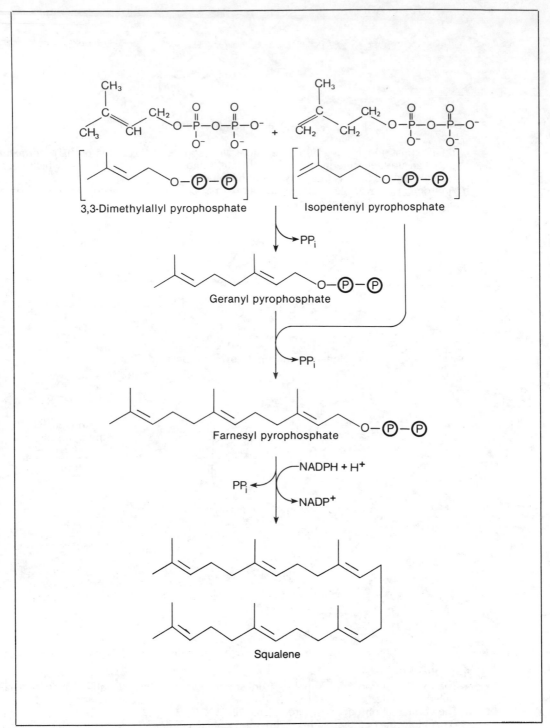

Figure 23-1. The pathway for the biosynthesis of squalene.

4. **Cyclization of squalene to form lanosterol.** The formation of lanosterol from squalene takes place in two steps.
   a. In the first step, **squalene 2,3-epoxide** is formed by a reaction requiring NADPH and molecular oxygen and is catalyzed by **squalene monooxygenase**.

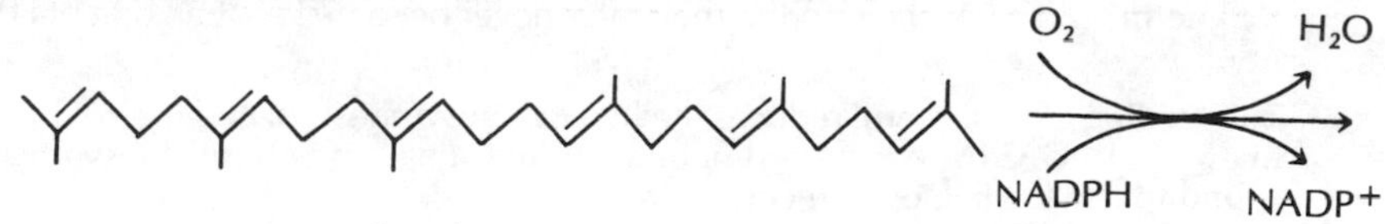

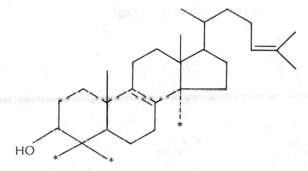

Squalene

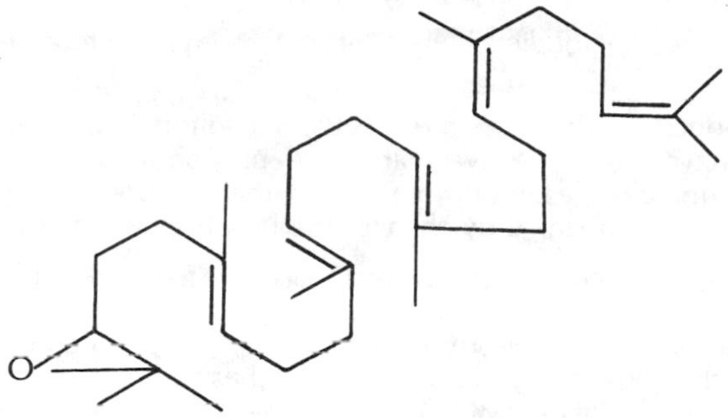

O_2 H_2O

NADPH $NADP^+$

Squalene 2,3-epoxide

b. In the next step, a cyclase catalyzes the cyclization of squalene to form **lanosterol**.

HO

Lanosterol

5. **Conversion of lanosterol to cholesterol**

   a. The three methyl groups starred in the above structural diagram for lanosterol are removed in a series of reactions leading from lanosterol, a C-30 compound, to cholesterol, a C-27 compound.

   b. Other reactions reduce the Δ^{24} double bond of lanosterol and bring about the migration of the Δ^8 double bond of lanosterol to the Δ^5 position found in cholesterol (see section I B).

   c. These final reactions taking lanosterol to cholesterol require NADPH, nicotinamide-adenine dinucleotide (NAD^+), and oxygen.

C. Esterification of cholesterol

1. The bulk of the cholesterol in tissues and about 65% of plasma cholesterol is esterified with long-chain fatty acids at C-3.

2. The synthesis of cellular cholesterol ester requires ATP in order to form fatty acyl CoA derivatives, which are then transferred to the 3-β-hydroxyl group of cholesterol.

3. Cholesterol associated with plasma lipoproteins can be esterified by **lecithin:cholesterol acyltransferase** (LCAT).

   Phosphatidylcholine + cholesterol →

   lysophosphatidylcholine + cholesterol fatty acyl ester

D. Control of cholesterol biosynthesis

1. The reduction of HMG CoA by HMG CoA reductase is regarded as the rate-limiting step in cholesterol biosynthesis. Although later steps may be affected by a prolonged stimulus (e.g.,

long-term feeding of cholesterol), their rates never become less than that of HMG CoA reductase.

2. The feeding of cholesterol reduces the hepatic biosynthesis of cholesterol by reducing the activity of HMG CoA reductase. Importantly, intestinal cholesterol biosynthesis does not respond to the feeding of high-cholesterol diets.

3. HMG CoA reductase activity is also reduced by fasting, which limits the availability of acetyl CoA and NADPH for cholesterol biosynthesis.

4. In contrast, the feeding of diets high in fat or carbohydrate tends to increase hepatic cholesterol biosynthesis.

5. HMG CoA reductase can undergo reversible phosphorylation–dephosphorylation; the phosphorylated enzyme is less active than the dephosphorylated form. However, there is no evidence that the changes in enzyme activity that are observed with cholesterol feeding or with fasting are accompanied by changes in phosphorylation of HMG CoA reductase.

6. **Hormonal effects on cholesterol biosynthesis**
   a. **Insulin** stimulates HMG CoA reductase activity. The hormone is required for the diurnal rhythm that occurs in cholesterol biosynthesis, a phenomenon that is probably related to feeding cycles and the need for bile acid synthesis (see section III). **Diabetes**, or lack of insulin, therefore, reduces HMG CoA reductase activity and abolishes the diurnal rhythm of hepatic cholesterol biosynthesis.
   b. **Glucagon** antagonizes the effect of insulin in intact animals.
   c. **Thyroid hormone** stimulates HMG CoA reductase activity.

III. CHOLESTEROL AS PRECURSOR OF BILE ACID

A. Structures

1. The predominant bile acids in humans are **cholic, chenodeoxycholic, deoxycholic**, and **lithocholic** acids.

2. The bile acids are C-24 steroids and are derived from cholesterol by scission of the side chain, which leaves
   a. A C-24 carboxyl group with the loss of three carbons
   b. Saturation of the Δ^5 double bond of cholesterol
   c. Hydroxylation of the steroid nucleus

3. The bile acids differ in the degree of hydroxylation. α-Hydroxyls occur in
   a. Cholic acid at C-3, C-7, and C-12
   b. Deoxycholic acid at C-3 and C-12
   c. Chenodeoxycholic acid at C-3 and C-7
   d. Lithocholic acid at C-3

Primary bile acids

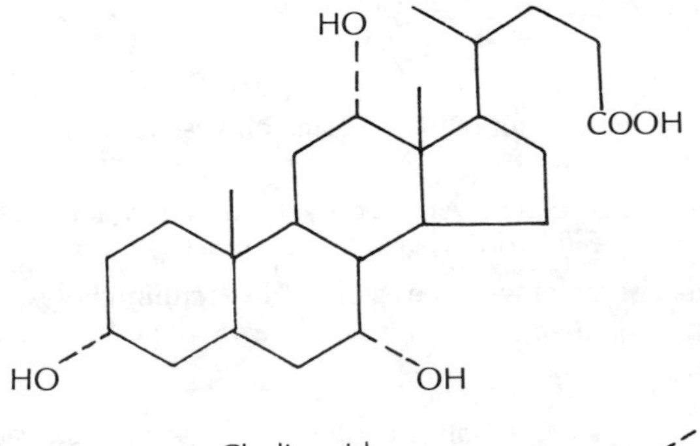

Cholic acid

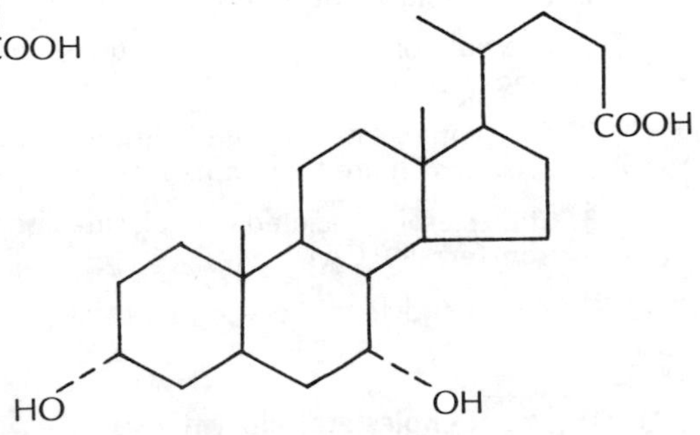

Chenodeoxycholic acid

Secondary bile acids

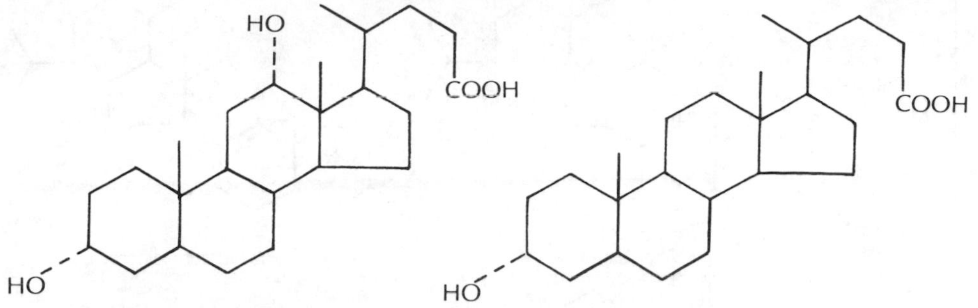

Deoxycholic acid

Lithocholic acid

4. Cholic and chenodeoxycholic acids are formed in the liver from cholesterol and are known as the **primary bile acids**. Deoxycholic acid and lithocholic acid are known as the **secondary bile acids** because they are formed from the primary bile acids; this occurs in the intestine through the action of intestinal bacterial enzymes.

5. The bile acids are conjugated to **glycine** or **taurine** in the liver. Cholic acid, for example, forms glyco- or taurocholate.

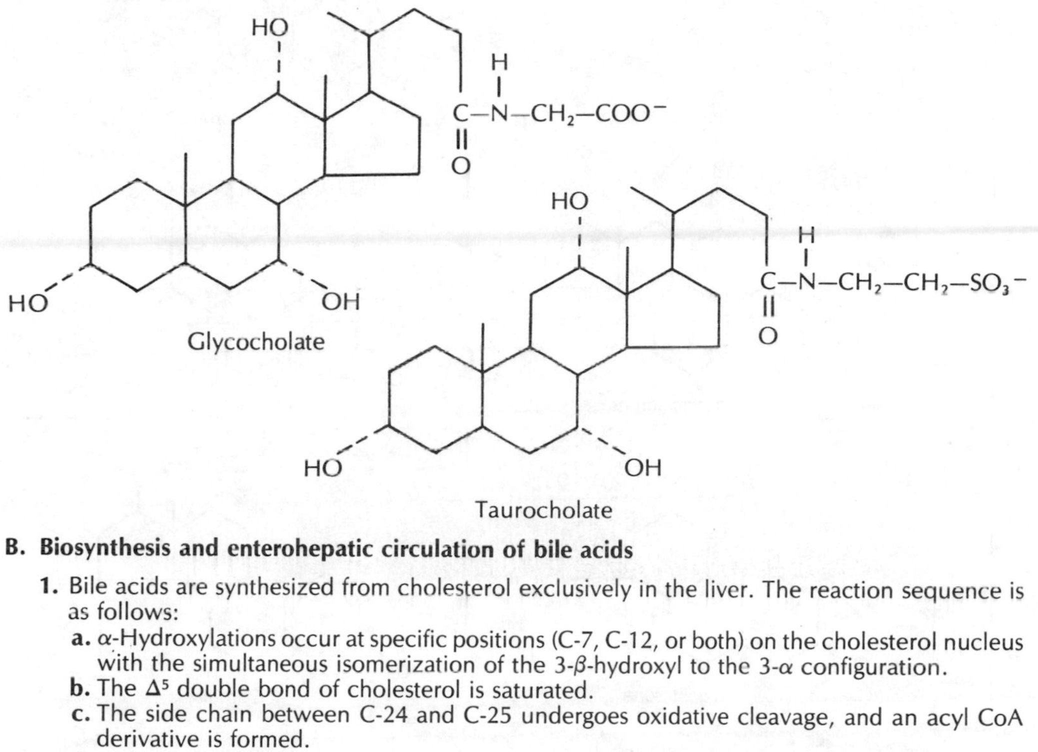

Glycocholate

Taurocholate

B. **Biosynthesis and enterohepatic circulation of bile acids**

1. Bile acids are synthesized from cholesterol exclusively in the liver. The reaction sequence is as follows:
   a. α-Hydroxylations occur at specific positions (C-7, C-12, or both) on the cholesterol nucleus with the simultaneous isomerization of the 3-β-hydroxyl to the 3-α configuration.
   b. The Δ^5 double bond of cholesterol is saturated.
   c. The side chain between C-24 and C-25 undergoes oxidative cleavage, and an acyl CoA derivative is formed.
   d. Taurine or glycine forms amide links at the C-24 carboxyl with the displacement of CoA.

2. The rate-limiting step in the sequence is the first hydroxylation, which occurs at C-7.

3. After release of the bile from storage in the gallbladder into the intestine, the two primary bile acids, cholic and chenodeoxycholic, are converted in part to the secondary bile acids, deoxycholate and lithocholate, by intestinal bacteria. The conversion involves the loss of the hydroxyls at C-7.

4. In the intestine, both primary and secondary bile acids are deconjugated, reabsorbed by the intestinal mucosa, and returned to the liver by the portal vein bound to serum albumin.

5. The liver takes up the bile acids from the portal circulation, reconjugates them with taurine or glycine, and secretes all four bile acids in the bile. The secretion of bile and its reabsorption,

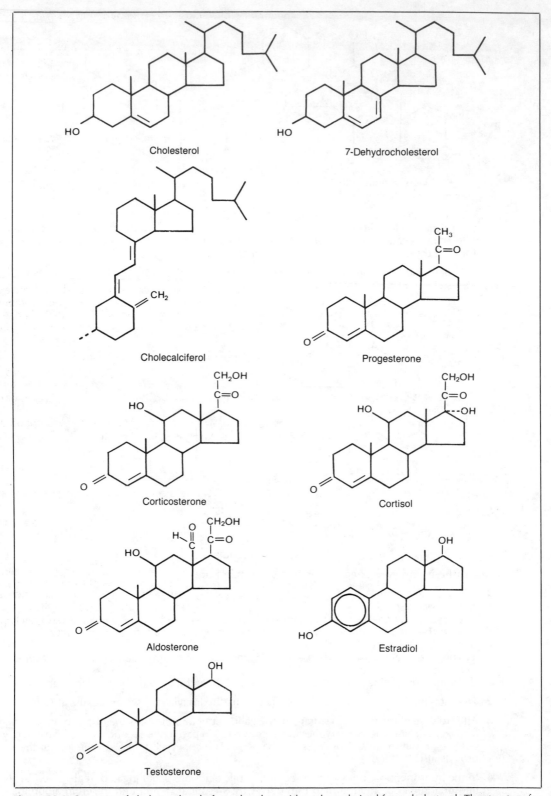

Figure 23-2. Structure of cholesterol and of sterol and steroid products derived from cholesterol. The structure for cholecalciferol is written in an extended form so that the hydroxy group at C-3 now lies below the plane of the molecule; it is, therefore, joined to the ring by a *dotted* line but is still termed a 3-β-hydroxyl group.

return to the liver, and resecretion into the gut form the **enterohepatic circulation of bile acids**.

6. The secretion of bile into the intestine is the main route for the excretion of cholesterol, as the steroid nucleus cannot be oxidized to carbon dioxide and water by human tissues.

7. The function of bile and the significance of the enterohepatic circulation are discussed in Chapter 24 (see sections I B 2 and IV D I).

C. Solubility of cholesterol in bile

1. Bile contains a considerable amount of cholesterol, which would be insoluble but for its association with bile salts and phospholipids (chiefly phosphatidylcholine).

2. Despite the presence of these solubilizing agents, an increased concentration of cholesterol in the bile can lead to the formation of **gallstones** in the gallbladder or duct. These are formed if cholesterol is precipitated out of solution around a core of protein and bilirubin (see Chapter 28, section II D), a condition known as **cholelithiasis**.

IV. STEROID HORMONES DERIVED FROM CHOLESTEROL

A. Structures

1. The steroid hormones produced in humans are formed and secreted by the **adrenal cortex**, the **testis**, the **ovary**, and the **placenta**. In addition, a steroid compound derived from 7-dehydrocholesterol, a precursor of cholesterol and related to vitamin D_3, is formed in the kidney (1,25-dihydroxycholecalciferol). The structures of these compounds are shown in Figure 23-2.

2. The **adrenal cortex** produces hormones with two kinds of physiologic activities.
   a. The zona fasciculata of the adrenal cortex primarily produces **cortisol** in humans. Cortisol regulates a number of key metabolic steps and also inhibits the inflammatory response (see Chapter 22, section III B 5). It is called a **glucocorticoid**, as is its 11-keto derivative **cortisone**, a clinically equivalent steroid.
   b. The adrenal zona glomerulosa produces **aldosterone**, which controls the reabsorption of Na^+ in the kidney. Aldosterone is called a **mineralocorticoid**.

3. Gonadal steroids
   a. The Leydig cells of the **testis** produce **testosterone**, the hormone responsible for the development of the male secondary sexual characteristics.
   b. The **ovary** produces **estradiol** in the thecal cells of the graafian follicle and **progesterone** in the corpus luteum; progesterone is also formed by the **placenta** in pregnancy.

B. Biosynthesis

1. Adrenocortical hormones
   a. In the adrenocortical cells, cholesterol is stored in lipid droplets, predominantly as cholesterol esters. It is hydrolyzed to free cholesterol and converted to **progesterone**, a key intermediate in the biosynthesis of adrenal and gonadal hormones.
   b. The conversion of cholesterol to progesterone involves side-chain cleavage between C-20 and C-22, followed by oxidation of the 3-hydroxyl to a keto, and then isomerization of the Δ^5 double bond to a Δ^4 configuration.

Cholesterol   Progesterone   + Isocapryl aldehyde

c. Progesterone then undergoes hydroxylations at C-17, C-21, and C-11 to form **cortisol**. The C-17 and C-21 hydroxylations occur in the microsomal fraction, while the final hydroxylation at C-11 occurs in the mitochondria. All the hydroxylations require NADPH and molecular oxygen.

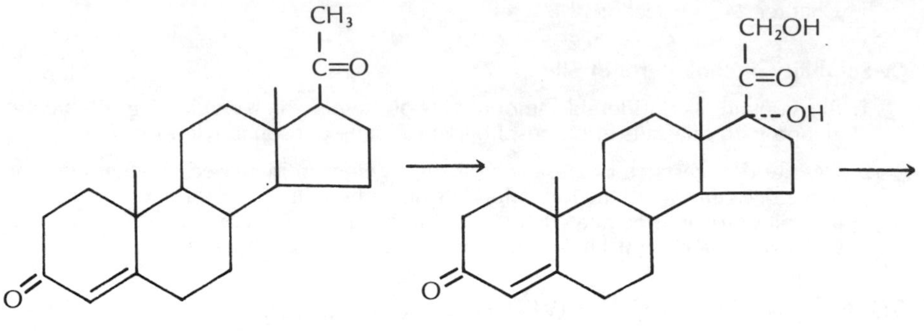

Progesterone   17α,21-Dihydroxyprogesterone

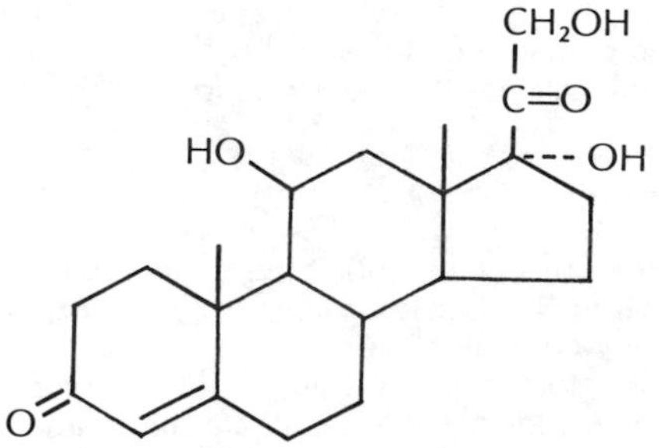

Cortisol

d. Aldosterone is formed from corticosterone (11β,17α,21-trihydroxyprogesterone) in the zona glomerulosa.

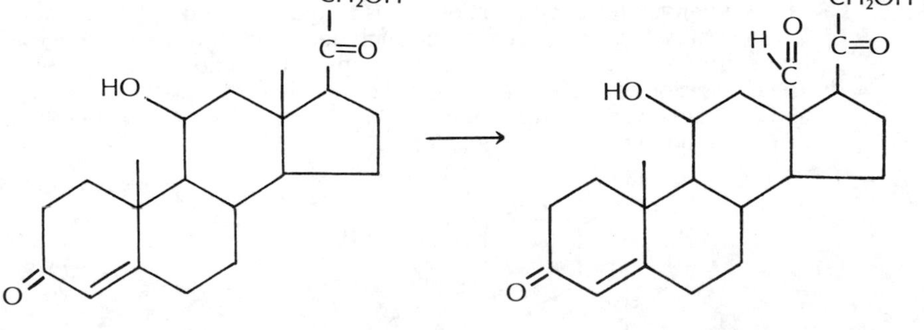

Corticosterone   Aldosterone

2. Gonadal steroids
   a. Testicular hormones. Progesterone is also an intermediary in the biosynthesis of **testosterone** from cholesterol in the Leydig cells of the testis. Progesterone is first hydroxylated at C-17, and the side chain is then removed to give **androstenedione**. This is then reduced to testosterone.

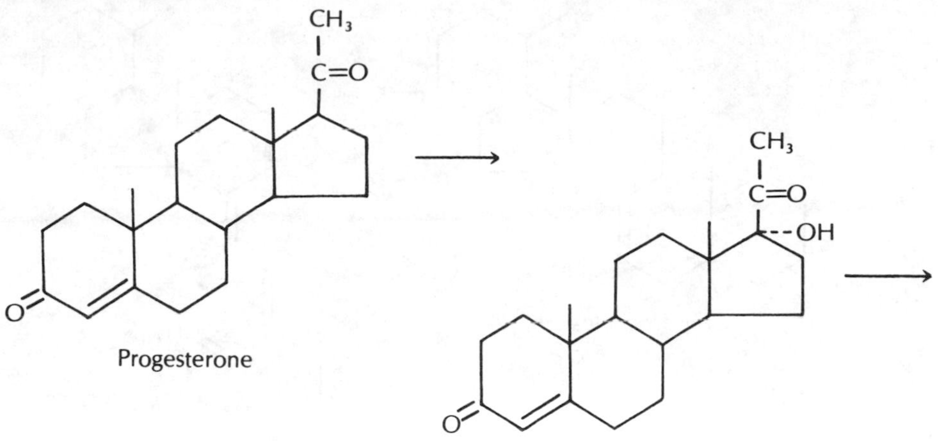

Progesterone

17α-Hydroxyprogesterone

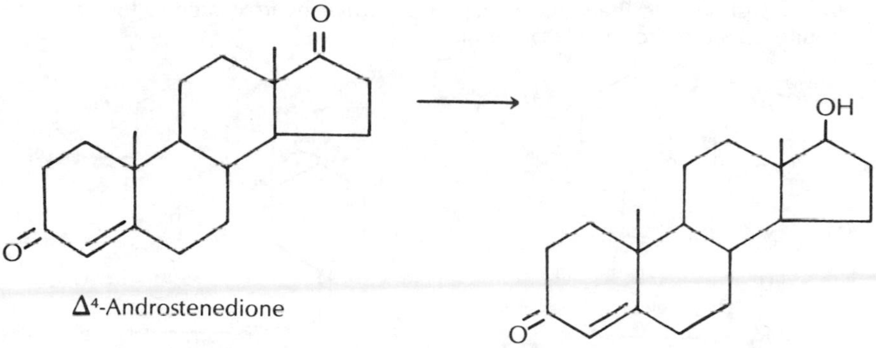

Δ⁴-Androstenedione

Testosterone

b. Ovarian hormones. The ovary produces two steroid hormones. 17β-Estradiol is formed in the thecal cells of the graafian follicle, and progesterone is formed by the cells of the corpus luteum and by the placenta in pregnancy. Estradiol formation involves aromatization of ring A and the consequent loss of the axial methyl group on carbon 10. Estradiol is, therefore, a C-18 steroid (see Fig. 23-2).

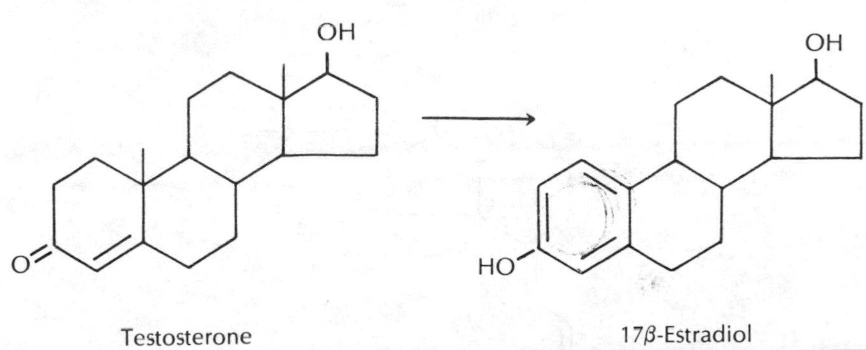

Testosterone

17β-Estradiol

3. Vitamin D
    a. 7-Dehydrocholestrol, an intermediate in the synthesis of cholesterol from lanosterol (see section II B 5), accumulates in the skin.
    b. Under the influence of ultraviolet light, 7-dehydrocholesterol is converted to cholecalciferol (vitamin D₃).

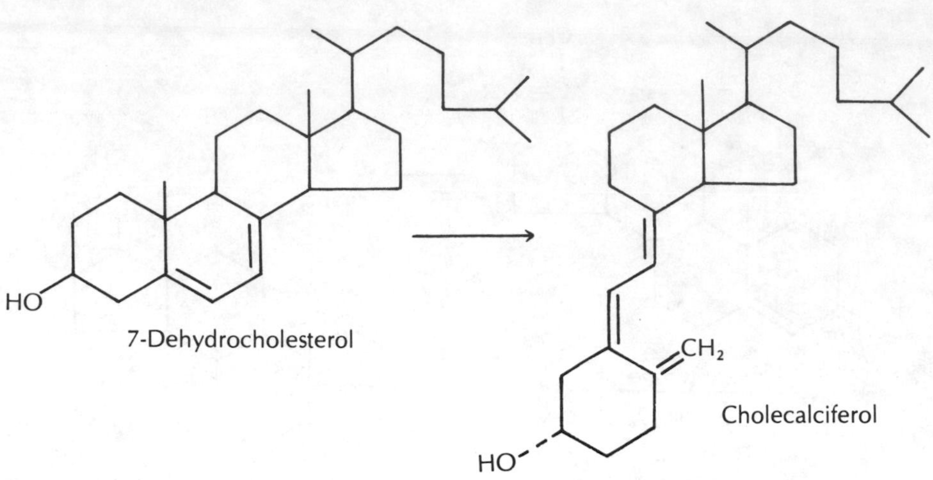

7-Dehydrocholesterol

Cholecalciferol

c. In the liver, cholecalciferol is hydroxylated by an enzyme that shows strong product inhibition (see Chapter 11, section VII E) to form 25-hydroxycholecalciferol. This compound is then transported to the kidney, where it is further hydroxylated to form the active compound 1,25-dihydroxycholecalciferol.

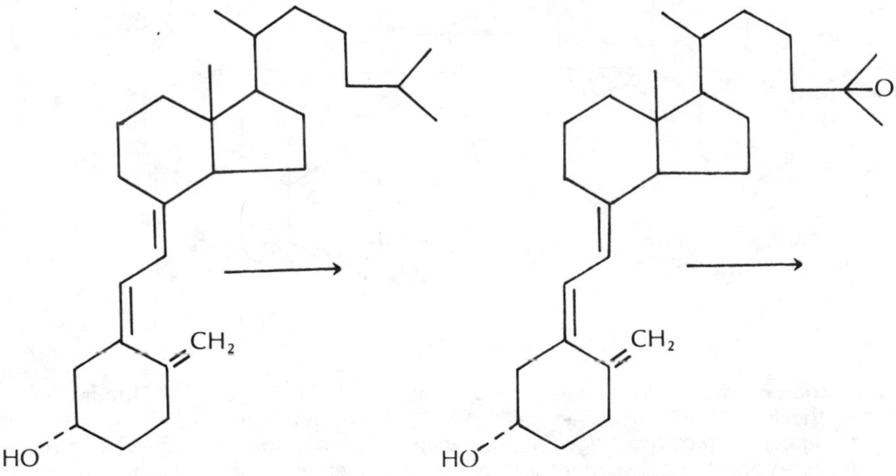

Cholecalciferol

25-Hydroxycholecalciferol

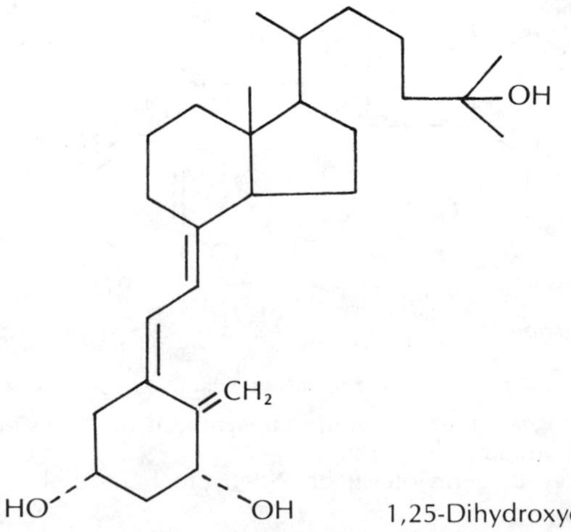

1,25-Dihydroxycholecalciferol

STUDY QUESTIONS

Directions: Each question below contains five suggested answers. Choose the **one best** response to each question.

1. Which of the following compounds is a precursor in the biosynthesis of both cholesterol and β-hydroxybutyrate (a "ketone" body)?

(A) Farnesyl pyrophosphate
(B) β-Hydroxy-β-methylglutaryl coenzyme A (HMG CoA)
(C) Methylmalonyl CoA
(D) Mevalonate
(E) None of the above

2. The committed step in the biosynthesis of cholesterol from acetyl coenzyme A (acetyl CoA) is

(A) the formation of acetoacetyl CoA from acetyl CoA
(B) the formation of β-hydroxy-β-methylglutaryl coenzyme A (HMG CoA) from acetyl CoA and acetoacetyl CoA
(C) the formation of mevalonic acid from HMG CoA
(D) the formation of squalene by squalene synthetase
(E) the cyclization of squalene to lanosterol

3. If the carboxyl carbon of mevalonic acid is labeled with ^{14}C and then incubated with a liver preparation capable of synthesizing cholesterol, which of the cholesterol carbons will be labeled?

(A) All carbons
(B) Axial methyl carbons, C-18 and C-19
(C) Every third carbon, starting with C-1
(D) Only the side-chain carbons
(E) None of the carbons

4. Bile acids are characterized by all of the following statements EXCEPT

(A) the rate-limiting step in the formation of cholic acid from cholesterol is the hydroxylation at C-7
(B) conjugation of cholic acid with glycine or taurine occurs in the intestine
(C) the secondary bile acids (deoxycholate and lithocholate) are deconjugated before they are reabsorbed by the intestinal mucosa prior to returning to the liver via the portal vein
(D) chenodeoxycholate differs from cholate in that it is not hydroxylated at C-12
(E) the primary bile acids lose their C-7 hydroxyl group in the intestine

5. Cholesterol can act as a precursor for each of the following compounds EXCEPT

(A) chenodeoxycholic acid
(B) 1,25-dihydroxycholecalciferol
(C) testosterone
(D) glycocholic acid
(E) cholecystokinin

6. The synthesis of 1,25-dihydroxycholecalciferol takes place

(A) in the skin under the action of ultraviolet light from 7-dehydrocholesterol
(B) in the liver from cholecalciferol
(C) in the kidney from 25-hydroxycholecalciferol
(D) in the intestine from cholecalciferol
(E) nowhere, as it is not synthesized in mammals

ANSWERS AND EXPLANATIONS

1. The answer is B. (*II B 1; Chapter 20 VII B 4*) The rate-limiting step in cholesterol biosynthesis is the reduction of β-hydroxy-β-methylglutaryl coenzyme A (HMG CoA) to mevalonate by HMG CoA reductase. This occurs in the cytoplasm, the HMG CoA being formed from cytoplasmic acetyl CoA. In the liver mitochondria, HMG CoA, formed from mitochondrial acetyl CoA when fatty acid oxidation is high, is a substrate for HMG CoA lyase, which cleaves it into acetyl CoA and the ketone body acetoacetate. β-Hydroxybutyrate is formed from acetoacetate and reduced nicotinamide-adenine dinucleotide by β-hydroxybutyrate dehydrogenase. Farnesyl pyrophosphate and mevalonate are intermediates in cholesterol biosynthesis but not in ketone body synthesis. Methylmalonyl CoA is derived from propionyl CoA, the end product of the β-oxidation of an odd-numbered carbon fatty acid.

2. The answer is C. (*II B 1 b*) The rate-limiting and committed step in cholesterol biosynthesis from acetyl CoA is the reduction of β-hydroxy-β-methylglutaryl coenzyme A (HMG CoA) by HMG CoA reductase to form mevalonate. This is also the controlled step: Hepatic cholesterol synthesis is reduced by the feeding of cholesterol, a feedback effect, or by fasting, which limits the supply of acetyl CoA and reduced nicotinamide-adenine dinucleotide phosphate for cholesterol synthesis. Diets high in fat, or in carbohydrate, increase hepatic cholesterol biosynthesis.

3. The answer is E. (*II B 2*) Mevalonic acid gives rise to the activated isoprenoid isomers, isopentenyl pyrophosphate and 3,3-dimethylallyl pyrophosphate by reactions requiring the expenditure of two high-energy phosphate bonds of adenosine triphosphate and the loss of the carboxyl carbon of mevalonate as carbon dioxide. Thus, none of the carbons of cholesterol would be labeled with ^{14}C.

4. The answer is B. (*III A, B*) The bile acids are formed in the liver from cholesterol. The reactions involved in their synthesis include α-hydroxylation at specific positions (C-7, C-12, or both) with the simultaneous isomerization of the 3-β-hydroxyl of cholesterol to the 3-α-isomer. The side chain between C-24 and C-25 undergoes oxidative cleavage with the formation of an acyl coenzyme A (acyl CoA) derivative. Taurine, or glycine, then forms amide links at the C-24 carboxyl with the displacement of CoA. The conjugated bile acids are then released from the liver for storage in the gallbladder. Thus, conjugation occurs in the liver, not in the intestine.

5. The answer is E. (*III A, B; IV B 2, 3; Chapter 24, I B 1 b*) Cholesterol is the precursor of bile acids in the liver. Cholic and chenodeoxycholic acids are formed by α-hydroxylation of the cholesterol nucleus at specific positions (cholic acid at C-7 and C-12, chenodeoxycholic acid at C-7) and oxidative cleavage of the side chain between C-24 and C-25. The acyl coenzyme A (acyl CoA) derivatives formed in the side-chain cleavage reaction react with taurine or glycine to form amide links at the C-24 carboxyl with the displacement of CoA. Glycocholic acid is a conjugate of cholic acid and glycine. The active form of vitamin D, 1,25-dihydroxycholecalciferol, is formed in the kidney from 25-hydroxycholecalciferol, which is itself formed in the liver from cholecalciferol. The latter is formed in the skin from 7-dehydrocholesterol by the action of ultraviolet light. 7-Dehydrocholesterol is a precursor of cholesterol biosynthesis from lanosterol and accumulates in the skin. Testosterone is formed in the Leydig cells of the testis from cholesterol. Scission of the side chain requires a prior α-hydroxylation at C-17, and the 3-β-hydroxyl and Δ^5 unsaturation of cholesterol is changed to the Δ^4-3-ketone of testosterone. Cholecystokinin is a peptide hormone secreted by the intestinal mucosa and is involved in the regulation of pancreatic secretion of digestive enzymes and in the contraction of the gallbladder.

6. The answer is C. (*IV B 3*) Vitamin D, cholecalciferol, arises from the action of ultraviolet light on 7-dehydrocholesterol in the skin. The latter compound is an intermediate in cholesterol biosynthesis and accumulates in the skin. Cholecalciferol is not active as a vitamin D and undergoes hydroxylations, first in the liver and then in the kidney. In the liver, cholecalciferol is hydroxylated to 25-hydroxycholecalciferol, and in the kidney, the 25-hydroxy derivative is further hydroxylated at C-1 to form the active compound, 1,25-dihydroxycholecalciferol.

24

Digestion, Absorption, and Transport of Lipids

I. DIGESTION

A. Dietary lipids

1. Triacylglycerol is the major dietary lipid of nutritional value, although many other lipid compounds are ingested.

2. Linoleic acid, a so-called **essential fatty acid**, which is a precursor for arachidonic acid synthesis (see Chapter 20, section II B 2), is also included in small amounts in an adequate diet.

B. Emulsification

1. Digestion of lipids begins in the duodenum, when the entrance of the acid chyme from the stomach stimulates the secretion of enteric hormones (small peptides) by the duodenal mucosa.

a. Gastric hydrochloric acid (HCl) stimulates the secretion of **secretin**, which in turn stimulates the release of HCO_3^- from the pancreas into the duodenum via the pancreatic duct.

b. Fats and amino acids stimulate the release of **pancreozymin–cholecystokinin**, which in turn stimulates

(1) The release of **zymogen granules** from the pancreas

(2) The release of **bile**, which leaves the gallbladder and enters the duodenum via the bile duct

2. The bile salts and phosphatidylcholine from the bile act as detergents in the duodenum due to their amphipathic structures; in this capacity, they aid in the formation of mixed micelles, which have a large surface area, from fat globules, which have a small surface area.

3. The micellar associations of lipids are the substrates for hydrolyzing enzymes.

C. Hydrolysis

1. Three lipid-specific proenzymes, which become active upon encountering the bile acids and the neutral pH, which is due to the neutralization of the acid chyme by HCO_3^-, are contained in the zymogen granules. The enzymes are

a. Pancreatic lipase, which cleaves triacylglycerols to 2-monoacylglycerol and two free fatty acids

$$
\begin{array}{c}
CH_2\!-\!O\!-\!\overset{\displaystyle O}{\overset{\|}{C}}\!-\!R_1 \\[4pt]
R_2\!-\!\overset{\displaystyle}{\underset{\|}{C}}\!-\!O\!-\!CH_2 \quad + 2\,H_2O \rightarrow \\
CH_2\!-\!O\!-\!\overset{\displaystyle}{\underset{\|}{C}}\!-\!R_3
\end{array}
\qquad
\begin{array}{c}
CH_2OH \\[4pt]
R_2\!-\!\overset{\displaystyle}{\underset{\|}{C}}\!-\!O\!-\!CH_2 \quad + R_1\!-\!COO^- + R_3\!-\!COO^- + 2\,H^+ \\
CH_2OH
\end{array}
$$

b. Cholesterol esterase, which hydrolyzes cholesterol esters

Cholesterol esters + H_2O → cholesterol + a fatty acid

c. Phospholipase A_2, which hydrolyzes phospholipid

Phospholipid + H_2O → lysophospholipid + a fatty acid

II. ABSORPTION AND RE-ESTERIFICATION

A. Absorption

1. Free cholesterol, fatty acids, and 2-monoacylglycerols passively diffuse into the columnar epithelial cells in the mucosa of the jejunum and ileum.

2. Within these cells, two processes occur: re-esterification and formation of chylomicrons.

B. Re-esterification

1. A thiokinase activates fatty acids to form fatty acyl coenzyme As (fatty acyl CoAs).

$$\text{Fatty acid } + \text{ CoASH } + \text{ ATP } \rightarrow \text{ fatty acyl CoA } + \text{ AMP } + \text{ PP}_i$$

2. Acyl CoA transferase esterifies the 2-monoacylglycerols at the 1 and 3 positions.

$$\text{2-Monoacylglycerol } + \text{ fatty acyl CoA } \rightarrow \text{ diacylglycerol } + \text{ CoASH}$$

$$\text{Diacylglycerol } + \text{ fatty acyl CoA } \rightarrow \text{ triacylglycerol } + \text{ CoASH}$$

3. Cholesterol and lysophospholipids are re-esterified in a similar manner.

C. Formation of chylomicrons

1. The re-esterification of triacylglycerols takes place in the cisternae of the endoplasmic reticulum of the mucosal cells.

2. The triacylglycerols are given a coat composed of protein, phospholipids, and cholesterol esters, forming particles, which are about 0.2 to 1.0 μ in diameter, called **chylomicrons**.

3. The chylomicrons are transported in membrane-bounded vesicles to the lateral cell membranes of the mucosal cells, where they are released by exocytosis into the extracellular space.

D. Very low-density lipoproteins (VLDLs), which are similar but smaller particles, are also synthesized by the mucosal cells. VLDLs contain more phospholipid in addition to the components contained in chylomicrons. (See Chapter 10, section IV B–D for the classification and structure of VLDLs, chylomicrons, and other lipoproteins.)

III. TRANSPORT

A. Movement and fate of chylomicrons

1. Chylomicrons enter the **lacteals** and are transported by the **lymphatic system** to the **thoracic duct**, which empties into the left **subclavian vein**.

2. Before the triacylglycerols of the chylomicrons can be taken up by tissues, they must be hydrolyzed to monoacylglycerols and free fatty acids.
  a. The enzyme responsible for triacylglycerol hydrolysis is **lipoprotein lipase**. This enzyme
    (1) Is localized in **endothelial cell** membranes
    (2) Also has phospholipase activity
    (3) Is activated by phospholipids and apoprotein C-II (see Chapter 10, section IV C), both of which are present in chylomicrons. The apoprotein C-II is obtained by chylomicrons from **high-density lipoprotein** particles (HDLs), which are synthesized in the liver (see section IV C).
  b. Most of the lipid in chylomicrons is removed (delipidated) by lipoprotein lipase, forming chylomicron remnants; the C-II apoprotein returns to the HDLs.
  c. A monoacylglycerol hydrolase completes the hydrolysis of monoacylglycerols to glycerol and fatty acids.

3. The chylomicron remnants that remain after delipidation are enriched in cholesterol and cholesterol ester. They are cleared by the liver, where most of the cholesterol is used for bile acid synthesis (see Chapter 23, section III B), while the fatty acids are used for phospholipid biosynthesis.

B. Transport of medium-chain fatty acids

1. Approximately 10% of the triacylglycerols in the diet contain fatty acids with 8 to 10 carbons. These medium-chain triacylglycerols are absorbed directly into the mucosal cells from the gut without prior hydrolysis.

2. In the mucosal cells, these triacylglycerols are hydrolyzed by endoplasmic reticulum lipases to free fatty acids and glycerol, which both enter the portal circulation.

C. Fate of the liberated fatty acids

1. Approximately 80% of the fatty acids liberated from triacylglycerols by lipoprotein lipase are taken up by muscle and adipose tissue, the remainder being cleared by the liver.

2. The fatty acids are used as an energy source in muscle or are re-esterified into triacylglycerols for storage in adipose tissue cells.

IV. LIPOPROTEIN SYNTHESIS BY THE LIVER

A. VLDLs

1. The main bulk of VLDLs is synthesized in the liver. A small amount is synthesized in the intestinal mucosal cells (see section II D).

2. VLDLs are composed primarily of triacylglycerols derived from
   a. Excess acetyl CoA obtained from carbohydrate
   b. The uptake of free fatty acids from the circulation

3. The transport and fate of VLDLs are very similar to the transport and fate of chylomicrons; the remnants from delipidation of VLDLs are called **low-density lipoproteins (LDLs)**.

B. LDLs

1. LDLs are cholesterol-rich remnants of VLDLs, which are bound to specific receptors on the cell surface of extrahepatic tissues.

2. LDLs enter the cell by pinocytosis and are degraded in lysosomes.

3. Cholesterol released from the degraded LDLs inhibits the formation of LDL receptors by the cell.

4. Most of the extrahepatic cholesterol is used in plasma membrane synthesis.

C. Formation and fate of HDLs

1. HDLs are synthesized in the liver as protein–cholesterol–phospholipid-rich particles containing **apoprotein A**.

2. During release of the HDLs into the circulation, **apoprotein C-II** is added to the particle.

3. HDLs are disk-shaped when first released into the circulation. An apoprotein A-1 activates plasma **lecithin:cholesterol acyltransferase (LCAT)**, which esterifies the cholesterol and converts the lecithin to lysophosphatidylcholine (see Chapter 23, section II C 3).

4. The completely hydrophobic cholesterol esters form a core within the HDL, and the discoid shape of the HDL converts to a fully activated, round shape.

5. HDLs exchange apoprotein C-II with chylomicrons and VLDLs for activation of lipoprotein lipase (see section III A 2 a).

6. HDLs may also shuttle excess cholesterol back from the peripheral tissues to the liver.

D. Cholesterol turnover. Cholesterol is cycled between extrahepatic sites and the liver by two processes.

1. The enterohepatic circulation of bile (see Chapter 23, section III B) returns most of the bile acids secreted into the gut back to the liver, together with a smaller proportion of cholesterol (either from the bile or ingested), the remainder being excreted in the feces after modification by bacterial action.

2. Intracellular lipoprotein hydrolysis and HDL transport
   a. Chylomicron remnants and LDLs are cholesterol-rich particles (because they have lost triacylglycerols), which are removed from the circulation by binding to specific LDL receptors that are localized on peripheral tissue cell membranes in areas called "coated pits." They are internalized, and the cholesterol esters are hydrolyzed to free cholesterol in the lysosomes.
   b. The cholesterol released within the extrahepatic tissue cells inhibits the neosynthesis of

Table 24-1. Classification of Primary Hyperlipidemias

Type	Genetic Classification	Genetic Form	Elevated Plasma Lipid	Elevated Plasma Lipoprotein	Risk of Atherosclerosis	Treatment
I	Familial lipoprotein lipase deficiency	Autosomal recessive	Triacylglycerols	Chylomicrons	Not clearly increased	Low-fat diet
IIa	Familial hypercholesterolemia (LDL receptor deficiency)	Autosomal dominant; common	Cholesterol	LDLs	Very high, especially in coronary artery	Low-cholesterol diet; cholestyramine; possible surgery
IIb	. . .	. . .	Triacylglycerol and cholesterol	LDLs and VLDLs	. . .	. . .
III	Familial dysbetalipoproteinemia	Uncertain	Triacylglycerol and cholesterol	β-VLDL	Very high, especially in peripheral vessels	Low-cholesterol, low-calorie diet; clofibrate
IV	Familial hypertriglyceridemia	Heterogeneous; common	Triacylglycerol	VLDL	Possible	Low-carbohydrate, low-calorie diet; no alcohol; niacin
V	. . .	Heterogeneous	Triacylglycerol	Chylomicrons and VLDLs	Not clearly increased	Low-fat, low-calorie diet; niacin

Note.—LDLs = low-density lipoproteins; VLDLs = very low-density lipoproteins; and β-VLDL = very low-density lipoprotein with an abnormal, or "floating β," band seen on electrophoresis isolated from the top fraction of centrifuged plasma (for normal electrophoretic mobilities, see Chapter 10, section IV B).

cholesterol by the cells. If the supply of cholesterol is still excessive, the cholesterol becomes associated with HDLs, which return to the liver where the cholesterol can either be used to synthesize bile acids or be disposed of in the bile as cholesterol.

V. CLINICAL DISORDERS OF LIPOPROTEIN METABOLISM. Familial disorders of lipid metabolism include both hyper- and hypolipidemias.

A. Hyperlipidemias

1. If hyperlipidemia is the principal manifestation of the disease, it is called a **primary hyperlipidemia**.
   a. The primary hyperlipidemias can be classified on the basis of plasma electrophoretic patterns after a 12-hr fast (Fig. 24-1).
   b. Table 24-1 gives some characteristics of the primary hyperlipidemias, which are classified according to the electrophoretic patterns of plasma proteins.

2. If the hyperlipidemia is due to an underlying disease process, as found in some cases of thyroid, liver, or kidney disease, it is called a **secondary hyperlipidemia**.

B. Hypolipidemias

1. **Hypobetalipoproteinemia (Bassen-Kornzweig syndrome)** is a rare genetic disorder with a recessive inheritance characterized by neurologic symptoms, including ataxia and mental retardation.
   a. Plasma triacylglycerol and cholesterol levels are decreased.
   b. There is a complete absence of β-lipoproteins (no chylomicrons or VLDLs).
   c. Absorption of fat is greatly reduced.
   d. There is no effective treatment to prevent the neurologic symptoms from appearing.

2. **α-Lipoprotein deficiency (Tangier disease)** is a rare familial disorder characterized by recurrent polyneuropathy, lymphadenopathy, tonsillar hyperplasia, and hepatosplenomegaly (from storage of cholesterol in reticuloendothelial cells).
   a. Plasma cholesterol levels are low; triacylglycerols are normal or increased.
   b. There is a marked decrease in plasma HDLs.
   c. There is no known treatment.

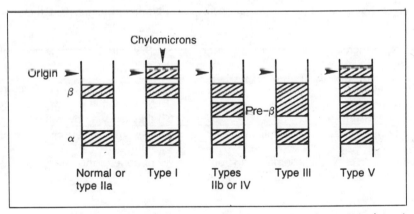

Figure 24-1. Plasma electrophoretic patterns (see Chapter 10, section IV B) in the various types of primary hyperlipidemia.

STUDY QUESTIONS

Directions: Each question below contains five suggested answers. Choose the **one best** response to each question.

1. Dietary triacylglycerols containing long-chain fatty acyl moieties are digested, absorbed from the small intestine, and stored in adipose tissue cells. The following processes occur at different times and sites:

1. Formation of chylomicrons
2. Hydrolysis by hormone-sensitive triacylglycerol lipase
3. Micellar formation with bile salts
4. Resynthesis from 2-acylglycerols
5. Hydrolysis by lipoprotein lipase
6. Transport in the circulation as free fatty acids bound to serum albumin

The correct order of these events is given by which of the following sequences?

(A) 3, 1, 5, 6, 4, 2
(B) 3, 4, 1, 2, 6, 5
(C) 1, 4, 3, 5, 6, 2
(D) 3, 4, 1, 5, 2, 6
(E) 1, 4, 5, 6, 2, 3

2. All of the following statements regarding type IIa hyperlipidemia (familial hypercholesterolemia) are correct EXCEPT

(A) there is an increased risk of coronary artery disease
(B) the serum low-density lipoprotein (LDL) levels are high
(C) it is due to a deficiency of LDL receptors in extrahepatic tissues
(D) cholesterol synthesis in extrahepatic cells is deregulated
(E) it is a rare, recessive, autosomal condition

3. Lipoprotein lipase is characterized by all of the following statements EXCEPT

(A) it is localized in endothelial cell membranes
(B) it has phospholipase activity
(C) it is activated by phospholipids and apolipoprotein C-II
(D) it hydrolyzes off fatty acids from the sn-2 carbon of the glycerol backbone
(E) it is missing or deficient in patients with type 1 hyperlipidemia

4. Which of the following statements regarding very low-density lipoproteins (VLDLs) is correct?

(A) VLDLs contain about 40% to 50% triacylglycerols and are formed chiefly in the liver
(B) VLDLs contain about 40% to 50% cholesterol and are formed chiefly in the mucosal cells of the small intestine
(C) VLDLs give rise to HDLs by delipidation in the circulation
(D) VLDLs contain lipid covalently bound to protein
(E) None of the above statements is correct

5. Bile salts act as detergents in the duodenum because

(A) of their amphipathic structures
(B) they contain hydrophobic groups
(C) they carry an overall positive charge
(D) they are zwitterions at the pH of the duodenum
(E) of none of the above reasons

Directions: Each question below contains four suggested answers of which **one or more** is correct. Choose the answer

A   if **1, 2, and 3** are correct
B   if **1 and 3** are correct
C   if **2 and 4** are correct
D   if **4** is correct
E   if **1, 2, 3, and 4** are correct

6. A malfunctioning pancreas would be expected to decrease the

(1) digestion of fat by decreasing the pH of the duodenal contents
(2) absorption of fat by increasing the pH of the duodenal contents
(3) absorption of fat by decreasing the secretion of lipid-digestive enzymes
(4) absorption of triacylglycerols containing medium-chain fatty acids (8 to 10 carbons)

7. Correct statements regarding pancreatic lipase include which of the following?

(1) The substrates for pancreatic lipase are triacylglycerols in micellar form
(2) The products of pancreatic lipase action are fatty acids and 2-acylglycerols
(3) Pancreatic lipase secretion is stimulated by pancreozymin–cholecystokinin
(4) Pancreatic lipase zymogen is activated by bile acids and neutral pH

8. The formation of chylomicrons takes place

(1) in the duodenum where they consist of lipid covalently bound to protein
(2) in specialized hepatic cells where they contain mainly cholesterol and protein
(3) in the blood from circulating lipoproteins where they contain about 60% triacylglycerols
(4) in the intestinal mucosal cells where they contain about 85% triacylglycerols

9. Correct statements regarding the absorption of triacylglycerols containing medium-chain fatty acids (8 to 10 carbons) include which of the following?

(1) They are absorbed by the mucosal cells of the gut following hydrolysis by pancreatic lipase
(2) They are absorbed by the mucosal cells of the gut as triacylglycerols
(3) They are hydrolyzed in the circulation to free fatty acids and glycerol by lipoprotein lipase
(4) They are hydrolyzed in the mucosal cells to free fatty acids and glycerol

ANSWERS AND EXPLANATIONS

1. The answer is D. (*I; II; III*) Dietary triacylglycerols containing long-chain fatty acids are first emulsified by bile salts, hydrolyzed by pancreatic lipase to 2-acylglycerols, which are absorbed by the mucosal cells of the small intestine and built into chylomicrons. Thus, the first three processes listed are 3, 4, and 1, and they occur in that order. Lipoprotein lipase hydrolyzes the triacylglycerols to monoacylglycerols and free fatty acids, which are taken up by the adipose tissue cells, re-esterified to triacylglycerols, and stored as such. Mobilization of fat stores is achieved by hydrolysis of the triacylglycerols by hormone-sensitive lipase with the release of free fatty acids and glycerol from the adipocytes. The free fatty acids are transported in the circulation bound to serum albumin. Thus, the order for the next three reactions listed is 5, 2, and 6.

2. The answer is E. (*IV B, D 2 a; V A b; Table 24-1*) Type IIa hyperlipidemia (familial hypercholesterolemia) is an autosomal dominant condition characterized by a deficiency of functional low-density lipoprotein (LDL) receptors on extrahepatic cell membranes. The serum LDL levels, and, therefore, the serum cholesterol levels, are high, and there is an increased risk of coronary artery disease. In addition, the failure of extrahepatic tissues to remove cholesterol from the LDLs results in loss of control over the rate of neocholesterol synthesis by these cells.

3. The answer is D. (*III A 2; Table 24-1*) Lipoprotein lipase is localized in the plasma membrane of endothelial cells. It hydrolyzes triacylglycerols in chylomicrons and in very low-density lipoproteins, removing fatty acids from either the sn-1 or sn-3 carbon of the glycerol moiety. A 2-acyl hydrolase completes the hydrolysis of triacylglyerols. Lipoprotein lipase is activated by phospholipids and apolipoprotein C-II, which is obtained by chylomicrons from high-density lipoprotein particles synthesized in the liver. Lipoprotein lipase also exhibits intrinsic phospholipase activity. It is missing or deficient in type I hyperlipidemia (familial lipoprotein lipase deficiency).

4. The answer is A. (*IV A*) The bulk of the very low-density lipoproteins (VLDLs) is formed in the liver, although a small amount is synthesized in the intestinal mucosal cells. VLDLs are composed primarily of triacylglycerols formed from excess acetyl coenzyme A, which is derived from carbohydrate catabolism and from the uptake by the liver of excess free fatty acids from the circulation. VLDLs contain only a small proportion of cholesterol (5% to 10% of total lipid), and on delipidation by lipoprotein lipase, they form cholesterol-rich remnants, the low-density lipoproteins. The lipid–protein interactions in the lipoproteins do not involve covalent bonding.

5. The answer is A. (*I B 2*) Bile salts, together with phosphatidylcholine (lecithin) from the bile, act as detergents in the duodenum due to their amphipathic structures; that is, they contain both hydrophilic groups (carboxylate and hydroxyl) and a hydrophobic region (the β side of the rather flat steroid ring structure). As detergents, they help to form micelles that have inner hydrophobic regions in which nonpolar substances can dissolve and that can exist in aqueous suspension due to their outer hydrophilic groups. None of the other choices are main reasons for the detergent-like action of the bile salts; in fact, the charge on the molecule is negative, not positive, and there is no possibility of zwitterion formation.

6. The answer is B (1, 3). (*I B, C; III B*) Disease of the exocrine tissues of the pancreas could interfere with the secretion of bicarbonate ions required to neutralize the acid chyme from the stomach as it enters the duodenum and with the release of zymogen granules of potential digestive enzymes. Both hydrolysis and absorption of fats would be impaired. Triacylglycerols containing medium-chain fatty acids (8 to 10 carbons) might be absorbed reasonably well, as they do not require prior hydrolysis in the duodenum.

7. The answer is E (all). (*I B, C*) The entry of fats and amino acids into the duodenum from the stomach causes the release of pancreozymin–cholecystokinin, which, in turn, stimulates the release of zymogen granules from the pancreatic acinar cells and evacuation of the gallbladder. The formation of mixed micelles containing triacylglycerols, phospholipids, cholesterol, and other lipids is due to the presence of bile salts and phosphatidylcholine in the duodenum. The pancreatic lipase zymogen is activated by bile acids and the neutral pH brought about by the secretion of bicarbonate by the pancreas. Pancreatic lipase requires the micellar form of its substrate for its catalytic function, which is to cleave triacylglycerols to 2-acylglycerol and fatty acids.

8. The answer is D (4). (*II C; Chapter 10, Table 10-3*) Re-esterification of absorbed fatty acids and 2-monoacylglycerols occurs in the cisternae of the endoplasmic reticulum of intestinal mucosal cells. The triacylglycerols thus formed receive a coat composed of protein, phospholipids, and cholesterol esters to form chylomicrons, which are particles of about 0.2 to 1 μ in diameter. Chylomicrons contain about 0.5% to 2.5% protein and 97.5% to 99.5% lipid, of which 84% to 89% is triacylglycerol. They are released from the lateral membranes of the mucosal cells and reach the lymphatic system. Entry from the lymph into the blood circulation occurs at the thoracic duct, which empties into the left subclavian vein.

9. The answer is C (2, 4). (*III B*) Fatty acids with 8 to 10 carbons, that is, medium-chain fatty acids, occur as triacylglycerols, constituting about 10% of the dietary triacylglycerol intake. They are absorbed into the mucosal cells as triacylglycerols, without undergoing prior hydrolysis in the intestine. In the intestinal mucosal cells, they are hydrolyzed to free fatty acids and glycerol by lipases of the endoplasmic reticulum. Both the free fatty acids and the glycerol enter the portal circulation.

I. MEMBRANE STRUCTURE

A. General aspects

1. **A limiting membrane** encompasses all cells. In higher animals, this membrane separates the cytoplasm of the cell from the interstitial fluid, which is in equilibrium with the plasma, and this limiting membrane is known as the **plasma membrane**.

2. **Intracellular membrane systems** in eukaryotes consist of
   a. **A nuclear membrane**, which encloses the genetic material
   b. **Cell organelle membranes**, such as the double membrane of mitochondria and the single membrane of lysosomes
   c. **The endoplasmic reticulum**, with its cisternae forming an enclosure
   d. **The Golgi apparatus**, which consists of a stack of flattened disk-shaped structures called a **Golgi stack** or **dictyosome**

B. Functions. Membranes serve several purposes.

1. They compartmentalize and segregate intracellular events, separate cells from one another, and segregate organ functions.

2. They mediate the regulation of cellular functions by acting as selective barriers.

3. They localize specific enzyme systems and provide a semisolid-state phase in an otherwise aqueous environment.

C. Methods of study

1. **Isolation of membranes**
   a. Erythrocytes (red blood cells) can be hemolyzed to yield "cell ghosts."
      (1) The red blood cells are isolated from blood by centrifugation.
      (2) Placing the red blood cells in distilled water lyses the cells with the hemoglobin going into solution in the water.
      (3) The red blood cell plasma membranes are collected by low-speed centrifugation and then washed. These preparations are called "ghosts" because most of the intracellular contents have disappeared.
   b. Membrane preparations from nucleated cells can be obtained by the homogenization of cells or tissues, using special media.
      (1) This is carried out in such a manner that the plasma membrane is broken open without destruction of the intracellular organelles.
      (2) Sequential differential centrifugation of the homogenate is then carried out. This is done at increasing forces to yield pellets enriched with various cellular components, the behavior of particles in a centrifugal field being influenced by the size, density, and shape of the particles.
      (3) Density-gradient centrifugations are used to separate organelles and membrane fractions according to their densities.

2. **Isolation of membrane components**
   a. Organic solvents are used to extract lipids; for example, chloroform–methanol mixtures are used to separate lipids from proteins and carbohydrates.
   b. Hydrophobic bonds can be disrupted by detergents. Examples include deoxycholic acid, which is a naturally occurring bile acid, and sodium dodecyl sulfate (SDS), a sulfated derivative of lauric acid that acts as a detergent.
   c. Hydrogen bonds can be disrupted by high concentrations of urea.

    d. Separation of membrane proteins, freed from lipid, can be carried out by disk electrophoresis or by SDS–gel electrophoresis [see Chapter 9, section IV B 2 d, 3 c (2)].

II. COMPOSITION OF MEMBRANES.
The reasons in sections II A and B led Davson and Danielli to postulate that plasma membranes are a bimolecular layer of polar lipids with hydrocarbon chains inward and polar groups outward, coated on each polar surface by monomolecular layers of proteins.

A. Lipids in membranes were deduced from

1. The low permeability for the entry of ions into cells and through artificial lipid membranes

2. The correlation between the lipid solubility of a compound and its rate of entry into cells. (This led Overton to postulate that the plasma membrane was entirely lipid, a continuous hydrocarbon phase.)

3. The ability of oils to coalesce with the plasma membrane of cells

4. The ability of lipolytic agents, such as benzene and carbon tetrachloride, to lyse plasma membranes

B. Proteins in, or on, membranes were deduced because

1. Plasma membranes have elasticity and mechanical strength.

2. Plasma membranes have a low surface tension.

3. Antibodies against cells often lyse plasma membranes.

C. Carbohydrates in membranes

1. Carbohydrates occur in membranes as **glycolipids** and **glycoproteins.**
     a. The glycolipids are derivatives of sphingosine (see Chapter 21, section II D).
     b. In the glycoproteins, chains of sugar residues are covalently linked to the side chains of serine, threonine, or asparagine residues of proteins (see Chapter 10, section II C).

2. The carbohydrate groups in membranes can be localized by labeling them with **lectins**. These are proteins formed by plants, which have a high affinity for specific sugar residues; for example
     a. Concanavalin A binds to nonreducing terminal α-mannosyl residues and to internal α-mannosyl residues.
     b. Wheat germ agglutinin binds to terminal N-acetylglucosamine residues.
     c. The lectin–sugar complex can be visualized in the electron microscope after its conjugation to the iron-containing protein, ferritin (see Chapter 28, section II E).

III. CHARACTERISTICS OF MEMBRANE COMPONENTS

A. Membrane lipids.
Phospholipids, glycolipids, and cholesterol are the major lipids in membranes.

1. They are all amphipathic molecules, having both hydrophobic and hydrophilic ends.

2. The membrane lipids spontaneously form **bilayers** in aqueous media, burying their hydrophobic tails and leaving their hydrophilic ends exposed to the water.

3. Membrane lipids undergo **lateral diffusion**, that is, side-to-side or **translational movement** and **flexing**.
     a. The exchange rate of near neighbors in the translational movement of membrane lipids is less than 1 μsec.
     b. The flexing motion of phospholipids is greatest at the hydrophobic end remote from the polar end group, as the fatty acid tails have some degree of rotational freedom. This is more pronounced with saturated fatty acids than with unsaturated fatty acids, as the latter have −*cis* configurations at the double bonds, which limit their ability to flex (see Chapter 20, section II A 2 b).
     c. Cholesterol decreases the ability of phospholipids to translate and to flex.

4. Phospholipids do not readily "flip-flop," that is, pass from one side of the bilayer to the other. If it occurs, it is a slow process.

5. Phospholipids are asymmetrically distributed in plasma membranes.
     a. Choline phospholipids constitute the outer half of the bilayer.
     b. The amino phospholipids constitute the inner, or cytoplasmic, half of the bilayer.

    6. Evidence for the asymmetric distribution of phospholipids is as follows:
      a. Reagents that react with phospholipid amino groups will only label proteins on the outer surface of *intact* red blood cells. The phospholipids are not labeled.
      b. The phospholipids of red blood cell *ghosts*, on the other hand, *are* labeled by reagents that react with amino groups.
      c. Intact red blood cells show a constant mobility in an electric field between pH 4.5 and pH 11. If amino phospholipid groups are present on the outer surface, they should titrate around pH 10.

    7. Glycolipids are found only in the outer half of the bilayer with their sugar residues exposed at the surface of the cell. It is believed, but not proven, that the complex oligosaccharides play a role in cell-to-cell interactions.

B. Membrane proteins

    1. Erythrocyte plasma membranes
      a. Membrane proteins have been investigated in a number of cell types, but the most detailed information has come from studies of erythrocyte membranes.
      b. Erythrocytes have no organelles, and hemolysis of the cells releases most of the cytoplasm, leaving "ghosts," which are almost pure plasma membrane.
      c. Erythrocytes can be prepared as
        (1) Sealed ghosts, which have their inner membrane surface protected from the medium
        (2) Leaky ghosts, in which the inner membrane surface is in contact with the medium

    2. Erythrocyte membrane proteins
      a. SDS–polyacrylamide gel electrophoresis of detergent-solubilized erythrocyte membranes, after staining with coomassie blue, shows a pattern of more than 10 bands. The major bands are referred to as bands 1, 2, 3, 4.1, 4.2, 5, 6, and 7.
      b. Staining with periodic acid Schiff reagent reveals four bands, referred to as PAS-1, PAS-2, PAS-3, and PAS-4, which are glycoproteins.
      c. Peripheral proteins are proteins that can be extracted from erythrocyte membranes by using relatively gentle procedures, such as a change in the salt concentration or the pH of the medium.
        (1) The peripheral proteins are not digested by proteases that are added to a medium in which intact erythrocytes or sealed ghosts are suspended. This is because the peripheral proteins are located on the cytoplasmic face of the erythrocyte membrane.
        (2) Bands 1, 2, 4.1, 4.2, 5, and 6 represent such proteins.
          (a) Glyceraldehyde 3-phosphate dehydrogenase, an enzyme of the glycolytic pathway (see Chapter 14, section III E), is represented by band 6.
          (b) Actin, a protein involved in muscle contraction and cell mobility, is represented by band 5.
          (c) Spectrin, composed of two complex polypeptides that form an extensive network that functions in stabilizing the shape of the erythrocyte membrane, is represented by bands 1 and 2.
      d. Integral proteins (bands 3 and 7, and all four PAS-positive bands) can only be extracted from erythrocyte membranes after treatment with detergents or organic solvents. These proteins are embedded in the hydrophobic phase of the membrane.
      e. Transmembrane proteins are exposed both on the outer half of the membrane and on the inner, or cytoplasmic, side. In other words, their peptide chains extend across the width of the membrane.

        (1) Glycophorin
          (a) Glycophorin is one of two proteins exposed on the outer surface of the human erythrocyte. It is a glycoprotein with most of its mass on the outer surface of the membrane, where all of the protein's carbohydrate residues are located.
          (b) It has an α-helix segment of about 20 amino acids, which spans the hydrophobic layer of the membranes.
          (c) Glycophorin is present in high concentration in the erythrocyte membrane and is only found in this cell type. However, its function remains unknown. Individuals who lack this protein appear to be normal.

        (2) Band 3 protein: a transport protein
          (a) The band 3 protein traverses the membrane in a more folded configuration than glycophorin shows with the peptide chain extending across the bilayer several times.
          (b) In contrast to glycophorin, band 3 protein is known to play an important role in erythrocyte function.

(i) Erythrocytes are concerned with the transport of oxygen from the lungs to the tissues and of carbon dioxide (CO_2) from the tissues to the lungs.

(ii) At the lungs, there is an exchange of CO_2 (from the red blood cell) with chloride ions (Cl^-) [from the plasma]. At the tissues, the reverse process takes place.

(iii) Band 3 protein appears to function as an anion channel, allowing the bicarbonate ion (HCO_3^-) [derived from the red blood cell CO_2] and Cl^- exchange to take place.

(c) The carbohydrate residues of this glycoprotein are also on the outer surface of the erythrocyte membrane.

3. Movement of proteins in plasma membranes

a. Because of the constancy of the orientation of proteins in membranes, as measured by labeling experiments, it is highly unlikely that membrane proteins "flip-flop."

b. **The lateral movement** of proteins is allowed for by the lipid bilayer. Direct evidence for the lateral movement of proteins in plasma membranes is provided by a study of **heterokaryons**, which are artificially produced hybrids of mouse and human cells.

(1) Human cells labeled with antibodies coupled to rhodamine (a red fluorescent dye) are fused to mouse cells labeled with antibodies coupled to fluorescein (a green fluorescent dye).

(2) Initially, the mouse and the human proteins remain confined to their own membrane halves, but within 1 hour, the two sets of proteins will spread over the entire surface of the cell, indicating rapid lateral diffusion of the proteins within the membrane bilayer.

c. **"Patching" of membrane proteins**

(1) When antibodies bind to specific plasma membrane surface proteins, the bivalent compounds cross-link some of the proteins, causing them to cluster in large complexes. This again indicates that the proteins are able to move laterally within the membrane.

(2) Partial digestion of the antibody with the formation of monovalent agents does not prevent the modified antibody from binding to the membrane protein, but cluster formation, that is, "patching," does not occur.

d. **"Capping"**

(1) After lymphocyte cell membrane proteins have been clustered by antibodies (or by lectins), the clusters rapidly collect over one pole of the cell to form a "cap."

(2) The mechanism of capping is not clear, but a lymphocyte being capped looks like a lymphocyte moving on a surface.

4. Restriction of membrane protein movement

a. The concept that biologic membranes are two-dimensional fluids (**the fluid-mosaic theory**) has been most helpful in understanding membrane structure and function. However, the picture of a biologic membrane as a lipid sea with proteins floating freely in it is oversimplified.

b. Many membrane proteins are restricted in their lateral mobility in order for them to fulfill specific cellular functions.

c. Devices used by cells to limit protein diffusion in the lipid bilayer include

(1) Confinement to limited areas

(2) Cell junctions

(3) Increases in mass by aggregation

(4) Cross-links by extrinsic elements

(5) Links to cytoplasmic components of the cytoskeleton

d. The need for glycoproteins to have their carbohydrate moiety on the outside in contact with the extracellular environment limits their movement.

IV. MEMBRANE FLUIDITY

A. The fact that some membrane components can move in the lipid bilayer of the membrane is very important for cell function. A number of factors influence the fluidity of membranes, which in turn influences the physiologic function.

B. The fluidity of membranes depends in large part on the nature of the packing and interaction of the fatty acyl chains in membrane phospholipids.

1. Long-chain saturated fatty acids pack closely and interact strongly, producing a relatively rigid structure (i.e., one with low fluidity).

2. With an increase in temperature, some of the *trans* C—C bonds become "gauche" (rotated 120°), and a more fluid structure results.

3. The change in fluidity is seen as the temperature rises above the T_m (melting temperature).

4. The longer the chain length of the fatty acids, the higher the T_m (see Chapter 20, Table 20-1).

5. Unsaturated fatty acids with their *cis* double bonds do not pack closely; this results in more fluid structures in which the T_m is lowered.

6. The greater the number of double bonds, the lower the T_m and the greater the fluidity of the membrane.

C. *Escherichia coli* can change the fluidity of its membrane in response to changes in temperature by modifying the nature of the fatty acids in its membrane phospholipids.

D. In higher animals, cholesterol reduces membrane fluidity by preventing the movement of fatty acyl chains.

STUDY QUESTIONS

Directions: Each question below contains five suggested answers. Choose the **one best** response to each question.

1. Molecular action within biologic membranes is best characterized by which of the following statements?

(A) Lipid molecules readily "flip-flop" from one side of the membrane to the other

(B) Lipid molecules exhibit lateral movement within the membrane bilayer

(C) Protein molecules in membranes are all situated on the cytoplasmic surface of the bilayer

(D) Protein molecules do not exhibit lateral movement within the membrane bilayer

(E) None of the above statements is true

2. In biologic membranes, integral proteins and lipids interact mainly by

(A) hydrophobic interactions

(B) hydrogen bonding

(C) covalent bonds

(D) ionic bonds

(E) both hydrophobic and covalent bonds

3. The fluid-mosaic model for membrane structure proposes that

(A) the outer and inner faces of the membrane are identical

(B) peripheral proteins are situated only on the outer face of the plasma membrane

(C) integral proteins are associated with the hydrophobic phase of the bilayer

(D) both polar and nonpolar ends of membrane phospholipids are within the hydrophobic phase of the bilayer

(E) membrane components cannot move laterally within the bilayer.

4. Lateral movement of membrane proteins can be visualized by following the movement of fluorescent-labeled proteins at 37°C in heterokaryons (artificial hybrids of mouse and human cells). If the incubation temperature is decreased by about 35°C, the lateral movement of the proteins is not observed. Which of the following statements best explains the temperature effect?

(A) The apparent lack of movement is an artifact due to the loss of fluorescence at the lower temperature

(B) The lateral movement of membrane proteins requires an expenditure of high-energy phosphate bonds, which are not available at the lower temperature

(C) The lateral movement of membrane proteins requires the rupture of covalent bonds, which does not occur at the lower temperature

(D) The lateral movement of membrane proteins requires the formation of new covalent bonds, which does not occur at the lower temperature

(E) The lateral movement of membrane proteins is influenced by membrane fluidity, which is low at the lower temperature

Directions: Each question below contains four suggested answers of which **one or more** is correct. Choose the answer

  A if **1, 2, and 3** are correct
  B if **1 and 3** are correct
  C if **2 and 4** are correct
  D if **4** is correct
  E if **1, 2, 3, and 4** are correct

5. Characteristics of peripheral membrane proteins include

(1) extractability using relatively gentle procedures
(2) location on the cytoplasmic or inner face of the red cell membrane
(3) interaction with polar ends of membrane phospholipids
(4) an invariable association with carbohydrate groups, which are covalently bound

6. Characteristics of the lateral movement of integral membrane proteins include which of the following?

(1) It can be influenced by membrane fluidity
(2) It can be restricted by cytoskeletal elements
(3) It can be restricted by cell junctions
(4) It is restricted by cholesterol in eukaryotic cells

ANSWERS AND EXPLANATIONS

1. The answer is B. (*III A, B 2–4*) Lipid molecules can move laterally within the membrane, and the fatty acyl chains of the phospholipids can show rotational flexing. However, they do not readily "flip-flop" from one side of the membrane to the other. If this process occurs, it is a slow one. Membrane proteins can also move laterally in the bilayer, unless their movement is restricted to allow the performance of a specific function, and they do not "flip-flop." Protein molecules in membranes are found associated with the inner, cytoplasmic, face of the membrane (peripheral proteins), or embedded in the hydrophobic phase of the membrane (integral proteins), or they may be found to traverse the membrane and appear on both membrane faces (transmembrane proteins).

2. The answer is A. (*III A 2, B 2 d*) Integral membrane proteins are embedded within the lipid bilayer and bond with the nonpolar chains by hydrophobic interactions. Peripheral membrane proteins or transmembrane proteins will interact with the lipid phase, where they are in contact with it, by similar hydrophobic interactions.

3. The answer is C. (*III B 3–4*) The fluid-mosaic model of membrane structure envisages the lipid bilayer as being fluid with lateral movement of membrane components, both lipid and protein, being possible. Integral proteins are those firmly associated with the hydrophobic phase of the bilayer. The phospholipids, being amphipathic compounds with polar and nonpolar ends, orient with the former on the face nearest the aqueous phase (either the outside or the cytoplasmic side) and the latter within the hydrophobic phase. The outer and inner faces of the membrane are not identical but show variation in the type of phospholipid with the amino phospholipids predominating in the inner face and the choline phospholipids predominating in the outer face. They also differ in the distribution of protein. Carbohydrate moieties associated with membrane glycoproteins are on the outer surface of the membrane.

4. The answer is E. (*III B 3 b; IV B*) The lateral movement of membrane proteins is influenced by the fluidity of the membrane, which in turn is a function of temperature. Reducing the temperature of the medium in which the fluorescent-labeled heterokaryons are incubated prevents the lateral movement of the proteins by reducing the fluidity of the membrane. The lateral movement of membrane proteins does not, in general, require the expenditure of high-energy phosphate bonds nor does it involve the formation of, nor rupture of, covalent bonds. The lower temperature does not inactivate the fluorescence reaction.

5. The answer is A (1, 2, 3). (*III B 2 c*) Some membrane proteins can be extracted in soluble form from the membrane by relatively gentle procedures, such as a change in salt concentration or pH. These are called peripheral proteins, and they are found on the cytoplasmic, or inner, face of the red cell membrane, interacting with the polar ends of membrane phospholipids. Peripheral proteins may also be found on the outer surface of other cell membranes. Peripheral proteins may or may not be glycoproteins. If they are glycoproteins, the carbohydrate moieties are usually on the outer surface of the membrane.

6. The answer is E (all). (*III B 4, IV D*) Some membrane proteins are restricted in their lateral mobility within the membrane bilayer as their physiologic function requires this. Decreasing membrane fluidity, by whatever mechanism, reduces the lateral mobility of the membrane proteins. Attachment to cytoskeletal elements, or to structures forming a cell-to-cell junction, will also restrict movement of the protein in the bilayer. Cholesterol abolishes the transition from rigid to fluid states in membranes and restricts the movement of fatty acyl–containing components. These restrictions on lipid movement also tend to reduce the lateral movement of integral membrane proteins.

26
Protein Turnover

I. PATTERN OF NITROGEN ECONOMY. Proteolysis is the breakdown of protein to free amino acids. The **metabolic pathways of amino acid degradation**, that is, the catabolism of amino acids to ammonia and carbon dioxide (CO_2), are discussed in Chapter 27. Figure 26-1 illustrates the major pathways of amino acid use.

 A. **Amino acid pool.** The amino acid "pool" depicted in Figure 26-1 is an abstraction used to simplify the presentation. In actuality, the pool has several compartments, which vary in their patterns of amino acids as well as in their concentrations.

  1. **The concentration of free amino acids** in the intracellular compartment is considerably higher than the concentration in the extracellular compartment, the gradient being maintained by active transport of amino acids into cells (see section II A 4). The size of the gradient varies with different amino acids, being greatest for glutamate and glutamine. The patterns in the intracellular compartment are also different from one tissue to another.

  2. **The *total* amount of free amino acid** in the body is about 100 g of which 50% is glutamate plus glutamine and about 10% is essential amino acids.

  3. **Essential amino acids** are those whose carbon skeletons cannot be synthesized by the organism; they are, therefore, essential dietary factors. There are nine essential amino acids (Table 26-1).

  4. **Inputs to the amino acid pool** shown in Figure 26-1 are from
   a. Dietary protein
   b. Proteolysis of cellular protein

  5. **Outputs from the amino acid pool** are from
   a. Protein synthesis, the major drain on the pool

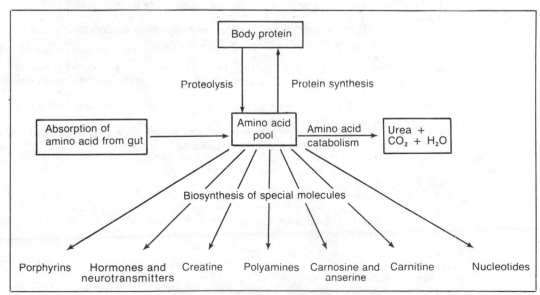

Figure 26-1. The major pathways of amino acid use.

Essential:
PUT
TIM HALL

Table 26-1. The Nine Essential Amino Acids

			Arginine
Histidine	Lysine	Threonine	
Isoleucine	Methionine	Tryptophan	
Leucine	Phenylalanine	Valine	

Note.—In growing children, arginine may not be synthesized in amounts adequate to fill the requirements for both protein synthesis and urea formation; under these circumstances, it would be considered an essential amino acid.

    b. Amino acid degradation (catabolism)
    c. Synthesis of special compounds

B. Protein content of the body

    1. Approximately half of the body protein (i.e., 6 kg to 7 kg) is associated with the skeleton and other supporting tissues, the major protein component being collagen, which turns over very slowly. Such protein is mainly extracellular.

    2. The other half of the body protein, mainly intracellular, is in a more dynamic state with continual protein synthesis being required to replace degraded protein.

    3. The plasma proteins, forming a small but vital part of the body protein, are extracellular but highly dynamic.

II. INPUTS TO THE AMINO ACID POOL

A. Gastrointestinal input

    1. About 70 g to 100 g of dietary protein and 35 g to 200 g of endogenous protein from sloughed-off intestinal mucosal cells and digestive enzymes are digested and absorbed daily.

    2. Polypeptides are not absorbed intact (except during a brief neonatal period) but must first be hydrolyzed to free amino acids.

    3. Proteolysis (digestion) of dietary protein is carried out by three groups of enzymes, classified according to their source.
      a. Gastric juice
        (1) The **hydrochloric acid** (pH 2) in gastric juice kills microorganisms, denatures proteins, and provides an acid environment for the action of pepsin.
        (2) The proteolytic enzyme **pepsin** works at acid pH. It is secreted into the stomach as the proenzyme **pepsinogen**, which is activated through the removal of 44 N-terminal amino acids by the high hydrogen ion concentration (**autoactivation**) or by pepsin itself (**autocatalysis**).
        (3) The major products of peptic hydrolysis of proteins are large peptides and some free amino acids.
        (4) The liberated peptides act as stimulants for the secretion of **cholecystokinin** in the duodenum, setting the stage for pancreatic proteolysis.
      b. Pancreatic juice
        (1) Endopeptidases and carboxypeptidases in pancreatic juice are secreted as proenzymes from the pancreatic acinar cells under the influence of **secretin** and **cholecystokinin–pancreozymin**.
        (2) Activation of the proenzymes occurs by the action of **enteropeptidase** (enterokinase), secreted by the duodenal epithelial cells.
        (3) Enteropeptidase activates **trypsinogen** by cleaving off six amino acids, and the activated **trypsin** in turn activates more trypsinogen. Trypsin also activates the proenzymes of the endopeptidases **chymotrypsin** and **elastase**, and the proenzymes of the exopeptidases **carboxypeptidase A and B** (Fig. 26-2).
      c. Intestinal proteases
        (1) Aminopeptidases and **dipeptidases** on the intestinal cell surface continue the digestion of protein to free amino acids.
        (2) Dipeptides and tripeptides are usually absorbed as such and are digested to free amino acids within the intestinal epithelial cells.

    4. Absorption takes place in the small intestine.
      a. Five separate systems have been identified for the transport of amino acids from the gut lumen into the intestinal epithelial cell.
        (1) One for neutral amino acids (alanine, valine, leucine, isoleucine, methionine, phenylalanine, tyrosine, and tryptophan)

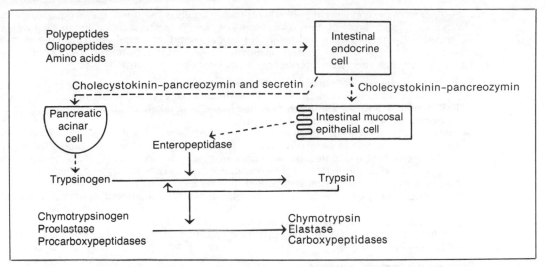

Figure 26-2. Secretion and activation of pancreatic enzymes. (Reprinted with permission from Freeman HJ, Kim YS: Digestion and absorption of protein. *Ann Rev Med* 29:102, 1978. ©Annual Reviews, Inc.)

    (2) One for basic amino acids (lysine, arginine, and cystine)
    (3) One for imino acids and glycine (proline, glycine, and hydroxyproline)
    (4) One for acidic amino acids (aspartic acid and glutamic acid)
    (5) One for β-amino acids (β-alanine and taurine)
  b. The transport is tightly coupled to the entry of Na^+ into the cell with the energy being derived from the electrochemical Na^+ gradient rather than from the adenosine triphosphate–dependent (ATP-dependent) transport of Na^+ out of the cell [i.e., the "downhill" entry of Na^+ can bring about the "uphill" concentration of amino acids in the brush border cells (see Chapter 14, section I C, sugar transport)].

B. Proteolysis of body protein. Body protein is continuously being broken down to free amino acids (proteolysis); the rate of degradation for individual proteins varies widely. Some liver enzymes have half-lives of only a few hours, whereas some structural proteins (e.g., collagen) have half-lives too long to be measured. Hemoglobin and the red blood cells have a total lifetime of 120 days.

  1. Factors affecting the rates of protein degradation
    a. The denaturation of protein (i.e., loss of its preferred native configuration) accelerates proteolytic breakdown.
    b. The activation of lysosomes increases the rate of intracellular proteolysis.
    c. Glucocorticoids increase protein degradation in muscle.
    d. Thyroid hormones, when in excess, increase protein turnover.
    e. Insulin reduces proteolysis and increases protein synthesis.

  2. The labile protein pool is the pool of proteins that turns over very rapidly.
    a. The carbon skeletons of amino acids from this pool can contribute to blood sugar formation (gluconeogenesis) and to energy (see Chapter 27, section II).
    b. When dietary protein is insufficient, the synthesis of proteins in this pool cannot match their breakdown. At such times, the excretion of the amino acid nitrogen as urea accounts for most of the increased nitrogen excretion seen early in starvation.

  3. Estimation of protein breakdown
    a. Muscle protein. Some of the histidine residues incorporated into the muscle protein complex, **actomyosin**, are methylated after their incorporation. When actomyosin breaks down, it liberates 3-methyl histidine, which is excreted unchanged in the urine. Thus, urinary **3-methyl histidine excretion** gives a measure of muscle protein breakdown.
    b. Plasma proteins. Turnover can be determined by measuring the exponential decline in protein-bound radioactivity in plasma after injection of a ^{125}I-labeled protein. For example, the normal half-life of plasma albumin is 20 days.

III. OUTPUTS FROM THE AMINO ACID POOL

  A. Protein synthesis. The use of amino acids as substrates for the synthesis of proteins represents a major output from the amino acid pool. This output in adults on a normal diet is equal to the input from proteolysis of body protein. The process of protein synthesis described in Chapter 4 is summarized in Figure 26-3. Four high-energy phosphate bonds are required to form the peptide bond, and if the cost of nucleic acid involvement is taken into account, the overall cost is approximately five high-energy phosphate bonds per peptide bond formed.

1. **Estimation of the rate of whole-body protein synthesis**
   a. Because the rate of protein synthesis in the steady state equals the rate of proteolysis, the **half-time ($t_{1/2}$) of the degradation rate**, that is, the time taken for 50% of the protein to be degraded, provides a measure of turnover.
   b. The turnover rate of whole-body protein must reflect the rates of synthesis both for the proteins that are turning over rapidly and for those turning over slowly. In most studies, the time scale, which is dictated by practical constraints, is too short to allow an adequate measure of the synthesis rate for the component turning over slowly and, therefore, overemphasizes that for the faster component.
   c. Protein turnover can be estimated by measuring the ^{15}N-enrichment in urea after the administration of ^{15}N-enriched amino acids. Because ^{15}N is not a radioactive isotope, the method is suitable for studies involving malnourished children.
   d. In adults on a normal diet, the rate of protein synthesis ranges in value from 200 g to 300 g of protein per 24 hr.

2. **Regulation of protein synthesis**
   a. *All* of the twenty common amino acids incorporated into protein must be present in adequate concentrations for intracellular protein synthesis to occur. Factors that influence amino acid uptake by cells will thus influence protein synthesis; for example, insulin increases amino acid transport into cells.
   b. Insulin must also be present because protein synthesis without it is severely inhibited in most tissues, such as
      (1) **Muscle**, primarily because of a decrease in the number of ribosomes and in their ability to carry out peptide chain initiation
      (2) **Liver**, primarily because of a decrease in the mRNA that codes for the synthesis of secretory proteins
   c. Regulation also occurs by control of gene expression with consequent formation of specific mRNA, which is translated at the ribosome during the process of protein synthesis.

B. **Catabolism of amino acids**

1. **Degradation of part of the amino acid pool** to urea and CO_2 is a continuous drain on the pool. Amino acid degradation is reduced in starvation but is never turned off; there must be a daily intake of essential amino acids in order to replace the amino acids lost from the body pool. The metabolic pathways involved in the degradation of amino acids are discussed in Chapter 27. At this time, a quantitative picture of the nitrogen loss is necessary in order to strike a balance between nitrogen intake (dietary protein) and nitrogen loss.

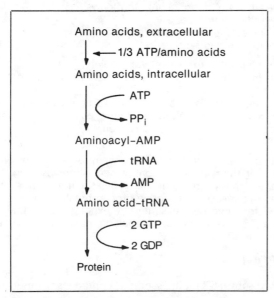

Figure 26-3. Summary of the process of protein synthesis (see Chapter 4 for details). *AMP, ATP* = adenosine mono- and triphosphates; *GDP, GTP* = guanosine di- and triphosphates; *PP$_i$* = inorganic pyrophosphate; and *tRNA* = transfer RNA.

2. Routes of nitrogen loss
   a. Excretion in the urine disposes of about 90% of the nitrogen lost from the body with urinary urea accounting for 70% to 85% of the total urinary nitrogen, depending upon nitrogen intake. Table 26-2 gives typical values for urinary nitrogen losses at three different levels of protein intake.
   b. Losses also occur through the skin, mainly as sweat, and via the feces (small amounts). With normal sweating, the 70-kg man loses about 2 g of nitrogen daily through the skin and feces.

IV. NITROGEN BALANCE. If the total daily nitrogen losses in urine, skin, and feces are equal to the total daily nitrogen intake, the subject is said to be in **nitrogen balance**, as a healthy, adequately fed adult should be.   *WHICH IS O.*

A. Positive nitrogen balance. If nitrogen losses are *less* than intake, the subject is in positive nitrogen balance, as healthy, growing children and convalescing adults should be.

B. Negative nitrogen balance. If nitrogen losses are *greater* than intake, the subject is in negative nitrogen balance, as in diseases involving tissue wasting or in starvation. Prolonged periods of negative balance are dangerous and are fatal if the loss of body protein reaches about one-third of the total body protein.

Table 26-2. Variations in Some Nitrogenous Urinary Constituents with Different Protein Intakes

	Normal Protein Intake		Protein-Rich Diet		Protein-Poor Diet	
	g	%N	g	%N	g	%N
Total urinary nitrogen	13.2	100.0	23.28	100.0	4.2	100.0
Protein represented by total urinary nitrogen	82.5	. . .	145.5	. . .	26.25	. . .
Urea nitrogen	11.36	86.1	20.45	87.9	2.9	69.1
Ammonia nitrogen	0.4	3.0	0.82	3.5	0.17	4.0
Creatinine nitrogen	0.61	4.6	0.64	2.7	0.6	14.3
Uric acid nitrogen	0.21	1.6	0.3	1.3	0.11	2.6
Undetermined nitrogen	0.62	4.7	1.07	4.6	0.42	10.0

STUDY QUESTIONS

Directions: Each question below contains five suggested answers. Choose the **one best** response to each question.

1. All of the following statements regarding the digestion of protein are correct EXCEPT

(A) pepsinogen is activated by autoactivation (pH 2) or autocatalysis
(B) the major products of peptic hydrolysis are large peptides and some free amino acids
(C) trypsin and chymotrypsin are secreted by the pancreas as inactive zymogens
(D) enteropeptidase (enterokinase) activates pancreatic trypsinogen
(E) most of the digested protein is absorbed in the intestine in the form of polypeptides

2. The rate of excretion of certain substances in the urine on a 24-hr basis can be used to estimate the rate of muscle protein loss. Which of the following substances falls into this category?

(A) Urea
(B) Uric acid
(C) 3-Methyl histidine
(D) Ammonia
(E) None of the above

3. A patient undergoing convalescence is in positive nitrogen balance if

(A) the amount of nitrogen ingested equals the amount of nitrogen excreted in the urine, feces, and sweat
(B) the amount of nitrogen excreted in the urine, feces, and sweat is less than the nitrogen ingested
(C) the amount of nitrogen excreted in urine, feces, and sweat is greater than the amount of nitrogen ingested
(D) the patient is fed a protein-free diet
(E) none of the above are true

Directions: Each question below contains four suggested answers of which **one or more** is correct. Choose the answer

A if **1, 2, and 3** are correct
B if **1 and 3** are correct
C if **2 and 4** are correct
D if **4** is correct
E if **1, 2, 3, and 4** are correct

4. Which of the following constitute essential amino acids; that is, which must be ingested on a daily basis by humans?

(1) Leucine
(2) Cysteine
(3) Phenylalanine
(4) Serine

5. Actions or substances that increase protein breakdown include

(1) denaturation of protein
(2) glucocorticoids
(3) excess thyroid hormone
(4) activation of lysosomes

6. A lack of insulin restricts protein synthesis because of a decrease in the

(1) number of ribosomes in muscle
(2) ability of muscle ribosomes to carry out peptide initiation
(3) liver mRNA coding for secretory proteins
(4) rate of entry of amino acids into muscle cells

ANSWERS AND EXPLANATIONS

1. The answer is E. (*II A*) Protein digestion is carried out by proteases found in gastric, pancreatic, and intestinal secretions. In the stomach, pepsinogen is activated by the low pH (due to secreted hydrochloric acid) to yield active pepsin. The major products of peptic hydrolysis are large peptides and some free amino acids. In the duodenum, trypsinogen from the pancreas is activated first by enterokinase and then by trypsin itself, which also activates chymotrypsinogen. In the intestine, proteolysis continues until the protein is digested to free amino acids and some dipeptides. Apart from a brief neonatal period, polypeptides are not absorbed as such but have to be broken down to free amino acids or, in some cases, to dipeptides.

2. The answer is C. (*II B 3*) Some of the histidine residues incorporated into muscle actomyosin are methylated after their incorporation (post-translational modification). When these proteins are broken down to free amino acids (proteolysis), 3-methyl histidine is released and excreted in the urine unchanged. The 3-methyl histidine output in the urine per 24 hr is, thus, a measure of muscle protein loss. Urea is the form in which most of the nitrogen in the amino acids that are degraded to carbon dioxide and ammonia (NH_3) is excreted in the urine. It is, thus, not a specific measure of muscle protein loss. Uric acid is the waste product of purine catabolism. NH_3 excretion is normally low, as most NH_3 (which is toxic) is transformed into urea (which is nontoxic).

3. The answer is B. (*IV A*) If the total daily losses of nitrogen in urine, feces, and sweat are equal to the total daily intake of nitrogen, a subject is in nitrogen balance. A patient recovering from an illness involving the loss of tissue protein (i.e., there was a period of negative nitrogen balance) should be in positive nitrogen balance during convalescence. During this time, the total nitrogen losses would be less than the daily nitrogen intake. A protein-free diet is essentially a nitrogen-free diet.

4. The answer is B (1, 3). (*I A 3; Table 26-1*) The carbon skeletons of the essential amino acids cannot be synthesized by human tissues. These amino acids have to be included in the diet on a daily basis, as there is little storage of free amino acids. The nine amino acids that are essential for humans are leucine, isoleucine, and valine (the three so-called branched-chain amino acids), methionine, threonine, lysine, histidine, phenylalanine, and tryptophan.

5. The answer is E (all). (*II B 1*) Protein degradation occurs more readily if a protein is denatured (i.e., has lost its preferred native configuration). The activation of lysosomes in cells increases the rate of proteolysis, and the administration of glucocorticoids in muscle does likewise. Excess thyroid hormone levels increase protein turnover and thus proteolysis. Insulin, on the other hand, increases protein synthesis and reduces proteolysis.

6. The answer is E (all). (*III A 2*) Insulin enhances amino acid transport into cells, and a lack of the hormone reduces protein synthesis because of this. Insulin deficiency is also responsible for a reduction in the number and the functional capacity of muscle ribosomes, and for the reduced formation of mRNA in the liver.

27

Amino Acid Degradation

I. DISPOSAL OF AMINO ACID NITROGEN

A. Enzymatic removal of nitrogen from amino acids

1. **Transamination.** This type of reaction involves the transfer of an amino group from an amino acid to an α-keto acid to form a new amino acid and a new α-keto acid.

   a. The enzymes, **transaminases (aminotransferases)**, have equilibrium constants between 1 and 10 and are widely distributed. For example, aspartate aminotransferase catalyzes the following reaction:

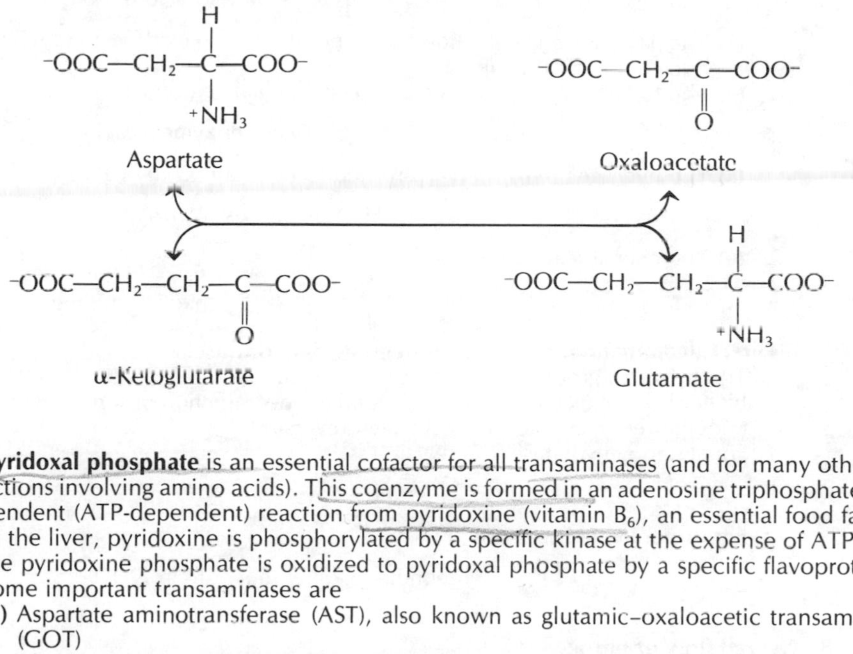

   b. **Pyridoxal phosphate** is an essential cofactor for all transaminases (and for many other reactions involving amino acids). This coenzyme is formed in an adenosine triphosphate–dependent (ATP-dependent) reaction from pyridoxine (vitamin B_6), an essential food factor. In the liver, pyridoxine is phosphorylated by a specific kinase at the expense of ATP, and the pyridoxine phosphate is oxidized to pyridoxal phosphate by a specific flavoprotein.

   c. Some important transaminases are

   (1) Aspartate aminotransferase (AST), also known as glutamic–oxaloacetic transaminase (GOT)

   (2) Alanine aminotransferase (ALT), also known as glutamic–pyruvic transaminase (GPT)

   (3) α-Amino acid–α-ketoglutarate aminotransferase, which is specific for α-ketoglutarate and glutamate and nonspecific for the other α-amino acid–α-keto acid pair

   (4) Amino acid–pyruvate aminotransferase, which is specific for pyruvate and alanine and nonspecific for the other α-amino acid–α-keto acid pair

2. **Oxidative deamination**

   a. **Glutamate dehydrogenase**

   (1) This enzyme catalyzes the following reaction:

   $$\text{Glutamate} + \text{NAD(P)}^+ + H_2O \rightleftharpoons \alpha\text{-ketoglutarate} + \text{NAD(P)H} + H^+ + NH_3$$

   (2) Glutamate dehydrogenase is located in the mitochondrial matrix space.

   (3) The enzyme is polymorphic, its base form being a hexamer made up of identical peptides.

(4) The hexamer can use oxidized nicotinamide-adenine dinucleotide (NAD^+) or NAD phosphate ($NADP^+$) as cofactor and can bind guanosine triphosphate (GTP) and adenosine diphosphate (ADP) as well as substrates.

(5) Under normal conditions in the liver, the reduced NADP (NADPH):$NADP^+$ ratio is high and the reduced NAD^+ (NADH):NAD^+ ratio is low. Thus, there is a pyridine nucleotide coenzyme available in its oxidized state to participate in the reaction in either direction.

(6) The reaction is controlled, therefore, by the relative levels of glutamate, α-ketoglutarate, and ammonia (NH_3).

(7) The allosteric inhibitors, NADH and GTP, prevent the depletion of mitochondrial α-ketoglutarate when the NADH:NAD^+ ratio is high as it is during rapid β-oxidation of fatty acids.

(8) When GTP and NADH are bound, the enzyme changes its substrate specificity and is active as a dehydrogenase toward alanine and other monocarboxylic amino acids with a concomitant lowering of glutamate dehydrogenase activity. However, the change in substrate specificity chiefly affects α-ketoglutarate levels, as the alanine dehydrogenase activity is only 2% of the glutamate dehydrogenase activity.

(9) The binding of ADP by the enzyme enchances glutamate dehydrogenase activity.

b. Amino acid oxidases

(1) Although most of the activity of liver and kidney homogenates toward L-amino acids is due to the coupled action of transaminases plus L-glutamate dehydrogenase, both L- and D-amino acid oxidase activities do occur in mammalian liver and kidney; their physiologic significance, however, is not clear.

(2) The reaction is

$$\text{Amino acid} + H_2O + \text{enzyme—flavin} \rightarrow \alpha\text{-keto acid} + NH_3 + \text{enzyme—flavin } H_2$$

(3) The flavin is riboflavin phosphate for L-amino acid oxidase and FAD (flavin-adenine dinucleotide) for D-amino acid oxidase.

(4) The reduced flavin is reoxidized by molecular oxygen.

$$\text{Enzyme—flavin } H_2 + O_2 \rightarrow \text{enzyme—flavin} + H_2O_2$$

(5) The hydrogen peroxide is converted to H_2O and O_2 by catalase.

$$H_2O_2 \rightarrow H_2O + \tfrac{1}{2} O_2$$

(6) The overall reaction is

$$\text{Amino acid} + \tfrac{1}{2} O_2 \rightarrow \text{keto acid} + NH_3$$

3. Direct deamination: serine and threonine dehydratase

a. The hydroxyl group in the side chain of serine and threonine gives these amino acids a higher level of oxidation than most, and the deamination of serine and threonine liberates enough free energy to be essentially irreversible.

b. For serine dehydratase, the reaction is

$$\text{Serine} \rightarrow \text{pyruvate} + NH_4^+$$

c. For threonine dehydratase, it is

$$\text{Threonine} \rightarrow \alpha\text{-ketobutyrate} + NH_4^+$$

B. Overall flow of nitrogen

1. Liver

a. Pathways of nitrogen flow in the liver are

$$\begin{array}{ccccc} \text{Alanine} & & \alpha\text{-Ketoglutarate} & NH_3 & \text{Carbon dioxide } (CO_2) \\ \\ \text{Pyruvate} & & \text{Glutamate} & & \text{Urea} \end{array}$$

b. Role of glutamate dehydrogenase

(1) High levels of ADP are associated with a low-energy charge, and this effector serves to increase the activity of the tricarboxylic acid (TCA) cycle via an increase in α-ketoglutarate.

(2) The low-energy charge is probably accompanied by an increased protein breakdown and the need for glutamate dehydrogenase activity in the direction of NH_3 production with the latter rapidly removed in liver mitochondria as urea.

(3) High levels of GTP and NADH, reflecting a high-energy charge and a limited protein breakdown, favor a slowdown of the glutamate dehydrogenase reaction.

2. Skeletal muscle. Pathways of nitrogen flow in skeletal muscle are

$$\alpha\text{-Amino acid} \quad \alpha\text{-Ketoglutarate} \quad \text{Alanine}$$
$$\alpha\text{-Keto acid} \quad \text{Glutamate} \quad \text{Pyruvate}$$

3. Transport from muscle to liver. The major transporters of amino acid nitrogen from muscle to liver are, by far, **glutamine** and **alanine**.
   a. Alanine
      (1) In skeletal muscle, the distribution of transaminases favors the collection of nitrogen from amino acids (largely the branched-chain amino acids) into **alanine**, which leaves the muscle and is taken up by the liver.
      (2) In the liver, alanine is a major gluconeogenic precursor (see Chapter 17, section IV B), with the amino group being transaminated to α-ketoglutarate to form glutamate.
   b. Glutamine
      (1) Glutamine is formed by glutamine synthetase in an ATP-dependent reaction.

$$\text{Glutamate} + NH_3 + ATP \rightarrow \text{glutamine} + ADP + P_i$$

      (2) Role of glutamine in other tissues
         (a) Glutamine is the major mode of disposing of NH_3 from the brain, and it is also formed and excreted by the kidney as part of the mechanism of acid–base control.
         (b) In the liver and in the kidney, glutamine can lose its amide group as NH_3 by the glutaminase reaction.

$$\text{Glutamine} + H_2O \rightarrow \text{glutamate} + NH_4^+$$

C. Urea cycle. The liver is the exclusive site of synthesis of urea, although the enzymatic machinery for producing arginine, the precursor of urea, is present elsewhere; for example, in brain, kidney, and skin.

1. Reactions of the urea cycle
   a. The formation of carbamoyl phosphate provides entry into the cycle.

$$NH_4^+ + HCO_3^- + 2\,ATP \rightarrow H_2N-\underset{\underset{O}{\|}}{C}-O-\underset{\underset{O}{\|}}{\overset{\overset{O^-}{|}}{P}}-O^- + 2\,ADP + P_i$$

Carbamoyl phosphate

      (1) The enzyme responsible is carbamoyl phosphate synthetase (ammonia), which is located in the mitochondrial matrix.
      (2) Two molecules of ATP are hydrolyzed, and the reaction is virtually irreversible.
      (3) N-acetylglutamate is a required positive allosteric effector.
   b. Ornithine transcarbamoylase transfers the carbamoyl group to the δ-amino group of ornithine to form citrulline. This is also a mitochondrial matrix reaction.

$$H_2N-\underset{\underset{O}{\|}}{\overset{\overset{O}{\|}}{C}}-O-\underset{\underset{O}{\|}}{\overset{\overset{O^-}{|}}{P}}-O^- + H_3N^+-(CH_2)_3-\overset{\overset{+NH_3}{|}}{CH}-COO^- \rightarrow H_2N-\underset{\underset{}{\overset{\overset{O}{\|}}{C}}}-NH-(CH_2)_3-\overset{\overset{+NH_3}{|}}{CH}-COO^- + P_i$$

Carbamoyl phosphate   Ornithine   Citrulline

   c. Citrulline is released from the mitochondria into the cytosol where it combines with aspartate and ATP to form argininosuccinate.

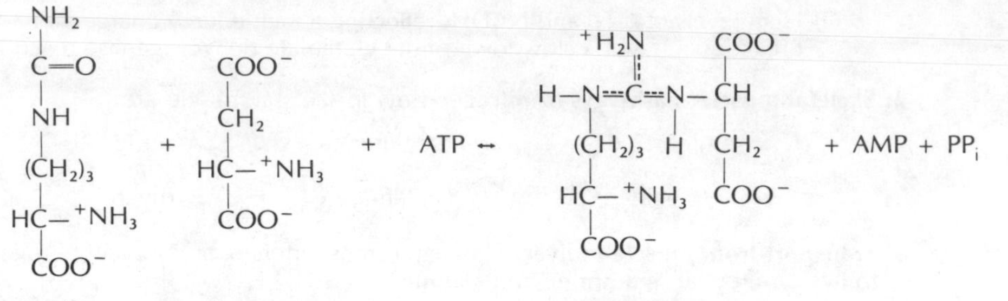

Citrulline   Aspartate   Argininosuccinate

(1) The catalyst for this reaction is argininosuccinate synthetase.
(2) The pyrophosphate formed in the reaction is a strong inhibitor of the enzyme, but it is rapidly cleaved to inorganic phosphate (P_i), rendering the reaction irreversible under usual cellular conditions.
(3) The aspartate for the reaction arises from oxaloacetate by transamination with cytosolic glutamate.

d. Argininosuccinate is cleaved by argininosuccinate lyase, forming arginine and fumarate.

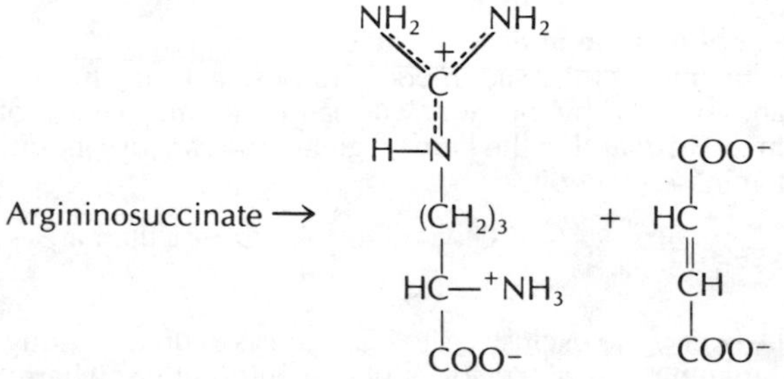

Argininosuccinate →

Arginine   Fumarate

e. The fumarate is converted to malate, which may enter the mitochondria or, which, by the action of cytosolic malate dehydrogenase, may give rise to oxaloacetate for further aspartate generation by transamination.

f. Arginine is then cleaved by arginase to yield urea and ornithine.

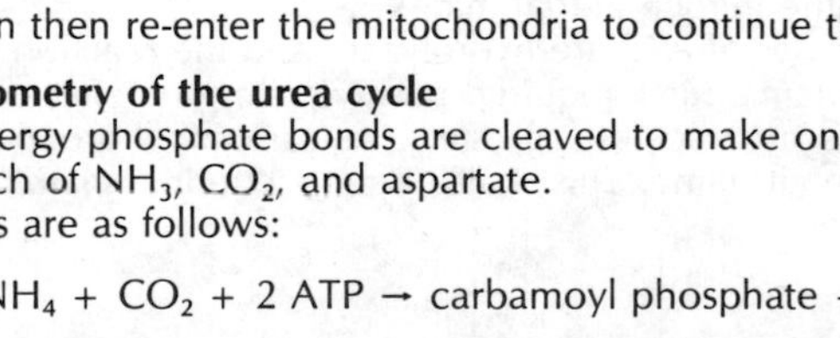

Arginine + H_2O →

Urea   Ornithine

g. Ornithine can then re-enter the mitochondria to continue the cycle.

2. Overall stoichiometry of the urea cycle
   a. Four high-energy phosphate bonds are cleaved to make one molecule of urea from one molecule each of NH_3, CO_2, and aspartate.
   b. The reactions are as follows:

$$^+NH_4 + CO_2 + 2\ ATP \rightarrow carbamoyl\ phosphate + 2\ ADP + P_i$$

$$Carbamoyl\ phosphate + ornithine \rightarrow citrulline + P_i$$

$$Citrulline + aspartate + ATP \rightarrow argininosuccinate + AMP + PP_i$$

$$AMP + ATP \rightarrow 2\ ADP$$

$$PP_i + H_2O \rightarrow 2\ P_i$$

$$Argininosuccinate \rightarrow arginine + fumarate$$

$$Arginine + H_2O \rightarrow urea + ornithine$$

c. The sum of the reactions is

$$^+NH_4 + CO_2 + 4\ ATP + aspartate \rightarrow urea + fumarate + 4\ ADP + 4\ P_i$$

3. Compartmentalization of the urea-cycle enzymes

    a. The fumarate formed in the cytosol at the argininosuccinate lyase step can be converted to aspartate by both cytoplasmic and mitochondrial enzyme systems.

        (1) These reactions involve the hydration of fumarate to malate, followed by oxidation of the malate to oxaloacetate. The latter will form aspartate on transamination with glutamate.

        (2) Because aspartate, malate, and glutamate can be transported in and out of the mitochondria, both mitochondrial and cytoplasmic amino acids can contribute to the maintenance of the urea cycle.

    b. Figure 27-1 summarizes the reactions supporting urea synthesis and the pathway of the carbons to glucose synthesis by way of oxaloacetate.

4. Interrelationships between the urea and TCA cycles (see Fig. 27-1)

    a. Both nitrogens of urea are ultimately derived from glutamate, which is present in both the mitochondrial matrix and the cytosol.

    b. In the cytosol, aspartate is formed by transamination of glutamate with cytosolic oxaloacetate.

    c. In the mitochondrial matrix, NH_3 can arise from glutamate via glutamate dehydrogenase and from glutamine via mitochondrial glutaminase.

5. Regulation of the urea cycle

    a. The levels of the urea-cycle enzymes fluctuate with changes in feeding patterns.

        (1) With an essentially protein-free diet, urea excretion accounts for only 60% of the total urinary nitrogen (as opposed to 80% with a normal diet), and the levels of all urea-cycle enzymes decline.

        (2) With a high-protein diet or in starvation (when gluconeogenesis from amino acids is high), the levels of the urea-cycle enzymes increase several-fold.

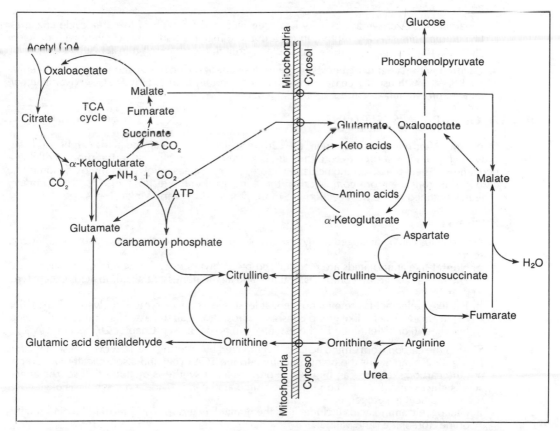

Figure 27-1. Interconversions between the urea cycle and the tricarboxylic acid cycle. (Adapted from Devlin TM: *Textbook of Biochemistry with Clinical Correlations,* 2nd ed. New York, John Wiley, 1986, p 448.)

b. The supplying of carbamoyl phosphate to the urea cycle is regulated by the positive allosteric effector, N-acetylglutamate. The activity of acetylglutamate synthetase, the enzyme that synthesizes N-acetylglutamate, is markedly increased by amino acids, particularly arginine.

D. NH₃ toxicity: hyperammonemia

1. Hyperammonemia is associated with comatose states such as may occur in hepatic failure.
   a. Acquired hyperammonemia is usually the result of cirrhosis of the liver with the development of a collateral circulation, which shunts the portal blood around the organ, thereby severely reducing the synthesis of urea.
   b. Inherited hyperammonemia results from genetic defects in the urea-cycle enzymes (see section I E).

2. Failure of the liver to remove NH_3 from the portal blood is particularly dangerous, as the portal blood levels of NH_3 are normally much higher than peripheral blood levels because bacterial urease activity in the gut hydrolyzes urea to NH_3 and CO_2.

3. The induction of the comatose state is believed to be due to the hyperammonemia itself, although there is no fully accepted rationale for the toxicity of NH_3.
   a. One possibility is that α-ketoglutarate in the brain is depleted during the removal of NH_3 by glutamate dehydrogenase with a consequent reduction in the respiratory activity of the brain cells.
   b. Another possibility is the transamination of glutamine in the brain to form a presumably toxic compound, α-ketoglutaramate, which normally is not a significant process because glutamine levels in the brain are usually low. However, there is no evidence of a relationship between α-ketoglutaramate levels and the clinical symptoms.

E. Genetic defects of urea-cycle enzymes

1. Patients with a total lack of one of the enzymes involved in urea synthesis do not survive the neonatal period.

2. A partial lack of enzymatic activity has been recorded for each of the urea-cycle enzymes. Hyperammonemia is a symptom in most of these patients and is most severe with deficiencies of carbamoyl phosphate synthetase and ornithine carbamoyl transferase.

3. Children with partial deficiencies of the urea-synthesizing enzymes show episodic encephalopathies, which usually cease on restriction of protein intake. However, survival is uncertain, as protein intake cannot be totally reduced.

II. FATE OF CARBON SKELETONS

A. Overview.
After removal of nitrogen, the carbon skeletons of the amino acids can be oxidized, providing an energy source even if they are first used to make glucose in the process of gluconeogenesis. There are 20 amino acids; therefore, the processes involved in the oxidation of the carbon skeletons are more complex than those involved in the disposal of NH_3. The summary schemes outlined in the next section are those proposed by Van Winkle.*

B. Pathways

1. Amino acids can be arranged in two groups based on their general pathways of use as energy sources.
   a. Nonfat-like amino acids are metabolized by pathways of carbohydrate metabolism only. The nonfat-like group comprises histidine and all the nonessential amino acids except tyrosine.
   b. Fat-like amino acids are metabolized, at least in part, by enzymes involved in β-oxidation of fatty acids. The fat-like group consists of tyrosine and all the essential amino acids except histidine. (For a definition of the essential amino acids, see Chapter 26, section I A 3.)

2. Most of the carbon from amino acids is metabolized by entering the TCA cycle at some point. Figure 27-2 summarizes the points of entry into the TCA cycle and also indicates whether or not the carbon skeleton can be used to provide a net synthesis of glucose. If so, the amino acid is **glucogenic**. If no carbon from the amino acid can provide a net synthesis of glucose, the amino acid is **ketogenic**.
   a. Glucogenic amino acids include all of the nonfat-like group plus three fat-like ones (methionine, threonine, and valine).

---

*Van Winkle LJ: A framework for teaching amino acid metabolism. *Biochem Educ* 8:72-75, 1980.

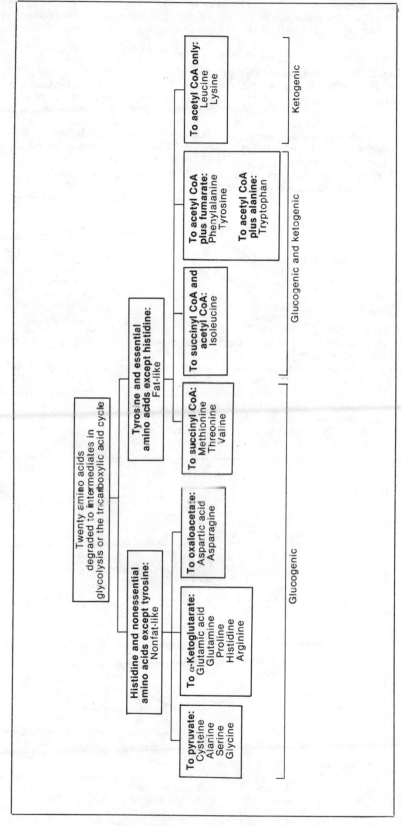

Figure 27-2. The degradation of nonfat-like and fat-like amino acids. (Reprinted with permission from Van Winkle LJ: A framework for teaching amino acid metabolism. *Biochem Educ* 8:72–75, 1980.)

b. The ketogenic amino acids are leucine and lysine.

c. Four are both glucogenic and ketogenic: isoleucine, phenylalanine, tyrosine, and tryptophan (some also include lysine in this category).

C. Points of entry of amino acid carbon into the TCA cycle

1. Degradation of the nonfat-like group. The catabolic pathways for this group of amino acids are summarized in Figure 27-3. All of the amino acids of this group are glucogenic, as they either form pyruvate or enter the TCA cycle as four-carbon compounds. Two of the reactions shown in the figure are **anaplerotic reactions**; that is, they replace intermediates in the TCA cycle (see Chapter 15, section IV B). These reactions are glutamate to α-ketoglutarate and aspartate to oxaloacetate. These pathways are also **amphibolic**; that is, they are involved in both the catabolic and anabolic pathways for these nonessential amino acids.

a. Entry at α-ketoglutarate

   (1) Proline is broken down to glutamate (the pyrrole ring is opened). (Proline synthesis is by a similar but not identical route.)

   (2) Arginine is cleaved by arginase to yield urea and ornithine, after which ornithine (if not reused for urea synthesis) can be converted to glutamate via glutamate semialdehyde.

$$\underset{\text{Ornithine}}{^+H_3N-(CH_2)_3-\underset{\underset{^+NH_3}{|}}{CH}-COO^-} \xrightleftharpoons[\text{α-ketoglutarate}]{\text{glutamate}} \underset{\text{Glutamate semialdehyde}}{HC-(CH_2)_2-\underset{\underset{^+NH_3}{|}}{\overset{\overset{O}{||}}{CH}}-COO^-}$$

   (3) Histidine [see section III B 5 b (2)]

   (4) Glutamine enters at α-ketoglutarate after hydrolysis by glutaminase to glutamate.

b. Entry at pyruvate

   (1) Alanine transaminates with α-ketoglutarate (through the action of ALT and other aminotransferases) to yield pyruvate.

   (2) Cysteine is metabolized to pyruvate by three different pathways in humans; the major route is uncertain. The sulfur ends up as inorganic sulfate.

   (3) Serine yields pyruvate through direct deamination by serine dehydratase (see section I A 3). **Glycine** and serine are interconverted in an important reaction described in sections III B 5 a and III C 3 a (8).

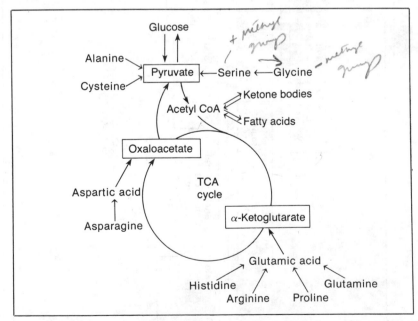

Figure 27-3. Points of entry of the carbons from the nonfat-like amino acids into the tricarboxylic acid (TCA) cycle. (Reprinted with permission from Van Winkle LJ: A framework for teaching amino acid metabolism. *Biochem Educ* 8:72–75, 1980.)

    c. Entry at oxaloacetate
       (1) <u>Aspartate transaminates with α-ketoglutarate to yield oxaloacetate.</u>
       (2) **Asparagine** is converted to aspartate by asparaginase to form NH_3.

2. Degradation of the fat-like group. The catabolic pathways for this group of amino acids are summarized in Figure 27-4. Some of the amino acids in this group are glucogenic, some are ketogenic, and some are both. The amino acids that are ketogenic have some of their carbons entering at acetyl coenzyme A (acetyl CoA) or acetoacetyl CoA (see Fig. 27-4). Two of the reactions are anaplerotic, namely, the entry of fumarate from tyrosine degradation and of succinyl CoA from the degradation of valine, methionine, threonine, and isoleucine. In contrast to the nonfat-like group, all but one of the fat-like amino acids are essential food factors because man does not have the enzymes required for the synthesis of their carbon skeletons. Their biosynthesis will, therefore, not be considered here.

    a. Entry at succinyl CoA
       (1) The carbons of four glucogenic amino acids—valine, methionine, threonine, and isoleucine—enter the TCA cycle at succinyl CoA. Isoleucine also gives rise to acetyl CoA and is, therefore, both ketogenic and glucogenic.
       (2) Both **threonine** and **methionine** are converted to α-ketobutyrate, threonine by direct deamination and methionine via cystathionine (see section III C 4).
       (3) α-Ketobutyrate is first *decarboxylated* to propionyl CoA by the pyruvate dehydrogenase complex.

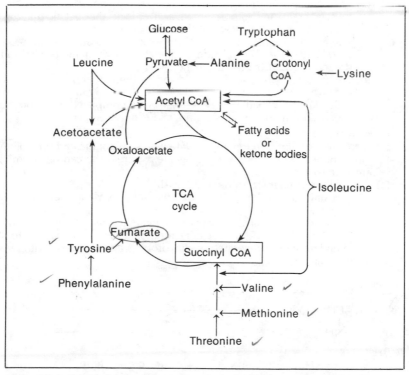

Figure 27-4. Points of entry of the carbons from the fat-like amino acids into the tricarboxylic acid (TCA) cycle. (Reprinted with permission from Van Winkle LJ: A framework for teaching amino acid metabolism. *Biochem Educ* 8:72–75, 1980.)

(4) Propionyl CoA is then *carboxylated* by a biotin-requiring carboxylase to form (S)-methylmalonyl CoA.

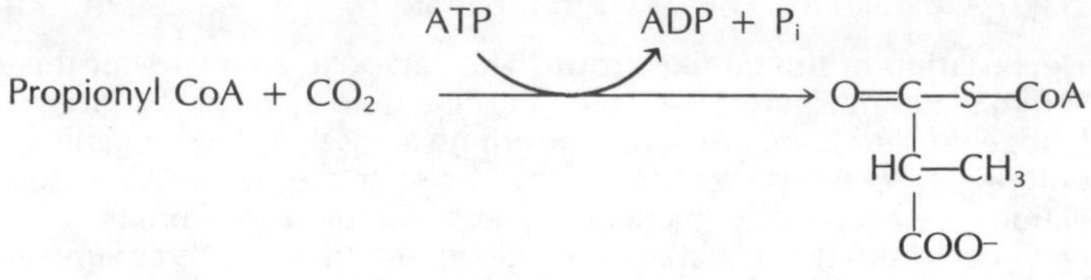

$$\text{Propionyl CoA} + CO_2 \xrightarrow[\text{ATP} \quad\quad \text{ADP} + P_i]{} \begin{array}{c} O{=}C{-}S{-}CoA \\ | \\ HC{-}CH_3 \\ | \\ COO^- \end{array}$$

(S)-Methylmalonyl coenzyme A

(5) (S)-Methylmalonyl CoA is racemized to the (R) form, which is then acted on by a vitamin B_{12}–dependent mutase to form succinyl CoA.

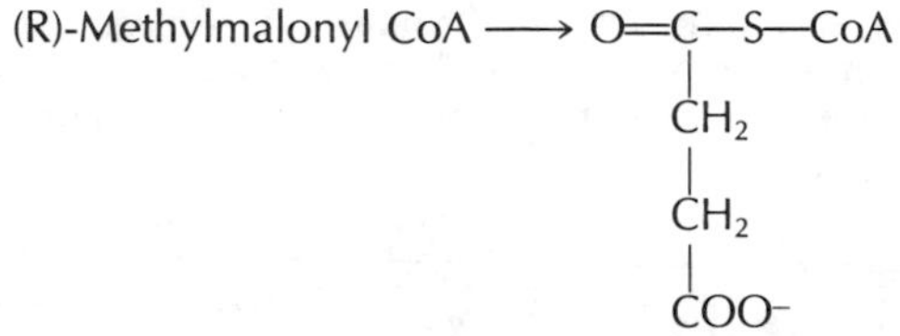

$$\text{(R)-Methylmalonyl CoA} \longrightarrow \begin{array}{c} O{=}C{-}S{-}CoA \\ | \\ CH_2 \\ | \\ CH_2 \\ | \\ COO^- \end{array}$$

Succinyl coenzyme A

(6) **Valine** and **isoleucine** are two of the branched-chain amino acids, and their catabolism is discussed next as a group. In terms of the fate of the carbon skeletons, valine is metabolized to propionyl CoA and isoleucine to propionyl CoA and acetyl CoA.

(7) **Catabolism of the branched-chain amino acids**
  (a) The branched-chain amino acids are **leucine**, **isoleucine**, and **valine**.
  (b) The transamination of the branched-chain amino acids occurs mainly in skeletal muscle, brain, and adipose tissue, as liver lacks the specific α-ketoglutarate–linked aminotransferase required for their transamination.
  (c) The α-keto acids, which result from the transamination reactions, are oxidized by an α-keto acid dehydrogenase complex similar in many respects to pyruvate dehydrogenase (see Chapter 15, section II A).
  (d) The fatty acyl CoAs are then oxidized by the mitochondrial β-oxidation enzymes, although the fatty acyl CoA derivative of leucine requires a carboxylation step in order for the β-oxidation sequence to be completed. Leucine carbons enter at acetyl CoA; therefore, this amino acid is only ketogenic.

b. Entry at fumarate
  (1) **Tyrosine** is catabolized to acetoacetate with some of the carbons splitting off as fumarate. These latter carbons are glucogenic, whereas the carbons forming acetoacetate are ketogenic.
  (2) **Catabolism of phenylalanine and tyrosine**
    (a) **Tyrosine** is formed by the hydroxylation of **phenylalanine** and is, therefore, a non-essential amino acid.
    (b) The reaction is catalyzed by phenylalanine hydroxylase, which requires as a cofactor **tetrahydrobiopterin**. The cofactor is oxidized in the reaction to dihydrobiopterin and must be regenerated by dihydrobiopterin reductase with NADPH as the reductant.

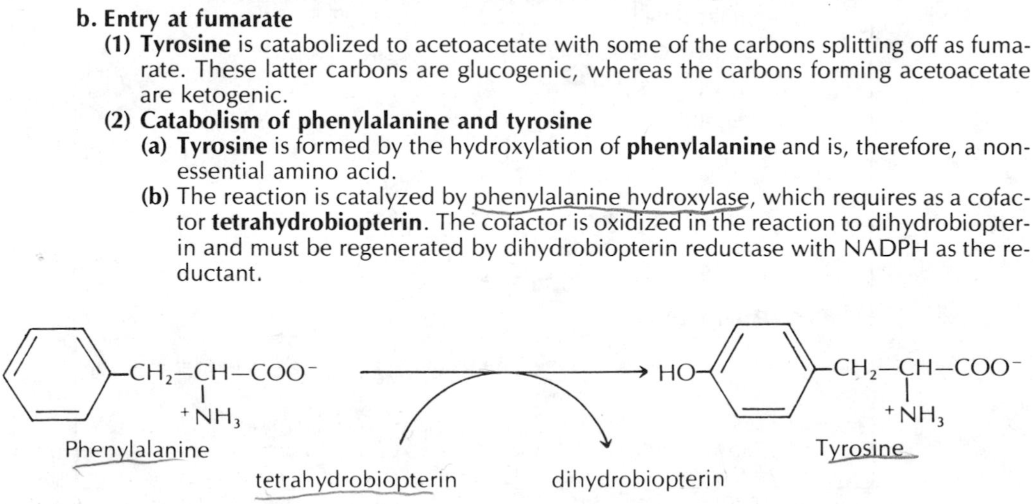

Phenylalanine   tetrahydrobiopterin   dihydrobiopterin   Tyrosine

$NADP^+ \longleftarrow \quad \longrightarrow NADPH + H^+$

(c) The next step is the transamination of tyrosine with α-ketoglutarate by a specific aminotransferase to form p-hydroxyphenylpyruvate.

$$\text{Tyrosine} \xrightleftharpoons[\alpha\text{-ketoglutarate}]{\text{glutamate}} \text{HO}-\bigcirc-CH_2-\underset{O}{\overset{}{\underset{\|}{C}}}-COO^-$$

p-Hydroxyphenylpyruvate

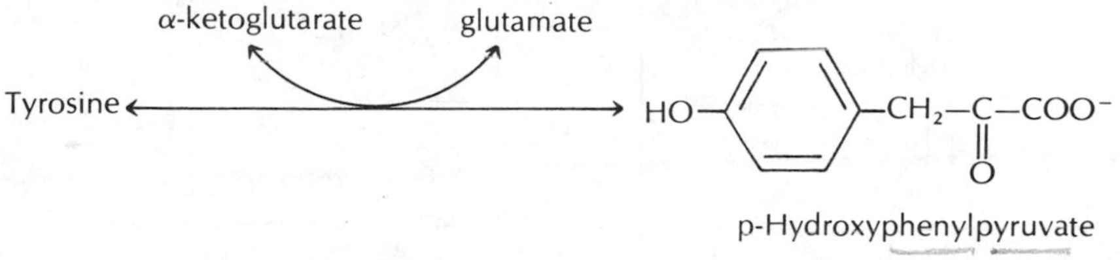

(d) Two oxidations using molecular oxygen and an isomerase convert p-hydroxyphenylpyruvate to fumarylacetoacetate, which is split to form fumarate and acetoacetate.

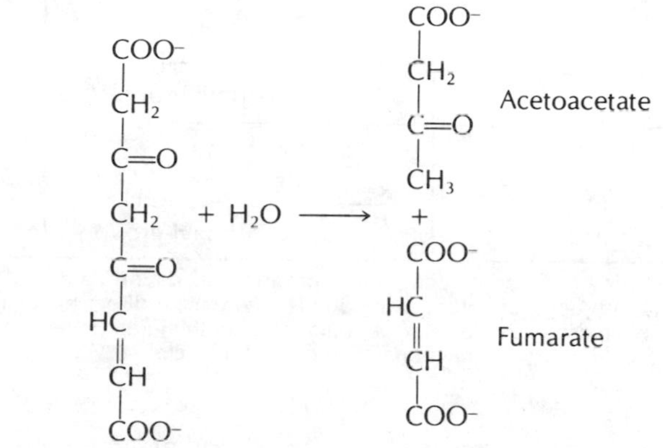

$$\begin{array}{l} COO^- \\ | \\ CH_2 \\ | \\ C=O \\ | \\ CH_2 \quad + H_2O \longrightarrow \\ | \\ C=O \\ | \\ HC \\ \| \\ CH \\ | \\ COO^- \end{array}$$

Fumarylacetoacetate

$$\begin{array}{l} COO^- \\ | \\ CH_2 \\ | \\ C=O \qquad \text{Acetoacetate} \\ | \\ CH_3 \\ + \\ COO^- \\ | \\ HC \\ \| \qquad \text{Fumarate} \\ CH \\ | \\ COO^- \end{array}$$

c. **Catabolism of tryptophan**
   (1) Tryptophan is used only sparingly as a component of protein, and its catabolism is not significant as an energy source nor as a precursor of blood glucose. However, tryptophan is an important precursor of the neurotransmitter serotonin and can contribute to the pool of pyridine nucleotide coenzymes.
   (2) The catabolism of tryptophan yields crotonyl CoA, formate, and alanine. Crotonyl CoA is an intermediate in the β-oxidation of fatty acids, and thus, tryptophan is ketogenic. It is also glucogenic by virtue of the carbons appearing as alanine.

d. **Catabolism of lysine**
   (1) Lysine is catabolized via saccharopine, a condensation product of lysine and α-ketoglutarate. The α-ketoglutarate is removed as glutamate (but none of these carbons is derived from lysine) to form, eventually, α-aminoadipate.
   (2) The catabolism of α-aminoadipate is not completely clear, but it is believed to give rise to crotonyl CoA. Thus, lysine is a ketogenic amino acid.

III. AMINO ACIDS AS MAJOR INPUTS TO THE ONE-CARBON POOL

A. Concept of a one-carbon pool

1. The transfer of single-carbon (one-carbon) groups occurs frequently in metabolism, very often as the loss or gain of CO_2, which is the highest oxidation state of the one-carbon groups (Chapter 12, section IV A 3). Table 27-1 illustrates the various oxidation levels for one-carbon groups.

2. Transfers of single-carbon groups of a lower state of oxidation are also needed, as in the formation of the purine ring (see Chapter 29, section I B 2), or in methylations involving nucleic acid bases (see Chapter 1, section IV C), or in the synthesis of compounds, such as creatine, choline, or epinephrine (see Chapter 21, section I B 4 and Chapter 28, sections I A 4 and III A 2).

3. These one-carbon fragments can be thought of as existing in a readily available pool with interchange between different oxidation levels, the fragments being borne on **carrier compounds**.

Table 27-1. Oxidation Levels of One-Carbon Groups

Parent Compound	Condensed Forms			
CH_4 Methane	$\xrightarrow{+\ R-H}$	None		
CH_3OH Methanol	$\xrightarrow[-\ H_2O]{+\ R-H}$	$R-CH_3$ Methyl group	$\xrightarrow{+\ R'-H}$	None
$\overset{\displaystyle O}{\overset{\displaystyle \|}{H-C-H}}$ Formaldehyde	$\xrightarrow{+\ R-H}$	$R-CH_2OH$ Hydroxymethyl group	$\xrightarrow[-\ H_2O]{+\ R'-H}$	$R-CH_2-R'$ Methylene group
$\overset{\displaystyle O}{\overset{\displaystyle \|}{H-C-OH}}$ Formic acid	$\xrightarrow[-\ H_2O]{+\ R-H}$	$\overset{\displaystyle O}{\overset{\displaystyle \|}{R-C-H}}$ Formyl group	$\xrightarrow[-\ H_2O]{+\ R'-H_2}$	$\overset{\displaystyle H}{\overset{\displaystyle \|}{R-C=R'}}$ Methylidyne group
$\overset{\displaystyle O}{\overset{\displaystyle \|}{HO-C-OH}}$ Carbonic acid	$\xrightarrow[-\ H_2O]{+\ R-H}$	$\overset{\displaystyle O}{\overset{\displaystyle \|}{R-C-OH}}$ Carboxyl group	$\xrightarrow[-\ H_2O]{+\ R'-H}$	$\overset{\displaystyle O}{\overset{\displaystyle \|}{R-C-R'}}$ Carbonyl group

Note.—The table shows one-carbon compounds of various oxidation states and their formal relationship to various groups found as substituents of natural compounds. There is no direct biologic interconversion of methanol and methyl groups. Otherwise examples may be found of all of the interconversions. (Reprinted with permission from McGilvery RW: *Biochemistry: A Functional Approach*. Philadelphia, WB Saunders, 1970, p 414.)

   a. Different carriers are employed for one-carbon groups of different levels of oxidation.
   (1) Biotin is a carrier of CO_2, at least in many carboxylation reactions.
   (2) Tetrahydrofolate carries groups at the level of oxidation of formic acid, formaldehyde, and methanol.
   (3) The major carrier of methyl groups is S-adenosylmethionine.
   b. The CO_2 transfers are excluded from this concept of a one-carbon pool mainly because animal cells cannot reduce CO_2, whereas the one-carbon fragments of lower oxidation level that are carried by tetrahydrofolate can undergo oxidation–reduction reactions.

B. Folate carrier. Tetrahydrofolate can carry one-carbon groups at various levels of oxidation; the oxidation state can be changed by specific dehydrogenases acting on the one-carbon groups while they are borne on the carrier.

   1. The structures of tetrahydrofolate (H_4-folate, more formally, **tetrahydropteroylglutamate**) and its precursors are shown in Figure 27-5.

   2. The synthesis of tetrahydrofolate is summarized in Figure 27-6.
   a. Humans cannot synthesize the pterin ring, and the formation of tetrahydrofolate is dependent upon the dietary intake of a precursor, **folate** (pteroylglutamate, folic acid).
   b. Folate in the diet is usually in the form of polyglutamates with the glutamate moieties attached in γ-peptidyl linkages, which are cleaved by conjugase, an enzyme in the intestinal mucosal cells, leaving a single glutamate moiety on the absorbed vitamin.
   c. The vitamin is usually in an oxidized state prior to absorption, and the intestinal mucosal cells also reduce the folic acid to tetrahydrofolic acid in two steps.

   3. Inhibition
   a. The folic acid antagonists **aminopterin** and **methotrexate** are analogues of dihydrofolic acid. As extremely potent inhibitors of the reductase reaction, dihydrofolate to tetrahydrofolate, they are widely used in cancer chemotherapy (see Chapter 29, section VIII F).
   b. The inhibition of bacterial synthesis of tetrahydrofolate is the basis of the action of the so-called **sulfa drugs** (e.g., sulfanilamide), which is an analogue of p-aminobenzoic acid, a precursor and constituent of tetrahydrofolate (see Fig. 27-5).

   4. One-carbon groups on tetrahydrofolate
   a. Methylene tetrahydrofolate carries the one-carbon group at the oxidation level of formaldehyde.

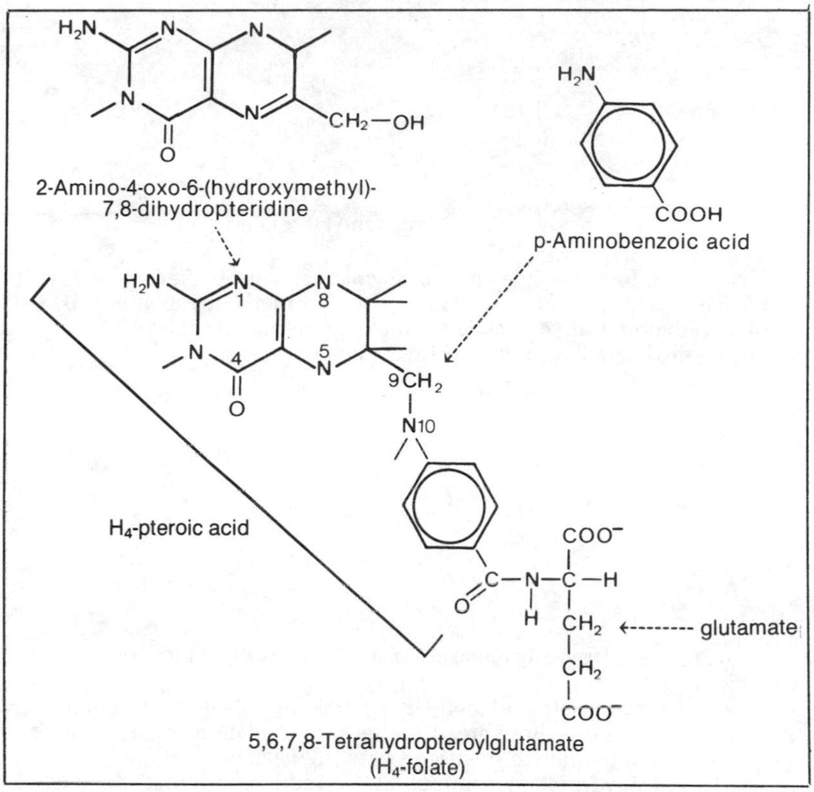

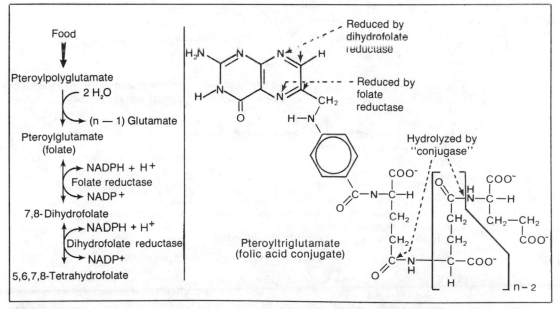

Figure 27-5. Tetrahydrofolate (H₄-folate, more formally, tetrahydropteroylglutamate) and its precursors. (Reprinted with permission from McGilvery RW: *Biochemistry: A Functional Approach,* 2nd ed. Philadelphia, WB Saunders, 1979, p 591.)

Figure 27-6. The formation of tetrahydrofolate. The tetrahydrofolates in foodstuffs often become oxidized before absorption. The intestinal mucosa removes all but one glutamyl group from pteroylpolyglutamates and reduces the resultant folate to its tetrahydro form in two steps. NADP⁺ = nicotinamide-adenine dinucleotide phosphate; NADPH = reduced NADP⁺. (Reprinted with permission from McGilvery RW: *Biochemistry: A Functional Approach,* 2nd ed. Philadelphia, WB Saunders, 1979, p 592.)

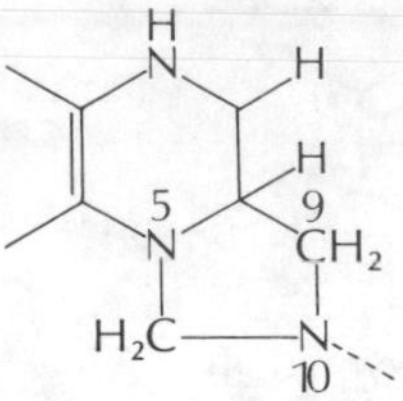

5,10-Methylene H$_4$-folate

It is used to form the side chain of thymidine (see Chapter 29, section II B) and can be a source of methyl groups by reduction and of methylidyne groups by oxidation.
b. Methylidyne tetrahydrofolate (or methenyl tetrahydrofolate) and the equivalent **formyl tetrahydrofolate** have the following structures:

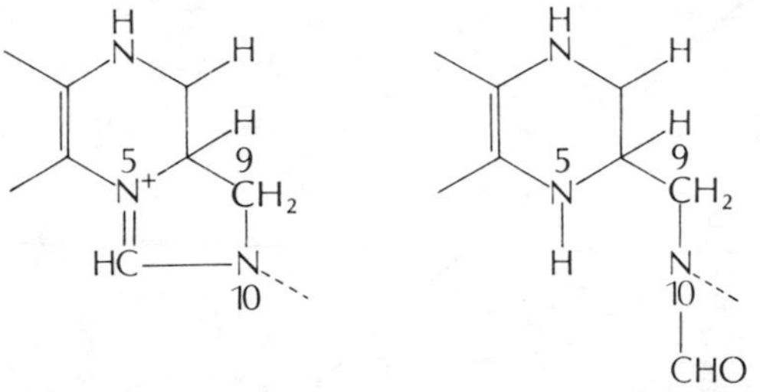

5,10-Methylidyne H$_4$-folate   10-Formyl H-folate

They arise by oxidation of methylene tetrahydrofolate and by combination of tetrahydro-folate with formate. Both are at the level of oxidation of formic acid and are obligatory precursors of the purine ring (see Chapter 29, section I B 2).
c. Methyl tetrahydrofolate is formed by the reduction of methylene tetrahydrofolate.

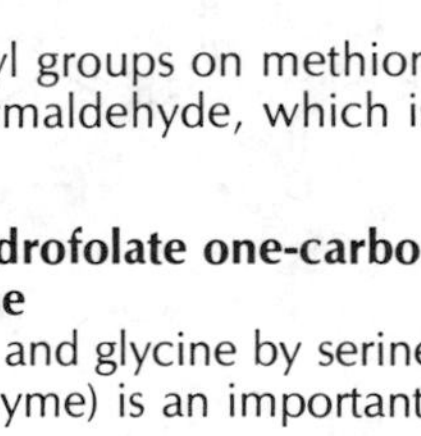

5-Methyl H$_4$-folate

It is used to regenerate methyl groups on methionine. The degradation of methyl groups results in the formation of formaldehyde, which is recovered by combination with tetra-hydrofolate.

5. Primary sources for the tetrahydrofolate one-carbon pool
  a. Methylene groups from serine
   (1) The equilibration of serine and glycine by serine hydroxymethyl transferase (a pyridoxal phosphate–dependent enzyme) is an important reaction of nitrogen metabolism.

$$\text{Serine} + \text{H}_4\text{-folate} \leftrightarrow \text{glycine} + \text{5,10-methylene H}_4\text{-folate}$$

It is a mitochondrial reaction and can satisfy man's needs for glycine, which is thus not an essential amino acid.
   (2) Glycine is also degraded in the mitochondria.

$$\text{Glycine} + \text{NAD}^+ + \text{H}_4\text{-folate} \rightarrow \text{NH}_4^+ + \text{CO}_2 + \text{5,10-methylene H}_4\text{-folate} + \text{NADH}$$

   (3) When necessary, the one-carbon pool can be maintained by one or both of these reactions involving glycine, as both produce 5,10-methylene tetrahydrofolate. If the pool is replete, glycine can be converted to serine, which is then metabolized to pyruvate.
   (4) Serine, a nonessential amino acid, is synthesized from glucose by two pathways: via phosphorylated intermediates from D-3-phosphoglycerate and via nonphosphorylated intermediates and hydroxypyruvate. Thus, glucose can function as a major source of carbon for the one-carbon pool.

b. Methylidyne groups

(1) These can be formed through oxidation of 5,10-methylene tetrahydrofolate by an NAD^+-coupled dehydrogenase. The reverse reaction is probably catalyzed by an NADPH-coupled dehydrogenase.

NAD+   NADH

NADP+   NADPH

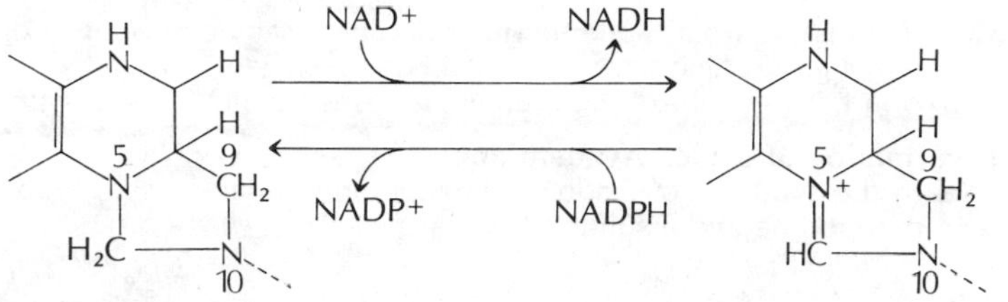

5,10-Methylene H_4-folate   5,10-Methylidyne H_4-folate

(2) Formation from histidine

(a) The opening of the imidazole ring of histidine during degradation forms N-formiminoglutamate (FIGLU), a nitrogen analogue of a formyl group. This formyl group is transferred to tetrahydrofolate; then, by removal of NH_4^+ and cyclization, the product is transformed to 5,10-methylidyne tetrahydrofolate.

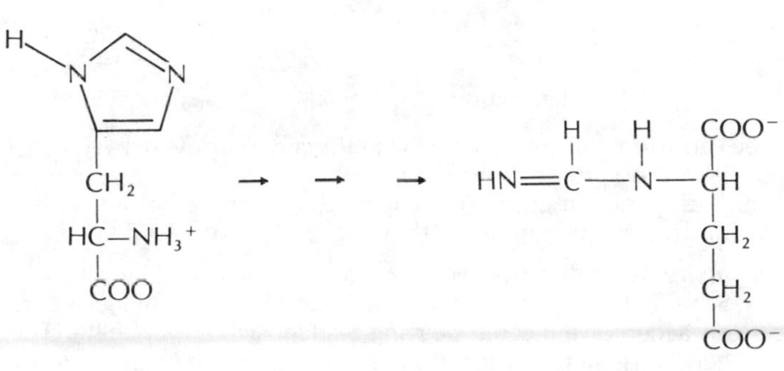

Histidine   N-Formiminoglutamate

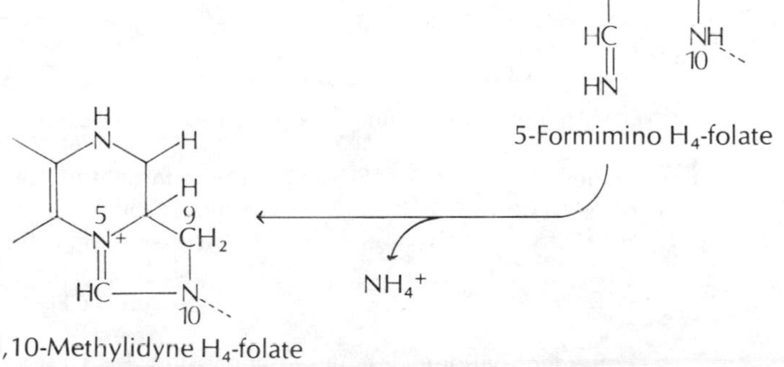

N-Formiminoglutamate   Glutamate

(FIGLU)

H_4-Folate

5-Formimino H_4-folate

NH_4^+

5,10-Methylidyne H_4-folate

(b) A quantitative estimation of urinary FIGLU has been used as a measure of folic acid deficiency in man. Although widespread in foods, folic acid is rather readily destroyed by overcooking, and this, combined with absorption difficulties, may account for the fact that folic acid deficiency is the most common vitamin deficiency worldwide.

(3) **Formation from formate.** The major source of formate is normally the oxidation of methyl groups. Formate is attached to tetrahydrofolate by a synthetase, forming 10-formyl tetrahydrofolate, which is cyclized enzymatically to form 5,10-methylidyne tetrahydrofolate.

C. **Methyl transfers from S-adenosylmethionine.** Many compounds of biologic importance contain methyl groups that have been attached to oxygen, nitrogen, or carbon atoms of a precursor compound. The donor of the methyl groups is usually S-adenosylmethionine.

1. **Formation of S-adenosylmethionine**
   a. In the synthesis of S-adenosylmethionine, an adenosyl group is transferred from ATP to methionine at the sulfur.

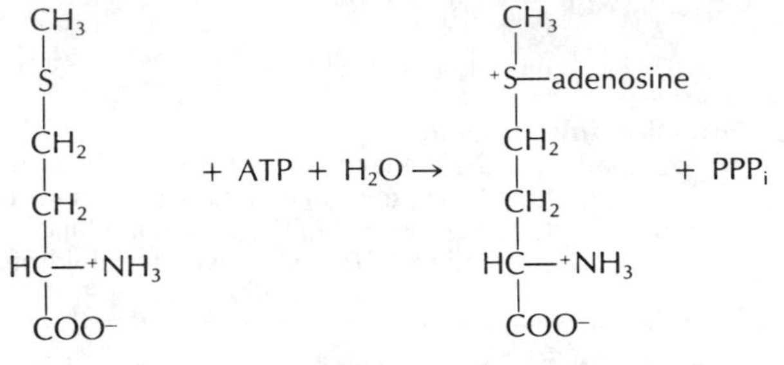

$$
\begin{array}{ccc}
CH_3 & & CH_3 \\
| & & | \\
S & & {}^+S-adenosine \\
| & & | \\
CH_2 & +\ ATP\ +\ H_2O \rightarrow & CH_2 \qquad +\ PPP_i \\
| & & | \\
CH_2 & & CH_2 \\
| & & | \\
HC-{}^+NH_3 & & HC-{}^+NH_3 \\
| & & | \\
COO^- & & COO^-
\end{array}
$$

$$\qquad\text{Methionine} \qquad\qquad\qquad \text{S-Adenosylmethionine}$$

   b. The formation of S-adenosylmethionine is driven by the hydrolysis of the triphosphate (PPP_i) to P_i by phosphohydrolases.
   c. The sulfonium group of S-adenosylmethionine has a standard free-energy change for hydrolysis to adenosine, methionine, and H^+ of -5.7 kcal.

2. **Transfer of methyl groups**
   a. Specific transferase enzymes can transfer the methyl group from S-adenosylmethionine to an acceptor molecule, leaving S-adenosylhomocysteine. The transfer may be to a nitrogen, oxygen, or carbon atom of the acceptor molecule, depending upon the specificity of the transferase.

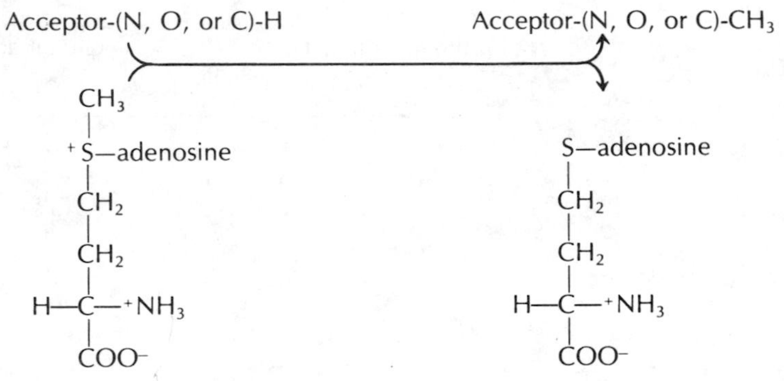

$$
\text{Acceptor-(N, O, or C)-H} \qquad\qquad \text{Acceptor-(N, O, or C)-CH}_3
$$

$$
\begin{array}{cc}
CH_3 & \\
| & \\
{}^+S-adenosine & S-adenosine \\
| & | \\
CH_2 & CH_2 \\
| & | \\
CH_2 & CH_2 \\
| & | \\
H-C-{}^+NH_3 & H-C-{}^+NH_3 \\
| & | \\
COO^- & COO^-
\end{array}
$$

$$\text{S-Adenosylmethionine} \qquad\qquad \text{S-Adenosylhomocysteine}$$

   b. After the methyl transfer from S-adenosylmethionine, the S-adenosylhomocysteine that remains is hydrolyzed to adenosine and homocysteine.

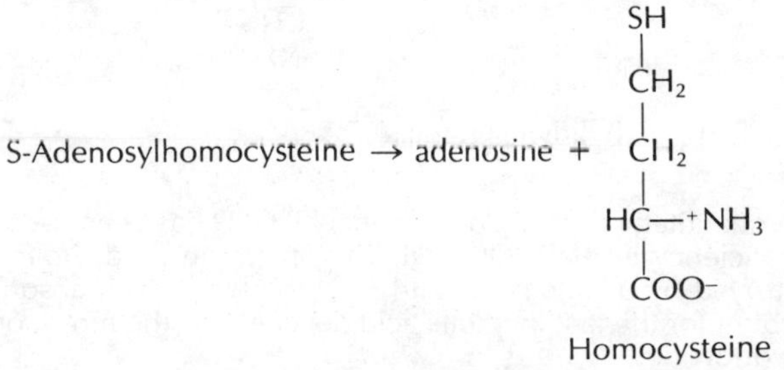

$$
\begin{array}{c}
SH \\
| \\
CH_2 \\
\text{S-Adenosylhomocysteine} \rightarrow \text{adenosine} + \quad CH_2 \\
| \\
HC-{}^+NH_3 \\
| \\
COO^-
\end{array}
$$

$$\text{Homocysteine}$$

3. Regeneration of methionine
   a. Recovery of a methyl group from choline
      (1) Choline, in the form of phosphatidylcholine, is formed from phosphatidylethanol-amine by the transfer of three methyl groups from S-adenosylmethionine (see Chapter 21, section I B 4).
      (2) One of the methyl groups of choline can be used to regenerate methionine from homocysteine.
      (3) Choline, after its release from phosphatidylcholine by phospholipase D, is degraded in a mitochondrial oxidative system to glycine betaine in a two-step reaction, requiring FAD-linked and NAD$^+$-linked dehydrogenases.

$$H_3C—^+N(CH_3)(CH_3)—CH_2—CH_2—OH \longrightarrow \longrightarrow H_3C—^+N(CH_3)(CH_3)—CH_2—COO^-$$

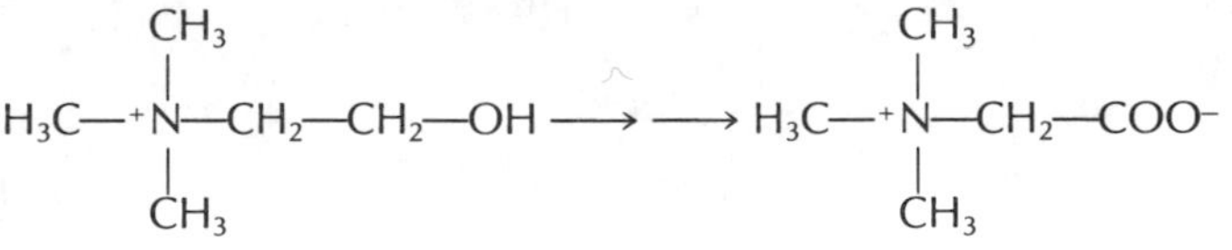

Choline   Glycine betaine

   (4) Glycine betaine can then donate one methyl group to homocysteine to regenerate methionine.

$$\text{Glycine betaine + homocysteine} \rightarrow \text{methionine} + H_3C—^+N(H)(CH_3)—CH_2—COO^-$$

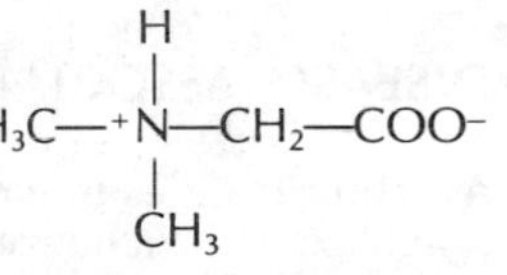

N,N-Dimethylglycine

   (5) Two further FAD-linked dehydrogenases oxidize the remaining two methyl groups on N,N-dimethylglycine to formaldehyde, leaving glycine.
   (6) All the electrons from this series of oxidations are fed into the mitochondrial electron transport system via the reduced cofactors.
   (7) The formaldehyde, using the intramitochondrial tetrahydrofolate pool, can form 5,10-methylene tetrahydrofolate, and this can react with glycine to form serine.
   (8) The serine can leave the mitochondria and, using the cytoplasmic tetrahydrofolate pool, can form glycine and 5,10-methylene tetrahydrofolate. Thus, one-carbon fragments can be readily ferried across the inner mitochondrial membrane even though it is poorly permeable to tetrahydrofolate.

   b. De novo synthesis from methylene tetrahydrofolate
      (1) 5,10-Methylene tetrahydrofolate is reduced by a flavoprotein, at the expense of NADH, to 5-methyl tetrahydrofolate in a virtually irreversible reaction.

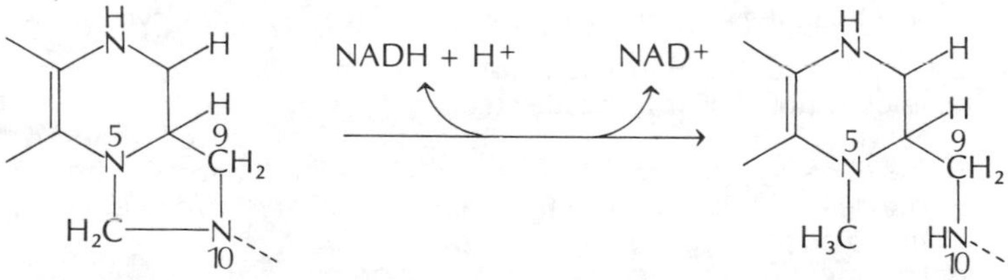

$$\text{NADH} + H^+ \qquad \text{NAD}^+$$

5,10-Methylene H$_4$-folate   5-Methyl H$_4$-folate

   (2) 5-Methyl tetrahydrofolate can be used to methylate homocysteine to methionine. This methylation reaction requires **methylcobalamin** as a cofactor. The 5-methyl tetrahydrofolate in fact methylates cobalamin to form the methyl donor methylcobalamin.
   (3) A deficiency of cobalamin (vitamin B$_{12}$) leads to a buildup, or trapping, of part of the folate pool in the form of 5-methyl tetrahydrofolate. This **methyl folate trap hypothesis** has found favor as an explanation of the effect of vitamin B$_{12}$ deficiency on DNA synthesis.

4. Degradation of homocysteine and the synthesis of cysteine
    a. Cystathionine synthetase, a pyridoxal phosphate–dependent enzyme, catalyzes the condensation of homocysteine with serine to form cystathionine.

$$
\begin{array}{c}
SH \\
| \\
CH_2 \\
| \\
CH_2 \\
| \\
HC\!-\!{}^+NH_3 \\
| \\
COO^-
\end{array}
\;+\;
\begin{array}{c}
COO^- \\
| \\
HC\!-\!{}^+NH_3 \\
| \\
CH_2 \\
| \\
OH
\end{array}
\;\longrightarrow\;
\begin{array}{c}
COO^- \\
| \\
HC\!-\!{}^+NH_3 \\
| \\
CH_2 \\
| \\
S \\
| \\
CH_2 \\
| \\
CH_2 \\
| \\
HC\!-\!{}^+NH_3 \\
| \\
COO^-
\end{array}
\;+\; H_2O
$$

              Homocysteine   Serine   Cystathionine

    b. Cystathionase, also a pyridoxal phosphate–dependent enzyme, then cleaves cystathionine to cysteine and α-ketobutyrate. Thus, cysteine can be formed from the essential amino acid methionine and is, therefore, not itself an essential amino acid. Cysteine that is present in the diet will, however, reduce the requirement for methionine.

IV. DISEASES ASSOCIATED WITH ABNORMAL AMINO ACID METABOLISM

  A. Abnormalities in amino acid transport mechanisms. The various mechanisms for the uptake, efflux, and exchange of amino acids are heterogeneously distributed among the different tissues and subcellular compartments of the body. The importance of the amino acid transport mechanisms is illustrated by several amino acid transport defects and their associated disease states, all of which are characterized by **aminoaciduria**. The identification of these heritable disorders has greatly amplified an awareness of the complexity of amino acid transport even though the molecular basis of the defects is not known.

  1. Hartnup's disease causes a massive generalized aminoaciduria due to a defect in the renal tubular and intestinal systems handling monoamino–monocarboxylic acids. In addition, the transport mechanisms for lysine and possibly for histidine are also affected.
    a. Diminished renal tubular reabsorption and abnormal intestinal (jejunal) absorption can lead to
      (1) A pellagra-like rash and psychiatric changes, ranging from emotional instability to delirium
      (2) A reduced ability to convert tryptophan to kynurenine and the pyridine nucleotide coenzymes
    b. The reduced ability to form kynurenine from tryptophan is probably due to an inhibition of the breakdown of tryptophan by indole, which arises by the action of gut bacteria on unabsorbed tryptophan.
    c. Nicotinamide administration usually relieves all of the symptoms except aminoaciduria.

  2. Cystinuria causes aminoaciduria of dibasic amino acids (lysine, ornithine, and cystine). Cysteine does not appear to be affected.
    a. As with Hartnup's disease, the disorder involves a specific transport system in the transepithelial cells of the renal tubules and in the intestine.
    b. The defect leads to urinary tract deposition of crystals of cystine, one of the most insoluble amino acids.
    c. Treatment is to reduce cystine deposition by
      (1) Increasing urine volume
      (2) Increasing cystine solubility by alkalizing the urine
      (3) Treating with the drug D-penicillamine (β,β-dimethylcysteine), which forms the mixed disulfide of cysteine-penicillamine, a compound that is significantly more soluble than cystine. (See Chapter 10, section III C 3 c for complications of this treatment.)

  3. Iminoglycinuria produces an aminoaciduria of glycine, proline, and hydroxyproline.
    a. The defect appears to be a deletion or inactivation of a membrane transport protein of the renal tubule, which selectively binds L-proline, hydroxyproline, and glycine.

        b. There is no consistent or typical illness associated with this disorder.

B. Hyperphenylalaninemias. Nine forms of hyperphenylalaninemia are now recognized; two are associated with primary tyrosinemia. Most cases are due to classic phenylketonuria.

  1. Classic phenylketonuria (PKU) is an inherited disorder due to the absence of activity of a component of phenylalanine hydroxylase. It is an autosomal recessive defect with a frequency of occurrence of 1 in 10,000 births.

    a. The inability to hydroxylate phenylalanine to tyrosine [see section II C 2 b (2)] leads to an increase in minor pathways of phenylalanine metabolism.

    b. In affected children, the levels of blood phenylpyruvate, the transamination product of phenylalanine, can reach 0.1–0.5 mM.

    c. Severe mental retardation occurs in untreated infants with this disease but can be avoided by early diagnosis and the feeding of a low-phenylalanine diet.

    d. The cause of the mental retardation is not known, but phenylpyruvate is a potent inhibitor of brain mitochondrial pyruvate translocase.

  2. Atypical phenylketonuria with dihydrobiopterin reductase deficiency

    a. Because of the enzyme deficiency, biopterin is not reduced to its active form, tetrahydrobiopterin.

    b. The symptoms of hyperphenylalaninemia appear soon after birth and do not respond to the standard low-phenylalanine dietary treatment.

    c. In one case, treatment with intravenous tetrahydrobiopterin lowered the serum phenylalanine levels.

    d. Tetrahydrobiopterin is also a cofactor in the synthesis of neurotransmitters (see Chapter 28, sections I A, D), which may underlie the severe neurologic changes that can occur. Substitution therapy with levodopa, carbidopa, and 5-hydroxytryptophan, in addition to dietary treatment, may be beneficial if started early in life.

C. Abnormalities of branched-chain amino acid metabolism

  1. Hypervalinemia causes high plasma levels of valine, but not of leucine or isoleucine, and no indication of increased α-ketoisovalerate excretion.

    a. The block is at the transamination of valine to α-ketoisovalerate.

    b. Only one case has been described, but this is of considerable interest as it implies the existence of a separate transaminase for valine, rather than a common enzyme for handling all three branched-chain amino acids, which is the widely held view.

  2. Maple syrup urine disease characteristically causes the urine and sweat to smell like maple syrup. Plasma (and urinary) levels of all three branched-chain amino acids and their corresponding α-keto acids are increased far above normal.

    a. The biochemical defect is the lack of the α-keto acid decarboxylase, which converts the branched-chain amino acids to acyl CoA esters and CO_2. Considerable evidence is now accruing to indicate that there is more than one enzyme system handling the three branched-chain amino acids.

    b. Untreated children do not usually live longer than 1 year.

    c. Treatment includes restricted intake of branched-chain amino acids, exchange transfusion, and peritoneal dialysis for acute episodes. Some cases respond to large doses of thiamine, the precursor vitamin for thiamine pyrophosphate, a cofactor of the keto acid decarboxylase.

  3. Intermittent branched-chain ketonuria is a variant of maple syrup urine disease in which the α-keto acid decarboxylase deficiency is much less severe.

    a. Symptoms appear later in life and occur only intermittently.

    b. Dietary restriction of the branched-chain amino acids is more successful in these cases.

  4. Isovaleric acidemia is due to a deficiency of isovaleryl CoA dehydrogenase, an enzyme involved in leucine catabolism.

    a. There is a "cheesy" odor on the breath and in urine. Vomiting, acidosis, and coma are induced by feeding of protein. Mild mental retardation occurs.

    b. Treatment is administration of glycine and reduction of dietary leucine.

D. Abnormalities of tyrosine metabolism

  1. Hereditary tyrosinemia is an inability to metabolize tyrosine and p-hydroxyphenylpyruvate along the main oxidative pathway.

    a. There appear to be two forms of the disease.

      (1) One is described as "acute," in which symptoms appear about 6 months after birth with death resulting from liver failure in about 90% of the untreated cases.

       (2) The other, described as "chronic," occurs in later childhood and causes severe nodular hepatic cirrhosis and a high frequency of hepatic carcinoma.

    b. Methionine levels may be high, particularly in the acute form of the disease, but the biochemical basis for this is not known.

    c. Treatment aims at reducing the dietary levels of both tyrosine and methionine, as both of these amino acids are toxic in high concentration.

2. Alkaptonuria is due to a hereditary lack of homogentisate oxidase, an enzyme in the catabolic pathway for tyrosine. (This is the genetic disease upon which Garrod* based his concepts of heritable metabolic disorders.)

    a. Homogentisate is excreted in large amounts and causes the urine to darken on standing in air.

    b. Eventually, there may be pigmentation of connective tissue and a form of arthritis.

---

*Garrod AE: *Inborn Errors of Metabolism* (reprinted with supplement by H. Harris). Fairlawn, NJ, Oxford University Press, 1963.

STUDY QUESTIONS

Directions: Each question below contains five suggested answers. Choose the **one best** response to each question.

1. All of the following statements concerning the actions of transaminases are true EXCEPT

(A) transaminases catalyze readily reversible reactions with an equilibrium constant (K_e) between 1 and 10
(B) transaminases interact with four substrates
(C) all require pyridoxal phosphate as a cofactor
(D) all require adenosine triphosphate as an energy source
(E) they appear in both cytosolic and mitochondrial compartments

2. The two enzymatic activities that catalyze the bulk of the nitrogen flow from amino acids to ammonia are

(A) transaminases and glutaminase
(B) transaminases and glutamate dehydrogenase
(C) transaminases and amino acid oxidases
(D) glutaminase and glutamate dehydrogenase
(E) glutaminase and amino acid oxidases

3. Correct statements regarding the urea cycle include all of the following EXCEPT

(A) the supply of carbamoyl phosphate to the urea-cycle enzymes is regulated by N-acetylglutamate
(B) the immediate precursor of urea in the cycle is ornithine
(C) the formation of carbamoyl phosphate requires the expenditure of two molecules of adenosine triphosphate per molecule of ammonia incorporated
(D) fumarate is formed in the urea cycle
(E) aspartate used in the urea cycle can arise from oxaloacetate by transamination with cytosolic α-ketoglutarate

4. The two nitrogen atoms in urea arise from

(A) ammonia (NH_3) and glutamine
(B) NH_3 and aspartic acid
(C) glutamine and aspartic acid
(D) glutamine and glutamic acid
(E) glutamic acid and alanine

5. Which of the following groups of amino acids are catabolized along common pathways?

(A) Alanine, leucine, isoleucine, and methionine
(B) Glycine, serine, glutamine, and asparagine
(C) Lysine, leucine, serine, and cysteine
(D) Proline, glutamate, glutamine, and arginine
(E) Tryptophan, tyrosine, phenylalanine, and histidine

6. Carbon from threonine and methionine can enter the tricarboxylic acid cycle as

(A) fumarate
(B) succinyl coenzyme A (succinyl CoA)
(C) α-ketoglutarate
(D) oxaloacetate
(E) acetyl CoA

7. Which of the following essential dietary factors is a precursor for a compound that can act as a carrier of one-carbon fragments at different levels of oxidation?

(A) Methionine
(B) Thiamine
(C) Folic acid
(D) Biotin
(E) Pyridoxine

8. The transformation of serine to glycine is dependent upon adequate absorption of which of the following pairs of vitamins from the gut?

(A) Niacin and B_{12}
(B) Thiamine and B_{12}
(C) B_6 and B_{12}
(D) NAD^+ and B_{12}
(E) Folic acid and B_{12}

9. Maple syrup urine disease is due to a deficiency of

(A) the enzymes that specifically transaminate the branched-chain amino acids
(B) the α-keto acid decarboxylase that converts branched-chain amino acids to acyl coenzyme A (acyl CoA) esters and carbon dioxide
(C) the mitochondrial enzyme system that oxidizes acyl CoA esters derived from the branched-chain amino acids
(D) mitochondrial CoA required for the oxidation of branched-chain amino acids
(E) amino acid transport systems

Directions: Each question below contains four suggested answers of which **one or more** is correct. Choose the answer

A   if **1, 2, and 3** are correct
B   if **1 and 3** are correct
C   if **2 and 4** are correct
D   if **4** is correct
E   if **1, 2, 3, and 4** are correct

10. Which of the following amino acids can undergo deamination by dehydration?

(1) Serine
(2) Leucine
(3) Threonine
(4) Valine

11. Correct statements regarding deficiencies of the urea-cycle enzymes include which of the following?

(1) Patients with a total lack of one of the enzymes of the urea cycle do not survive the neonatal period
(2) Restriction of protein intake reduces the encephalopathies seen in children with a deficiency of a urea-cycle enzyme
(3) Acquired hyperammonemia is usually due to cirrhosis of the liver, rather than to lack of an enzyme of the urea cycle
(4) Hyperammonemia is most severe in patients with arginase deficiency

12. Glucogenic amino acids include which of the following?

(1) Proline
(2) Phenylalanine
(3) Isoleucine
(4) Leucine

13. Primary sources for the one-carbon pool include which of the following amino acids?

(1) Alanine
(2) Threonine
(3) Leucine
(4) Serine

14. Hartnup's disease can cause which of the following conditions?

(1) A massive aminoaciduria
(2) A pellagra-like rash
(3) Psychiatric changes
(4) A reduced ability to convert tryptophan to the pyrimidine nucleotide coenzymes

15. A deficiency of tetrahydrobiopterin, or its abnormal metabolism, would affect the

(1) biosynthesis of a neurotransmitter
(2) catabolism of methionine
(3) biosynthesis of tyrosine
(4) catabolism of branched-chain amino acids

ANSWERS AND EXPLANATIONS

1. The answer is D. (*I A 1*) Transaminases catalyze the transfer of an amino group from an amino acid to an α-keto group to form a new amino acid and a new α-keto acid. The reactions are readily reversible as the equilibrium constant (K_e) is between 1 and 10. Each transaminase accepts two amino acids and two α-keto acids as substrates, and all require pyridoxal phosphate as a cofactor in the transfer of amino groups from amino acids to the α-keto acids. They are found within the inner mitochondrial membrane (in the matrix space) and in the cytosol. The transamination reaction is not coupled to the hydrolysis of a high-energy phosphate bond of adenosine triphosphate.

2. The answer is B. (*I B 1–3*) The greater part of the nitrogen from amino acid degradation in skeletal muscle is collected by transamination reactions as alanine, which is exported to the liver. The liver uses the alanine as a gluconeogenic substrate after the amino group is transferred to α-ketoglutarate to form glutamate. Glutamate dehydrogenase removes the amino group as ammonia, which is a substrate for urea synthesis in liver mitochondria.

3. The answer is B. (*I C*) The mitochondrial carbamoyl phosphate synthetase, which is involved in urea synthesis, requires N-acetylglutamate as a positive allosteric regulator, and two molecules of adenosine triphosphate are expended per molecule of ammonia incorporated. The level of the synthetase that catalyzes the formation of N-acetylglutamate is markedly increased by amino acids, particularly arginine. The immediate precursor of urea in the cycle is arginine, which is cleaved by arginase to form urea and ornithine. Fumarate arises from the urea cycle by the action of argininosuccinate lyase, which cleaves argininosuccinate to arginine and fumarate. The latter can be converted to malate by fumarase and then to oxaloacetate by cytosolic malate dehydrogenase. The oxaloacetate is then available for transamination with α-ketoglutarate to form aspartate for the urea cycle.

4. The answer is B. (*I C 2*) Mitochondrial carbamoyl phosphate synthetase uses ammonia as the nitrogen source in the formation of carbamoyl phosphate. The cytosolic carbamoyl phosphate synthetase involved in pyrimidine synthesis uses glutamine as the nitrogen source (see Chapter 29, section I C 2). The other nitrogen in urea arises from aspartate when citrulline is released from the mitochondria into the cytosol and combines with aspartate to form argininosuccinate. The latter is cleaved to arginine and fumarate, and urea arises from the hydrolysis of arginine by arginase.

5. The answer is D. (*II C 1 a*) Carbon from proline, arginine, glutamate, and glutamine can enter the tricarboxylic acid cycle via α-ketoglutarate. Histidine is also in this category. Proline is directly broken down to glutamate, which yields α-ketoglutarate by the action of glutamate dehydrogenase or by transamination. Arginine is cleaved by arginase to urea and to ornithine, which can be converted to glutamate via glutamate semialdehyde. Glutamine gives rise to glutamate by the action of glutaminase.

6. The answer is B. (*II C 2 a; III C*) Threonine is deaminated by threonine dehydratase to yield ammonia and α-ketobutyrate. Methionine is also catabolized to α-ketobutyrate, in this case via cystathionine. Methionine, an essential amino acid, is the precursor of the methyl donor S-adenosylmethionine, which, when it donates its methyl group to an acceptor, becomes S-adenosylhomocysteine. The latter is hydrolyzed to adenosine and to homocysteine, which condenses with serine to form cystathionine. The α-ketobutyrate formed in the catabolism of threonine and methionine is **decarboxylated** to propionyl coenzyme A (propionyl CoA), which is, in turn, **carboxylated** to form (S)-methylmalonyl CoA. This is then racemized to the (R) form and converted to succinyl CoA by a mutase that requires a cofactor derived from vitamin B_{12}.

7. The answer is C. (*III B*) Folic acid is a vitamin precursor of tetrahydrofolic acid, a carrier of one-carbon fragments at the level of oxidation of methanol, formaldehyde, or formic acid. Biotin is also a vitamin and carrier of a one-carbon fragment, but it is specific for carbon dioxide. Thiamine is the vitamin precursor of thiamine pyrophosphate, a carrier of the acetate group (two-carbon fragment). Methionine is an essential amino acid that is the precursor of S-adenosylmethionine, the methyl donor. It is thus a carrier of a one-carbon fragment but only at the level of oxidation of methanol. Pyridoxine is the vitamin precursor of pyridoxal phosphate, a cofactor involved in many reactions using amino acids as substrates.

8. The answer is E. (*III C 3 b; Chapter 31 II D 3 a, b*) The reversible transformation of serine and glycine requires tetrahydrofolate as cofactor. This is a derivative of the vitamin folic acid. However, with inadequate vitamin B_{12} uptake from the gut, the metabolism of the folate coenzymes is impaired. In particular, there is an accumulation of the inactive form 5-methyl tetrahydrofolate. This is because the only way in which 5-methyl tetrahydrofolate can be used is in the reaction converting homocysteine to methionine, which requires methylcobalamin, derived from vitamin B_{12}, as a cofactor.

9. The answer is B. (*IV C 2*) Maple syrup urine disease, so-called because of the characteristic odor of urine and sweat, is due to a deficiency of the α-keto acid decarboxylase that converts branched-chain amino acids to acyl coenzyme A (acyl CoA) esters and carbon dioxide. The accumulation of the acyl CoA derivatives of the branched-chain amino acids in body fluids is the cause of the maple syrup odor and probably the accompanying mental retardation and early death. The α-keto acid decarboxylase is part of a multienzyme complex similar to pyruvate and α-ketoglutarate dehydrogenases. These complexes all require thiamine pyrophosphate as one of the cofactors, and it is of interest that some cases of maple syrup urine disease respond to large doses of thiamine.

10. The answer is B (1, 3). (*I A 3*) The hydroxyl group in the side chain of serine and threonine gives these amino acids a higher level of oxidation than most, and the deamination of these amino acids liberates enough free energy to be essentially irreversible. The reaction involves the elimination of the elements of water; hence, the names of these enzymes are serine and threonine dehydratase.

11. The answer is A (1, 2, 3). (*I D, E*) Patients with a total lack of one of the enzymes of the urea cycle do not survive the neonatal period because of blood ammonia (NH_3) levels that reach toxic proportions. Hyperammonemia is a common feature in patients with a partial deficiency of a urea-cycle enzyme with the problem being most acute in carbamoyl phosphate synthetase deficiency. Restriction of protein intake helps to keep NH_3 levels down in patients with urea-cycle enzyme deficiencies, but the protein intake cannot be reduced to zero. Acquired hyperammonemia is usually due to cirrhosis of the liver when the development of a collateral circulation bypasses the liver.

12. The answer is A (1, 2, 3). (*II C; Figures 27-3 and 27-4*) The catabolism of proline involves its conversion to glutamate, which, upon transamination or oxidative deamination, yields α-ketoglutarate. Three carbons of α-ketoglutarate can provide a net synthesis of glucose and are, therefore, glucogenic. None of the carbon from leucine is glucogenic, but isoleucine is metabolized via succinyl coenzyme A (succinyl CoA), which is glucogenic, and acetyl CoA, which is ketogenic. Phenylalanine is also both glucogenic and ketogenic, some of the carbon entering the tricarboxylic acid cycle as fumarate (glucogenic) and some forming acetoacetate (ketogenic).

13. The answer is D (4). (*III B 5*) Serine reacts with tetrahydrofolate to form glycine and 5,10-methylene tetrahydrofolate. This is a mitochondrial system catalyzed by serine transhydroxymethylase. The reaction supplies humans with sufficient glycine, so that glycine is is not an essential amino acid, and also charges a carrier of the tetrahydrofolate pool with a one-carbon fragment at the level of oxidation of formaldehyde. Alanine, threonine, and leucine are not involved in reactions that result in the input of carbon to the tetrahydrofolate one-carbon pool.

14. The answer is E (all). (*IV A 1*) Hartnup's disease is due to a deficiency in amino acid transport. There is a defect (not clearly defined) in the renal tubular and intestinal transport of monoamino–monocarboxylic acids in addition to lysine and possibly histidine. A pellagra-like rash (i.e., similar to that seen in niacin deficiency) and psychiatric changes may be seen. The pellagra-like rash, which may be relieved by the administration of niacin, is due to a reduced ability to convert tryptophan to the pyrimidine nucleotide coenzymes. Specifically, indole, formed by gut bacteria from tryptophan, which cannot be efficiently absorbed, inhibits the formation of kynurenine from tryptophan, a step in the formation of pyridine nucleotide coenzymes.

15. The answer is B (1, 3). (*IV B 2; Chapter 28, I D*) Tetrahydrobiopterin is a cofactor in the hydroxylation of phenylalanine to tyrosine and in the biosynthesis of the neurotransmitter serotonin (5-hydroxytryptamine) from tryptophan. The abnormal metabolism of the cofactor can be due to a deficiency of the reductase, which reduces biopterin to its active form, tetrahydrobiopterin.

Amino Acids as Precursors of Biologically Important Compounds

I. AMINO ACIDS AS PRECURSORS OF NEUROTRANSMITTERS AND HORMONES

A. Biosynthesis of catecholamines from tyrosine. Each of the three closely related catecholamines—dopamine, norepinephrine, and epinephrine—has a specific function as a neurotransmitter or hormone.

1. Formation of dopa

a. Tyrosine is hydroxylated to 3,4-dihydroxyphenylalanine (dopa) by tyrosine hydroxylase, a mixed-function oxidase, which requires tetrahydrobiopterin as a cofactor.

b. Reduced nicotinamide-adenine dinucleotide phosphate (NADPH) is the reductant in the regeneration of tetrahydrobiopterin from dihydrobiopterin.

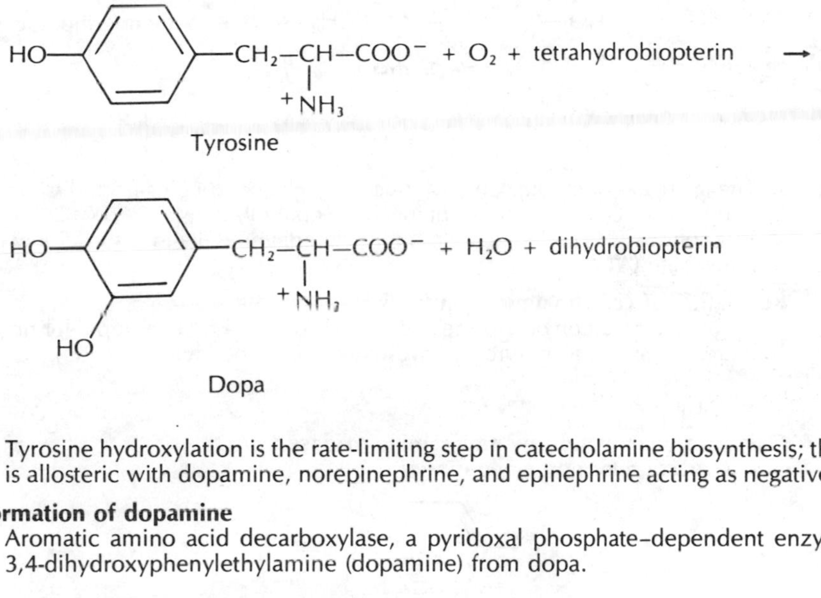

$$HO-\langle\rangle-CH_2-CH-COO^- + O_2 + \text{tetrahydrobiopterin} \longrightarrow$$
$$\overset{|}{{}^+NH_3}$$

Tyrosine

$$HO-\langle\rangle-CH_2-CH-COO^- + H_2O + \text{dihydrobiopterin}$$
$$\overset{|}{{}^+NH_3}$$
$$HO$$

Dopa

c. Tyrosine hydroxylation is the rate-limiting step in catecholamine biosynthesis; the enzyme is allosteric with dopamine, norepinephrine, and epinephrine acting as negative effectors.

2. Formation of dopamine

a. Aromatic amino acid decarboxylase, a pyridoxal phosphate–dependent enzyme, forms 3,4-dihydroxyphenylethylamine (dopamine) from dopa.

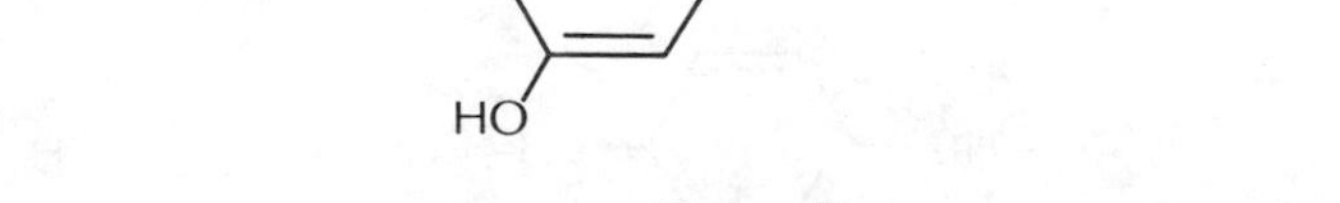

$$\text{Dopa} + H^+ \longrightarrow HO-\langle\rangle-CH_2-CH_2-{}^+NH_3 + CO_2$$
$$HO$$

Dopamine

b. Dopamine serves as a neurotransmitter, particularly in some neurons of the basal ganglia.

3. Formation of norepinephrine

a. Another mixed-function oxidase, dopamine β-hydroxylase, hydroxylates dopamine on the side chain to yield norepinephrine.

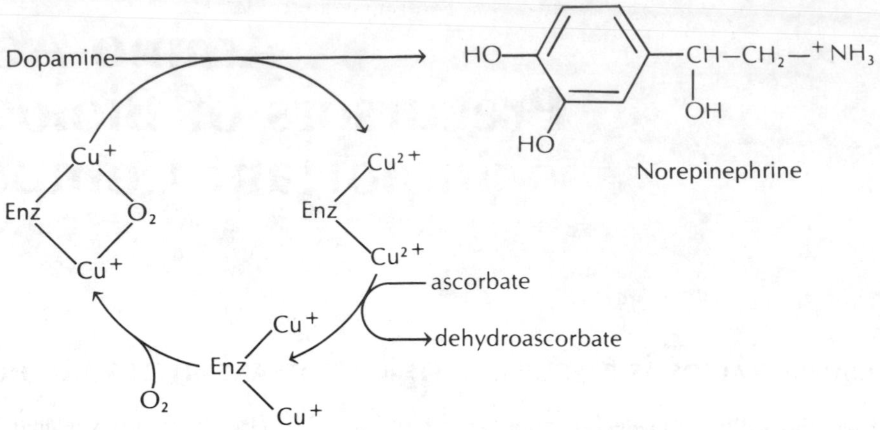

Norepinephrine

b. The enzyme contains copper and requires ascorbate and molecular oxygen.
c. Norepinephrine is the major neural transmitter of the sympathetic nervous system.

4. Formation of epinephrine
  a. In the adrenal medulla, norepinephrine is methylated by phenylethanolamine N-methyl transferase, using S-adenosylmethionine as the methyl donor.

Norepinephrine + S-adenosylmethionine ⟶

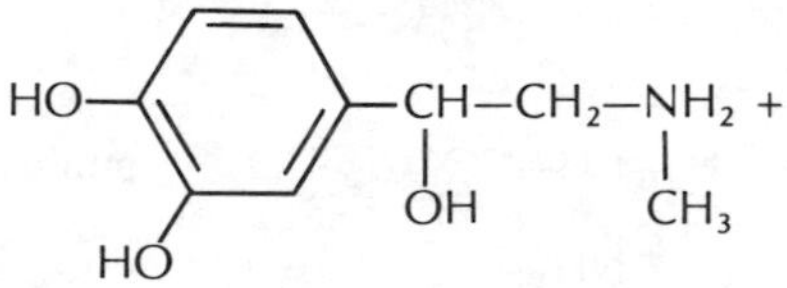

Epinephrine

  b. The synthesis of the enzyme is induced by glucocorticoids from the surrounding adrenal cortex, and its production is inhibited by epinephrine.
  c. Epinephrine is stored in granules in the adrenal medulla in association with adenosine triphosphate (ATP).

5. Regulation of catecholamine synthesis is exerted by
  a. Feedback inhibition of tyrosine hydroxylase by dopamine, norepinephrine, or epinephrine
  b. Increased synthesis of tyrosine hydroxylase by repetitive nerve firing

B. Catabolism of catecholamines

1. Oxidative deamination by monoamine oxidase (MAO)
  a. The reaction for norepinephrine is

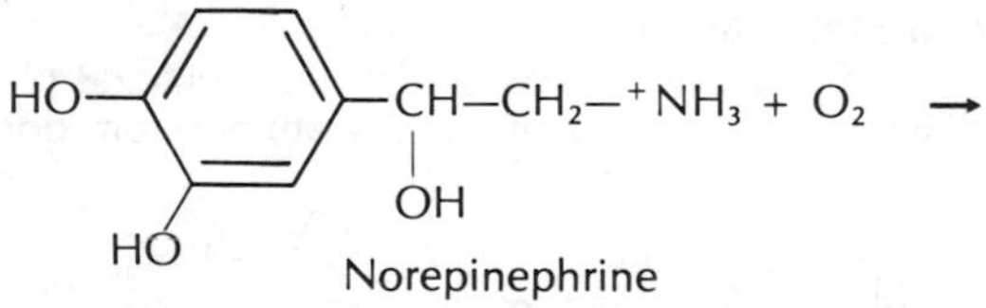

Norepinephrine

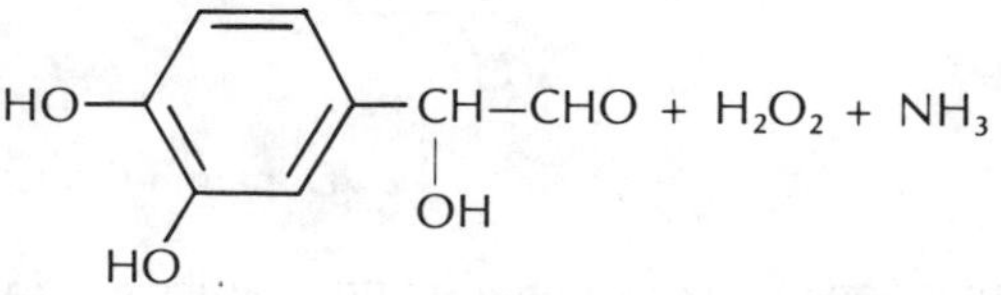

3,4-Dihydroxyphenylglycoaldehyde

  b. The aldehyde formed is then oxidized, in this case to 3,4-dihydroxymandelate, by aldehyde dehydrogenase.

2. Methylation by catechol-O-methyl transferase (COMT)

a. The reaction for norepinephrine is

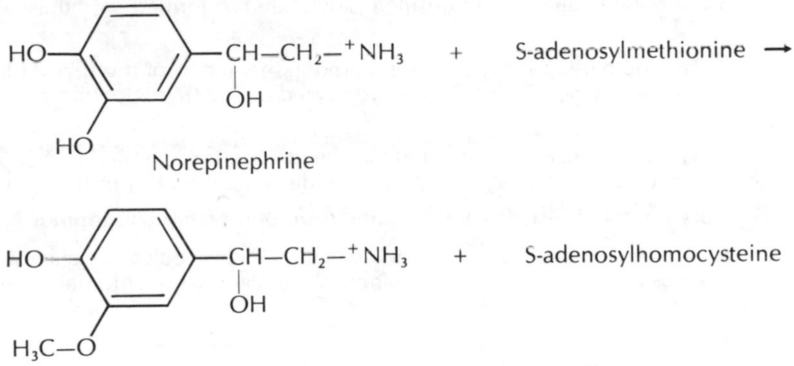

Norepinephrine + S-adenosylmethionine →

3-Methoxynorepinephrine (normetanephrine) + S-adenosylhomocysteine

b. This is followed by oxidation by MAO and aldehyde dehydrogenase to give the major excretory product, 3-methoxy-4-hydroxymandelic acid (vanillylmandelic acid, VMA).

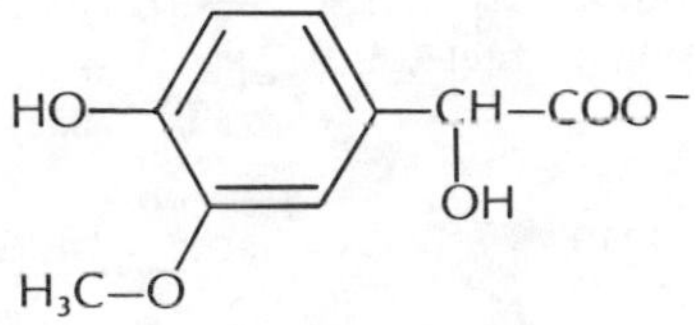

Vanillylmandelic acid

3. Catabolism may also occur with the reaction sequence of MAO and COMT reversed. Epinephrine, norepinephrine, and their methylated derivatives are usually excreted as conjugates with sulfate or glucuronic acid.

4. VMA excretion levels are used as an aid in the diagnosis of adrenal pheochromocytomas, tumors that produce huge amounts of catecholamines.

C. Biosynthesis of thyroxine and triiodothyronine from tyrosine

1. Two iodinated derivatives of tyrosine are secreted by the thyroid gland and function as hormones. They are

a. 3,3′,5,5′-Tetraiodothyronine (thyroxine or T_4)

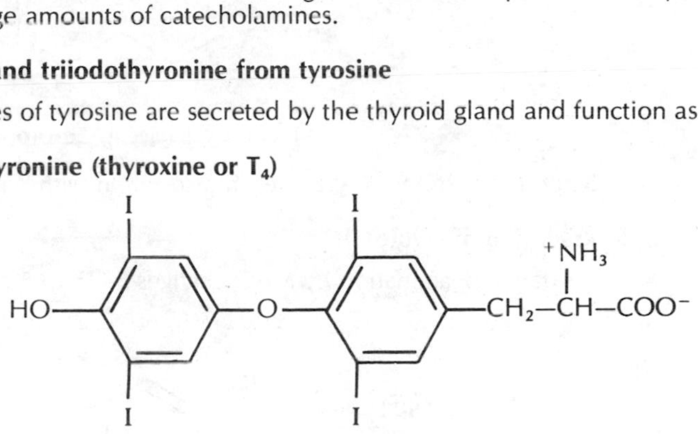

Thyroxine (T_4)

b. 3,3′,5-Triiodothyronine (T_3)

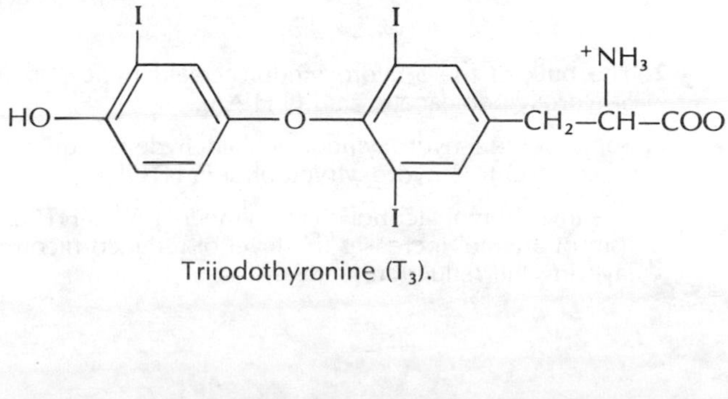

Triiodothyronine (T_3).

2. The hormones are formed in the follicle cells of the thryoid gland by iodination of tyrosine residues, which are in peptide linkage in chains of the protein thyroglobulin.

3. Monoiodo- and diiodotyrosine residues are first formed, and these then react to form T_3 and T_4.

4. The iodinated thyroglobulin is stored in the lumen of the thyroid follicles until required for secretion when T_3 and T_4 are hydrolyzed off the thyroglobulin and secreted into the circulation.

5. T_3 is the more active form and may be the only form to bind to receptors in cell nuclei. About two-thirds of the plasma T_3 arises by deiodination of T_4 in the liver.

D. Biosynthesis of 5-hydroxytryptamine (serotonin) from tryptophan

1. Serotonin is synthesized by neurons whose cell bodies are chiefly located in the hypothalamus and brain stem, by the pineal gland, and by the chromaffin cells of the gastrointestinal tract.

2. The rate-limiting step is catalyzed by tryptophan hydroxylase, which requires tetrahydrobiopterin as the reductant and which forms 5-hydroxytryptophan (5-HTP).

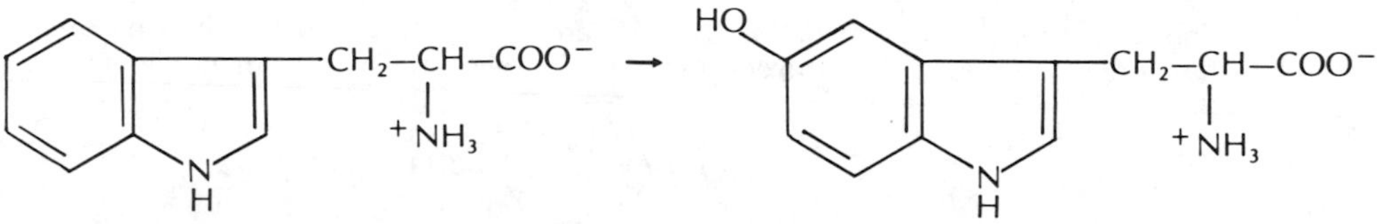

Tryptophan   5-Hydroxytryptophan

3. Decarboxylation occurs by the same enzyme that decarboxylates dopa to dopamine (see section I A 2), yielding serotonin (5-hydroxytryptamine, 5-HT).

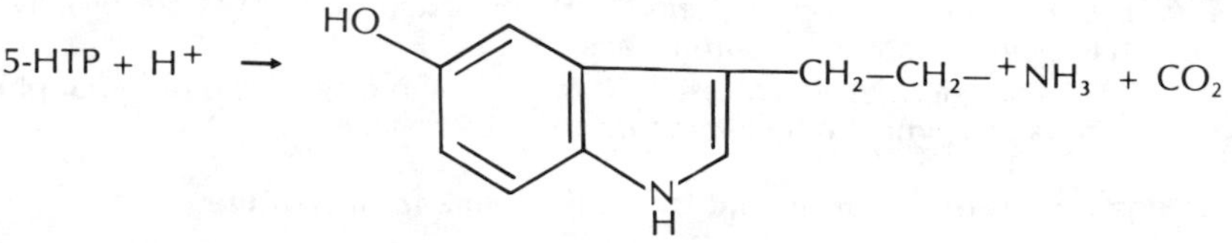

Serotonin

4. Serotonin is stored in granules in association with ATP.

E. Degradation of serotonin

1. Oxidative deamination by MAO changes 5-HT to 5-hydroxyindoleacetaldehyde.

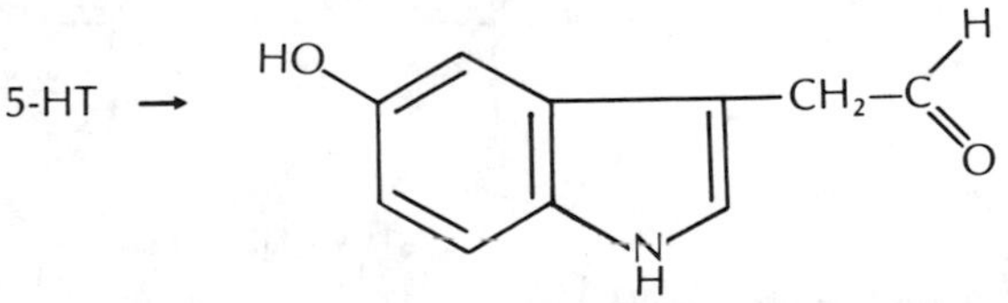

5-Hydroxyindoleacetaldehyde

2. The bulk of the 5-hydroxyindoleacetaldehyde is oxidized by aldehyde dehydrogenase to 5-hydroxyindoleacetic acid (5-HIAA).

3. Some of the 5-hydroxyindoleacetaldehyde is reduced by aldehyde reductase (alcohol dehydrogenase) to 5-hydroxytryptophol (5-HTOL).

4. The ingestion of alcohol changes the 5-HIAA:5-HTOL ratio in the urine because the metabolism of alcohol increases the level of reduced nicotinamide-adenine dinucleotide (NADH), favoring the reduction pathway.

F. Biosynthesis of melatonin

1. Melatonin is a hormone produced by the pineal gland, which has effects on the hypothalamic–pituitary system.
2. 5-HT is formed in the pineal gland and converted to 5-hydroxy-N-acetyl tryptamine by N-acetyl transferase.

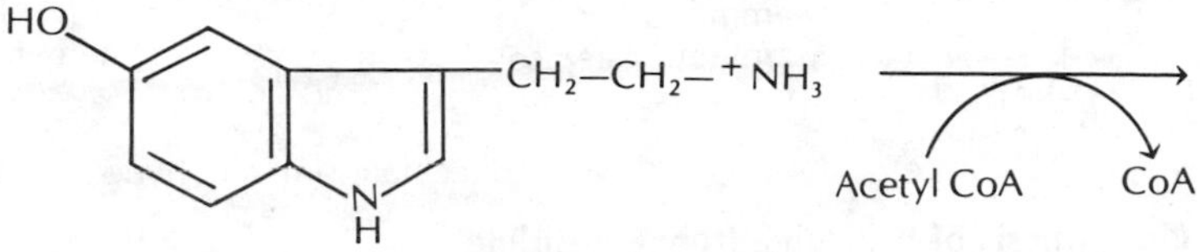

5-Hydroxytryptamine

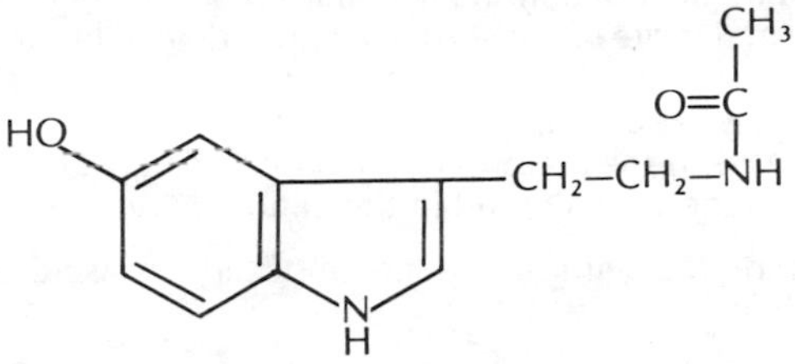

5-Hydroxy-N-acetyl tryptamine

3. 5-Hydroxy-N-acetyl tryptamine is then methylated by an O-methyl transferase and S-adenosylmethionine to melatonin.

5-Hydroxy-N-acetyl tryptamine + S-adenosylmethionine →

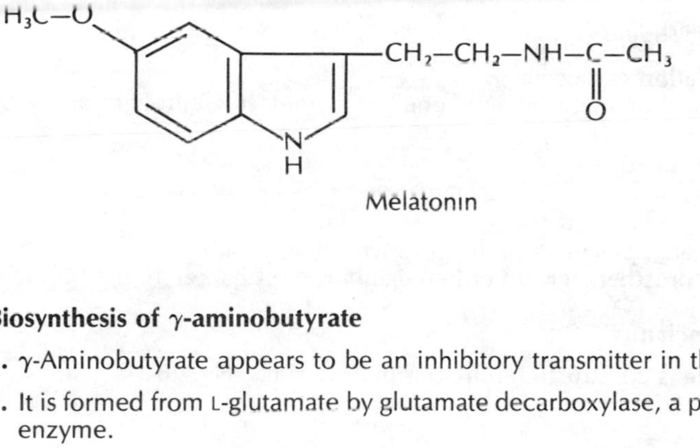

\+ S-adenosylhomocysteine

Melatonin

G. Biosynthesis of γ-aminobutyrate

1. γ-Aminobutyrate appears to be an inhibitory transmitter in the brain and spinal cord.
2. It is formed from L-glutamate by glutamate decarboxylase, a pyridoxal phosphate–dependent enzyme.

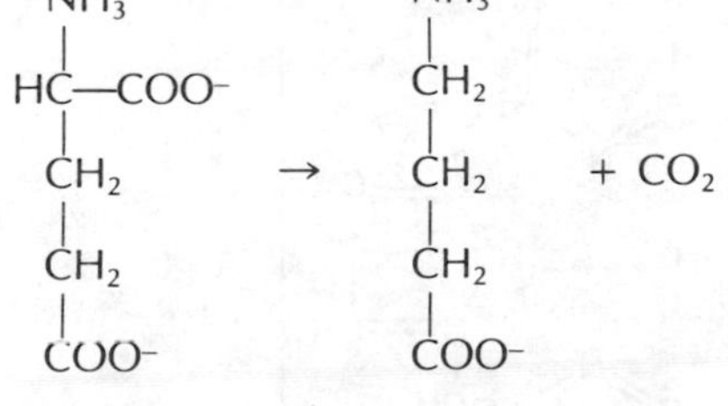

L-Glutamate γ-Aminobutyrate

3. γ-Aminobutyrate is metabolized within the neurons to succinate, bypassing the α-ketoglutarate dehydrogenase step of the Krebs cycle.

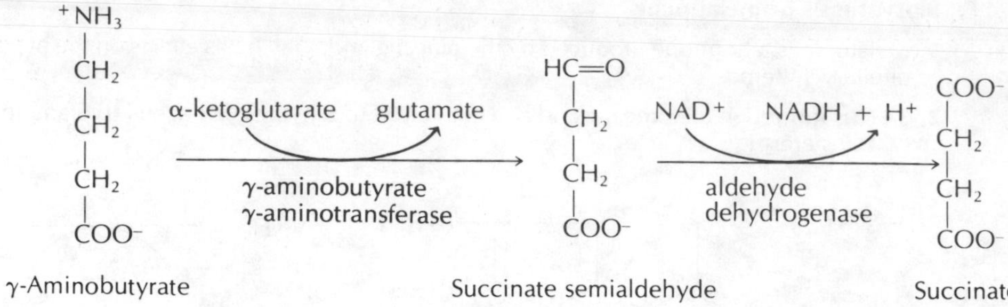

γ-Aminobutyrate   Succinate semialdehyde   Succinate

H. Biosynthesis of histamine from L-histidine

1. **Histamine distribution.** Most of the body histamine "pool" is in the skin, lungs, and stomach.
   a. The major site of conversion of L-histidine to histamine is in the mast cells found in connective tissue near blood vessels and in the pleural membrane of the lung.
   b. The release of histamine from gastric mucosal cells (whose identity is not clear) causes an increase in the secretion of hydrochloric acid by the oxyntic gastric mucosal cells.
   c. Histamine is also found in the central nervous system and may be a neural transmitter.

2. **Histamine biosynthesis.** Histidine decarboxylase catalyzes the formation of histamine from L-histidine.

L-Histidine   Histamine

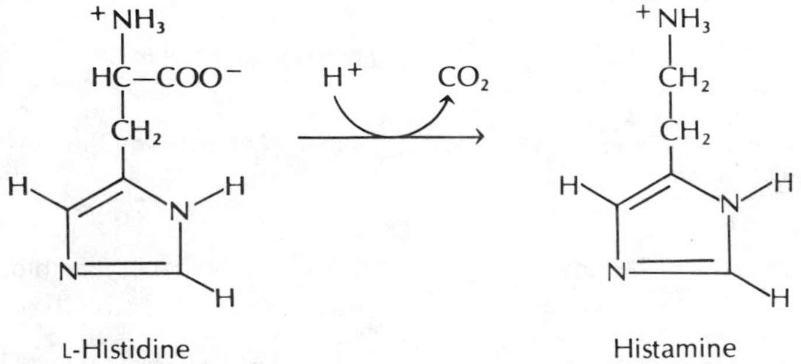

3. **Histamine degradation** can occur by
   a. S-adenosylmethionine methylation of one of the imidazole nitrogens and subsequent oxidation by MAO
   b. Direct oxidation by MAO

II. PORPHYRINS

A. **Glycine** is the amino acid precursor of the porphyrin ring, which plays an essential role as the nucleus of the **heme prosthetic group of hemoglobin** (see Chapter 10, section I B 1 d)

B. **Structure and nomenclature**

1. The **porphyrin ring** is constructed from four **pyrrole rings** (Fig. 28-1).

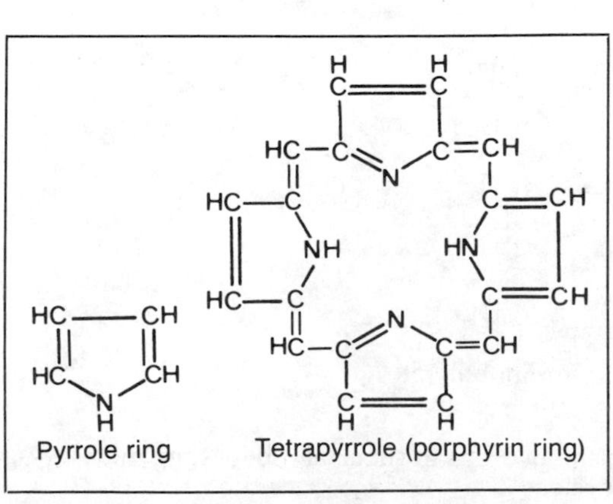

Pyrrole ring   Tetrapyrrole (porphyrin ring)

Figure 28-1. The pyrrole ring as the building block of the porphyrins.

Table 28-1. Kinds of Porphyrins and Their Constituent Pyrroles

Porphyrin	Constituent Pyrroles					
Uroporphyrin	4					
Coproporphyrin		4				
Protoporphyrin		2	2			
Etioporphyrin				4		
Hematoporphyrin		2			2	
Mesoporphyrin		2		2		
Deuteroporphyrin		2				2

Note.—Reprinted with permission from McGilvery RW: *Biochemistry: A Functional Approach*. Philadelphia, WB Saunders, 1970, p 632.

2. The porphyrins are named and classified by the side chains of their constituent pyrroles.
   a. Each of the four pyrrole rings has two side chains, and the name of the porphyrin reflects the type of side chains present (Table 28-1).
   b. The side chains of the pyrrole rings can be different among porphyrins of a given name, and these differences are indicated by a Roman numeral (I through XV).
   c. All naturally occurring porphyrins are derived from uroporphyrinogen I and uroporphyrinogen III. The porphyrin in heme is designated protoporphyrin IX.

C. **Biosynthetic pathway.** An outline of the synthesis of protoporphyrin IX and its relation to hemoglobin synthesis is shown in Figure 28-2.

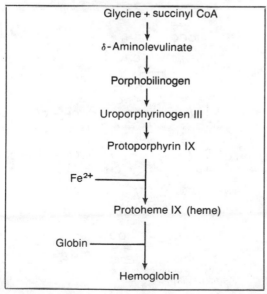

Figure 28-2. An outline of the synthesis of protoporphyrin IX in relation to hemoglobin synthesis.

1. δ-Aminolevulinate synthesis. The first and the rate-controlling step of porphyrin biosynthesis occurs in the mitochondria; it is the condensation of succinyl coenzyme A (succinyl CoA) and glycine to form δ-aminolevulinate, catalyzed by δ-aminolevulinate synthetase.

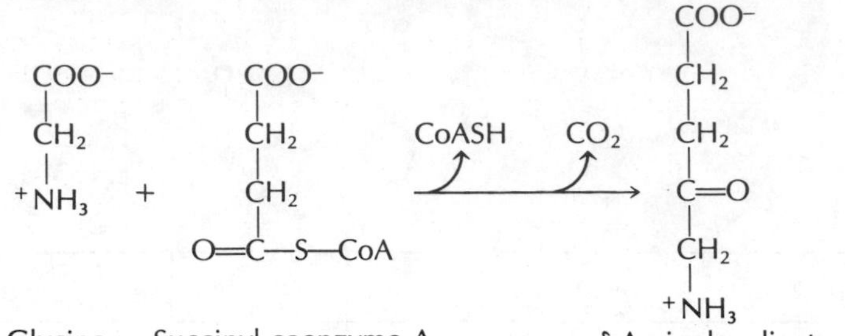

Glycine   Succinyl coenzyme A   δ-Aminolevulinate

2. Porphobilinogen synthesis. A cytoplasmic enzyme, porphobilinogen synthetase (δ-aminolevulinate dehydratase), forms porphobilinogen from two molecules of δ-aminolevulinate.

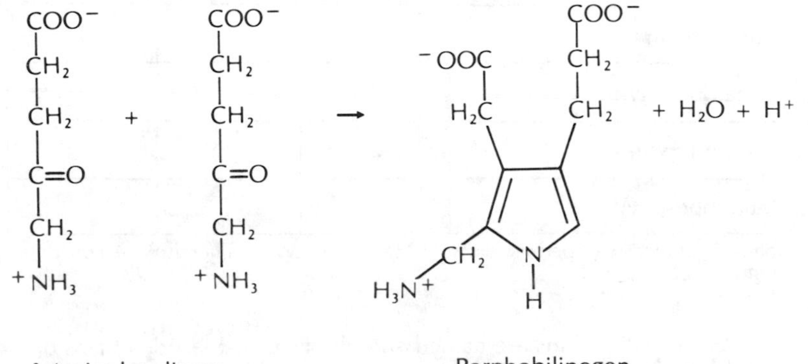

δ-Aminolevulinate   Porphobilinogen

3. Tetrapyrrole synthesis. The condensation of four porphobilinogen units to form uroporphyrinogen III requires two proteins.

   a. The enzyme uroporphyrinogen synthetase on its own catalyzes the *head-to-tail* condensation of porphobilinogen molecules to form uroporphyrinogen I.

   b. A noncatalytic protein, uroporphyrinogen III cosynthetase, alters the specificity of uroporphyrinogen I synthetase, so that one of the porphobilinogen molecules is condensed *head-to-head* to give uroporphyrinogen III (Fig. 28-3).

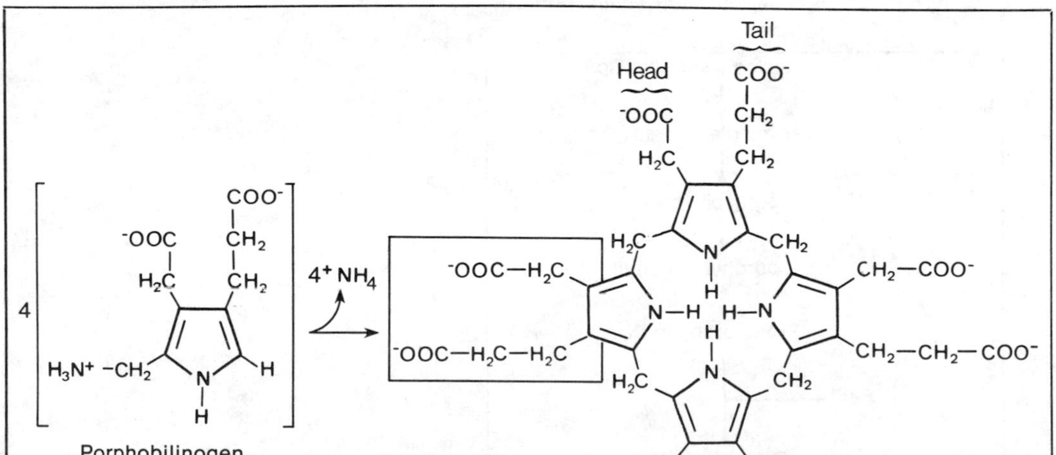

Figure 28-3. Formation of uroporphyrinogen III from four porphobilinogen molecules. In the presence of the cosynthetase, the pyrrole in the *box* is joined head-to-head, rather than head-to-tail, as the others are.

4. Heme synthesis

a. All aerobic mammalian cells can synthesize heme, as they all must synthesize the heme-containing cytochrome respiratory enzymes.

b. Enzymes for heme synthesis are found in relatively high concentration in the liver, bone marrow, intestinal mucosa, nucleated erythrocytes (reticulocytes), and kidney.

c. Conversion of uroporphyrinogen III to protoheme IX (heme)

(1) Uroporphyrinogen III is decarboxylated at its aceto side chains by a cytosolic enzyme to form coproporphyrinogen III with methyl side chains (Fig. 28-4).

(2) Two mitochondrial enzymes then oxidatively decarboxylate two side chains and oxidize the methylene bridges to form protoporphyrin IX (see Fig. 28-4).

(3) The final step in the formation of protoheme IX, or **heme**, is the addition of Fe^{2+} by a mitochondrial enzyme (Fig. 28-5).

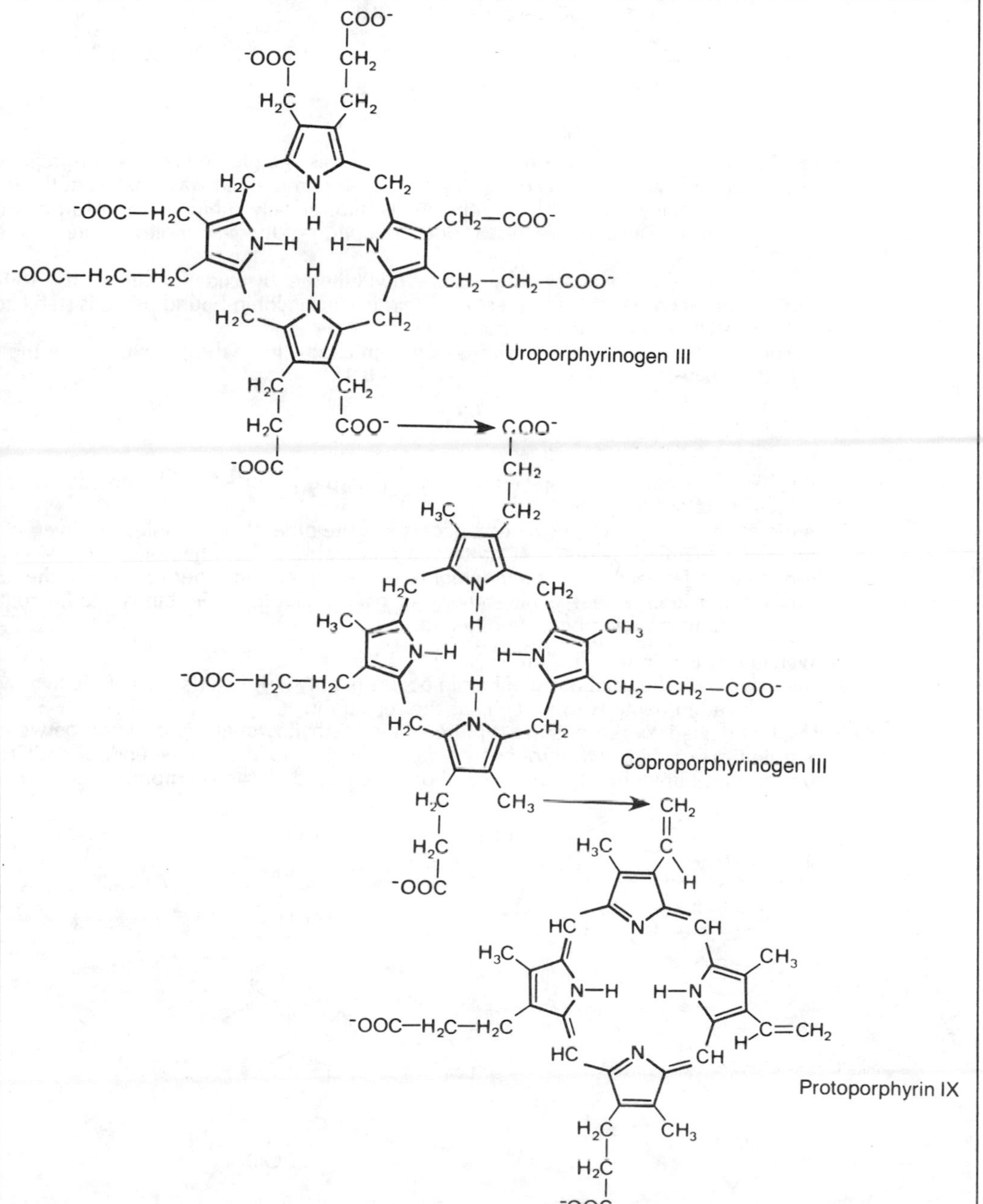

Figure 28-4. The sequence from uroporphyrinogen III to protoporphyrin IX.

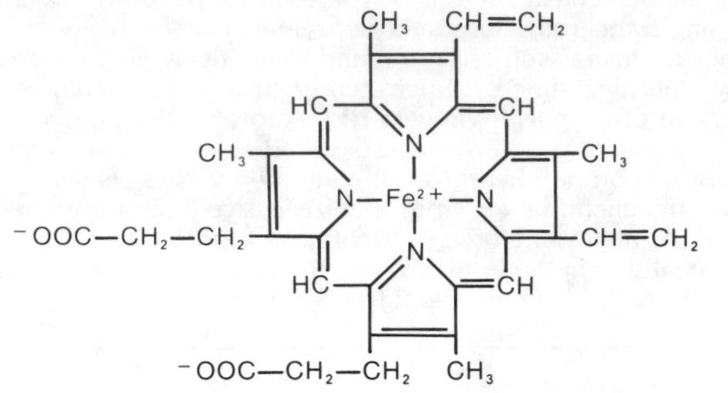

Figure 28-5. Protoheme IX, also known as heme.

5. **Hemoglobin synthesis regulation.** Both heme synthesis and globin synthesis are regulated.
    a. Regulation of **heme synthesis** occurs at the first step in the pathway, that is, at the formation of δ-aminolevulinate. The synthetase is allosterically inhibited by **protohemin IX (hemin)**, which is also a corepressor of transcription of the δ-aminolevulinate synthetase gene.
    b. Protohemin IX (hemin) arises from protoheme IX (heme) by oxidation of the Fe^{2+} to Fe^{3+}. This occurs when heme is in excess of globin because globin-bound heme is stabilized in the Fe^{2+} state, but free heme is not.
    c. Protohemin IX regulates the **synthesis of globin chains** in erythropoietic cells at the level of mRNA translation (see Chapter 8, section I B 1).

D. Porphyrin degradation

1. **Red cell destruction**
    a. This usually occurs in the spleen, which is thus the major site of heme catabolism, although some occurs in the liver.
    b. In the event that red cell destruction occurs at a site other than the spleen or liver (e.g., in hemolytic anemias), two carrier proteins are available to bind hemoglobin in which the iron is in the ferric state (methemoglobin), or free hemin, in order to prevent the loss of iron via the kidney that could otherwise occur. **Haptoglobin** binds methemoglobin dimers, and **hemopexin** binds free hemin.

2. **Conversion of hemin to bilirubin**
    a. The first step in the catabolism of hemin occurs mainly in Kupffer's cells of the liver where it is reduced to heme with NADPH as the reductant.
    b. The first degradative step is the rupture of the methylidyne group of heme between the pyrrole rings carrying vinyl groups by a mixed-function oxidase in the endoplasmic reticulum to form **biliverdin**. The bridge carbon is removed as carbon monoxide.

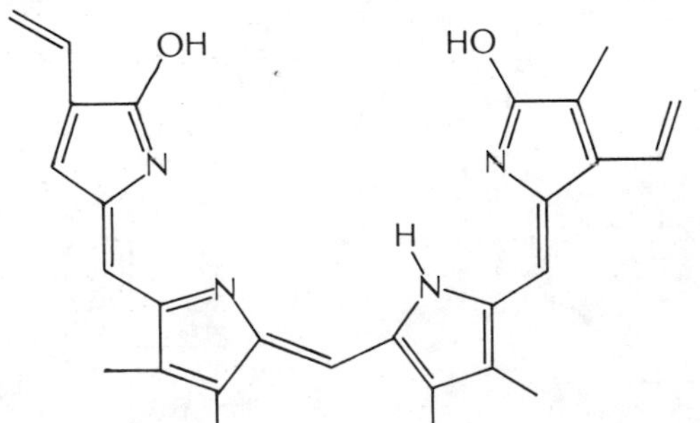

Biliverdin

c. The iron dissociates, yielding the free tetrapyrrole biliverdin, which has a green color.
d. Biliverdin is reduced (with NADPH as the reductant) to **bilirubin**, which is reddish-brown in color.

Bilirubin

3. Excretion of bilirubin
a. Bilirubin, which is poorly soluble in water, is transferred to the liver bound to plasma albumin.
   (1) In the liver cell, it is transported through the cytoplasm to the endoplasmic reticulum bound to a protein called ligandin.
   (2) Here it is conjugated to glucuronic acid by a specific glucuronyl transferase that transfers the glucuronyl moiety from uridine diphosphate glucuronate (UDP-glucuronate) [see Chapter 19, section II B 3 a] to one of the propionyl side chains of bilirubin.
   (3) At the plasma membrane, the diglucuronide of bilirubin is formed by the reaction of two molecules of bilirubin monoglucuronide to form bilirubin diglucuronide and free bilirubin. This last reaction is catalyzed by bilirubin glucuronidoside glucuronyl transferase.
   (4) The water-soluble bilirubin diglucuronide is secreted into the bile.
b. Failure to remove bilirubin as the glucuronide leads to jaundice in which the yellowing of the skin and sclera are due to the deposition of bilirubin and bilirubin diglucuronide.
   (1) In biliary tract obstruction, plasma levels of bilirubin diglucuronide may be increased.
   (2) If the conjugation system is damaged, as in liver disease, or is nonfunctional due to abnormal development or genetic absence, free bilirubin accumulates.
   (3) **The van den Bergh reaction** is a means of determining whether jaundice is due to the accumulation of free bilirubin or the glucuronide. Samples of plasma are reacted with diazotized sulfanilic acid both in aqueous solution and in an organic solvent.
      (a) The aqueous solution ("direct bilirubin") measures only the conjugated bilirubin diglucuronide.
      (b) The organic solvent ("total bilirubin") measures both free and conjugated bilirubin.
      (c) "Indirect bilirubin," a measure of the amount of unconjugated bilirubin, is obtained by taking the difference between the total and the direct measurements.
   (4) **Phototherapy**
      (a) Currently, the irradiation of jaundiced infants during neonatal life by fluorescent lights is the most common treatment of neonatal hyperbilirubinemia.
      (b) Exposure of infants to fluorescent light therapy was first shown to decrease serum bilirubin levels in 1958.
      (c) The precise mechanism by which phototherapy induces reductions of serum bilirubin concentrations remains unclear, but in vitro studies and studies with the Gunn rat (hereditary lack of the conjugation system) have suggested that absorbance of light energy results in a solubilization of the bilirubin.
      (d) The products from the irradiation of bilirubin are more soluble than bilirubin and can be excreted by the liver into the bile without conjugation with glucuronic acid. These products have been known collectively as "photobilirubin."
c. Bilirubin diglucuronide is hydrolyzed to free bilirubin in the bowel, where microorganisms reduce the bilirubin to **urobilinogens** and **stercobilinogens**. Part of the urobilinogen is absorbed from the bowel and excreted in the urine.

d. The reduced bilinogens are colorless compounds, which become colored by partial oxidation on exposure to air and are responsible for the color of urine and feces.

E. Transport and storage of iron (Fe^{3+})

1. Fe^{3+} is transported in the blood bound to a protein synthesized in the liver, **transferrin**, which has a very high affinity for Fe^{3+}, binding two atoms per molecule.

2. Fe^{3+} is stored in the cells in combination with the protein **ferritin**. Ferritin can bind a very large number of Fe^{3+} atoms because its 24 subunits, arranged in a sphere, form channels by which the Fe^{3+} enters the core to be deposited as a hydroxyphosphate.

3. With very high Fe^{3+} intakes, some of the Fe^{3+} is found in granules of **hemosiderin**, a complex of Fe^{3+}, protein, and polysaccharide, which is 3% Fe^{3+} by weight.

F. Abnormal porphyrin biosynthesis. Accumulations of porphyrins may occur under conditions of genetically defective porphyrin synthesis in the liver (**hepatic porphyrias**) or in the erythropoietic tissues (**erythropoietic porphyrias**), as these systems are under separate genetic control.

1. **Acute intermittent porphyria** is a hepatic porphyria.
   a. It is due to a deficiency of uroporphyrinogen I synthetase and is an autosomal dominant condition.
   b. Excessive amounts of porphobilinogen and δ-aminolevulinate are formed.
   c. The symptoms are acute abdominal pain and neurologic symptoms (e.g., the "madness" of George III).

2. **Congenital erythropoietic porphyria** is a rare disease characterized by the urinary excretion of large amounts of uroporphyrin I.
   a. It is an autosomal recessive condition, causing mutilating skin lesions and hemolytic anemia.
   b. A decrease of uroporphyrinogen III cosynthetase to about one-tenth to one-third of the normal level may be the basis of the disease.

III. MISCELLANEOUS COMPOUNDS

A. Creatine from glycine, arginine, and S-adenosylmethionine

1. Creatine phosphate is a high-energy phosphate-storage compound found in muscle and formed from parts of three amino acids.

2. **Synthesis**
   a. The guanidinium group of arginine is transferred to glycine to form **guanidinoacetate**.

b. N-Methylation of the guanidinoacetate by S-adenosylmethionine forms creatine.

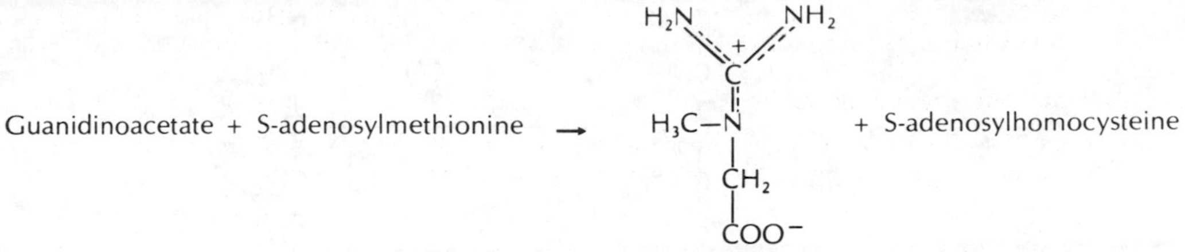

$$H_2N \diagdown \quad \diagup NH_2$$

Guanidinoacetate + S-adenosylmethionine $\longrightarrow$ H_3C-N ... $+$ S-adenosylhomocysteine

$$CH_2$$
$$COO^-$$

Creatine

c. Phosphorylation of creatine to creatine phosphate occurs at the expense of ATP.

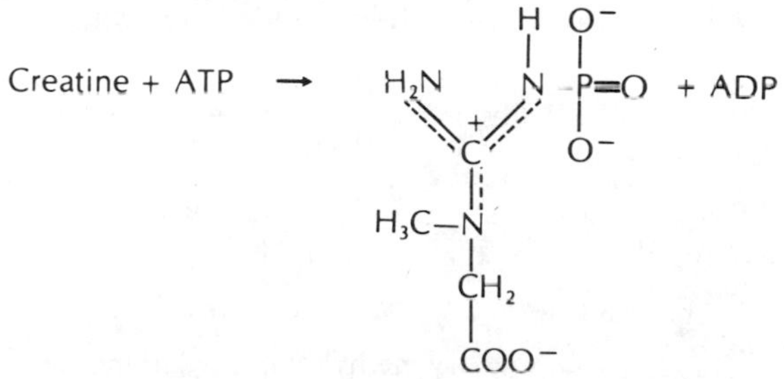

Creatine + ATP $\longrightarrow$ $H_2N \diagdown \quad \diagup N - P=O + ADP$

Creatine phosphate

3. Formation and excretion of creatinine

a. A constant fraction of the muscle pool of creatine phosphate spontaneously cyclizes to creatinine, which is excreted in the urine.

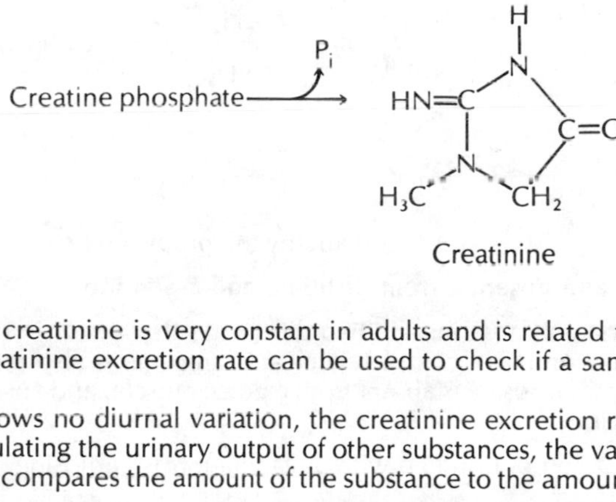

Creatine phosphate $\xrightarrow{P_i}$ Creatinine structure

Creatinine

b. The excretion of creatinine is very constant in adults and is related to muscle mass.

(1) The 24-hr creatinine excretion rate can be used to check if a sample of urine is complete.

(2) Because it shows no diurnal variation, the creatinine excretion rate can be used as a basis for calculating the urinary output of other substances, the values being expressed as a ratio that compares the amount of the substance to the amount of creatinine in the urine sample.

(3) Urinary creatinine levels also give a measure of body cell mass (i.e., muscle mass) by relating the milligrams of creatinine excreted in 24 hr to height, the creatinine–height index being the ratio of these parameters to the values for subjects of normal body composition.

(4) Lastly, blood creatinine levels are a good indicator of kidney function, as creatinine is normally cleared very efficiently from the plasma by the kidney.

B. Polyamines, spermidine, and spermine from ornithine

1. The polyamines are polycations at physiologic pH and associate with negatively charged cell components, such as membranes and nucleic acids.

2. Ornithine decarboxylase, an inducible enzyme with an extremely short half-life ($\sim$ 20 min), causes the decarboxylation of ornithine to form **putrescine**.

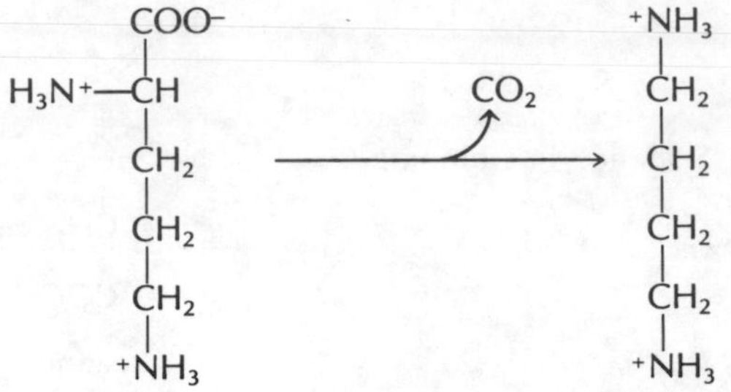

$$
\begin{array}{c}
\text{COO}^- \\
| \\
\text{H}_3\text{N}^+\!\!-\!\!\text{CH} \\
| \\
\text{CH}_2 \\
| \\
\text{CH}_2 \\
| \\
\text{CH}_2 \\
| \\
^+\text{NH}_3
\end{array}
\qquad
\xrightarrow{\;\;\text{CO}_2\;\;}
\qquad
\begin{array}{c}
^+\text{NH}_3 \\
| \\
\text{CH}_2 \\
| \\
\text{CH}_2 \\
| \\
\text{CH}_2 \\
| \\
\text{CH}_2 \\
| \\
^+\text{NH}_3
\end{array}
$$

Ornithine   Putrescine

3. Condensation of putrescine with an aminopropyl residue donated by S-adenosylmethylthio-propylamine, the latter derived by decarboxylation of S-adenosylmethionine, yields spermidine.

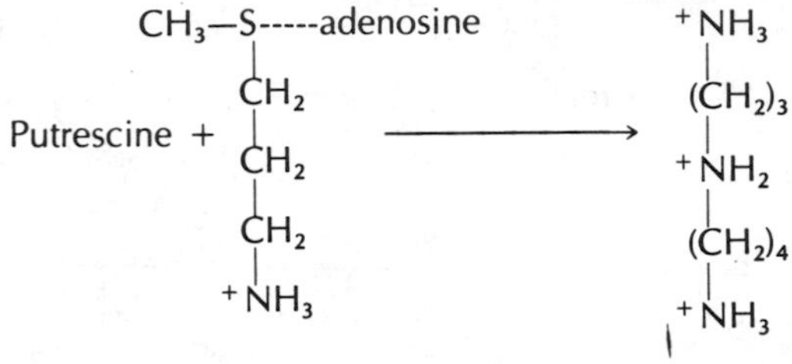

$$
\text{Putrescine} +
\begin{array}{c}
\text{CH}_3\!\!-\!\!\text{S}\text{-----adenosine} \\
| \\
\text{CH}_2 \\
| \\
\text{CH}_2 \\
| \\
\text{CH}_2 \\
| \\
^+\text{NH}_3
\end{array}
\qquad \longrightarrow \qquad
\begin{array}{c}
^+\text{NH}_3 \\
| \\
(\text{CH}_2)_3 \\
| \\
^+\text{NH}_2 \\
| \\
(\text{CH}_2)_4 \\
| \\
^+\text{NH}_3
\end{array}
$$

S-Adenosylmethylthiopropylamine   Spermidine

4. Condensation of spermidine with another aminopropyl residue yields spermine.

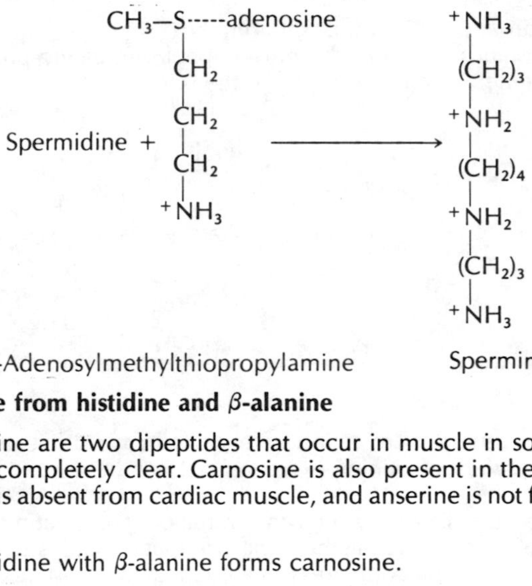

$$
\text{Spermidine} +
\begin{array}{c}
\text{CH}_3\!\!-\!\!\text{S}\text{-----adenosine} \\
| \\
\text{CH}_2 \\
| \\
\text{CH}_2 \\
| \\
\text{CH}_2 \\
| \\
^+\text{NH}_3
\end{array}
\qquad \longrightarrow \qquad
\begin{array}{c}
^+\text{NH}_3 \\
| \\
(\text{CH}_2)_3 \\
| \\
^+\text{NH}_2 \\
| \\
(\text{CH}_2)_4 \\
| \\
^+\text{NH}_2 \\
| \\
(\text{CH}_2)_3 \\
| \\
^+\text{NH}_3
\end{array}
$$

S-Adenosylmethylthiopropylamine   Spermine

C. Carnosine and anserine from histidine and β-alanine

1. Carnosine and anserine are two dipeptides that occur in muscle in some species, although their function is not completely clear. Carnosine is also present in the olfactory pathway in the brain. Carnosine is absent from cardiac muscle, and anserine is not found in human skeletal muscle.

2. Condensation of histidine with β-alanine forms carnosine.

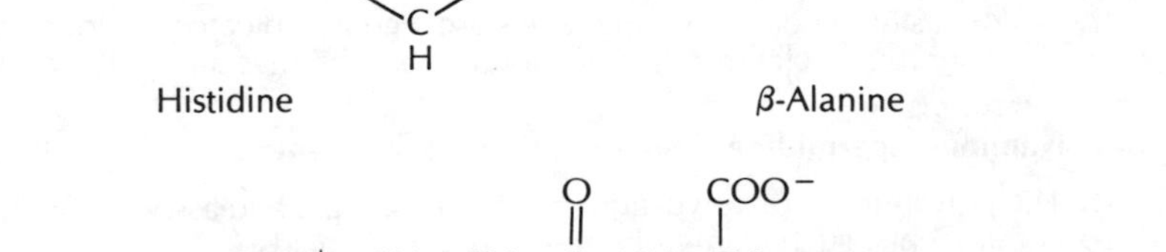

$$
\begin{array}{c}
^+\text{NH}_3 \\
| \\
^-\text{OOC}\!\!-\!\!\text{CH}\!\!-\!\!\text{CH}_2\!\!-\!\!\text{C}\!\!=\!\!\!=\!\!\text{CH} \\
\qquad\quad \text{HN} \quad\;\; \text{NH}^+ \\
\qquad\qquad\;\; \diagdown\text{C}\diagup \\
\qquad\qquad\qquad | \\
\qquad\qquad\qquad \text{H}
\end{array}
\;+\;
\text{H}_3\text{N}^+\!\!-\!\!\text{CH}_2\!\!-\!\!\text{CH}_2\!\!-\!\!\text{COO}^-
\;\longrightarrow
$$

Histidine   β-Alanine

$$
\text{H}_3\text{N}^+\!\!-\!\!\text{CH}_2\!\!-\!\!\text{CH}_2\!\!-\!\!\overset{\displaystyle \text{O}}{\overset{\|}{\text{C}}}\!\!-\!\!\text{NH}\!\!-\!\!\overset{\displaystyle \text{COO}^-}{\underset{}{\text{CH}}}\!\!-\!\!\text{CH}_2\!\!-\!\!\langle\text{imidazole}\rangle
$$

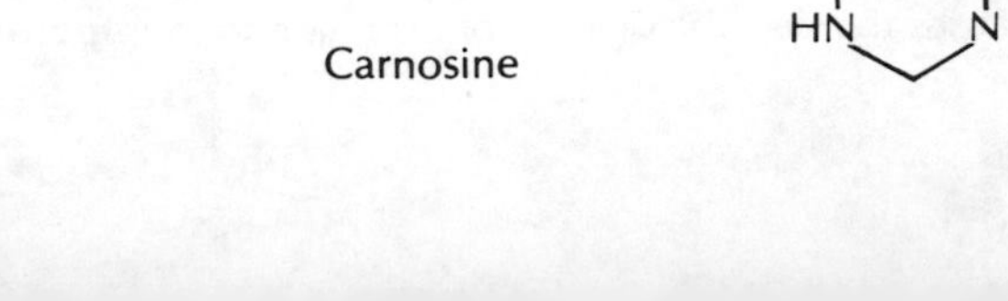

Carnosine

3. Methylation of carnosine by S-adenosylmethionine forms anserine.

Carnosine + S-adenosylmethionine $\longrightarrow$

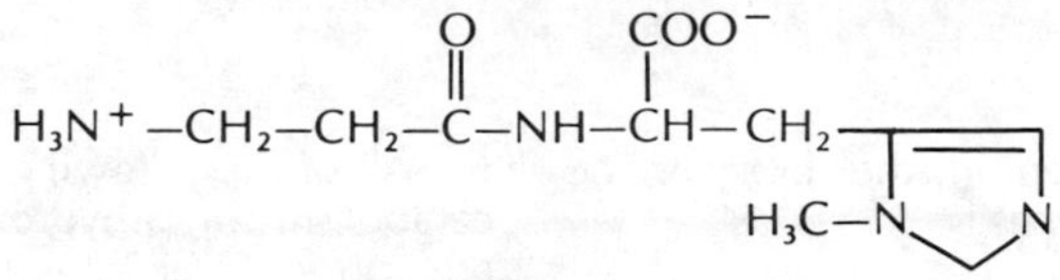

$$H_3N^+ - CH_2 - CH_2 - \overset{\overset{\displaystyle O}{\|}}{C} - NH - \overset{\overset{\displaystyle COO^-}{|}}{CH} - CH_2 \underset{H_3C - N \diagdown \diagdown N}{\boxed{}} \quad + \text{ S-adenosylhomocysteine}$$

Anserine

D. The use of amino acids in nucleotide biosynthesis is described in Chapter 29.

STUDY QUESTIONS

Directions: Each question below contains five suggested answers. Choose the **one best** response to each question.

1. The rate-limiting step in the biosynthetic pathway for catecholamines is

(A) the hydroxylation of tyrosine
(B) the hydroxylation of phenylalanine to tyrosine
(C) the formation of dopamine
(D) the reduction of biopterin
(E) none of the above

2. Epinephrine is formed from norepinephrine by

(A) hydroxylation
(B) decarboxylation
(C) oxidative deamination
(D) O-methylation
(E) N-methylation

3. Which of the following substances is present in high concentration in the urine of patients with pheochromocytomas?

(A) Dopamine
(B) Epinephrine
(C) 3-Methoxy-4-hydroxymandelic acid
(D) Norepinephrine
(E) None of the above

4. Which of the following amino acids is the precursor for the thyroid hormones T_3 and T_4?

(A) Tryptophan
(B) Lysine
(C) Tyrosine
(D) Histidine
(E) Threonine

5. Which of the following enzymes catalyzes reactions in the biosynthesis of *both* catecholamines and indoleamines (serotonin)?

(A) Dopamine β-hydroxylase
(B) Phenylethanolamine N-methyl transferase
(C) Aromatic amino acid decarboxylase
(D) Tryptophan hydroxylase
(E) Tyrosine hydroxylase

6. All of the following statements regarding the catabolism of 5-hydroxytryptamine (serotonin) are correct EXCEPT

(A) degradation products of serotonin are excreted in the urine
(B) oxidative deamination by monoamine oxidase changes serotonin to 5-hydroxyindole-acetaldehyde
(C) the bulk of the 5-hydroxyindoleacetaldehyde is oxidized by nicotinamide-adenine dinucleotide (NAD^+)–linked aldehyde dehydrogenase to 5-hydroxyindoleacetic acid (5-HIAA)
(D) some of the 5-hydroxyindoleacetaldehyde is reduced by reduced NAD^+–linked aldehyde reductase to 5-hydroxytryptophol (5-HTOL)
(E) the ingestion of large amounts of alcohol elevates the ratio of 5-HIAA:5-HTOL in the urine

7. Which of the following reactions in the biosynthesis of porphyrins is the rate-controlling step?

(A) δ-Aminolevulinate synthetase
(B) Porphobilinogen synthetase
(C) Uroporphyrinogen synthetase
(D) Decarboxylation of uroporphyrinogen III
(E) Conversion of protoporphyrin IX to protoheme IX

8. All of the following statements regarding the degradation of porphyrins are correct EXCEPT

(A) the formation of biliverdin is the first step in the degradation of heme
(B) reduction of biliverdin yields bilirubin
(C) bilirubin is conjugated to glucuronic acid in the liver
(D) failure to conjugate bilirubin to glucuronic acid leads to the condition known as jaundice
(E) free (unconjugated) bilirubin may be reduced to urobilinogen in the liver

9. Creatine is formed from which of the following amino acids?

(A) Glycine, lysine, and methionine
(B) Glycine, arginine, and S-adenosylmethionine
(C) Glycine, ornithine, and S-adenosylmethionine
(D) Glycine and S-adenosylmethionine
(E) None of the above

10. Correct statements regarding creatinine include all of the following EXCEPT

(A) creatinine is formed by the spontaneous cyclization of a constant fraction of muscle creatine phosphate
(B) the excretion of creatinine in the urine of adults is very constant from day to day
(C) although the 24-hr excretion rate is very constant, there is a marked diurnal fluctuation
(D) urinary creatinine levels give a measure of muscle mass
(E) blood levels of creatinine are a good indicator of kidney function

Directions: Each question below contains four suggested answers of which **one or more** is correct. Choose the answer

A if **1, 2, and 3** are correct
B if **1 and 3** are correct
C if **2 and 4** are correct
D if **4** is correct
E if **1, 2, 3, and 4** are correct

11. Which of the following compounds can serve as neurotransmitters?

(1) Norepinephrine
(2) 5-Hydroxytryptamine
(3) Dopamine
(4) γ-Aminobutyrate

12. Catecholamine catabolism involves which of the following enzymes?

(1) Monoamine oxidase
(2) Aldehyde dehydrogenase
(3) Catechol-O-methyl transferase
(4) Phenylethanolamine N-methyl transferase

13. Characteristics of γ-aminobutyrate include which of the following?

(1) It is an inhibitory neurotransmitter in the brain and spinal cord
(2) It is formed from L-glutamate by decarboxylation
(3) It is metabolized within the neurons to succinate
(4) It is a five-carbon amino acid

14. Characteristics of acute intermittent porphyria include which of the following?

(1) It is a hepatic porphyria
(2) It is due to partial deficiency of uroporphyrinogen III cosynthetase
(3) It causes excessive amounts of porphobilinogen and δ-aminolevulinate to form
(4) It is an autosomal recessive condition

ANSWERS AND EXPLANATIONS

1. The answer is A. (*I A 1 c*) Tyrosine hydroxylation is the rate-limiting step in catecholamine biosynthesis that forms 3,4-dihydroxyphenylalanine (dopa). The enzyme, tyrosine hydroxylase, requires tetrahydrobiopterin as a cofactor, but the reduction of biopterin (the oxidized form of the cofactor) would not normally be the slow step in catecholamine biosynthesis.

2. The answer is E. (*I A 4*) Epinephrine is formed in the adrenal medulla from norepinephrine by N-methylation by the methyl donor S-adenosylmethionine in a reaction catalyzed by phenylethanolamine N-methyl transferase. O-Methylation and oxidative deamination are steps in the degradation of catecholamines, and hydroxylation and decarboxylation are steps in the synthesis of norepinephrine.

3. The answer is C. (*I B 4*) 3-Methoxy-4-hydroxymandelic acid (vanillylmandelic acid, VMA) occurs in the urine in large amounts when a medullary pheochromocytoma is present. These tumors produce huge amounts of catecholamines, which, by oxidative deamination followed by oxidation of the aldehyde formed, and then O-methylation, yield VMA. The catecholamines—epinephrine, norepinephrine, and dopamine—are not readily detectable in urine as unchanged compounds, as they are rapidly catabolized to VMA or related products.

4. The answer is C. (*I C 1*) Tyrosine is the precursor of the thyroid hormones. It is iodinated on the ring while it is in peptide linkage in the protein thyroglobulin. Monoiodo- and diiodotyrosines are formed, which then react to form the double-ring thyronine compounds, 3,3',5,5'-tetraiodothyronine (T_4) and 3,3',5-triiodothyronine (T_3).

5. The answer is C. (*I D 3*) The neurotransmitter dopamine is formed from 3,4-dihydroxyphenylalanine (dopa) by the action of aromatic amino acid decarboxylase. The same enzyme decarboxylates 5-hydroxytryptophan to yield 5-hydroxytryptamine (serotonin), an indoleamine. The three hydroxylases listed are specific for their substrates, and phenylethanolamine N-methyl transferase is responsible for the conversion of norepinephrine to epinephrine.

6. The answer is E. (*I E*) Serotonin undergoes oxidative deamination catalyzed by monoamine oxidase to form 5-hydroxyindoleacetaldehyde. Most of this compound is oxidized to 5-hydroxyindoleacetic acid (5-HIAA), but some is reduced to 5-hydroxytryptophol (5-HTOL). The oxidation pathway requires nicotinamide-adenine dinucleotide (NAD^+) and the reduction pathway, NADH. Because the metabolism of alcohol increases the level of cellular NADH in the liver, the reduction pathway is favored, and the ratio of 5-HIAA:5-HTOL *decreases* as more of the 5-HTOL and less of the 5-HIAA is formed.

7. The answer is A. (*II C 1*) The synthesis of δ-aminolevulinate is the first and the rate-controlling step of porphyrin biosynthesis, the synthetase that catalyzes the condensation of succinyl coenzyme A and glycine to form δ-aminolevulinate being allosterically inhibited by protohemin IX (hemin).

8. The answer is E. (*II D 2, 3 a–c*) The first degradative step in porphyrin catabolism is the rupture of the methylidyne group of heme between the pyrrole rings carrying vinyl groups to form biliverdin (green in color). The iron dissociates and the free tetrapyrrole is reduced with reduced nicotinamide-adenine dinucleotide phosphate as the reductant to bilirubin (reddish-brown in color). Normally, the insoluble bilirubin is rendered water soluble by conjugation to glucuronic acid in the liver. Failure to conjugate bilirubin to glucuronic acid leads to jaundice. Bilirubin diglucuronide is excreted in the bile, and in the gut, it is cleaved to the free bilirubin. Here, gut bacterial enzymes reduce the bilirubin to urobilinogen and stercobilinogen.

9. The answer is B. (*III A*) Creatine is formed from parts of three amino acids. In the first step, the guanidinium group of arginine is transferred to glycine to form guanidinoacetate. N-Methylation follows with S-adenosylmethionine acting as the methyl donor to form creatine.

10. The answer is C. (*III A 3*) Creatinine is formed by the spontaneous cyclization of a constant fraction of the muscle pool of creatine phosphate. Thus, it gives a measure of muscle mass. There is no diurnal fluctuation in creatinine excretion; therefore, the rate can be used as a basis for measuring the excretion of other substances. As it is constant from day to day in the adult, the creatinine excretion gives a measure of the completeness of a urine collection. Creatinine is normally cleared very efficiently from the plasma by the kidney; thus, the blood levels of creatinine are a good indicator of kidney function.

11. The answer is E (all). (*I A 2 b, 3 c, D 1, G 1*) All four compounds serve as neurotransmitters. Dopamine, a precursor of norepinephrine, is formed in some neurons of the basal ganglia. Norepinephrine is the major neural transmitter of the sympathetic nervous system. 5-Hydroxytryptamine (serotonin) is synthesized by neurons whose cell bodies are chiefly located in the hypothalamus and brain stem, by the pineal gland, and by the chromaffin cells of the gastrointestinal tract. γ-Aminobutyrate appears to be an inhibitory transmitter in the brain and spinal cord.

12. The answer is A (1, 2, 3). *(I A 4, B 1, 2)* The degradation of catecholamines may be initiated by oxidative deamination or by O-methylation. If oxidative deamination catalyzed by monoamine oxidase (MAO) occurs first, the aldehyde formed is oxidized by aldehyde dehydrogenase. If O-methylation by catechol-O-methyl transferase occurs first, it is followed by oxidative deamination by MAO to the aldehyde, which is then oxidized by aldehyde dehydrogenase. Phenylethanolamine N-methyl transferase is the enzyme that is responsible for the methylation of norepinephrine to epinephrine and is, thus, involved in the biosynthesis of catecholamines rather than in their catabolism.

13. The answer is A (1, 2, 3). *(I G)* γ-Aminobutyrate is a four-carbon amino acid (but not an α-amino acid), which plays an inhibitory role in neural transmission in the brain and spinal cord. It is formed from L-glutamate by decarboxylation and is metabolized within the neurons to succinate via succinate semialdehyde.

14. The answer is B (1, 3). *(II F 1)* Acute intermittent porphyria has an autosomal dominant inheritance pattern and is a hepatic porphyria that is due to a deficiency of uroporphyrinogen I synthetase. Excessive amounts of porphobilinogen (the substrate of the deficient enzyme) and δ-aminolevulinate are formed. An erythropoietic porphyria is known, which is a congenital condition due to a deficiency of uroporphyrinogen III cosynthetase; in this condition, the urine contains large amounts of uroporphyrin I.

29
Nucleotide Turnover

I. DE NOVO SYNTHESIS OF PURINE AND PYRIMIDINE NUCLEOTIDES

A. Formation of 5'-phosphoribosyl-1-pyrophosphate (PRPP)

1. An intermediate of major significance in nucleotide metabolism, PRPP is formed from ribose 5-phosphate and adenosine triphosphate (ATP).

$$\text{Ribose 5-phosphate} + \text{ATP} \rightarrow \text{PRPP} + \text{AMP}$$

2. Ribose 5-phosphate can arise from
   a. Glucose metabolism via glucose 6-phosphate (G6P) and the hexose monophosphate shunt (see Chapter 19, section I)
   b. Ribose 1-phosphate derived from nucleoside degradation (see section VI A 4), the reaction being catalyzed by phosphoglucomutase (see Chapter 18, section II A 2)

3. PRPP is involved in
   a. The de novo synthesis of both pyrimidine and purine nucleotides
   b. The salvage pathways for both purine and pyrimidine bases
   c. Nucleotide coenzyme biosynthesis

4. The cellular concentrations of PRPP are closely regulated and are usually low.

5. The formation of PRPP is catalyzed by PRPP synthetase, an allosteric enzyme with an absolute requirement for inorganic phosphate ions (P_i).
   a. At the usual low cellular levels of P_i, the sigmoidal kinetics toward $[P_i]$ displayed by PRPP synthetase keep the activity of the enzyme low. If the cellular level of P_i increases significantly, there is a marked increase in enzyme activity.
   b. PRPP synthetase is allosterically inhibited by nucleoside di- and triphosphates. In the red blood cell, 2-3-diphosphoglycerate (2,3-DPG) [see Chapter 10, section I C 3 c] is also an inhibitor.

B. Purine nucleotides

1. Origin of the carbons and nitrogens of the purine ring
   a. The purine ring is built up on a molecule of PRPP. The precursors of the ring are PRPP, glutamine, glycine, carbon dioxide (CO_2), aspartate, and two one-carbon fragments from the one-carbon folate pool (see Chapter 27, section III A 3).

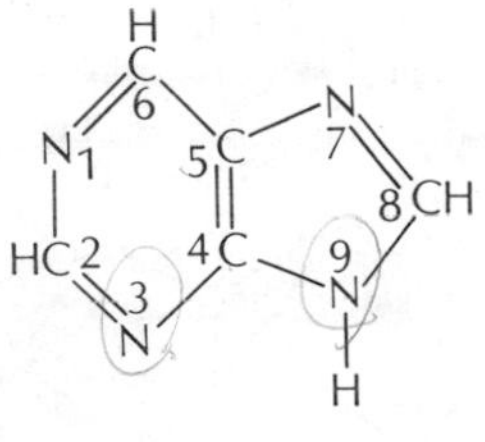

   b. Sources
      (1) N-3 and N-9 are from the amide nitrogen of glutamine.
      (2) C-4, C-5, and N-7 are from glycine.
      (3) C-2 and C-8 are from formate via tetrahydrofolate derivatives.
      (4) C-6 is from CO_2.
      (5) N-1 is from aspartate.

2. Synthesis of inosine monophosphate (IMP)

a. The first purine nucleotide formed is IMP in a ten-step pathway that requires the expenditure of six high-energy phosphate bonds, counting the pyrophosphate (PP_i) bond split at the committed step described next. It is, therefore, an energetically expensive process.

b. Pathway

(1) The **committed step** is the formation of 5′-phosphoribosylamine from PRPP and glutamine.

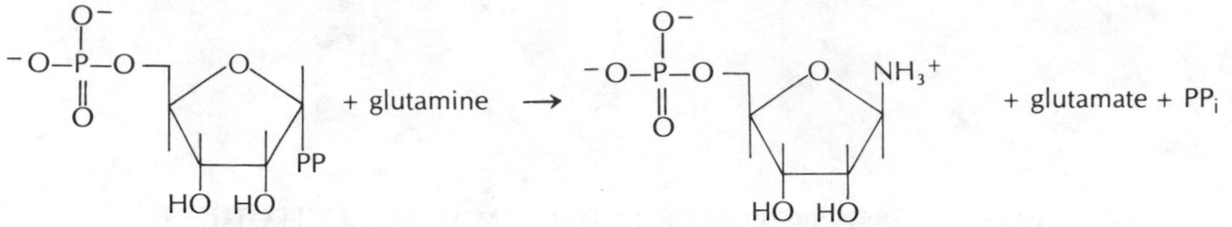

5′-Phosphoribosyl-1-pyrophosphate   + glutamine →   5′-Phosphoribosylamine   + glutamate + PP_i

(a) The reaction is catalyzed by PRPP-amidotransferase and is a reversible reaction but is driven to the right by the hydrolysis of inorganic pyrophosphate ions (PP_i) to P_i by cellular pyrophosphatases.

(b) Note the inversion at C-1 of the ribose ring as the amine group is in the β-configuration, which is the orientation of the bases in nucleotides, whereas the PP in PRPP is in the α-configuration.

(2) A molecule of glycine is added in its entirety to form 5′-phosphoribosylglycinamide.

5′-Phosphoribosylamine + glycine + ATP →

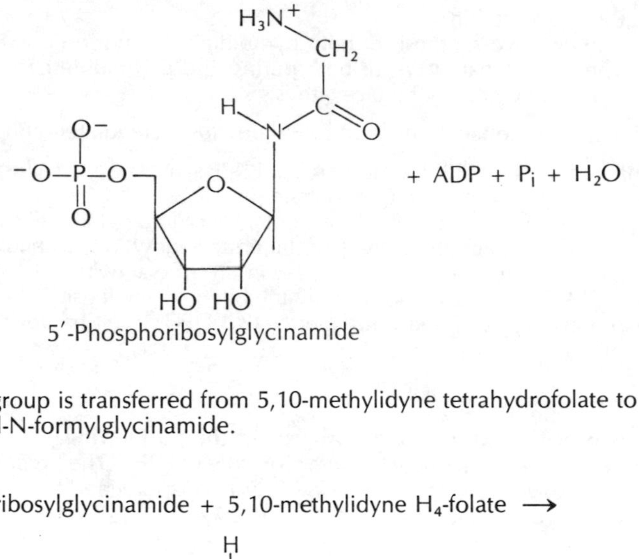

+ ADP + P_i + H_2O

5′-Phosphoribosylglycinamide

(3) A formyl group is transferred from 5,10-methylidyne tetrahydrofolate to form 5′-phosphoribosyl-N-formylglycinamide.

5′-Phosphoribosylglycinamide + 5,10-methylidyne H_4-folate →

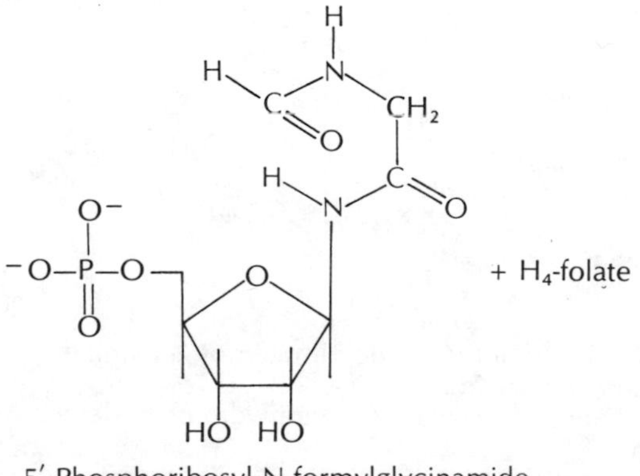

+ H_4-folate

5′-Phosphoribosyl-N-formylglycinamide

(4) Another glutamate transfers its amide group with the accompanying hydrolysis of ATP to form 5′-phosphoribosyl-N-formylglycinamidine.

5′-Phosphoribosyl-N-formylglycinamide + glutamine + ATP $\longrightarrow$

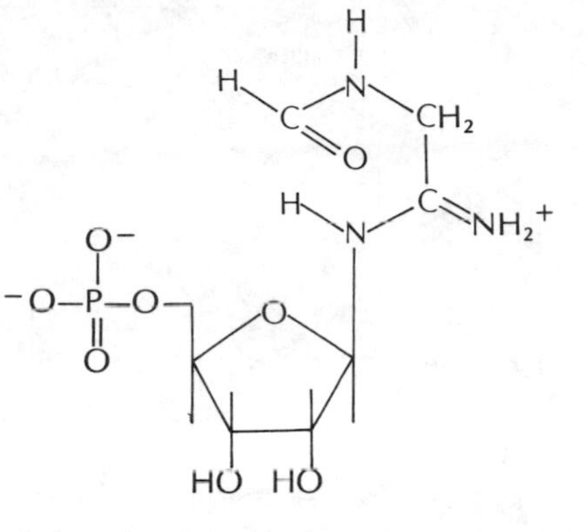

+ glutamate + ADP + P_i

5′-Phosphoribosyl-N-formylglycinamidine

(5) With ring closure, the imidazole ring is formed in a reaction driven by the hydrolysis of ATP, forming 5′-phosphoribosyl-5-aminoimidazole.

5′-Phosphoribosyl-N-formylglycinamidine + ATP $\longrightarrow$

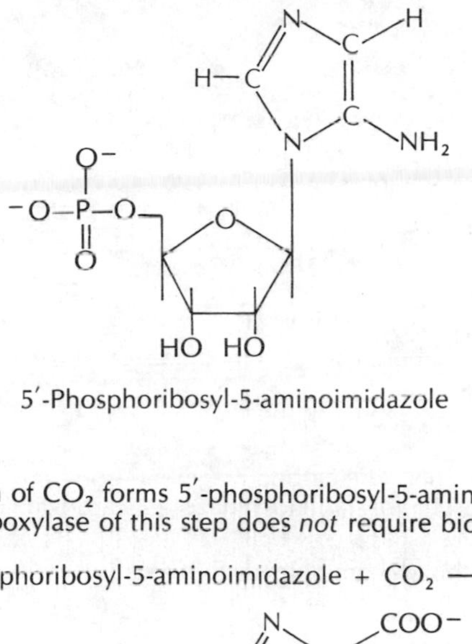

+ ADP + P_i

5′-Phosphoribosyl-5-aminoimidazole

(6) The addition of CO_2 forms 5′-phosphoribosyl-5-aminoimidazole-4-carboxylate. (Note that the carboxylase of this step does *not* require biotin as a cofactor.)

5′-Phosphoribosyl-5-aminoimidazole + CO_2 $\longrightarrow$

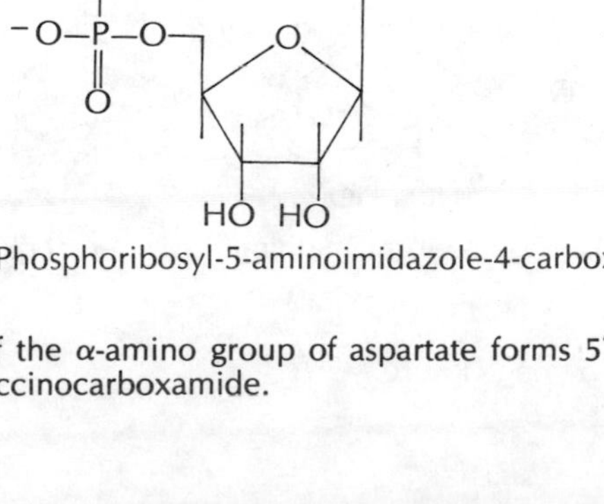

+ H^+

5′-Phosphoribosyl-5-aminoimidazole-4-carboxylate

(7) Transfer of the α-amino group of aspartate forms 5′-phosphoribosyl-5-aminoimidazole-4-N-succinocarboxamide.

5′-Phosphoribosyl-5-aminoimidazole-4-carboxylate + aspartate + ATP⟶

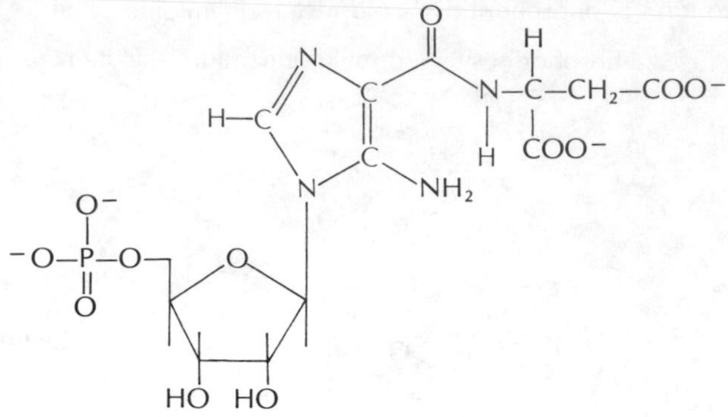

+ ADP + P_i

5′-Phosphoribosyl-5-aminoimidazole-4-N-succinocarboxamide

(8) Cleavage of the aspartate moiety forms 5′-phosphoribosyl-5-aminoimidazole-4-carboxamide.

5′-Phosphoribosyl-5-aminoimidazole-4-N-succinocarboxamide ⟶

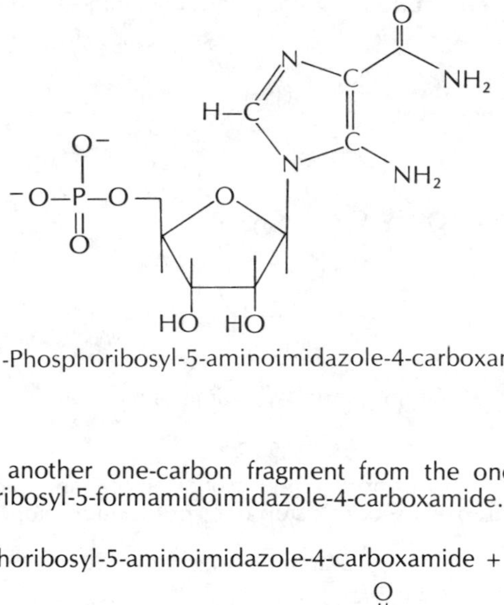

+ fumarate

5′-Phosphoribosyl-5-aminoimidazole-4-carboxamide

(9) Transfer of another one-carbon fragment from the one-carbon folate pool forms 5′-phosphoribosyl-5-formamidoimidazole-4-carboxamide.

5′-Phosphoribosyl-5-aminoimidazole-4-carboxamide + 10-formyl H_4-folate ⟶

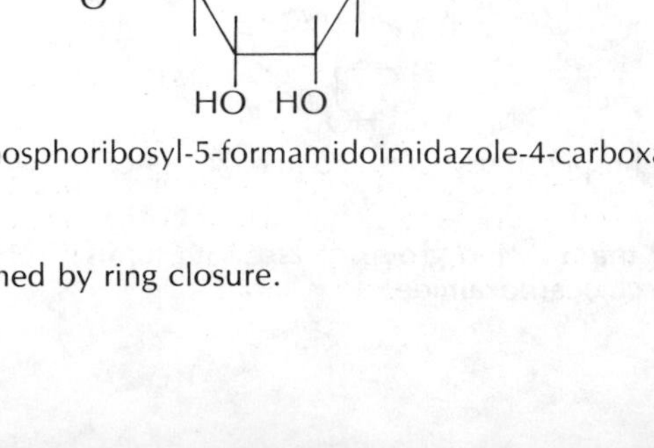

+ H_4-folate

5′-Phosphoribosyl-5-formamidoimidazole-4-carboxamide

(10) IMP is formed by ring closure.

5′-Phosphoribosyl-5-formamidoimidazole-4-carboxamide $\longrightarrow$

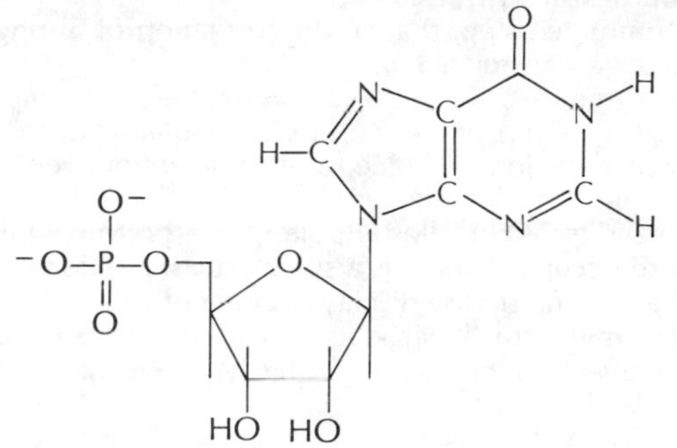

Inosine monophosphate

$+ \ H_2O$

3. Synthesis of adenosine and guanosine monophosphates (AMP and GMP) from IMP
a. Formation of AMP from IMP
(1) The first step is the formation of adenylosuccinate (N^6-succinoadenosine-5′-phosphate).

IMP + aspartate + GTP $\longrightarrow$

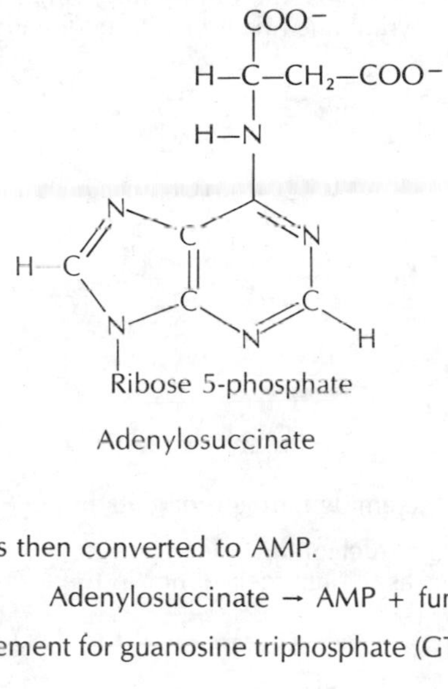

Ribose 5-phosphate

Adenylosuccinate

$+ \ GDP + P_i$

(2) Adenylosuccinate is then converted to AMP.

Adenylosuccinate $\rightarrow$ AMP + fumarate

(a) Note the requirement for guanosine triphosphate (GTP) in the formation of adenylosuccinate.
(b) Adenylosuccinate is converted to AMP by the same enzyme that is involved in the cleavage of the aspartate moiety in 5′-phosphoribosyl-5-aminoimidazole-4-N-succinocarboxamide [see section I B 2 b (8)].

b. Formation of GMP from IMP
(1) Step one is the formation of xanthosine 5′-monophosphate (XMP).

IMP + NAD^+ $\rightarrow$ XMP + NADH + H^+

(2) In step two, XMP is converted to GMP.

XMP + glutamine + ATP $\rightarrow$ GMP + glutamate + AMP + PP_i

(3) Note the expenditure of two high-energy phosphate bonds in step two.

4. Cost of purine synthesis in terms of high-energy phosphate bonds
a. The synthesis of IMP requires the expenditure of five high-energy phosphate bonds.
b. IMP to AMP requires one high-energy bond.
c. IMP to GMP requires two high-energy bonds.

5. Regulation of the de novo synthesis of purine nucleotides

a. Regulation of IMP synthesis

(1) The committed step, that is, the formation of 5′-phosphoribosylamine by PRPP-amidotransferase, is regulated.

(2) The enzyme exists in two forms, a smaller active molecule and a larger inactive one.

(3) IMP, GMP, and AMP are negative modulators of the enzyme and cause a shift to the larger inactive form. Hence, these nucleotides regulate the cellular concentration of 5′-phosphoribosylamine.

(4) The enzyme has two allosteric sites, one for IMP or GMP and the other for AMP. If both sites are occupied, there is a synergistic inhibition.

(5) PRPP-amidotransferase displays sigmoidal kinetics towards PRPP, whereas the kinetics for glutamine are hyperbolic. Thus, changes in the concentration of PRPP have a greater effect on the rate of purine synthesis than do equivalent changes in glutamine concentration.

b. Regulation of IMP conversion to AMP and GMP

(1) The two enzymes that use IMP as substrate are IMP dehydrogenase at the first step in GMP synthesis and adenylosuccinate synthetase at the first step in AMP synthesis; both enzymes have similar Michaelis constants for IMP.

(2) AMP is a competitive inhibitor of adenylosuccinate synthetase, and GMP is a competitive inhibitor of IMP dehydrogenase.

(3) In addition, GTP is an energy source for adenylosuccinate formation, and ATP is an energy source for the conversion of XMP to GMP.

C. Pyrimidine nucleotides

1. Origin of the carbons and nitrogens of the pyrimidine ring

a. The pyrimidine ring, unlike the purine ring, is not built up on a ribose phosphate molecule. Rather, the pyrimidine ring is first formed, and then PRPP is used to form the nucleotide.

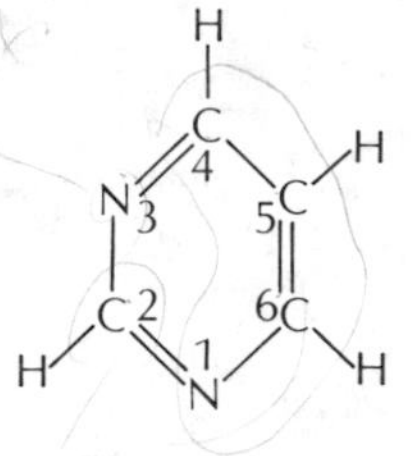

b. Sources

(1) N-1, C-4, C-5, and C-6 are from aspartate.

(2) C-2 is from CO_2.

(3) N-3 is from the amide nitrogen of glutamine.

2. Formation of uridine nucleotides

a. Glutamine serves as the nitrogen donor in the formation of carbamoyl phosphate.

$$\text{Glutamine} + CO_2 + 2\ \text{ATP} \longrightarrow$$

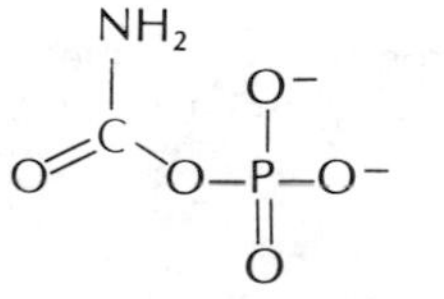

+ glutamate + 2 ADP + P_i

Carbamoyl phosphate

(1) The reaction is catalyzed by carbamoyl phosphate synthetase (glutamine), a cytosolic enzyme distinct from the hepatic mitochondrial enzyme involved in the urea cycle (see Chapter 27, section I C).

(2) Unlike the mitochondrial enzyme, the cytosolic enzyme does not require N-acetylglutamate as a cofactor.

b. The **committed step** is the addition of the entire aspartate molecule to form N-carbamoyl aspartate.

Carbamyol phosphate + aspartate →

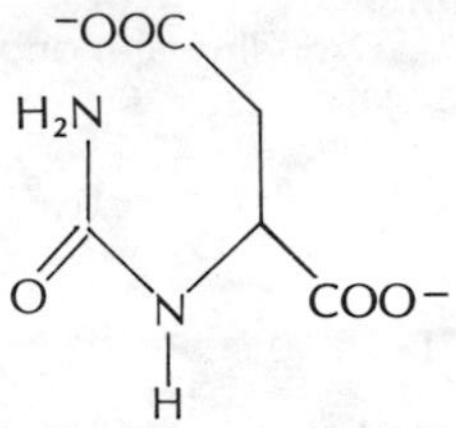

+ P$_i$

N-Carbamyol aspartate

c. Ring closure forms the first pyrimidine, dihydroorotate.

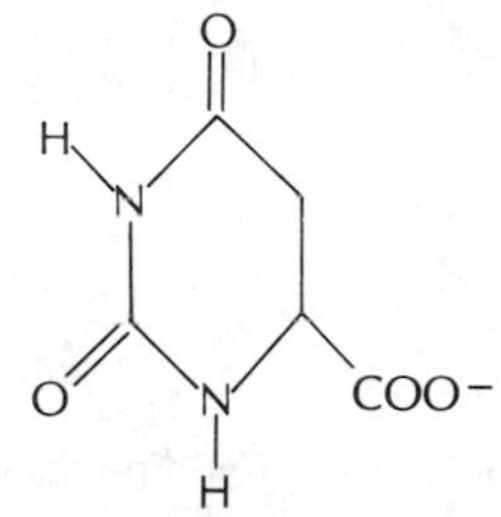

Dihydroorotate

d. Orotate is formed by dehydrogenation.

Dihydroorotate + NAD$^+$ ⟶

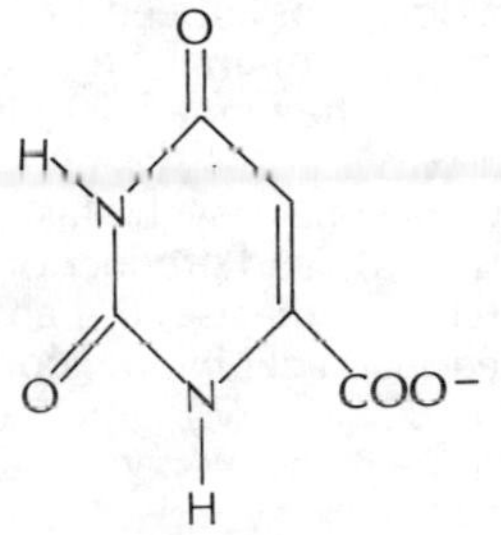

+ NADH + H$^+$

Orotate

Dihydroorotate dehydrogenase is a mitochondrial enzyme, whereas the other enzymes of pyrimidine synthesis are cytosolic.

e. Orotate acquires a ribose phosphate group from PRPP in the formation of the first pyrimidine nucleotide, orotidine 5′-monophosphate.

Orotate + PRPP →

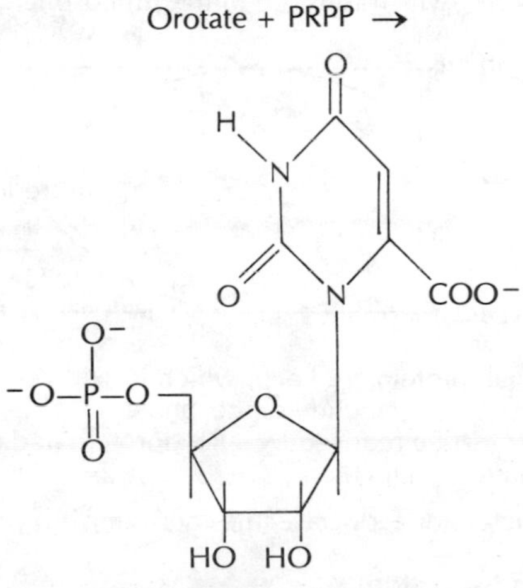

+ PP$_i$

Orotidine 5′-monophosphate

f. Decarboxylation of orotidine 5′-monophosphate forms uridine 5′-monophosphate.

Orotidine 5′-monophosphate $\longrightarrow$

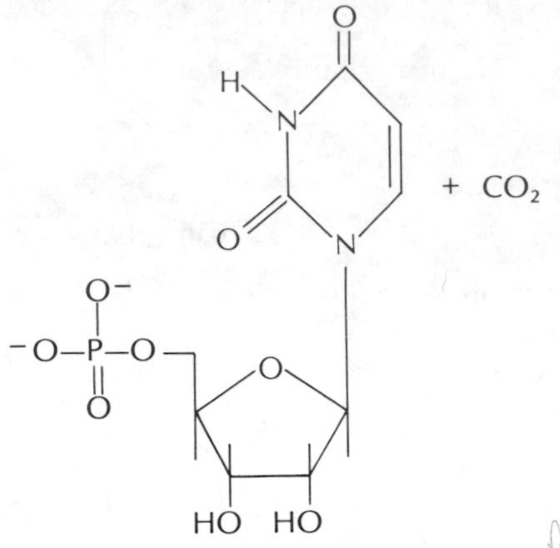

$+$ CO_2

Uridine 5′-monophosphate

3. Formation of cytidine nucleotides. Cytidine nucleotides are formed from uridine nucleotides at the triphosphate level.

Uridine 5′-triphosphate + ATP + glutamine →
$$\text{cytidine 5′-triphosphate} + ADP + P_i + \text{glutamate}$$

4. Regulation of pyrimidine synthesis
   a. In mammals, the pyrimidine biosynthetic pathway is regulated at the carbamoyl phosphate synthetase (glutamine) step, unlike the case for *Escherichia coli* and other bacteria, in which the regulated step is the formation of N-carbamoyl aspartate, catalyzed by aspartate carbamoyl transferase.
   b. Mammalian carbamoyl phosphate synthetase (glutamine) is inhibited allosterically by uridine triphosphate (UTP), the end product of the pathway.
   c. Cytidine triphosphate (CTP) synthetase, the enzyme catalyzing the formation of cytidine nucleotides, is stimulated markedly by low concentrations of GTP. In addition, CTP synthetase, in the presence of CTP, shows sigmoidal kinetics with respect to [UTP], which prevents all of the UTP from being converted to CTP.

II. DEOXYRIBONUCLEOTIDES

   A. Formation of deoxyribonucleotides. The cellular levels of deoxyribonucleotides are usually very low, only being increased at the time of DNA replication (i.e., in the S phase of the cell cycle; see Chapter 1, section VI C 3). Deoxyribonucleotides are formed by reduction of the corresponding ribonucleotides, which must be in the diphosphate form.

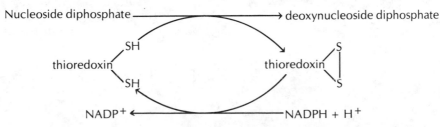

   1. Thioredoxin is a small protein (12 kdal), which is oxidized during the reduction of the ribose 2′-hydroxyl group of the nucleotide diphosphate. In order to maintain the reaction, the oxidized thioredoxin must be reduced by a flavoprotein and reduced nicotinamide-adenine dinucleotide phosphate (NADPH).

   2. Mammalian ribonucleotide reductase, the catalyzing enzyme, contains two nonidentical subunits.
   a. One subunit binds effectors.
   b. The other subunit is the catalytic subunit and contains nonheme iron.

3. Regulation is by

   a. Modulation of the cellular levels of the enzyme

   b. Allosteric effectors

   (1) The reduction of a specific nucleoside diphosphate requires a specific nucleoside triphosphate as a positive regulator.

   (2) Other nucleoside triphosphates act as negative regulators.

B. Synthesis of deoxythymine nucleotide

1. Deoxythymidylate (dTMP) is formed from deoxyuridylate (dUMP) in a reaction in which a methylene group is both transferred from 5,10-methylene tetrahydrofolate and, at the same time, reduced to a methyl group; in the reaction, the folate coenzyme is oxidized to inactive dihydrofolate.

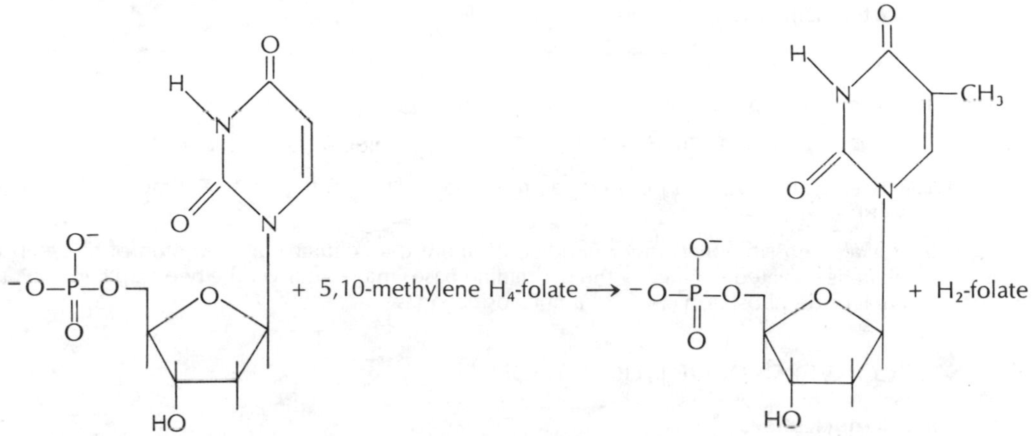

2. The inactive dihydrofolate must be reduced again in order to function as a carrier of one-carbon moieties (see Chapter 27, section III D, Figure 27-6).

3. The source of dUMP for the synthesis of deoxythymidine nucleotide is probably from the direct deamination of deoxycytidylate (dCMP).

III. SALVAGE PATHWAYS FOR PURINE AND PYRIMIDINE BASES

A. General comments

1. The de novo synthesis of nucleotides is expensive in terms of the use of high-energy phosphate bonds (particularly for purine biosynthesis), and many cells also have pathways that "salvage" purine and pyrimidine bases for the formation of the corresponding nucleotides.

2. In the case of purines, some cells cannot carry out de novo synthesis of purine nucleotides, for example, red blood cells, which lack PRPP-amidotransferase.

3. Nucleotides do not enter cells directly but are converted to nucleosides by cell-membrane nucleotidase. After entering the cell, the nucleoside is either converted to the nucleotide by a kinase (see section V A) or degraded to the base by nucleoside phosphorylase.

B. Purine salvage pathways

1. Hypoxanthine-guanine phosphoribosyl transferase (HGPRT) catalyzes the one-step formation of the nucleotides from either hypoxanthine or guanine, using PRPP as the donor of the ribosyl moiety.

$$\text{Guanine} + \text{PRPP} \rightarrow \text{GMP} + \text{PP}_i$$

$$\text{Hypoxanthine} + \text{PRPP} \rightarrow \text{IMP} + \text{PP}_i$$

The enzyme is regulated by the competitive inhibition (with respect to PRPP) of IMP and GMP.

2. Adenine phosphoribosyl transferase (APRT) catalyzes the following reaction:

$$\text{Adenine} + \text{PRPP} \rightarrow \text{AMP} + \text{PP}_i$$

AMP inhibits the reaction.

3. Nucleoside phosphorylase–nucleoside kinase pathway

a. Under some conditions, it is theoretically possible for purine bases to be salvaged by a two-step process.

(1) Nucleoside phosphorylase is an enzyme that functions in nucleotide breakdown (see section VI A 4). The reaction is readily reversible and, in fact, the *synthesis* of the nucleoside is the favored direction.

$$\text{Base + ribose 1-phosphate} \rightarrow \text{nucleoside} + P_i$$

(2) Once the nucleoside is formed, a kinase may phosphorylate it to the 5'-nucleotide.

b. However, neither guanosine nor inosine kinases have been detected in mammalian cells, so that adenine is the only purine that might be salvaged by the two-step pathway.

c. Under usual cellular conditions, it is also unlikely that the cellular concentrations of base and ribose 1-phosphate will be sufficiently high for efficient salvage of the base by the two-step pathway.

C. Pyrimidine base salvage

1. Pyrimidine phosphoribosyl transferase catalyzes the following reaction:

$$\text{Pyrimidine base + PRPP} \rightarrow \text{pyrimidine nucleotide} + PP_i$$

2. The enzyme (obtained from human red blood cells) can use orotate, uracil, and thymine as substrates.

3. It is also possible that under conditions where the cellular concentration of ribose 1-phosphate is elevated, salvage of the pyrimidine bases may occur via the two-step pathway, using nucleoside phosphorylase and a nucleoside kinase.

IV. INTERCONVERSION OF NUCLEOTIDES

A. Purine nucleotides

1. AMP deaminase catalyzes the reductive deamination of AMP to IMP.

$$\text{AMP} + H_2O \rightarrow \text{IMP} + \text{ammonia (NH}_3\text{)}$$

GMP can then be formed from IMP via conversion to XMP (see section I B 3).

2. GMP reductase catalyzes the reduction of GMP to IMP.

$$\text{GMP} + \text{NADPH} + H^+ \rightarrow \text{IMP} + \text{NADP}^+$$

AMP can then be formed from IMP via conversion to adenylosuccinate (see section I B 3)

3. Regulation of the interconversion of AMP and GMP provides a balance of cellular nucleotide concentrations for nucleic acid synthesis. This regulation involves several mechanisms.
a. XMP competitively inhibits GMP reductase.
b. GTP activates GMP reductase.
c. ATP activates AMP deaminase.
d. P_i, GDP, and GTP inhibit AMP-deaminase.

B. Pyrimidine nucleotides. The interconversion of pyrimidine nucleotides is summarized in Figure 29-1; the dashed lines in the figure indicate the mechanisms for regulation.

V. NUCLEOSIDE AND NUCLEOTIDE KINASES

A. Kinases for the salvage of nucleosides

1. The reaction is

$$\text{Nucleoside} + \text{ATP} \rightarrow \text{nucleotide} + \text{ADP}$$

2. Such kinases exist for adenosine, uridine, cytidine, deoxycytidine, and thymidine, but not for guanosine or inosine.

B. Kinases for the conversion of the monophosphates to the diphosphates

1. Pyrimidine nucleoside monophosphate kinase uses cytidine and uridine monophosphates and dCMP as substrates.

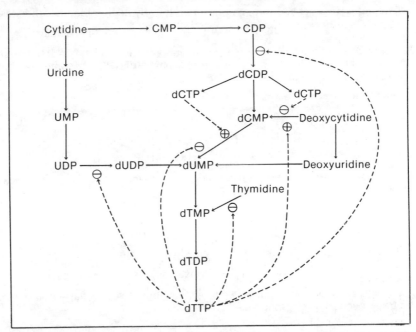

Figure 29-1. Pyrimidine interconversions. The *solid arrows* represent enzyme-catalyzed reactions. The *dashed lines* represent the regulatory effects of certain nucleotides on the various steps (+, activation; −, inhibition). UMP, UDP = uridine mono-(uridylate) and diphosphates; CMP, CDP = cytidine mono- and diphosphates; dCMP, dCDP, dCTP = deoxycytidine mono-(deoxycytidylate), di-, and triphosphates; dTMP, dTDP, dTTP = deoxythymidine mono-(deoxythymidylate), di-, and triphosphates; dUMP, dUDP = deoxyuridine mono-(deoxyuridylate) and diphosphates. (Reprinted with permission from Devlin TM: *Textbook of Biochemistry with Clinical Correlations*, 2nd ed. New York, John Wiley, 1986, p 512.)

2. Thymidylate kinase catalyzes the reaction

$$dTMP + ATP \rightarrow \text{deoxythymidine diphosphate (dTDP)} + ADP$$

3. AMP kinase (myokinase) catalyzes the readily reversible reaction

$$AMP + ATP \leftrightarrow 2\ ADP$$

4. GMP kinase catalyzes the reaction

$$GMP + ATP \rightarrow GDP + ADP$$

C. Nucleoside diphosphate kinase converts nucleoside diphosphates to triphosphates.

$$NDP + ATP \rightarrow NTP + ADP$$

The enzyme is nonspecific for the purine or pyrimidine base or for the ribose or deoxyribose sugar.

VI. DEGRADATION OF PURINE AND PYRIMIDINE BASES

A. Nucleic acid breakdown

1. The purine and pyrimidine bases arise from the hydrolysis of cellular nucleic acids and from ingested nucleoprotein, which is degraded during digestion to nucleosides or bases before absorption.

2. A variety of nucleases, both exo- and endonucleases showing specificity toward RNA or DNA and toward the nucleic acid bonds cleaved, break down nucleic acids to nucleotides.

3. The nucleotides are hydrolyzed to nucleosides by nucleotidases.

4. The bases are liberated through phosphorolysis by nucleoside phosphorylase.

$$\text{Nucleoside} + P_i \rightarrow \text{base} + \text{ribose 1-phosphate}$$

5. Mammalian cells have a very active deoxyuridine triphosphatase (dUTPase), which does not allow cellular levels of deoxyuridine triphosphate (dUTP) to become high enough for dUMP to be incorporated in error into DNA.

B. Degradation of purine bases

1. AMP is hydrolyzed to adenosine by nucleotidase or deaminated to IMP by AMP deaminase.

$$\text{AMP} + H_2O \rightarrow \text{adenosine} + P_i$$

$$\text{AMP} + H_2O \rightarrow \text{IMP} + NH_4^+$$

2. Adenosine deaminase then converts adenosine to inosine, and nucleotidases convert IMP to inosine.

$$\text{Adenosine} + H_2O \rightarrow \text{inosine} + NH_4^+$$

$$\text{IMP} + H_2O \rightarrow \text{inosine} + P_i$$

3. Inosine undergoes phosphorolysis to give hypoxanthine.

$$\text{Inosine} + P_i \rightarrow \text{hypoxanthine} + \text{ribose 1-phosphate}$$

4. Guanine nucleotides are converted to guanine by the sequential action of nucleotidases and purine nucleoside phosphorylase.

$$\text{GMP} + H_2O \rightarrow \text{guanosine} + P_i$$

$$\text{Guanosine} + P_i \rightarrow \text{guanine} + \text{ribose 1-phosphate}$$

5. Guanine is deaminated to xanthine by the enzyme guanase.

$$\text{Guanine} + H_2O \rightarrow \text{xanthine} + NH_4^+$$

6. The final steps of purine degradation in humans are carried out by the enzyme xanthine oxidase.

a. Xanthine oxidase oxidizes hypoxanthine to xanthine and then xanthine to uric acid.

Hypoxanthine	Xanthine	Uric acid

b. Xanthine oxidase contains molybdenum, which accounts for the trace requirement for molybdenum in animals, and also contains Fe^{3+} and oxidized flavin-adenine dinucleotide (FAD). Molecular oxygen is required, and hydrogen peroxide (H_2O_2) is formed.

C. Degradation of pyrimidine bases. Uracil and thymine are degraded by similar pathways.

1. An NADPH-dependent reduction converts uracil to dihydrouracil and thymine to dihydrothymine.

2. Ring opening has the following effects:

$$\text{Dihydrouracil} \rightarrow \beta\text{-ureidopropionic acid}$$

$$\text{Dihydrothymine} \rightarrow \beta\text{-ureidoisobutyric acid}$$

3. Uracil (and cytosine) degradation finally gives rise to β-alanine.

4. Thymine is degraded finally to β-aminoisobutyrate which, being excreted in the urine, can provide a measure of DNA turnover.

VII. BIOSYNTHESIS OF NUCLEOTIDE COENZYMES

A. Pyridine nucleotide coenzymes

1. Synthesis from tryptophan

a. Tryptophan is present in body proteins and in dietary protein, to only a limited extent, but if there is an excess of this essential amino acid, part of the tryptophan pool can be metabolized to quinolinate.

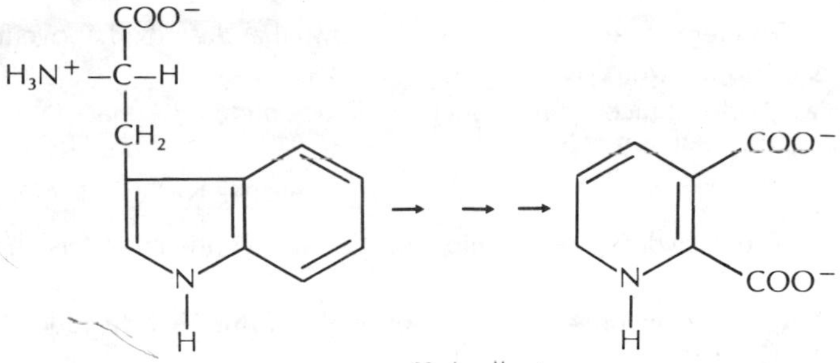

Quinolinate

b. PRPP then donates a ribose phosphate moiety to form nicotinate mononucleotide.

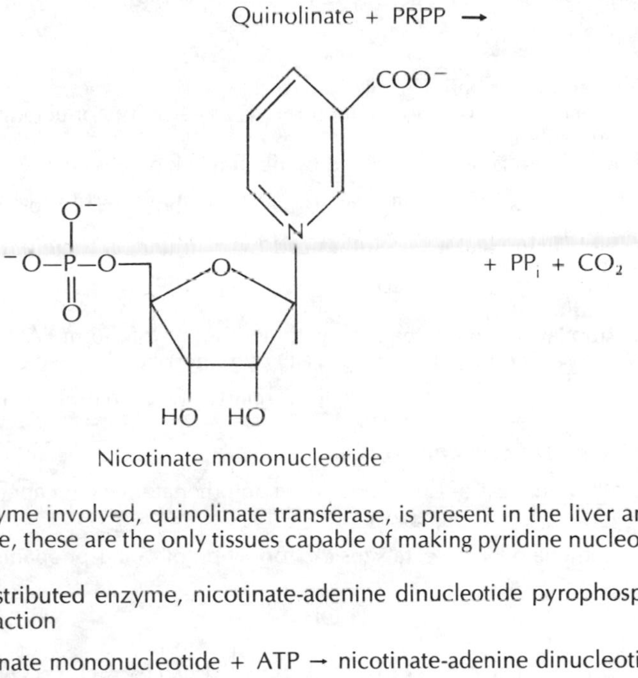

$$\text{Quinolinate} + \text{PRPP} \longrightarrow$$

$$+ \text{PP}_i + \text{CO}_2$$

Nicotinate mononucleotide

    (1) The enzyme involved, quinolinate transferase, is present in the liver and kidney only.
    (2) Therefore, these are the only tissues capable of making pyridine nucleotides from tryptophan.

c. A widely distributed enzyme, nicotinate-adenine dinucleotide pyrophosphorylase, catalyzes the reaction

$$\text{Nicotinate mononucleotide} + \text{ATP} \rightarrow \text{nicotinate-adenine dinucleotide} + \text{PP}_i$$

    (1) In nucleated cells, nicotinate-adenine dinucleotide pyrophosphorylase is exclusively found in the nucleus.
    (2) In erythrocytes, which have no nucleus, NAD^+ can be synthesized from nicotinate or nicotinamide.

d. Glutamine then donates its amide group to the nicotinate moiety to form NAD^+, a reaction catalyzed by NAD^+ synthetase.

$$\text{Nicotinate-adenine dinucleotide} + \text{glutamine} + \text{ATP} \rightarrow NAD^+ + \text{glutamate} + \text{ADP} + P_i$$

2. Synthesis from nicotinate or nicotinamide

a. Nicotinic acid (niacin) or nicotinamide are equally effective in supplying the requirement for this vitamin in humans.

    (1) The vitamin is an essential dietary factor, as tryptophan can probably only supply about 10% of the body's requirements for pyridine nucleotides.
    (2) There is no known direct conversion of nicotinate to nicotinamide.

b. Nicotinate phosphoribosyl transferase is a widely distributed enzyme that catalyzes the reaction

$$\text{Nicotinate} + \text{PRPP} \rightarrow \text{nicotinate mononucleotide} + \text{PP}_i$$

c. Nicotinamide phosphoribosyl transferase, the same enzyme that uses nicotinate as substrate, catalyzes the reaction

$$\text{Nicotinamide} + \text{PRPP} \rightarrow \text{nicotinamide mononucleotide} + \text{PP}_i$$

d. The remainder of the pathway follows that described above for quinolinate to NAD^+.

3. Formation of $NADP^+$
  a. Oxidized nicotinamide-adenine dinucleotide phosphate ($NADP^+$) is derived from NAD^+ by the action of NAD^+ kinase.

$$NAD^+ + ATP \rightarrow NADP^+ + ADP$$

  b. The $2'$-hydroxyl on the nicotinamide nucleotide moiety is phosphorylated.
  c. NADPH is a negative modulator of NAD^+ kinase.

4. NAD^+ glycohydrolase is a microsomal enzyme that degrades NAD^+ to nicotinamide and ADP-ribose.

5. Poly (ADP) ribose synthetase catalyzes the polymerization of the ADP-ribose moiety of NAD^+ onto nuclear proteins with the release of nicotinamide. It is exclusively a nuclear enzyme with its highest activity in the G2 phase of the cell cycle and its lowest in the S phase.

B. FAD biosynthesis

1. Formation of riboflavin phosphate
  a. Riboflavin is also known, incorrectly, as flavin mononucleotide (FMN), although it is not a nucleotide.
  b. Riboflavin is phosphorylated by riboflavin kinase with ATP as the phosphate donor.

$$\text{Riboflavin} + ATP \rightarrow \text{riboflavin phosphate} + ADP$$

  c. Riboflavin cannot be synthesized by man and is, therefore, an accessory food factor or vitamin.

2. Formation of FAD
  a. Riboflavin phosphate reacts with another ATP to form FAD.
  b. The reaction is catalyzed by FAD pyrophosphorylase.

$$\text{Riboflavin phosphate} + ATP \rightarrow FAD + \text{PP}_i$$

C. Biosynthesis of coenzyme A (CoA)

1. The biosynthesis of CoA starts with pantothenate, which cannot be made by man and must be ingested as an accessory food factor or vitamin.

2. Pantothenate kinase catalyzes the formation of 4-phosphopantothenate.

$$\text{HO—CH}_2\text{—}\overset{\overset{\displaystyle CH_3}{|}}{\underset{\underset{\displaystyle CH_3}{|}}{C}}\text{—}\overset{\overset{\displaystyle CH_3OH}{|}}{CH}\text{—}\overset{\overset{\displaystyle}{}}{\underset{\underset{\displaystyle O}{\|}}{C}}\text{—NH—CH}_2\text{—CH}_2\text{—COO}^- + \text{ATP} \rightarrow$$

$$^{2-}\text{O}_3\text{PO—CH}_2\text{—}\overset{\overset{\displaystyle CH_3}{|}}{\underset{\underset{\displaystyle CH_3}{|}}{C}}\text{—}\overset{\overset{\displaystyle CH_3OH}{|}}{CH}\text{—}\overset{\overset{\displaystyle}{}}{\underset{\underset{\displaystyle O}{\|}}{C}}\text{—NH—CH}_2\text{—CH}_2\text{—COO}^- + \text{ADP}$$

4-Phosphopantothenate

3. Cysteine is added in a reaction catalyzed by phosphopantothenoylcysteine synthetase, forming 4-phosphopantothenoylcysteine.

4-Phosphopantothenate + cysteine + ATP →

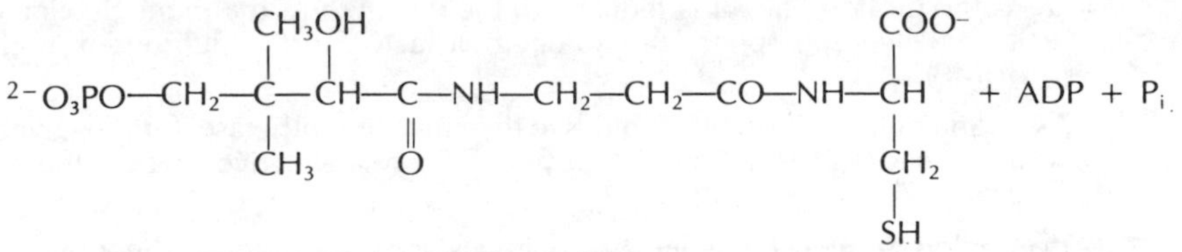

```
                CH₃OH                                    COO⁻
                 |  |                                     |
²⁻O₃PO—CH₂—C—CH—C—NH—CH₂—CH₂—CO—NH—CH    + ADP + Pᵢ
                 |     ||                                 |
                CH₃    O                                 CH₂
                                                          |
                                                          SH
```

4-Phosphopantothenoylcysteine

4. The removal of the α-carboxyl group from cysteine yields 4-phosphopantotheine.

4-Phosphopantothenoylcysteine →

```
                CH₃OH
                 |  |
²⁻O₃PO—CH₂—C—CH—C—NH—CH₂—CH₂—CO—NH—CH₂—CH₂—SH  + CO₂
                 |     ||
                CH₃    O
```

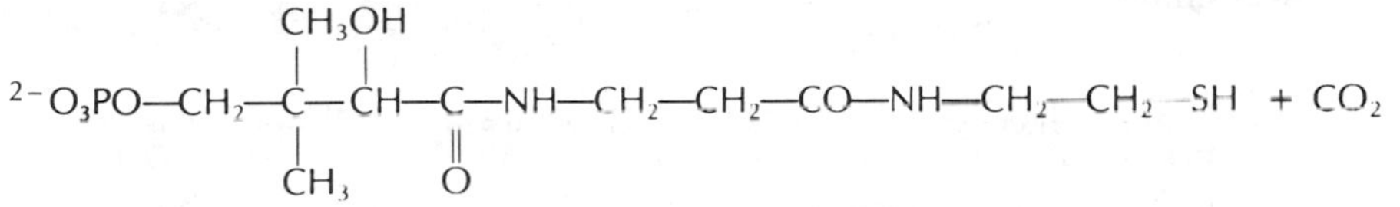

4-Phosphopantotheine

5. ATP is then added by dephospho-CoA pyrophosphorylase to form dephospho-CoA.

4-Phosphopantotheine + ATP →

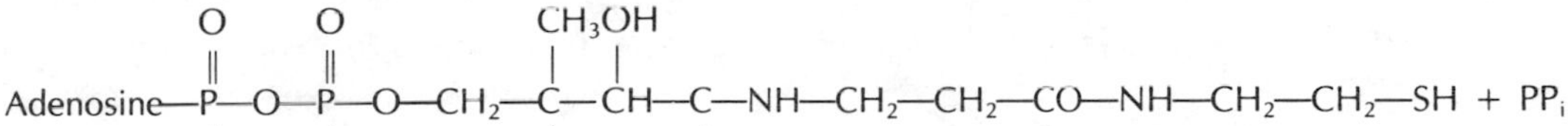

```
             O    O       CH₃OH
             ||   ||       |  |
Adenosine—P—O—P—O—CH₂—C—CH—C—NH—CH₂—CH₂—CO—NH—CH₂—CH₂—SH + PPᵢ
             |    |         |     ||
             O⁻   O⁻       CH₃    O
```

Dephosphocoenzyme A

6. ATP is again used, this time to phosphorylate dephospho-CoA at the 3′-hydroxyl on the adenosine moiety to give CoA.

VIII. INHIBITORS OF PURINE AND PYRIMIDINE METABOLISM.
The turnover of nucleic acids in malignant tissues is very high. For this reason, much attention has been paid to the therapeutic possibility of inhibiting the supply of precursors to the nucleic acid synthesizing systems. Many compounds have been screened as inhibitors of the de novo pathways of purine and pyrimidine synthesis, or as compounds that could be converted to inhibitors by purine or pyrimidine salvage reactions.

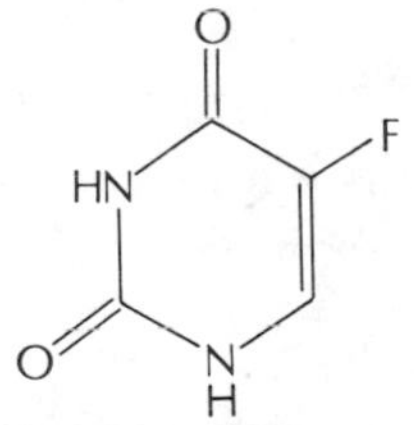

5-Fluorouracil

A. Fluorouracil (5-FU)

1. 5-FU has been used in the treatment of solid tumors.

2. This compound must be extensively metabolized to be an effective inhibitor.
  a. The first metabolite is 5-fluorouridine, formed via nucleoside phosphorylase.
  b. It is then phosphorylated to the monophosphate by uridine kinase. Steps a and b are those of the two-step salvage pathway.

     c. Further phosphorylation to the diphosphate is next.

     d. As the diphosphate, it is reduced to the deoxy form by ribonucleotide reductase.

     e. It is finally dephosphorylated to one of its active forms, fluorodeoxyuridylate (5-fluoro-2'-dUMP).

   3. As an analogue of 2'-dUMP, it binds to thymidylate synthetase, forming, with the enzyme and with tetrahydrofolate, a stable ternary complex that effectively blocks the formation of thymidine.

   4. In addition, some of the 5-fluoro-2'-dUMP is phosphorylated to the triphosphate level and incorporated into RNA, where it has an inhibitory effect on the maturation of the 45S precursor of the 28S and 18S rRNA species.

   5. The dual inhibitory effect is evident from the need for both thymidine and uridine in the "rescue" of treated cells.

B. 5-Iodouracil

   1. This inhibitor functions as an analogue of thymidine.

   2. In the triphosphate form, it can be incorporated into DNA, where it tautomerizes into a form that bonds with G rather than A, thus causing misreading of the DNA sequence.

   3. It has been used in the treatment of herpes infections.

C. 6-Mercaptopurine

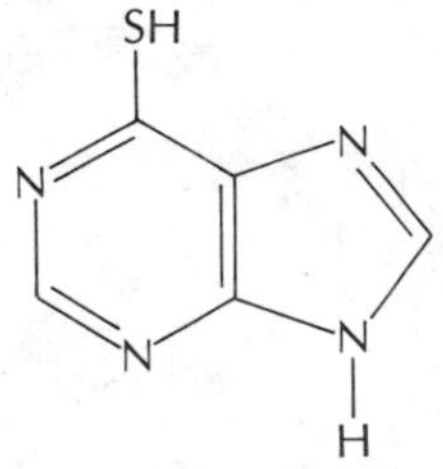

6-Mercaptopurine

   1. 6-Mercaptopurine, a useful antitumor drug, is an inhibitor of the de novo synthesis of purines.

   2. It is metabolized to 6-mercaptopurine ribonucleotide by the APRT salvage enzyme (see section III B 2).

   3. It inhibits IMP conversion to GMP and AMP.

   4. It also has a feedback inhibitory effect on PRPP-amidotransferase, the controlling step of purine synthesis.

   5. 6-Mercaptopurine is also a substrate for xanthine oxidase, which oxidizes it to 6-thiouric acid, deactivating it as an inhibitor. Concomitant administration of allopurinol, an inhibitor of xanthine oxidase, therefore, potentiates the inhibitory action of 6-mercaptopurine.

D. Cytosine and adenine arabinoside

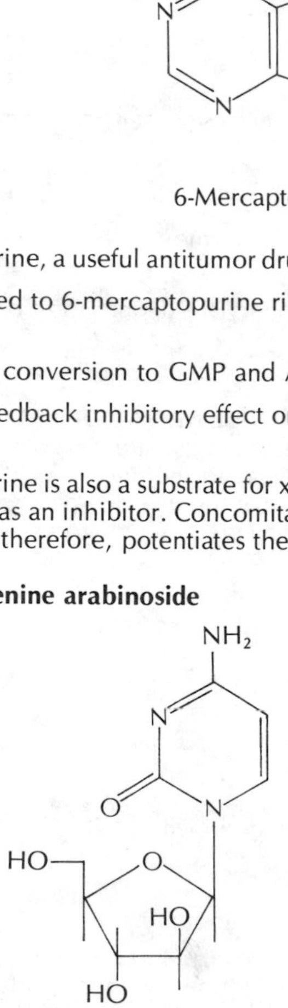

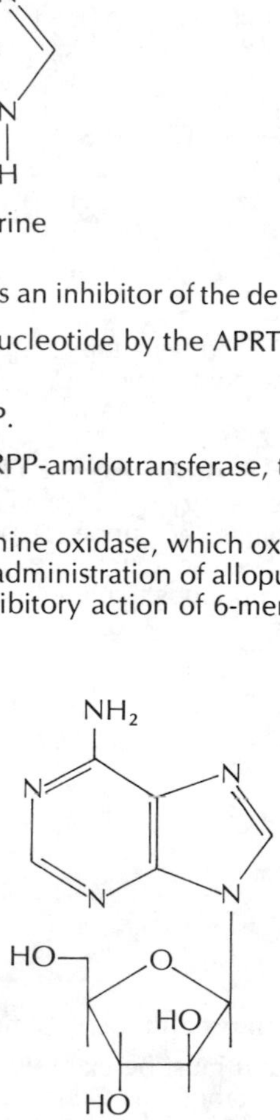

Cytosine arabinoside   Adenine arabinoside

1. Cytosine arabinoside has been used in the treatment of some acute leukemias. It is rapidly degraded by cytosine deaminase and has, therefore, a very short half-life.

2. Adenine arabinoside has been used as an antiviral and antitumor agent in man.

3. Both compounds are active inhibitors of DNA polymerase after they have been metabolized to the triphosphate forms.

E. Azaserine

$$
\begin{array}{c}
\overset{+}{\text{HC}}=\text{N}=\overset{-}{\text{N}} \\
| \\
\text{C}=\text{O} \\
| \\
\text{O} \\
| \\
\text{CH}_2 \\
| \\
\text{H}_3\text{N}^+\!\!-\!\text{CH} \\
| \\
\text{COO}^-
\end{array}
$$

Azaserine

1. Azaserine inhibits all reactions involving the transfer of the amide group of glutamine, including the entry of N-3 and N-9 into the purine ring. It inhibits the synthesis of GMP from IMP and the synthesis of CTP from UTP.

2. It irreversibly inactivates the enzymes and is too toxic for clinical use but has been extensively used in metabolic studies of purine synthetic pathways.

F. Methotrexate and aminopterin

1. Methotrexate and aminopterin are analogues of folic acid, which inhibit both folic acid and dihydrofolic acid reductase (see Chapter 27, section III B 3).

2. They are widely used therapeutically in various forms of cancer because they reduce the intracellular levels of the active cofactor tetrahydrofolate, preventing the reduction of the oxidized form (dihydrofolate), which is formed each time a uracil is methylated to thymine.

IX. DISEASES ASSOCIATED WITH DEFECTS OF NUCLEOTIDE METABOLISM

A. Lesch-Nyhan syndrome

1. A hereditary X-linked recessive condition, Lesch-Nyhan syndrome is due to a very severe or complete deficiency of hypoxanthine-guanine phosphoribosyl-transferase (HGPRT) activity.

2. The symptoms include excessive uric acid production, hyperuricemia, increased excretion of hypoxanthine, and, in severe cases, neurologic problems, with spasticity, mental retardation, and self-mutilation.

3. Several forms of the deficiency have been identified.
   a. Some patients have a normal level of HGPRT *protein*, as judged by titration with antibodies prepared against the purified enzyme, yet no measurable HGPRT *activity*.
   b. Some patients have a form of the enzyme that appears unstable, as its activity is much higher in new red blood cells than in old.
   c. One patient with the full spectrum of Lesch-Nyhan symptoms had a normal level of HGPRT when the enzyme was assayed under standard conditions of saturating substrate (PRPP), but had an activity in the Lesch-Nyhan range if the enzyme was assayed with intracellular levels of PRPP.

4. Because of the lack of HGPRT activity, hypoxanthine and guanine are not salvaged and the intracellular levels of PRPP increase while those of IMP and GMP decrease. Both of these factors lead to an increased de novo synthesis of purines.

5. The basis of the neurologic symptoms is unknown, but the brain (frontal lobe, basal ganglia, cerebellum) normally has levels of purine salvage enzymes 20 times higher than the levels in most cells.

6. Allopurinol treatment reduces the uric acid formation but does not shut down purine production, nor does it help with the neurologic problems.

B. Primary gout

1. **Causes.** Primary gout is caused by an excessive formation of uric acid due to an overproduction of purine nucleotides by the de novo pathway.
   a. The serum levels of uric acid exceed the solubility limit (6.8 mg/dl), and crystals of sodium urate are deposited in the joints of the extremities. These deposits may be so severe that they form deformities on the joint. These protuberances, known as **tophi**, are filled with sodium urate crystals.
   b. The deposition of sodium urate crystals can also occur in the kidney, causing extensive impairment of renal function. This is a common cause of death in Lesch-Nyhan patients.
   c. Hyperuricemia due to a primary overproduction of purines can be distinguished from hyperuricemia due to kidney disease (inability to excrete uric acid), or due to increased cellular breakdown during radiation therapy, by feeding ^{15}N-glycine. If there is an enrichment of ^{15}N in the urinary uric acid, the problem is one of excessive purine synthesis. If the enrichment is insignificant, the problem is due to secondary causes.

2. **Molecular basis for the hyperuricemia in primary gout**
   a. **PRPP synthetase.** Patients have been identified in whom the PRPP synthetase is abnormal, in that it does not respond to regulation by cellular levels of P_i or to feedback inhibition by purine nucleoside diphosphates.
   b. **Partial deficiency of HGPRT.** As in the Lesch-Nyhan syndrome, this condition leads to increased cellular levels of PRPP and hence to overproduction of purines. The deficiency in gout is not severe enough to provoke the neurologic symptoms associated with the Lesch-Nyhan syndrome.
   c. **Glycogen storage disease, type I (von Gierke's disease)**
      (1) This hereditary disorder is due to a deficiency of glucose 6-phosphatase activity (see Chapter 18, section IV A).
      (2) The increased levels of G6P enhance pentose shunt activity and the intracellular levels of ribose phosphates are increased.
      (3) The elevated ribose 1-phosphate increases cellular levels of PRPP with consequent overproduction of purine nucleotides.
   d. **Hyperactivity of glutathione reductase** has also been identified in cases of hyperuricemia, the increased cellular levels of glutathione reductase boosting cellular levels of $NADP^+$. $NADP^+$ is a major controlling factor in the pentose phosphate shunt pathway (see Chapter 19, section I).

3. **Treatment**
   a. Allopurinol is the major drug for the treatment of gout.
   b. It is oxidized by xanthine oxidase to oxypurinol.

Allopurinol   Oxypurinol

   c. The oxypurinol binds very tightly to reduced xanthine oxidase, effectively inhibiting its ability to catalyze the oxidation of xanthine or hypoxanthine. This inhibition is an example of **"suicide" inhibition**.
   d. The advantage of blocking uric acid formation is that hypoxanthine will not precipitate in the joints, and xanthine (solubility in plasma 10 mg/dl) may not do so.
   e. A secondary effect of treatment with allopurinol is a lowering of the rate of purine synthesis due to the formation of allopurinol ribonucleotide by the HGPRT salvage enzyme. This serves to reduce PRPP levels and to increase the feedback inhibition of PRPP-amidotransferase by the nucleotide. This secondary action does not occur in the Lesch-Nyhan syndrome.

C. Immunodeficiency diseases associated with abnormal purine metabolism

1. **Adenosine deaminase deficiency** is associated with severe combined immunodeficiency involving T-cell and B-cell dysfunction.

      a. There is a marked increase in cellular concentrations of deoxyadenosine triphosphate (dATP), a strong inhibitor of ribonucleotide reductase.

      b. The increase in [dATP] is due to the lack of conversion of excess deoxyadenosine to deoxyinosine and hypoxanthine.

   2. Purine nucleoside phosphorylase deficiency is associated with impaired T-cell function but with a normal B-cell function.

      a. There is a considerable reduction in uric acid production and increased levels of guanosine, deoxyguanosine, inosine, and deoxyinosine.

      b. It has been suggested that deoxyguanosine triphosphate is the toxic agent. It is a known inhibitor of CDP reductase and has been found in high concentration in the red blood cells of a patient with this condition.

D. Orotic aciduria

   1. This hereditary condition is characterized by high levels of urinary orotic acid, retarded growth, and severe anemia.

   2. The high levels of urinary orotic acid are due to the absence of orotate phosphoribosyl transferase, orotate monophosphate decarboxylase, or both.

   3. Patients with this condition can be successfully treated with uridine or cytidine, either of which will provide the pyrimidine nucleotides required for nucleic acid synthesis, and will also exert a feedback inhibition on carbamoyl phosphate synthetase (glutamine) to reduce pyrimidine production.

STUDY QUESTIONS

Directions: Each question below contains five suggested answers. Choose the **one best** response to each question.

1. All of the following reactions require 5′-phosphoribosyl-1-pyrophosphate as a substrate EXCEPT

(A) formation of 5′-phosphoribosylamine
(B) conversion of hypoxanthine to inosine monophosphate
(C) formation of nicotinate mononucleotide
(D) formation of orotic acid in pyrimidine biosynthesis
(E) conversion of adenine to adenosine monophosphate

4. The committed step in the de novo synthesis of purine nucleotides is the formation of

(A) ribose 5-phosphate from ribose 1-phosphate
(B) 5′-phosphoribosyl-1-pyrophosphate (PRPP) from ribose 5-phosphate
(C) 5′-phosphoribosylamine from PRPP
(D) 5′-phosphoribosylglycinamide from 5′-phosphoribosylamine
(E) none of the above

2. The control of purine nucleotide synthesis is characterized by all of the following statements EXCEPT

(A) 5′-phosphoribosyl-1-pyrophosphate (PRPP) synthetase is allosterically inhibited by nucleoside di- and triphosphates
(B) changes in PRPP concentration have a greater effect on the rate of purine synthesis than do equivalent changes in glutamine concentration
(C) inosine, guanosine, and adenosine monophosphates (IMP, GMP, and AMP) are negative modulators of PRPP-amidotransferase
(D) guanosine triphosphate (GTP) is an energy source for adenylosuccinate formation, and adenosine triphosphate is an energy source for the conversion of xanthosine 5′-monophosphate to GMP
(E) GMP is a competitive inhibitor of adenylosuccinate synthetase, and AMP is a competitive inhibitor of IMP dehydrogenase

5. The synthesis of carbamoyl phosphate associated with all of the following EXCEPT

(A) de novo synthesis of pyrimidines located in the cytosol
(B) de novo synthesis of pyrimidines, using glutamine as the nitrogen source
(C) de novo synthesis of pyrimidines, requiring N-acetylglutamate as an allosteric positive modulator
(D) urea synthesis in the liver, using ammonia as the nitrogen source
(E) urea synthesis located in mitochondria in the liver

6. Deoxyribonucleotides are formed by reduction of

(A) ribonucleosides
(B) ribonucleoside monophosphates
(C) ribonucleoside diphosphates
(D) ribonucleoside triphosphates
(E) none of the above

3. How many carbons of the purine ring are contributed by the folate one-carbon pool during purine biosynthesis?

(A) 0
(B) 1
(C) 2
(D) 3
(E) 4

7. The end product of purine catabolism in normal humans is

(A) urea
(B) uric acid
(C) creatinine
(D) xanthine
(E) hypoxanthine

8. The biosynthesis of pyridine nucleotide coenzymes is characterized by all of the following statements EXCEPT

(A) nicotinamide-adenine dinucleotide (NAD$^+$) can be formed from tryptophan in the human liver and kidney
(B) most of the requirement for NAD$^+$ and its phosphate (NADP$^+$) is met by synthesis from ingested niacin (nicotinate or nicotinamide)
(C) Nicotinate-adenine dinucleotide pyrophosphorylase is found exclusively in the cell nucleus
(D) nicotinamide can be formed from nicotinate by a reaction using glutamine as the amine group donor
(E) NADP$^+$ is formed by phosphorylation of the 2′ hydroxyl of the nicotinamide moiety of NAD$^+$

9. What is the chemical difference between nicotinamide-adenine dinucleotide (NAD$^+$) and its phosphate (NADP$^+$)?

(A) NADP$^+$ has an additional phosphate group at the 2′ position on the pentose adjacent to adenine
(B) NADP$^+$ has a nicotinate moiety rather than a nicotinamide
(C) NADP$^+$ has an additional pyrophosphate group
(D) NADP$^+$ has an additional phosphate linked directly to the adenine ring
(E) NADP$^+$ accepts two protons and two electrons, whereas NAD$^+$ accepts one proton and two electrons or a hydride ion

10. The antimetabolite 5-fluorouracil is characterized by all of the following statements EXCEPT

(A) it has to be extensively metabolized to be an effective inhibitor
(B) it is converted to 5-fluorouridine monophosphate by the two-step salvage pathway
(C) 5-fluorouridine diphosphate is reduced to the deoxyribose form
(D) the deoxyribose diphosphate form is dephosphorylated to one active form, 5-fluoro-2′-dUMP
(E) 5-fluoro-2′-dUMP can be dephosphorylated to 5-fluoro-2′-uridine, which is another active inhibitory form

11. The hereditary X-linked recessive condition known as the Lesch-Nyhan syndrome is due to a severe, or complete, deficiency of

(A) 5′-phosphoribosyl-1-pyrophosphate amidotransferase
(B) ribonucleotide reductase
(C) adenine phosphoribosyl transferase
(D) hypoxanthine-guanine phosphoribosyl transferase
(E) purine nucleoside phosphorylase

12. The hyperuricemia seen in cases of von Gierke's disease (glucose 6-phosphatase deficiency) is due to

(A) decreased inhibition of 5′-phosphoribosyl-1-pyrophosphate amidotransferase by nucleotides
(B) decreased conversion of inosine monophosphate to adenosine and guanosine monophosphates
(C) increased levels of PRPP due to increased hexose monophosphate shunt activity
(D) increased levels of glutamine
(E) increased activity of xanthine oxidase

13. Allopurinol reduces purine biosynthesis by

(A) inhibiting xanthine oxidase
(B) inhibiting 5′-phosphoribosyl-1-pyrophosphate (PRPP) amidotransferase
(C) virtue of its conversion to allopurinol ribonucleotide
(D) increasing the solubility of uric acid in plasma
(E) inhibiting the formation of PRPP

Directions: Each question below contains four suggested answers of which **one or more** is correct. Choose the answer

A if **1, 2, and 3** are correct
B if **1 and 3** are correct
C if **2 and 4** are correct
D if **4** is correct
E if **1, 2, 3, and 4** are correct

14. The cellular concentration of 5'-phospho-ribosyl-1-pyrophosphate (PRPP) is controlled by modulators of PRPP synthetase activity, including

(1) inorganic phosphate
(2) nucleoside diphosphates
(3) 2,3-diphosphoglycerate
(4) nucleoside triphosphates

15. Steps in the biosynthesis of deoxythmidylate (dTMP) include which of the following?

(1) dTMP is formed by methylation of deoxy-uridylate (dUMP)
(2) The donor of the methyl group is 5,10-methy-lene tetrahydrofolate
(3) The tetrahydrofolate coenzyme is oxidized to inactive dihydrofolate in the reaction
(4) The inactive dihydrofolate is reduced twice in order to function as a carrier of one-carbon moieties

16. Correct statements regarding the *interconversion* of purine nucleotides include which of the following?

(1) Adenosine monophosphate (AMP) can be formed by reduction of guanosine monophosphate (GMP)
(2) Inosine monophosphate (IMP) can be formed by deamination of AMP
(3) GMP can be formed by deamination of IMP
(4) IMP can be formed by reduction of GMP

17. Mammalian kinases are able to convert which of the following nucleosides to nucleotides?

(1) Adenosine
(2) Inosine
(3) Cytidine
(4) Guanine

ANSWERS AND EXPLANATIONS

1. The answer is D. (*I A 3, B 2 b (1), C 1 a, 2 e; III B 1; VII A 1 b*) 5′-Phosphoribosyl-1-pyrophosphate (PRPP) is involved in the biosynthesis of purines, pyrimidines, and pyridine nucleotide coenzymes. It is also a substrate for the one-step salvage pathways for purine and pyrimidine bases. Specifically, PRPP is a substrate with glutamine in the formation of 5′-phosphoribosylamine, the committed step in the de novo synthesis of purines. It is *not* required for the formation of orotic acid, an intermediate in pyrimidine synthesis, but is a substrate for the conversion of orotic acid to orotidine 5′-monophosphate. PRPP is a substrate for hypoxanthine-guanine phosphoribosyl transferase, the salvage enzyme, which converts hypoxanthine to inosine monophosphate or guanine to guanine monophosphate, and for adenine phosphoribosyl transferase, which converts adenine to adenosine monophosphate. PRPP is required for the conversion of quinolinate (from tryptophan) to nicotinate mononucleotide in the biosynthesis of pyridine nucleotide coenzymes.

2. The answer is E. (*I A 5, B 5*) Increases in 5′-phosphoribosyl-1-pyrophosphate (PRPP) concentration can drive purine nucleotide biosynthesis to abnormally high levels. PRPP levels are normally well controlled, the synthetase being inhibited by nucleoside di- and triphosphates. The committed step in purine nucleotide synthesis, catalyzed by PRPP-amidotransferase, displays sigmoidal kinetics toward [PRPP], whereas the kinetics for [glutamine] are hyperbolic. Thus, changes in the concentration of PRPP have a greater effect on the rate of purine nucleotide biosynthesis than equivalent changes in glutamine concentration. Inosine, guanosine, and adenosine monophosphates (IMP, GMP, and AMP) are negative modulators of PRPP-amidotransferase with AMP being synergistic with IMP or GMP. GTP is an energy source for AMP formation from IMP, and adenosine triphosphate is an energy source for GMP formation from IMP. AMP is a competitive inhibitor of adenylosuccinate synthetase, a step in the path to GMP, and GMP is a competitive inhibitor of IMP dehydrogenase, a step in the path to AMP. This complex of checks and balances ensures that approximately equal amounts of each purine nucleotide are formed for the synthesis of nucleic acids.

3. The answer is C. (*I B 1*) Two carbons from the folate one-carbon pool are used in the formation of the purine nucleotides. The first reaction to use a one-carbon group from the folate pool transfers a formyl group from 5,10-methylidyne tetrahydrofolate to 5′-phosphoribosylglycinamide to form 5′-phosphoribosyl-N-formylglycinamide. Another one-carbon group is transferred from 10-formyl tetrahydrofolate to form 5′-phosphoribosyl-5-formamidoimidazole-4-carboxamide, the immediate precursor of inosine monophosphate.

4. The answer is C. (*I B 2 b*) The committed step in the de novo synthesis of purine nucleotides is the formation of 5′-phosphoribosylamine from 5′-phosphoribosyl-1-pyrophosphate (PRPP), catalyzed by PRPP-amidotransferase. It is a reversible reaction, but in the cell the hydrolysis of pyrophosphate, one of the products, by pyrophosphatases renders the reaction irreversible.

5. The answer is C. (*I C 2 a*) The formation of carbamoyl phosphate from glutamine and carbon dioxide (CO_2) is the regulated step of pyrimidine synthesis in higher animals. It is catalyzed by a cytoplasmic enzyme and does *not* require N-acetylglutamate as an allosteric regulator. The enzyme is referred to as carbamoyl phosphate synthetase (glutamine). An enzyme located in liver mitochondria catalyzes the formation of carbamoyl phosphate from ammonia and CO_2. This enzyme is involved in urea synthesis and *does* require N-acetylglutamate as an allosteric positive regulator. It is referred to as carbamoyl phosphate synthetase (ammonia).

6. The answer is C. (*II A*) Deoxyribonucleotides are formed by reduction of the corresponding ribonucleosides, which must be in the diphosphate form. The reduction of the 2′-hydroxyl group of ribose requires a 12-kdal protein, thioredoxin, as electron donor. A flavoprotein and reduced nicotinamide-adenine dinucleotide phosphate are required to reduce the oxidized form of thioredoxin.

7. The answer is B. (*VI B 6 a*) The purine bases are converted to hypoxanthine (adenosine and inosine) or xanthine (guanine). The enzyme xanthine oxidase oxidizes hypoxanthine to xanthine and also xanthine to uric acid, which, as the final product of purine catabolism, is excreted in the urine.

8. The answer is D. (*VII A*) The vitamin niacin (nicotinic acid or nicotinamide) is the major source of pyridine nucleotide coenzymes in the human. About 10% of the total pyridine nucleotide coenzymes required can be met from excess tryptophan. Although nicotinic acid and nicotinamide are both equally effective as precursors of nicotinamide-adenine dinucleotide (NAD^+) and its phosphate ($NADP^+$), there is no known conversion of nicotinate to nicotinamide. The nicotinamide moiety of NAD^+ arises by glutamine-dependent amidation of the nicotinate moiety when it is in the form of the dinucleotide. NAD^+ is formed exclusively in the nucleus by NAD^+-pyrophosphorylase.

9. The answer is A. (*VII A 3*) Nicotinamide-adenine dinucleotide (NAD$^+$) and its phosphate (NADP$^+$) are chemically identical except for a phosphoryl group borne by NADP$^+$ on the 2$'$ position of the ribose attached to the adenine. They both accept one proton and two electrons or a hydride ion (H$^-$).

10. The answer is E. (*VIII A*) The antimetabolite 5-fluorouracil has been used in the treatment of solid tumors. It can block the biosynthesis of thymidine, required for DNA synthesis, and can interfere with the formation of rRNA species. These two inhibitory actions of cell growth and division are due to the formation of different inhibitory forms of the antimetabolite. 5-Fluorouracil is first converted to 5-fluorouridine monophosphate by the sequential action of nucleoside phosphorylase and uridine kinase. This is the two-step salvage pathway. Further phosphorylation to the diphosphate is followed by reduction to the 2$'$-deoxyribose form and dephosphorylation to fluorodeoxyuridylate (5-fluoro-2$'$-dUMP). This is the compound that can bind to thymidylate synthetase, forming a stable ternary complex with the enzyme and tetrahydrofolate that effectively blocks the formation of thymidine. Some of the 5-fluoro-2$'$-dUMP is phosphorylated to the triphosphate derivative and incorporated into RNA, where it is an inhibitor of the maturation of the 45S precursor of the 28S and 18S rRNA species.

11. The answer is D. (*IX A*) The Lesch-Nyhan syndrome is due to severe, or complete, deficiency of hypoxanthine-guanine phosphoribosyl transferase. The adenine phosphoribosyl transferase is not affected. This deficiency in its more severe forms leads to hyperuricemia and gout-like symptoms, increased excretion of hypoxanthine, and neurologic problems of spasticity, mental retardation, and self-mutilation. A deficiency of 5$'$-phosphoribosyl-1-pyrophosphate amidotransferase would lead to a deficiency in the de novo synthesis of purines and not to excessive purine synthesis and hyperuricemia. The inability to form deoxyribonucleotides (ribonucleotide reductase deficiency) would probably be lethal even if the deficiency was only partial. A deficiency of purine nucleoside phosphorylase is associated with impaired T-cell function, with a normal B-cell function, in the immune system.

12. The answer is B. (*IX B 2 c*) The glycogen storage disease due to an inherited deficiency of glucose 6-phosphatase (von Gierke's disease) is characterized by a high plasma level of uric acid. The failure of the liver to convert glucose 6-phosphate (G6P) to glucose due to the deficiency of the phosphatase causes a buildup of G6P and an increased rate of ribose 5-phosphate formation by the hexose monophosphate shunt pathway. This increases 5$'$-phosphoribosyl-1-pyrophosphate levels, which drives purine biosynthesis to abnormally high rates. The excessive uric acid production is, therefore, due to the abnormally high purine synthesis and not to enhanced activity of xanthine oxidase.

13. The answer is C. (*IX B 3 e*) Allopurinol is used to reduce the formation of uric acid in patients with gout. In this condition, crystals of sodium urate are deposited in the joints. Allopurinol inhibits uric acid formation by inhibiting xanthine oxidase, which oxidizes both hypoxanthine to xanthine and xanthine to uric acid. A secondary effect of allopurinol in reducing purine synthesis is due to the formation of allopurinol ribonucleotide by the hypoxanthine-guanine phosphoribosyltransferase salvage enzyme. This serves to reduce 5$'$-phosphoribosyl-1-pyrophosphate (PRPP) levels and to increase the feedback inhibition of PRPP-amidotransferase by mononucleotides.

14. The answer is E (all). (*I A*) The cellular concentration of PRPP is controlled by modulators of PRPP synthetase activity. This allosteric enzyme has an absolute requirement for inorganic phosphate (P$_i$) and displays sigmoidal kinetics towards [P$_i$]. Normally the P$_i$ level in the cell is low enough to keep the enzyme relatively inactive, but if [P$_i$] increases, due perhaps to excessive nucleotide catabolism, the activity of PRPP synthetase is markedly increased. The enzyme is inhibited by nucleoside di- and triphosphates. In the erythrocyte, 2$'$,3$'$-diphosphoglycerate is also inhibitory.

15. The answer is A (1, 2, 3). (*II B*) Deoxythymidylate (dTMP) is formed by the methylation of deoxyuridylate (dUMP). A methylene group and a hydrogen are transferred from 5,10-methylene tetrahydrofolate to the C-5 position of dUMP. In this reaction, the tetrahydrofolate coenzyme is oxidized to inactive dihydrofolate. The inactive dihydrofolate must be reduced by dihydrofolate reductase in order to function again as a carrier of one-carbon moieties.

16. The answer is C (2, 4). (*IV A*) Adenosine monophosphate (AMP) deaminase catalyzes the deamination of AMP, a 6-amino purine derivative, to inosine monophosphate (IMP), a 6-keto purine derivative. Guanosine monophosphate (GMP) reductase catalyzes the reduction of GMP, a 2-amino-6-keto purine derivative, to IMP. AMP is not formed by the reduction of GMP, nor is GMP formed by deamination of IMP.

17. The answer is B (1, 3). (*V A 2*) The known mammalian kinases can phosphorylate the nucleosides adenosine, uridine, cytidine, deoxycytidine, and thymidine to their corresponding nucleoside monophosphates. Kinases that can phosphorylate guanine or inosine to the corresponding nucleoside monophosphate have not been identified in higher animals.

I. OVERVIEW. The **endocrine system** embraces all the hormone-producing tissues. However, in this review, only those hormones that are primarily involved in metabolic regulation will be considered.

A. Definitions

1. A **hormone** was defined almost 80 years ago by Starling and Bayliss as "a chemical agent which is released from one group of cells and travels, via the bloodstream, to affect one or more different groups of cells."

2. That definition has been expanded by others to include the functional aspect of **information transfer**. Thus, in Huxley's concept (1905), as modified by Robison, Butcher, and Sutherland (1971), a hormone became "an information-transferring molecule, the essential function of which is to transfer information from one set of cells to another for the good of the cell population as a whole."

B. Metabolic regulation. Hormones that regulate the flux of metabolic fuels can be regarded as being either

1. **Storage hormones**, which activate systems involved in the storage of metabolic fuels, such as glycogenesis, lipogenesis, and, in one sense, protein synthesis. The major storage hormone is the peptide **insulin**, secreted by the β cells of the islets of Langerhans in the pancreas.

2. **Mobilizing hormones**, which control the rate at which glycogen is broken down to glucose and triacylglycerols to free fatty acids or which result in the increased use of amino acids as fuels. The major mobilizing hormones are
   a. **Glucagon**, a peptide secreted by the α cells of the islets of Langerhans
   b. **The catecholamines**, epinephrine and norepinephrine
   c. **Cortisol**, produced by the cortex of the adrenal glands
   d. **Somatotropin** (growth hormone), produced by the anterior lobe of the pituitary gland (**adenohypophysis**)

C. Mechanisms of action. Hormones interact with their target cells in different ways.

1. By binding to a receptor molecule on the outer surface of the cell membrane with the concomitant activation of membrane-bound adenylate cyclase

2. By binding to receptor molecules on the outer surface of the cell membrane with the transmission of the signal within the cell achieved by mechanisms not yet elucidated

3. By binding to intracellular receptors with the hormone–receptor complex modulating gene expression. The initial hormone–receptor interaction may occur in the cytoplasm with subsequent transfer to the nucleus, or it may occur initially in the nucleus.

II. HORMONES THAT ACTIVATE ADENYLATE CYCLASE

A. Hormone–receptor interaction

1. The receptor molecules on the outer surface of the membrane are large integral membrane proteins (see Chapter 25, section III B 2 d) with a very high affinity for a given hormone.

2. The binding between hormone and receptor is reversible, and the hormonal action declines as the plasma levels of hormone decline

3. The hormone does not enter the cell.

B. Receptor–adenylate cyclase complex

1. The enzyme **adenylate cyclase** catalyzes the formation of cyclic adenosine monophosphate (cAMP) and inorganic pyrophosphate (PP_i) from adenosine triphosphate (ATP), the reaction being virtually irreversible owing to the hydrolysis of PP_i to inorganic orthophosphate (P_i).

2. The enzyme is a large integral membrane protein and is coupled to the receptor by the **G protein**, so-called because it binds guanine nucleotides.

3. The G protein exists in two forms.
   a. As a complex with guanosine triphosphate (GTP), in which form it can activate adenylate cyclase
   b. As a complex with guanosine diphosphate (GDP), in which form it cannot activate adenylate cyclase

4. The conversion of the GDP form of the G protein to the GTP form occurs by nucleotide exchange and is catalyzed by the hormone–receptor complex but not by the receptor alone.

5. The active adenylate cyclase is inactivated by the hydrolysis of the bound GTP to GDP by a guanosine triphosphatase (GTPase), which is an integral part of the G protein molecule.

6. The state of activity of adenylate cyclase thus depends upon the relative rates of GTP–GDP exchange and GTP hydrolysis. In the presence of hormone, the hormone–receptor complex markedly accelerates the exchange reaction.

C. cAMP–phosphodiesterase activity

1. The increased cellular concentrations of cAMP due to the activation of adenylate cyclase can be reduced to resting levels by the action of a cytoplasmic enzyme, **cyclic nucleotide phosphodiesterase (PDEase)**, which catalyzes the reaction

$$cAMP + H_2O \longrightarrow 5'\text{-}AMP$$

2. Thus, cellular cAMP concentrations are the result of the activities of adenylate cyclase and PDEase.

3. The activity of PDEase is inhibited by methyl xanthines, a group of compounds that are found in coffee and tea, including caffeine and theophylline.

D. cAMP-dependent protein kinase

1. The effect of cAMP on cellular function is promulgated via the activation of a protein kinase, which phosphorylates specific cellular proteins with ATP donating its γ-phosphate group.

2. The protein kinase is specific for cAMP and is thus called the cAMP-dependent protein kinase. (There are other protein kinase activities that do not require cAMP as an activator.)

3. The cAMP-dependent protein kinase is a tetramer having two kinds of subunits.
   a. Regulatory (R) subunits
   b. Catalytic (C) subunits

4. The R_2C_2 tetramer is inactive.

5. Two molecules of cAMP bind to each R subunit, whereupon the R_2C_2 complex dissociates into an R_2 subunit and two C units that are now catalytically active.

6. The active protein kinase transfers the γ-phosphate group from ATP to a specific cell protein.

E. Phosphorylation of cellular proteins

1. Phosphorylation of a cell protein is a covalent modification (see Chapter 11, section VII G). If the cell protein is an enzyme, phosphorylation may either activate it or inactivate it, depending upon the particular cell protein. For example, phosphorylation
   a. Activates glycogen phosphorylase and phosphorylase kinase
   b. Inactivates glycogen synthetase, pyruvate kinase, and pyruvate dehydrogenase

2. Note that some phosphorylations are carried out by other protein kinases besides the cAMP-dependent protein kinase. For example, glycogen phosphorylase is phosphorylated by phosphorylase kinase (see Chapter 18, section III B 2).

F. Hormone activators

1. Glucagon

a. Biosynthesis
(1) Glucagon, a peptide containing 29 residues, is formed in the α cells of the islets of Langerhans in the pancreas.
(2) There are two precursor forms, **preproglucagon** and **proglucagon**, the former bearing a signal peptide sequence at the N-terminal end (see Chapter 4, section IX B 2).
(3) Proglucagon is converted to glucagon by the removal of eight residues from the N-terminal end.
(4) A peptide similar to α-cell glucagon is produced by cells in the gastrointestinal tract.

b. Release
(1) The major signal for the release of glucagon from the α cells is a lowered blood sugar level, although increases in plasma amino acid levels or sympathetic nervous system activity also stimulate the release of glucagon.
(2) Elevated levels of blood glucose inhibit glucagon release.

c. Functions
(1) Glucagon serves to combat hypoglycemia by several means.
    (a) It activates liver glycogenolysis by binding to specific glucagon receptors on the liver cell membrane, thereby activating the cAMP–adenylate cyclase cascade (see Chapter 18, section III D).
    (b) It inhibits glycolysis by promoting cAMP-dependent phosphorylation and by inhibiting the actions of pyruvate kinase and phosphofructokinase.
    (c) It stimulates gluconeogenesis by increasing the passage of fuel through the gluconeogenic pathway from pyruvate to glucose, particularly at the pyruvate carboxylase and phosphoenolpyruvate carboxykinase steps.
(2) **Insulin** (see section IV) counteracts most actions of glucagon, and it is the insulin:glucagon ratio that determines the direction of metabolic flux: A low ratio directs it toward carbohydrate use; a high ratio toward storage.

2. Catecholamines
a. Biosynthesis
(1) For the formation of the catecholamines—dopamine, norepinephrine, and epinephrine—from tyrosine, see Chapter 28, section I A.
(2) A synthetic catecholamine, **isoproterenol**, has been widely used in studies of catecholamine action. It has the following structure:

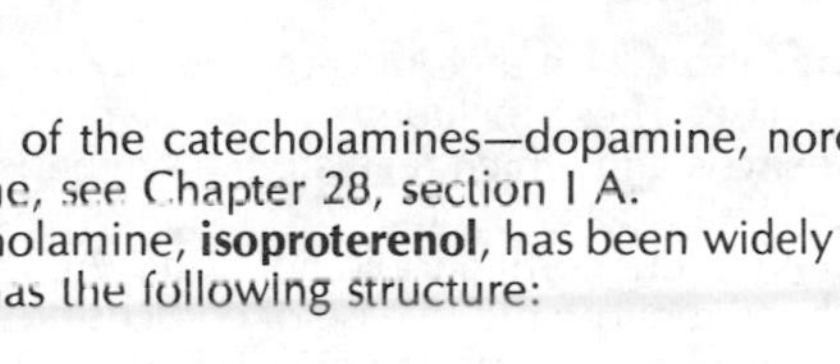

Isoproterenol

(3) **Norepinephrine** is the major neural transmitter in the sympathetic nervous system. **Epinephrine** is a hormone secreted from the adrenal medulla in times of stress, including exercise.

b. Functions
(1) The role of catecholamines in fuel metabolism is to serve as mobilizing agents.
(2) The release of norepinephrine from nerve endings on target tissues, such as adipose tissue, has the same kind of effect as the arrival of epinephrine via the circulation.
(3) The catecholamines also have numerous other actions in the body that affect the cardiovascular system, the visceral smooth muscle, and many exocrine glands.

c. Cellular receptors
(1) Intensive studies of the relative potencies of norepinephrine, epinephrine, and isoproterenol on a wide variety of cellular responses, coupled with studies using specific inhibitors (blockers), have led to the concept that there are at least three kinds of cell receptors for the catecholamines.
    (a) α receptors, for which the order of potency in activating the cell response is epinephrine > norepinephrine > isoproterenol
    (b) β_1 receptors, where isoproterenol > norepinephrine > epinephrine
    (c) β_2 receptors, where isoproterenol > epinephrine > norepinephrine
(2) Both the β_1 and β_2 receptors are coupled to adenylate cyclase by the G protein, and the catecholamine stimulus is relayed via increased cAMP formation, which is in effect acting as a second messenger.
(3) The α receptor is not coupled to adenylate cyclase, and the hormonal signal may be re-

layed by an increased influx of calcium into the cell, which is triggered in some manner by the hormone–receptor interaction.

d. Cellular responses to catecholamines

(1) α **Responses**
  (a) Increased glycogenolysis in the liver
  (b) Increased gluconeogenesis

(2) β_1 **Responses**
  (a) Increased glycogenolysis in cardiac muscle
  (b) Increased lipolysis in adipose tissue
  (c) Increased thermogenesis in brown adipose tissue
  (d) Increased heart rate (chronotropic effect)
  (e) Increased force of contraction of the heart (inotropic effect)

(3) β_2 **Responses**
  (a) Increased glycogenolysis in skeletal muscle
  (b) Relaxation of bronchial smooth muscle

3. Epinephrine and regulation of cardiac muscle contraction

a. Epinephrine has both inotropic and chronotropic effects, stimulating both the force and the rate of cardiac muscle contractions.

b. The action of epinephrine on β_1 receptors of the muscle cell membrane results in activation of the membrane-bound adenylate cyclase with a consequent rise in the cellular levels of cAMP.

c. The elevated levels of cAMP stimulate the cAMP-dependent protein kinase, which is responsible for the phosphorylation of specific proteins in cardiac muscle cells.

d. The phosphorylation (covalent modification) of these proteins alters their function and eventually allows Ca^{2+} ion concentrations in the sarcoplasm to increase and potentiate myocardial contraction.

4. Other hormones also act on their target tissues via a cell membrane receptor that is coupled to adenylate cyclase. These include several polypeptide hormones, which are produced in the adenohypophysis and which regulate the synthesis and release of other hormones from peripheral endocrine organs, such as

a. **Corticotropin** (adrenocorticotropic hormone, ACTH), which acts on the adrenal cortex, regulating the formation and release of corticosteroids (see section III B 3)

b. **Thyrotropin** (thyroid-stimulating hormone, TSH), which regulates the production of triiodothyronine (T_3) and thyroxine (T_4) by the thyroid gland (see section III C 2)

c. **Luteinizing hormone** and **follicle-stimulating hormone**, which control the function of male and female gonads

III. HORMONES THAT MODULATE GENE EXPRESSION

A. General description

1. Two classes of hormones act on their target tissues to modulate gene function so that the gene products are changed in kind or in quantity.
   a. One set of hormones are steroids derived from cholesterol.
   b. The other hormones discussed here are the thyroid hormones, T_3 and T_4, which are derived from tyrosine.

2. Both classes of hormones enter the cell cytoplasm before interacting with specific receptor molecules. This is in contrast to the hormones discussed in section II, which interact with receptors on the outside of the plasma membrane.

B. Steroid hormones

1. **Classification.** The steroid hormones include
   a. **Cortisol**, the glucocorticoid produced in the fasciculata region of the adrenal cortex
   b. **Aldosterone**, the mineralocorticoid produced by cells in the glomerulosa region of the adrenal cortex
   c. **Testosterone**, the male gonadal hormone, and **estradiol** and **progesterone**, the female gonadal hormones, which will not be considered here
   d. **1,25-Dihydroxycholecalciferol**, the active form of vitamin D, produced by the kidney from 25-hydroxycholecalciferol

2. **Biosynthesis** (See Chapter 23, section IV B.)

3. **Release and functions**
   a. **Cortisol**
     (1) The signal for the synthesis and release (there is no storage) of cortisol is provided by

the pituitary secretion of ACTH, which acts on the adrenal cortex via adenylate cyclase activation.

(2) The release of adrenocorticotropin is part of the body's response to stress of various kinds.

(3) Cortisol circulates bound to a plasma globulin, **transcortin**.

(4) Cortisol has a number of important **actions**, which include

    (a) Effects on the liver, such as

        (i) Increased production of glucose and deposition of glycogen

        (ii) Increased synthesis of RNA and protein, particularly enzymes involved in gluconeogenesis from amino acids

    (b) Effects on skeletal muscle, such as

        (i) Decreased glucose use

        (ii) Increased degradation of amino acids

    (c) Effects on adipose tissue, such as

        (i) Decreased use of glucose

        (ii) Increased release of free fatty acids

    (d) Inhibition of the inflammatory response (see Chapter 22, section III B 5).

b. Aldosterone

(1) The rate of aldosterone synthesis is increased by both ACTH and **angiotensin II**, an octapeptide formed in the blood by a kidney enzyme, **renin**, acting on a protein substrate formed in the liver.

(2) The synthesis and secretion of aldosterone are modulated by Na^+ and K^+ with a deficit of Na^+ accelerating synthesis and a deficit of K^+ inhibiting it.

(3) Aldosterone acts on most cells in the body to regulate the exchange of Na^+, K^+, and H^+. Its effects, however, are particularly marked on the cells of the ascending loop of Henle in the kidney, where it stimulates the reabsorption of Na^+ from the tubule and the transfer of K^+ and/or H^+ from the blood to the tubule.

c. 1,25-Dihydroxycholecalciferol

(1) 1,25-Dihydroxycholecalciferol (1,25-DHC) is the active hormonal form of vitamin D.

(2) It is released from cells in the kidney, which hydroxylate 25-hydroxycholecalciferol at the C-1 position.

(3) The formation and release of 1,25-DHC is regulated by a peptide hormone, **parathormone**, which is produced by the parathyroid glands and is believed to act on the 1,25-DHC–producing cells via the stimulation of adenylate cyclase.

(4) 1,25-DHC acts on the mucosa of the small intestine, where it increases the absorption of Ca^{2+}, and on bone, where it increases the deposition of Ca^{2+}.

3. Mechanism of action

a. A model for steroid hormone action was outlined in Chapter 8, section I C in connection with the role of estradiol and progesterone in eukaryotic gene regulation.

b. The model proposes that

(1) Cytoplasmic receptors interact with the hormone

(2) The hormone–receptor complex enters the nucleus, where it interacts with the chromatin

c. Receptors for all the major classes of steroid hormones have been found, and although the details vary from hormone to hormone, the sequence of events appears to be the same in all cases.

d. The change in gene expression induced by steroid hormones is complex, involving changes in the expression of multiple genes, and only in a few cases have gene products been clearly identified.

C. Thyroid hormones. Triiodothyronine (T_3) and thyroxine (T_4), the thyroid hormones, circulate bound to plasma proteins, chiefly thyroxine-binding globulin (TBG). T_3 is three to five times more active biologically than T_4.

1. Biosynthesis (See Chapter 28, section I C.)

2. Release

a. The formation and release of the thyroid hormones are under the control of pituitary **TSH**, which is itself regulated by **thyrotropin-releasing hormone (TRH)**, a factor produced in the hypothalamus.

b. Feedback inhibition of thyroid hormone release occurs at both the hypothalamic and pituitary levels.

3. Functions

a. The thyroid hormones appear to affect all cells in the body with the exception of adult brain cells and the testis.

b. Binding to a receptor in the chromatin appears to be the first step in thyroid hormone action.

c. An increase in thermogenesis and in oxygen consumption is a common cellular response.

d. **Metabolic effects** are seen in most phases of fuel metabolism.

(1) In the carbohydrate sector, the overall effect is hyperglycemia.

(2) In the fat sector, cholesterol levels increase in hypothyroidism and decrease in hyperthyroidism. The response to administration of thyroid hormones is an overall increase in lipid use.

(3) In the protein sector, hyperthyroidism increases protein catabolism; a normal thyroid hormone level (euthyroid state) is a necessity for normal growth and development.

IV. INSULIN. Most hormones that function in the control of metabolism can be classified as mobilizing hormones, in terms of their action on metabolic fuel reserves. Insulin, on the other hand, is the major hormone concerned with the **storage** of metabolic fuels.

A. **Biosynthesis**

1. Insulin is produced in the β cells of the islets of Langerhans in the pancreas.

2. It is a polypeptide consisting of two chains connected by disulfide bridges. The A chain is 21 residues in length; the B chain has 30 residues.

3. The mRNA for human insulin specifies a peptide of 110 residues, called **preproinsulin**.

4. Twenty-four residues at the N-terminal end are hydrophobic in nature and form the signal peptide, which indicates that the protein is to be exported (see Chapter 4, section IX B 2). The 24 residues are removed in the cisternae of the endoplasmic reticulum to yield **proinsulin** (86 residues).

5. The proinsulin is transferred to the Golgi apparatus, where it is packaged into granules.

6. Within the secretory granules, an endopeptidase and an exopeptidase hydrolyze specific bonds to form active **insulin** (51 residues, 2 chains), the inactive **C peptide** (31 residues), and four basic amino acids.

7. Both insulin and the C peptide are secreted into the circulation on demand.

B. **Release of insulin from the β cells**

1. The signals for the release of insulin from the β cells are a high blood sugar level and, to a lesser extent, high plasma levels of some amino acids.

2. The exact mechanism for the release signal is not known, but it appears to involve an increase in intracellular Ca^{2+} concentration.

3. The release of insulin in response to a glucose load shows two phases.

a. An early rapid burst of insulin release

b. A later slower output of hormone

C. **Functions**

1. Insulin counterbalances the mobilizing hormones, setting the stage for the storage of glucose as glycogen and free fatty acids as triacylglycerols and enhancing protein synthesis.

2. Insulin is an absolute requirement for the adequate uptake of glucose by some tissues, notably skeletal and cardiac muscle and adipose tissue.

3. In the absence of insulin or in cases where the amount of insulin released is inadequate to control blood glucose levels, **diabetes mellitus** develops. The increased blood glucose levels in diabetes result from

a. The failure of muscle to take up glucose

b. Unchecked gluconeogenesis due to the unopposed action of glucagon

c. Impairment of carbohydrate use, particularly at the level of pyruvate dehydrogenase

D. **Mechanism of action**

1. Insulin interacts with a receptor on the outside of the plasma membrane of target cells. The insulin–receptor complex is eventually internalized, but it is unclear whether or not this is part of the mechanism of action.

2. The manner by which the hormone–receptor interaction is coupled to the physiologic response is not known.

STUDY QUESTIONS

Directions: Each question below contains five suggested answers. Choose the **one best** response to each question.

1. The role of guanine nucleotide–binding protein (G protein) in adenylate cyclase activation is best described by which of the following statements?

(A) The G protein forms a complex with different hormones, and the hormone–G protein complex is responsible for activation of adenylate cyclase

(B) The G protein, when complexed with guanosine triphosphate (GTP), is able to activate adenylate cyclase

(C) Active adenylate cyclase is inactivated by the creation of the GTP form of the G protein in a reaction catalyzed by the hormone–receptor complex

(D) Complex formation between the G protein and adenylate cyclase is sufficient to activate the latter

(E) None of the above

2. Insulin is released from the β cells of the pancreatic islets when

(A) the concentration of amino acids in blood is below normal

(B) the concentration of free fatty acids in blood is below normal

(C) the rate of free fatty acid release from adipose tissue is elevated

(D) the rate of ketone body formation by the liver is increased

(E) blood sugar levels are elevated above normal

3. All of the following statements regarding steroid hormones are correct EXCEPT

(A) the synthesis and secretion of aldosterone by adrenal glomerulosa cells is modulated by Na^+ and K^+

(B) cortisol increases the production of glucose in liver, the deposition of glycogen in liver, and protein synthesis in liver

(C) 1,25-dihydroxycholecalciferol acts on mucosal cells of the small intestine to increase the rate of Ca^{2+} absorption

(D) steroid hormones interact with receptors on the outside of the plasma membrane of target cells, and the complex is then transferred to the nucleus, where it modulates gene expression

(E) cortisol release from the fasciculata region of the adrenal cortex is regulated by pituitary adrenocorticotropin

4. All of the following statements regarding insulin are correct EXCEPT

(A) it is produced by the β cells of the islets of Langerhans in the pancreas

(B) preproinsulin is clipped to form proinsulin in the cisterna of the endoplasmic reticulum

(C) proinsulin is converted to insulin in storage vesicles in the β cell

(D) the active peptide contains 51 amino acids

(E) the active hormone is made up of one B chain linked by disulfide bonds to a C chain

5. All of the following statements are true for both the β_1 and β_2 catecholamine receptors EXCEPT

(A) they are linked to adenylate cyclase by the G protein

(B) they are located on the outer surface of the plasma membrane

(C) the order of potency for activating the cell response is isoproterenol > norepinephrine > epinephrine

(D) the catecholamine stimulus is relayed via increased cAMP formation

(E) they are associated with increased glycogenolysis in cardiac and skeletal muscles

6. Which of the following statements best describes the mechanism of action of insulin on target cells?

(A) Insulin binds to cytoplasmic receptor molecules and is transferred as a hormone–receptor complex to the nucleus, where it acts to modulate gene expression

(B) Insulin binds to a receptor molecule on the outer surface of the plasma membrane, and the hormone–receptor complex activates adenylate cyclase via the guanine nucleotide-binding protein (G protein)

(C) Insulin binds to a receptor molecule on the outer surface of the plasma membrane, with the hormone–receptor complex initiating the cellular response in an unknown fashion

(D) Insulin binds to the enzyme hexokinase, thereby stimulating the transfer of glucose into the cell

(E) None of the above

Directions: Each question below contains four suggested answers of which **one or more** is correct. Choose the answer

A   if **1, 2, and 3** are correct
B   if **1 and 3** are correct
C   if **2 and 4** are correct
D   if **4** is correct
E   if **1, 2, 3, and 4** are correct

7.   Epinephrine acts on the liver to

(1) increase blood sugar levels
(2) decrease glycogen synthesis
(3) activate glycogen phosphorylase
(4) activate adenylate cyclase

8.   Which of the following substances can interact with target cells by binding to a receptor on the outer side of the plasma membrane?

(1) Triiodothyronine
(2) Glucagon
(3) 17β-Estradiol
(4) Epinephrine

9.   Glucagon acts to combat hypoglycemia by

(1) activation of liver glycogenolysis
(2) inhibition of liver glycolysis
(3) stimulation of gluconeogenesis
(4) inhibition of glucose uptake by skeletal muscle

ANSWERS AND EXPLANATIONS

1. The answer is B. (*II B*) Hormones that act on cells via the activation of membrane-bound adenylate cyclase react first with a receptor molecule on the outside of the plasma membrane of target cells. The hormone–receptor complex, but not the receptor alone, catalyzes a nucleotide exchange reaction, whereby a protein, the G protein, exchanges a bound guanosine diphosphate (GDP) for a guanosine triphosphate (GTP) molecule. The GTP form of the G protein is capable of activating adenylate cyclase.

2. The answer is E. (*IV B*) Insulin is the major storage hormone, and the most effective signal for its release from the β cells in the pancreatic islets is an elevated blood sugar level such as occurs during feeding. Free fatty acids provide only a relatively weak stimulus, but high plasma levels of certain amino acids effectively augment the glucose stimulus. Increased plasma levels of ketone bodies are not a signal for insulin release. In fact, insulin suppresses ketone body formation.

3. The answer is D. (*III A 2, B*) The synthesis and release of aldosterone from the adrenocortical glomerulosa cells are modulated by Na^+ and K^+, with a deficit of Na^+ accelerating synthesis and a deficit of K^+ inhibiting it. Cortisol is released from the fasciculata region of the adrenal cortex under the control of adrenocorticotropin, and it acts on the liver to increase glucose formation and glycogen deposition. Protein synthesis in the liver is increased by cortisol mainly because it promotes the induction of a number of enzymes concerned with gluconeogenesis from amino acid precursors. 1,25-Dihydroxycholecalciferol, the hormonal form of vitamin D, acts on the mucosal cells of the small intestine to accelerate calcium transport. Steroids interact with intracellular receptors in their target cells, not with receptor molecules on the outer surface of the cell. The hormone–receptor complex is transferred to the nucleus where it acts by modulating gene expression.

4. The answer is E. (*IV*) Because insulin is a protein that the β cells of the islets of Langerhans in the pancreas are going to secrete, it is first formed by ribosomes bound to the endoplasmic reticulum as preproinsulin, which contains a leader, or signal, sequence of 23 amino acids at its amino end. This signal sequence directs the protein to the cisterna of the endoplasmic reticulum. In the cisterna, the signal sequence is clipped off to give proinsulin (81 residues), which is stored in vesicles in the β cell. Prior to secretion, two peptidases clip out a sequence of 30 residues which is called the C peptide, leaving active insulin (51 residues), which is secreted. Insulin contains two chains, a B chain and an A chain, linked by disulfide bonds. The inactive C peptide is also secreted.

5. The answer is C. (*II F 2 c*) At least three kinds of receptor molecules (α, β_1, and β_2) appear to be responsible for the primary interaction between catecholamines and target cells. These receptors are located on the outer face of the plasma membrane. Two kinds of β receptors are linked to adenylate cyclase by the G protein and differ in the relative affinities shown for different catecholamines. The β_1 receptors show the highest affinity for isoproterenol and the least for epinephrine, whereas with β_2 receptors the order of affinity is isoproterenol > epinephrine > norepinephrine. β_1 responses are associated with increased glycogenolysis in cardiac muscle, and β_2 responses are associated with increased glycogenolysis in skeletal muscle.

6. The answer is C. (*IV D*) Insulin binds to receptor molecules on the outer surface of the plasma membrane of target cells. The hormone–receptor interaction triggers the cellular responses, such as increased glucose entry into skeletal muscle cells, by an unknown linking mechanism. Insulin may be internalized, together with receptor protein, but it is not clear how, if at all, this is related to the mechanism of action.

7. The answer is E (all). (*II A, B, D, E, F 3*) Epinephrine, acting on receptors on the outer surface of the liver cell, activates adenylate cyclase, which forms the second messenger cyclic adenosine monophosphate (cAMP) from adenosine triphosphate (ATP). cAMP, in turn, activates cAMP-dependent protein kinase. This enzyme, by phosphorylating glycogen synthetase, converts it to its glucose 6-phosphate–dependent (G6P-dependent) form, which is inactive at normal cellular levels of G6P. Thus, glycogen synthesis is reduced. At the same time, the activated cAMP-dependent protein kinase also phosphorylates phosphorylase b kinase and in so doing activates this enzyme. Phosphorylase b kinase, in turn, phosphorylates glycogen phosphorylase, converting it to the active phosphorylase a. This results in an increased breakdown of glycogen, providing a source of blood glucose.

8. The answer is C (2, 4). (*II A, F*) A group of peptide hormones, including glucagon, bind to receptors on the outer side of the plasma membrane of their target cells. This is also true of the catecholamines, such as epinephrine. The receptor molecules are specific for a given substance, and in a cell, such as the hepatocyte, which responds to both glucagon and epinephrine, both kinds of receptor are present on the cell surface. Steroid hormones, such as 17β-estradiol, interact with *intracellular* receptors, which carry the hormone to the nucleus. Triiodothyronine, a thyroid hormone, also interacts with intracellular receptors, which in this case are sited in the nucleus.

9. The answer is A (1, 2, 3). *(II F 1 c)* Glucagon activates liver glycogenolysis via the cyclic adenosine monophosphate (cAMP)–adenylate cyclase cascade, thereby providing glucose for export to the blood. The increased cellular levels of cAMP, due to the glucagon action, also inhibit the flux of carbohydrate through glycolysis by the inhibition of pyruvate kinase and phosphofructokinase. Glucagon also provides for an increase in hepatic glucose production by stimulating gluconeogenesis, particularly at the pyruvate carboxylase and phosphoenolpyruvate carboxykinase steps. Skeletal muscle does not have cellular receptors for glucagon, so that this hormone is without direct effect on this tissue.

31
Vitamins

I. OVERVIEW

A. Definition. Vitamins are organic compounds that human tissues cannot synthesize but that are required for normal growth and development. Also known as **accessory food factors**, vitamins must be included in the diet.

1. Only small quantities are required, amounts that are insufficient in mass to be suppliers of carbon or nitrogen or to act as energy sources.

2. Nonetheless, when intake in the diet is below the needed level, deficiency symptoms appear in humans (and in other animals).

3. The time of onset of symptoms is a function of the size and the daily flux of the body reserves.

B. Sources of vitamins. Humans have two sources of vitamins.

1. All can be supplied by foods, but no single food source is a *rich* source of *all* vitamins.

2. Some can also be synthesized by intestinal microorganisms, but these microorganisms may not provide humans with the total requirement.

C. Classification. Vitamins are conveniently classified as **water-soluble** or **fat-soluble** vitamins.

1. **Water-soluble vitamins** form a very heterogeneous group chemically.

2. **Fat-soluble vitamins**
   a. Fat-soluble vitamins can all be regarded as being built up from isoprene units (see Chapter 23, section II B 2).

$$-CH_2-C(CH_3)=CH-CH_2-$$

   b. Human tissues can manufacture vitamin D but not the other fat-soluble vitamins, although they can make and handle isoprene units in the synthesis of cholesterol and ubiquinone (coenzyme Q).

D. Causes of deficiency

1. **Inadequate dietary intake** can be due to famine, poverty, food faddism, or alcoholism.

2. **Inadequate absorption** may result from
   a. Biliary obstruction, since lack of bile leads to decreased absorption of the fat-soluble vitamins (A, E, K, and D)
   b. Regional ileitis
   c. Diseases such as tropical sprue or celiac disease (nontropical sprue), the latter induced by wheat gluten in the diet of susceptible persons
   d. Pernicious anemia, owing to lack of the intrinsic factor

3. **Inadequate use** may result from
   a. Lack of a transport protein for a particular vitamin in the serum
   b. Failure to synthesize the active form of a vitamin that is ingested in an inactive precursor form

4. **Increased requirements.** In some instances, an increased caloric requirement can turn a borderline vitamin deficiency into a frank deficiency. These increased requirements occur during
   a. Growth

b. Pregnancy
c. Lactation
d. Wound healing and convalescence

5. **Increased excretion**, for example in kidney malfunction, may lead to a demand not fulfilled by an unchanging intake.

6. **Drug-induced deficiency** may also occur, such as
   a. Loss of microbial synthesis in the intestine because of antibiotic therapy
   b. Pyridoxine deficiency in patients receiving isoniazid (isonicotinic acid hydrazide) for the treatment of tuberculosis. This compound is an antagonist of pyridoxal phosphate, a cofactor derived from the vitamin pyridoxine (B_6).

II. THE WATER-SOLUBLE VITAMINS.
The water-soluble vitamins are, in most cases, cofactors in enzyme systems or precursors of cofactors. Table 31-1 lists the water-soluble vitamins and some of the major enzyme systems that require a cofactor derived from a particular vitamin.

A. Thiamine (vitamin B_1)

1. **Structure**

Thiamine

2. **Nutritional requirements**
   a. The recommended dietary allowance (RDA)* is 1.5 mg per day. The requirement depends primarily on carbohydrate intake and may be stated as 0.5 mg per 1000 calories consumed.
   b. Sources include pork, organ meats, yeast, lean meats, eggs, green vegetables, whole-grain cereals, nuts, and legumes. There is no significant supply from synthesis by intestinal microorganisms.

3. **Metabolism**
   a. Thiamine is the vitamin precursor of thiamine pyrophosphate, which is formed, at the expense of adenosine triphosphate (ATP), in the liver and to a lesser extent in muscle, brain, heart, and nucleated red cells (reticulocytes).
   b. Storage is limited, and the liver stores can be depleted in 12 to 14 days.

4. **Functions.** Thiamine pyrophosphate is a cofactor involved in the enzyme complexes that dehydrogenate α-keto acids [see Chapter 15, sections II B 1 and III D 3; Chapter 27, section II C 2 a (7)]. It is also a cofactor for transketolase (see Chapter 19, section I B 3).

5. **Deficiency**
   a. Biochemical sequelae include
      (1) Disturbances in carbohydrate metabolism
      (2) Low levels of transketolase activity, particularly in erythrocytes and leukocytes
   b. Clinical symptoms
      (1) Deficiency leads to cardiovascular and neurologic lesions and to emotional disturbances.
      (2) "Dry" beriberi develops when the diet chronically contains slightly less than the thiamine requirement. Symptoms include peripheral neuropathy with the extremities of greatest use most affected, fatigue, and an impaired capacity to work.
      (3) "Wet" beriberi develops when the deficiency is more severe. In addition to the neurologic manifestations, cardiovascular symptoms are more apparent. The heart shows a right-sided enlargement, tachycardia occurs, and cardiac failure is common after stress; edema and anorexia are characteristic.

*The RDA is based on a 70-kg adult consuming 3000 calories per day.

Table 31-1. Water-Soluble Vitamins as Cofactors or Precursors of Cofactors in Enzymatic Reactions

Vitamin	Cofactor	Examples of Enzyme Systems
Thiamine	Thiamine pyrophosphate (cocarboxylase)	Pyruvate dehydrogenase (pyruvate → acetyl CoA) Pyruvate decarboxylase (pyruvate → acetaldehyde) α-Ketoglutarate dehydrogenase (α-Ketoglutarate → succinyl CoA) Oxidative decarboxylation of α-keto acids derived from leucine, isoleucine, valine, threonine, and serine Transketolase
Biotin	ε-N-Biotinyl-L-lysine enzyme complex	Acetyl CoA carboxylase (acetyl CoA → malonyl CoA) 3-Methylcrotonyl CoA carboxylase (methylcrotonyl CoA → methylglutaconyl CoA in degradative pathway of leucine) Propionyl CoA carboxylase (propionyl CoA → 5-methylmalonyl CoA) Pyruvate carboxylase (pyruvate → oxaloacetate)
Pyridoxine (pyridoxol)	Pyridoxal phosphate	All transaminases (aminotransferases) Serine-threonine dehydratase Amino acid decarboxylation (e.g., histidine → histamine) Cystathionase (cystathionine → cysteine + α-ketobutyrate) Cystathionine synthetase (homocysteine + serine → cystathionine) Kynureninase (3-hydroxykynurenine → alanine + 3-hydroxyanthranilate in degradative pathway of tryptophan) Glycine dehydrogenase (glycine → NH$_4^+$ + CO$_2$ + 5,10-methylene H$_4$-folate) Serine transhydroxymethylase H$_4$-folate (Serine ⟷ glycine) Diamine oxidase (e.g., histamine → imidazole-4-acetaldehyde) Glycogen phosphorylase (glycogen → n-glucose-1-phosphate) 5-Aminolevulinate synthetase (succinyl CoA + glycine → 5-aminolevulinate)
Riboflavin	Flavin mononucleotide and flavin-adenine dinucleotide	Electron carriers in oxidoreduction (usually tightly bound to the apoenzyme)
Niacin	Nicotinamide-adenine dinucleotide and its phosphate	Electron carriers in oxidoreduction (mobile electron carriers)
Pantothenic acid	Coenzyme A	Pyruvate dehydrogenase α-Ketoglutarate dehydrogenase Oxidative decarboxylation of α-keto acids Fatty acid activation Generally, a carrier of acyl groups

Table 31-1. Continued

Vitamin	Cofactor	Examples of Enzyme Systems
Folic acid (pteroylglutamic acid)	Tetrahydrofolic acid	Glutamate–H_4-folate formimino transferase Glycine dehydrogenase Serine transhydroxymethylase 5-Methyl H_4-folate homocysteine transmethylase 10-Formyl H_4-folate synthetase Thymidylate synthetase *In purine synthesis:* Phosphoribosyl-glycinamide formyl transferase Phosphoribosylaminoimidazole-carboxyamide formyl transferase
Vitamin B_{12} (cyanocobalamin)	5-Deoxyadenosylcobalamin, methylcobalamin	Methylmalonyl CoA isomerase 5-Methyl H_4-folate homocysteine transmethylase
Vitamin C	Ascorbic acid	Prolyl and lysyl hydroxylases

Note.—The table gives the major enzyme systems with which each of the water-soluble vitamins is associated.

  (4) Wernicke's encephalopathy develops in the most acute deficiencies and is seen primarily in alcoholics. Hemorrhagic polioencephalopathies accompany neurologic manifestations, including ocular paralysis; coma and death ensue.

B. Riboflavin (vitamin B₂)

  1. Structure

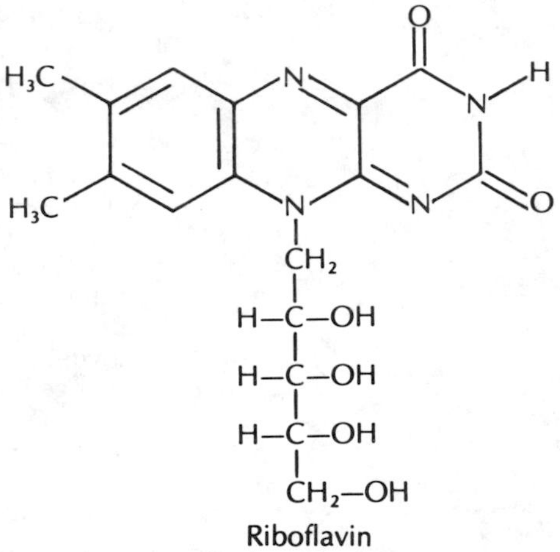

Riboflavin

  2. Nutritional requirements
  a. The RDA is 1.8 mg per day. The requirement for riboflavin does not appear to be related to caloric requirements or to muscular activity.
  b. Need is related to protein use and is increased during growth, pregnancy, lactation, and wound healing.
    (1) The urinary excretion of riboflavin is affected by alterations in nitrogen balance.
    (2) In positive nitrogen balance (see Chapter 26, section IV A), there is a decrease in urinary riboflavin.
    (3) This suggests the correlation with protein use and the increased need during growth, pregnancy, lactation, wound healing, and convalescence.
  c. Sources include milk, liver, kidneys, and green vegetables. The supply from intestinal microorganisms is limited.

3. Metabolism
  a. Riboflavin is converted to riboflavin phosphate in the intestinal mucosal cells and then to flavin-adenine dinucleotide (FAD) in the liver (see Chapter 29, section VII B).
  b. It is excreted primarily as riboflavin.

4. Functions
  a. Riboflavin is the precursor of coenzymes required by several oxidative enzymes (see Table 31-1).
  b. It is needed for maintenance of mucosal, epithelial, and ocular tissues.

5. Deficiency. Ariboflavinosis is characterized by
  a. Lesions of the lips, mouth, skin, and genitalia, especially angular stomatitis, cheilosis, glossitis, and seborrheic dermatitis
  b. Vascularization of the cornea

C. Pyridoxine (vitamin B6)

1. Structure
  a. Several closely related compounds—pyridoxine, pyridoxamine, and pyridoxal—can function as vitamin B6.

Pyridoxine   Pyridoxamine   Pyridoxal

  b. Phosphorylation yields the biologically active form.

Pyridoxal phosphate
(cofactor form)

2. Nutritional requirements
  a. The RDA is ordinarily 1.8 mg per day; requirements increase during pregnancy and lactation.
  b. Sources include liver, fish, nuts, and whole-grain cereals; some can come from intestinal bacterial synthesis.

3. Metabolism
  a. The nonphosphorylated forms are absorbed in the upper intestinal tract.
  b. These are converted to phosphate esters primarily in the brain, liver, and kidneys by pyridoxal kinase, which uses ATP as the phosphate donor.
  c. The vitamin is stored in the brain, liver, and muscle. About one-half of the pyridoxal phosphate in the body is bound to muscle glycogen phosphorylase.
  d. The vitamin is extensively metabolized. The major metabolite is pyridoxic acid, which is formed in the liver and excreted in the urine.

4. Functions
  a. Vitamin B6 serves as cofactor for many enzymes that have amino acids as substrates (see Table 31-1).
  b. It is found in glycogen phosphorylase.

5. Deficiency
   a. Biochemical sequelae include
   (1) Decreased pyridoxal phosphate levels in the blood
   (2) Decreased transaminase activity in erythrocytes
   (3) Increased excretion of xanthurenic acid in urine following a tryptophan load
   (4) Decreased urinary excretion of pyridoxic acid and increased urinary excretion of oxalic acid
   (5) Increased urinary excretion of cystathionine
   b. Clinical symptoms include
   (1) Lesions of the skin and mucosa
   (2) Anemia
   (3) Neuronal dysfunction, including convulsions
   (4) Personality changes

D. Cobalamin (vitamin B_{12})

1. **Structure.** See Figure 31-1.

2. **Nutritional requirements**
   a. The RDA is 3 μg per day; the need is increased during pregnancy.
   b. Sources include liver, kidney, other meats, and milk.

3. **Metabolism**
   a. Vitamin B_{12} is absorbed in the ileum as a complex with **intrinsic factor**, a glycoprotein formed in the gastric mucosa. The complex is cleaved to yield free vitamin B_{12} in the mucosal cells. In addition, up to 1% of a given dose is absorbed by passive diffusion along the entire small intestine.
   b. The vitamin is transported in serum bound to globulin and is converted to the metabolically active forms, adenosylcobalamin and methylcobalamin, in the liver, bone marrow cells, and reticulocytes.
   c. There is no significant catabolism.
   d. Small amounts of the vitamin are excreted in the bile, but most of this is reabsorbed in the ileum.
   e. The enterohepatic circulation of B_{12} provides almost total conservation of the vitamin, and a deficiency would take decades to develop if the vitamin were to be removed from the

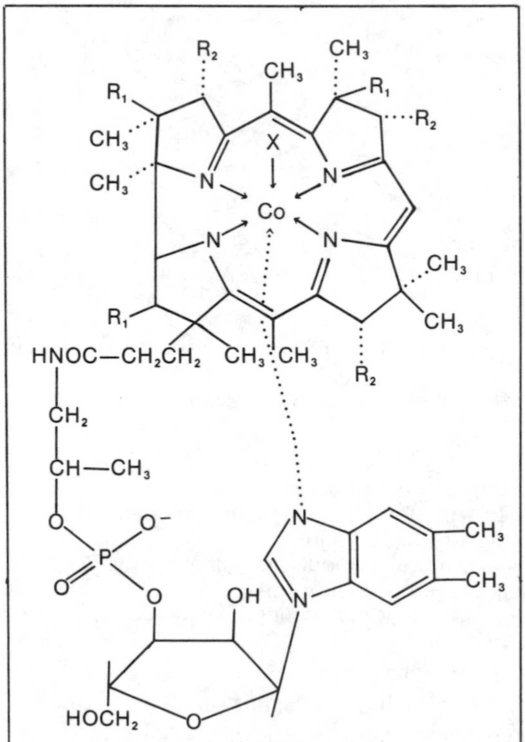

Figure 31-1. Structure of methylcobalamin (vitamin B_{12}). In another cofactor form, adenosylcobalamin, the methyl group attached to the cobalt is replaced by an adenosyl group. Commercial vitamin B_{12} has a cyanide moiety in place of the methyl group. (Reprinted with permission from Devlin TM: *A Textbook of Biochemistry with Clinical Correlations,* 2nd ed. New York, John Wiley, 1986, p 980).

diet. However, damage to the stomach or the ileum causes deficiency to occur more rapidly (3–6 years).

4. Functions

  a. There are only two human enzyme systems that are known to require cobalamin cofactors.

   (1) Methylmalonyl coenzyme A (methylmalonyl CoA) isomerase, which is involved in the catabolism of isoleucine and valine and in the use of propionyl CoA, requires adenosylcobalamin as its cofactor [see Chapter 27, section II C 2 a (5)].

   (2) Homocysteine:H_4-folate methyltransferase, which catalyzes the methylation of homocysteine to methionine, requires methylcobalamin as its cofactor (see Chapter 27, section III C 3 b).

  b. The role of B_{12} in hemopoiesis is related to folate metabolism, as lack of methylcobalamin leads to a deficiency in the folate coenzyme pool (see Chapter 27, section III C 3 b).

  c. Vitamin B_{12} is also involved in an unknown manner in the maintenance of the myelin sheath and epithelial cells.

5. Deficiency. Pernicious anemia is the classic consequence of vitamin B_{12} deficiency.

  a. It is due to an absence of intrinsic factor.

  b. Consequences include macrocytic anemia, megaloblastosis of bone marrow, degeneration of axis cylinders of spinal cord neurons, lesions of mucous surfaces, glossitis, and methylmalonic aciduria.

E. Folic acid (pteroylglutamic acid)

1. Structure

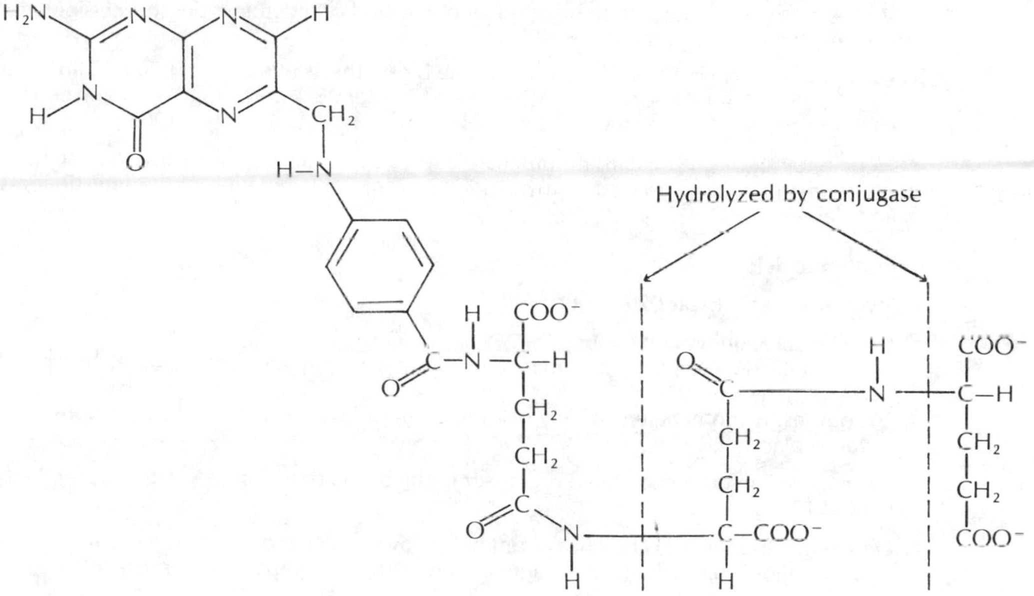

2. Nutritional requirements

  a. The RDA is 0.4 mg per day; requirements increase during pregnancy and lactation.

  b. Sources include synthesis by intestinal bacteria as well as liver, yeast, and green vegetables. Folates are easily destroyed by cooking.

3. Metabolism

  a. Folate in food is primarily present in polyglutamate form. The glutamate residues are split off by an intestinal conjugase prior to absorption. Unconjugated folate is absorbed primarily from the proximal third of the small intestine.

  b. In serum, liver, and other tissues, the predominant form of the vitamin is 5-methyl tetrahydrofolate.

  c. Total body stores (12–15 mg) will last 4–6 months after cessation of folate ingestion.

4. Functions

  a. Folate coenzymes act as carriers of one-carbon fragments at different levels of oxidation (see Chapter 27, section III).

b. Deficiency affects purine nucleotide and deoxythymidylate synthesis (see Chapter 29, sections I B and II B).

5. Deficiency. Clinical symptoms include
   a. Macrocytic anemia
   b. Megaloblastosis of bone marrow
   c. Gastrointestinal disturbances

F. Niacin

1. Structure. Nicotinic acid (niacin) and nicotinamide (nicotinic acid amide) are equally effective in supplying human needs.

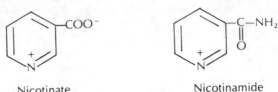

Nicotinate   Nicotinamide

2. Nutritional requirements
   a. The RDA is 20 mg per day.
   b. Tryptophan is a precursor; 60 mg is equivalent to 1 mg of nicotinic acid (see Chapter 29, section VII A).
   c. Nicotinic acid derived from the catabolism of tryptophan can only provide about 10% of the pyridine nucleotide cofactor pool.
   d. The remainder comes from ingested nicotinic acid or nicotinamide; sources include yeast, liver, legumes, and meats.

3. Functions. Niacin is an essential factor in the biosynthesis of nicotinamide-adenine dinucleotide (NAD^+) and its phosphate ($NADP^+$) [see Chapter 29, VII A], cofactors involved in biologic oxidation–reduction reactions (see Table 31-1).

4. Deficiency of dietary niacin in humans is the cause of the disease known as **pellagra**. The symptoms include dermatitis, diarrhea from chronic inflammation of the intestinal mucosa, and dementia.

G. Pantothenic acid

1. Structure. See Chapter 29, section VII C.

2. Nutritional requirements
   a. Although requirements in humans are unknown, they are probably met by an intake of 5–10 mg per day.
   b. Sources. Pantothenate is widely distributed in foods. Some synthesis by intestinal bacteria takes place.

3. Functions. Pantothenic acid is a precursor in the biosynthesis of CoA (see Chapter 29, section VII C).

4. Deficiency in humans is practically unknown. Symptoms produced by experimental feeding of an antagonist include nausea, fatigue, and burning cramps in the limbs.

H. Biotin

1. Structure

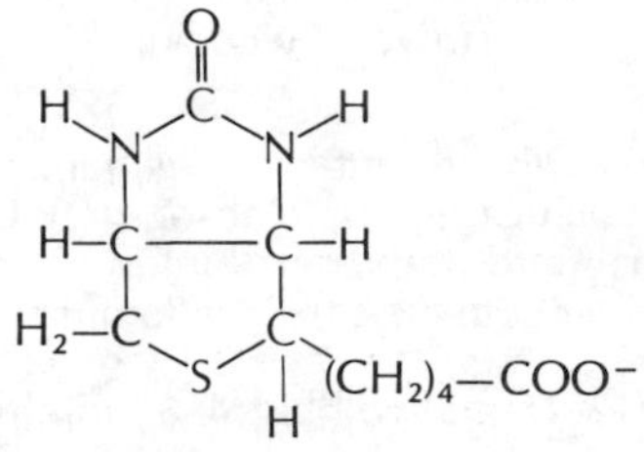

Biotin

2. Nutritional requirements
   a. Biotin is synthesized by intestinal microorganisms in such large quantities that a dietary

source is probably not necessary. This is suggested by the finding that the amount of biotin excreted is two to five times greater than the dietary intake.

   b. Sources in the diet include liver, kidneys, vegetables, and egg yolk.

3. Functions
   a. Biotin acts as a coenzyme in carboxylation reactions, where it is the carrier of carbon dioxide (CO_2) [see Table 31-1].
   b. ATP is required for the formation of the CO_2–biotin complex.

4. Deficiency
   a. Biotin deficiency can be produced by
      (1) Drugs that inhibit the growth of intestinal bacteria (e.g., antibiotics and sulfonamides)
      (2) The oral administration of avidin, a protein extracted from raw egg white, which binds to biotin and prevents its absorption
   b. Symptoms include seborrheic dermatitis, anorexia, nausea, and muscular pain.

I. L-Ascorbic acid (vitamin C)

   1. Structure. See Chapter 19, section II B 4.

   2. Nutritional requirements
      a. The RDA is 45 mg per day; about 60 to 100 mg per day maintains saturated plasma levels.
      b. Sources include citrus fruits and their juices, strawberries, cantaloupes, and raw or minimally cooked vegetables.

   3. Metabolism. Ascorbic acid is readily oxidized by heavy metals (see Chapter 19, section II B 4 b for the reaction).

   4. Functions. Vitamin C is required in the hydroxylation of proline and lysine during the biosynthesis of procollagen by the fibroblast [see Chapter 10, section III B 1 b (3)].

   5. Deficiency leads to the condition known as **scurvy** in human subjects.
      a. Scurvy causes impairment of wound healing and bone formation and a lessening of the integrity of blood vessels, which leads to a hemorrhagic diathesis.
      b. Anemia is often present.

III. FAT-SOLUBLE VITAMINS

   A. Retinol (vitamin A)
      1. Structure. Carotenoids (e.g., β-carotene) are provitamins; retinal and retinol are the forms used by the body.

β-Carotene

Retinal

Retinol

   2. Nutritional requirements
      a. The RDA is 1000 retinol equivalents (5000 IU, 1 mg retinol, 6 mg β-carotene) per day.
      b. Sources
         (1) Carotenoids are found in green leafy and yellow vegetables.
         (2) All-*trans*-retinyl esters occur in fish liver oils, liver, kidney, and butter fat.

   3. Absorption and metabolism
      a. In the upper intestine, retinyl esters are hydrolyzed and retinol is absorbed.
      b. In the intestinal mucosal cells, the retinol is re-esterified with palmitate and transported to the liver, where 90% of the body's vitamin A is stored.

    c. Carotenoids may be directly absorbed (bile is required for micelle formation), but most of the carotenoids are cleaved to retinal and then converted to retinol, which is absorbed.

  4. Functions
    a. Role in vision
      (1) On entering the retina, retinol is esterified to a fatty acid, providing a means of concentrating the retinol within the cell.
      (2) The fatty acid esters are hydrolyzed, and the retinol is oxidized to retinal by a specific NAD^+-linked dehydrogenase.
      (3) The retinal forms complexes with proteins called **opsins** in the rods and cones. Oxidation of retinol continues until all the opsins are saturated with retinal.
      (4) The opsins preferentially bind the 11-*cis* isomer of retinal, as opposed to the all-*trans* form.
      (5) The absorption of light is accompanied by a conformational change in the opsin and a change of the 11-*cis* isomer of retinal to the all-*trans* isomer, which is only weakly bound to the opsin.
      (6) In some manner, the conformational change in the opsin results in the generation of an action potential.
    b. Other functions of vitamin A are not clear, but administration of retinoic acid prevents the appearance of many deficiency symptoms (but not those associated with vision), so that a retinol-retinal oxidation–reduction reaction is probably not involved.

  5. Deficiency
    a. Insufficient vitamin A causes defects in vision, progressing from night blindness to xeromalacia and total blindness.
    b. There is also an abnormal formation of keratin in mucous membranes and a failure of bone remodeling, leading to thick, solid bones in the skull with an increase in cerebrospinal fluid pressure. Gonadal dysfunction occurs in males and abortion in females.
    c. Deprivation of vitamin A ultimately results in death.

B. Cholecalciferol (vitamin D₃)

  1. Structure. See Chapter 23, section IV B 3.

  2. Nutritional requirements
    a. The RDA is 400 IU (10 µg cholecalciferol) per day.
    b. Vitamin D is only required as an accessory food factor when human beings are deprived of sunlight, as it is synthesized from 7-dehydrocholesterol in the skin by the action of ultraviolet light.

  3. Functions. Cholecalciferol (vitamin D₃), which is formed from 7-dehydrocholesterol, is a precursor of the hormone 1,25-dihydroxycholecalciferol, a regulator of calcium metabolism (see Chapter 23, section IV B 3).

  4. Deficiency
    a. In children, vitamin D deficiency leads to a condition known as **rickets**, which manifests as a malformation of the long bones.
    b. In adults, whose bony growth is complete, bone deformities are rare, but there is an increased radiolucency of bones and an increased tendency for fractures to occur. The condition is known as **osteomalacia.**

C. Vitamin K

  1. Structure. Menaquinone is the form of vitamin K found in mammalian liver. It is a member of a large family of compounds with vitamin K activity, including one form made by plants (phylloquinone).

Menaquinone

2. Nutritional requirements
a. The daily requirement for vitamin K has been difficult to establish. Approximately 0.03 mg per day given intravenously is required to obtain normal blood clotting in vitamin K–depleted human adults.
b. Sources include green vegetables; synthesis by intestinal bacteria takes place.
c. Bile is required for absorption.

3. Functions. Vitamin K plays a role in blood coagulation.
a. Several enzyme factors involved in the blood clotting process become active by binding Ca^{2+}.
b. These proenzymes bind Ca^{2+} strongly because their amino terminal segments contain glutamate residues that have been carboxylated to dicarboxylate forms, which have a high affinity for Ca^{2+}.
c. Vitamin K acts as a cofactor in the carboxylation reaction, which also requires oxygen.

4. Deficiency
a. A primary deficiency of vitamin K is uncommon in humans due to its extensive distribution in plant and animal tissue products and to synthesis by intestinal microorganisms.
b. Deficiency may show up in newborn children, particularly if the gut is sterile, as the placenta is relatively weak in the transmission of lipids.

D. α-Tocopherol (vitamin E)

1. Structure

α-Tocopherol

2. Nutritional requirements
a. The RDA is 15 IU per day, but the requirement is related to the intake of unsaturated fatty acids.
b. Sources include vegetable and seed oils.
c. Bile is required for absorption. Approximately 60% of ingested tocopherols are excreted in the feces.

3. Functions
a. The tocopherols have the ability to prevent free radical formation in polyunsaturated fatty acids by donating electrons, although the mechanism is not clear.
b. This property enables them to participate with glutathione peroxidase in the removal of peroxides formed in the polyunsaturated fatty acids.

4. Deficiency of vitamin E in humans is associated with hemolysis of erythrocytes due to lack of protection against peroxides and with creatinuria due to increased muscle breakdown.

STUDY QUESTIONS

Directions: Each question below contains five suggested answers. Choose the **one best** response to each question.

1. Which of the following characteristics would be seen in a patient with a severe deficiency of thiamine?

(A) A decreased level of blood pyruvate and lactate
(B) An increased clotting time of blood
(C) A low level of red cell transaminase activity
(D) An increased urinary excretion of xanthurenic acid following a tryptophan load
(E) A decreased level of transketolase activity in red blood cells

2. Increased protein use is accompanied by an increased dietary requirement for

(A) ascorbic acid (vitamin C)
(B) riboflavin (vitamin B_2)
(C) cobalamin (vitamin B_{12})
(D) pyridoxine (vitamin B_6)
(E) thiamine (vitamin B_1)

3. Pyridoxal phosphate is a cofactor for which of the following enzymatic reactions?

(A) Fixation of carbon dioxide
(B) Oxidation–reduction
(C) Aminotransferase
(D) Phosphate group transfer
(E) None of the above

4. The disease pellagra is due to a deficiency of

(A) vitamin B_6
(B) biotin
(C) pantothenic acid
(D) folic acid
(E) niacin

5. The absorption of light by cells in the retina of the eye results in the conversion of

(A) β-carotene to retinal
(B) *cis*-retinal to all-*trans*-retinal
(C) all-*trans*-retinal to *cis*-retinal
(D) retinal to retinol
(E) retinol to retinal

Directions: Each question below contains four suggested answers of which **one or more** is correct. Choose the answer

A   if **1, 2, and 3** are correct
B   if **1 and 3** are correct
C   if **2 and 4** are correct
D   if **4** is correct
E   if **1, 2, 3, and 4** are correct

6. Which of the following enzymes require biotin as a cofactor?

(1) Acetyl coenzyme A (acetyl CoA) carboxylase
(2) Pyruvate carboxylase
(3) Propionyl CoA carboxylase
(4) Carbamoyl phosphate synthetase I (mitochondrial)

7. The symptoms of scurvy are due to

(1) a failure to incorporate ascorbic acid into the procollagen molecule
(2) failure to remove the N-terminal fragments from procollagen molecules
(3) inadequate cross-linking of the α-chains in tropocollagen
(4) inadequate formation of hydroxyproline and hydroxylysine

ANSWERS AND EXPLANATIONS

1. The answer is E. (*II A 5*) Transketolase activity in erythrocytes and leukocytes is a sensitive indicator of body thiamine levels. In severe thiamine deficiency, transketolase activity is depressed. Blood levels of pyruvate and lactate would not be expected to be low; rather, pyruvate levels are increased in thiamine deficiency due to a reduced use of pyruvate by the thiamine pyrophosphate-dependent pyruvate dehydrogenase complex. Thiamine deficiency should not significantly influence the clotting time of blood, the catabolism of tryptophan, or the transaminase activity in red blood cells.

2. The answer is B. (*II B 2*) The need for riboflavin (vitamin B_2) is related to the increased protein deposition that occurs during growth and pregnancy and to increased protein synthesis during lactation. During these times the dietary requirement for the vitamin is increased.

3. The answer is C. (*II C 4; Table 31-1*) Pyridoxal phosphate is a cofactor derived from the water-soluble vitamin pyridoxine. It is involved in many reactions in which an amino acid is a substrate, and all aminotransferases (transaminases) require it as a cofactor. The fixation of carbon dioxide (CO_2) in carboxylation reactions usually requires biotin as a cofactor, which acts as the carrier of CO_2. Oxidation–reduction reactions often use nicotinamide-adenine dinucleotide and its phosphate, flavin mononucleotide, and flavin-adenine dinucleotide as the electron acceptors. Phosphotransferases do not usually require a cofactor but nearly always show a dependency on a metal such as Mg^{2+} or Mn^{2+}.

4. The answer is E. (*II F 4*) Pellagra is a disease due to a deficiency of niacin. The symptoms include dermatitis, diarrhea due to chronic inflammation of the mucosa of the intestinal tract, and dementia. Niacin (nicotinic acid or nicotinamide) is a precursor of nicotinamide-adenine dinucleotide and its phosphate, which are electron carriers in many oxidation–reduction reactions.

5. The answer is B. (*III A 4*) The *cis* isomer of retinal is preferentially bound by proteins called opsins in the cells of the retina. The absorption of a photon of light by the opsin-*cis*-retinal complex is accompanied by a change in the configuration of the opsin and the conversion of *cis*-retinal to all-*trans*-retinal, which then dissociates from the opsin. In some manner, the conformational change in the opsin results in the generation of an action potential. Most of the ingested β-carotene is cleaved in the intestine to retinal, which is absorbed after its conversion to retinol.

6. The answer is A (1, 2, 3). (*II H 3; Table 31-1*) Most carboxylation reactions require biotin as the carrier of carbon dioxide (CO_2) The reaction of the biotinyl–enzyme complex with CO_2 requires adenosine triphosphate. Acetyl coenzyme A (acetyl CoA) carboxylase, pyruvate carboxylase, and propionyl CoA carboxylase are all in this category. The formation of carbamoyl phosphate from CO_2 and NH_4^+ by mitochondrial carbamoyl phosphate synthetase does not require biotin.

7. The answer is D (4). (*II I 4, 5*) Scurvy develops from a deficiency of dietary ascorbic acid. The symptoms include impairment of wound healing, convalescence, bone formation, and a lessening of the integrity of blood vessels, which leads to a hemorrhagic diathesis, anemia being often present. The symptoms are due to the improper synthesis of collagen, which is not cross-linked in a normal fashion because of a reduction in the formation of hydroxyproline and hydroxylysine during the synthesis of procollagen. Ascorbic acid acts as a cofactor for prolyl and lysyl hydroxylases.

32
Integration of Metabolism

I. INTRODUCTION. The various anabolic and catabolic pathways by which carbohydrates, fats, and proteins are processed as metabolic fuels for energy supply or as precursors in the biosynthesis of compounds required by the cell for maintenance or growth are intimately coordinated. In this chapter, some aspects of the integration and regulation of these pathways are considered.

II. METABOLIC FUELS

A. Definition

1. Metabolic fuels are substances that are used by the body as sources of carbon or oxidized to release free energy, which is used to support anabolic processes and other cellular functions.

2. The four kinds of molecules used as major metabolic fuels are sugars, fatty acids, ketone bodies, and amino acids.

B. Caloric value of metabolic fuels

1. If an organic substance is reacted with molecular oxygen in a bomb calorimeter and all of the carbon is converted to carbon dioxide (CO_2), the heat evolved is a measure of the potential free energy available in this substance.

2. The **caloric value** of a metabolic fuel is expressed in terms of **kilocalories (kcal) per gram**.

3. Useful approximations for the caloric value of the major metabolic fuels are shown in Table 32-1.

C. Body stores of metabolic fuels

1. **Carbohydrate**
   a. Circulating glucose in the blood is a major metabolic fuel, and a number of mechanisms are used to maintain adequate blood glucose levels.
   b. Carbohydrate is stored primarily as glycogen in the liver and skeletal muscle and to some extent in most other tissues. Brain stores of glycogen are very small.
   c. Only those cells that contain glucose 6-phosphatase (i.e., liver, kidney, and small intestine; see Chapter 17, section I A) can use glycogen to maintain blood glucose levels.
   d. Triacylglycerols form a potential source of glucose, insofar as the glycerol derived from hydrolysis of triacylglycerols can be converted to glucose by gluconeogenic tissues (see Chapter 17, section IV C).
   e. The catabolism of glucogenic amino acids (see Chapter 27, section II B 2) can provide glucose, 100 g of protein yielding 60 g of glucose.
   f. The amounts of carbohydrate stored, and thus potentially available, are given in Table 32-2.

Table 32-1. Caloric Values of the Major Metabolic Fuels

Metabolic Fuel	Caloric Value (kcal/g)
Carbohydrate (glucose, glycogen, etc.)	4
Amino acids (average)	4
Ketone bodies (hydroxybutyrate, acetoacetate)	4
Fatty acids	9

Table 32-2. Amounts of Carbohydrate Stored in Body Tissues

	Tissue Vol or Wt	Concentration	Carbohydrate — Total Amount Low	High
Blood glucose	6 L	80 mg/dl	5 g	
Glucose in extracellular water	19 L	80 mg/dl	15 g	
Liver glycogen	1.6 kg	2%–10%	30 g	150 g
Muscle glycogen	21 kg	0.3%–2%	60 g	400 g
Carbohydrate in other BCM* tissues	5–6 kg	Unknown	(25 g)[†]	

Potential carbohydrate
   From triacylglycerol: 1.7 kg glucose from 17 kg triacylglycerol
   From protein: 5.5 kg glucose from 9 kg BCM (approximately ⅓ of BCM protein can be mobilized)

*BCM = body cell mass.
[†]Very approximate.

2. **Fat**
   a. In the average 70-kg male, about 25% of body weight, or 17.5 kg, is adipose tissue. In the average 60-kg female, the figure is about 35%, or a total of 21 kg.
   b. Adipose tissue contains about 10% water (5% intracellular; 5% extracellular). Therefore, the 17 kg of adipose tissue in the adult male provides 8 kcal/g, rather than the 9 kcal/g of pure triacylglycerol. Thus, the stores of fat yield approximately 135,000 kcal in the male and 165,000 kcal in the female.

3. **Protein**
   a. The tissues of the body cell mass (BCM) contain 25% water (H_2O) and 75% protein; thus, the caloric value of the BCM is 1 kcal/g.
   b. One-third of the BCM, the maximum that can be mobilized, thus yields 40% × 70 kg × ⅓ × 1, which gives approximately 9000 kcal.

III. USE OF METABOLIC FUELS

A. Feeding–fasting cycle

1. Humans are intermittent feeders. Although in affluent societies the time periods between intakes of food may be rather short during waking hours, most individuals undergo a 6–8 hr fast overnight. Any consideration of metabolism and the use of metabolic fuels has to take into account this feeding–fasting cycle.

2. During feeding and shortly thereafter, the metabolic fuels used by tissues may be quite different from those used during a fast. In the first case, the fuel may be derived directly from the ingested, digested, and absorbed food molecules, and in the second case, from mobilized stores of fuel molecules.

3. To facilitate discussions of the integration of metabolism and the use of metabolic fuels, it is useful to define the different states of the feeding–fasting cycle and states of extended fasting.
   a. Postprandial state. Just after a meal, plasma substrate levels are still elevated above fasting levels.
   b. Postabsorptive state. After an overnight fast
   c. Starvation. After 2 to 3 days without food
   d. Prolonged starvation. When substrate levels in plasma are stable
   e. Terminal phase. When reserves of fuels are exhausted

B. Metabolic fuels and the respiratory quotient (RQ)

1. The use of metabolic fuels for the production of energy involves oxidation reactions and results in transformation of the carbon and hydrogen of the fuel into CO_2 and H_2O. For carbohydrates and fatty acids, these reactions may be expressed quantitatively by writing equations for the oxidation of glucose or palmitic acid to CO_2 and H_2O.

$$C_6H_{12}O_6 + 6 O_2 \rightarrow 6 CO_2 + 6 H_2O$$

$$\frac{\text{Glucose}}{1 \text{ mol}} \quad \frac{CO_2 \text{ produced}}{O_2 \text{ consumed}} = \frac{6 \text{ mol}}{6 \text{ mol}} = 1 \longrightarrow$$

$$C_{16}H_{32}O_2 + 23\ O_2 \rightarrow 16\ CO_2 + 16\ H_2O$$

$$\frac{\text{Palmitic acid}}{1\ \text{mol}} \quad \frac{CO_2\ \text{produced}}{O_2\ \text{consumed}} = \frac{16\ \text{mol}}{23\ \text{mol}} = 0.7$$

2. The measurement of O_2 uptake and CO_2 output by humans, experimental animals, or surviving tissues allows the calculation of the proportion of fats and carbohydrates that are being oxidized. This measurement is called **indirect calorimetry**, as opposed to the direct measurement of heat production by **direct calorimetry**.

3. The ratio of CO_2 produced to O_2 consumed is called the **respiratory quotient (RQ)**. When calculated from CO_2 production and O_2 consumption only, it is referred to as the **nonprotein RQ**.

4. The oxidation of protein is complex, but an approximate correction can be made for protein oxidation by using calculations based on the urinary nitrogen output. The RQ of protein is taken as 0.8. (One gram of urinary nitrogen represents 5.93 L O_2 used and 4.754 L of CO_2 produced.)

C. Hormonal regulation of storage and mobilization of metabolic fuels

1. **Postprandial state**
   a. Blood levels of insulin are elevated due to stimulation of its release by increased portal vein glucose levels.
   b. The secretion of glucagon by the α cells of the pancreatic islets of Langerhans is depressed by the high blood glucose levels.
   c. In this state, insulin is dominant, and the **storage phase** of the feeding–fasting cycle is operative. Thus
      (1) Glucose is stored in liver and muscle as glycogen.
      (2) Free fatty acids are stored as triacylglycerols, primarily in adipose tissue.
      (3) Protein synthesis surpasses protein breakdown.
      (4) Gluconeogenesis from excess dietary amino acids will occur, primarily due to substrate pressure.

2. **Postabsorptive state**
   a. Substrate levels in plasma have declined because of the activities during the storage phase, and insulin levels are back to normal.
   b. The glycogen reserve in liver is drawn on to maintain blood glucose levels, a step that requires the increased output of mobilizing hormones, particularly glucagon. The signal for this mobilization is a declining blood glucose level, which triggers glucagon release.
   c. If the fast is prolonged, gluconeogenesis from amino acids is increased by continued glucagon action in the liver in order to maintain blood glucose levels.
   d. The mobilization of free fatty acids from adipose tissue is induced by the action of norepinephrine released from sympathetic nerve endings, providing an alternative to glucose for muscle fuel and, thus, sparing glucose for use by the brain, which cannot use free fatty acids.

D. Competition between metabolic fuels for oxidation

1. All metabolic fuels are eventually degraded to CO_2 and H_2O in the tricarboxylic acid (TCA) cycle with the concomitant formation of adenosine triphosphate (ATP) via the electron transport system and oxidative phosphorylation.

2. The rate at which the TCA cycle provides the cell with ATP is directly geared to the rate of use of ATP by the cell.

3. If the rate of use of ATP declines, then the rate of oxidation of all fuels also declines.

4. Because there is a common pathway for the final oxidation of all metabolic fuels, namely the TCA cycle, there is competition between fuels for common cofactors, namely coenzyme A, (CoA), nicotinamide-adenine dinucleotide (NAD^+), and flavin-adenine dinucleotide (FAD).

5. With the exception of the brain, tissues that use a large amount of metabolic fuel will use free fatty acids preferentially because the high levels of acetyl CoA from the β-oxidation of the free fatty acids shut down carbohydrate oxidation at the pyruvate dehydrogenase (PDH) step.

6. The block at PDH by acetyl CoA also reduces the use of carbon from amino acids as a metabolic fuel.

E. Ability to survive starvation

1. By the end of an overnight fast, gluconeogenesis is already providing glucose from amino acids.

2. As judged by the output of urinary nitrogen, gluconeogenesis from amino acids increases during the first 3 days of starvation and then declines, until the urinary nitrogen is close to the obligatory output at about day 30 of starvation.

3. The decline in the use of protein as a metabolic fuel is essential for prolonged survival, as the protein represents the "fabric" and "enzymatic machinery" of the body and can only be depleted to a certain extent (see section II C 3).

4. Gluconeogenesis from amino acids declines as starvation continues because free fatty acids mobilized under conditions of low plasma insulin levels are oxidized preferentially.

5. The mobilization of free fatty acids is unchecked and lasts as long as the reserves of triacylglycerols will allow. The high levels of acetyl CoA in the liver drive the formation of ketone bodies, which are produced at the maximum rate as early as day 3 of starvation.

6. The ketone bodies are readily used by cardiac and skeletal muscle as fuel and by the brain when the plasma levels of ketone bodies are high enough.

7. After about 20 days of starvation, muscle, for reasons that are not clear, turns wholly to free fatty acids for fuel, sparing the ketone bodies for oxidation by the brain, which then is obtaining about 60% of its energy from ketone bodies.

IV. METABOLISM OF SPECIALIZED TISSUES.
Cells and tissues are specialized for specific functions, and different patterns of gene products are present in the different cell types. In consequence, different cell types show different patterns of metabolic activity and use fuel molecules to different degrees. The metabolism of many tissues changes radically with changes in the rate of function (e.g., in muscle at rest versus at work) and also in relation to changes in states of the feeding–fasting cycle.

A. Skeletal muscle.
Skeletal muscle tissue is a heavy consumer of metabolic fuel and oxygen. Owing to its great mass, it is the major consumer of the body, accounting for over 60% of the total oxygen consumption in the basal resting state.

1. **Types of muscle fibers**
   a. The main function of muscles is to contract and relax.
   b. The contractile fibers of striated muscle are of two kinds: white and red.
      (1) Red fibers have a large amount of sarcoplasm (cytoplasm), more nuclei, more mitochondria, more myoglobin, and more lipid droplets than the white fibers.
      (2) They twitch more slowly and for a longer time than the white fibers.
   c. Metabolically, three fiber types are distinguished:
      (1) **Fast-twitch white fibers** show low respiration, high glycogenolysis, and high myosin adenosine triphosphatase (ATPase) activity (an indication of contractile ability).
      (2) **Fast-twitch red fibers** show high respiration, high glycogenolysis, and high myosin ATPase activity.
      (3) **Slow-twitch red fibers** show high respiration, low glycogenolysis, and low myosin ATPase activity.

2. **Postprandial metabolism**
   a. The main fuel of skeletal muscle is glucose.
   b. Plasma insulin levels are high, which
      (1) Facilitate the entry of glucose into muscle cells
      (2) Stimulate oxygen consumption (by activation of PDH)
      (3) Stimulate amino acid uptake
      (4) Promote the accumulation of protein by inhibiting protein breakdown
   c. Glucose is metabolized mainly by glycolysis with only about 2% of the glucose being processed by the hexose monophosphate shunt.
   d. Fast-twitch red and white fibers probably synthesize glycogen from lactate. This process is reduced in low-twitch red fibers, owing to a low level of fructose 1,6-diphosphatase (F1,6DPase).
   e. **Resting muscle**
      (1) Glycolysis and the TCA cycle are operating at about 10% of maximum. There is little accumulation of lactate.
      (2) Muscle anoxia stimulates glucose uptake and lactate production, probably because phosphofructokinase (PFK) activity is increased as ATP levels decrease.

f. Contracting muscle
  (1) During contraction, the uptake of glucose and the use of oxygen increase about 20-fold.
  (2) During prolonged heavy contraction, the rate of pyruvate formation exceeds the oxidative capacity of the TCA cycle, and large amounts of lactate are formed.

3. Fasting metabolism
  a. In the early stages of fasting, free fatty acids are the preferred fuel for muscle cells.
    (1) The free fatty acids suppress the uptake and oxidation of glucose, indirectly promoting the conversion of PDH to its inactive form (see Chapter 15, section II C 3).
    (2) The use of free fatty acids by muscle cells does not depend upon insulin, which is low in the fasting state.
    (3) The free fatty acids supply at least 50% of the energy requirements of skeletal muscle both at rest and during contraction.
  b. Later in fasting (starvation), ketone bodies are readily used as energy sources by muscle.
  c. The use of free fatty acids and ketone bodies by muscle during starvation spares glucose for other tissues.
  d. The branched-chain amino acids—leucine, isoleucine, and valine—are degraded in peripheral tissues, particularly in skeletal muscle and cardiac muscle.
    (1) The carbon skeletons of these amino acids provide energy for muscular contraction.
    (2) Muscle is a major site for branched-chain amino acid degradation (the liver is inactive in this respect), and the first step is transamination with pyruvate. The resulting alanine leaves the muscle cell and is an important gluconeogenic precursor in the liver (see Chapter 17, section IV B; the **alanine cycle**).

B. Cardiac muscle

  1. Characteristics. Cardiac muscle fibers are striated and rich in mitochondria. They are, therefore, like the red fibers of skeletal muscle except that they are branched. The oxidation of free fatty acids makes up 60% to 90% of the total fuel oxidized by cardiac muscle.

  2. Postprandial metabolism
    a. Glucose, pyruvate, and lactate are used at this time when the plasma levels of free fatty acids are low.
    b. The uptake and oxidation of free fatty acids and branched-chain amino acids are also both inhibited by pyruvate and lactate.

  3. Fasting metabolism
    a. The major fuel of cardiac muscle in fasting is free fatty acids derived from adipose tissue stores.
    b. When plasma levels of ketone bodies rise in starvation, they form an important fuel for cardiac muscle.

C. Adipose tissue

  1. Types of adipose tissue. There are two kinds of adipose tissue, white and brown. Both white and brown adipose tissue are well supplied with blood capillaries, and both are well innervated.
    a. White adipose tissue
      (1) Triacylglycerols comprise about 80% of the wet weight of the tissue.
      (2) In humans, the fatty acids present in the triacylglycerols are mainly oleic (45%), palmitic (20%), linoleic (10%), stearic (6%), and myristic (4%).
      (3) The adipocytes are spherical, containing a single huge vacuole filled with triacylglycerol, which occupies most of the cell.
      (4) The cytoplasm is a thin film around the fat vacuole, and the nucleus is flattened.
      (5) The function of white adipose tissue is the storage of metabolic fuel in the form of triacylglycerols.
    b. Brown adipose tissue
      (1) Brown adipose tissue adipocytes contain more cytoplasm than those of white adipose tissue.
      (2) They are richly endowed with mitochondria, and the nucleus is not flattened.
      (3) The lipid stores are contained in many small droplets.
      (4) The main function of brown adipose tissue in the human is heat production in newborns, although some tissue may persist in some adults.

  2. Metabolism of white adipose tissue. The function of white adipose tissue is the storage of metabolic fuels; the quantity used for energy supply to the tissue forms only a minor fraction of the metabolic flux.

a. Postprandial metabolism
(1) Chylomicrons and very low-density lipoproteins (VLDL) are hydrolyzed to free fatty acid and glycerol by lipoprotein lipase, and the free fatty acid enters the adipocyte (see Chapter 24, sections III and IV).
(2) Re-esterification of the free fatty acid to triacylglycerols in the adipocyte depends upon ongoing glycolysis for the supply of glycerol 3-phosphate, as adipose tissue lacks a glycerol kinase.
(3) In turn, glycolysis depends upon insulin for the entry of glucose into the adipocyte. Thus, insulin promotes the synthesis of triacylglycerols.
(4) Insulin also suppresses the activity of the enzyme system that mobilizes free fatty acid from triacylglycerols.
(5) In the postprandial state, both glucose and insulin levels are elevated in the plasma, and all signals are set for fat storage.

b. Fasting metabolism
(1) Adipose tissue taken from rodents responds to a number of hormones in in vitro systems with increased rates of lipolysis.
(2) In humans, the regulation of lipolysis in adipose tissue is probably achieved mainly by norepinephrine released from the innervating autonomic nervous system and by the level of plasma insulin.
  (a) Norepinephrine activates the hormone-sensitive lipase via adenylate cyclase and cyclic adenosine monophosphate (cAMP).
  (b) In the postprandial state, the high insulin levels in plasma inhibit this system.
  (c) Growth hormone (pituitary somatotropin) is also a mobilizing hormone in adipose tissue.
(3) In an emergency situation or under stress, catecholamine released from the adrenal medulla stimulates lipolysis in the adipocytes.

3. Metabolism of brown adipose tissue. The free fatty acids are used by adipocyte mitochondria as an oxidizable substrate with the free fatty acids uncoupling oxidative phosphorylation in some manner, so that heat, rather than ATP, is the end product.

D. Liver
1. General characteristics
a. Blood supply
(1) The liver has two afferent blood supplies.
  (a) The portal vein, which carries blood from the alimentary tract, spleen, and pancreas
  (b) The hepatic artery
(2) The parenchymal cells of the liver are penetrated by a network of sinusoids, which drain into the central veins.
(3) The sinusoids are lined by cells called **Kupffer's cells**, which function as phagocytes.
b. Specialized functions of the liver include
(1) The maintenance of blood glucose levels
(2) The production of ketone bodies in starvation
(3) The removal of metabolic waste products and potentially dangerous compounds, such as drugs, hormones, and poisons from the blood. These compounds may be converted to metabolic fuels or transformed to compounds that may safely be excreted in the urine or feces.
(4) The synthesis of plasma proteins

2. Postprandial metabolism
a. The liver is the first organ to receive the substrate-laden blood from the intestine after a meal and can extract what it needs before the blood reaches any other tissue.
b. With insulin dominant, glucose is directed toward glycogen synthesis with gluconeogenesis from dietary amino acids contributing to this process.
c. Protein turnover in the liver is rapid, and a considerable part of the incoming amino acid pool is used by the liver for protein synthesis both of its enzymatic machinery and also of proteins for export (plasma albumin, transferrin, etc.).
d. There is some conversion of glucose into fatty acids, which can be used for phospholipid synthesis. The phospholipids can be exported to other tissues in the form of lipoproteins.

3. Fasting metabolism
a. In fasting, a major role of the liver is the maintenance of blood sugar levels.
b. With the insulin:glucagon ratio decreasing during the fast, glycogen is mobilized to provide glucose for export.

    c. Gluconeogenesis from lactate, glycerol, or the major amino acids entering the liver at this time (i.e., alanine and glutamine) becomes the major source of blood glucose as glycogen stores are depleted.

    d. *Excess* free fatty acids from increased lipolysis may be re-esterified in the liver, built into lipoproteins, and returned to adipose tissue for storage when insulin levels rise again.

E. Kidney

1. General characteristics

    a. The kidney has two well-defined regions, the outer cortex and the inner medulla.

    b. The cortex contains the glomeruli, where the filtration of plasma occurs, and the bulk of the nephrons is found in the medulla.

    c. Among the major functions of the kidney are

        (1) Filtration of plasma by the glomeruli

        (2) Selective reabsorption by the tubules of some substances required for the maintenance of the internal environment (e.g., water, salts, sugars, and amino acids)

        (3) Secretion by the tubules of certain unwanted substances into the tubular lumen for excretion in the urine

2. Metabolism

    a. The kidney can use all major fuels—free fatty acids, glucose, lactate, glutamine, citrate, glycerol, and ketone bodies—although the cortex and the medulla have very different capabilities in this respect.

    b. The **renal cortex** has a high oxygen consumption because of its high concentrations of enzymes from the TCA cycle and the electron transport chain. The RQ of about 0.75 indicates that free fatty acid is the major fuel. Ketone bodies can also be used by the renal cortex as fuel molecules, and they are also produced in this tissue.

    c. The **renal medulla** is low in TCA and electron transport enzymes but has high glycolytic activity. Oxygen uptake is low. The sources of glucose for glycolysis are

        (1) Plasma glucose under postprandial conditions

        (2) Gluconeogenesis by the renal cortex in the fasting state

3. Gluconeogenesis in the kidney

    a. The kidney has the capacity to carry out gluconeogenesis at a greater rate per tissue weight than the liver has, but usually little of the glucose thus formed is released to the circulation.

    b. The glucose that is formed by gluconeogenesis in the renal cortex under fasting conditions is believed to be used by the renal medulla.

    c. The rate of gluconeogenesis in the kidney, which is regulated by glucagon and epinephrine via adenylate cyclase and cAMP, is increased during fasting.

    d. The rate of gluconeogenesis in the kidney is also increased in acidosis, the key enzyme, phosphoenolpyruvate carboxykinase, being induced by an unknown mechanism.

F. Brain

1. Energy source. Under usual conditions, that is, postprandially or after an overnight fast, the brain uses glucose as its sole fuel. The glucose is almost completely oxidized to CO_2 and H_2O, there being very little storage as glycogen.

2. Oxidative activity. The brain has a high oxygen consumption rate. Approximately, 25% of an adult's total oxygen consumption is due to brain oxidations. In infants and children under age 4 years, the brain's share is 50% of the total oxygen consumed.

3. Glycolysis in the brain normally proceeds at about 20% of its maximum capacity, but the TCA cycle proceeds at a rate near its maximum.

4. The reserve capacity for glycolysis is an important part of the defense against cerebral anoxia.

    a. Glycolysis (and lactate formation) is increased fivefold to eightfold by the end of one minute of anoxia.

    b. The mechanism for increasing glycolysis in anoxia is the **Pasteur effect**, whereby declining ATP levels allow the activation of phosphofructokinase. Although the mechanism is not clear, anoxia also causes activation of brain hexokinase.

5. Long-term starvation with somewhat reduced plasma glucose levels is tolerated by brain tissue because the plasma level of ketone bodies is high, providing metabolic fuel that the brain can use.

6. Insulin cannot cross the blood–brain barrier and is, therefore, without a direct effect on brain metabolism.

STUDY QUESTIONS

Directions: Each question below contains five suggested answers. Choose the **one best** response to each question.

1. A 70-kg human male with 25% of his body weight as adipose tissue has stored as triacylglycerols approximately

(A) 35,000 kcal
(B) 42,000 kcal
(C) 70,000 kcal
(D) 140,000 kcal
(E) 157,000 kcal

2. The postprandial state is characterized by

(A) high blood levels of glucose and low levels of insulin
(B) high blood levels of glucose and high levels of insulin
(C) high blood levels of free fatty acids and low levels of glucagon
(D) high blood levels of free fatty acids and high levels of glucagon
(E) high blood levels of insulin and glucagon

3. What is meant by the respiratory quotient?

(A) The ratio of the volume of expired air to inspired air when at rest
(B) The oxygen (O_2) consumption of an average 70-kg human male or an average 60-kg human female over a period of 1 hr under resting conditions
(C) The volume of carbon dioxide (CO_2) expired by a normal human over a period of 1 hr under resting conditions
(D) The ratio of the volume of CO_2 expired to the volume of O_2 consumed
(E) None of the above is the correct meaning

4. All of the following statements regarding brown and white adipose tissue are correct EXCEPT

(A) brown adipose tissue adipocytes contain more cytoplasm than those of white adipose tissue
(B) brown adipose tissue adipocytes contain numerous small droplets of triacylglycerols
(C) white adipose tissue contains many more mitochondria than brown adipose tissue because energy is required to store triacylglycerols
(D) brown adipose tissue produces heat by uncoupled oxidative phosphorylation
(E) the largest deposits of brown adipose tissue are found in the newborn

5. The source of glucose for glycolysis by renal medullary cells in the fasting state is believed to be

(A) plasma glucose
(B) glycogen in medullary cells
(C) gluconeogenesis by the renal cortex
(D) gluconeogenesis by the medullary cells
(E) conversion of fructose to glucose

Directions: Each question below contains four suggested answers of which **one or more** is correct. Choose the answer

A if **1, 2, and 3** are correct
B if **1 and 3** are correct
C if **2 and 4** are correct
D if **4** is correct
E if **1, 2, 3, and 4** are correct

6. In the postprandial state, the triacylglycerol stores in adipose tissue are built up because

(1) high levels of blood insulin allow glucose entry into adipocytes
(2) there are high levels of triacylglycerols in the blood, carried by chylomicrons and very low-density lipoproteins
(3) cyclic adenosine monophosphate levels in adipocytes are low and, therefore, the rate of intracelluar lipolysis of triacylglycerols is low
(4) there are high levels of glycerol 3-phosphate in adipose tissue because of increased glycolysis

E

7. Which of the following statements can explain, in part, the decline in nitrogen excretion seen during starvation?

(1) Gluconeogenesis from amino acids in the liver declines as the breakdown of protein in muscle is slowed
(2) Muscle protein catabolism is reduced when alternative fuels such as free fatty acids and ketone bodies are used by muscle
(3) Nitrogen excretion, in the form of urea, declines as amino acid degradation decreases
(4) The reduced excretion of nitrogen in starvation is due to the preferential synthesis of enzymes vital for survival

A

8. The action of insulin on muscle cells under postprandial conditions includes

(1) facilitated entry of glucose
(2) reduced entry of amino acids
(3) stimulation of oxygen consumption
(4) inhibition of pyruvate dehydrogenase activity

B

*

ANSWERS AND EXPLANATIONS

1. The answer is D. *(II C 2)* Adipose tissue contains about 10% water, half of which is intracellular and half extracellular. Thus, the caloric value of adipose tissue is 8 kcal/g, rather than 9 kcal/g as in the case of triacylglycerols. The 70-kg male contains about 17.5 kg of adipose tissue (25% of 70 kg) and a store of about 17.5 kg x 8 kcal/g = 140,000 kcal.

2. The answer is B. *(III A, C 1)* The postprandial state is characterized by high blood levels of glucose arising from the ingested meal and, in consequence, high blood levels of insulin, which is secreted in response to rising blood glucose levels. Free fatty acids in blood are at minimal levels in the postprandial state, most of the circulating fat being triacylglycerols in the form of chylomicrons or very low-density lipoproteins. Insulin and glucagon are antagonistic in action, and opposing signals preclude their concomitant release from the islet tissue in the pancreas. For example, high blood glucose levels promote insulin release and inhibit the release of glucagon.

3. The answer is D. *(III B 3)* The ratio of carbon dioxide produced to oxygen consumed is called the respiratory quotient (RQ). The value of the RQ depends upon the metabolic fuel being oxidized by the body, a ratio of 1 indicating that carbohydrate is the predominant fuel and a ratio of 0.7 indicating that free fatty acids are the major fuel being oxidized.

4. The answer is C. *(IV C 1, 3)* The main function of the adipocytes in white adipose tissue is the storage of triacylglycerols. These are contained in a single large vacuole, which fills most of the cell. There is little cytoplasm, and mitochondria are relatively few. Brown adipose tissue adipocytes contain more cytoplasm and more mitochondria than white adipose tissue, and the triacylglycerols are stored in numerous small droplets. Brown adipose tissue carries out oxidation of triacylglycerols in a relatively uncoupled manner, so that the major product is heat, rather than ATP.

5. The answer is C. *(IV E 2, 3)* The renal medullary cells are low in enzymes of the tricarboxylic acid cycle and the electron transport system and depend upon an active glycolytic pathway for energy production. In the fasting state, when plasma glucose levels are low and glycogen stores are depleted, the source of glucose for glycolysis in the renal medulla appears to be gluconeogenesis in the renal cortex, which is more active on a weight basis than liver in this process.

6. The answer is E (all). *(III C, IV C 2)* In the postprandial state, blood insulin levels are high, and the stores of adipocyte triacylglycerols are built up from ingested lipid, which in this state is present in blood as chylomicrons or very low-density lipoproteins. These are hydrolyzed by lipoprotein lipase, and the free fatty acids enter the adipocytes. The free fatty acids are immediately reesterified to triacylglycerols because the insulin-induced glucose uptake and increased glycolysis provide adequate glycerol 3-phosphate for the re-esterification. Insulin also reduces cyclic adenosine monophosphate levels, possibly by increasing cyclic nucleotide phosphodiesterase activity, thus minimizing lipolysis of triacylglycerols.

7. The answer is A (1, 2, 3). *(III E)* For the first 3 days of starvation, urinary nitrogen excretion increases as gluconeogenesis from amino acids is used to maintain blood glucose levels. As the fast continues, free fatty acids from mobilized reserves of triacylglycerols and ketone bodies derived from them become the preferred fuels for muscle, reducing the oxidation of amino acids, thus having a protein-sparing effect. This slowdown in amino acid oxidation results in less urea being excreted in the urine. The catabolism of amino acids is never zero, and when the fat reserves are exhausted, fatal inroads into the enzymatic protein of the body will occur.

8. The answer is B (1, 3). *(IV A 2)* In the postprandial state, insulin levels in blood are high, which facilitates the entry of both glucose and amino acids into muscle cells. Oxygen uptake by muscle cells is increased by insulin, due mainly to activation of pyruvate dehydrogenase.

Appendix:
Water and Acid–Base
Equilibria

I. WATER

A. Introduction

1. In living cells, most chemical reactions occur in an aqueous environment, approximately 75% of the mass of a living cell being ascribable to water.

2. Water is a reactant in many systems in the cell and is often a factor in determining the properties of macromolecules in cells.

B. Molecular structure of water

1. Two hydrogen atoms share their electrons with an unshared pair of electrons of an oxygen atom.

2. If the structure (Fig. A-1) is looked at as an irregular tetrahedron with an oxygen atom at its center, then
   a. The two bonds with hydrogen point toward two corners of the tetrahedron.
   b. The unshared electrons on the two hybridized sp^3 orbitals of the oxygen atom occupy the two other corners.
   c. The angle between the two hydrogen atoms is 105°, which is less than the regular tetrahedral angle of 109.5°.

3. The side of the oxygen that is opposite to the two hydrogens is electron-rich and, therefore, electronegative, while the other side is opposite to the hydrogen nuclei, which are electropositive. The water molecule, therefore, has a **dipole** nature.

C. Associations between water molecules

1. **Molecular ordering of water molecules** occurs both in liquid water and in ice.
   a. In the solid state, each water molecule is associated with four other water molecules.
   b. In the liquid state, the association is transient, and the number of water molecules in association is less.
   c. The half-life for the association–dissociation of liquid water molecules is about 1 μsec.

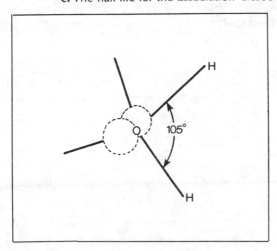

Figure A-1. Structure of the water molecule.

d. The formation of these ordered arrays of water molecules is favored by the dipole nature of the water molecules with the possibility of forming hydrogen bonds (Fig. A-2).

2. **Hydrogen bonds.** The electrostatic interaction between the hydrogen nucleus of one water molecule and the unshared electron pair of another is called a hydrogen bond (see Fig. A-2).
   a. Note that hydrogen bonds are not restricted to water molecules.
   (1) They tend to form between a small, highly electronegative atom and a hydrogen atom that is covalently bonded to another electronegative atom.
   (2) Thus, oxygen, fluorine, and, in particular, nitrogen atoms can participate in hydrogen bonding; for example,

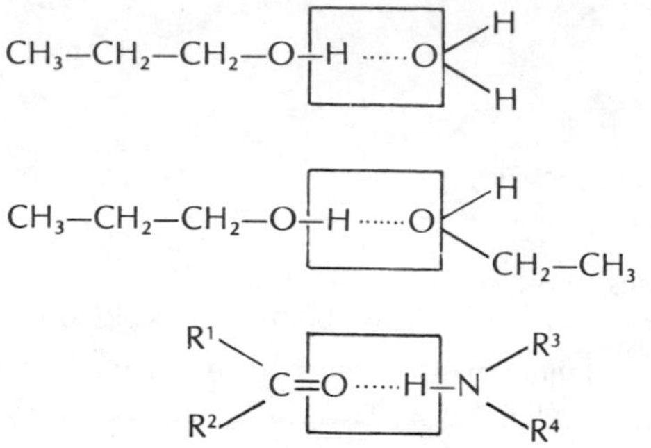

$$CH_3-CH_2-CH_2-O\text{—}H \cdots\cdots O\Big\langle {}^H_H$$

$$CH_3-CH_2-CH_2-O\text{—}H \cdots\cdots O\Big\langle {}^H_{CH_2-CH_3}$$

$${}^{R^1}_{R^2}\Big\rangle C=O \cdots\cdots H\text{—}N\big\langle {}^{R^3}_{R^4}$$

b. Hydrogen bonds are weaker than covalent bonds with only about 4.5 kcal of energy being required to break 1 mol of hydrogen bonds between molecules in liquid water, compared to 110 kcal/mol for the O—H bond *within* the water molecule.
c. However, although individually weak, when taken together many hydrogen bonds confer significant structure on dipolar molecules as diverse as water, alcohols, DNA, and proteins.

D. Dissociation of water

1. Water molecules have a limited tendency to dissociate (**ionize**) into H^+ and OH^-.

$$H_2O \leftrightarrow H^+ + OH^- \tag{1}$$

Note that the dissociation is reversible.

2. A possibly more accurate description of the dissociation is given by

$$2 H_2O \leftrightarrow H_3O^+ + OH^-$$

a. Even this representation, using the hydronium ion (H_3O^+), may not be as chemically correct as the hydrogen-bonded $H_9O_4^+$ or some other species.
b. Therefore, in what follows, the expression given in equation (1) is used, as it is the simplest form, and for the purposes of this discussion, it presents no problem of loss of rigor.

3. The tendency of water to dissociate is given by

$$K = \frac{[H^+][OH^-]}{[H_2O]}$$

where the terms in brackets represent concentrations of hydrogen ions, hydroxyl ions, and

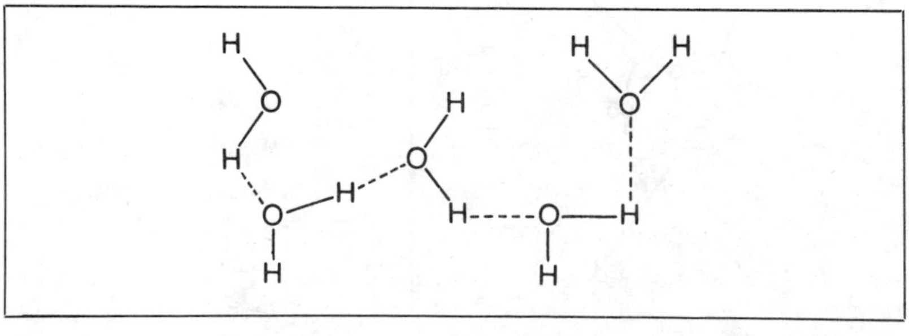

Figure A-2. Hydrogen bonding between water molecules.

undissociated water molecules at equilibrium; that is, when the rate of the forward reaction in equation (1) [dissociation] equals the rate of the backward reaction (association), and K is the equilibrium constant, or the **dissociation constant**, of water.

4. At 25°C, the value of K is 1.8×10^{-16} mol/L.

5. Ion product of water

a. Since undissociated water is present in great excess, its concentration is virtually constant at 55.56 M. (The molecular weight of water equals 18; therefore, in 1 L, or 1000 g, there are 1000/18 mol.)

b. This "constant" value for the concentration of water can be incorporated into the dissociation constant to give a new constant, the so-called **ion product of water**, or K_w.

$$K_w = [H^+][OH^-]$$

c. At 25°C,

$$K_w = 1 \times 10^{-14} \text{ mol}^2/L^2$$

d. It is important to realize that the ion product of water, namely $[H^+][OH^-]$, is constant for *all* aqueous solutions, even those that contain dissolved acids (proton donors) or dissolved bases (proton acceptors).

(1) If a large number of hydrogen ions (protons) are added to pure water, the concentration of OH^- ions must decrease in order that the product $[H^+][OH^-]$ will remain 10^{-14} at 25°C.

(2) Conversely, if a large number of hydroxyl ions are added, the $[H^+]$ will have to decrease.

II. ACID–BASE EQUILIBRIUM

A. Dissociation of a weak acid

1. Strong and weak electrolytes

a. Strong electrolytes are regarded as being fully dissociated in aqueous solution, at least when dilute solutions are used.

b. Weak electrolytes are only partially dissociated (usually less than 5%) in aqueous solution.

2. Bronsted-Lowry definition of acids and bases. Acids are defined as **proton donors** and **bases** as **proton acceptors**, the equilibrium system of undissociated acid and conjugate base being referred to as a **conjugate acid–base pair. The general form of the reaction** is

$$HA \leftrightarrow H^+ + A^-$$

where HA is the undissociated acid and A^- is its conjugate base. A specific example is

$$CH_3COOH \leftrightarrow H^+ + CH_3COO^-$$

Likewise,

$$BH^+ \leftrightarrow H^+ + B$$

For example,

$$CH_3NH_3^+ \leftrightarrow H^+ + CH_3NH_2$$

3. The dissociation of a weak acid is given by

$$K_a = \frac{[A^-][H^+]}{[HA]} \qquad (2)$$

where K_a is the dissociation constant for a weak acid.

a. In biochemical and medical applications, it is usual to designate the concentration in terms of moles per liter in the dissociation expression. In this case, the constant, K_a, becomes the **apparent dissociation constant**, K_a'.

b. The true, thermodynamic, dissociation constant requires the use of ion activities in place of molar concentrations for its calculation.

B. Hydrogen ion concentration of a solution of a weak acid

1. Equation (2) above can be rearranged to give

$$[H^+] = K_a' \times \frac{[HA]}{[A^-]} \qquad (3)$$

Note that the ratio of undissociated acid to conjugate base ($[HA]/[A^-]$) refers to the value obtaining at *equilibrium*.

2. It is usual to express the variables in equation (3) in terms of their negative logarithms.

$$-\log H^+ \ = \ -\log K_a{'} \ - \log \ \frac{[HA]}{[A^-]} \tag{4}$$

C. pH scale and $pK_a{'}$

1. pH is defined as the negative logarithm of $[H^+]$; that is,

$$pH \ = \ -\log_{10} [H^+]$$

2. $pK_a{'}$ is defined as the negative logarithm of the dissociation constant for a weak acid.

3. Henderson-Hasselbalch equation. Thus, equation (4) becomes

$$pH = pK_a{'} \ + \ \log \ \frac{[A^-]}{[HA]} \tag{5}$$

This relationship [equation (5)] is known as the Henderson-Hasselbalch equation and is the most useful formulation in acid-base equilibrium applications.

4. Measurement of $[H^+]$ and pH
   a. The **hydrogen gas electrode** provides the primary standard for the measurement of $[H^+]$.
   b. However, the simpler pH-sensing **glass electrode** is now used in nearly all laboratory and clinical applications.
      (1) The glass pH-sensing electrode is the nearest approach to a universal pH-indicating electrode known at present.
      (2) The glass electrode establishes a voltage across a glass membrane separating two solutions. The voltage magnitude depends on the difference in pH of the two solutions.
      (3) In practice, a solution of constant, known pH is put inside a chamber within the electrode, and the electrode is then immersed in the solution to be tested. The voltage developed is proportional to the pH of the external solution.

D. Behavior of weak acids in the presence of their salts

1. Titration of a weak acid with a strong alkali
   a. If a strong alkali [e.g., sodium hydroxide (NaOH)] is added in known increments to the solution of a weak acid (e.g., lactic acid) and the pH is measured with a glass electrode after each addition, a plot of the pH against the degree of neutralization (or amount of alkali added) can be prepared, as shown in Figure A-3, which illustrates the plot obtained from a titration of 50 ml 0.1 N lactic acid with 0.1 N NaOH. Lactic acid has a $pK_a{'}$ of 3.86.

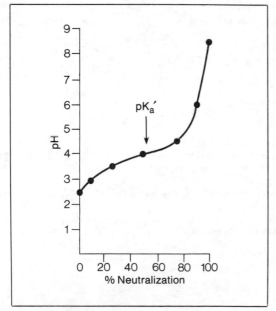

Figure A-3. Titration curve for 50 ml of 0.1 N lactic acid ($pK_a{'}$ = 3.86) titrated with increments of 0.1 N NaOH.

b. As the strong alkali is added, several reactions occur.
   (1) The proton in lactic acid is displaced by the Na^+ from NaOH.

$$H\ lactate + NaOH \leftrightarrow Na\ lactate + H_2O$$

   (2) The sodium lactate formed during the titration is, like most salts, fully dissociated into positively and negatively charged ions.

$$Na\ lactate \leftrightarrow Na^+ + lactate^-$$

c. At 10% neutralization, there is sufficient lactate ion present to suppress the dissociation of the remaining lactic acid. Thus, the lactate ion that is present is due entirely to the amount of alkali added.

d. At 100% neutralization, there is only lactate ion present and, of course, Na^+.
   (1) This lactate ion, being the conjugate base of a weak acid, is a relatively strong base.
   (2) Therefore, it will accept protons from water.

$$Lactate^- + H_2O \leftrightarrow H\ lactate + OH^-$$

   (3) The resulting increase in the concentration of OH^- explains why the pH of the solution at the end of the titration is greater than pH 7 (see Fig. A-3).

2. pH calculations

a. pH at the midpoint of the titration
   (1) At that time, the concentration of undissociated lactic acid and that of lactate ion are equal.
   (2) Expressing this equivalence in terms of equation (5) above, $[A^-]/[HA] = 1$.
   (3) If one substitutes the ratio of 1 for $[A^-]/[HA]$ in equation (5), the pH is found to be equal numerically to the pK_a'.
   (4) Thus, the pK_a' of a weak acid is the pH at the midpoint of its titration with a strong alkali.

b. Calculating the initial pH
   (1) The initial pH of the lactic acid solution, before the addition of alkali, cannot be calculated from equation (5).
   (2) It can be calculated, however, if equation (3) is modified so that the initial concentration of undissociated acid can be used rather than the equilibrium value. Adjusting equation (3)

$$as\ [H^+] = [A^-],\ then\ K_a' = \frac{[H^+]^2}{[HA]}$$

   (3) If the initial concentration of acid is C mol/L, then the equilibrium concentration of undissociated acid ($[HA]$) is given by $C - [H^+]$
Thus,

$$K_a' = \frac{[H^+]^2}{C - [H^+]}$$

or

$$[H^+]^2 + K_a' \times [H^+] + K_a' \times C = 0$$

from which $[H^+]$ can be calculated.

c. Calculating the pH at other points
   (1) The pH at points on the plot lying between 10% and 90% neutralization can be calculated, using equation (5), the Henderson-Hasselbalch equation.
     (a) For example, at 40% neutralization, the ratio of $[A^-]/[HA]$ is 40/60, or 2/3.
     (b) Therefore,

$$\begin{aligned} pH &= 3.86 + \log 2/3 \\ &= 3.86 + 0.3 - 0.48 \\ &= 3.68 \end{aligned}$$

   (2) Below 10% and above 90% neutralization, equation (5) cannot be used.
     (a) Below 10% neutralization, the dissociation of the weak acid is significant.
     (b) Above 90% neutralization, hydrolysis of the conjugate base, the lactate ion, is significant and cannot be ignored.

3. Buffering

a. The **behavior of the pH around the midpoint** region in the titration plot in Figure A-3 is also of interest.

(1) Close to the 50% neutralization point, the pH changes very slowly with each increment of alkali added.

(2) In this region, the solution is said to be exhibiting **buffering** or resistance to change in pH upon the addition of hydroxyl or hydrogen ions.

(3) The buffering effect is maximal at the midpoint of the titration; that is, when the pH is at the pK_a' of the conjugate acid-base pair.

   b. The extent to which a solution containing a weak acid and its salt can resist a change in pH on the addition of acid or alkali depends upon two factors.

   (1) The closeness of the buffer pH to the pK_a' of the weak acid: the closer the values, the greater the **buffer capacity.**

   (2) The concentration of the buffer solution: the higher the buffer concentration, the greater the buffer capacity.

   (3) The capacity of a buffer to resist change in pH upon the addition of protons or hydroxyl ions can be quantitated by measuring the moles of protons or hydroxyl ions required to change the pH of 1 L of the solution by 1.0 pH unit.

   c. As the pH of a buffer solution depends upon the ratio $[A^-]/[HA]$, one would not expect the pH of a buffer solution to change upon dilution of the buffer. However, it does change. The reason for the instability of the pH upon dilution is that the ion activities of the free conjugate base and undissociated acid change to different degrees upon dilution.

E. Uses of the Henderson-Hasselbalch equation. In using the Henderson-Hasselbalch equation— that is, equation (5)—to calculate pH values at different points on the titration curve, the object, in section II D, was to stress the consequences that occur when the dissociation of a weak acid is suppressed by an increasing level of free conjugate base because the phenomenon has several practical applications.

1. One use of the Henderson-Hasselbalch equation is in the **preparation of solutions of known pH**.

   a. If one wants to obtain a solution with a given pH, it is sufficient to mix a solution of a weak acid with a solution of a salt of that weak acid.

   b. The only requirements are that the
   (1) pK_a' of the weak acid be close to the desired pH
   (2) Ratio of salt to acid be the ratio required to give the pH wanted; this ratio can be calculated using equation (5).

   c. It is important to realize that the concentration of undissociated acid and the concentration of the conjugate base derived from the salt **do not change** from their initial values. The equilibrium concentrations are, therefore, known and can be substituted directly into equation (5).

2. The Henderson-Hasselbalch relationship also allows one to **calculate the degree of dissociation of ionizing groups** in relation to the prevailing $[H^+]$.

   a. Given the pH of the medium and the pK_a' of the ionizing group in question, the degree of dissociation can be calculated by substituting these for pH and pK_a' in equation (5) and then solving for $\log([A^-]/[HA])$.

   b. For example, if the pK_a' of carboxyl side chain groups on a protein is given as 3.0, then at pH 7, the degree of dissociation can be calculated as follows:

$$pH = pK_a' + \log[COO^-]/[COOH]$$

$$7 = 3 + \log[COO^-]/[COOH]$$

$$\log[COO^-]/[COOH] = 4$$

Therefore,

$$[COO^-]/[COOH] = 10,000/1$$

or virtually 100% dissociation.

STUDY QUESTIONS

Directions: Each question below contains five suggested answers. Choose the **one best** response to each question.

1. Which of the following statements best describes what is meant by the ion product of water?

(A) The total number of negatively and positively charged ions in 1 L of an aqueous solution of an electrolyte

(B) The product of the concentrations of hydrogen ions and hydroxyl ions that are derived only from water molecules in aqueous solutions of electrolytes

(C) The number of ionized molecules of H_2O in 1 mol of pure water

(D) The product of the concentrations of hydrogen ions and hydroxyl ions in water or in aqueous solutions of electrolytes

(E) None of the above statements describes the ion product of water

2. If a buffer solution contains a mixture of a weak acid at 10 mM concentration and its potassium salt at 1 mM, the pH of the buffer solution would be approximately

(A) pK_a'

(B) $pK_a' - 1$

(C) $pK_a' - 2$

(D) $pK_a' + 1$

(E) $pK_a' + 2$

3. A 0.1 M solution of lactic acid has a pH of 2.4. How much sodium hydroxide would be required to titrate 100 ml of the solution to the end point?

(A) 100 Eq

(B) 100 mEq

(C) 100 ml of 0.1 M

(D) 2.4 mEq

(E) 0.1 mmol

4. If 0.1 M solutions of sodium dihydrogen phosphate (NaH_2PO_4) and disodium hydrogen phosphate (Na_2HPO_4) are mixed together in equal proportions, what would the pH of the mixture be? (Note that the pK_a' values of orthophosphoric acid (H_3PO_4) are 2.0, 6.8, and 12.0.)

(A) 2.0

(B) 4.4

(C) 6.8

(D) 9.4

(E) 12.0

5. A sample of urine from a patient with ketosis contained 30 mmol of total acetoacetic acid–acetoacetate ($pK_a = 4.8$). The pH of the urine was 4.8. How many millimoles of sodium ion were excreted with this keto acid?

(A) 5

(B) 10

(C) 15

(D) 20

(F) 25

6. The pK_a' of carbonic acid in plasma is 6.1. The pH of a sample of plasma was found to be 7.1. What is the ratio of bicarbonate ion to carbonic acid?

(A) 100:1

(B) 50:1

(C) 20:1

(D) 10:1

(E) 5:1

Directions: The question below contains four suggested answers of which **one or more** is correct. Choose the answer

    A if **1, 2, and 3** are correct
    B if **1 and 3** are correct
    C if **2 and 4** are correct
    D if **4** is correct
    E if **1, 2, 3, and 4** are correct

7.   Among the characteristics of the buffer capacity of a buffer solution are that it

(1) is maximal at a pH equal to the pK_a' of the weak acid present

(2) decreases upon dilution of the buffer solution

(3) can be quantitated by measuring the moles of hydrogen or hydroxyl ions required to change the pH of 1 L of buffer solution by 1.0 pH unit

(4) can be evaluated by measuring the resistance of the buffer solution to a change in the pH on dilution of the buffer

ANSWERS AND EXPLANATIONS

1. The answer is D. (*I D 5*) The product of the concentrations of hydrogen ions and hydroxyl ions is a constant, at a given temperature, for water *and* for all aqueous solutions of electrolytes, even when the dissociation of the electrolyte leads to an increase in the concentration of hydrogen or hydroxyl ions. The ion product has a value of 10^{-14} mol ^{2}/L^2 at 25°C.

2. The answer is B. (II C 3, D 2) The pH of a buffer solution composed of a weak acid and a salt formed from the weak acid and the strong base depends upon the ratio of dissociated conjugate base to un-dissociated weak acid. The relationship between this ratio, the pK_a of the weak acid, and the pH of the solution is given by the Henderson-Hasselbalch equation. If the ratio is 0.1, the pH will be approx-imately equal to the $pK_a - 1$, which can be determined by substituting the log of 0.1 into the Henderson-Hasselbalch equation.

3. The answer is C. (*II D 1*) The complete neutralization of the dissociable hydrogen of a lactic acid solution with a sodium hydroxide solution requires an amount of hydroxyl ions equivalent to the hy-drogen ions potentially available from the lactic acid. Lactic acid is a weak acid with one dissociable proton, and 100 ml of a 0.1 M solution contains 10 mEq of potential hydrogen ions. Therefore, 10 mEq of sodium hydroxide are required. This amount is contained in 100 ml of 0.1 M sodium hydroxide solu-tion.

4. The answer is C. (*II D 1, 2*) Orthophosphoric acid (H_3PO_4) has three dissociable protons with pK_a' values of 2.0, 6.8, and 12.0. The dissociation of $H_2PO_4^{1-}$ to the HPO_4^{2-} species has a pK_a' of 6.8. When sodium dihydrogen phosphate (NaH_2PO_4) and disodium hydrogen phosphate (Na_2HPO_4) are mixed in equal proportions, the two ion species $H_2PO_4^{1-}$ and HPO_4^{2-} are present in equal concen-trations; this is equivalent to the condition where the $H_2PO_4^{1-}$ species has donated 50% of its dissoci-able proton. The pH is, therefore, at the pK_a value, namely, 6.8.

5. The answer is C. (*II D 2 a*) The pH of the urine is the same as the pK_a' of the acetoacetic acid con-jugate acid-base pair, that is, 4.8. At this pH, 50% of the acetoacetic acid is present as acetoacetate ion, which carries a single negative charge. Sodium will be excreted as the positively charged counter-ion in equivalent amount, that is, 50% of 30 mmol, or 15 mmol.

6. The answer is D. (*II D 2 c*) Carbonic acid (H_2CO_3), which is a weak acid, partially dissociates into its conjugate base, bicarbonate ion (HCO_3^-), and a proton. In plasma, the pK_a', or the pH at 50% dis-sociation, is 6.1. The Henderson-Hasselbalch equation describes the relationship between pH, pK_a', and the ratio of conjugate base to conjugate acid.

$$pH = pK_a' + \log \frac{[base]}{[acid]}$$

At pH 7.1, the log of the ratio of base to acid is 7.1 − 6.1, which is 1. Therefore, since 1 is the log of 10, the ratio is 10:1.

7. The answer is A (1, 2, 3). (*II D 3*) The buffer capacity of a solution of a weak acid and its conjugate base (a buffer solution) depends upon (1) the closeness of the pH of the buffer to the pK_a' of the weak acid and (2) the concentration of the buffer. The buffer capacity is maximal at the pK_a', and it increases with increasing buffer concentration. Buffer capacity can be quantitated by measuring the moles of hydrogen or hydroxyl ions required to change the pH of 1 L of buffer solution by 1.0 pH unit. A buffer pH depends upon the ratio of conjugate base to conjugate acid. Although on dilution of a buffer, the ratio of conjugate base to conjugate acid does not change in terms of molar concentrations, it does in terms of ion activities. This change with dilution is not a measure of buffer capacity.

Challenge
Exam

Introduction

One of the least attractive aspects of pursuing an education is the necessity of being examined on what one has learned. Instructors do not like preparing tests, and students do not like taking them.

Students are required to take many examinations during their learning careers, but little if any time is spent acquainting them with the positive aspects of tests and with systematic and successful methods for approaching them. Students perceive tests as punitive and sometimes feel that they are merely opportunities for the instructor to discover what the student has forgotten or has never learned. Students need to view tests as opportunities to display their knowledge and to use tests as tools for developing prescriptions for further study and learning.

A brief history and discussion of the National Board of Medical Examiners (NBME) examinations (i.e., Parts I, II, and III and FLEX) are presented in this preface, along with some ideas concerning psychological preparation for the examinations. Also presented are some general considerations and test-taking tips as well as how practice exams can be used as educational tools. (The literature provided by the various examination boards contains detailed information concerning the construction and scoring of specific exams.)

National Board of Medical Examiners Examinations

Before the various NBME exams were developed, each state attempted to license physicians through its own procedures. Differences between the quality and testing procedures of the various state examinations resulted in the refusal of some states to recognize the licensure of physicians licensed in other states. This obviously made it difficult for physicians to move freely from one state to another and produced an uneven quality of medical care in the United States.

To remedy this situation, the various state medical boards decided they would be better served if an outside agency prepared standard exams to be given in all states; this would allow each state to meet its own needs and have a common standard by which to judge the educational preparation of individuals applying for licensure.

One misconception concerning these outside agencies is that they are licensing authorities. This is not the case; they are examination boards only. The individual states retain the power to grant and revoke licenses. The examination boards are charged only with designing and scoring valid and reliable tests. They are primarily concerned with providing the states with feedback on how examinees have performed and with making suggestions about the interpretation and usefulness of scores. The states use this information as partial fulfillment of qualifications upon which they grant licenses.

Students should remember that these exams are administered nationwide and, although the general medical information is similar, educational methodologies and faculty areas of expertise differ from institution to institution. It is unrealistic to expect the students to know all the material presented in the exams. Students may face questions on the exams in areas that were only superficially covered in their classes. The testing authorities recognize this situation, and their scoring procedures take it into account.

Scoring the Exams

The diversity of curriculum necessitates that these tests be scored using a criteria-based normal curve. An individual score is based not only on how many questions were answered correctly by a specific student but also on how this one performance relates to the distribution of all scores of the criteria group. In the case of NBME, Part I, the criteria group consists of those students who have completed 2 years of medical training in the United States and are taking the test for the first time and those students who took the test during the previous four June sittings.

Since this test has been constructed to measure a wide range of educational situations, the mean, or average, score generally can be achieved by answering 64% to 68% of the questions correctly. Passing the exam requires answering correctly 55% to 60% of the questions. The competition for acceptance into medical school and the performance levels necessary to stay in school are at such high levels that many students who have always achieved these high levels naturally assume they must perform in a similar fashion and attain equivalent scores on the NBME exams. This is not the case. In fact, among students who are accustomed to performing at levels exceeding 80% to 90%, fewer than 4% taking these tests perform at that high level. Unrealistically high personal expectations leave students psychologically unprepared for these tests, and the anxiety of the moment renders them incapable of doing their best work.

Actually, most students have learned quite well, but they fail to display this learning when they are tested because they do not understand the construction, purpose, or scoring procedures of board exams. It is imperative that they understand that they are **not** expected to score as well as they have in the past and that the measurement criteria is group performance, not only individual performance.

While preparing for an exam, it is important that students learn as much as they can about the subject they will be tested on as well as prepare to discover just how much they may not know. Students should study to acquire knowledge, not just to prepare for tests. **For the well-prepared candidate, the chances of passing far exceed the chances of failing**.

Test Preparation

In preparation for the tests, many students collect far too much study material only to find that they simply do not have the time to go through all of it. They are defeated before they begin because either they cannot get through all the material leaving areas unstudied, or they race through the material so quickly that they cannot benefit from the activity.

It is generally more efficient for the student to use materials already at hand, that is, class notes, one good outline to cover or strengthen areas not locally stressed and for quick review of the whole topic, and one good text as a reference for looking up complex material needing further explanation.

Also, many students attempt to memorize far too much information, rather than learning and understanding less material and then relying on that learned information to determine the answers to questions at the time of the examination. Relying too heavily on memorized material causes anxiety, and the more anxious students become during a test, the less learned knowledge they are likely to use.

Positive Attitude

A positive attitude and a realistic approach are essential to successful test taking. If concentration is placed on the negative aspects of tests or on the potential for failure, anxiety increases and performance decreases. A negative attitude generally develops if the student concentrates on "I must pass" rather than "I can pass." "What if I fail?" becomes the major factor motivating the student to run from failure rather than toward success. This results from placing too much emphasis on scores rather than understanding that scores have only slight relevance to future professional performance.

The score received is only one aspect of test performance. Test performance also indicates the student's ability to use information during evaluation procedures and reveals how this ability might be used in the future. For example, when a patient enters the physician's office with a problem, the physician begins by asking questions, searching for clues, and seeking diagnostic information. Hypotheses are then developed, which will include several potential causes for the problem. Weighing the probabilities, the physician will begin to discard those hypotheses with the least likelihood of being correct. Good differential diagnosis involves the ability to deal with uncertainty, to reduce potential causes to the smallest number, and to use all learned information in arriving at a conclusion.

This same thought process can and should be used in testing situations. It might be termed **paper-and-pencil differential diagnosis**. In each question with five alternatives, of which one is correct, there are four alternatives that are incorrect. If deductive reasoning is used, as in solving a clinical problem, the choices can be viewed as having possibilities of being correct. The elimination of wrong choices increases the odds that a student will be able to recognize the correct choice. Even if the correct choice does not become evident, the probability of guessing correctly increases. Just as differential diagnosis in a clinical setting can result in a correct diagnosis, eliminating incorrect choices on a test can result in choosing the correct answer.

Answering questions based on what is incorrect is difficult for many students since they have had nearly 20 years experience taking tests with the implied assertion that knowledge can be displayed only by knowing what is correct. It must be remembered, however, that students can display knowledge by knowing something is wrong, just as they can display it by knowing something is right. **Students should begin to think in the present as they expect themselves to think in the future**.

Paper-and-Pencil Differential Diagnosis

The technique used to arrive at the answer to the following question is an example of the paper-and-pencil differential diagnosis approach.

> A recently diagnosed case of hypothyroidism in a 45-year-old man may result in which of the following conditions?

- **(A)** Thyrotoxicosis
- **(B)** Cretinism
- **(C)** Myxedema
- **(D)** Graves' disease
- **(E)** Hashimoto's thyroiditis

It is presumed that all of the choices presented in the question are plausible and partially correct. If the student begins by breaking the question into parts and trying to discover what the question is attempting to measure, it will be possible to answer the question by using more than memorized charts concerning thyroid problems.

- The question may be testing if the student knows the difference between "hypo" and "hyper" conditions.
- The answer choices may include thyroid problems that are not hypothyroid problems.
- It is possible that one or more of the choices are "hypo" but are not "thyroid" problems, that they are some other endocrine problems.
- "Recently diagnosed in a 45-year-old man" indicates that the correct answer is not a congenital childhood problem.
- "May result in" as opposed to "resulting from" suggests that the choices might include a problem that **causes** hypothyroidism rather than **results from** hypothroidism, as stated.

By applying this kind of reasoning, the student can see that choice **A**, thyroid toxicosis, which is a disorder resulting from an overactive thyroid gland ("hyper") must be eliminated. Another piece of knowledge, that is, Graves' disease is thyroid toxicosis, eliminates choice **D**. Choice **B**, cretinism, is indeed hypothyroidism, but it is a childhood disorder. Therefore, **B** is eliminated. Choice **E** is an inflammation of the thyroid gland—here the clue is the suffix "itis". The reasoning is that thyroiditis, being an inflammation, may **cause** a thyroid problem, perhaps even a hypothyroid problem, but there is no reason for the reverse to be true. Myxedema, choice **C**, is the only choice left and the obvious correct answer.

Preparing for Board Examinations

1. **Study for yourself**. Although some of the material may seem irrelevant, the more you learn now, the less you will have to learn later. Also, do not let the fear of the test rob you of an important part of your education. If you study to learn, the task is less distasteful than studying solely to pass a test.

2. **Review all areas.** You should not be selective by studying perceived weak areas and ignoring perceived strong areas. This is probably the last time you will have the time and the motivation to review **all** of the basic sciences.

3. **Attempt to understand, not just to memorize, the material.** Ask yourself: To whom does the material apply? When does it apply? Where does it apply? How does it apply? Understanding the connections among these points allows for longer retention and aids in those situations when guessing strategies may be needed.

4. Try to **anticipate questions that might appear on the test.** Ask yourself how you might construct a question on a specific topic.

5. **Give yourself a couple days of rest before the test.** Studying up to the last moment will increase your anxiety and cause potential confusion.

Taking Board Examinations

1. In the case of NBME exams, be sure to **pace yourself** to use the time optimally. As soon as you get your test booklet, go through and circle the questions numbered 40, 80, 120, and 160. The test is constructed so that you will have approximately 45 seconds for each question. If you are at a circled number every 30 minutes, you will be right on schedule. A 2-hour test will have 150–170 questions and a 2½-hour test will have approximately 200 questions. You should use all of your allotted time; if you finish too early, you probably did so by moving too quickly through the test.

2. **Read each question and all the alternatives carefully** before you begin to make decisions. Remember the questions contain clues, as do the answer choices. As a physician, you would not make a clinical decision without a complete examination of all the data; the same holds true for answering test questions.

3. **Read the directions for each question set carefully.** You would be amazed at how many students make mistakes in tests simply because they have not paid close attention to the directions.

4. It is not advisable to leave blanks with the intention of coming back to answer the questions later. Because of the way board examinations are constructed, you probably will not pick up any new information that will help you when you come back, and the chances of getting numerically off on your answer sheet are greater than your chances of benefiting by skipping around. If you feel that you must come back to a question, mark the best choice and place a note in the margin. Generally speaking, it is best not to change answers once you have made a decision, unless you have learned new information. Your intuitive reaction and first response are correct more often than changes made out of frustration or anxiety. **Never turn in an answer sheet with blanks.** Scores are based on the number that you get correct; you are not penalized for incorrect choices.

5. **Do not try to answer the questions on a stimulus–response basis.** It generally will not work. Use all of your learned knowledge.

6. **Do not let anxiety destroy your confidence.** If you have prepared conscientiously, you know enough to pass. Use all that you have learned.

7. **Do not try to determine how well you are doing as you proceed.** You will not be able to make an objective assessment, and your anxiety will increase.

8. **Do not expect a feeling of mastery** or anything close to what you are accustomed. Remember, this is a nationally administered exam, not a mastery test.

9. **Do not become frustrated or angry** about what appear to be bad or difficult items. You cannot know everything.

Specific Test-Taking Strategies

Read the entire question carefully, regardless of format. Test questions have multiple parts. Concentrate on picking out the pertinent key words that might help you begin to problem solve. Words such as "always," "all," "never," "mostly," "primarily," and so forth play significant roles in questions. In all types of questions, distractors with terms like "always" or "never" most often are incorrect. Adjectives and adverbs can completely change the meaning of questions—pay close attention to them. Also, medical prefixes and suffixes (e.g., "hypo-," "hyper-," "-ectomy," "-itis") are sometimes at the root of the question. Normal English grammar often can be valuable in dissecting questions.

Multiple-Choice Questions

Read the question and the choices carefully to become familiar with the data as given. Remember, in multiple-choice questions there is one correct answer and there are four distractors, or incorrect answers. (Distractors are plausible and possible or they would not be called distractors.) They are generally correct for part of the question but not for the entire question. Dissecting the question into parts sometimes aids in discerning these distractors.

If the correct answer is not immediately evident, begin eliminating the distractors. (Many students feel that they must always start at option A and make a decision before they move to B, thus forcing decisions they are not ready to make.) Your first decisions should be made on those choices you feel the most confident about.

Compare the choices to each part of the question. **To be wrong**, a choice needs to be incorrect for only part of the question. **To be correct**, it must be **totally** correct. If you believe a choice is partially incorrect, tentatively eliminate that choice. Make notes next to the choices regarding tentative decisions. One method is to place a minus sign next to the choices you are certain are incorrect and a plus sign next to those that potentially are correct. Finally, place a zero next to any choice you do not understand or need to come back to for further inspection. Do not feel that you must make final decisions until you have examined all choices carefully.

When you have eliminated as many choices as you can, decide which of those that are left has the highest probability of being correct. Remember to use paper-and-pencil differential diagnosis. Above all, be honest with yourself. If you do not know the answer, eliminate as many choices as possible and choose reasonably.

Multiple True–False Questions

Multiple true–false questions are not as difficult as some students make them. These are the questions in which you must mark:

- **A** if **1, 2, and 3** are correct,
- **B** if **1 and 3** are correct,
- **C** if **2 and 4** are correct,
- **D** if only **4** is correct, or
- **E** if **all** are correct.

You should remember that the name for this type of question is multiple true–false and use this concept. You should become familiar with each choice and make notes. Then concentrate on the one choice you feel is definitely incorrect. If you can find one incorrect alternative, you can eliminate three choices immediately and be down to a fifty–fifty probability of guessing the correct answer. In this format, if choice 1 is incorrect, so is choice 3; they go together. Alternatively, if 1 is correct, so is 3. The combinations of alternatives are constant; they will not be mixed. You will not find a situation where choice 1 is correct, but 3 is incorrect.

After eliminating the choices you are sure are incorrect, concentrate on the choice that will make your final decision. For instance, if you discard choice 1, you have eliminated alternatives A, B, and E. This leaves C (2 and 4) and D (4 only). Concentrate on choice 2, and decide if it is true or false. Rereading and concentrating on choice 4 only wastes time; choice 2 will be the decision maker. (Take the path of least resistance and concentrate on the smallest possible number of items while making a decision.) Obviously, if none of the choices is found to be incorrect, the answer is E (all).

Remember you have a better chance of knowing enough to be certain that one choice is incorrect than you have of knowing all there is to know about all the choices.

Comparison-Matching Questions

Comparison-matching questions are also easier to address if you concentrate on one alternative at a time. Choose the option:

A if the question is associated with **(A) only**,
B if the question is associated with **(B) only**,
C if the question is associated with **both (A) and (B)**, or
D if the question is associated with **neither (A) nor (B)**.

Here again, the elimination of obvious wrong alternatives helps clear away needless information and can help you make a clearer decision.

Single Best Answer–Matching Sets

Single best answer–matching sets consist of a list of words or statements followed by several numbered items or statements. Be sure to pay attention to whether the choices can be used more than once, only once, or not at all. Consider each choice individually and carefully. Begin with those with which you are the most familiar. It is important always to break the statements and words into parts, as with all other question formats. **If a choice is only partially correct, then it is incorrect**.

Guessing

Although nothing takes the place of a firm knowledge base, with little to work with, even after playing paper-and-pencil differential diagnosis, you may find it necessary to guess at the correct answer. Some simple rules can help increase your guessing accuracy. Always guess consistently if you have no idea what is correct; that is, after eliminating all that you can, make the choice that agrees with your intuition or choose the option closest to the top of the list that has not been eliminated as a potential answer.

When guessing at questions that present with choices in numerical form, you will often find the choices listed in an ascending or descending order. It is generally not wise to guess the first or last alternative, since these are usually extreme values and are most likely incorrect.

Using the Challenge Exam to Learn

All too often, students do not take full advantage of practice exams. There is a tendency to complete the exam, score it, look up the correct answers to those questions missed, and then forget the entire thing.

In fact, great educational benefits could be derived if students would spend more time using practice tests as learning tools. As mentioned earlier, incorrect choices in test questions are plausible and partially correct or they would not fulfill their purpose as distractors. This means that it is just as beneficial to look up the incorrect choices as the correct choices to discover specifically why they are incorrect. In this way, it is possible to learn better test-taking skills as the subtlety of question construction is uncovered. Additionally, it is advisable to go back and attempt to restructure each question to see if all the choices can be made correct by modifying the question. By doing this, four times as much will be learned. By all means, look up the right answer and explanation. Then, focus on each of the other choices and ask yourself under what conditions they might be correct? For example, the entire thrust of the sample question concerning hypothyroidism could be changed by rewording it to read:

> Hyperthyroidism recently discovered in
>
> Hypothyroidism prenatally occurring in
>
> Hypothyroidism resulting from

This question can be used to learn and understand the area of thyroid problems in general, not only to memorize answers to specific situations.

The Challenge Exam that follows contains 180 questions and explanations. Every effort has been made to simulate the types of questions and the degree of question difficulty in the various licensure and qualifying exams (i.e., NBME Parts I, II, and III and FLEX). While taking this exam, the student should attempt to create the testing conditions that might be experienced during actual testing situations. Approximately 1 minute should be allowed for each question, and the entire test should be finished before it is scored.

Summary

Ideally, examinations are designed to determine how much information students have learned and how that information is used in the successful completion of the examination. Students will be successful if they employ the following suggestions.

- Develop a positive attitude and maintain that attitude.
- Be realistic in determining the amount of material they attempt to master and in the score they hope to attain.
- Read the directions for each type of question and the questions themselves closely and follow the directions carefully.
- Guess intelligently and consistently when guessing strategies must be used.

- Bring the paper-and-pencil differential diagnosis approach to each question in the examination.
- Use the test as an opportunity to display their knowledge and as a tool for developing prescriptions for further study and learning.

National Board examinations are not easy. They may be almost impossible for those who have unrealistic expectations or for those who allow misinformation concerning the exams to produce anxiety out of proportion to the task at hand. They are manageable if they are approached with a positive attitude and with consistant use of all the information the student has learned.

Directions: Each question below contains five suggested answers. Choose the **one best** response to each question.

1. The high-energy phosphate bond of adenosine diphosphate is used in biologic systems by

(A) hydrolysis of the terminal phosphate to give adenosine monophosphate
(B) coupling of the hydrolysis of the terminal phosphate to another reaction via a common intermediate
(C) transfer of the terminal phosphate to glucose or similar substrates
(D) transfer of the terminal phosphate of one adenosine diphosphate (ADP) to another ADP to form adenosine triphosphate
(E) none of the above as it is not active in phosphoryl transfer reactions

2. In one turn of a "closed" tricarboxylic acid cycle, all of the following steps occur EXCEPT

2 H₂° moles used

(A) an acetate group is consumed
(B) a net of 2 molecules of carbon dioxide are produced for each turn of the cycle, *+ 2 co₂ used*
(C) a net of 1 molecule of oxaloacetate is produced for each turn of the cycle
(D) 1 molecule of guanosine triphosphate is generated via a substrate level phosphorylation
(E) reduced nicotinamide-adenine dinucleotide is formed in three of the four oxidative steps

3. The rate of oxygen consumption in an intact mammal would be controlled primarily by the

(A) insulin:glucagon ratio in the blood
(B) level of reduced nicotinamide-adenine dinucleotide in the cell
(C) nature of the substrates metabolized
(D) rate of pulmonary ventilation
(E) respiratory control in the mitochondria

4. All of the following statements describe the properties of histones EXCEPT

(A) they are important constituents of chromatin
(B) they form ionic interactions with DNA
(C) they are found in nucleosomes
(D) they are highly species-specific
(E) they are found in repressed chromatin

5. Which of the following reactions requires the hydrolysis of guanosine triphosphate?

(A) Binding of mRNA to the 40S initiation complex
(B) Activation of alanine by alanine-tRNA synthetase
(C) Peptide bond formation catalyzed by peptidyl transferase
(D) Dissociation of the ribosome into subunits
(E) Binding of aminoacyl-tRNA to the A site on the ribosomes

6. The number of hydrogen bonds between the two strands of the duplex oligonucleotide illustrated below is

AGCTC
TCGAG

(A) 5
(B) 7
(C) 10
(D) 11
(E) 13

7. Nicotinamide-adenine dinucleotide phosphate may be produced in reactions catalyzed by all of the following enzymes EXCEPT

(A) glucose 6-phosphate dehydrogenase
(B) glutamate dehydrogenase
(C) malate dehydrogenase
(D) malic enzyme
(E) 6-phosphogluconate dehydrogenase

8. The activation of chymotrypsinogen requires

(A) the incorporation of dietary lipids into micelles
(B) the presence of a specific lipase
(C) small amounts of bile acid
(D) the action of a proteolytic enzyme
(E) the action of pepsin in the stomach

Questions 9 and 10

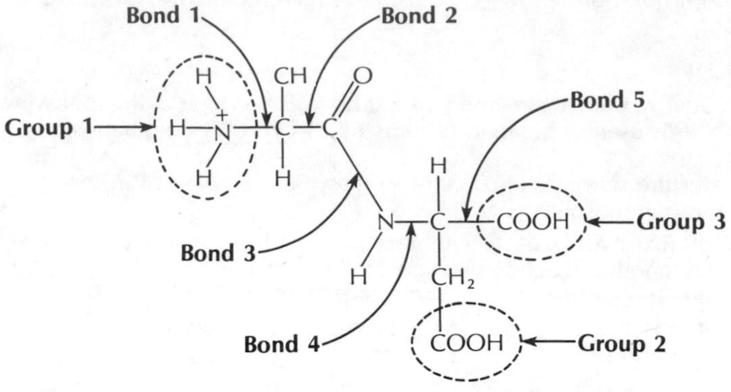

Bond 1   Bond 2

Group 1 →

Bond 5

Group 3

Bond 3

Bond 4

Group 2

9. Which of the bonds in the structure above is not free to rotate?

(A) Bond 1
(B) Bond 2
(C) Bond 3
(D) Bond 4
(E) Bond 5

10. The structure above would not migrate in an electric field when the pH is

(A) equal to pK_a of group 1
(B) equal to pK_a of group 2
(C) equal to pK_a of group 3
(D) midway between the pK_a's of groups 1 and 2
(E) midway between the pK_a's of groups 2 and 3

(end of group question)

11. Proper initiation of transcription in prokaryotes requires all of the following factors EXCEPT

(A) DNA melting
(B) DNA template
(C) promoters
(D) rho factor
(E) sigma factor

12. The major products of the hydrolysis of amylopectin by pancreatic amylase are

(A) maltose, lactose, and limit dextrins
(B) maltotriose, limit dextrins, and glucose
(C) glucose, maltose, and maltotriose
(D) maltose, maltotriose, and limit dextrins
(E) lactose, sucrose, and maltose

13. Glycogenolysis in muscle does not contribute directly to blood glucose concentration because muscle lacks the enzyme

(A) phosphorylase
(B) phosphoglucomutase
(C) glucose-6-phosphatase
(D) glucokinase
(E) phosphoglucoisomerase

14. All of the following tissues are capable of using ketone bodies EXCEPT the

(A) brain
(B) heart
(C) red blood cells
(D) renal cortex
(E) skeletal muscle

15. In eukaryotes, DNA is best differentiated from RNA in that DNA (but not RNA) has which of the following characteristics?

(A) It is confined to the nucleus
(B) It is double-stranded
(C) It contains a deoxyribosyl group rather than a ribosyl group
(D) It requires a DNA template for synthesis
(E) It contains equimolar amounts of G and C

16. Correct statements about ketone bodies include which of the following?

(A) Acetoacetyl coenzyme A (acetoacetyl CoA) is the most abundant ketone body
(B) β-hydroxybutyrate yields more energy equivalents per mole than the other ketone bodies
(C) The liver is the only organ capable of metabolizing acetoacetate
(D) Acetoacetate enters the tricarboxylic acid cycle in the form of acetyl CoA
(E) High levels of reduced nicotinamide-adenine dinucleotide favor a high ratio of acetoacetate to β-hydroxybutyrate

17. De novo purine nucleotide synthesis is regulated by all of the following actions EXCEPT

(A) activation of 5'-phosphoribosyl-1-pyrophosphate (PRPP) amidotransferase by guanosine monophosphate (GMP)
(B) inhibition of PRPP-amidotransferase by adenosine monophosphate (AMP)
(C) inhibition of the condensation of inosine monophosphate (IMP) and aspartate by AMP
(D) inhibition by GMP of the oxidation of IMP
(E) stimulation of the PRPP-amidotransferase reaction by increasing PRPP concentration

18. An inhibitor of glucose 6-phosphate dehydrogenase in the liver might result in an

(A) accumulation of glucose 6-phosphate
(B) impaired conservation of energy and synthesis of adenosine triphosphate
(C) impaired fatty acid synthesis
(D) inability to catabolize glucose
(E) inability to synthesize ribose 5-phosphate

19. A product of the series of reactions that converts carbamoyl phosphate to urea is

(A) arginine
(B) aspartate
(C) adenosine triphosphate
(D) citrulline
(E) fumarate

20. Ornithine has all of the following properties EXCEPT

(A) it is an intermediate of the urea cycle
(B) it is not found in proteins
(C) it is a monoamino–monocarboxylic acid
(D) it is glucogenic
(E) it can be formed from either arginine or glutamic acid

21. An individual who consumes 100 g of protein loses 13.5 g of nitrogen in the urine, 2 g in the feces, and 0.5 g by other routes. This individual is most apt to be

(A) a woman in her eighth month of pregnancy
(B) a 6-year-old child
(C) recovering from major surgery
(D) consuming a diet consistently deficient in lysine
(E) a normal, healthy adult

22. All of the following steps in the catabolism of long-chain fatty acids involve an alteration of a coenzyme–prosthetic group or a coenzyme–cosubstrate EXCEPT

(A) activation of the fatty acid
(B) transport into the mitochondrion
(C) dehydrogenation of the acyl form to the enoyl form (unsaturated)
(D) conversion of the enoyl form to the hydroxyl acyl form
(E) thiolysis of the keto acyl form, liberating acetyl coenzyme A

23. Metals found as natural constituents of proteins include all of the following EXCEPT

(A) cobalt
(B) copper
(C) iron
(D) mercury
(E) molybdenum

24. Lipid micelles in the intestine have which of the following properties?

(A) They contain mostly triacylglycerols
(B) They are absorbed intact in the duodenum
(C) They are secreted into the capillary system by cells lining the intestine
(D) They are absorbed mainly in the lower portion of the small intestine
(E) They facilitate the transfer of hydrolysis products of triacylglycerol to the cells lining the intestine

25. Which of the following titration curves would be expected for glycylarginine, a dipeptide?

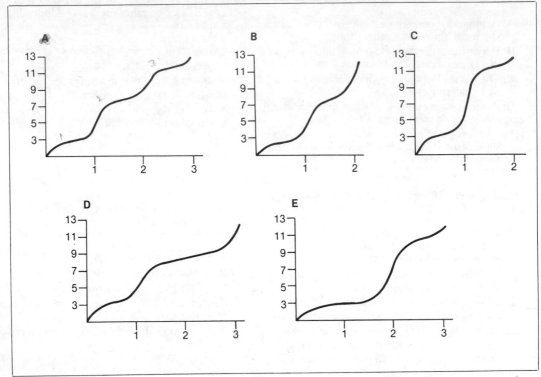

Note.—pH is plotted on the ordinate, and mEq of sodium hydroxide added per mEq of dipeptide is plotted on the abscissa.

26. The phosphate:oxygen (P:O) ratio is defined as

(A) the moles of phosphate consumed divided by the moles of oxygen consumed
(B) the moles of adenosine triphosphate (ATP) synthesized divided by the atom equivalents of oxygen consumed
(C) the moles of ATP formed divided by the milligrams of protein
(D) the moles of oxygen consumed in the presence of adenosine diphosphate (ADP) divided by the moles of oxygen consumed in the absence of ADP
(E) the moles of carbon dioxide produced divided by the moles of oxygen consumed

27. A rate-limiting step for an anabolic or catabolic process is controlled by all of the following enzymes EXCEPT

(A) acetyl coenzyme A (acetyl CoA) carboxylase
(B) cyclooxygenase
(C) β-hydroxy-β-methylglutaryl CoA synthetase
(D) hormone-sensitive lipase
(E) phospholipase A

28. Quantitatively, the most important enzyme involved in generating ammonia from amino acids in humans is

(A) L-amino acid oxidase
(B) arginase
(C) glutamate dehydrogenase
(D) glutamine synthetase
(E) serine dehydratase

29. To initiate translation of mRNA in eukaryotes, all of the following factors are required EXCEPT

(A) adenosine and guanosine triphosphates
(B) eukaryotic initiation factors
(C) formylmethionine-tRNA
(D) mRNA with the triplet AUG near the 5′ end
(E) 40S and 60S ribosomal subunits

30. To prevent rickets associated with renal disorders, administration of which of the following substances is needed?

(A) 1,25-Dihydroxycholecalciferol
(B) Ergocalciferol
(C) High dietary calcium
(D) 25-Hydroxycholecalciferol
(E) Cholecalciferol (vitamin D_3)

31. Tryptophan is best described by which of the following statements?

(A) It is an acidic amino acid
(B) It is converted to thyroid hormones
(C) It is a precursor of the pineal hormone melatonin
(D) It is glucogenic only
(E) It is abundantly distributed in plasma proteins

32. Degradation and excretion of bioactive amines involve all of the following reactions EXCEPT

(A) conjugation to asparagine
(B) conjugation to glucuronic acid
(C) conjugation to sulfate
(D) methylation
(E) oxidation of a carbon bearing an amino group

33. The three major lipid components of animal cell membranes are

(A) glycolipids, prostaglandins, and cholesterol esters
(B) triacylglycerols, free fatty acids, and cholesterol
(C) phospholipids, sphingolipids, and cholesterol
(D) triacylglycerols, phospholipids, and cholesterol esters
(E) phospholipids, dolichol, and free fatty acids

34. Dietary triacylglycerols are absorbed from the intestinal lumen mainly after hydrolysis to

(A) fatty acids and 1,2-diacylglycerol
(B) fatty acids and glycerol
(C) fatty acids and glycerol phosphate
(D) fatty acids and 2-monoacylglycerol
(E) fatty acyl CoA and glycerol

35. Generally accepted features of biologic membranes include all of the following EXCEPT

(A) asymmetric arrangement of lipids
(B) rapid diffusion of inorganic ions across the lipid bilayer
(C) lateral diffusion of lipids
(D) lateral diffusion of integral and peripheral proteins
(E) infrequent transverse ("flip-flop") movement of lipids from one face of the bilayer to the other

36. Diabetic individuals frequently have type IV hyperlipidemia, which is characterized by elevated concentrations of very low-density lipoprotein (VLDL). The most likely explanation for this condition is that

(A) elevated ketone bodies cause allosteric inhibition of lipoprotein lipase
(B) fatty acids delivered to the liver in excess of those used for β-oxidation are incorporated into triglycerides of VLDL
(C) decreased glucose utilization causes cortisol secretion, which leads to induction of acetyl coenzyme A carboxylase
(D) decreased insulin levels lead to increased protein synthesis, including the B-100 protein of VLDL
(E) reduced dependence upon fatty acid oxidation in muscle causes decreased catabolism of VLDL

37. The effect of reduced nicotinamide-adenine dinucleotide on isocitrate dehydrogenase is an example of *act all sites other than active sites of enzyme.*

(A) competitive inhibition
(B) homotropic activation
(C) heterotropic inhibition
(D) homotropic inhibition
(E) none of the above

38. All of the following are rate-limiting steps in a pathway EXCEPT

(A) acetyl coenzyme A (acetyl CoA) + bicarbonate ion + adenosine triphosphate (ATP) → malonyl CoA + adenosine diphosphate + inorganic phosphate
(B) palmitate + CoA + ATP → palmitoyl CoA + adenosine monophosphate + inorganic pyrophosphate
(C) 3-hydroxyl-methylglutaryl CoA → mevalonic acid
(D) 1-stearoyl-2-arachidonoyl phosphatidylcholine → 1-stearoyl lysophosphatidylcholine + arachidonic acid
(E) stearoyl CoA + carnitine → stearoyl carnitine + CoA

carbamoyle phosphate ✓

39. If an individual had a genetic defect in the enzyme that produces N-acetyl glutamate, the most likely clinical findings would be hyperammonemia with

(A) elevated levels of argininosuccinate (the condensation product of citrulline and aspartate)
(B) no detectable citrulline
(C) elevated levels of arginine
(D) elevated levels of urea
(E) no detectable ornithine

40. Which of the following reactions is catalyzed by a nucleoside diphosphate kinase?

(A) Adenine + 5′-phosphoribosyl-1-pyrophosphate → adenosine monophosphate (AMP) + inorganic pyrophosphate
(B) AMP + adenosine triphosphate (ATP) ⇌ 2 adenosine diphosphate (ADP)
(C) deoxythymine + deoxyribose 1-phosphate → deoxythymidine + inorganic phosphate
(D) uridine diphosphate + ATP → uridine triphosphate + ADP
(E) uridine + ATP → uridine monophosphate + ADP

41. Requirements for protein synthesis include all of the following EXCEPT

(A) 20 different amino acids in the form of aminoacyl-tRNAs
(B) ribosomes
(C) guanosine triphosphate
(D) mRNA
(E) lysosomes

42. The net yield of high-energy bonds from the complete oxidation of acetoacetate in the brain is

(A) 11
(B) 12
(C) 23
(D) 24
(E) 25

2 mol of Acetyl coA
↓
TCA
↓ 24 ATP -1 =23

ATP

43. The adenosine triphosphate molecule contributes one or more atoms in the synthesis of each of the following compounds EXCEPT

(A) carbamoyl phosphate
(B) formyl tetrahydrofolate
(C) 5′-phosphoribosyl-1-pyrophosphate
(D) 3′-phosphoadenosine-5′-phosphosulfate
(E) S-adenosylmethionine

44. An enzyme involved in the catabolism of fructose to pyruvate in the liver is

(A) hexokinase
(B) phosphofructokinase
(C) 6-phosphogluconate dehydrogenase
(D) glyceraldehyde 3-phosphate dehydrogenase
(E) phosphoglucomutase

45. Low levels of cellular β-hydroxy-β-methylglutaryl coenzyme A reductase activity in humans is most likely to result from

(A) a vegetarian diet
(B) the administration of a bile acid sequestering resin
(C) a "prudent" diet
(D) familial hypercholesterolemia
(E) a long-term high cholesterol diet

46. Sphingolipid degradation is described by all of the following statements EXCEPT

(A) it is catalyzed by hydrolytic enzymes contained in lysosomes
(B) the degradative pathways require the expenditure of high-energy phosphate bonds
(C) it is a sequential, stepwise removal of constituents
(D) in the sphingolipidoses (lysosomal storage diseases), it is abnormal
(E) it is catalyzed by enzymes that are specific for a type of linkage rather than for a particular compound

PD *protein kinase not AMP P dependent*

47. All of the following liver enzymes are phosphorylated by a cyclic adenosine monophosphate–dependent protein kinase EXCEPT *PD*

(A) acetyl CoA carboxylase
(B) glycogen synthetase
(C) phosphofructokinase-2
(D) pyruvate dehydrogenase
(E) pyruvate kinase

48. The 3′-terminal sequence of the mRNA for the α-chain of globin is

$$\ldots\ldots AACUAU\overset{140}{C}ACUAAGCUAGUUAAUAA\ldots 3'$$

141

The 140th and 141st amino acid residues of the globin are tyrosine and histidine, respectively. An abnormal α-globin isolated from a patient exhibits an elongated chain with leucine and serine at positions 142 and 143. (Codons for leucine are UUA, UUG, CUU, CUC, CUA, and CUG; codons for serine are AGC and AGU). What is the mutation responsible for this elongated globin?

(A) Deletion of a base
(B) Insertion of a base
(C) Missense mutation
(D) Nonsense mutation
(E) Reverse terminator mutation

49. Thioredoxin is described best by which of the following statements?

(A) It is an enzyme that reduces xanthine to hypoxanthine
(B) It is a component of the salvage pathway for pyrimidine bases
(C) It is the source of reducing equivalents for the conversion of a nucleoside diphosphate to a deoxyribonucleotide
(D) It is the source of reducing equivalents for the conversion of deoxyuridine monophosphate to deoxythymidine monophosphate
(E) It is the source of reducing equivalents for certain mixed-function oxidases, such as phenylalanine hydroxylase

50. During long-term fasting, the major control of the rate of gluconeogenesis is the

(A) cyclic adenosine monophosphate level in the liver
(B) adenosine triphosphate level in the liver
(C) availability of free fatty acids to the liver
(D) availability of alanine in the blood
(E) insulin:glucagon ratio

51. A sudden fall in adenosine triphosphate in muscle cells results in a transient increase in the concentration of all of the following metabolites EXCEPT

(A) citrate
(B) fructose 1,6-diphosphate
(C) glucose 1-phosphate
(D) α-ketoglutarate
(E) 3-phosphoglycerate

52. A competitive inhibitor of an enzyme has which of the following properties?

(A) It becomes covalently attached to an enzyme
(B) It is frequently a feedback inhibitor
(C) It causes irreversible inactivation of the enzyme
(D) It interferes with substrate binding to the enzyme
(E) It decreases the V_{max}

53. In the transamination of amino acids all of the following statements are true EXCEPT

(A) pyridoxal phosphate is a coenzyme/prosthetic group
(B) the reaction is tightly controlled by allosteric effectors
(C) the amino acid acceptor is typically an α-keto acid
(D) ammonia is neither consumed nor produced
(E) in the catabolism of most, but not all, amino acids, transamination precedes all catabolic changes in the carbon skeleton

54. A compound that transfers reducing equivalents from mitochondria to the cytosol during gluconeogenesis is

(A) phosphoenolpyruvate
(B) glycerol 3-phosphate
(C) aspartate
(D) malate
(E) oxaloacetate

Directions: Each question below contains four suggested answers of which **one or more** is correct. Choose the answer

A if **1, 2, and 3** are correct
B if **1 and 3** are correct
C if **2 and 4** are correct
D if **4** is correct
E if **1, 2, 3, and 4** are correct

55. Most patients with gout who are treated with allopurinol, an inhibitor of xanthine oxidase, respond with which of the following changes?

(1) A decreased level of de novo purine nucleotide synthesis
(2) A decreased level of uric acid in the urine
(3) An increased level of hypoxanthine in the blood
(4) An increased level of xanthine in the blood

56. Synthesis of glycogen from glucose 6-phosphate requires

(1) phosphoglucomutase
(2) glycogen phosphorylase
(3) uridine triphosphate
(4) cyclic adenosine monophosphate

57. Antibiotics that inhibit bacterial protein synthesis include

(1) erythromycin
(2) chloramphenicol
(3) streptomycin
(4) puromycin

58. Excision repair of DNA involves which of the following substances?

(1) RNA primer
(2) DNA polymerase I
(3) DNA polymerase III
(4) DNA ligase

59. The structure of tRNA includes one or more specific regions, the properties of which include

(1) containing a triplet of nucleotides known as the anticodon
(2) binding specifically to amino acids by hydrogen bonding
(3) allowing a specific secondary structure
(4) being recognized by a specific amino acid

60. The degradation of heme to bilirubin is characterized by

(1) occurrence only in the liver
(2) the production of carbon monoxide
(3) the reduction of the propionic side groups of heme
(4) the requirement for reduced nicotinamide-adenine dinucleotide phosphate

61. Efficient expression of the tryptophan operon requires

(1) the dissociation of the cyclic adenosine monophosphate–catabolite activator protein complex from the gene
(2) low tryptophan levels
(3) completed transcription and processing of the mRNA
(4) mRNA with a Shine-Dalgarno sequence

62. According to the fluid mosaic model of a membrane, characteristics of membrane proteins include which of the following?

(1) Proteins may be embedded in the lipid bilayer
(2) Transverse movement ("flip-flop") of a protein in the membrane is thermodynamically favorable
(3) Individual proteins may be degraded and resynthesized
(4) Proteins are distributed symmetrically across the membrane

63. The structural stability of collagen depends on

(1) the triple helical structure
(2) the high content of proline and hydroxyproline
(3) a glycyl residue at every third position
(4) glycosylation of hydroxylysine residues

64. Correct statements concerning the initiation of transcription in *Escherichia coli* include which of the following?

(1) RNA polymerase holoenzyme binds to a specific region of the gene
(2) The association of histones with DNA must be disrupted
(3) There is localized melting of DNA in the region of the initiation site
(4) Both a DNA template and a DNA primer are required

65. The metabolism of a 17:1 fatty acid and valine is expected to have which of the following points in common?

(1) Neither can contribute to gluconeogenesis
(2) Both require a vitamin B_{12}–dependent enzyme for conversion of an intermediate to succinyl CoA
(3) Both require an aminotransferase for their complete metabolism
(4) Both require the bicarbonate ion (or carbon dioxide) for their complete metabolism

66. In DNA replication, DNA polymerase I is responsible for

(1) removal of RNA primers by its 5',3'-exonuclease activity
(2) addition of nucleotides by its DNA polymerase activity
(3) editing by its 3',5'-exonuclease activity
(4) driving the reaction to completion by its pyrophosphatase activity

67. Correct statements about fatty acid transport into the mitochondria include which of the following?

(1) Long-chain fatty acids are transported by a sodium-dependent cotransport process, which is driven by the sodium ion gradient
(2) Carnitine acyl transferases participate in fatty acyl group transport across the inner mitochondrial membrane
(3) Fatty acyl group transfer across the mitochondrial membrane requires the expenditure of adenosine triphosphate
(4) Carnitine concentrations are higher in heart muscle than in the brain

68. Respiratory control has which of the following characteristics?

(1) It is an example of allosterism
(2) It is lacking in uncoupled mitochondria
(3) It is dependent on the ratio of nicotinamide-adenine dinucleotide (NAD^+) to reduced NAD
(4) It is dependent on the ratio of adenosine diphosphate to adenosine triphosphate

69. Intrachain hydrogen bonds stabilize the secondary structure of

(1) human hemoglobin A_1
(2) tropocollagen
(3) tRNA
(4) triacylglycerols

70. In mammals, the reduced nicotinamide-adenine dinucleotide (NADH) produced by the glyceraldehyde 3-phosphate dehydrogenase step of glycolysis may be reoxidized by

(1) lactate dehydrogenase
(2) malate dehydrogenase
(3) glycerol 3-phosphate dehydrogenase
(4) NADH dehydrogenase

71. A patient who has glycogen storage disease type IV (i.e., deficiency of liver branching enzyme) should show

(1) normal rapid glycogenolysis
(2) formation of glycogen with short branches (i.e., limit dextrans)
(3) greatly reduced glucose uptake from the blood
(4) an abnormally high ratio of glucose 1-phosphate to glucose in the immediate products of glycogen breakdown

72. Correct statements regarding the de novo synthesis of pyrimidine nucleotides in man include which of the following?

(1) Aspartate is the source of several atoms of the pyrimidine ring
(2) The pyrimidine base, orotic acid, is an intermediate
(3) A primary site of control is at the level of the enzyme carbamoyl phosphate synthetase II
(4) The pathway branches at the level of orotidine monophosphate to generate either uridine monophosphate or cytidine monophosphate

SUMMARY OF DIRECTIONS

A	B	C	D	E
1, 2, 3 only	1, 3 only	2, 4 only	4 only	All are correct

73. Type I glycogen storage disease (i.e., deficiency of glucose 6-phosphatase and type VI (i.e., deficiency of liver glycogen phosphorylase) can be distinguished by differences in blood glucose levels in response to the administration of

(1) glucagon
(2) epinephrine
(3) fructose
(4) glucose

74. There is an increase in "direct" bilirubin of the blood in which of the following conditions?

(1) Hemolytic jaundice
(2) Decreased hepatic uptake of bilirubin
(3) Familial deficiency of uridine diphosphoglucuronyl transferase for bilirubin
(4) Obstruction of the bile duct

75. The degradation of epinephrine and norepinephrine is described by which of the following statements?

(1) It includes a methylation reaction followed by an oxidation of the amino group
(2) It proceeds by reduction of the aromatic ring
(3) It includes a reaction catalyzed by monoamine oxidase followed by a reaction catalyzed by catechol-O-methyl transferase
(4) It requires 5-methyl tetrahydrofolate as a methyl donor

76. The activity that takes place in the oxidative phase of the hexose monophosphate shunt is described by which of the following statements?

(1) Glucose 6-phosphate is converted to ribose 5-phosphate
(2) Nicotinamide-adenine dinucleotide phosphate is reduced
(3) Glucose 6-phosphate oxidation is stimulated by fatty acid synthesis
(4) Carbon dioxide and adenosine triphosphate are produced

77. Carbons in dihydroxyacetone phosphate could eventually be incorporated into

(1) liver glycogen
(2) testosterone
(3) trioleoglycerol
(4) linoleic acid

78. In *Escherichia coli*, the RNA polymerase holoenzyme has which of the following properties?

(1) It is required for correct initiation of transcription
(2) It contains four different kinds of subunits
(3) It binds to the promoter region of a gene
(4) It is required for elongation of RNA

79. During collagen synthesis, intracellular events that take place include

(1) hydroxylation of selected prolines and lysines
(2) glycosylation of selected hydroxylysines
(3) formation of interchain and intrachain disulfide bonds
(4) cleavage of extension peptides

80. Observations that are consistent with the chemiosmotic coupling theory include which of the following?

(1) The mitochondrial inner membrane is impermeable to protons
(2) Electron transport through complex IV (cytochrome oxidase) results in proton transport from the matrix to the intermembrane space
(3) Uncouplers increase the permeability of the mitochondrial inner membrane to protons
(4) Adenosine triphosphate synthesis occurs in the process of moving protons from the matrix to the intermembrane space

81. A deficiency in plasma lecithin:cholesterol acyltransferase (LCAT) is likely to have which of the following consequences?

(1) A decrease in the percentage of cholesteryl ester in high-density lipoproteins (HDLs)
(2) An increase in the percentage of phosphatidylcholine in HDLs
(3) An increase in the percentage of free cholesterol in HDLs
(4) An increase in the apo C-II content of HDLs

82. An α-keto acid dehydrogenase catalyzes which of the following reactions?

(1) Conversion of pyruvate to acetyl coenzyme A (acetyl CoA)
(2) The reaction catalyzed by the enzyme that is defective in maple syrup urine disease
(3) Conversion of α-ketobutyrate to propionyl CoA in the metabolism of threonine
(4) Conversion of α-ketoglutarate to succinyl CoA in the tricarboxylic acid cycle

83. In DNA replication, RNA functions in which of the following ways?

(1) It forms a phosphodiester bond with a deoxyribonucleotide
(2) It forms hydrogen bonds with the template strand
(3) It is necessary for synthesis of both the leading and lagging strands
(4) It is a substrate for DNA ligase

84. The synthesis of δ-aminolevulinate is characterized by which of the following statements?

(1) It requires succinyl coenzyme A as a precursor
(2) It requires vitamin B_6
(3) It is inhibited by hemin
(4) It is inhibited by lead

85. A eukaryotic gene, cloned into a vector and replicated in bacteria, might not be expressed as a functional protein because the gene

(1) contains introns
(2) is not properly oriented to a promoter
(3) contains no information for a Shine-Dalgarno sequence
(4) cannot code for a polyadenylated mRNA

86. The decreased rate of fatty acid oxidation in the postprandial liver (1–4 hours after eating) is the result of

(1) the effect of a high insulin:glucagon ratio on adipose tissue
(2) activation of acetyl coenzyme A (acetyl CoA) carboxylase by citrate
(3) low levels of free fatty acids in blood
(4) inhibition of acyl CoA–carnitine transferase by acetyl CoA

87. From the following diagram, it can be concluded that

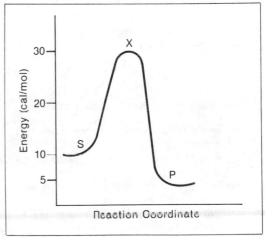

(1) the reaction is endergonic
(2) the activation energy for the reaction S → P is 20 cal/mol
(3) an enzyme might accelerate this reaction by raising the energy of S from 10 cal/mol to 20 cal/mol
(4) X corresponds to the transition state of the reaction

88. Nonheme iron is found in association with

(1) reduced nicotinamide-adenine dinucleotide dehydrogenase
(2) succinate dehydrogenase
(3) cytochromes b and c
(4) an equimolar amount of inorganic sulfur

89. An uncoupler of oxidative phosphorylation has which of the following effects?

(1) It inhibits respiration
(2) It decreases adenosine triphosphate (ATP) synthesis
(3) It increases the standard free-energy change ($\Delta G°'$) of hydrolysis for ATP
(4) It increases heat production

SUMMARY OF DIRECTIONS

A	B	C	D	E
1, 2, 3 only	1, 3 only	2, 4 only	4 only	All are correct

90. Enzymes that are allosterically inhibited by reduced nicotinamide-adenine dinucleotide include

(1) 3-phosphoglycerate kinase
(2) pyruvate dehydrogenase kinase
(3) pyruvate dehydrogenase
(4) isocitrate dehydrogenase

91. Study of the genetic code reveals that

(1) some triplets can code for more than one amino acid
(2) some amino acids can be coded for by more than one triplet
(3) initiation and termination codons are "nonsense" codons
(4) the specificity of base-pairing is less stringent for the 5'-end base of the anticodon

92. Cellular membrane proteins have which of the following properties?

(1) They may have catalytic activity
(2) If peripheral, they can be removed from the membrane without disrupting the membrane structure
(3) If integral, they can span the width of the membrane
(4) They can move laterally in the plane of the membrane

93. When a coenzyme acts as a prosthetic group, it experiences which of the following activities?

(1) The coenzyme undergoes modification during the reaction
(2) The coenzyme cycles between at least two chemical forms during the catalytic process
(3) Removal of the coenzyme inactivates the enzyme
(4) Removal of the coenzyme yields the holoenzyme

94. From the following Lineweaver-Burk plot, one can conclude that

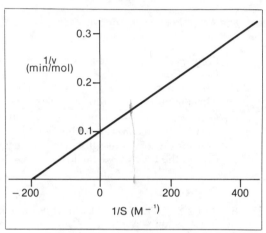

(1) $K_m/V_{max} = 100$
(2) $K_m = 0.005$ M
(3) turnover number = 100 min
(4) $V_{max} = 100$ mol/min

95. If the sulfur of methionine is labeled with ^{35}S and administered to an animal, labeled sulfur would be found in

(1) the urine as inorganic sulfate ion
(2) the urine conjugated with catecholamine catabolites
(3) coenzyme A
(4) cysteine residues in protein

96. Antibiotics that can be used clinically to inhibit bacterial ribosome-mediated translation include

(1) tetracycline
(2) puromycin
(3) erythromycin
(4) penicillin

97. Correct statements about the synthesis of alanine in man, using ammonia as the source of nitrogen, include which of the following?

(1) Glucose could be the source of carbons
(2) Glutamate dehydrogenase is involved
(3) Glutamate–alanine aminotransferase is involved
(4) Both reduced nicotinamide-adenine dinucleotide phosphate and pyridoxal phosphate serve as coenzymes

E

PGLA

98. Substances that are important for gluconeogenesis during fasting are

(1) pyruvate
(2) glycerol
(3) lactate
(4) amino acids

E

99. The proteins required for both DNA replication and DNA repair include

(1) DNA ligase
(2) topoisomerase
(3) DNA polymerase I
(4) primase

B

100. The effects of increased blood glucose levels after a high carbohydrate meal include which of the following?

(1) The liver increases glucose utilization due to decreased product inhibition of glucokinase
(2) The muscle increases glucose uptake if the insulin levels rise in the blood
(3) The brain increases glucose utilization due to an increased saturation of hexokinase
(4) The red blood cells do not increase glucose utilization appreciably

C

101. The "signal" hypothesis involves which of the following actions?

(1) Removal of one or more C-terminal amino acid residues from the protein being synthesized
(2) A sequence of hydrophilic amino acids at the N-terminal end of a secretory protein, which is cleaved off after synthesis
(3) Synthesis of cytosolic proteins by free polysomes
(4) Provision of a signal for polysome attachment to the endoplasmic reticulum to facilitate export of secretory proteins

D

102. The following sequence

$$5'\text{-GTAAG-}3'$$
$$3'\text{-CAUUC-}5'$$

could appear during

(1) transcription
(2) replication
(3) hybridization
(4) reverse transcription

E

103. Synthesis of 1 mol of glucose from pyruvate will require exactly

(1) 2 mol of guanosine triphosphate
(2) 2 mol of adenosine triphosphate
(3) 2 mol of reduced nicotinamide-adenine dinucleotide (NADH)
(4) 2 mol of NADH phosphate

B

104. Rotenone, which inhibits complex I, also inhibits the mitochondrial oxidation of

(1) succinate
(2) malate
(3) glycerol 3-phosphate
(4) α-ketoglutarate

105. Iron–sulfur–protein centers have which of the following properties?

(1) They take part in the one-electron transfers of the reduced nicotinamide-adenine dinucleotide dehydrogenase complex
(2) They contain inorganic sulfur in amounts equimolar to that of nonheme iron
(3) They are found in association with the succinate dehydrogenase complex
(4) They are involved in the transfer of electrons from reduced ubiquinone to cytochrome c

E

106. The formation of alanine from valine requires which of the following enzymes?

(1) Branched-chain α-keto acid dehydrogenase
(2) Succinate dehydrogenase
(3) Phosphoenolpyruvate carboxykinase
(4) Pyruvate kinase

E

107. Factors favoring glucose mobilization include

(1) low plasma glucose concentration
(2) glucagon
(3) an adequate supply of substrate for gluconeogenesis
(4) insulin

SUMMARY OF DIRECTIONS

A	B	C	D	E
1, 2, 3 only	1, 3 only	2, 4 only	4 only	All are correct

108. Thiamine deficiency (beriberi) can cause

(1) diarrhea
(2) heart failure
(3) dermatitis
(4) polyneuropathy

109. Chemical examination of reticulocyte DNA, mRNA, tRNA, and ribosomes from an individual with sickle cell anemia reveals differences from normal in

(1) DNA
(2) tRNA
(3) mRNA
(4) ribosomes

110. An increase in the ratio of nicotinamide-adenine dinucleotide (NAD$^+$) to its reduced form (NADH) in the mitochondria has which of the following effects?

(1) The reduction potential, $E_{NAD^+/NADH}$, becomes, less negative
(2) Isocitrate dehydrogenase is inhibited
(3) The malate:oxaloacetate ratio decreases
(4) The succinate:fumarate ratio increases

111. Catabolism and excretion of catecholamines involves

(1) oxidation by monoamine oxidase in the liver
(2) methylation catalyzed by catechol-O-methyl transferase
(3) conjugation of catabolites with a charged molecule
(4) excretion of catabolites in the urine

112. The peptide bond has which of the following characteristics?

(1) It has a partial double-bond character
(2) It is polar
(3) It cannot rotate freely
(4) It is trans in the α-helix

113. All enzymes have which of the following characteristics?

(1) They form complexes with their substrates
(2) They exhibit sigmoidal kinetics
(3) They decrease the activation energy of the catalytic reaction
(4) They increase the equilibrium constant of the catalytic reaction

114. Attenuation in gene regulation involves

(1) translation of the leader peptide
(2) a repressor bound to the operator
(3) changes in the RNA secondary structure
(4) activation by a cyclic adenosine monophosphate–catabolite activator protein complex

115. The rate of glycogenolysis in muscle is increased by

(1) the phosphorylation of phosphorylase kinase
(2) the dephosphorylation of phosphorylase by a protein phosphatase
(3) the allosteric effects of adenosine monophosphate on phosphorylase
(4) product inhibition of glycogen synthetase by glucose 6-phosphate

116. Ubiquinone (coenzyme Q) has which of the following properties?

(1) It contains a B vitamin as a part of its structure
(2) It can be either a one- or two-electron acceptor
(3) It is water soluble and freely dissociable from the membrane
(4) It is reduced by either complex I or complex II

117. The bacterial transforming principle has which of the following characteristics?

(1) Resistance to heating
(2) Strong absorption of light of about 260 nm wavelength
(3) Resistance to ribonuclease
(4) Resistance to protease

Directions: The groups of questions below consist of lettered choices followed by several numbered items. For each numbered item, select the one lettered choice with which it is most closely associated. Each lettered choice may be used once, more than once, or not at all. Choose the answer

   A   if the item is associated with **(A) only**
   B   if the item is associated with **(B) only**
   C   if the item is associated with **both (A) and (B)**
   D   if the item is associated with **neither (A) nor (B)**

Questions 118–120

For each description of activation listed below, select the amino acid that is most closely associated with it.

(A) Methionine
(B) Choline
(C) Both
(D) Neither

C 118. After suitable activation, it can be a methyl group donor in a transmethylation reaction

A 119. Activation to a methyl group donor requires adenosine triphosphate

A 120. In activated form, it transfers a methyl group to many different acceptors

Questions 121 and 122

For each operon listed below, select the process by which it is regulated.

(A) Negative control
(B) Attenuation
(C) Both
(D) Neither

A 121. Regulates the lactose operon

C 122. Regulates the tryptophan operon

Questions 123–125

For each description listed below, select the enzyme most likely to be associated with it.

(A) Carbamoyl phosphate synthetase I (mitochondrial)
(B) Carbamoyl phosphate synthetase II (cytosolic)
(C) Both
(D) Neither

A 123. N-Acetyl glutamate is a required allosteric activator

B 124. Glutamine is the preferred source of nitrogen

A 125. A genetic deficiency results in hyperammonemia

Questions 126 and 127

Match the following.

(A) Adenosine triphosphate hydrolysis is required
(B) Guanosine triphosphate hydrolysis is required
(C) Both
(D) Neither

C 126. Initiation of protein synthesis in eukaryotes

B 127. Termination of protein synthesis in prokaryotes

Questions 128 and 129

For each activity listed below, select the defect in porphyrin synthesis most likely to cause it.

(A) Hereditary coproporphyria
(B) Hereditary protoporphyria
(C) Both
(D) Neither

D 128. Increases the excretion of protoporphyrinogen I

B 129. Decreases the activity of the enzyme ferrochelatase

Questions 130–132

Match the following.

(A) De novo synthesis of purine nucleotides
(B) Salvage of purine bases
(C) Both
(D) Neither

C 130. 5-Phosphoribosyl-1-pyrophosphate (PRPP) is a substrate

B 131. A deficiency of hypoxanthine-guanine phosphoribosyl transferase is detrimental

C 132. Inosine monophosphate is an intermediate in the formation of adenosine monophosphate

Questions 133 and 134

For each description listed below, select the structure that is most closely associated with it.

(A) α-Helix
(B) β-pleated sheet
(C) Both
(D) Neither

133. It may be found in globular proteins

134. The sequence Gly-Pro-Hyp is a common and important structural element

Questions 135 and 136

Match the following.

(A) Amino acid activation
(B) Fatty acid activation
(C) Both
(D) Neither

135. An enzyme-bound acyl–adenosine monophosphate is an intermediate species

136. The energy for the process is supplied by the hydrolysis of adenosine triphosphate to adenosine diphosphate and inorganic pyrophosphate

Questions 137 and 138

Match the following.

(A) Ferritin
(B) Transferrin
(C) Both
(D) Neither

137. Binds iron in the ferric (oxidized) state

138. Is complexed with polysaccharide

Questions 139–143

For each property of protein synthesis listed below, select the type of organism in which it is most apt to be found.

(A) Protein synthesis in prokaryotes
(B) Protein synthesis in eukaryotes
(C) Both
(D) Neither

139. The first step in the formation of the initiation complex involves the binding of mRNA

140. Elongation factors Tu, Ts, and G are required during elongation

141. Translation of mRNA is initiated while the mRNA is being transcribed

142. Gene expression may be controlled at the level of translation of mRNA

143. Termination of translation is effected by the peptidyl transferase of the ribosome

Questions 144–146

Match the following.

(A) Collagen
(B) Typical globular protein
(C) Both
(D) Neither

144. Its structure is stabilized by hydrogen bonding

145. α-Helix is an important structural element

146. Its biosynthesis involves both intra- and extracellular reactions

Questions 147 and 148

For each statement listed below, select the compound with which it is most apt to be associated.

(A) S-Adenosylmethionine
(B) 5-Methyl tetrahydrofolate
(C) Both
(D) Neither

B 147. The immediate precursor of the methyl group in the compound is a more oxidized form of carbon than a methyl group

C 148. If serine is uniformly labeled with carbon-14, the label appears in the methyl group

Questions 149–151

Match the following.

(A) Substrate level phosphorylation
(B) Oxidative phosphorylation
(C) Both
(D) Neither

C 149. Occurs in mitochondria

A 150. Occurs in the cytosol

B 151. Is uncoupled by 2,4-dinitrophenol

Questions 152–154

For each description that follows, select the type of glycosidic linkage most apt to be described by it.

D (A) Type I glycosidic linkage
(B) Type II glycosidic linkage
(C) Both
(D) Neither

A 152. The carbohydrate chain is initially synthesized attached to dolichol – I

D 153. Attachment of carbohydrate to protein is at the Gln residue of the sequeon, Gln-X-Thr, where X is any amino acid *I & II*

C 154. A threonine residue may be involved in determining the site of linkage to the protein

Questions 155–160

For each description listed below, select the physiologic response that is most apt to be associated with it.

(A) Bohr effect
(B) Blood buffering
(C) Both
(D) Neither

C 155. Participates in the control of pH changes due to physiologic carbon dioxide transport

A 156. Made possible by the conformational change that hemoglobin experiences as it binds and releases oxygen

A 157. Requires a change in pK of weakly acidic groups

B 158. The combined effects of blood proteins and bicarbonate:carbonic acid ratio

B 159. Participation by histidyl residues of both the α-chains and the β-chains of hemoglobin

D 160. Due primarily to the —SH groups of cysteinyl residues

Questions 161–167

For each property of mRNA listed below, select the organism in which it is most likely to be found.

(A) Eukaryotic mRNA
(B) Prokaryotic mRNA
(C) Both
(D) Neither

B 161. Translation may begin before transcription is completed

A 162. It has a poly A tail

C 163. AUG is an initiation codon for translation

B 164. It contains a Shine-Dalgarno sequence

B 165. It is polycistronic *C template for several*

D 166. It contains thymine bases *polypeptide chains.*

A 167. It contains a cap

Questions 168–172

For each characteristic of a vitamin deficiency listed below, select the disease with which it is most likely to be associated.

(A) Pellagra
(B) Beriberi
(C) Both
(D) Neither

A 168. Results from a nicotinic acid (niacin) deficiency

B 169. Results from a thiamine hydrochloride (vitamin B₁) deficiency

D 170. Results from a riboflavin (vitamin B₂) deficiency

A 171. Presents with diarrhea and dermatitis

C 172. Progresses to degenerative changes in the central nervous system

Directions: The groups of questions below consist of lettered choices followed by several numbered items. For each numbered item select the **one** lettered choice with which it is **most** closely associated. Each lettered choice may be used once, more than once, or not at all.

Questions 173–175

Match the following.

(A) Bicarbonate ion
(B) Deoxyhemoglobin
(C) Carbonic acid
(D) Dissolved carbon dioxide (CO_2)
(E) Myoglobin

C 173. It does not significantly contribute to physiologic (CO_2) transport

B 174. Its contribution to CO_2 transport would increase in the absence of 2,3-diphosphoglycerate

A 175. It is a major factor in the buffering of plasma

Questions 176–180

For each vitamin listed below, select the metabolic process with which is most likely to be associated.

(A) Synthesis of amino acids
(B) Synthesis of DNA
(C) Calcium metabolism
(D) Electron transport
(E) Pentose phosphate pathway

A 176. Pyridoxal phosphate (vitamin B₆)

B 177. Folic acid (pteroylglutamic acid)

C 178. Cholecalciferol (vitamin D₃)

D 179. Niacin (nicotinic acid)

E 180. Thiamine (vitamin B₁)

ANSWERS AND EXPLANATIONS

1. The answer is D. (*Chapter 29 V B 3*) The enzyme adenosine monophosphate kinase (myokinase) catalyzes the reaction:

Adenosine triphosphate (ATP) + AMP ⇌ adenosine diphosphate (ADP) + ADP

It is a readily reversible reaction with an equilibrium constant near 1. In the backward reaction, the high-energy terminal phosphate bond of one ADP is transferred to another ADP to form ATP. The terminal phosphate group of ADP is not used to phosphorylate glucose or similar substrates, nor is its transfer coupled to another reaction via a common intermediate.

2. The answer is C. (*Chapter 15 I C; III E*) The tricarboxylic acid (TCA) cycle provides the degradation of two-carbon acetyl residues derived from carbohydrate, fatty acids, and amino acids, to carbon dioxide (CO_2) and water. For each turn of the cycle, a net of 2 molecules of CO_2 are produced, 1 molecule of guanosine triphosphate is produced via a substrate level phosphorylation, and 3 molecules of reduced nicotinamide-adenine dinucleotide (NADH) are formed. The cycle does not produce, or consume, any intermediate of the cycle; thus, a net of 1 molecule of oxaloacetate is not produced for each turn of the cycle.

3. The answer is E. (*Chapter 32 III D; Chapter 16 VIII*) The amount of energy derived from the oxidation of metabolic fuels is geared to the energy expenditure of the organism. The rate of oxidation of all metabolic fuels in the mitochondria, the main site of energy formation and oxygen consumption, is governed by the rate of phosphorylation of adenosine diphosphate (ADP) to adenosine triphosphate (ATP). It is the inhibitory action of accumulating ATP (the obverse being the shortage of ADP acceptor) that controls the rate of oxidation of metabolic fuels (respiratory control). The insulin:glucagon ratio influences the mobilization of stored metabolic fuel, and the rate of pulmonary ventilation influences the oxygen supply; however, it is the respiratory control exercised in the mitochondria that is the primary control.

4. The answer is D. (*Chapter 1 VI D 1*) The DNA in eukaryotes (but not prokaryotes) is associated with basic proteins called histones and other proteins called nonhistone chromosomal proteins. Chromatin contains about equal amounts (by weight) of DNA and protein. The histones are small basic proteins that carry a considerable positive charge and form ionic interactions with the negatively charged phosphate backbone of DNA. They are among the most highly conserved proteins known. They form the basis of the structure of nucleosomes, which have a core that consists of an octomer of histones noncovalently bonded to 140 base-pairs of DNA; these base-pairs are wrapped around the histone in a left-handed superhelix. Histones are found both in euchromatin and heterochromatin. The former represents *dispersed* chromatin, which can be transcribed into RNA, and the latter represents *repressed* chromatin, which cannot be transcribed into RNA and is found mainly in the centromere of chromosomes.

5. The answer is E. (*Chapter 4 IV B 4; VII A 3, B 1, 2; VIII B 5 e*) The delivery of an aminoacyl-tRNA to the A site on the ribosome is effected by the elongation factor Tu (EF-Tu) in prokaryotes. Guanosine triphosphate (GTP) bound to EF-Tu is hydrolyzed to guanosine diphosphate at this time. The eukaryotic elongation factor 1_ψ (eEF-1_α) is analogous to the prokaryote EF-Tu. Binding of mRNA to the 40S initiation complex in eukaryotes does not require the hydrolysis of GTP. The activation of alanine by alanine-tRNA synthetase requires adenosine triphosphate for the formation of the high-energy link between the amino acid and the tRNA, which is an amino acid–ester bond. There is sufficient energy in this bond for the formation of a peptide bond catalyzed by peptidyl transferase. Dissociation of ribosomes into small and large subunits is effected by initiation factors 1 and 3 (IF-1 and IF-3) in prokaryotes and does not involve the hydrolysis of GTP.

6. The answer is E. (*Chapter 5 Figure 5-1*) There are two hydrogen (H) bonds between the bases of an AT pair and three H bonds between a GC pair. Thus, in the duplex oligonucleotide shown, there are two AT pairs and three GC pairs, making a total of thirteen H bonds between the two strands.

7. The answer is C. (*Chapter 15 III H; Chapter 19 I B 1 a, b; Chapter 20 III B 2 a; Chapter 27 I A 2 a*) The dehydrogenases listed in the question that use nicotinamide-adenine dinucleotide phosphate as the electron acceptor cofactor include glucose 6-phosphate dehydrogenase, glutamate dehydrogenase, malic enzyme, and 6-phosphogluconate dehydrogenase. Malate dehydrogenase, on the other hand, is specific for nicotinamide-adenine dinucleotide. Malic enzyme catalyzes the oxidative decarboxylation of malate to pyruvate and carbon dioxide.

8. The answer is D. (*Chapter 24 I B, C; Chapter 26 II A 3 b*) Chymotrypsinogen is a proenzyme involved in the digestion of dietary proteins. It is transformed into an active enzyme in the duodenum by the proteolytic action of trypsin, not pepsin. The activation of chymotrypsinogen does not involve the incorporation of its substrates (proteins) into micelles, the presence of bile acid, nor a specific lipase, all of which are requirements for the digestion of dietary fats.

9. The answer is C. (*Chapter 9 III A 1 c*) The figure that accompanies the question represents the structure of a peptide bond. Bond 1 links the α-amino group of amino acid 1 to the α-carbon. Bond 2 links the α-carboxyl carbon of amino acid 1 to the α-carbon. Bond 3 links the α-carboxyl carbon of amino acid 1 to the α-amino nitrogen of amino acid 2. This bond has a partial double-bond character, which prevents rotation about the bond axis. Bond 4 links the nitrogen of the α-amino group of amino acid 2 to its α-carbon. Bond 5 links the α-carboxyl carbon to the α-carbon of amino acid 2.

10. The answer is E. (*Chapter 9 II C; Appendix II D 2*) Group 1 is an α-amino group with a pK_a of approximately 8–9. Groups 2 and 3 are carboxyl groups, which in peptides would have pK_a's of approximately 3. At a pH equal to the pK_a of the α-amino group, the two carboxyl groups would be negatively charged, and the α-amino group would bear a half positive charge. At this pH, the compound would move towards the anode in an electric field. At a pH midway between the pK_a's of groups 2 and 3, the carboxyls would have a negative charge of 1, and the α-amino group a positive charge of 1, so that the electrically neutral molecule would not move in an electric field.

11. The answer is D. (*Chapter 3 I C 1*) Transcription can be divided into four stages: binding of RNA polymerase to the DNA template at specific sites, initiation of polymerization, chain elongation, and chain termination/release. Promoters are specific regions contained in DNA that are recognized by the σ subunit (sigma factor) of RNA polymerase. The core enzyme, that is, the RNA polymerase minus the σ subunit, cannot recognize the promoter region. After RNA polymerase has bound to the mRNA, a complex is formed called the open–promoter complex within which localized melting of the DNA has occurred, starting about 10 bases from the upstream side of the Pribnow box, one of the binding sites for RNA polymerase on the DNA template. A protein, the rho factor, is involved in chain termination in some instances, but it is not involved in the initiation of transcription.

12. The answer is D. (*Chapter 14 I B 1*) α-Amylase, which exists in different isozyme forms, attacks α-1,4 linkages except for those that serve as branch points in hydrated (i.e., heated or cooked) starch and in glycogen; it does not attack α-1,6 linkages. Its action produces maltose [α-glucose (1,4) glucose], maltotriose [α-glucose (1,4) α-glucose (1,4) glucose], and limit dextrins (i.e., highly branched molecules composed of about eight glucose units joined by one or more α-1,6 bonds).

13. The answer is C. (*Chapter 17 II F 2 a*) Glucose-6-phosphatase is a gluconeogenic enzyme, occurring in the liver and kidney. Thus, muscle cannot synthesize glucose, and glycogenolysis in exercising muscle leads to an elevation of blood lactate. Indirectly, muscle contributes significantly to the level of blood glucose by providing the liver with an important gluconeogenic substance, lactate.

Muscle and liver contain glycogen phosphorylase, which is necessary for the depolymerization of glucose to form glucose-1-phosphate. These two tissues also contain phosphoglucomutase and phosphoglucoisomerase, which convert glucose-1-phosphate to glucose-6-phosphate and glucose-6-phosphate to fructose-6-phosphate, respectively.

Muscle, unlike liver, also lacks the glucokinase enzyme, which phosphorylates glucose to form glucose-6-phosphate. The phosphorylation of glucose in muscle is catalyzed by hexokinase, an enzyme with broad specificity and distribution in most tissues.

14. The answer is C. (*Chapter 14 VI B 1; Chapter 20 VII C 1*) Ketone bodies can be oxidized as metabolic fuel by all tissues with a functional tricarboxylic acid (TCA) cycle except for the liver. Thus, the brain, heart muscle, skeletal muscle, and renal cortex are all capable of using ketone bodies. The red blood cells cannot oxidize ketone bodies as they have no mitochondria and no TCA cycle.

15. The answer is C. (*Chapter 1 II A 2; IV A, B 1; Chapter 3 I B 2*) In eukaryotes, although DNA is confined to the nucleus, RNA is also found there. Both DNA and RNA may be double-stranded, although RNA may be only partially double-stranded. Being double-stranded, DNA contains equimolar amounts of G and C, whereas RNA does not have to contain equimolar amounts of G and C (and usually does not). RNA requires a DNA template for synthesis. However, the most clear-cut difference between DNA and RNA is the presence of a deoxyribosyl group in DNA and a ribosyl group in RNA.

16. The answer is B. (*Chapter 20 VII B 1, 4 e, C 3–5*) Of the three ketone bodies, β-hydroxybutyrate, acetoacetate, and acetone, only the first two are significantly metabolized as acetone is lost in expired air. The ratio of β-hydroxybutyrate to acetoacetate is determined by the ratio of reduced nicotinamide-adenine dinucleotide (NADH) to NAD^+, which is usually high in the liver during ketone body formation (high fatty acid oxidation) and which accounts for the predominance of β-hydroxybutyrate. Acetoacetyl coenzyme A (acetoacetyl CoA) is not a ketone body. The ketone bodies are formed in the liver during high rates of fatty acid oxidation, but the liver cannot use them as energy sources. β-hydroxybutyrate and acetoacetate enter the tricarboxylic acid cycle as acetyl CoA after activation by 3-keto acid CoA transferase. The CoA donor is succinyl CoA, and the acetoacetyl CoA formed is cleaved to two acetyl CoAs by thiolase.

17. The answer is A. (*Chapter 29 I A 5 b, B 5*) The committed and rate-limiting step of de novo purine nucleotide synthesis is catalyzed by 5′-phosphoribosyl-1-pyrophosphate (PRPP) amidotransferase. It is inhibited, allosterically, by adenosine, guanosine, and inosine monophosphates (AMP, GMP, and IMP). The enzyme has two allosteric sites, one for IMP and GMP and one for AMP. If both sites are occupied, there is synergistic inhibition. PRPP-amidotransferase displays sigmoidal kinetics towards PRPP, whereas the kinetics for glutamine are hyperbolic. Thus, changes in the concentration of PRPP have a greater effect on the rate of purine biosynthesis than do equivalent changes in glutamine concentration. The condensation of IMP and aspartate to form adenylosuccinate requires GTP as a cofactor and is inhibited by AMP. The oxidation of IMP to xanthine 5′-monophosphate requires oxidized nicotinamide-adenine dinucleotide as electron acceptor and is inhibited by IMP.

18. The answer is C. (*Chapter 19 I B 1; II B 6 a; Chapter 20 III B 2 c*) An inhibitor of glucose 6-phosphate dehydrogenase (G6PD) in the liver prevents the generation of reduced nicotinamide-adenine dinucleotide phosphate (NADPH) by glucose 6-phosphate (G6P) and 6-phosphogluconate dehydrogenase in the pentose phosphate pathway. The NADPH formed in these reactions is an important source of reducing equivalents for reductive syntheses, particularly in the biosynthesis of fatty acids. The 6-phosphogluconate dehydrogenase does not function in the absence of an active G6PD as the formation of its substrate depends on the activity of G6PD. G6P does not accumulate if the pentose phosphate pathway is inhibited as there are other pathways of metabolism for G6P in the liver, such as conversion to glucose, glycolysis to pyruvate, or storage as glycogen. The pentose phosphate pathway does not generate high-energy phosphate bonds, and its inhibition does not impair conservation of energy or synthesis of adenosine triphosphate. Glucose is metabolized to glycogen or metabolized by the glycolytic pathway in the absence of G6PD. Ribose 5-phosphate, in the absence of G6PD, can be synthesized from triose phosphates via transaldolase and transketolase activities or by the uronic acid pathway.

19. The answer is E. (*Chapter 27 I C 1*) In liver mitochondria, carbamoyl phosphate is formed from ammonia, carbon dioxide, and adenosine triphosphate. The carbamoyl group is then transferred to ornithine by ornithine transcarbamoylase. The product is citrulline, which leaves the mitochondria and combines with aspartate in the cytosol to form argininosuccinate. The latter is cleaved by argininosuccinate lyase to form arginine and fumarate. Arginine is then cleaved by arginase to yield urea and ornithine. Thus, of the choices given, only fumarate would be a product of this series of reactions.

20. The answer is C. [*Chapter 27 I C 1 b, f; II C 1 a (2)*] Ornithine is a basic α-amino acid with a $H_3N^+ —(CH_2)_3 —$ side chain, which is not found in proteins. It can be formed from either arginine or glutamic acid. In the former case, arginase cleaves arginine to urea and ornithine in the cytoplasm of the liver with ornithine entering the mitochondria to continue the urea cycle pathway. In the latter case, glutamate can be converted to ornithine via glutamate semialdehyde with the α-amino group of ornithine being donated by a transamination from glutamate.

21. The answer is E. (*Chapter 26 III B 2; IV; Table 26-2*) The 100 g of protein consumed by the individual described in the question provided 16 g of nitrogen (see Table 26-2 for this relationship). The total nitrogen loss was $13.5 + 2 + 0.5 = 16$ g; thus, the individual is in nitrogen balance with intake and losses being equal. Because this individual is in nitrogen balance, this individual is not a growing child, a woman in the late stages of pregnancy, or someone who is convalescing, because in all three cases, the individual would be in positive nitrogen balance with nitrogen losses being less than intake. Also, the diet could not be consistently deficient in lysine, because this is an essential amino acid and general protein synthesis would be greatly reduced. Under these conditions, the individual would be in negative nitrogen balance with nitrogen losses exceeding intake.

22. The answer is D. (*Chapter 20 VI A 1 a, 2, 3; Figure 20-4*) Prior to oxidation, long-chain fatty acids are activated by an endoplasmic reticulum acyl coenzyme A (acyl CoA) synthetase (thiokinase), which uses adenosine triphosphate (ATP) and CoA. The ATP is split to adenosine monophosphate and inorganic pyrophosphate. Transport of the long-chain fatty acyl CoAs across the mitochondrial inner membrane involves the transfer of the acyl group from CoA to carnitine at the outer side of the mitochondrial inner membrane and the transfer of the acyl group at the inner side to CoA drawn from the matrix pool. Dehydrogenation of the acyl group to form the enoyl compound is the first step of a quartet of reactions that are repeated until the fatty acid has been degraded, two carbons at a time, to acetyl CoA. This acyl CoA dehydrogenase step requires flavin-adenine dinucleotide (FAD) as the electron acceptor with the formation of reduced FAD ($FADH_2$). The second step is the conversion of the enoyl form to the β-hydroxyacyl CoA by enoyl CoA hydratase; this step does not involve an alteration of a coenzyme or prosthetic group. The third step is the oxidation of the β-hydroxyacyl CoA to the β-ketoacyl form by the nicotinamide-adenine dinucleotide (NAD^+)–linked β-hydroxyacyl CoA dehydrogenase with the formation of the reduced cofactor NADH. Lastly, the β-ketoacyl CoA is split by thiolase, which uses CoA to split the two-carbon acetyl CoA from the remainder of the fatty acyl moiety of the β-ketoacyl CoA.

23. The answer is D. [*Chapter 10 III B 5 a; Chapter 16 III A 3, E 1; Chapter 28 II C 4 c (3); Chapter 29 VI B 6 b; Chapter 31 II D 1; Figure 31-1*] Cobalt is required for the formation of cobalamin, a vitamin for humans and other animals, which is synthesized by plants. Copper is a required element for cytochrome oxidase, the terminal electron transport complex in the mitochondrial electron transport chain. This complex interacts directly with molecular oxygen. Copper is also an essential part of lysyl oxidase, an enzyme involved in the maturation of collagen. Iron is required for a number of enzymatic activities. It is present in the form of heme in the cytochrome electron transporters of the mitochondrial electron transport chain and in the oxygen-binding proteins (i.e., myoglobin and hemoglobin). It is also found in the electron transport entities known as the iron–sulphur–protein centers. Molybdenum is found in xanthine oxidase, an enzyme involved in the oxidation of purine degradation products to uric acid, the disposal form of purine catabolism. This requirement for molybdenum by xanthine oxidase accounts for the human requirement for molybdenum. Mercury is not a required element in humans.

24. The answer is E. (*Chapter 24 I B; II A*) Lipid micelles in the intestine are formed from fat globules (all kinds of dietary fat) by the action of bile salts and phosphatidylcholine, which act as detergents. The lipid micelles have a much larger surface area than the fat globules and act as substrates for the lipid hydrolyzing enzymes. The products of hydrolysis pass from the micelles into the columnar epithelial cells in the mucosa of the intestine. The micelles are not absorbed as such, nor are they secreted into the circulation.

25. The answer is A. (*Chapter 9 II C; III B; Appendix II D 1*) The dipeptide glycylarginine has three dissociable protons. In its fully protonated state, it has a positive charge on the N-terminal amino group (glycine), a positive charge on the side chain of arginine (the guanidinium group), and a zero charge on the protonated C-terminal carboxyl (arginine). The ψ-carboxyl of glycine and the α-amino of arginine are lost in the formation of the dipeptide. Titration with alkali gives a titration curve with three points of inflexion. The first is due to the dissociation of the carboxyl proton of arginine at about pH 3; the second is due to the dissociation of a proton from the α-amino group of glycine; and the third is due to the dissociation of a proton from the guanidinium group on the side chain of arginine, which will occur at about pH 12.

26. The answer is B. (*Chapter 16 VIII C*) The phosphate:oxygen (P:O) ratio is a measure of the degree of coupling of adenosine triphosphate (ATP) formation to electron transport in mitochondria. It is also known as the adenosine diphosphate:oxygen (ADP:O) ratio. It is usually measured as the number of moles of ADP (or inorganic phosphate) that disappear per gram atom of oxygen used. The moles of ATP formed per atom equivalent of oxygen give the same measure. The calculation of the P:O ratio does not employ moles as the unit of oxygen used, which eliminates choices A, D, and E.

27. The answer is B. (*Chapter 20 III B 3; V B 2; Chapter 22 III C 1; Chapter 23 II B 1; Chapter 24 I C 1 c*) Acetyl coenzyme A (acetyl CoA) carboxylase is the rate-limiting step in the biosynthesis of fatty acids by the cytoplasmic system. In this step, acetyl CoA is activated by carboxylation to form malonyl CoA in a reaction that requires carbon dioxide, biotin, and adenosine triphosphate. β-Hydroxy-β-methylglutaryl CoA synthetase (HMG CoA synthetase) is the committed step, and the rate-limiting step, in the synthesis of cholesterol. In this reaction, mevalonic acid is formed from acetyl CoA via HMG CoA. Hormone-sensitive lipase is involved in the hydrolysis of phospholipids. This catabolic enzyme requires activation by phosphorylation by the cyclic adenosine monophosphate–dependent protein kinase. The hormone-sensitive designation is due to the activation of the protein kinase in this system by catecholamines, either from neural activity or from hormonal catecholamine released from the adrenal medulla. Phospholipase A_2, also a catabolic enzyme, is concerned with the hydrolysis of fatty acids from phospholipids. Specifically, it cleaves a fatty acid from the sn-2 position of the glycerol moiety of a phospholipid. Cyclooxygenase is not involved with pathways that handle metabolic fuels, but it is the rate-limiting step in the synthesis of prostaglandins from arachidonic acid.

28. The answer is C. [*Chapter 27 I A 2 a, b, 3 b, B 3 b (1), C 1 f*] Glutamate dehydrogenase is the major enzyme involved in the production of ammonia in humans; the rate of the reaction is controlled by the relative levels of glutamate, α-ketoglutarate, and ammonia. L-Amino acid oxidase activity is present in mammalian liver and kidney, but their physiologic significance is not clear, as most of the activity by tissue homogenates toward L-amino acids is due to the combined action of transaminases and gluta-mate dehydrogenase. Arginase cleaves arginine to urea and ornithine; glutamine synthetase catalyzes the formation of glutamine from glutamate and ammonia in a reaction that requires adenosine triphos-phate; and serine dehydratase oxidizes serine to pyruvate and ammonia.

29. The answer is C. (*Chapter 4 VIII B 2 a, c, e, 5 e, 7*) Initiation of protein synthesis in eukaryotes re-quires free 40S and 60S ribosomal subunit particles, mRNA with the start codon AUG near the 5′ end, guanosine triphosphate, and, in contrast to prokaryotic initiation, adenosine triphosphate as well. Pro-tein factors, called eukaryotic initiation factors (i.e., eIF-1, eIF-2, and eIF-3), and at least seven other factors are also required. Again, in contrast to prokaryotes, eukaryotic initiation requires methionine-tRNA, not formylmethionine-tRNA; that is, the methionine is not formylated.

30. The answer is A. (*Chapter 23 IV B 3 c; Chapter 31 III B 3*) The active form of vitamin D is 1,25-dihy-droxycholecalciferol. It is formed in the kidney from 25-hydroxycholecalciferol, which is itself formed in the liver from cholecalciferol (vitamin D_3, or ergocalciferol). In certain renal disorders, the hydroxy-lase forming 1,25-dihydroxycholecalciferol is ineffective, and substitution therapy with 1,25-dihydroxy-cholecalciferol is indicated. High dietary calcium would not be effective as the transport of calcium from the intestine would be reduced because of the lack of 1,25-dihydroxycholecalciferol.

31. The answer is C. (*Chapter 9 Table 9-1; Chapter 27 II C 2 c; Chapter 28 I D–F*) Tryptophan is a monocarboxylic–monoamino acid with an aromatic ring, which is not charged. It is used in very limited amounts in protein structures. It is also a precursor of the neurotransmitter serotonin and the pineal hormone melatonin. Some of its carbon can contribute to the net formation of glucose via glu-coneogenesis as a fumarate moiety is split off from the tryptophan molecules during its catabolism. Tryptophan is also ketogenic as other carbons from this amino acid give rise to crotonyl coenzyme A (crotonyl CoA), which is metabolized to acetyl CoA by the β-oxidation pathway of fatty acids.

32. The answer is A. (*Chapter 21 II D 2; Chapter 28 I B*) Two enzymes are active in the degradation of bioactive amines, such as the catecholamines, norepinephrine and epinephrine. One is monoamine oxidase (MAO), which oxidizes the carbon bearing an amino group to yield an aldehyde, and the other is catechol-O-methyl transferase (COMT), which is active in the methylation of a ring hydroxyl of the catecholamines. After both MAO and COMT have acted upon the catecholamine, there is further ca-tabolism, resulting in the formation of a glucuronide by transfer of the glucuronyl moiety from uridine diphosphate glucuronide or the formation of a sulfate by transfer of the sulfate moiety from 3′-phos-phoadenosine-5′-phosphosulfate. Conjugation to asparagine does not occur.

33. The answer is C. (*Chapter 25 III A*) The major lipid components of animal cell membranes are phospholipids, glycolipids (glycosphingolipids), and cholesterol. The phospholipids are asymmetrically distributed in plasma membranes with choline phospholipids found in the outer half of the membrane and aminophospholipids in the inner, or cytoplasmic, half of the membrane.

34. The answer is D. (*Chapter 24 I C*) Triacylglycerols, a major dietary lipid, are incorporated into mixed micelles by the action of bile salts and phosphatidylcholine. These micelles form the substrates for hydrolyzing enzymes. The triacylglycerols are hydrolyzed by pancreatic lipase, which cleaves them to two free fatty acids and 2-monoacylglycerol.

35. The answer is B. (*Chapter 25 II A 1; III A 3–5, B 3 b, 4 b*) Lipids, particularly phospholipids, are asymmetrically arranged in membranes with choline phospholipids found in the outer half of the mem-brane and aminophospholipids in the inner, or cytoplasmic, half. Biologic membranes are not readily permeable to inorganic ions, a fact taken to indicate the presence of lipids in membranes. Membrane lipids undergo lateral diffusion, that is, side-to-side or translational movement and flexing, but they do not readily "flip-flop," that is, pass from one side of the bilayer to the other. The lateral diffusion of pro-teins in membranes occurs, although in many cases proteins are restricted in their lateral mobility in order for them to fulfill specific cellular functions.

36. The answer is B. (*Chapter 20 III E 1 b; Chapter 24 Table 24-1; Chapter 30 IV C 1*) In diabetic in-dividuals, the lack of insulin leads to an excessive mobilization of fat reserves far in excess of energy re-quirements. Excess fatty acids may be converted in the liver to very low-density lipoprotein (VLDL) par-ticles. The elevated levels of circulating ketone bodies do not inhibit lipoprotein lipase nor does the decreased utilization of glucose lead to an induction of acetyl coenzyme A carboxylase. In insulin defi-ciency, all protein synthesis is reduced, not increased.

37. The answer is C. (*Chapter 11 VII H; Chapter 15 III C 5*) Isocitrate dehydrogenase is a nicotinamide-adenine dinucleotide (NAD^+)–linked enzyme of the tricarboxylic acid cycle that catalyzes the conver-sion of isocitrate to α-ketoglutarate and carbon dioxide. It is an allosteric regulatory enzyme, which is inhibited by adenosine triphosphate or reduced NAD^+ (NADH), both indicators of a high-energy charge, and which is stimulated by adenosine diphosphate, an indicator of a low-energy charge. The action of NADH (or ATP) is an example of heterotropic inhibition with ATP or NADH acting as small molecule modulators at sites other than the active site of the enzyme.

38. The answer is B. (*Chapter 20 III B 3; VI A 2; Chapter 22 III B 3 a; Chapter 23 II B 1 b*) The formation of malonyl coenzyme A (malonyl CoA) by the carboxylation of acetyl CoA is the rate-limiting step in the cytosolic fatty acid synthesizing system. The rate-limiting step of cholesterol biosynthesis is the reduction of β-hydroxy-β-methylglutaryl (HMG) CoA to mevalonate by HMG CoA reductase. The release of arachidonic acid from membrane phospholipids is the rate-limiting step in the biosynthesis of the eicosanoids (i.e., prostaglandins, thromboxanes, and leukotrienes). In fatty acid β-oxidation, long-chain fatty acids are transported across the mitochondrial inner membrane by carnitine palmitoyl transferase I into the matrix space where the oxidation enzymes are located. This is the slow step of β-oxidation of fatty acids. The activation of palmitate by CoASH and adenosine triphosphate to form palmitoyl CoA is not a rate-limiting step in a pathway.

39. The answer is B. (*Chapter 27 I C 1 a, b*) The nitrogens in urea, the disposal form of ammonia, come from ammonia and aspartate. Ammonia reacts with carbon dioxide and adenosine triphosphate to form carbamoyl phosphate in a reaction catalyzed by carbamoyl phosphate synthetase (ammonia). This enzyme requires N-acetyl glutamate as a positive allosteric effector. Without this reaction, urea would not be formed, and ammonia levels would be high. The next step in the urea cycle is the reaction of carbamoyl phosphate with ornithine to form citrulline. In the absence of carbamoyl phosphate, ornithine levels are high, citrulline is undetectable, and neither argininosuccinate nor arginine is formed.

40. The answer is D. (*Chapter 29 III B 2, C 3; V A, B 3, C*) The reaction of adenine with $5'$-phosphoribosyl-1-pyrophosphate (PRPP) forms the nucleotide adenosine monophosphate (AMP) and is called the salvage pathway for adenine. It is catalyzed by adenine phosphoribosyl transferase. The enzyme myokinase (or AMP kinase) catalyzes the readily reversible reaction AMP + adenosine triphosphate (ATP) $\rightleftharpoons$ 2 adenosine diphosphate (ADP). It is not a nucleoside diphosphate kinase. The reaction of deoxythymidine and deoxyribose 1-phosphate is an example of the first step in the two-step salvage reaction for a pyrimidine base, catalyzed by nucleoside phosphorylase. The reaction of uridine diphosphate with ATP to form uridine triphosphate and ADP is an example of a nucleoside diphosphate kinase catalyzed reaction. The reaction of uridine and ATP to form uridine monophosphate is a kinase reaction for the salvage of a nucleoside. Such kinases exist for adenosine, uridine, cytidine, deoxycytidine, and thymine, but not for guanosine or inosine.

41. The answer is E. (*Chapter 4 IV A, B; VI A; VII A 1, 2*) An mRNA molecule specifies a specific peptide or, in the case of the polycistronic message of prokaryotes, a group of peptides. In general, the translation of the message of mRNA into the amino acid sequence of a polypeptide is achieved by the binding of the mRNA to ribosomes with the amino acids being delivered to the ribosome attached covalently to tRNAs (aminoacyl-tRNAs). This process in prokaryotes is initiated by the formation of a 30S initiation complex, consisting of the 30S ribosomal subunit, formylmethionine-tRNA, initiation factors (i.e., IF-1, IF-2, and IF-3), guanosine triphosphate, and mRNA. Lysosomes contain degradative enzymes, including proteases, which hydrolyze peptide bonds. These enzymes are not involved in protein synthesis.

42. The answer is C. (*Chapter 20 VII C 5*) The ketone body, acetoacetate, has to be activated in order to be used as an energy substrate. Activation occurs by the formation of the coenzyme A (CoA) thioester of acetoacetate, catalyzed by 3-keto acid CoA transferase. The CoA donor is succinyl CoA. This use of succinyl CoA bypasses the formation of 1 mol of guanosine triphosphate, which is formed when 1 mol of succinyl CoA is converted to succinate in the tricarboxylic acid (TCA) cycle. Thus, activation of acetoacetate costs 1 mol high-energy phosphate bond. The acetoacetyl CoA is cleaved by thiolase to 2 mol of acetyl CoA, which can be oxidized in the TCA cycle. Each mole of acetyl CoA yields 12 mol of adenosine triphosphate (ATP) via the TCA cycle, electron transport, and oxidative phosphorylation. Thus, the yield of high-energy phosphate bonds from acetoacetate in the brain (or any tissue that can metabolize acetoacetate) is $-1 + 24 = 23$ mol of ATP per mole of acetoacetate.

43. The answer is B. (*Chapter 21 II D 2 a; Chapter 27 I C 1 a; III B 4 b, C 1; Chapter 29 I A, C 2 a*) Carbamoyl phosphate is synthesized in two different metabolic pathways. In a mitochondrial reaction, its synthesis is catalyzed by carbamoyl phosphate synthetase I (ammonia) from carbon dioxide, ammonia, and 2 molecules of adenosine triphosphate (ATP). This is the beginning of the urea cycle for the disposal of ammonia. In a cytosolic reaction, carbamoyl phosphate synthetase II (glutamine) catalyzes the reaction of carbon dioxide, glutamine, and 2 molecules of ATP to form carbamoyl phosphate in the first and regulated step of pyrimidine nucleotide biosynthesis. The formation of $5'$-phosphoribosyl-1-pyrophosphate (PRPP), an intermediate of major significance in nucleotide metabolism, is catalyzed by

PRPP synthetase from ribose 5-phosphate and ATP with a pyrophosphate group from ATP ending up in PRPP. "Active sulfur," or 3'-phosphoadenosine-5'-phosphosulfate, is a sulfate group donor, which is formed from inorganic sulfate and ATP. S-adenosylmethionine, a methyl group donor, is formed from methionine and ATP with the adenosyl group of ATP being transferred to the sulfur of methionine with the triphosphate moiety as the leaving group. Formyl tetrahydrofolate is formed by oxidation of methylene tetrahydrofolate, a reaction in which ATP is not involved.

44. The answer is D. (*Chapter 19 III A*) Fructose is phosphorylated by fructokinase to fructose 1-phosphate in the liver, kidney, and the small intestine. Hexokinase does not phosphorylate fructose. Fructose 1-phosphate is metabolized by cleavage to dihydroxyacetone phosphate and D-glyceraldehyde catalyzed by phosphofructoaldolase (aldolase B), an isozyme of the aldolase in the glycolytic pathway. The dihydroxyacetone phosphate can equilibrate with glyceraldehyde 3-phosphate (G3P) in a reaction catalyzed by triose phosphate isomerase with the G3P being converted to 1,3-diphosphoglycerate (1,3-DPG) by the nicotinamide-adenine dinucleotide–linked glyceraldehyde 3-phosphate dehydrogenase. The D-glyceraldehyde formed at the aldolase B step can be metabolized along several pathways. One of these may be phosphorylated to G3P at the expense of adenosine triphosphate with subsequent conversion to 1,3-DPG. 6-Phosphogluconate is an intermediate in the pentose phosphate pathway, which arises from glucose 6-phosphate (G6P) and not from fructose 1 phosphate. Fructose 1-phosphate derived from fructose does not give rise to G6P and, therefore, does not involve the activity of phosphoglucomutase.

45. The answer is E. (*Chapter 23 II B 1 c*) The rate-limiting and regulated step of cholesterol biosynthesis is the formation of mevalonate from β-hydroxy-β-methylglutaryl coenzyme A (HMG CoA) catalyzed by HMG CoA reductase. This enzyme is inhibited by both dietary cholesterol as well as by endogenously synthesized cholesterol. A vegetarian diet, a "prudent" diet (i.e., one low in cholesterol), and the administration of bile acid sequestering resin all result in a reduced intake of cholesterol, which will not reduce HMG CoA reductase activity. Familial hypercholesterolemia is a result of a deficiency of low-density lipoprotein receptors.

46. The answer is B. (*Chapter 21 II E; Figure 21-1*) The degradation (catabolism) of sphingolipids occurs by the stepwise hydrolysis of the component moieties, starting at the hydrophilic terminal of the molecule, and is catalyzed by hydrolytic enzymes located in the lysosomes. A number of inherited defects, involving the lysosomal enzymes that degrade sphingolipids, are known, the most common of which are Tay-Sachs disease (i.e., deficiency of lysosomal hexosaminidase) and G_{M1} gangliosidosis (i.e., deficiency of β-galactosidase). The lysosomal enzymes that degrade sphingolipids (gangliosides) show specificity toward a type of linkage rather than toward a specific compound.

47. The answer is D. [*Chapter 14 III C 3 a, f; I 5 c; Chapter 15 II C 3 b, d; Chapter 18 III A 2; Chapter 20 III E 1 a (3)*] All of the enzymes listed in the question (i.e., acetyl coenzyme A carboxylase, glycogen synthetase, phosphofructokinase, and pyruvate kinase) are phosphorylated by a cyclic adenosine monophosphate–dependent protein kinase except pyruvate dehydrogenase. The latter is phosphorylated and inactivated by a protein kinase that is itself stimulated by acetyl coenzyme A (acetyl CoA) and reduced nicotinamide-adenine dinucleotide (NADH) and inhibited by free CoA (CoASH), NAD^+, or pyruvate. In all of the other enzymes listed, the phosphorylation of the enzyme protein results in inactivation of the enzyme activity.

48. The answer is B. (*Chapter 4 III C; Table 4-1*) The mutant α-globin chain is longer than the normal chain, which occurs when a mutation has eliminated a stop codon. In the normal chain, the codon following that specifying the 141st residue is a stop codon (UAA), so that the C-terminal residue of this 141-residue chain is histidine. The 142nd residue in the mutated chain is leucine, which arises from the stop codon by the insertion of the base C (at the 5' side of the stop codon) to give CUA, which codes for leucine, followed by AGC (coding for serine) and UAG, a stop codon. The mutated chain, thus, becomes 143 residues in length, or two residues longer than the normal chain with a C-terminal residue of serine. The deletion of the 5' base of the stop codon in the normal chain does not give a mutant structure with 142 residues, but one with at least 145 residues.

49. The answer is C. (*Chapter 29 II A 1*) Thioredoxin is a small protein, which is oxidized during the reduction of the ribose 2'-hydroxyl group of a nucleoside diphosphate to a deoxyribonucleotide in a reaction catalyzed by ribonucleotide reductase. The oxidized thioredoxin is reduced by a flavoprotein, using electrons transferred from reduced nicotinamide-adenine dinucleotide phosphate in order to maintain the ribonucleotide reductase catalyzed reaction. Thioredoxin is not involved in the conversion of xanthine to hypoxanthine, in the salvage pathways for pyrimidine bases, in the conversion of deoxyuridine monophosphate to deoxythymidine monophosphate, or in providing reducing equivalents for mixed-function oxidases.

50. The answer is D. (*Chapter 17 IV B; Chapter 32 III E*) During long-term fasting the insulin:glucagon ratio is low, and mobilization of metabolic fuels predominates. The cyclic adenosine monophosphate levels in the liver are high due to glucagon dominance, and the adenosine triphosphate levels in the liver are always adequate. Free fatty acids from mobilized triacylglycerol stores in adipose tissue are available to the liver and other tissues. However, the main controlling factor for the rate of gluconeogenesis is substrate availability. The substrates for gluconeogenesis in prolonged fasting are amino acids, although after 3 to 4 days of fasting, the rate of supply of amino acids to the liver from peripheral tissues declines. Alanine (and glutamine) is a major form by which amino acids arrive at the liver from muscle and other tissues.

51. The answer is A. (*Chapter 14 III C 3, F, G; V B; Chapter 15 III A, D; Chapter 18 II C 1*) A sudden fall in the concentration of adenosine triphosphate (ATP) in muscle cells initiates signals to speed up glycolysis and glycogenolysis to provide fuel for the tricarboxylic acid (TCA) cycle. Thus, the concentration of fructose 1,6 diphosphate increases as the regulatory signals turn on phosphofructokinase. Glucose 1-phosphate increases in concentration as glycogen is broken down by glycogen phosphorylase. The mitochondrial citrate level falls as it is drawn on by the TCA cycle. α-Ketoglutarate increases in concentration as the rate of the TCA cycle increases, and anaplerotic reactions increase the level of the TCA cycle intermediates. 3-Phosphoglycerate increases transiently as the rate of the glycolytic pathway increases.

52. The answer is D. (*Chapter 11 V A 1; VII H 2 b*) Competitive inhibition is the competition between the substrate of an enzyme and another substance that can reversibly bind to the active site of the enzyme. The competitive inhibitor does not become covalently bonded to the enzyme, but it interferes with the binding of the substrate to the active site of the enzyme. It increases the Michaelis constant (K_m) of the enzyme for its substrate, but not the V_{max}, as increasing the substrate concentration overcomes the inhibition by a competitive inhibitor.

53. The answer is B. (*Chapter 27 I A 1 a, b*) Transamination involves the transfer of an amino group from an amino acid to an α-keto acid to form a new amino acid and a new α-keto acid. Pyridoxal phosphate is a coenzyme for all aminotransferases. The equilibrium constants of the catalyzed reactions are between 1 and 10, and they are regulated entirely by stoichiometric control. Ammonia is not taken up, nor produced, by aminotransferases as the amino group is transferred from one molecule to another. Most, but not all, amino acids undergo transamination before oxidative degradation of the carbon skeletons occurs.

54. The answer is D. [*Chapter 17 II C 2 a (1)*] In mitochondria, oxaloacetate, which cannot cross the mitochondrial inner membrane, can be converted to malate by mitochondrial malate dehydrogenase during gluconeogenesis when the level of reduced nicotinamide-adenine dinucleotide (NADH) is high. Malate can then pass through the inner membrane of the mitochondria and, in the cytoplasm, give rise to oxaloacetate and NADH by the action of cytoplasmic malate dehydrogenase. Thus, both oxaloacetate carbon and reducing equivalents are transferred from the mitochondria to the cytosol. In the mitochondria in humans, phosphoenolpyruvate can arise from oxaloacetate, aspartate can be formed from oxaloacetate by transamination with α-ketoglutarate, and both can pass into the cytosol. However, reducing equivalents are not formed in these reactions. The glycerol 3-phosphate shuttle transfers reducing equivalents from the cytosol to the mitochondria.

55. The answer is E (all). (*Chapter 29 VI B 6; IX B 3*) In the final steps of purine degradation in humans, xanthine oxidase oxidizes hypoxanthine to xanthine, and then xanthine to uric acid, which is excreted as such in the urine. Inhibition of xanthine oxidase by allopurinol reduces uric acid output in the urine with a concomitant increase in blood levels of xanthine and hypoxanthine. Xanthine is somewhat more soluble than uric acid and hypoxanthine considerably so. Thus, neither of these compounds are as likely to precipitate in the joints in gouty patients as uric acid. Some of the allopurinol is converted to allopurinol ribonucleotide by the purine salvage enzyme hypoxanthine-guanine phosphoribosyl transferase. This nucleotide acts as a feedback inhibitor of phosphoribosyl pyrophosphate amidotransferase, the rate-limiting step of purine biosynthesis, thereby reducing the rate of uric acid formation.

56. The answer is B (1, 3). (*Chapter 18 II A 1, 2, C*) The immediate precursor of glycogen is uridine diphosphate glucose (UDP-glucose), which is formed from glucose 1-phosphate (G1P) and uridine triphosphate in a reaction catalyzed by UDP-glucose pyrophosphorylase. Hydrolysis of the pyrophosphate formed by cellular pyrophosphatases renders the reaction essentially irreversible. G1P is derived from glucose 6-phosphate by the action of phosphoglucomutase. Glycogen phosphorylase is the rate-limiting step in the breakdown of glycogen, and although this enzyme can synthesize glycogen in vitro, under cellular conditions, it does not. Cyclic adenosine monophosphate is involved in the inactivation of glycogen synthetase, not in the synthesis of glycogen.

57. The answer is E (all). (*Chapter 4 X A 2, B 1 a, c, d*) All four of the compounds listed in the question (i.e., erythromycin, chloramphenicol, streptomycin, and puromycin) are inhibitors of protein synthesis. Puromycin inhibits protein synthesis in both prokaryotes and eukaryotes as an analogue of the aminoacyl–adenosine moiety of an aminoacyl-tRNA. It binds to the A site on the ribosome and blocks the entry of aminoacyl-tRNAs. Erythromycin, chloramphenicol, and streptomycin inhibit prokaryotic, but not eukaryotic, protein synthesis. Erythromycin inhibits translocation by binding to 50S ribosomal subunits. Chloramphenicol inhibits the peptidyl transferase activity of 50S ribosomal subunits. Streptomycin inhibits the initiation of protein synthesis by preventing the binding of formylmethionine-$tRNA_f$, to the P site of the initiation complex. It also causes a misreading of the mRNA sequence.

58. The answer is C (2, 4). (*Chapter 5 II A*) In the excision repair of DNA, involving the removal of thymine dimers, or repair of cytosine deamination, specific enzymes are involved in the initiation of the repair process. However, in both cases of DNA repair, DNA polymerase I fills in gaps, using the intact complementary strand as a template. The gaps between the newly synthesized segment and the main chain are sealed by DNA ligase. An RNA primer is not required by DNA polymerase I nor is DNA polymerase III involved in DNA repair.

59. The answer is B (1, 3). (*Chapter 3 II B 2, 3 b; Chapter 4 IV B*) Virtually all tRNA shares similarities in base composition and general primary and secondary structure. tRNA has a basic two-dimensional cloverleaf construction, consisting of three (sometimes four) arms with three loops. The last three nucleotides at the 3' terminus are always CAA, and the aminoacyl-tRNA synthetases attach the amino acid to the 2'- or 3'-hydroxyl of the terminal A. A specific amino acid is not recognized by tRNA, but solely by the aminoacyl-tRNA synthetase. One arm of the tRNA, called the anticodon loop, contains the triplet of bases, the anticodon, which has complementarity to the triplet codon on mRNA.

60. The answer is C (2, 4). (*Chapter 28 II D 2*) Red blood cells are usually destroyed in the spleen, which is the major site of heme catabolism, although some destruction occurs in Kupffer's cells of the liver. Heme removed from its carrier globin is rapidly oxidized to hemin (Fe^{3+}), and the first step in hemin catabolism is its reduction back to heme (Fe^{2+}) by an enzyme system that uses reduced nicotinamide-adenine dinucleotide phosphate (NADPH) as the electron donor. The first degradative step is rupture of the methylidyne group of heme between the pyrrole rings carrying vinyl groups by a mixed-function oxidase in the endoplasmic reticulum with the bridge carbon being lost as carbon monoxide. The iron dissociates, yielding the free tetrapyrrole, biliverdin, which has a greenish color. Biliverdin is reduced (with NADPH as the reductant) to bilirubin, which is reddish-brown in color.

61. The answer is C (2, 4). [*Chapter 4 VII A 6 b (2); Chapter 7 II B 3 a, b (2)*] The prokaryote tryptophan (*trp*) operon is an example of a negatively controlled repressible operon, which is regulated in two ways: by an operator and by an attenuator. Operator control is achieved by a protein, the *trp* repressor, coded for by the *trpR* gene. The repressor forms a complex with tryptophan, which acts as a corepressor and which binds to the operator site, overlapping the promoter site in the process. Thus, the complex prevents the binding of RNA polymerase at the promoter site. Low levels of tryptophan prevent the repressor complex from forming and allow transcription to occur. Cyclic adenosine monophosphate and the catabolite activator protein are not involved in this control. The polycistronic mRNA bearing the tryptophan structural genes has a Shine-Dalgarno sequence on the 5' side of the initiating codon of the cistron that allows the 30S ribosomal subunit to bind to it for the initiation of protein synthesis. As with most prokaryotic cistrons, transcription does not have to be completed before translation of the *trp* operon genes commences. Processing of the mRNA is not required.

62. The answer is B (1, 3). (*Chapter 25 III B 2–4*) Membrane proteins may be found on the outer or inner surfaces of membranes or embedded in the lipid bilayer (i.e., in the hydrophobic phase). The so-called peripheral proteins can be extracted from the membrane, using relatively gentle procedures, such as a change in the salt concentration or pH of the medium. These proteins are located on the inner, or cytoplasmic face, of the membrane. The integral proteins can only be extracted from cell membranes after treatment with detergents or organic solvents. They are embedded in the hydrophobic phase of the membrane. It is very unlikely that membrane proteins "flip-flop" as they appear from labeling experiments to maintain a constant orientation in the membrane. Membrane proteins are degraded and resynthesized, as are all proteins, although the rates of turnover may vary very widely. The membrane proteins are distributed in the membrane in accord with their function and are not necessarily symmetrically arranged.

63. The answer is E (all). (*Chapter 10 III A, B*) A unique distribution of amino acid residues distinguishes collagen in which about 33% of the total residues are glycine (i.e., a Gly at every third position), 10% proline, 10% hydroxyproline, and 1% hydroxylysine. The proline-rich chains of tropocollagen form a triple helical structure quite different from the α-helix in that it is more open, there is no intrachain peptide hydrogen bonding, and it is a left-handed (counterclockwise) helix. In addition to the formation of hydroxyproline and hydroxylysine by enzyme systems in the cisternae of the endoplasmic reticulum, the formation of a stable triplex helix requires glycosylation of specific hydroxylysyl residues.

64. The answer is B (1, 3). (*Chapter 1 VI D 1; Chapter 3 I C 1*) Transcription of RNA in *Escherichia coli* begins with RNA polymerase holoenzyme, that is, the complete enzyme containing the σ subunit, binding to specific sites on the gene. Two sites are involved: the Pribnow box, a sequence of bases in the DNA located 5–10 bases upstream from the first base that will be copied into RNA, and the "-35" sequence, which is upstream from the Pribnow box and thought to be the initial site of σ subunit binding. Prior to initiation of polymerization, the open–promoter complex is formed where there is localized melting of the DNA, starting about 10 bases from the upstream side of the Pribnow box. Prokaryotic DNA is not associated with histones, and although RNA polymerase does require a DNA template, it does not require a primer.

65. The answer is C (2, 4). [*Chapter 20 VI B; Chapter 27 II C 2 a (6)*] A 17:1 fatty acid has an odd number of carbon atoms and, thus, ends up in the β-oxidation system as the three-carbon compound propionyl coenzyme A (propionyl CoA). The amino acid valine on degradation also gives rise to propionyl CoA. The latter enters the tricarboxylic acid cycle after conversion to succinyl CoA. This process involves carboxylation of propionyl CoA to the S form of methylmalonyl CoA, which, after racemization to the R form, is converted to succinyl CoA by a vitamin B_{12}–dependent mutase. Three carbons of succinyl CoA can contribute to a net synthesis of glucose and are, therefore, glucogenic. Thus, both valine and the odd-carbon chain fatty acid are glucogenic, both require the vitamin B_{12}–dependent mutase for conversion to succinyl CoA, and both require bicarbonate ion (or carbon dioxide) for the carboxylation of propionyl CoA. The first step in the catabolism of valine is the transamination of valine to yield the 2-keto acid derivative. However, an aminotransferase is not involved in the metabolism of odd-chain length fatty acids.

66. The answer is A (1, 2, 3). [*Chapter 2 III A 1 a (1), (3) (a), (b), 3 b*] In prokaryotic DNA replication, at the replication fork, DNA polymerase I removes the RNA primer via nick translation (5′,3′-exonuclease activity) and replaces it with DNA by its polymerase and 3′,5′-exonuclease editing activities. DNA polymerase does not have pyrophosphatase activity; the polymerase reaction is driven to completion by hydrolysis of the pyrophosphate leaving group by cellular pyrophosphatases.

67. The answer is C (2, 4). (*Chapter 20 VI A 1, 2; Chapter 32 III D 5*) Long-chain fatty acids are transported across the inner mitochondrial membrane by carnitine acyl transferases. After activation of the fatty acids by formation of fatty acyl coenzyme As (fatty acyl CoAs) by an endoplasmic reticulum enzyme system, an enzyme on the outer surface of the inner mitochondrial membrane, carnitine palmitoyl transferase I, catalyzes the transfer of the acyl group from CoA to carnitine (β-hydroxy-γ-trimethylammonium butyrate), which is in the inner membrane. The fatty acyl group is translocated across the membrane to the inner surface, where the enzyme carnitine palmitoyl transferase II catalyzes the transfer of the acyl group to CoA drawn from the matrix CoA pool. This process does not require the concomitant expenditure of adenosine triphosphate. Carnitine concentrations are higher in heart muscle than in the brain because circulating free fatty acids are a major source of metabolic fuel for heart muscle, but they are not used for this purpose by the brain.

68. The answer is C (2, 4). (*Chapter 16 VIII A–C*) Mitochondria that oxidize a substrate in the presence of adequate levels of adenosine diphosphate (ADP), inorganic phosphate, and oxygen consume oxygen at a rate limited only by the rate at which the electron transport chain can transfer electrons. Such a respiratory rate is called rate III respiration and occurs if ADP is not limiting and oxygen uptake is obligatorily coupled to the phosphorylation of ADP to adenosine triphosphate. If the concentration of ADP becomes limiting, phosphorylation of ADP to ATP is reduced, and the level of oxygen utilization is reduced to a level called rate IV respiration. In uncoupled mitochondria, when the oxygen uptake is not obligatorily coupled to the phosphorylation of ADP, the rate of respiration increases markedly over the level of state III respiration. Thus, respiratory control is dependent upon the ATP:ADP ratio and not upon the ratio of nicotinamide-adenine dinucleotide (NAD^+) to reduced NAD (NADH) as the rate of formation of NADH is linked to the rate of oxidation but not to phosphorylation in uncoupled mitochondria.

69. The answer is B (1, 3). (*Chapter 3 II B 2 b; Chapter 9 V B 2 c; Chapter 10 I B 1 b, C 1 c; III A 3 a*) Human hemoglobin A_1 is a tetrameric protein that contains two α subunits and two β subunits. Both of these subunits closely resemble the monomeric myoglobin as far as tertiary structure is concerned. Myoglobin contains eight major α-helix segments in which the secondary structure is stabilized by intrachain hydrogen bonds (H bonds) between the peptide bonds. tRNAs have a basic two-dimensional cloverleaf construction; the single chain shows a pairing of bases along the arms of the cloverleaf. The pairing is effected by H bonds between base-pairs, which serves to stabilize the two-dimensional cloverleaf structure of tRNA. Tropocollagen, the fundamental unit of collagen structure, forms from three proline-rich peptides, which form a helical structure quite different from the α-helix in that it is more open, there is no intrachain peptide H-bonding, and it is a left-handed (counterclockwise) helix. Triacylglycerols are not polymers and, therefore, do not have intrachain H bonds.

70. The answer is A (1, 2, 3). [*Chapter 14 III E 5; IV B 3; Chapter 17 II C 2 a (1)*] The reduced nicotin-amide-adenine dinucleotide (NADH) produced by the glyceraldehyde 3-phosphate dehydrogenase step of glycolysis has to be reoxidized to NAD^+ for the pathway to continue functioning. Under anaer-obic conditions, the NADH is reoxidized by the reduction of pyruvate formed during glycolysis to lac-tate in a reaction catalyzed by lactate dehydrogenase. Under aerobic conditions, the NADH formed can be oxidized by either of two systems. One is the glycerol 3-phosphate shuttle, and the other is the malate shuttle. In the former system, a cytoplasmic glycerol 3-phosphate dehydrogenase equilibrates glycerol 3-phosphate, dihydroxyacetone phosphate, NAD^+, and NADH. The glycerol 3-phosphate formed enters the mitochondria and is oxidized to dihydroxyacetone by the flavin-adenine dinucleo-tide (FAD)–linked mitochondrial glycerol 3-phosphate dehydrogenase. The reduced FAD ($FADH_2$) formed transfers its electrons to complex II of the electron transport chain; thus, 2 molecules of adeno-sine triphosphates are produced for each NADH oxidized. In the malate shuttle, the cytoplasmic NADH is oxidized to NAD^+ by the functioning of a cytoplasmic malate dehydrogenase, which con-verts cytoplasmic oxaloacetate to malate. The malate can enter the mitochondria, where it is con-verted to oxaloacetate by mitochondrial NAD^+-linked malate dehydrogenase.

71. The answer is D (4). (*Chapter 18 II B 2, C 2; Table 18-1*) Glycogen synthetase catalyzes the forma-tion of amylose chains in which glucose residues are in 1,4-glycosidic linkages. Branch chains are formed by transferring segments of the amylose chain onto the C-6 hydroxyl of neighboring chains for α-1,6 linkages. The enzyme responsible is glycosyl-4:6 transferase (branching enzyme). It is missing or deficient in glycogen storage disease type IV. The lack of branching enzyme means that the debranch-ing enzyme (amylo-1,6-glucosidase) has only very few triplets of glucosyl residues to transfer and very few single residues on C-6, which its amylo-6-glucosidase activity removes as free glucose. Thus, in this condition, there is an abnormally high ratio of glucose 1-phosphate (due to phosphorylase action) to free glucose derived from glycogen. Also in this condition, glycogenolysis proceeds at a reduced rate with glycogen accumulation, leading to progressive cirrhosis in the liver in juvenile cases. The deficit is in the number of branches formed, not in their length. There should not be any gross change in the rate of uptake of glucose from the blood.

72. The answer is A (1, 2, 3). (*Chapter 29 I C 2 b, d; 3; 4 a, b*) The de novo synthesis of pyrimidine nu-cleotides starts with the synthesis of carbamoyl phosphate from glutamine, carbon dioxide, and adeno-sine triphosphate catalyzed by carbamoyl phosphate synthetase II. This is the controlled step of pyrimi-dine biosynthesis in man. In the committed step of pyrimidine biosynthesis, carbamoyl phosphate re-acts with aspartate to form N-carbamoyl aspartate in which all of the atoms of aspartate are retained. Ring closure then forms dihydroorotate, which is oxidized to orotate by a nicotinamide-adenine dinu-cleotide–linked dehydrogenase. Orotidine monophosphate (OMP) is formed by transfer of a ribose phosphate group from 5'-phosphoribosyl-1-pyrophosphate, and uridine monophosphate (UMP) is formed by decarboxylation. Cytidine nucleotides arise from uridine triphosphate, thus the pathway does **not** branch at the level of OMP to generate either UMP or cytidine monophosphate.

73. The answer is B (1, 3). [*Chapter 18 Table 18-1; Chapter 19 III A; Chapter 30 II F 1 c, (1), d (1); IV B*] In type I glycogen storage disease, the liver cannot release glucose into the circulation as a result of a deficiency of glucose 6-phosphatase (G6Pase). Thus, agents that stimulate glycogenolysis, such as glu-cagon or epinephrine, do not increase the blood sugar level even though glycogen phosphorylase has been activated and glycogen is broken down to form glucose 1-phosphate. The administration of fruc-tose to a normal subject causes an increase in blood glucose as the liver, kidney, and small intestine convert the fructose to glucose. However, in patients with a deficiency of G6Pase, glucose cannot be formed from the glucose 6-phosphate derived from the fructose. In type VI glycogen storage disease in which liver phosphorylase is deficient, neither glucagon nor epinephrine increases blood glucose levels as glycogenolysis, the source of the glucose, is not active. However, fructose in these patients in-creases blood glucose levels as G6Pase is active in these patients.

74. The answer is D (4). (*Chapter 28 II D 3*) The van den Bergh reaction is a means of determining whether jaundice is due to the accumulation of free bilirubin or of bilirubin diglucuronide. Samples of plasma are reacted with diazotized sulfanilic acid in aqueous solution and in an organic solvent. The aqueous solution ("direct bilirubin") measures only the conjugated bilirubin diglucuronide. The or-ganic solvent ("total bilirubin") measures both free and conjugated bilirubin. "Indirect bilirubin," a measure of the amount of unconjugated bilirubin, is obtained by taking the difference between the total and the direct measurements. In the case of an obstruction of the bile duct, there is a buildup of conjugated bilirubin, which is reflected in increased blood levels (i.e., an increased "direct bilirubin").

In cases of a decreased hepatic uptake of bilirubin, there is a reduced formation of bilirubin diglucuronide as in the genetic absence of the conjugation system. Hemolytic jaundice occurs if the rate of bilirubin formation outstrips the capacity of the hepatic bilirubin conjugation system. In all three cases, there is an increase in the blood level of "indirect bilirubin."

75. The answer is B (1, 3). (*Chapter 28 I B 1, 2*) Two enzymes play important roles in the degradation of catecholamines. Monoamine oxidase (MAO) catalyzes the oxidative deamination of norepinephrine or epinephrine to yield, in the case of norepinephrine, 3,4-dihydroxyphenylglycoaldehyde. Catechol-O-methyl transferase (COMT) catalyzes the O-methylation of catecholamines or the products of MAO action with S-adenosylmethionine as the methyl donor. If the catecholamines are acted on by COMT, the product undergoes oxidative deamination by MAO.

76. The answer is A (1, 2, 3). (*Chapter 19 I B; Chapter 20 III B 2*) In the oxidative phase of the hexose monophosphate shunt, glucose 6-phosphate (G6P) is converted to ribose 5-phosphate by a pathway that includes two nicotinamide-adenine dinucleotide phosphate ($NADP^+$)–linked dehydrogenases. The first is glucose 6-phosphate dehydrogenase and the second is 6-phosphogluconate dehydrogenase. Thus, oxidation of 1 mol of G6P to 1 mol of ribose 5-phosphate yields 2 mol of reduced $NADP^+$ (NADPH). The rate of oxidation of G6P in this pathway is regulated by the concentration of $NADP^+$. In fatty acid synthesis, NADPH is required, and, in this process, is oxidized to $NADP^+$. The increased ratio of $NADP^+$ to NADPH, which occurs during active fatty acid synthesis, stimulates the hexose mono phosphate pathway to produce more NADPH.

77. The answer is A (1, 2, 3). (*Chapter 14 III D; Chapter 15 II A; Chapter 17 II D, E; Chapter 18 II; Chapter 20 III B 1; IV B; Chapter 23 II A*) Carbons from dihydroxyacetone phosphate are incorporated into pyruvate during glycolysis and hence into acetate by the action of pyruvate dehydrogenase. Carbons from acetyl coenzyme A (acetyl CoA) are incorporated into fatty acids (e.g., oleic acid) and then into triacylglycerols (e.g., trioleoglycerol). Carbons from acetyl CoA are also incorporated into cholesterol, and all 27 carbons of cholesterol arise from this source. Thus, carbons from dihydroxyacetone phosphate also appear in testosterone as this steroid is derived from cholesterol. Carbons from this compound are incorporated into glucose 6-phosphate (G6P) by the gluconeogenic pathway, using the enzyme fructose 1,6-diphosphatase in gluconeogenic tissues (e.g., liver, kidney, and small intestine) to bypass the irreversible phosphofructokinase step of glycolysis. Carbons from G6P are incorporated into glycogen. No carbons from dihydroxyacetone phosphate would be incorporated into linoleic acid, which is an essential fatty acid and cannot be synthesized from acetyl CoA by humans.

78. The answer is A (1, 2, 3). (*Chapter 3 I B 6, C 1, 3*) The RNA polymerase holoenzyme of *Escherichia coli* contains four different kinds of subunits (five subunits in all) and is required for the correct initiation of transcription, that is, the initiation stage of synthesis of RNA on a DNA template. It binds to the promoter region of a gene, which is a stretch of DNA that is recognized by the σ subunit of the RNA polymerase. In prokaryotes, such as *Escherichia coli*, one RNA polymerase synthesizes all cellular RNA. The holoenzyme is not required, however, for the elongation stage of RNA synthesis once initiation has occurred, and the σ subunit dissociates from the core enzyme after about eight bases have been polymerized.

79. The answer is A (1, 2, 3). (*Chapter 10 III B 1*) In the formation of mature collagen, there are several stages. The synthesis of preprocollagen takes place on ribosomes attached to the endoplasmic reticulum, and the signal peptide sequences allow the molecule to enter into the cisternae of the endoplasmic reticulum. Here, a number of events take place before the tropocollagen is secreted into the extracellular space. From then on, the maturation of collagen involves extracellular events. The intracellular events in the cisternae of the endoplasmic reticulum include hydroxylation of selected proline and lysine residues, glycosylation of selected hydroxylysines, and formation of interchain and intrachain disulfide bonds. The cleavage of the extension peptides (where the disulfide bonds are located) at the N- and C-terminal ends of the tropocollagen molecule is an extracellular event.

80. The answer is A (1, 2, 3). (*Chapter 16 IX A*) The chemiosmotic coupling theory of oxidative phosphorylation proposes that an electrochemical gradient of protons (H^+) across the mitochondrial inner membrane serves as the means of coupling the energy flow of electron transport to the formation of adenosine triphosphate (ATP). The electron carriers are hypothesized to act as pumps, which cause vectorial (directional) pumping of H^+ across the membrane, which is otherwise impermeable to protons. Thus, electron transport through complex I, complex III, or complex IV results in the transport of protons from the matrix to the intermembrane space. ATP synthesis occurs when the protons in the intermembrane space pass through the inner membrane and back into the matrix at a special site, or "pore," where ATP synthetase resides. The dissipation of energy that occurs as the protons pass down the concentration gradient to the matrix allows the phosphorylation of adenosine diphosphate to ATP by the synthetase. Uncouplers of oxidative phosphorylation, such as 2,4-dinitrophenol, create holes in the mitochondrial inner membrane, rendering it permeable to protons so that a proton gradient cannot form.

81. The answer is A (1, 2, 3). (*Chapter 24 IV C 3*) High-density lipoproteins (HDLs) are synthesized in the liver as particles rich in protein, cholesterol, and phospholipids. During release of the HDLs into the circulation, apoprotein C-II is added to the particle. An apoprotein A-1 activates plasma lecithin:cholesterol acyltransferase (LCAT), which esterifies the cholesterol and converts the lecithin (phosphatidylcholine) to lysophosphatidylcholine. Thus, with a fully active LCAT, the percentage of cholesteryl ester in the HDLs increases, the percentage of cholesterol declines by an equivalent amount, and the percentage of phosphatidylcholine decreases. In the case of LCAT deficiency, these changes do not occur, and compared to a normal situation, the percentage of cholesterol esters in the HDLs decreases, the percentage of cholesterol increases, and the percentage of phosphatidylcholine increases. There is no change in the apoprotein C-II content of HDLs in LCAT deficiency.

82. The answer is E (all). [*Chapter 15 II A; III D; Chapter 27 II C 2 a (3); IV C 2*] An α-keto acid dehydrogenase catalyzes all four of the reactions listed in the question. The conversion of pyruvate to acetyl coenzyme A (acetyl CoA) is catalyzed by the mitochondrial enzyme pyruvate dehydrogenase (PDH), a multienzyme complex. A different α-keto acid dehydrogenase complex catalyzes the oxidative decarboxylation of the branched-chain amino acids—valine, leucine, and isoleucine—to acyl CoA esters. There may be more than one enzyme involved in the decarboxylation of the three amino acids. Lack of this activity in infants leads to maple syrup urine disease. Threonine undergoes direct deamination to yield α-ketobutyrate, a reaction catalyzed by threonine dehydratase. The α-ketobutyrate undergoes oxidative decarboxylation to propionyl CoA catalyzed by the PDH complex. Finally, the conversion of α-ketoglutarate to succinyl CoA in the tricarboxylic acid cycle is catalyzed by α-ketoglutarate dehydrogenase, a multienzyme complex very similar to PDH.

83. The answer is A (1, 2, 3). (*Chapter 2 III B; IV D*) Replication of DNA in prokaryotes is carried out by the concerted action of a number of proteins. Polymerization of incoming deoxyribonucleotides by DNA polymerase III requires a DNA template and a short strand of RNA as a primer. Thus, the first incoming deoxyribonucleotide forms a phosphodiester bond with the RNA primer. The RNA primer is synthesized on the DNA template by primase on the lagging strand and RNA polymerase on the leading strand; in each case, complementary base-pairing involves the formation of hydrogen bonds with the incoming ribonucleotides and the template strand. RNA is not a substrate for DNA ligase, which joins DNA strands, preferably of duplex DNA molecules.

84. The answer is A (1, 2, 3). (*Chapter 28 II C 1, 5; Chapter 31 Table 31-1*) δ-Aminolevulinate synthetase catalyzes the formation of δ-aminolevulinate from glycine and succinyl CoA. It is the first and rate-controlling step of porphyrin biosynthesis. It is a pyridoxal phosphate (vitamin B_6)–dependent enzyme. The synthetase is allosterically inhibited by hemin (i.e., protoporphyrin IX with its iron in the oxidized ferric state). Lead interferes with porphyrin metabolism at more than one site, but not at the step catalyzed by δ-aminolevulinate synthetase.

85. The answer is A (1, 2, 3). [*Chapter 3 II A 3 d, e; Chapter 4 VII A 6 b (2); Chapter 6 II C 4*] A typical eukaryotic gene contains introns, which are DNA sequences that interrupt the expressed DNA sequences, or exons, which are spliced out after transcription in eukaryotes. A cDNA molecule formed from mRNA does not contain introns. However, the bacterial system does not carry out splicing of introns, and if the message is translated, a nonfunctional protein is formed. Depending on where in the bacterial genome the vector ended up, it might not be oriented properly to an effective bacterial promoter. If the cloned gene did not contain a prokaryote-like Shine-Dalgarno sequence, the bacterial RNA polymerase might not be able to initiate polymerization of RNA to form a functional mRNA. The gene does not have to code for a polyadenylated mRNA as it is replicated and expressed in a bacterium in which the mRNAs are not polyadenylated.

86. The answer is A (1, 2, 3). (*Chapter 20 III E 1 a; V C 3; VI A 3*) The insulin:glucagon ratio is high in the postprandial state, increasing the rate of glucose entry into adipocytes so that the formation of dihydroxyacetone phosphate and glycerol 3-phosphate is increased. The availability of these products of glycolysis increases the rate of re-esterification of free fatty acids to triacylglycerols, thus reducing the rate of release of free fatty acids from the adipocytes. This results in reduced blood levels of free fatty acids. The high insulin:glucagon ratio also activates liver coenzyme A (CoA) carboxylase, thereby increasing the rate of fatty acid synthesis from acetyl CoA and reducing the amount of acetyl CoA oxidized to carbon dioxide and water.

87. The answer is C (2, 4). (*Chapter 11 III A; Chapter 12 II D 4*) If the change in free energy of a reaction is more than zero ($\Delta G > 0$), the reaction cannot proceed spontaneously unless there is an input of energy to drive the reaction forward. In the example shown, ΔG is from 10 to < 5 cal/mol (i.e., > -5

cal/mol). Thus, the reaction is not endergonic. The activation energy for the reaction S to P is given by 30 cal/mol − 10 cal/mol, or 20 cal/mol, and X corresponds to the transition state of the reaction in which there is a high probability that a chemical bond will be made or broken to form the product. Enzymes catalyze reactions by decreasing the energy of activation of a reaction (i.e., by lowering X, not by raising S, in terms of cal/mol).

88. The answer is E (all). (*Chapter 16 III A 3, B 1 b, C 3, D 1*) The nonheme iron–sulfur–protein (FeS-protein) centers are one-electron carriers in the mitochondrial electron transport system. The nonheme iron is associated with an equimolar amount of inorganic sulfur. These centers are associated with reduced nicotinamide-adenine dinucleotide dehydrogenase (or NADH coenzyme Q reductase) in complex I; with succinate-coenzyme Q reductase in complex II; and with QH_2–cytochrome c reductase in complex III, which consists of cytochrome b, an FeS-protein center, and cytochrome c_1.

89. The answer is C (2, 4). (*Chapter 16 VIII D; Chapter 32 IV C 3*) Uncouplers of oxidative phosphorylation are compounds that allow mitochondria to use oxygen regardless of whether or not there is any phosphate acceptor [e.g., adenosine diphosphate (ADP)] available. Addition of an uncoupler to a suspension of mitochondria that are oxidizing a substrate in the presence of adequate ADP and oxygen results in an increased rate of oxygen uptake, a cessation of the phosphorylation of ADP to adenosine triphosphate (ATP), and a release of the energy of electron transport as heat. The standard free-energy change ($\Delta G°'$) of hydrolysis is a property of the reaction and is not influenced by a change in reaction conditions.

90. The answer is D (4). (*Chapter 14 III F; Chapter 15 II A, C 3 d; III C*) Of the four enzymes listed in the question only isocitrate dehydrogenase is allosterically inhibited by reduced nicotinamide-adenine dinucleotide (NADH), indicating that the energy charge of the mitochondria is high and that the tricarboxylic acid cycle activity should be reduced. Pyruvate dehydrogenase is inhibited by NADH (and by acetyl coenzyme A), but the effect of NADH in this case is via the stimulation of a protein kinase (pyruvate dehydrogenase kinase), which inhibits pyruvate dehydrogenase by phosphorylating it. 3-Phosphoglycerate kinase catalyzes the transfer of phosphate from 1,3-diphosphoglycerate to adenosine diphosphate to form adenosine triphosphate. This enzyme activity is not allosterically inhibited by NADH, although a high-energy charge means a slow down of this activity as glycolysis is reduced.

91. The answer is C (2, 4). (*Chapter 4 III A–C*) The genetic code consists of 64 codons, which are triplets of bases, 61 triplets coding for an amino acid and 3 coding for stop or chain termination signals. The code is degenerate in the sense that most amino acids are coded for by more than one codon. It is not normally ambiguous as a given codon designates only one amino acid. The initiation codons (i.e., AUG or GUG) and the termination codons (i.e., UAA, UAG, or UGA) are not "nonsense" codons but are required for the initiation or termination of translation.

92. The answer is E (all). (*Chapter 25 III B 2 c–e, 3 b; Chapter 30 II A 1, B 1*) Cellular membrane proteins are of two main types: peripheral proteins, which can be extracted from cellular membrane preparations using relatively gentle procedures, and integral proteins, which can only be extracted by treatments that disrupt membrane structure, such as detergents or organic solvents. Integral proteins can span the width of the membrane, for example, glycophorin in the erythrocyte cell membrane. Lateral movement of proteins in the plane of the membrane can occur as has been shown by a study of heterokaryons of mouse and human cells. Cellular membrane proteins can have catalytic activity as in the hormone receptor–adenylate cyclase complex found in certain cell membranes.

93. The answer is A (1, 2, 3). (*Chapter 11 I C*) Enzymes that require a coenzyme for activity that is tightly bound to the enzyme protein are conjugated proteins, and the coenzyme is called a prosthetic group. The activity of such an enzyme depends upon the protein's conformation and the proper binding of the prosthetic group. Apoenzyme + prosthetic group = holoenzyme, the active form. Removal of the cofactor inactivates the enzyme. Coenzymes can undergo modification during the reaction cycling between at least two chemical forms during the catalytic process.

94. The answer is C (2, 4). (*Chapter 11 IV D 2 a*) The Lineweaver-Burk plot is a graph of a linear transform of the Michaelis-Menten equation. This transform is given by:

$$\frac{1}{v} = \frac{1}{V_{max}} + \frac{K_m}{V_{max}} \times \frac{1}{[S]},$$

where v = the reaction velocity; V_{max} = the maximum reaction velocity; K_m = the Michaelis constant; and [S] = the substrate concentration. The intercept on the x axis gives a value for $1/V_{max}$ of − 200 from which it can be calculated that the value of V_{max} is 100 mol/min. The intercept on the y axis gives a value for $-1/K_m$ from which it can be calculated that K_m has a value of 0.005 M. In this example, $K_m/V_{max} = 0.005/100 = 5 \times 10^{-5}$, not 100. The turnover number cannot be obtained from the Michaelis-Menten equation.

95. The answer is E (all). [*Chapter 21 II D 2; Chapter 27 II C 1 b (2); III C 2 a, 4; Chapter 28 I B 3; Chapter 29 VII C 3*] Methionine labeled with ^{35}S can be degraded to cysteine ^{35}S. When methionine donates its methyl group, S-adenosylhomocysteine is a product, which is hydrolyzed to adenosine and homocysteine. The latter is catabolized by cystathionine synthetase, a pyridoxal phosphate–dependent enzyme, which catalyzes the condensation of homocysteine and serine to form cystathionine. Cystathionase, also a pyridoxal phosphate–dependent enzyme, then cleaves cystathionine into cysteine and α-ketobutyrate. The cysteine, which will have an ^{35}S label, can be used in protein synthesis or in the formation of coenzyme A. It may also be degraded to pyruvate by one of three different pathways with the ^{35}S ending up as inorganic sulfate ion ($^{35}SO_4^{2-}$), which is excreted in the urine as a $^{35}SO_4^{2-}$. The $^{35}SO_4^{2-}$ could also be used to form "active sulfate," (3'-phosphoadenosine-5'-phosphosulfate), which can be used to form sulfates of the catabolic products of the catecholamines, which are partially excreted in the urine as sulfates.

96. The answer is B (1, 3). (*Chapter 4 X A, B*) Inhibition of bacterial ribosome-mediated translation is an effective antibiotic strategy as it prevents the growth of bacterial populations by inhibiting protein synthesis. Clinically, it is desirable to use antibiotics that inhibit prokaryotic (bacterial) protein synthesis but do not affect eukaryotic (human) protein synthesis. Puromycin inhibits protein synthesis in both prokaryotes and eukaryotes and has, in consequence, only a limited clinical use. Penicillin does not directly inhibit protein synthesis. Tetracycline and erythromycin are both effective antibiotics in humans, inhibiting protein synthesis in prokaryotes but not in eukaryotes. Tetracycline inhibits the binding of aminoacyl-tRNAs to the 30S subunit of prokaryote ribosomes, and erythromycin inhibits translation in prokaryotes by binding to the 50S ribosomal subunits.

97. The answer is E (all). (*Chapter 14 II A; Chapter 27 I A 1, 2*) Alanine can arise from pyruvate by transamination with glutamate catalyzed by the pyridoxal phosphate–dependent glutamate pyruvate aminotransferase. The pyruvate is derived from glucose via the glycolytic pathway, and the glutamate is formed from α-ketoglutarate in the glutamate dehydrogenase reaction, using ammonia as the nitrogen source and reduced nicotinamide-adenine dinucleotide phosphate as the electron acceptor.

98. The answer is E (all). (*Chapter 17 I A 1, 2*) The substrates for gluconeogenesis are noncarbohydrate precursors, which include all the intermediates of glycolysis and the citric acid cycle. These precursors are: lactate from erythrocytes and exercising skeletal muscle; glycerol from triacylglycerols in adipose tissue; and α-keto acids, such as pyruvate, oxaloacetate, and α-ketoglutarate from glycogenic (glucogenic) amino acids. Gluconeogenesis occurs primarily in liver, that is, 90% of the glucose is made here; however, during prolonged starvation and metabolic acidosis, the kidney may contribute up to 50% of the glucose formed.

99. The answer is B (1, 3). (*Chapter 2 III D; IV F; Chapter 5 II A*) The replication of DNA in *Escherichia coli* requires the participation of approximately 15 proteins, including the following enzymes: DNA polymerases I and III (DNA pol I and III), DNA ligase, primase, and RNA polymerase. A type of DNA repair mechanism, exemplified by the repair of thymine dimers that form during exposure of the skin to ultraviolet light (UV light), requires the participation of a UV-specific endonuclease, DNA pol I, and DNA ligase. Thus, the enzymes that participate in both DNA replication and repair are DNA ligase and DNA pol I. Topoisomerase and primase are both required for DNA replication, but not for DNA repair.

100. The answer is C (2, 4). [*Chapter 14 III A 2 c (1), (2); Chapter 30 IV C; Chapter 32 IV A 2 b (1)*] A high carbohydrate meal stimulates the release and production of insulin by the β cells of the islets of Langerhans in the pancreas. The increase in blood levels of insulin stimulates uptake of glucose from the blood by muscle cells but not by the brain, liver, or red blood cells. In the liver, insulin affects glucose utilization by increasing the rate of synthesis of glucokinase, the main enzyme phosphorylating glucose in this tissue. The high Michaelis constant (K_m) of glucokinase allows it to phosphorylate effectively the high levels of glucose found in the portal system after a meal. Glucokinase is not inhibited by its product glucose 6-phosphate. Brain utilization of glucose is not increased by rising levels of blood glucose as the brain hexokinase has a very low K_m and is saturated at fasting levels of blood glucose.

101. The answer is D (4). (*Chapter 4 IX B 2*) The signal hypothesis refers to the synthesis of proteins that are exported from the cell and that are synthesized on ribosomes attached to the endoplasmic reticulum. These proteins contain a special amino terminal sequence of approximately 20 amino acids called signal sequences. These sequences contain a stretch of hydrophobic amino acids and function in the binding of ribosomes to the endoplasmic reticulum, allowing the growing peptide chain to pass into the cisternae of the endoplasmic reticulum.

102. The answer is E (all). (*Chapter 2 IV D; Chapter 3 I B; Chapter 6 II C 1; III A 4*) The sequence, 3'-CAUUC-5', represents RNA as it contains uracil (U). It could arise during transcription of the DNA strand 5'-GTAAG-3', that is, during the synthesis of an RNA by DNA-dependent RNA polymerase. It could also arise during replication of DNA in the form of RNA primers synthesized on the separated parental strands by primase on the lagging strand and by RNA polymerase on the leading strand. The duplex shown in the question could also represent a hybrid of a single DNA strand with an RNA strand, formed during annealing of a melted mix of DNA and RNA. Reverse transcriptase catalyzes the formation of DNA strands, using an RNA template. Synthesis is in a 5' to 3' direction; thus, the hybrid segment should be written as

5'-CUUAC-3'
3'-GAATG-5'

with the RNA template written in the 5' to 3' direction.

103. The answer is B (1, 3). (*Chapter 17 II D 4; III A*) The transformation of 2 mol of pyruvate to 1 mol of glucose requires the expenditure of 6 mol of high-energy phosphate bonds. The formation of 1 mol of 1,3-diphosphoglycerate (1,3-DPG) from pyruvate requires expending 1 mol of adenosine triphosphate (ATP) at the pyruvate to oxaloacetate step, 1 mol of guanosine triphosphate (GTP) at the oxaloacetate to phosphoenolpyruvate step, and 1 mol of ATP at the 3-phosphoglycerate to 1,3-DPG step, which makes a total of 2 mol of ATP and 1 mol of GTP for each pyruvate molecule, or 4 mol of ATP and 2 mol of GTP per mol of glucose. The conversion of 2 mol of 1,3-DPG to glyceraldehyde 3-phosphate requires 2 mol of reduced nicotinamide-adenine dinucleotide (NADH). NADH phosphate is not required in the conversion of pyruvate to glucose.

104. The answer is C (2, 4). (*Chapter 16 II C 1, 2; III A, B; V A*) Complex I is an association of reduced nicotinamide-adenine dinucleotide coenzyme Q reductase (NADH-Q reductase) and nonheme iron–sulfur–protein (FeS-protein) centers. Electrons from NADH enter the electron transport chain at the level of complex I. Both malate and α-ketoglutarate dehydrogenases of the tricarboxylic acid cycle use nicotinamide-adenine dinucleotide (NAD^+) as the electron acceptor; thus, inhibition by rotenone of complex I prevents the reoxidation of NADH and, thus, inhibits the oxidation of these substrates. On the other hand, both succinate and glycerol 3-phosphate dehydrogenases use flavin-adenine dinucleotide as the electron acceptor, and this cofactor transfers its electrons to the succinate coenzyme Q reductase association with FeS-protein centers, called complex II. Complex II is not inhibited by rotenone, and electron transport to the next electron carrier in the electron transport chain can proceed. Thus, both succinate and glycerol 3-phosphate can be oxidized in the presence of rotenone.

105. The answer is E (all). (*Chapter 16 III A–C; IX A 4 a; Figure 16-7*) The iron–sulfur–protein (FeS-protein) centers are one-electron carriers like the cytochromes and as opposed to the two-electron carriers, nicotinamide-adenine dinucleotide, flavin-adenine dinucleotide, and coenzyme Q (ubiquinone). The nonheme iron and inorganic sulfur (FeS) are present in equimolar amounts. They are found in association with both the reduced nicotinamide-adenine dinucleotide dehydrogenase complex (complex I) and the succinate dehydrogenase complex (complex II), taking part in the one-electron transfers of both complexes. They are also involved with the transfer of electrons from reduced coenzyme Q to cytochrome c.

106. The answer is E (all). (*Chapter 15 Figure 15-1; Chapter 17 II C; Chapter 27 II C 2 a*) Valine is one of three branched-chain amino acids and is transaminated to the α-keto acid derivative with α-ketoglutarate as the acceptor of the α-amino group. The α-keto derivative is oxidized by a branched-chain α-keto acid dehydrogenase, and the acyl coenzyme A (acyl CoA) derivative is oxidized by the mitochondrial oxidation system to methylmalonyl CoA. The latter is converted to succinyl CoA by the adenosylcobalamin-dependent methylmalonyl CoA mutase. The metabolism of succinyl CoA continues in the Kreb's cycle, of which it is an intermediate, and succinyl dehydrogenase is required en route to oxaloacetate. The latter gives rise to phosphoenolpyruvate by the action of guanosine triphosphate–linked phosphoenolpyruvate carboxykinase. Pyruvate kinase is required to form pyruvate from phosphoenolpyruvate, and transamination of the pyruvate with α-ketoglutarate yields alanine.

107. The answer is A (1, 2, 3). (*Chapter 17 V B 1, 2*) Hypoglycemia evokes the secretion of epinephrine and glucagon. In turn, epinephrine stimulates glycogenolysis, especially in muscle, thereby providing lactate for hepatic gluconeogenesis. Similarly, glucagon stimulates hepatic gluconeogenesis. In contrast, these two hormones have opposite effects on insulin secretion in that epinephrine inhibits, while glucagon stimulates, insulin secretion. Insulin tends to decrease blood glucose concentrations by increasing glucose utilization and promoting glycogen formation.

108. The answer is C (2, 4). (*Chapter 15 II B 1 a; Chapter 31 II A 1, 5 b, F 4; Table 31-1*) Severe dietary deficiency of thiamine (vitamin B_1) causes beriberi, and in alcoholics, Wernicke's encephalopathy. The cardiac disease of beriberi is characterized by an enlargement, usually dilatation, of the heart, absence of arrhythmia, predominant failure of the right ventricle, bounding arterial pulsations, and a high-output failure. The manifestations of polyneuropathy include weakness, paresthesias, and sometimes pain. The ocular disturbances consist of weakness or paralysis (ophthalmoplegia) of the external recti, nystagmus, and palsies of conjugate gaze. The majority of patients are apathetic, listless, and severely confused. Diarrhea and dermatitis are associated with pellagra, which is due to a deficiency of nicotinic acid (niacin).

109. The answer is B (1, 3). [*Chapter 3 II A 1 c; Chapter 10 I C 4 b (1)*] Individuals who are homozygous with respect to sickle cell anemia, that is, who have both allelic chromosomes bearing the mutant hemoglobin S β-chain gene, show a difference in one base from the gene in normal (nonsickle cell) individuals. If mRNA is examined, a single base difference is also found. The tRNA and ribosomes of individuals with sickle cell anemia appear normal.

110. The answer is B (1, 3). (*Chapter 12 IV B 3; Chapter 15 III C 5, F 1, H 1*) The standard reduction potential for the half-reaction nicotinamide-adenine dinucleotide (NAD^+) → reduced NAD^+ (NADH) + $H^+ + 2 e^-$ is $- 0.320$ volts. Increasing the ratio of mitochondrial NAD^+:NADH would make the reduction potential, E, less negative. The increased NAD^+:NADH ratio does not inhibit isocitrate dehydrogenase, which is allosterically inhibited by NADH, but it decreases the malate:oxaloacetate ratio as the conversion of malate to oxaloacetate by the NAD^+-linked malate dehydrogenase reaction is driven to the right. The succinate:fumarate ratio is not directly influenced by a change in the NAD^+:NADH ratio as succinate dehydrogenase, which converts succinate to fumarate, is linked to flavin-adenine dinucleotide, not to NAD^+.

111. The answer is E (all). (*Chapter 19 II A 1; Chapter 21 II D 2 a; Chapter 28 I B*) The catecholamines, dopamine, norepinephrine, and epinephrine, are catabolized by monoamine oxidase (MAO) and catechol-O-methyl transferase (COMT). The former enzyme catalyzes an oxidative deamination to form aldehyde derivatives, and the latter enzyme forms 3-methoxy compounds. The sequence order of the reactions does not matter; that is, the COMT reaction may precede the MAO reaction and vice versa. The catabolites are further metabolized and excreted in the urine after conjugation with sulfate, transferred from 3-phosphoadenosine-5-phosphosulfate, or with glucuronate, transferred from uridine diphosphate glucuronide.

112. The answer is E (all). (*Chapter 9 III A 1 c; V B 2 b; Figure 9-1*) The peptide bond is a planar structure with the two adjacent α-carbons, a carbonyl oxygen, an α-amino-N and its associated hydrogen (H) atom, and the carbonyl carbon all lying in the same plane. The —CN— bond has a partial double-bond character that prevents rotation about the bond axis. The —CN— bond is trans in the α-helix. It is a polar group and enters into H-bonding with other peptide bonds or with water.

113. The answer is B (1, 3). (*Chapter 11 I A 1; III A 2, IV C, Figure 11-2*) Enzymes share some of the properties of chemical catalysts. They are neither consumed nor produced during the course of a reaction. They do not alter the equilibrium constants of reactions that they catalyze, but rather, they speed up reactions that would ordinarily proceed at a much slower rate. Enzymes bind reversibly to their substrates, thereby decreasing the activation energy of the reaction so that the reaction proceeds at a greatly increased rate over that of the uncatalyzed reaction rate. Most enzymes exhibit a rectangular hyperbolic plot when reaction velocity is plotted against substrate concentration. In contrast, certain regulatory enzyme (allosteric) reactions exhibit a sigmoidal plot when reaction rate is plotted against substrate concentration.

114. The answer is B (1, 3). (*Chapter 7 II B 3 b*) The tryptophan (*trp*) operon of prokaryotes is an example of a negatively controlled repressible operon, which may be controlled in two ways: by operator control and by attenuator control. Operator control involves a repressor protein, which forms a complex with tryptophan, the corepressor. Attenuator control is related to the level of tryptophan available to the cell. During the transcription process, some RNA molecules become detached from the DNA at a control point called the attenuator. This point is located in a stretch of DNA known as the leader sequence, which lies between the operator and the *trpE* gene. The lower the concentration of tryptophan in the cell, the greater the number of RNA polymerase molecules that can travel beyond the attenuator and reach the promoter. Sensing of the cellular tryptophan concentration by the *trp* operon is due to the presence of *trp* codons in the leader sequence. If tryptophan is scarce, the ribosome cannot pass the *trp* codons in the leader sequence during translation, and it comes to a halt. The arrested ribosome alters the secondary structure of the mRNA so that it cannot terminate transcription at the attenuator region, and RNA polymerase can continue to the DNA sequences of the structural genes. Thus, attenuation involves translation of the leader sequence and changes in the RNA secondary structure. The repressor belongs to the operator control system; the cyclic adenosine monophosphate–catabolite activator protein complex is not involved.

115. The answer is B (1, 3). [*Chapter 18 III A 2, B 1 b (1), 2*] The rate-limiting step in glycogenolysis is the phosphorolysis of glycogen by glycogen phosphorylase. The enzyme is a dimer with an inactive form (phosphorylase b) and an active form (phosphorylase a). The inactive phosphorylase b can be allosterically activated by high concentrations of adenosine monophosphate (AMP). The inactive phosphorylase b can also be activated by phosphorylation by phosphorylase kinase at the expense of adenosine triphosphate. Phosphorylase kinase also exists in inactive and active forms, which are interconverted by protein phosphorylation. The phosphorylation of phosphorylase kinase by cyclic AMP–dependent protein kinase activates the enzyme and allows the phosphorylation of phosphorylase. In muscle, phosphorylase kinase is partially activated by calcium ion. Glycogen synthetase is not product inhibited by glucose 6-phosphate (G6P). The latter is not a product of catalysis by this enzyme but is an allosteric activator of one form of glycogen synthetase. This form is the D or dependent form, which is inactive in the absence of G6P.

116. The answer is C (2, 4). (*Chapter 16 III A–D; Figure 16-2*) Ubiquinone (coenzyme Q) is a lipid soluble electron-carrying coenzyme. It is a reversibly reducible quinone with a long isoprenoid side chain (see Figure 16-2). It is not derived from a B vitamin. Ubiquinone can accept two electrons from reduced nicotinamide-adenine dinucleotide coenzyme Q reductase (complex I) or succinate coenzyme Q reductase (complex II) and donate single electrons to cytochrome c reductase (complex III).

117. The answer is E (all). (*Chapter 1 III B 1 b; V A 3*) The bacterial transforming factor was shown to be DNA in 1944 by Avery, McLeod, and McCarthy. It is resistant to heating, although increasing temperature reversibly dissociates the double helix into single strands. It shows a strong absorption of light at 260 nm, and this absorption increases as the duplex strands separate. DNA is degraded by deoxyribonuclease, but not by ribonuclease or protease, which hydrolyze RNA and protein, respectively.

118–120. The answers are: 118-C, 119-A, 120-A. (*Chapter 27 III C 1, 2, 3 a*) Both methionine and choline can be methyl group donors, the former requiring a prior adenosine triphosphate (ATP)–dependent activation to S-adenosylmethionine, which is the methyl group donor. Choline requires conversion to glycine betaine to act as a methyl donor, and it has a very limited capability in this direction in contrast to S-adenosylmethionine, which is a methyl group donor in a wide variety of reactions. Choline can donate a methyl group to homocysteine to form methionine. The products of a transfer reaction involving S-adenosylmethionine and an acceptor are a methylated acceptor and S-adenosylhomocysteine. The latter breaks down to adenosine and homocysteine. By methylating homocysteine to methionine, choline plays a role in regenerating methionine. The expenditure of another ATP provides another S-adenosylmethionine methyl donor.

121 and 122. The answers are: 121-A, 122-C. (*Chapter 7 I C; II B 3 a, b*) The lactose (*lac*) operon is inducible and negatively controlled by the *lac* repressor, the product of the regulatory gene, which binds to the operator region to block transcription of the structural genes, z, y, and a. The tryptophan (*trp*) operon exhibits negative repressible control with the controlling metabolite (tryptophan) acting as a corepressor to prevent transcription. However, the *trp* operon can also be controlled by an attenuator; this control is related to the level of tryptophan available to the cell.

123–125. The answers are: 123-A, 124-B, 125-A. [*Chapter 27 I C 1 a (1), E; Chapter 29 I C 2 a (1)*] Carbamoyl phosphate synthetase I occurs in mammalian liver mitochondria, uses ammonia as the source of nitrogen, and requires N-acetyl glutamate as a positive allosteric effector. It is the regulated step in the cycle of events that lead to urea formation and is a major factor in removing ammonia from amino acid degradation. A partial genetic deficiency in the enzyme leads to hyperammonemia in newborn children. A total deficiency is incompatible with life.

Carbamoyl phosphate synthetase II is a cytosolic enzyme that uses the amide group of glutamine as the source of nitrogen and does not require N-acetyl glutamate as an activator. This enzyme is the first step in the synthesis of pyrimidine nucleotides.

126 and 127. The answers are: 126-C, 127-B. (*Chapter 4 VII A; VIII B 2*) Initiation of protein synthesis in eukaryotes requires the 40S ribosomal subunit, methionine-tRNA, mRNA, guanosine triphosphate (GTP), eukaryotic initiation factors (at least 10), and adenosine triphosphate, which is required for this process in eukaryotes but not in prokaryotes. Termination of protein synthesis in prokaryotes requires one of the stop codons (i.e., UAA, UAG, or UGA) on mRNA, which is recognized by proteins called release factors (i.e., RF-1, RF-2, and RF-3). RF-3 promotes either RF-1 or RF-2 binding to the ribosome in a GTP-dependent manner. Binding of the release factor to the termination codon causes hydrolysis of the ester linkages of the peptidyl tRNA. This hydrolysis is probably effected by the peptidyl transferase of the ribosome working in reverse.

128 and 129. The answers are: 128-D, 129-B. (*Chapter 28 II C 4; Figure 28-4*) A number of different types of genetically defective porphyrin syntheses are known, leading to accumulations of intermediates in enzymatic reactions that precede the genetically deficient enzyme in the porphyrin biosynthetic pathway. In hereditary coproporphyria, there is an abnormal accumulation and excretion of coproporphyrinogen III, and in hereditary protoporphyria, there is an increased excretion of protoporphyrin IX, not protoporphyrinogen I. In the latter case, the activity of the mitochondrial enzyme ferrochelatase, which catalyzes the insertion of ferrous iron into the protoporphyrin IX molecule to form heme, decreases.

130–132. The answers are: 130-C, 131-B, 132-C. [*Chapter 29 I A 3, B 2 b (10), 3 a; IX A*] 5′-Phosphoribosyl-1-pyrophosphate (PRPP) is an intermediate of major significance in nucleotide metabolism. It is involved in the de novo synthesis of purine (and pyrimidine) nucleotides, in the salvage pathways for purine (and pyrimidine) bases, and in nucleotide coenzyme biosynthesis. The rate-limiting and committed step of the de novo synthesis of purine nucleotides is the reaction of PRPP with glutamine to form 5′-phosphoribosylamine catalyzed by PRPP-amidotransferase. This pathway forms inosine monophosphate (IMP) from which adenosine monophosphate (AMP) can be formed in two reactions. A hereditary deficiency of hypoxanthine-guanine phosphoribosyl transferase, a purine base salvage enzyme, leads to an X-linked recessive condition known as the Lesch-Nyhan syndrome. The symptoms include excessive uric acid production, hyperuricemia, increased excretion of hypoxanthine, and, in severe cases, neurologic problems, with spasticity, mental retardation, and self-mutilation.

133 and 134. The answers are: 133-C, 134-D. (*Chapter 9 V B 2 c, d; Chapter 10 III A 3*) The α-helix is a rod-like secondary structure with peptide bonds coiled tightly inside and the side chains of the residues protruding outward. Each —CO— group is hydrogen-bonded (H-bonded) to the —NH— of a peptide bond that is four residues away from it on the same chain. There are 3.6 amino acid residues per turn of the helix, which is right-handed. The β-pleated sheets are more extended secondary structures than the α-helix; they are "pleated" because the carbon–carbon bonds are tetrahedral and cannot be in straight lines. The β-pleated sheets lie side by side with H bonds forming between the —CO— group of one peptide bond and the —NH— of another peptide bond in the neighboring chain. The chains may run in the same direction, forming a parallel β-pleated sheet, or run in opposite directions, as in a globular protein in which an extended chain is folded back on itself, forming an antiparallel β-structure. Both kinds of secondary structures are found in globular proteins. The sequence Gly-Pro-Hyp is found specifically in the unit peptide chains of tropocollagen. This sequence dictates a special helical form, which is more open than the α-helix, which is a left-handed helix, and in which there is no intrachain H-bonding.

135 and 136. The answers are: 135-C, 136-D. (*Chapter 4 IV B 4; VII B 2; Chapter 20 IV A*) Amino acids are prepared for polymerization by activation by aminoacyl-tRNA synthetases. The reaction of an amino acid and tRNA requires adenosine triphosphate (ATP) and proceeds via the formation of an aminoacyl intermediate, which is bound to the enzyme until transfer of the aminoacyl group to the tRNA. Pyrophosphate (PP_i) is a product that is rapidly hydrolyzed to inorganic phosphate (P_i), rendering the reaction irreversible in the cell. The link between the amino acid and the tRNA is an amino acid–ester bond, which is a high-energy bond. This supplies sufficient energy for peptide formation during protein synthesis. Thus, the energy for amino acid activation comes from ATP but is preserved as the amino acid–ester bond rather than arising from the hydrolysis of ATP to adenosine diphosphate and P_i. Fatty acids are activated by the formation of a coenzyme A (CoA) thioester catalyzed by acyl CoA synthetase. This reaction requires ATP and proceeds via the formation of an enzyme-bound acyl–adenosine monophosphate (AMP) intermediate, followed by the release of AMP and PP_i. As with amino acid activation, hydrolysis of PP_i to P_i renders the reaction irreversible under cellular conditions.

137 and 138. The answers are: 137-C, 138-D. (*Chapter 28 II E*) Iron in its oxidized (ferric) state (Fe^{3+}) is transported in the blood bound to transferrin, a protein with a very high affinity for Fe^{3+}, which binds two atoms per molecule. Fe^{3+} is stored in the cells in combination with the protein ferritin, which can bind a very large number of Fe^{3+} atoms; these atoms are deposited as hydroxyphosphate in the channels formed by the spherical arrangement of the 24 subunits of ferritin. With very high intakes of iron, some of the Fe^{3+} is found in granules of hemosiderin, a complex of Fe^{3+}, protein, and polysaccharide, which is 3% Fe^{3+} by weight.

139–143. The answers are: 139-A, 140-A, 141-A, 142-B, 143-C. (*Chapter 4 VII A 6, B 1; VIII A, B 4, C 1, D*) Initiation of protein synthesis in prokaryotes involves the formation of an initiation complex, the first step being the binding of mRNA to the 30S ribosomal subunit, which has already bound the initiation factors (i.e., IF-1, IF-2, and IF-3). In contrast, the first step in the formation of an initiation complex in eukaryotes is the formation of a preinitiation complex, which is a ternary complex of eukaryotic initiation factor 2 (eIF-2), guanosine triphosphate, and methionine-tRNA. The next step is the binding of mRNA, a complex and incompletely understood process. Elongation in prokaryotes involves the participation of the elongation factors (i.e., EF-Tu, EF-Ts, and EF-G). In eukaryotic protein synthesis, the elongation factors are eEF-1$_\alpha$, eEF-1$_\beta$, and eEF-2; they are different proteins from the elongation factors of prokaryotes.

Translation of the polycistronic mRNA of prokaryotes starts before transcription is complete. In contrast, the monocistronic mRNA of eukaryotes has to be transported out of the nucleus into the cytoplasm where translation occurs on the ribosomes; thus, transcription has to be completed before translation starts. Gene expression in eukaryotes can be controlled at the level of translation, whereas prokaryotic gene expression is usually controlled at the level of transcription. Termination of translation is effected by peptidyl transferase acting in reverse in both prokaryotic and eukaryotic systems.

144–146. The answers are: 144-C, 145-B, 146-A. (*Chapter 9 V B 2; Chapter 10 III A 3, B 3*) The fundamental unit of collagen structure is tropocollagen, a molecule with three peptide chains (α-chains) of about 1000 residues each. Every third amino acid in the α-chains is glycine, and the sequences Gly-Pro-X$_1$ and Gly-X$_2$-Hyp (where X$_1$ and X$_2$ are any amino acids) are repeated about 100 times; that is, they form 60% of the chains. The proline-rich chains of tropocollagen form a tight left-handed helical structure quite different from the α-helix in that it is more open and there is no intrachain peptide hydrogen bonding (H-bonding). The closeness of the three helical chains in tropocollagen is made possible by the high glycine content (every third residue), and the structure is believed to be stabilized by interchain H bonds and steric repulsion of the pyrrolidone rings of proline residues. Tropocollagen is excreted from the cell and undergoes a series of reactions in the extracellular space, commencing with cleavage of the N-terminal and C-terminal regions. The α-helix is stabilized by intrachain peptide H bonds. Each carbonyl group in a peptide bond is H-bonded to the —NH of a peptide bond that is four residues away from it along the same chain. The α-helix may form an important structural element in a typical globular protein (e.g., hemoglobin) with the total length of the α-helix varying from zero to more than 75% of the total chain length.

147 and 148. The answers are: 147-B, 148-C. [*Chapter 27 III B 5 a (1), C 3 b (1)*] 5-Methyl tetrahydrofolate arises from the reduction of 5,10-methylene tetrahydrofolate by a flavoprotein at the expense of reduced nicotinamide-adenine dinucleotide phosphate; the methyl group is the reduced product of the 5,10-methylene group. Serine is a major input to the one-carbon folate pool. Thus, a carbon-14 label can appear in the one-carbon group carried by tetrahydrofolate. The reaction transferring a ^{14}C from ^{14}C-serine to tetrahydrofolate is catalyzed by serine hydroxymethyl transferase, which would result in the formation of 5,10-methylene tetrahydrofolate labeled with ^{14}C in the methylene carbon. Thus, on reduction, the ^{14}C label would appear in the methyl group of 5-methyl tetrahydrofolate. The latter compound can be used to methylate homocysteine to methionine. This reaction requires methylcobalamin as a cofactor. The 5-methyl tetrahydrofolate, in fact, methylates cobalamin (vitamin B$_{12}$) to form the methyl donor methylcobalamin.

149–151. The answers are: 149-C, 150-A, 151-B. (*Chapter 14 III E 4, F, I 2; Chapter 15 III E 1; Chapter 16 I B*) Both substrate level phosphorylation and oxidative phosphorylation occur in mitochondria. In the tricarboxylic acid cycle, the conversion of succinyl coenzyme A (succinyl CoA) to succinate is coupled to the phosphorylation of guanosine diphosphate to guanosine triphosphate; this is a substrate level phosphorylation. Oxidative phosphorylation is the process whereby the free energy that is released when electrons are transferred along the electron transport chain is coupled to the formation of adenosine triphosphate (ATP) from adenosine diphosphate (ADP) and inorganic phosphate (P$_i$). This coupling of phosphorylation and electron flow is uncoupled by 2,4-dinitrophenol. Oxidative phosphorylation does not occur in the cytosol. In the cytosol, there are three important substrate phosphorylation reactions in the glycolytic pathway. One is the reaction catalyzed by glyceraldehyde 3-phosphate dehydrogenase in which glyceraldehyde 3-phosphate is converted to 1,3-diphosphoglycerate (1,3-DPG) with the phosphate coming from the P$_i$ pool. Another reaction is the conversion of 1,3-DPG to 3-phosphoglycerate with the coupled phosphorylation of ADP to ATP. The third reaction is the conversion of phosphoenolpyruvate to pyruvate, a reaction that is also coupled to the phosphorylation of ADP to ATP.

152–154. The answers are: 152-A, 153-D, 154-C. (*Chapter 4 IX B 3; Chapter 10 II C 1, 2*) Glycoproteins contain oligosaccharide units attached to asparagine residues (type I glycosidic linkages). These carbohydrate units are transferred from an activated lipid carrier, dolicholpyrophosphate, the link being formed only when the asparagine residue is in the sequence Asn-X-Thr (the sequeon), where X is any amino acid. A threonine residue may be involved in both type I glycosidic linkages (i.e., it is part of the sequeon) and in type II linkages, which are formed between serine or threonine residues and a sugar.

155–160. The answers are: 155-C, 156-A, 157-A, 158-B, 159-B, 160-D. (*Chapter 10 I C 3 b; Appendix II D 3*) An increase in plasma P$_{CO_2}$ and a decrease in plasma pH are both characteristic of actively metabolizing cells. Both hydrogen ion (H$^+$) and carbon dioxide (CO$_2$) bind to hemoglobin. The protons bind to the deoxy (T) form of hemoglobin because it is a weaker acid than the R form; the subsequent stabilization of the T form and the concomitant release of oxygen is known as the Bohr effect.

The binding of protons by hemoglobin is an important part of the blood buffering system. CO_2 binds to hemoglobin as carbamate on reaction with deproteinized terminal NH_2 groups and is released at the lungs when the high oxygen concentration promotes the reformation of the T form of hemoglobin. The Bohr effect is made possible by the conformational change in hemoglobin, occurring in the R- to T-form change, when an increase in the pK_a of weakly acidic groups accompanies the structural changes of deoxygenation. The major buffering system of plasma is the bicarbonate:carbonic acid ratio where the concentration of carbonic acid is taken to be the sum of carbonic acid plus dissolved CO_2. This ratio is normally about 20:1.

The imidazole ring in the side chain of the basic amino acid histidine has a pK'_a that is close to the pH of blood. Thus, it participates in the buffering mechanisms employed by the red cell. The cysteinyl groups of hemoglobin do not participate directly in blood buffering or in the Bohr effect.

161–167. The answers are: 161-B, 162-A, 163-C, 164-B, 165-B, 166-D, 167-A. [*Chapter 1 II B 2, C 2; Chapter 3 II A 2, 3; Chapter 4 III C 2; VII A 6 b (2); VIII A 1, B 1 a*] Prokaryotic mRNA is polycistronic (and eukaryotic mRNA is monocistronic); that is, it is a template for several polypeptide chains. The 5′ end of the mRNA may contain a sequence that is never translated into protein called the leader sequence or 5′ untranslated sequence. This is not the cap structure that is found in eukaryotic mRNA, which is a 7-methylguanosine linked to the 5′ end of the mRNA by a 5′-5′ pyrophosphate linkage with the adjacent riboses methylated at the 2′-hydroxyl position. Most eukaryotic mRNA also has a poly A tail, consisting of 20–200 adenine nucleotides at the 3′ end. No equivalent structure is found in prokaryotic mRNAs, although they may have 3′ untranslated sequences. Both prokaryotic and eukaryotic mRNA contain an AUG initiation codon, although recognition of this codon by RNA polymerase depends upon a number of characteristics of the mRNAs, for example, the Shine-Dalgarno sequences on the 5′ side of the initiating codon in prokaryotes. The recognition of the initiation codon in eukaryotes is more complex and is still unclear. Thymine bases are only found in DNA, not RNA.

168–172. The answers are: 168-A, 169-B, 170-D, 171-A, 172-C. [*Chapter 27 IV A 1; Chapter 31 II A 5, B 5, D 5, F 4*] Pellagra, a disease affecting the skin (dermatitis), the gastrointestinal tract (diarrhea), and the central nervous system (dementia), is the result of a niacin deficiency. Because niacin can be formed from tryptophan, an essential amino acid, dietary treatment of pellagra must take into consideration daily allowances for both niacin and tryptophan. Endemic pellagra is no longer a common occurrence; however, it is a manifestation of two disorders of tryptophan metabolism, Hartnup disease and the carcinoid syndrome. Hartnup disease is an autosomal recessive defect in which patients have a reduced ability to convert tryptophan to niacin. In the carcinoid syndrome, dietary tryptophan is metabolized in the hydroxylation pathway (a minor pathway), leaving little tryptophan for the formation of niacin. Administration of large amounts of niacin can cure the pellagra associated with these conditions.

Beriberi is a severe thiamine deficiency syndrome associated with malnutrition endemic to areas where there is a high intake of highly milled (polished) rice. Clinical characteristics of this deficiency range from cardiovascular and neurologic lesions to emotional disturbances. Cardiovascular changes include right-sided enlargement (dilatation), tachycardia, and "high-output" cardiac failure. Neuromuscular manifestations include peripheral neuropathy (neuritis), weakness, fatigue, and an impaired capacity to do work. Edema and anorexia are also characteristic. In the United States, thiamine deficiency is seen primarily in association with chronic alcoholism, which leads to Wernicke's encephalopathy, which presents with the classic triad of confusion, ataxia and ophthalmoplegia. In thiamine deficiency, motor and sensory peripheral nerve lesions are marked by neuromuscular findings of numbness and tingling of the legs, atrophy and weakness of the muscles of the extremities compounded by the loss of reflexes. Mental depression may also accompany these findings. The dementia caused by niacin deficiency results from degeneration of the ganglion cells of the brain, accompanied by degeneration of the fibers of the spinal cord.

173–175. The answers are: 173-C, 174-B, 175-A. [*Chapter 10 I C 3 a (3), b (3); Appendix II C 3; explanation to study question 6*] The ratio of bicarbonate ion to carbonic acid in plasma is normally maintained at about 20:1, so that carbonic acid is only a minor contributor to physiologic carbon dioxide (CO_2) transport. Both CO_2 and 2,3-diphosphoglycerate (2,3-DPG) tend to stabilize the tight (T) form of hemoglobin (i.e., the deoxygenated form). As the same sites are involved in the binding of CO_2 and 2,3-DPG to hemoglobin, their effects are not fully additive, but in the absence of 2,3-DPG, there would be an increase in the transport of CO_2 by hemoglobin. Myoglobin is an oxygen transport and repository molecule in muscle. It is not involved with CO_2 transport.

176–180. The answers are: 176-A, 177-B, 178-C, 179-D, 180-E. (*Chapter 31 II A 4, 5, C 4, 5, E 4, 5, F 3, 4; III B 3, 4*) Phosphorylation of pyridoxine yields the biologically active form of vitamin B_6 (pyridoxal phosphate). This form serves as a coenzyme for a large number of enzymes, particularly those that catalyze reactions involving amino acids, for example, transamination, deamination, decarboxylation, and condensation.

The biologically active form of folic acid is tetrahydrofolic acid. This form transfers one-carbon fragments to appropriate metabolites in the synthesis of amino acids, purines, and thymidylic acid, which is the characteristic pyrimidine of DNA.

Vitamin D_3 is converted to the active form, 1,25-dihydroxycholecalciferol, by two sequential hydroxylation reactions. The active form of vitamin D_3 stimulates the intestinal absorption of calcium and phosphate, increases the mobilization of calcium and phosphate from bone, and promotes the renal reabsorption of calcium and phosphate (in physiologic amounts).

The biologically active forms of niacin, or nicotinic acid, are nicotinamide-adenine dinucleotide (NAD^+) and nicotinamide-adenine dinucleotide phosphate ($NADP^+$). These two cofactors serve as coenzymes in oxidation–reduction reactions in which the coenzyme undergoes reduction of the pyridine ring by accepting a hydride ion (hydrogen ion plus one electron). The reduced forms of these two dinucleotides are NADH and NADPH, respectively.

Thiamine pyrophosphate is the biologically active form of the vitamin. It serves as a cofactor in the oxidative decarboxylation of keto acids. It is a cofactor in the nonoxidative reactions of the pentose phosphate pathway because it is the prosthetic group of transketolase.

Index

Note: Page numbers followed by f denote figures; those followed by t denote tables; those followed by Q denote questions; and those followed by E denote explanations.

maybe 12

6/19-23, WMCA 281-2626

Protein
Carb
amino acids
nucleic acid

X-Roy 238 pg +